WORLD HEALTH ORGANIZATION

INTERNATIONAL AGENCY FOR RESEARCH ON CANCER

IARC MONOGRAPHS
ON THE
EVALUATION OF CARCINOGENIC RISKS TO HUMANS

Dry Cleaning, Some Chlorinated Solvents and Other Industrial Chemicals

VOLUME 63

This publication represents the views and expert opinions
of an IARC Working Group on the
Evaluation of Carcinogenic Risks to Humans,
which met in Lyon,

7–14 February 1995

1995

IARC MONOGRAPHS

In 1969, the International Agency for Research on Cancer (IARC) initiated a programme on the evaluation of the carcinogenic risk of chemicals to humans involving the production of critically evaluated monographs on individual chemicals. In 1980 and 1986, the programme was expanded to include evaluations of carcinogenic risks associated with exposures to complex mixtures and other agents.

The objective of the programme is to elaborate and publish in the form of monographs critical reviews of data on carcinogenicity for agents to which humans are known to be exposed and on specific exposure situations; to evaluate these data in terms of human risk with the help of international working groups of experts in chemical carcinogenesis and related fields; and to indicate where additional research efforts are needed.

This project is supported by PHS Grant No. 5-UO1 CA33193-13 awarded by the United States National Cancer Institute, Department of Health and Human Services. Additional support has been provided since 1986 by the European Commission.

©International Agency for Research on Cancer 1995

IARC Library Cataloguing in Publication Data

IARC Working Group on the Evaluation of Carcinogenic Risks to
  Humans (1995 : Lyon, France)
  Dry cleaning, some chlorinated solvents and other industrial chemicals : views and expert opinions of an IARC Working Group on the Evaluation of Carcinogenic Risks to Humans which met in Lyon, 7–14 February 1995.

  (IARC monographs on the evaluation of carcinogenic risks to humans ; 63)

  1. Carcinogens – toxicity – congresses 2. Chemical industries – congresses
  3. Neoplasms – chemically induced – congresses 4. Solvents – toxicity I. Series

  ISBN 92 832 1263 0   (NLM Classification: W1)
  ISSN 0250-9555

Publications of the World Health Organization enjoy copyright protection in accordance with the provisions of Protocol 2 of the Universal Copyright Convention.

All rights reserved. Application for rights of reproduction or translation, in part or in toto, should be made to the International Agency for Research on Cancer.

Distributed by the International Agency for Research on Cancer
and by the Secretariat of the World Health Organization, Geneva

PRINTED IN THE UNITED KINGDOM

CONTENTS

NOTE TO THE READER 1

LIST OF PARTICIPANTS 3

PREAMBLE 9
   Background 9
   Objective and Scope 9
   Selection of Topics for Monographs 10
   Data for Monographs 11
   The Working Group 11
   Working Procedures 11
   Exposure Data 12
   Studies of Cancer in Humans 14
   Studies of Cancer in Experimental Animals 17
   Other Data Relevant to an Evaluation of Carcinogenicity and Its Mechanisms ... 19
   Summary of Data Reported 21
   Evaluation 22
   References 26

THE MONOGRAPHS
   Dry cleaning 33

   Some chlorinated solvents and related chemicals
      Trichloroethylene 75
      Tetrachloroethylene 159
      1,2,3-Trichloropropane 223
      Chloral and chloral hydrate 245
      Dichloroacetic acid 271
      Trichloroacetic acid 291
      1-Chloro-2-methylpropene 315
      3-Chloro-2-methylpropene 325

   Other industrial chemicals
      Acrolein 337
      Crotonaldehyde 373
      Furan 393
      Furfural 409
      Benzofuran 431

    Vinyl acetate 443
    Vinyl fluoride 467

SUMMARY OF FINAL EVALUATIONS 477

APPENDIX 1. SUMMARY TABLES OF GENETIC AND RELATED EFFECTS 481

APPENDIX 2. ACTIVITY PROFILES FOR GENETIC AND RELATED EFFECTS 499

SUPPLEMENTARY CORRIGENDA TO VOLUMES 1–62 521

CUMULATIVE INDEX TO THE *MONOGRAPHS* SERIES 523

NOTE TO THE READER

The term 'carcinogenic risk' in the *IARC Monographs* series is taken to mean the probability that exposure to an agent will lead to cancer in humans.

Inclusion of an agent in the *Monographs* does not imply that it is a carcinogen, only that the published data have been examined. Equally, the fact that an agent has not yet been evaluated in a monograph does not mean that it is not carcinogenic.

The evaluations of carcinogenic risk are made by international working groups of independent scientists and are qualitative in nature. No recommendation is given for regulation or legislation.

Anyone who is aware of published data that may alter the evaluation of the carcinogenic risk of an agent to humans is encouraged to make this information available to the Unit of Carcinogen Identification and Evaluation, International Agency for Research on Cancer, 150 cours Albert Thomas, 69372 Lyon Cedex 08, France, in order that the agent may be considered for re-evaluation by a future Working Group.

Although every effort is made to prepare the monographs as accurately as possible, mistakes may occur. Readers are requested to communicate any errors to the Unit of Carcinogen Identification and Evaluation, so that corrections can be reported in future volumes.

IARC WORKING GROUP ON THE EVALUATION OF CARCINOGENIC RISKS TO HUMANS: DRY CLEANING, SOME CHLORINATED SOLVENTS AND OTHER INDUSTRIAL CHEMICALS

Lyon, 7–14 February 1995

LIST OF PARTICIPANTS

Members[1]

A. Abbondandolo, Laboratory of Mutagenesis, National Institute for Research on Cancer, viale Benedetto XV, 10, 16132 Genoa, Italy

O. Axelson, Department of Occupational Medicine and Industrial Ergonomics, University Hospital, 581 85 Linköping, Sweden

S. Cordier, INSERM, U 170, 16 avenue Paul Vaillant-Couturier, 94807 Villejuif Cédex, France

W. Dekant, Institute of Toxicology and Pharmacology, University of Würzburg, Versbacher Str. 9, 97078 Würzburg, Germany

E. Dybing, Department of Environmental Medicine, National Institute of Public Health, 0462 Oslo, Norway (*Chairman*)

J. Fajen, National Institute for Occupational Safety and Health, 4676 Columbia Parkway, Cincinnati OH 45226-1988, United States

G.R. Howe, NCIC Epidemiology Unit, Faculty of Medicine, McMurrich Building, University of Toronto, 13 Queen's Park Crescent West, Toronto, Ontario M5S 1A8, Canada

A. Huici-Montagud, National Institute of Security and Hygiene at the Workplace, National Center of Working Conditions, Dulcet 2-10, 08034 Barcelona, Spain

Y. Konishi, Department of Oncological Pathology, Cancer Center, Nara Medical University, 840 Shijo-cho, Kashihara, Nara 634, Japan

H. Kromhout, Department of Air Quality, Wageningen Agricultural University, PO Box 8129, 6700 EV Wageningen, Netherlands

---

[1] Unable to attend: W.T. Allaben, National Center for Toxicological Research, Jefferson, AR 72079-9502, United States; P.W.J. Peters, National Institute of Public Health and Environmental Protection and Medical Faculty, University of Utrecht, PO Box 1, BA Bilthoven, Netherlands

L.S. Levy, Institute of Occupational Health, University of Birmingham, Edgbaston, Birmingham B15 2TT, United Kingdom

E. Lynge, Danish Cancer Society, Strandboulevarden 49, 2100 Copenhagen Ø, Denmark (*Vice-Chairman*)

R.L. Melnick, National Institute of Environmental Health Sciences, PO Box 12233, Research Triangle Park NC 27709, United States

H. Norppa, Department of Industrial Hygiene and Toxicology, Institute of Occupational Health, Topeliuksenkatu 41 a A, 00250 Helsinki, Finland

S. Olin, Risk Science Institute, International Life Sciences Institute, 1126 Sixteenth Street NW, Washington DC 20036, United States

C. Rosenberg, Department of Industrial Hygiene and Toxicology, Institute of Occupational Health, Topeliuksenkatu 41 a A, 00250 Helsinki, Finland

N.H. Stacey, Worksafe Australia, GPO Box 58, Sydney NSW 2001, Australia

S. Vamvakas, Institute of Toxicology and Pharmacology, University of Würzburg, Versbacher Str. 9, 97078 Würzburg, Germany

Representatives/observers

Representative of the National Cancer Institute

J. Sontag, Epidemiology and Biostatistics Program, National Cancer Institute, Executive Plaza North, Room 543, Bethesda MD 20892, United States

American Industrial Health Council

N.S. Weiss, Department of Epidemiology, SC-36, University of Washington, Room F-263, Health Sciences Building, 1959 NE Pacific Street, Seattle, WA 98195, United States

European Center for Ecotoxicology and Toxicology of Chemicals

D.G. Farrar, ICI Chemicals and Polymer Ltd, Occupational Health, PO Box 13, The Heath Runcorn, Cheshire WA7 4QF, United Kingdom

European Commission

E. de Pauw, Industrial Medicine and Hygiene Unit, Health and Safety Directorate, European Commission, Bâtiment Jean Monnet, 2920 Luxembourg, Grand Duchy of Luxembourg

National Cancer Institute of Canada

D. Burch, NCIC Epidemiology Unit, Faculty of Medicine, McMurrich Building, University of Toronto, Toronto, Ontario M5S 1A8, Canada

United States Environmental Protection Agency

J.C. Parker, Human Health Assessment Group, Office of Research and Development, United States Environmental Protection Agency, Washington DC 20460, United States

PARTICIPANTS

IARC Secretariat

P. Boffetta, Unit of Environmental Cancer Epidemiology
A. Dufournet, Unit of Carcinogen Identification and Evaluation
M. Friesen, Unit of Environmental Carcinogenesis
M.-J. Ghess, Unit of Carcinogen Identification and Evaluation
E. Heseltine, 24290 St Léon-sur-Vézère, France
V. Krutovskikh, Unit of Multistage Carcinogenesis
M. Lang, Programme of Molecular Toxicology
D. McGregor, Unit of Carcinogen Identification and Evaluation
D. Mietton, Unit of Carcinogen Identification and Evaluation
H. Møller, Unit of Carcinogen Identification and Evaluation
A. Mylvaganam, Programme of Radiation and Cancer
C. Partensky, Unit of Carcinogen Identification and Evaluation
S. Ruiz, Unit of Carcinogen Identification and Evaluation
P. Webb, Unit of Carcinogen Identification and Evaluation
J. Wilbourn, Unit of Carcinogen Identification and Evaluation
H. Yamasaki, Unit of Multistage Carcinogenesis

Secretarial assistance

M. Lézère
J. Mitchell
S. Reynaud

PREAMBLE

IARC MONOGRAPHS PROGRAMME ON THE EVALUATION OF CARCINOGENIC RISKS TO HUMANS[1]

PREAMBLE

1. BACKGROUND

In 1969, the International Agency for Research on Cancer (IARC) initiated a programme to evaluate the carcinogenic risk of chemicals to humans and to produce monographs on individual chemicals. The *Monographs* programme has since been expanded to include consideration of exposures to complex mixtures of chemicals (which occur, for example, in some occupations and as a result of human habits) and of exposures to other agents, such as radiation and viruses. With Supplement 6 (IARC, 1987a), the title of the series was modified from *IARC Monographs on the Evaluation of the Carcinogenic Risk of Chemicals to Humans* to *IARC Monographs on the Evaluation of Carcinogenic Risks to Humans*, in order to reflect the widened scope of the programme.

The criteria established in 1971 to evaluate carcinogenic risk to humans were adopted by the working groups whose deliberations resulted in the first 16 volumes of the *IARC Monographs series*. Those criteria were subsequently updated by further ad-hoc working groups (IARC, 1977, 1978, 1979, 1982, 1983, 1987b, 1988, 1991a; Vainio *et al.*, 1992).

2. OBJECTIVE AND SCOPE

The objective of the programme is to prepare, with the help of international working groups of experts, and to publish in the form of monographs, critical reviews and evaluations of evidence on the carcinogenicity of a wide range of human exposures. The *Monographs* may also indicate where additional research efforts are needed.

The *Monographs* represent the first step in carcinogenic risk assessment, which involves examination of all relevant information in order to assess the strength of the available evidence that certain exposures could alter the incidence of cancer in humans. The second step is quantitative risk estimation. Detailed, quantitative evaluations of epidemiological data may be

---

[1]This project is supported by PHS Grant No. 5-UO1 CA33193-13 awarded by the United States National Cancer Institute, Department of Health and Human Services. Since 1986, the programme has also been supported by the European Commission.

made in the *Monographs*, but without extrapolation beyond the range of the data available. Quantitative extrapolation from experimental data to the human situation is not undertaken.

The term 'carcinogen' is used in these monographs to denote an exposure that is capable of increasing the incidence of malignant neoplasms; the induction of benign neoplasms may in some circumstances (see p. 18) contribute to the judgement that the exposure is carcinogenic. The terms 'neoplasm' and 'tumour' are used interchangeably.

Some epidemiological and experimental studies indicate that different agents may act at different stages in the carcinogenic process, and several different mechanisms may be involved. The aim of the *Monographs* has been, from their inception, to evaluate evidence of carcinogenicity at any stage in the carcinogenesis process, independently of the underlying mechanisms. Information on mechanisms may, however, be used in making the overall evaluation (IARC, 1991a; Vainio *et al.*, 1992; see also pp. 25-27).

The *Monographs* may assist national and international authorities in making risk assessments and in formulating decisions concerning any necessary preventive measures. The evaluations of IARC working groups are scientific, qualitative judgements about the evidence for or against carcinogenicity provided by the available data. These evaluations represent only one part of the body of information on which regulatory measures may be based. Other components of regulatory decisions may vary from one situation to another and from country to country, responding to different socioeconomic and national priorities. **Therefore, no recommendation is given with regard to regulation or legislation, which are the responsibility of individual governments and/or other international organizations.**

The *IARC Monographs* are recognized as an authoritative source of information on the carcinogenicity of a wide range of human exposures. A users' survey, made in 1988, indicated that the *Monographs* are consulted by various agencies in 57 countries. Each volume is generally printed in 4000 copies for distribution to governments, regulatory bodies and interested scientists. The Monographs are also available from the International Agency for Research on Cancer in Lyon and via the Distribution and Sales Service of the World Health Organization.

3. SELECTION OF TOPICS FOR MONOGRAPHS

Topics are selected on the basis of two main criteria: (a) there is evidence of human exposure, and (b) there is some evidence or suspicion of carcinogenicity. The term 'agent' is used to include individual chemical compounds, groups of related chemical compounds, physical agents (such as radiation) and biological factors (such as viruses). Exposures to mixtures of agents may occur in occupational exposures and as a result of personal and cultural habits (like smoking and dietary practices). Chemical analogues and compounds with biological or physical characteristics similar to those of suspected carcinogens may also be considered, even in the absence of data on a possible carcinogenic effect in humans or experimental animals.

The scientific literature is surveyed for published data relevant to an assessment of carcinogenicity. The IARC information bulletins on agents being tested for carcinogenicity (IARC, 1973-1994) and directories of on-going research in cancer epidemiology (IARC, 1976-1994) often indicate those exposures that may be scheduled for future meetings. Ad-hoc working groups convened by IARC in 1984, 1989, 1991 and 1993 gave recommendations as to which agents should be evaluated in the IARC Monographs series (IARC, 1984, 1989, 1991b, 1993).

As significant new data on subjects on which monographs have already been prepared become available, re-evaluations are made at subsequent meetings, and revised monographs are published.

4. DATA FOR MONOGRAPHS

The *Monographs* do not necessarily cite all the literature concerning the subject of an evaluation. Only those data considered by the Working Group to be relevant to making the evaluation are included.

With regard to biological and epidemiological data, only reports that have been published or accepted for publication in the openly available scientific literature are reviewed by the working groups. In certain instances, government agency reports that have undergone peer review and are widely available are considered. Exceptions may be made on an ad-hoc basis to include unpublished reports that are in their final form and publicly available, if their inclusion is considered pertinent to making a final evaluation (see pp. 25–27). In the sections on chemical and physical properties, on analysis, on production and use and on occurrence, unpublished sources of information may be used.

5. THE WORKING GROUP

Reviews and evaluations are formulated by a working group of experts. The tasks of the group are: (i) to ascertain that all appropriate data have been collected; (ii) to select the data relevant for the evaluation on the basis of scientific merit; (iii) to prepare accurate summaries of the data to enable the reader to follow the reasoning of the Working Group; (iv) to evaluate the results of epidemiological and experimental studies on cancer; (v) to evaluate data relevant to the understanding of mechanism of action; and (vi) to make an overall evaluation of the carcinogenicity of the exposure to humans.

Working Group participants who contributed to the considerations and evaluations within a particular volume are listed, with their addresses, at the beginning of each publication. Each participant who is a member of a working group serves as an individual scientist and not as a representative of any organization, government or industry. In addition, nominees of national and international agencies and industrial associations may be invited as observers.

6. WORKING PROCEDURES

Approximately one year in advance of a meeting of a working group, the topics of the monographs are announced and participants are selected by IARC staff in consultation with other experts. Subsequently, relevant biological and epidemiological data are collected by IARC from recognized sources of information on carcinogenesis, including data storage and retrieval systems such as MEDLINE and TOXLINE, and EMIC and ETIC for data on genetic and related effects and reproductive and developmental effects, respectively.

For chemicals and some complex mixtures, the major collection of data and the preparation of first drafts of the sections on chemical and physical properties, on analysis, on production and use and on occurrence are carried out under a separate contract funded by the United States National Cancer Institute. Representatives from industrial associations may assist in the preparation of sections on production and use. Information on production and trade is obtained from governmental and trade publications and, in some cases, by direct contact with industries. Separate production data on some agents may not be available because their publication could disclose confidential information. Information on uses may be obtained from published sources but is often complemented by direct contact with manufacturers. Efforts are made to supplement this information with data from other national and international sources.

Six months before the meeting, the material obtained is sent to meeting participants, or is used by IARC staff, to prepare sections for the first drafts of monographs. The first drafts are compiled by IARC staff and sent, prior to the meeting, to all participants of the Working Group for review.

The Working Group meets in Lyon for seven to eight days to discuss and finalize the texts of the monographs and to formulate the evaluations. After the meeting, the master copy of each monograph is verified by consulting the original literature, edited and prepared for publication. The aim is to publish monographs within six months of the Working Group meeting.

The available studies are summarized by the Working Group, with particular regard to the qualitative aspects discussed below. In general, numerical findings are indicated as they appear in the original report; units are converted when necessary for easier comparison. The Working Group may conduct additional analyses of the published data and use them in their assessment of the evidence; the results of such supplementary analyses are given in square brackets. When an important aspect of a study, directly impinging on its interpretation, should be brought to the attention of the reader, a comment is given in square brackets.

7. EXPOSURE DATA

Sections that indicate the extent of past and present human exposure, the sources of exposure, the people most likely to be exposed and the factors that contribute to the exposure are included at the beginning of each monograph.

Most monographs on individual chemicals, groups of chemicals or complex mixtures include sections on chemical and physical data, on analysis, on production and use and on occurrence. In monographs on, for example, physical agents, occupational exposures and cultural habits, other sections may be included, such as: historical perspectives, description of an industry or habit, chemistry of the complex mixture or taxonomy. Monographs on biological agents have sections on structure and biology, methods of detection, epidemiology of infection and clinical disease other than cancer.

For chemical exposures, the Chemical Abstracts Services Registry Number, the latest Chemical Abstracts Primary Name and the IUPAC Systematic Name are recorded; other synonyms are given, but the list is not necessarily comprehensive. For biological agents, taxonomy and structure are described, and the degree of variability is given, when applicable.

Information on chemical and physical properties and, in particular, data relevant to identification, occurrence and biological activity are included. For biological agents, mode of replication, life cycle, target cells, persistence and latency and host response are given. A description of technical products of chemicals includes trades names, relevant specifications and available information on composition and impurities. Some of the trade names given may be those of mixtures in which the agent being evaluated is only one of the ingredients.

The purpose of the section on analysis or detection is to give the reader an overview of current methods, with emphasis on those widely used for regulatory purposes. Methods for monitoring human exposure are also given, when available. No critical evaluation or recommendation of any of the methods is meant or implied. The IARC publishes a series of volumes, *Environmental Carcinogens: Methods of Analysis and Exposure Measurement* (IARC, 1978–93), that describe validated methods for analysing a wide variety of chemicals and mixtures. For biological agents, methods of detection and exposure assessment are described, including their sensitivity, specificity and reproducibility.

The dates of first synthesis and of first commercial production of a chemical or mixture are provided; for agents which do not occur naturally, this information may allow a reasonable estimate to be made of the date before which no human exposure to the agent could have occurred. The dates of first reported occurrence of an exposure are also provided. In addition, methods of synthesis used in past and present commercial production and different methods of production which may give rise to different impurities are described.

Data on production, international trade and uses are obtained for representative regions, which usually include Europe, Japan and the United States of America. It should not, however, be inferred that those areas or nations are necessarily the sole or major sources or users of the agent. Some identified uses may not be current or major applications, and the coverage is not necessarily comprehensive. In the case of drugs, mention of their therapeutic uses does not necessarily represent current practice nor does it imply judgement as to their therapeutic efficacy.

Information on the occurrence of an agent or mixture in the environment is obtained from data derived from the monitoring and surveillance of levels in occupational environments, air, water, soil, foods and animal and human tissues. When available, data on the generation, persistence and bioaccumulation of the agent are also included. In the case of mixtures, industries, occupations or processes, information is given about all agents present. For processes, industries and occupations, a historical description is also given, noting variations in chemical composition, physical properties and levels of occupational exposure with time and place. For biological agents, the epidemiology of infection is described.

Statements concerning regulations and guidelines (e.g. pesticide registrations, maximal levels permitted in foods, occupational exposure limits) are included for some countries as indications of potential exposures, but they may not reflect the most recent situation, since such limits are continuously reviewed and modified. The absence of information on regulatory status for a country should not be taken to imply that that country does not have regulations with regard to the exposure. For biological agents, legislation and control, including vaccines and therapy, are described.

8. STUDIES OF CANCER IN HUMANS

(a) Types of studies considered

Three types of epidemiological studies of cancer contribute to the assessment of carcinogenicity in humans—cohort studies, case–control studies and correlation (or ecological) studies. Rarely, results from randomized trials may be available. Case series and case reports of cancer in humans may also be reviewed.

Cohort and case–control studies relate individual exposures under study to the occurrence of cancer in individuals and provide an estimate of relative risk (ratio of incidence or mortality in those exposed to incidence or mortality in those not exposed) as the main measure of association.

In correlation studies, the units of investigation are usually whole populations (e.g. in particular geographical areas or at particular times), and cancer frequency is related to a summary measure of the exposure of the population to the agent, mixture or exposure circumstance under study. Because individual exposure is not documented, however, a causal relationship is less easy to infer from correlation studies than from cohort and case–control studies. Case reports generally arise from a suspicion, based on clinical experience, that the concurrence of two events—that is, a particular exposure and occurrence of a cancer—has happened rather more frequently than would be expected by chance. Case reports usually lack complete ascertainment of cases in any population, definition or enumeration of the population at risk and estimation of the expected number of cases in the absence of exposure. The uncertainties surrounding interpretation of case reports and correlation studies make them inadequate, except in rare instances, to form the sole basis for inferring a causal relationship. When taken together with case–control and cohort studies, however, relevant case reports or correlation studies may add materially to the judgement that a causal relationship is present.

Epidemiological studies of benign neoplasms, presumed preneoplastic lesions and other end-points thought to be relevant to cancer are also reviewed by working groups. They may, in some instances, strengthen inferences drawn from studies of cancer itself.

(b) Quality of studies considered

The Monographs are not intended to summarize all published studies. Those that are judged to be inadequate or irrelevant to the evaluation are generally omitted. They may be mentioned briefly, particularly when the information is considered to be a useful supplement to that in other reports or when they provide the only data available. Their inclusion does not imply acceptance of the adequacy of the study design or of the analysis and interpretation of the results, and limitations are clearly outlined in square brackets at the end of the study description.

It is necessary to take into account the possible roles of bias, confounding and chance in the interpretation of epidemiological studies. By 'bias' is meant the operation of factors in study design or execution that lead erroneously to a stronger or weaker association than in fact exists between disease and an agent, mixture or exposure circumstance. By 'confounding' is meant a situation in which the relationship with disease is made to appear stronger or to appear weaker than it truly is as a result of an association between the apparent causal factor and another factor that is associated with either an increase or decrease in the incidence of the disease. In evaluating the extent to which these factors have been minimized in an individual study, working groups consider a number of aspects of design and analysis as described in the report of the study. Most of these considerations apply equally to case–control, cohort and correlation studies. Lack of clarity of any of these aspects in the reporting of a study can decrease its credibility and the weight given to it in the final evaluation of the exposure.

Firstly, the study population, disease (or diseases) and exposure should have been well defined by the authors. Cases of disease in the study population should have been identified in a way that was independent of the exposure of interest, and exposure should have been assessed in a way that was not related to disease status.

Secondly, the authors should have taken account in the study design and analysis of other variables that can influence the risk of disease and may have been related to the exposure of interest. Potential confounding by such variables should have been dealt with either in the design of the study, such as by matching, or in the analysis, by statistical adjustment. In cohort studies, comparisons with local rates of disease may be more appropriate than those with national rates. Internal comparisons of disease frequency among individuals at different levels of exposure should also have been made in the study.

Thirdly, the authors should have reported the basic data on which the conclusions are founded, even if sophisticated statistical analyses were employed. At the very least, they should have given the numbers of exposed and unexposed cases and controls in a case–control study and the numbers of cases observed and expected in a cohort study. Further tabulations by time since exposure began and other temporal factors are also important. In a cohort study, data on all cancer sites and all causes of death should have been given, to reveal the possibility of reporting bias. In a case–control study, the effects of investigated factors other than the exposure of interest should have been reported.

Finally, the statistical methods used to obtain estimates of relative risk, absolute rates of cancer, confidence intervals and significance tests, and to adjust for confounding should have been clearly stated by the authors. The methods used should preferably have been the generally accepted techniques that have been refined since the mid-1970s. These methods have been reviewed for case–control studies (Breslow & Day, 1980) and for cohort studies (Breslow & Day, 1987).

(c) Inferences about mechanism of action

Detailed analyses of both relative and absolute risks in relation to temporal variables, such as age at first exposure, time since first exposure, duration of exposure, cumulative exposure and time since exposure ceased, are reviewed and summarized when available. The analysis of temporal relationships can be useful in formulating models of carcinogenesis. In particular, such analyses may suggest whether a carcinogen acts early or late in the process of carcinogenesis, although at best they allow only indirect inferences about the mechanism of action. Special attention is given to measurements of biological markers of carcinogen exposure or action, such as DNA or protein adducts, as well as markers of early steps in the carcinogenic process, such as proto-oncogene mutation, when these are incorporated into epidemiological studies focused on cancer incidence or mortality. Such measurements may allow inferences to be made about putative mechanisms of action (IARC, 1991a; Vainio et al., 1992).

(d) Criteria for causality

After the quality of individual epidemiological studies of cancer has been summarized and assessed, a judgement is made concerning the strength of evidence that the agent, mixture or exposure circumstance in question is carcinogenic for humans. In making their judgement, the Working Group considers several criteria for causality. A strong association (i.e. a large relative risk) is more likely to indicate causality than a weak association, although it is recognized that relative risks of small magnitude do not imply lack of causality and may be important if the disease is common. Associations that are replicated in several studies of the same design or using different epidemiological approaches or under different circumstances of exposure are more likely to represent a causal relationship than isolated observations from single studies. If there are inconsistent results among investigations, possible reasons are sought (such as differences in amount of exposure), and results of studies judged to be of high quality are given more weight than those of studies judged to be methodologically less sound. When suspicion of carcinogenicity arises largely from a single study, these data are not combined with those from later studies in any subsequent reassessment of the strength of the evidence.

If the risk of the disease in question increases with the amount of exposure, this is considered to be a strong indication of causality, although absence of a graded response is not necessarily evidence against a causal relationship. Demonstration of a decline in risk after cessation of or reduction in exposure in individuals or in whole populations also supports a causal interpretation of the findings.

Although a carcinogen may act upon more than one target, the specificity of an association (i.e. an increased occurrence of cancer at one anatomical site or of one morphological type) adds plausibility to a causal relationship, particularly when excess cancer occurrence is limited to one morphological type within the same organ.

Although rarely available, results from randomized trials showing different rates among exposed and unexposed individuals provide particularly strong evidence for causality.

When several epidemiological studies show little or no indication of an association between an exposure and cancer, the judgement may be made that, in the aggregate, they show evidence of lack of carcinogenicity. Such a judgement requires first of all that the studies giving rise to it meet, to a sufficient degree, the standards of design and analysis described above. Specifically, the possibility that bias, confounding or misclassification of exposure or outcome could explain the observed results should be considered and excluded with reasonable certainty. In addition, all studies that are judged to be methodologically sound should be consistent with a relative risk of unity for any observed level of exposure and, when considered together, should provide a pooled estimate of relative risk which is at or near unity and has a narrow confidence interval, due to sufficient population size. Moreover, no individual study nor the pooled results of all the studies should show any consistent tendency for relative risk of cancer to increase with increasing level of exposure. It is important to note that evidence of lack of carcinogenicity obtained in this way from several epidemiological studies can apply only to the type(s) of cancer studied and to dose levels and intervals between first exposure and observation of disease that are the same as or less than those observed in all the studies. Experience with human cancer indicates that, in some cases, the period from first exposure to the development of clinical cancer is seldom less than 20 years; latent periods substantially shorter than 30 years cannot provide evidence for lack of carcinogenicity.

9. STUDIES OF CANCER IN EXPERIMENTAL ANIMALS

All known human carcinogens that have been studied adequately in experimental animals have produced positive results in one or more animal species (Wilbourn *et al.*, 1986; Tomatis *et al.*, 1989). For several agents (aflatoxins, 4-aminobiphenyl, azathioprine, betel quid with tobacco, BCME and CMME (technical grade), chlorambucil, chlornaphazine, ciclosporin, coal-tar pitches, coal-tars, combined oral contraceptives, cyclophosphamide, diethylstilboestrol, melphalan, 8-methoxypsoralen plus UVA, mustard gas, myleran, 2-naphthylamine, nonsteroidal oestrogens, oestrogen replacement therapy/steroidal oestrogens, solar radiation, thiotepa and vinyl chloride), carcinogenicity in experimental animals was established or highly suspected before epidemiological studies confirmed the carcinogenicity in humans (Vainio *et al.*, 1995). Although this association cannot establish that all agents and mixtures that cause cancer in experimental animals also cause cancer in humans, nevertheless, **in the absence of adequate data on humans, it is biologically plausible and prudent to regard agents and mixtures for which there is sufficient evidence (see p. 24) of carcinogenicity in experimental animals as if they presented a carcinogenic risk to humans**. The possibility that a given agent may cause cancer through a species-specific mechanism which does not operate in humans (see p. 26) should also be taken into consideration.

The nature and extent of impurities or contaminants present in the chemical or mixture being evaluated are given when available. Animal strain, sex, numbers per group, age at start of treatment and survival are reported.

Other types of studies summarized include: experiments in which the agent or mixture was administered in conjunction with known carcinogens or factors that modify carcinogenic effects; studies in which the end-point was not cancer but a defined precancerous lesion; and experiments on the carcinogenicity of known metabolites and derivatives.

For experimental studies of mixtures, consideration is given to the possibility of changes in the physicochemical properties of the test substance during collection, storage, extraction, concentration and delivery. Chemical and toxicological interactions of the components of mixtures may result in nonlinear dose–response relationships.

An assessment is made as to the relevance to human exposure of samples tested in experimental animals, which may involve consideration of: (i) physical and chemical characteristics, (ii) constituent substances that indicate the presence of a class of substances, (iii) the results of tests for genetic and related effects, including genetic activity profiles, DNA adduct profiles, proto-oncogene mutation and expression and suppressor gene inactivation. The relevance of results obtained, for example, with animal viruses analogous to the virus being evaluated in the monograph must also be considered. They may provide biological and mechanistic information relevant to the understanding of the process of carcinogenesis in humans and may strengthen the plausibility of a conclusion that the biological agent that is being evaluated is carcinogenic in humans.

(a) Qualitative aspects

An assessment of carcinogenicity involves several considerations of qualitative importance, including (i) the experimental conditions under which the test was performed, including route and schedule of exposure, species, strain, sex, age, duration of follow-up; (ii) the consistency of the results, for example, across species and target organ(s); (iii) the spectrum of neoplastic response, from preneoplastic lesions and benign tumours to malignant neoplasms; and (iv) the possible role of modifying factors.

As mentioned earlier (p. 11), the *Monographs* are not intended to summarize all published studies. Those studies in experimental animals that are inadequate (e.g. too short a duration, too few animals, poor survival; see below) or are judged irrelevant to the evaluation are generally omitted. Guidelines for conducting adequate long-term carcinogenicity experiments have been outlined (e.g. Montesano *et al.*, 1986).

Considerations of importance to the Working Group in the interpretation and evaluation of a particular study include: (i) how clearly the agent was defined and, in the case of mixtures, how adequately the sample characterization was reported; (ii) whether the dose was adequately monitored, particularly in inhalation experiments; (iii) whether the doses and duration of treatment were appropriate and whether the survival of treated animals was similar to that of controls; (iv) whether there were adequate numbers of animals per group; (v) whether animals of both sexes were used; (vi) whether animals were allocated randomly to groups; (vii) whether the duration of observation was adequate; and (viii) whether the data were adequately reported. If available, recent data on the incidence of specific tumours in historical controls, as well as in concurrent controls, should be taken into account in the evaluation of tumour response.

When benign tumours occur together with and originate from the same cell type in an organ or tissue as malignant tumours in a particular study and appear to represent a stage in the progression to malignancy, it may be valid to combine them in assessing tumour incidence (Huff *et al.*, 1989). The occurrence of lesions presumed to be preneoplastic may in certain instances aid in assessing the biological plausibility of any neoplastic response observed. If an agent or mixture induces only benign neoplasms that appear to be end-points that do not readily undergo transition to malignancy, it should nevertheless be suspected of being a carcinogen and requires further investigation.

(b) Quantitative aspects

The probability that tumours will occur may depend on the species, sex, strain and age of the animal, the dose of the carcinogen and the route and length of exposure. Evidence of an increased incidence of neoplasms with increased level of exposure strengthens the inference of a causal association between the exposure and the development of neoplasms.

The form of the dose–response relationship can vary widely, depending on the particular agent under study and the target organ. Both DNA damage and increased cell division are important aspects of carcinogenesis, and cell proliferation is a strong determinant of dose–

response relationships for some carcinogens (Cohen & Ellwein, 1990). Since many chemicals require metabolic activation before being converted into their reactive intermediates, both metabolic and pharmacokinetic aspects are important in determining the dose–response pattern. Saturation of steps such as absorption, activation, inactivation and elimination may produce nonlinearity in the dose–response relationship, as could saturation of processes such as DNA repair (Hoel *et al.*, 1983; Gart *et al.*, 1986).

(c) Statistical analysis of long-term experiments in animals

Factors considered by the Working Group include the adequacy of the information given for each treatment group: (i) the number of animals studied and the number examined histologically, (ii) the number of animals with a given tumour type and (iii) length of survival. The statistical methods used should be clearly stated and should be the generally accepted techniques refined for this purpose (Peto *et al.*, 1980; Gart *et al.*, 1986). When there is no difference in survival between control and treatment groups, the Working Group usually compares the proportions of animals developing each tumour type in each of the groups. Otherwise, consideration is given as to whether or not appropriate adjustments have been made for differences in survival. These adjustments can include: comparisons of the proportions of tumour-bearing animals among the effective number of animals (alive at the time the first tumour is discovered), in the case where most differences in survival occur before tumours appear; life-table methods, when tumours are visible or when they may be considered 'fatal' because mortality rapidly follows tumour development; and the Mantel-Haenszel test or logistic regression, when occult tumours do not affect the animals' risk of dying but are 'incidental' findings at autopsy.

In practice, classifying tumours as fatal or incidental may be difficult. Several survival-adjusted methods have been developed that do not require this distinction (Gart *et al.*, 1986), although they have not been fully evaluated.

10. OTHER DATA RELEVANT TO AN EVALUATION OF CARCINOGENICITY AND ITS MECHANISMS

In coming to an overall evaluation of carcinogenicity in humans (see p. 25), the Working Group also considers related data. The nature of the information selected for the summary depends on the agent being considered.

For chemicals and complex mixtures of chemicals such as those in some occupational situations and involving cultural habits (e.g. tobacco smoking), the other data considered to be relevant are divided into those on absorption, distribution, metabolism and excretion; toxic effects; reproductive and developmental effects; and genetic and related effects.

Concise information is given on absorption, distribution (including placental transfer) and excretion in both humans and experimental animals. Kinetic factors that may affect the dose–response relationship, such as saturation of uptake, protein binding, metabolic activation, detoxification and DNA repair processes, are mentioned. Studies that indicate the metabolic fate of the agent in humans and in experimental animals are summarized briefly, and comparisons of data from humans and animals are made when possible. Comparative information on the relationship between exposure and the dose that reaches the target site may be of particular importance for extrapolation between species. Data are given on acute and chronic toxic effects (other than cancer), such as organ toxicity, increased cell proliferation, immunotoxicity and endocrine effects. The presence and toxicological significance of cellular receptors is described. Effects on reproduction, teratogenicity, fetotoxicity and embryotoxicity are also summarized briefly.

Tests of genetic and related effects are described in view of the relevance of gene mutation and chromosomal damage to carcinogenesis (Vainio *et al.*, 1992). The adequacy of the reporting of sample characterization is considered and, where necessary, commented upon; with regard to complex mixtures, such comments are similar to those described for animal carcinogenicity tests on p. 17. The available data are interpreted critically by phylogenetic group according to the end-points detected, which may include DNA damage, gene mutation, sister chromatid exchange, micronucleus formation, chromosomal aberrations, aneuploidy and cell transformation. The concentrations employed are given, and mention is made of whether use of an exogenous metabolic system *in vitro* affected the test result. These data are given as listings of test systems, data and references; bar graphs (activity profiles) and corresponding summary tables with detailed information on the preparation of the profiles (Waters *et al.*, 1987) are given in appendices.

Positive results in tests using prokaryotes, lower eukaryotes, plants, insects and cultured mammalian cells suggest that genetic and related effects could occur in mammals. Results from such tests may also give information about the types of genetic effect produced and about the involvement of metabolic activation. Some end-points described are clearly genetic in nature (e.g. gene mutations and chromosomal aberrations), while others are to a greater or lesser degree associated with genetic effects (e.g. unscheduled DNA synthesis). In-vitro tests for tumour-promoting activity and for cell transformation may be sensitive to changes that are not necessarily the result of genetic alterations but that may have specific relevance to the process of carcinogenesis. A critical appraisal of these tests has been published (Montesano *et al.*, 1986).

Genetic or other activity manifest in experimental mammals and humans is regarded as being of greater relevance than that in other organisms. The demonstration that an agent or mixture can induce gene and chromosomal mutations in whole mammals indicates that it may have carcinogenic activity, although this activity may not be detectably expressed in any or all species. Relative potency in tests for mutagenicity and related effects is not a reliable indicator of carcinogenic potency. Negative results in tests for mutagenicity in selected tissues from animals treated *in vivo* provide less weight, partly because they do not exclude the possibility of an effect in tissues other than those examined. Moreover, negative results in short-term tests with genetic end-points cannot be considered to provide evidence to rule out carcinogenicity of agents or mixtures that act through other mechanisms (e.g. receptor-mediated effects, cellular toxicity with regenerative proliferation, peroxisome proliferation) (Vainio *et al.*, 1992). Factors that may lead to misleading results in short-term tests have been discussed in detail elsewhere (Montesano *et al.*, 1986).

When available, data relevant to mechanisms of carcinogenesis that do not involve structural changes at the level of the gene are also described.

The adequacy of epidemiological studies of reproductive outcome and genetic and related effects in humans is evaluated by the same criteria as are applied to epidemiological studies of cancer.

Structure–activity relationships that may be relevant to an evaluation of the carcinogenicity of an agent are also described.

For biological agents—viruses, bacteria and parasites—other data relevant to carcinogenicity include descriptions of the pathology of infection, molecular biology (integration and expression of viruses, and any genetic alterations seen in human tumours) and other observations, which might include cellular and tissue responses to infection, immune response and the presence of tumour markers.

11. SUMMARY OF DATA REPORTED

In this section, the relevant epidemiological and experimental data are summarized. Only reports, other than in abstract form, that meet the criteria outlined on p. 11 are considered for evaluating carcinogenicity. Inadequate studies are generally not summarized: such studies are usually identified by a square-bracketed comment in the preceding text.

(a) *Exposures*

Human exposure to chemicals and complex mixtures is summarized on the basis of elements such as production, use, occurrence in the environment and determinations in human tissues and body fluids. Quantitative data are given when available. Exposure to biological agents is described in terms of transmission, and prevalence of infection.

(b) *Carcinogenicity in humans*

Results of epidemiological studies that are considered to be pertinent to an assessment of human carcinogenicity are summarized. When relevant, case reports and correlation studies are also summarized.

(c) *Carcinogenicity in experimental animals*

Data relevant to an evaluation of carcinogenicity in animals are summarized. For each animal species and route of administration, it is stated whether an increased incidence of neoplasms or preneoplastic lesions was observed, and the tumour sites are indicated. If the agent or mixture produced tumours after prenatal exposure or in single-dose experiments, this is also indicated. Negative findings are also summarized. Dose–response and other quantitative data may be given when available.

(d) *Other data relevant to an evaluation of carcinogenicity and its mechanisms*

Data on biological effects in humans that are of particular relevance are summarized. These may include toxicological, kinetic and metabolic considerations and evidence of DNA binding, persistence of DNA lesions or genetic damage in exposed humans. Toxicological information, such as that on cytotoxicity and regeneration, receptor binding and hormonal and immunological effects, and data on kinetics and metabolism in experimental animals are given when considered relevant to the possible mechanism of the carcinogenic action of the agent. The results of tests for genetic and related effects are summarized for whole mammals, cultured mammalian cells and nonmammalian systems.

When available, comparisons of such data for humans and for animals, and particularly animals that have developed cancer, are described.

Structure–activity relationships are mentioned when relevant.

For the agent, mixture or exposure circumstance being evaluated, the available data on endpoints or other phenomena relevant to mechanisms of carcinogenesis from studies in humans, experimental animals and tissue and cell test systems are summarized within one or more of the following descriptive dimensions:

(i) Evidence of genotoxicity (i.e. structural changes at the level of the gene): for example, structure–activity considerations, adduct formation, mutagenicity (effect on specific genes), chromosomal mutation/aneuploidy

(ii) Evidence of effects on the expression of relevant genes (i.e. functional changes at the intracellular level): for example, alterations to the structure or quantity of the product of a proto-oncogene or tumour suppressor gene, alterations to metabolic activation/inactivation/DNA repair

(iii) Evidence of relevant effects on cell behaviour (i.e. morphological or behavioural changes at the cellular or tissue level): for example, induction of mitogenesis, compensatory cell proliferation, preneoplasia and hyperplasia, survival of premalignant or malignant cells (immortalization, immunosuppression), effects on metastatic potential

(iv) Evidence from dose and time relationships of carcinogenic effects and interactions between agents: for example, early/late stage, as inferred from epidemiological studies; initiation/promotion/progression/malignant conversion, as defined in animal carcinogenicity experiments; toxicokinetics

These dimensions are not mutually exclusive, and an agent may fall within more than one of them. Thus, for example, the action of an agent on the expression of relevant genes could be summarized under both the first and second dimension, even if it were known with reasonable certainty that those effects resulted from genotoxicity.

12. EVALUATION

Evaluations of the strength of the evidence for carcinogenicity arising from human and experimental animal data are made, using standard terms.

It is recognized that the criteria for these evaluations, described below, cannot encompass all of the factors that may be relevant to an evaluation of carcinogenicity. In considering all of the relevant scientific data, the Working Group may assign the agent, mixture or exposure circumstance to a higher or lower category than a strict interpretation of these criteria would indicate.

(a) *Degrees of evidence for carcinogenicity in humans and in experimental animals and supporting evidence*

These categories refer only to the strength of the evidence that an exposure is carcinogenic and not to the extent of its carcinogenic activity (potency) nor to the mechanisms involved. A classification may change as new information becomes available.

An evaluation of degree of evidence, whether for a single agent or a mixture, is limited to the materials tested, as defined physically, chemically or biologically. When the agents evaluated are considered by the Working Group to be sufficiently closely related, they may be grouped together for the purpose of a single evaluation of degree of evidence.

(i) *Carcinogenicity in humans*

The applicability of an evaluation of the carcinogenicity of a mixture, process, occupation or industry on the basis of evidence from epidemiological studies depends on the variability over time and place of the mixtures, processes, occupations and industries. The Working Group seeks to identify the specific exposure, process or activity which is considered most likely to be responsible for any excess risk. The evaluation is focused as narrowly as the available data on exposure and other aspects permit.

The evidence relevant to carcinogenicity from studies in humans is classified into one of the following categories:

Sufficient evidence of carcinogenicity: The Working Group considers that a causal relationship has been established between exposure to the agent, mixture or exposure circumstance and human cancer. That is, a positive relationship has been observed between the exposure and cancer in studies in which chance, bias and confounding could be ruled out with reasonable confidence.

Limited evidence of carcinogenicity: A positive association has been observed between exposure to the agent, mixture or exposure circumstance and cancer for which a causal interpretation is considered by the Working Group to be credible, but chance, bias or confounding could not be ruled out with reasonable confidence.

Inadequate evidence of carcinogenicity: The available studies are of insufficient quality, consistency or statistical power to permit a conclusion regarding the presence or absence of a causal association, or no data on cancer in humans are available.

Evidence suggesting lack of carcinogenicity: There are several adequate studies covering the full range of levels of exposure that human beings are known to encounter, which are mutually consistent in not showing a positive association between exposure to the agent, mixture or exposure circumstance and any studied cancer at any observed level of exposure. A conclusion of 'evidence suggesting lack of carcinogenicity' is inevitably limited to the cancer sites, conditions and levels of exposure and length of observation covered by the available studies. In addition, the possibility of a very small risk at the levels of exposure studied can never be excluded.

In some instances, the above categories may be used to classify the degree of evidence related to carcinogenicity in specific organs or tissues.

(ii) *Carcinogenicity in experimental animals*

The evidence relevant to carcinogenicity in experimental animals is classified into one of the following categories:

Sufficient evidence of carcinogenicity: The Working Group considers that a causal relationship has been established between the agent or mixture and an increased incidence of malignant neoplasms or of an appropriate combination of benign and malignant neoplasms in (a) two or more species of animals or (b) in two or more independent studies in one species carried out at different times or in different laboratories or under different protocols.

Exceptionally, a single study in one species might be considered to provide sufficient evidence of carcinogenicity when malignant neoplasms occur to an unusual degree with regard to incidence, site, type of tumour or age at onset.

Limited evidence of carcinogenicity: The data suggest a carcinogenic effect but are limited for making a definitive evaluation because, e.g. (a) the evidence of carcinogenicity is restricted to a single experiment; or (b) there are unresolved questions regarding the adequacy of the design, conduct or interpretation of the study; or (c) the agent or mixture increases the incidence only of benign neoplasms or lesions of uncertain neoplastic potential, or of certain neoplasms which may occur spontaneously in high incidences in certain strains.

Inadequate evidence of carcinogenicity: The studies cannot be interpreted as showing either the presence or absence of a carcinogenic effect because of major qualitative or quantitative limitations, or no data on cancer in experimental animals are available.

Evidence suggesting lack of carcinogenicity: Adequate studies involving at least two species are available which show that, within the limits of the tests used, the agent or mixture is not carcinogenic. A conclusion of evidence suggesting lack of carcinogenicity is inevitably limited to the species, tumour sites and levels of exposure studied.

(b) *Other data relevant to the evaluation of carcinogenicity and its mechanisms*

Other evidence judged to be relevant to an evaluation of carcinogenicity and of sufficient importance to affect the overall evaluation is then described. This may include data on preneoplastic lesions, tumour pathology, genetic and related effects, structure–activity relationships, metabolism and pharmacokinetics, physicochemical parameters and analogous biological agents.

Data relevant to mechanisms of the carcinogenic action are also evaluated. The strength of the evidence that any carcinogenic effect observed is due to a particular mechanism is assessed, using terms such as weak, moderate or strong. Then, the Working Group assesses if that particular mechanism is likely to be operative in humans. The strongest indications that a particular mechanism operates in humans come from data on humans or biological specimens obtained from exposed humans. The data may be considered to be especially relevant if they show that the agent in question has caused changes in exposed humans that are on the causal pathway to carcinogenesis. Such data may, however, never become available, because it is at least conceivable that certain compounds may be kept from human use solely on the basis of evidence of their toxicity and/or carcinogenicity in experimental systems.

For complex exposures, including occupational and industrial exposures, the chemical composition and the potential contribution of carcinogens known to be present are considered by the Working Group in its overall evaluation of human carcinogenicity. The Working Group also determines the extent to which the materials tested in experimental systems are related to those to which humans are exposed.

(c) Overall evaluation

Finally, the body of evidence is considered as a whole, in order to reach an overall evaluation of the carcinogenicity to humans of an agent, mixture or circumstance of exposure.

An evaluation may be made for a group of chemical compounds that have been evaluated by the Working Group. In addition, when supporting data indicate that other, related compounds for which there is no direct evidence of capacity to induce cancer in humans or in animals may also be carcinogenic, a statement describing the rationale for this conclusion is added to the evaluation narrative; an additional evaluation may be made for this broader group of compounds if the strength of the evidence warrants it.

The agent, mixture or exposure circumstance is described according to the wording of one of the following categories, and the designated group is given. The categorization of an agent, mixture or exposure circumstance is a matter of scientific judgement, reflecting the strength of the evidence derived from studies in humans and in experimental animals and from other relevant data.

Group 1—The agent (mixture) is carcinogenic to humans.
The exposure circumstance entails exposures that are carcinogenic to humans.

This category is used when there is *sufficient evidence* of carcinogenicity in humans. Exceptionally, an agent (mixture) may be placed in this category when evidence in humans is less than sufficient but there is *sufficient evidence* of carcinogenicity in experimental animals and strong evidence in exposed humans that the agent (mixture) acts through a relevant mechanism of carcinogenicity.

Group 2

This category includes agents, mixtures and exposure circumstances for which, at one extreme, the degree of evidence of carcinogenicity in humans is almost sufficient, as well as those for which, at the other extreme, there are no human data but for which there is evidence of carcinogenicity in experimental animals. Agents, mixtures and exposure circumstances are assigned to either group 2A (probably carcinogenic to humans) or group 2B (possibly carcinogenic to humans) on the basis of epidemiological and experimental evidence of carcinogenicity and other relevant data.

Group 2A—The agent (mixture) is probably carcinogenic to humans.
The exposure circumstance entails exposures that are probably carcinogenic to humans.

This category is used when there is *limited evidence* of carcinogenicity in humans and sufficient evidence of carcinogenicity in experimental animals. In some cases, an agent (mixture) may be classified in this category when there is inadequate evidence of carcinogenicity in humans and *sufficient evidence* of carcinogenicity in experimental animals and strong evidence that the carcinogenesis is mediated by a mechanism that also operates in humans. Exceptionally, an agent, mixture or exposure circumstance may be classified in this category solely on the basis of limited evidence of carcinogenicity in humans.

Group 2B—The agent (mixture) is possibly carcinogenic to humans.
The exposure circumstance entails exposures that are possibly carcinogenic to humans.

This category is used for agents, mixtures and exposure circumstances for which there is *limited evidence* of carcinogenicity in humans and less than *sufficient evidence* of carcinogenicity in experimental animals. It may also be used when there is *inadequate evidence* of carcinogenicity in humans but there is *sufficient evidence* of carcinogenicity in experimental animals. In some instances, an agent, mixture or exposure circumstance for which there is *inadequate evidence* of carcinogenicity in humans but *limited evidence* of carcinogenicity in experimental animals together with supporting evidence from other relevant data may be placed in this group.

Group 3—The agent (mixture or exposure circumstance) is not classifiable as to its carcinogenicity to humans.

This category is used most commonly for agents, mixtures and exposure circumstances for which the evidence of carcinogenicity is inadequate in humans and inadequate or limited in experimental animals.

Exceptionally, agents (mixtures) for which the evidence of carcinogenicity is inadequate in humans but sufficient in experimental animals may be placed in this category when there is strong evidence that the mechanism of carcinogenicity in experimental animals does not operate in humans.

Agents, mixtures and exposure circumstances that do not fall into any other group are also placed in this category.

Group 4—The agent (mixture) is probably not carcinogenic to humans.

This category is used for agents or mixtures for which there is *evidence suggesting lack of carcinogenicity* in humans and in experimental animals. In some instances, agents or mixtures for which there is *inadequate evidence* of carcinogenicity in humans but *evidence suggesting lack of carcinogenicity* in experimental animals, consistently and strongly supported by a broad range of other relevant data, may be classified in this group.

References

Breslow, N.E. & Day, N.E. (1980) *Statistical Methods in Cancer Research*, Vol. 1, *The Analysis of Case–Control Studies* (IARC Scientific Publications No. 32), Lyon, IARC

Breslow, N.E. & Day, N.E. (1987) *Statistical Methods in Cancer Research*, Vol. 2, *The Design and Analysis of Cohort Studies* (IARC Scientific Publications No. 82), Lyon, IARC

Cohen, S.M. & Ellwein, L.B. (1990) Cell proliferation in carcinogenesis. *Science*, **249**, 1007–1011

Gart, J.J., Krewski, D., Lee, P.N., Tarone, R.E. & Wahrendorf, J. (1986) *Statistical Methods in Cancer Research*, Vol. 3, *The Design and Analysis of Long-term Animal Experiments* (IARC Scientific Publications No. 79), Lyon, IARC

Hoel, D.G., Kaplan, N.L. & Anderson, M.W. (1983) Implication of nonlinear kinetics on risk estimation in carcinogenesis. *Science*, **219**, 1032–1037

Huff, J.E., Eustis, S.L. & Haseman, J.K. (1989) Occurrence and relevance of chemically induced benign neoplasms in long-term carcinogenicity studies. *Cancer Metastasis Rev.*, **8**, 1–21

IARC (1973–1994) *Information Bulletin on the Survey of Chemicals Being Tested for Carcinogenicity/Directory of Agents Being Tested for Carcinogenicity*, Numbers 1–16, Lyon

    Number 1 (1973)   52 pages
    Number 2 (1973)   77 pages
    Number 3 (1974)   67 pages
    Number 4 (1974)   97 pages
    Number 5 (1975)   8 pages
    Number 6 (1976)   360 pages
    Number 7 (1978)   460 pages
    Number 8 (1979)   604 pages
    Number 9 (1981)   294 pages
    Number 10 (1983)   326 pages
    Number 11 (1984)   370 pages
    Number 12 (1986)   385 pages
    Number 13 (1988)   404 pages
    Number 14 (1990)   369 pages
    Number 15 (1992)   317 pages
    Number 16 (1994)   293 pages

IARC (1976–1994)

    Directory of On-going Research in Cancer Epidemiology 1976. Edited by C.S. Muir & G. Wagner, Lyon

    Directory of On-going Research in Cancer Epidemiology 1977 (IARC Scientific Publications No. 17). Edited by C.S. Muir & G. Wagner, Lyon

    Directory of On-going Research in Cancer Epidemiology 1978 (IARC Scientific Publications No. 26). Edited by C.S. Muir & G. Wagner, Lyon

    Directory of On-going Research in Cancer Epidemiology 1979 (IARC Scientific Publications No. 28). Edited by C.S. Muir & G. Wagner, Lyon

    Directory of On-going Research in Cancer Epidemiology 1980 (IARC Scientific Publications No. 35). Edited by C.S. Muir & G. Wagner, Lyon

    Directory of On-going Research in Cancer Epidemiology 1981 (IARC Scientific Publications No. 38). Edited by C.S. Muir & G. Wagner, Lyon

    Directory of On-going Research in Cancer Epidemiology 1982 (IARC Scientific Publications No. 46). Edited by C.S. Muir & G. Wagner, Lyon

    Directory of On-going Research in Cancer Epidemiology 1983 (IARC Scientific Publications No. 50). Edited by C.S. Muir & G. Wagner, Lyon

    Directory of On-going Research in Cancer Epidemiology 1984 (IARC Scientific Publications No. 62). Edited by C.S. Muir & G. Wagner, Lyon

    Directory of On-going Research in Cancer Epidemiology 1985 (IARC Scientific Publications No. 69). Edited by C.S. Muir & G. Wagner, Lyon

Directory of On-going Research in Cancer Epidemiology 1986 (IARC Scientific Publications No. 80). Edited by C.S. Muir & G. Wagner, Lyon

Directory of On-going Research in Cancer Epidemiology 1987 (IARC Scientific Publications No. 86). Edited by D.M. Parkin & J. Wahrendorf, Lyon

Directory of On-going Research in Cancer Epidemiology 1988 (IARC Scientific Publications No. 93). Edited by M. Coleman & J. Wahrendorf, Lyon

Directory of On-going Research in Cancer Epidemiology 1989/90 (IARC Scientific Publications No. 101). Edited by M. Coleman & J. Wahrendorf, Lyon

Directory of On-going Research in Cancer Epidemiology 1991 (IARC Scientific Publications No.110). Edited by M. Coleman & J. Wahrendorf, Lyon

Directory of On-going Research in Cancer Epidemiology 1992 (IARC Scientific Publications No. 117). Edited by M. Coleman, J. Wahrendorf & E. Démaret, Lyon

Directory of On-going Research in Cancer Epidemiology 1994 (IARC Scientific Publications No. 130). Edited by R. Sankaranarayanan, J. Wahrendorf & E. Démaret, Lyon

IARC (1977) *IARC Monographs Programme on the Evaluation of the Carcinogenic Risk of Chemicals to Humans*. Preamble (IARC intern. tech. Rep. No. 77/002), Lyon

IARC (1978) *Chemicals with Sufficient Evidence of Carcinogenicity in Experimental Animals*—IARC Monographs *Volumes 1–17* (IARC intern. tech. Rep. No. 78/003), Lyon

IARC (1978–1993) *Environmental Carcinogens. Methods of Analysis and Exposure Measurement*:

Vol. 1. *Analysis of Volatile Nitrosamines in Food* (IARC Scientific Publications No. 18). Edited by R. Preussmann, M. Castegnaro, E.A. Walker & A.E. Wasserman (1978)

Vol. 2. *Methods for the Measurement of Vinyl Chloride in Poly(vinyl chloride), Air, Water and Foodstuffs* (IARC Scientific Publications No. 22). Edited by D.C.M. Squirrell & W. Thain (1978)

Vol. 3. Analysis of Polycyclic Aromatic Hydrocarbons in Environmental Samples (IARC Scientific Publications No. 29). Edited by M. Castegnaro, P. Bogovski, H. Kunte & E.A. Walker (1979)

Vol. 4. *Some Aromatic Amines and Azo Dyes in the General and Industrial Environment* (IARC Scientific Publications No. 40). Edited by L. Fishbein, M. Castegnaro, I.K. O'Neill & H. Bartsch (1981)

Vol. 5. *Some Mycotoxins* (IARC Scientific Publications No. 44). Edited by L. Stoloff, M. Castegnaro, P. Scott, I.K. O'Neill & H. Bartsch (1983)

Vol. 6. *N-Nitroso Compounds* (IARC Scientific Publications No. 45). Edited by R. Preussmann. I.K. O'Neill, G. Eisenbrand, B. Spiegelhalder & H. Bartsch (1983)

Vol. 7. *Some Volatile Halogenated Hydrocarbons* (IARC Scientific Publications No. 68). Edited by L. Fishbein & I.K. O'Neill (1985)

Vol. 8. *Some Metals: As, Be, Cd, Cr, Ni, Pb, Se, Zn* (IARC Scientific Publications No. 71). Edited by I.K. O'Neill, P. Schuller & L. Fishbein (1986)

Vol. 9. *Passive Smoking* (IARC Scientific Publications No. 81). Edited by I.K. O'Neill, K.D. Brunnemann, B. Dodet & D. Hoffmann (1987)

Vol. 10. *Benzene and Alkylated Benzenes* (IARC Scientific Publications No. 85). Edited by L. Fishbein & I.K. O'Neill (1988)

Vol. 11. *Polychlorinated Dioxins and Dibenzofurans)* (IARC Scientific Publications No. 108). Edited by C. Rappe, H.R. Buser, B. Dodet & I.K. O'Neill (1991)

Vol. 12. *Indoor Air* (IARC Scientific Publications No. 109). Edited by B. Seifert, H. van de Wiel, B. Dodet & I.K. O'Neill (1993)

IARC (1979) *Criteria to Select Chemicals for* IARC Monographs (IARC intern. tech. Rep. No. 79/003), Lyon

IARC (1982) *IARC Monographs on the Evaluation of the Carcinogenic Risk of Chemicals to Humans, Supplement 4, Chemicals, Industrial Processes and Industries Associated with Cancer in Humans* (IARC Monographs, Volumes 1 to 29), Lyon

IARC (1983) *Approaches to Classifying Chemical Carcinogens According to Mechanism of Action* (IARC intern. tech. Rep. No. 83/001), Lyon

IARC (1984) *Chemicals and Exposures to Complex Mixtures Recommended for Evaluation in IARC Monographs and Chemicals and Complex Mixtures Recommended for Long-term Carcinogenicity Testing* (IARC intern. tech. Rep. No. 84/002), Lyon

IARC (1987a) *IARC Monographs on the Evaluation of Carcinogenic Risks to Humans, Supplement 6, Genetic and Related Effects: An Updating of Selected* IARC Monographs *from Volumes 1 to 42*, Lyon

IARC (1987b) *IARC Monographs on the Evaluation of Carcinogenic Risks to Humans, Supplement 7, Overall Evaluations of Carcinogenicity: An Updating of* IARC Monographs *Volumes 1 to 42*, Lyon

IARC (1988) *Report of an IARC Working Group to Review the Approaches and Processes Used to Evaluate the Carcinogenicity of Mixtures and Groups of Chemicals* (IARC intern. tech. Rep.No. 88/002), Lyon

IARC (1989) *Chemicals, Groups of Chemicals, Mixtures and Exposure Circumstances to be Evaluated in Future IARC Monographs, Report of an ad hoc Working Group* (IARC intern. tech. Rep. No. 89/004), Lyon

IARC (1991a) *A Consensus Report of an IARC Monographs Working Group on the Use of Mechanisms of Carcinogenesis in Risk Identification* (IARC intern. tech. Rep. No. 91/002), Lyon

IARC (1991b) *Report of an Ad-hoc* IARC Monographs *Advisory Group on Viruses and Other Biological Agents Such as Parasites* (IARC intern. tech. Rep. No. 91/001), Lyon

IARC (1993) *Chemicals, Groups of Chemicals, Complex Mixtures, Physical and Biological Agents and Exposure Circumstances to be Evaluated in Future* IARC Monographs, *Report of an ad-hoc Working Group* (IARC intern. Rep. No. 93/005), Lyon

Montesano, R., Bartsch, H., Vainio, H., Wilbourn, J. & Yamasaki, H., eds (1986) *Long-term and Short-term Assays for Carcinogenesis—A Critical Appraisal* (IARC Scientific Publications No. 83), Lyon, IARC

Peto, R., Pike, M.C., Day, N.E., Gray, R.G., Lee, P.N., Parish, S., Peto, J., Richards, S. & Wahrendorf, J. (1980) Guidelines for simple, sensitive significance tests for carcinogenic effects in long-term animal experiments. In: *IARC Monographs on the Evaluation of the Carcinogenic Risk of Chemicals to Humans*, Supplement 2, *Long-term and Short-term Screening Assays for Carcinogens: A Critical Appraisal*, Lyon, pp. 311–426

Tomatis, L., Aitio, A., Wilbourn, J. & Shuker, L. (1989) Human carcinogens so far identified. *Jpn. J. Cancer Res.*, **80**, 795–807

Vainio, H., Magee, P., McGregor, D. & McMichael, A., eds (1992) *Mechanisms of Carcinogenesis in Risk Identification* (IARC Scientific Publications No. 116), Lyon, IARC

Vainio, H., Wilbourn, J. & Tomatis, L. (1995) Identification of environmental carcinogens: the first step in risk assessment. In: Mehlman, M.A. & Upton, A., eds, *The Identification and Control of Environmental and Occupational Diseases*, Princeton, Princeton Scientific Publishing Company, pp. 1–19

Waters, M.D., Stack, H.F., Brady, A.L., Lohman, P.H.M., Haroun, L. & Vainio, H. (1987) Appendix 1. Activity profiles for genetic and related tests. In: *IARC Monographs on the Evaluation of Carcinogenic Risks to Humans*, Suppl. 6, *Genetic and Related Effects: An Updating of Selected IARC Monographs from Volumes 1 to 42*, Lyon, IARC, pp. 687–696

Wilbourn, J., Haroun, L., Heseltine, E., Kaldor, J., Partensky, C. & Vainio, H. (1986) Response of experimental animals to human carcinogens: an analysis based upon the IARC Monographs Programme. *Carcinogenesis*, **7**, 1853–1863

THE MONOGRAPHS

DRY CLEANING

1. Exposure Data

1.1 Historical overview

Dry cleaning is believed to have originated in France in 1825: The French Federation of Dyeing and Cleaning gave an account of a servant in the household of Jean-Baptiste Jolly in Paris in 1825 who spilled the contents of a lamp on a soiled tablecloth. When the tablecloth dried, the spots had disappeared. The liquid in the lamp was probably camphene, produced from pinene, which is the main ingredient of turpentine. Camphene was initially used as a spotting agent, but it was later used for complete cleaning by hand. Use of volatile organic liquids to clean cloths became known as 'French cleaning' in Scotland and 'chemical cleaning' in Germany (Michelsen, 1957). Eventually, such processes were called 'dry cleaning' because they do not involve the use of water.

The common cleaning practice of early dyers and cleaners was to scour, using soap and water, and to add various acids and alkalis to the rinse. In the 1840s, garments were sometimes sponged inside and out with water and ox gall; dark clothes were sponged with water and logwood. A formulation for removing grease spots comprised fuller's earth, French chalk, yellow soap, pearlash and turpentine. When white fabrics became too soiled to be cleaned, they were dyed a darker colour. The early dry cleaning operation consisted of four steps: dusting (with rattan whips and brushes), hand brushing, solvent washing and drying (Michelsen, 1957).

Camphene was later replaced by benzene soap (1 lb soap for 2 gallons of benzene [60 g/L]), benzene (see IARC, 1987a), benzine (light petroleum naphtha), benzoline (a low-boiling subfraction of benzine), naphtha (see IARC, 1989a) and gasoline (see IARC, 1989b). The agents used early on for removing persistent spots on fabric (spotting agents) were benzine, turpentine, water, alcohol, cream of tartar, diethyl ether, chloroform (see IARC, 1987b), ox gall, egg yolks, boiling milk, honey, fuller's earth and spirits of wine. Attempts were made to provide mechanical agitation by adding sawdust, crushed marble and bran; however, use of these substances was soon discontinued because they were difficult to remove from the folds of the fabric (Michelsen, 1957).

The main hazard associated with the increasing use of volatile organic solvents in dry cleaning was fire. In 1928, W.J. Stoddard, president of the United States National Institute of Drycleaning (now the International Fabricare Institute) introduced Stoddard solvent, a petroleum-based solvent (see IARC, 1989c; Reich & Cormany, 1979) that was less inflammable and odour-free, which rapidly gained acceptance in the industry. Around 1917, many companies used 'cleaners' naphtha', which was similar to the light naphtha used in the paint industry. Subsequently, other petroleum solvents were marketed, including mineral turpentine, mineral spirits and oleum spirits. Use of these materials in the war effort in the early 1940s greatly curtailed their availability for dry cleaning (Michelsen, 1957).

Carbon tetrachloride (see IARC, 1987c) was the first chlorinated hydrocarbon solvent used for dry cleaning. It was introduced in the United States of America and Europe as an alternative to the expensive petroleum-based solvents, and by 1940 it was estimated that 20 thousand tonnes were used annually for dry cleaning in the United States. Widespread use of carbon tetrachloride in this context was discontinued in the 1950s because of its toxicity and corrosiveness, and tetrachloroethylene (see monograph, this volume) became the most widely used solvent. Trichloroethylene (see monograph, this volume) was also used, but on only a limited basis because it caused bleeding of many acetate dyes (Reich & Cormany, 1979).

Figure 1 shows the evolution of solvent use in dry cleaning over time.

1.2 Description of the industry

Cleaning is the process of removing dirt from fabric through the use of aqueous and non-aqueous solvents. Aqueous solvents swell the hydrophilic textile fibres, so that the dimensions of the fibre change, which can cause wrinkles or shrinkage of the fabric. Nonaqueous dry cleaning solvents do not swell the fibres and can remove oily stains at low temperatures, whereas a high-temperature, colloidal suspension is needed in aqueous (laundering) processes (Reich & Cormany, 1979).

Dry cleaning is a batch process that involves washing fabrics in a solvent solution, extracting the solvent from the fabrics during a spin cycle and tumble-drying the damp fabrics with warm air. In many commercial facilities where tetrachloroethylene is used, washing and extraction are done in a single machine. The damp fabric is then transferred manually to a tumbler for drying. This is known as the 'transfer' process (Stricoff, 1983).

A more recent process, in which all three steps are carried out in one machine, is referred to as the 'dry-to-dry' process. After the wash cycle, the solvent is drained, and residual solvent is removed by centrifugal force. This method has eliminated the handling of solvent-laden fabrics. All coin-operated dry cleaning equipment is dry-to-dry, while industrial processes may be either transfer or dry-to-dry (Stricoff, 1983).

Dry cleaning involves three basic steps: washing, solvent extraction and drying. The common tetrachloroethylene-based process is described below (Reich & Cormany, 1979).

1.2.1 The wash cycle

Before the wash cycle, detergent, a small amount of water and solvent are introduced into the tank. Water and detergent are added because they increase the solubility of dirt. The optimal temperature of solvent for cleaning cloth is 75–80 °F [23–27 °C]. Clothes are normally inspected first for stains. If they are stained, they are 'spotted' with a chemical (see below); they are then weighed and placed in a drum. About 1 gallon [3.8 L] of tetrachloroethylene is added to the drum for every 2 pounds [0.91 kg] of clothing.

The wash cycle lasts about 5–10 min. While the clothes are being washed, the solvent is passed continuously through a series of wire-mesh strips coated with diatomaceous earth and

Fig. 1. Use of solvents in dry cleaning over time

Solvent	Time							
	Before 1920	1920	1930	1940[a]	1950	1960[b]	1970[c]	1980	1990[d]
Camphene								
Benzene								
Naphtha								
Benzene soap								
Gasoline								
Stoddard solvent								
140–F solvent[e]								
Carbon tetrachloride								
Trichloroethylene								
Tetrachloroethylene								
1,1,1-Trichloroethane								
Chlorofluorocarbons[f]								
Dyeing operations[g]								

From Michelsen, 1957; Blair et al., 1990; Linak et al., 1992. The pattern generally reflects current world use, with some exceptions:
- in Japan, 70% petroleum-based, 20% tetrachloroethylene, 5% chlorofluorocarbons, 5% 1,1,1-trichloroethane (Kubota, 1992)
- in Oklahoma (USA), 50% petroleum-based (Duh & Asal, 1984); trichloroethylene used for cleaning industrial cloths and chlorofluorocarbons for fur coats
- in Denmark, petroleum spirit was used after the Second World War until the 1950s, when tetrachloroethylene became the primary solvent, with some trichloroethylene (industrial clothing) and chlorofluorocarbons (fur)

[a] Use of chlorinated hydrocarbons greatly reduced during Second World War, with increased use of petroleum-based solvents (Michelsen, 1957)
[b] Introduction of synthetic textile fibres resulted in a reduction in the volume of clothes dry cleaned (Linak et al., 1992)
[c] Environmental and engineering controls introduced (Linak et al., 1992)
[d] Worldwide environmental concern and regulatory action on chlorinated hydrocarbons
[e] Petroleum distillate similar to Stoddard solvent, with a flash-point of 60 °C (Reich & Cormany, 1979)
[f] Phased out by 1996 under the Montreal Protocol on ozone-depleting substances (Rice & Weinberg, 1994)
[g] Dyeing performed in dry cleaning shops until the mid-1950s in the USA (Michelsen, 1957)

activated carbon or through disposable cartridge filters, in order to remove fugitive dyes and insoluble material. At the end of the wash cycle, the solvent is drained from the drum through a strainer that traps buttons, coins, pins and other items that could damage the pumps, and is then pumped to the still or solvent storage tank. After the solvent has been drained from the drum, some still remains in the clothes. It is estimated that 100 pounds [45 kg] of clothes will contain 11–12 gallons [42–45 L] of tetrachloroethylene at this stage. In order to remove the residual solvent, the drum spins at 350–450 revolutions per minute for 3–5 min. At the end of the extraction process, 2–3 gallons [7.6–11 L] of tetrachloroethylene still remain per 100 pounds [45 kg] of clothes.

The filter medium must be replaced periodically. When cartridge filters are used, they must be drained and discarded. When diatomaceous earth is used, the filter medium is removed from its carrier and either discarded or heated to recover solvent from the filter muck. Most dry cleaners also distil the solvent to clean it (Douglas *et al.*, 1993).

1.2.2 *The drying and reclaiming process*

Once the extraction step is completed, the clothes are dried. Air is blown across coils heated by steam or electrically and then into the drum, to evaporate the tetrachloroethylene that is retained in the tumbling clothes. The temperature of the air is 120–150 °F [49–65 °C]. During the drying cycle, the exhaust from the drum is directed through a lint trap to filter out fabric particles; the trap is usually cleaned twice daily. The exhaust air is directed to a water-cooled or refrigerated condenser, where the vapours are condensed. The drying cycle usually lasts 10–25 min, depending on the type of fabric. Activated carbon adsorbers may be used to recover tetrachloroethylene from washer and dryer exhaust lines (Douglas *et al.*, 1993).

The final step in the drying cycle is either aeration or cooling. In aeration, heated ambient air is blown through the drum to remove the residual tetrachloroethylene. In the cooling cycle, the air in the drum is cooled in a refrigerated condenser and recirculated to the drum. Cooler air retains less tetrachloroethylene than warm air and causes fewer wrinkles in fabrics.

1.2.3 *Finishing and pressing*

At the end of the drying cycle, the clothes are removed from the dryer, examined for any remaining stains and spot cleaned. They are then pressed with machines that use steam and physical pressure to remove wrinkles. In the steam process, residual tetrachloroethylene may be volatilized. After final inspection, the garments are tagged and placed on racks.

Many dry cleaning operations provide other services, such as repair, dyeing and moth- and waterproofing. Optical brighteners, antistatic agents and sizing may be added to improve the brightness and the feel of garments.

1.2.4 *Spotting*

The removal of stains or spots from clothing before or after dry cleaning is both an art and a science. Historically, 'spotters' used recipes and methods that were carefully guarded secrets; today, companies provide proprietary products to the dry cleaning industry. 'Spotting' involves the selective application of chemicals, steam, detergent and/or water to loosen or remove specific stains from soiled garments. Depending on the size of the dry cleaning plant and the nature of the dry cleaning process, spotting can require a full time employee; however, this step is usually handled by the operator of a dry cleaning machine. Industrial dry cleaning plants seldom employ spotting operators.

The chemicals used typically include chlorinated solvents, amyl acetate, bleaching agents, acetic acid, aqueous ammonia, oxalic acid, hydrogen peroxide (see IARC, 1987d) and dilute hydrogen fluoride solutions. These chemicals are generally applied from plastic squeeze-bottles and are then either rubbed into the fabric with a brush, a spatula or by hand, allowed to soak into the fabric (which is subsequently handled) or flushed with steam from a steam gun. Thus, employees engaged in spotting may be exposed to toxic materials through both skin contact with liquids and inhalation of airborne vapours and mists (Stricoff, 1983).

1.3 Exposures in the workplace

1.3.1 Current practices

Today, commercial dry cleaners make up the largest segment of the industry (by number of establishments and quantity of garments cleaned). These firms handle primarily clothing and household products. Coin-operated dry cleaning machines are used directly by consumers for cleaning clothes and household furnishings. Industrial dry cleaners supply businesses, industrial plants and institutions with high-volume cleaning of uniforms, towels, linen and cleaning cloths, which are sometimes heavily soiled with industrial chemicals, oil and other materials.

In Europe, tetrachloroethylene comprises about 90% of the solvents used in dry cleaning; the main chlorofluorocarbon used is 1,1,2-trichloro-1,2,2-trifluoroethane (CFC-113) (Vieths *et al.*, 1988). There are an estimated 6500 dry cleaning establishments in the United Kingdom (Edmondson & Palin, 1993). In 1985, there were an estimated 2000 dry cleaners in Finland (Rantala *et al.*, 1992).

In 1970, there were 2886 laundries and dry cleaning shops in Denmark. White spirit (a petroleum-based solvent) was the most commonly used solvent in these establishments after the Second World War, but consumption decreased when automatic machines were introduced in the late 1950s. Tetrachloroethylene then became the most prevalent solvent, supplemented by trichloroethylene for the cleaning of work clothes and the fluorocarbons trichlorofluoromethane (CFC-11) and CFC-113 for cleaning fur coats (Lynge & Thygesen, 1990).

In Switzerland in 1991, 25% of dry cleaning shops used CFC-113 (Grob *et al.*, 1991). In France, 71 spotting agents used in dry cleaning shops were analysed. Benzene was not found; toluene (see IARC, 1989d) was found in 7% of the shops, xylenes (see IARC, 1989e) in 34%, methanol in 3% and chloroform in 1%. Chlorofluorocarbons were used in four of 30 shops studied and tetrachloroethylene in 26 (Davezies *et al.*, 1983).

In Japan, both dry cleaning and laundering are done in dry cleaning shops. Of the dry cleaning machines used, 70% (35 500) use petroleum-based solvents, 20% (10 700) use tetrachloroethylene, 5% (1900) use CFC-113 and 5% (1800) use 1,1,1-trichloroethane. In 1989, the industry consumed 7700 tonnes of petroleum solvent, 18 000 tonnes of tetrachloroethylene, 2700 tonnes of 1,1,1-trichloroethane and 4100 tonnes of CFC-113 (Kubota, 1992). In Nagoya, Japan, 50% of dry cleaning shops were reported to use petroleum solvents, 46% tetrachloroethylene, 4% CFC-113 and 2% a combination of tetrachloroethylene and petroleum solvents (Takeuchi *et al.*, 1981). Almost all of the machines involving use of petroleum-based solvents are transfer types, whereas in those that use the other three types of solvents the equipment is dry-to-dry. There are relatively few coin-operated machines in operation (Kubota, 1992).

By 1991, tetrachloroethylene was used in an estimated 28 121 dry cleaning plants in the United States, including 24 947 commercial plants and 130 industrial and 3044 coin-operated establishments (United States Environmental Protection Agency, 1991a). About two-thirds had dry-to-dry machines and one-third, transfer machines. Of commercial dry cleaning establishments in the United States in 1991, 82% used tetrachloroethylene, 14% used Stoddard solvent, 3% used CFC-113 and less than 1% used 1,1,1-trichloroethane (United States Environmental Protection Agency, 1991b; Earnest *et al.*, 1994). Of industrial cleaners, about 50% used tetrachloroethylene and 50% used Stoddard solvent. Coin-operated facilities used only tetrachloroethylene (Linak *et al.*, 1992). In Oklahoma, about 50% of dry cleaning involves use of Stoddard solvent (Duh & Asal, 1984). In Canada, 90% of commercial dry cleaning is based on tetrachloroethylene (Rice & Weinberg, 1994).

In December 1991, the United States Environmental Protection Agency (1991c) began regulating tetrachloroethylene as a hazardous air pollutant under Section 112 of the Clean Air Act. Regulatory action with regard to tetrachloroethylene is also being taken on an international level (Linak *et al.*, 1992). The international dry cleaning industry has responded with increased research into alternative solvents (Kirschner, 1994) and cleaning methods: a shift from transfer machines to closed-loop dry-to-dry machines and innovations in vapour recovery equipment and other devices to reduce occupational exposures and environmental emissions. Many of the exposure problems identified during studies in the late 1970s and early 1980s still exist, however, because transfer machines continue to be used, many of the controls that have been developed are prohibitively expensive and some work practices are inadequate.

The several distinct job titles and duties involved in dry cleaning are summarized in Table 1. It is estimated that several million people are employed in dry cleaning worldwide.

1.3.2 Sources of exposure

Occupational exposure to tetrachloroethylene occurs through inhalation, skin absorption and ingestion, although most occurs by inhalation whenever tetrachloroethylene vapours escape from a dry cleaning machine in the form of process or fugitive emissions.

Process emissions are those vented during steps such as aeration, transfer of clothing and distillation. Process emissions from vented machines can account for 50% or more of the total emitted (United States Environmental Protection Agency, 1991c). From the point of view of occupational health, process emissions are not a major concern as long as they are vented outside of the work environment; however, they may then become an environmental concern. Inadequate local exhaust or general ventilation can result in exposure to process emissions. In a study conducted in the early 1980s at 67 dry cleaning facilities in the United States, 28% had non-functioning local exhaust systems and 32% of the transfer machines had inadequate local exhaust ventilation, on the basis of face velocity at the washer door (Materna, 1985).

Table 1. Job titles and duties in the dry cleaning industry

Job title	Duties
Operator	Load and retrieve garments from dry cleaning machines, apply spotters (usually solvents or detergents) to stains; may also include routine maintenance tasks, such as cleaning lint and button traps or draining the still used to recover solvent
Presser	Operate pressing machine on garments that have been dry cleaned
Clerk	Counter duties (customer service), sort and bag dirty clothing, inspect finished clothing
Managerial	Administrative duties such as bookkeeping, purchasing, supervising
Seamstress	Mend and alter garments
Other	Includes disparate categories with small numbers of individuals, including driver, wet laundry worker, leather cleaner/dryer, maintenance worker

From Solet *et al.* (1990)

Fugitive emissions are those from leaks, rather than from venting as part of the process. Fugitive emissions can result from leaky seals and fittings, residual solvent in clothing, transfer of wet clothing, solvent evaporation in storage, improper operating practices, poor maintenance and housekeeping and system malfunctions. In one study conducted by the International Fabricare Institute, 25–30% of dry cleaning machines had visible solvent leaks; however, improvements in machine design have significantly reduced this problem (Earnest *et al.*, 1994). The United States Environmental Protection Agency estimated that about 87 000 tonnes (67%) of the 130 100 tonnes of tetrachloroethylene used in the United States in 1991 were lost through emissions to the atmosphere (Cantin, 1992).

Proper equipment and operating procedures are especially important around transfer machines, which are the most readily apparent source of exposure. Transfer involves holding solvent-saturated items directly in the breathing zone and in direct contact with the skin. Premature removal of wet garments from the drying cycle can be another cause of exposure.

Maintenance is an important part of any industrial process, and dry cleaning is no exception. A number of machine components must be maintained on a regular basis, daily, weekly, monthly or less often. Some maintenance procedures can result in high exposure to tetrachloroethylene; these include changing filters, cleaning stills and removing residue, cleaning lint and button traps and maintaining water separators.

Improper storage of hazardous waste, including still residues, dirty filters, contaminants in lint and button traps and water separator run-off, can increase background levels of tetrachloroethylene if these wastes are not stored in air-tight containers. When they are stored in a shop, they contribute to occupational exposure to background levels of solvent.

Excess tetrachloroethylene may be present in a cylinder if the extraction system fails, if the vapour recovery system fails or if improper operating procedures are followed. The balance between the water temperature of refrigerated condensers and the drying air temperature is crucial: If the air is not cool enough, the solvent will not be recovered; if it is too cold, the air will not dry the garments effectively. If the carbon beds in carbon adsorbers are not properly regenerated, solvent can be discharged through vents, which may be inside the shop (Stricoff, 1983).

1.3.3 Measurements of exposure

Occupational exposure to tetrachloroethylene has been measured in the dry cleaning industry throughout the world (Table 2). Because tetrachloroethylene has been the most commonly used solvent during the last two or three decades, when most of the measurements were made, only data on that compound are included in Table 2. Data on other exposures to tetrachloroethylene, for example, in the vicinity of dry cleaning establishments and in dry cleaned clothes carried home, are included in the monograph on tetrachloroethylene (this volume).

Differences in personal exposures to tetrachloroethylene between plants and shops are often many times larger than the differences between machine operators and other staff within dry cleaning premises. Such differences between staff have decreased in time from a factor of roughly 6 in the early 1980s to a factor of 2 in the early 1990s. As can be seen from Table 2, this coincides with decreasing average exposure levels, from 50–100 ppm [339–678 mg/m^3] in the 1970s to 10–50 ppm [67.8–339 mg/m^3] in the 1980s. Applying fixed multipliers for job titles in the calculation of cumulative exposure to tetrachloroethylene will introduce severe misclassification when factors that modify exposures in specific plants and shops are not taken into account.

In a study of dry cleaners and their families in Italy, a significant correlation was observed between the concentration of tetrachloroethylene in alveolar air samples collected at the end of the work day and 8-h time-weighted average (TWA) levels in ambient air ($r = 0.758$; $p < 0.001$; $n = 49$). A similar relationship was found between tetrachloroethylene concentrations in alveolar air samples collected from dry cleaners at home ($n = 33$), about 2 h after the end of the exposure, and the 8-h TWA concentration during the work day ($r = 0.665$; $p < 0.001$). Alveolar samples collected the next day still showed a significant correlation with the 8-h TWA value of the day before ($r = 0.549$; $p < 0.001$). The concentrations of tetrachloroethylene in alveolar air in this population were 0.49–353 mg/m^3, with a geometric mean of 15.4 mg/m^3 (Aggazzotti *et al.*, 1994).

In a study in Japan, the personal TWA levels of tetrachloroethylene to which dry cleaning workers were exposed were 0.6–100.8 ppm [4.07–683 mg/m^3]. The concentrations were 0.3–87 ppm [2.03–590 mg/m^3] in expired air and 0.01–0.73 mg/L in blood; the level of total trichlorinated compounds in urine was 0.6–19.2 mg/L (Hayashi *et al.*, 1990). In another study in Japan, the average annual consumption of solvents in dry cleaning shops was estimated to be 1280 kg of petroleum solvent, 1450 kg of tetrachloroethylene and 275 kg of CFC-113. The concentration of total trichloro compounds in urine samples was 3.2 ± 2.0 mg/L in the morning and 2.6 ± 1.9 mg/L in the afternoon (Takeuchi *et al.*, 1981).

Table 2. Exposures to tetrachloroethylene in dry cleaning shops

Country (year)	No. of plants	Job/task/industry	No. of samples	Air concentration				Reference
				Mean		Range		
				ppm	mg/m^3	ppm	mg/m^3	
Belgium	6	Dry cleaning	26 subjects (P) TWA	20.8	141	8.9–37.5	60–254	Lauwerys et al. (1983)
Finland (1982–85)	6	Dry cleaning	10 (A) TWA	13	88	3–29	20–197	Rantala et al. (1992)
France	26	Dry cleaning	> 100 per shop			0–100	0–678	Davezies et al. (1983)
France (1989, 1990)	36	Dry cleaning operator	5–10 per shop^a	17.5	120	NR–100	NR–678	Anon. (1991)
Germany		Dry cleaning	19 workers 55 (P)		62 (end of week) 43 (following Monday)		16–672	Pannier et al. (1986)
Germany		Dry cleaners	101 (P) workers TWA		205			Seeber (1989)
Germany (1987, 1989)	15	Dry cleaning	75 (A)			45% > 50 33% > 100 9% > 200	3.1–331	Gulyas & Hemmerling (1990)
Germany (1993, 1994)	21	Dry cleaning operator	100		7.4		< 0.02–27	Klein & Kurz (1994)
Italy	47	Dry cleaning	143 workers	11.3	77	1–80.8	7–548	Missere et al. (1988)
Italy (1992–1993)	28	Dry cleaning	60 workers^b (P) TWA (A)		NR 36		2.6–221.5 0.19–308	Aggazzotti et al. (1994)

Table 2 (contd)

Country (year)	No. of plants	Job/task/industry	No. of samples	Air concentration					Reference
				Mean			Range		
				ppm	mg/m^3		ppm	mg/m^3	
Netherlands (1976)	1	Dry cleaning	48 5 workers	6.7	45		3.7–25.9	25–176	van der Tuin & Hoevers (1977a)
(1977)	1	Dry cleaning	86 10 workers	41.3	280		11–101	75–685	van der Tuin & Hoevers (1977b)
(1978)	1	Dry cleaning	80 9 workers	59.7	405		10.0–250	68–1695	van der Tuin (1979)
Switzerland	10	Dry cleaning	1 week	18.5	125				Boillat et al. (1986)
United Kingdom	90	Dry cleaning shops	333 (P)	74% < 30 88% < 50 97% < 100	203 339 678				Shipman & Whim (1980)
	41	Dry cleaning factories	160 (P)	53% < 30 76% < 50 93% < 100	203 339 678				
United Kingdom (1990–91)	81	Dry cleaning operator	405 (P) TWA	22.5	153		0–360	0–2441	Edmondson & Palin (1993)
Japan	3	Dry cleaning	56 workers TWA	20 g	136		3.8–94.4	26–640	Cai et al. (1991)
USA		Dry cleaners, commercial	19 (A) 12 workers	91.5	621		31–270	210–1831	Kerr (1972)
		Dry cleaners, coin-operated	11 (A) 4 workers	125	848		87–264	590–1790	
USA		Dry cleaning area Spotter and dry cleaners	3 (A) 4 (P)	33 62	222 420		21–48 10–171	142–325 68–1160	Eddleston & Polakoff (1974)

Table 2 (contd)

Country (year)	No. of plants	Job/task/industry	No. of samples	Air concentration				Reference
				Mean		Range		
				ppm	mg/m³	ppm	mg/m³	
USA (1982)	17	Transfer	(P) TWA	86.6	587	28.5–302.7	193–2052	Materna (1985)
			(P) 5-min peak	135.9	921	11.3–533.8	77–3619	
USA (1984)	3	Dry-to-dry	(P) TWA	28.2	192	3.0–75.9	20–515	Pryor (1985)
	1	Dry cleaning (transfer process)	2 (P) TWA	54.5	370	45–64	305–434	
			10 (P) 15-min ceiling	306	2075	68–597	461–3899	
USA (1985)	1	Dry cleaning (dry-to-dry process)	4 (A) TWA		119		79–135	Burr & Todd (1986)
USA (1980s)		Transfer	TWA					International Fabricare Institute (1987)
	471	Machine operator		49.8	338			
	43	Counter person, etc.		16.7	113			
	57	Spotter, finisher		18.1	123			
		Dry-to-dry						
	157	Machine operator		23.2	157			
	16	Counter person, etc		10.8	73			
	24	Spotter, finisher		12.1	82			
USA		Dry cleaning	34 workers (P)		7.9		0.002–55	Eskenazi et al. (1991a)
USA	10	Operator/presser in dry cleaning (transfer and dry-to-dry)	60 (P) 13 workers	10	68	0.17–44.1	1.2–300	Petreas et al. (1992)

P, personal air sample; A, area air sample; NR, not reported; TWA, time-weighted average
[a] Worst-case sampling (highest exposed worker)
[b] Six to eight spot samples taken at each plant
[c] Arithmetic mean
[d] Geometric mean

Table 2 (contd)

Country (year)	No. of plants	Job/task/industry	No. of samples	Air concentration Mean ppm	Mean mg/m^3	Range ppm	Range mg/m^3	Reference
USA		Dry cleaners	96 (P)	41	278	5–125	34–848	Center for Chemical Hazard Assessment (1985)
USA (1975)		5 machine operators	(A)	37.2	252	12.2–62.2	83–422	Tuttle et al. (1977)
		2 pressers		11.4	77	4.6–18.3	31–124	
		5 counters		1.3	9	0.4–2.3	3–16	
		7 miscellaneous		3	20	0.9–5.1	6–35	
		5 machine operators	(P)	20.5	139	2.3–38.7	16–262	
		2 pressers		4.48	30	4.1–4.9	28–33	
		5 counters		0.95	6.4	0.3–1.6	2–11	
		7 miscellaneous		2	14	0–4.3	0–29	
USA (1977–79)	44	Machine operator	45 (P) TWA	31^c	210	4.0–149	27–1010	Ludwig (1981); Ludwig et al. (1983)
				22^d	149			
	35	Presser	52 (P) TWA	5.7^c	39	0.1–37	0.7–250	
				3.3^d	22			
	12	Seamstress	12 (P) TWA	6.6^c	45	0.6–29	4–197	
				3^d	20			
	31	Counter area	31 (P) TWA	5.9^c	40	0.3–26	2–176	
				3.1^d	21			
	39	Machine operator during transfer	134 (P) 5-min peak	76^c	515	3.3–366	22–2482	
				44^d	298			
	30	Machine operator during transfer	49 (P) 15-min peak	55^c	373	1–269	7–1824	
				33^d	224			
USA	44	Washer to dryer transfer	175 (A)	95.8	650	1.0–775	7–5255	US National Institute for Occupational Safety and Health (1981)
		Counter area	39 (A)	4.8	33	0.3–26.4	2–179	
		Dry cleaning area	36 (A)	32.8	222	0.5–177	3.4–1200	
		Washer area	25 (A)	22	169	2.0–91	14–617	
		Pressing area	26 (A)	6	41	0.2–40	1.4–271	
		Spotting area	14 (A)	12.3	83.4	0.9–35	6.1–237	

In one dry cleaning shop in Switzerland, the concentrations of tetrachloroethylene and CFC-113 were 10 mg/m^3 and 150 mg/m^3 inside the shop and 25 mg/m^3 and 1000 mg/m^3 in the machine room (Grob *et al.*, 1991).

Analysis of 138 blood samples taken from workers in 55 dry cleaning shops in Finland in 1987 showed a geometric mean tetrachloroethylene concentration of 0.75 µmol/L (range, < 0.1–14.0 µmol/L) [0.12 (0.02–2.3) mg/L]; the arithmetic mean was 1.25 µmol/L [0.21 mg/L] (Rantala *et al.*, 1992).

In a study conducted by the United States National Institute for Occupational Safety and Health in 1988, occupational exposures to tetrachloroethylene and trichloroethylene in dry cleaning and exposure to 2-butoxyethanol, hexylene glycol, *n*-butyl acetate and methyl isobutyl ketone during spotting were evaluated. The concentrations of spotting agents in air were low: 2-Butoxyethanol was not detectable ($n = 10$), and the mean concentrations were 1.2 ppm [5.8 mg/m^3] (range, 0–6.9 ppm [0–33 mg/m^3]; $n = 17$) for hexylene glycol, 0.32 ppm [1.3 mg/m^3] (range, 0–4.2 ppm [0–17.2 mg/m^3]; $n = 57$) for methyl isobutyl ketone and 1.2 ppm [5.7 mg/m^3] (range, 0–5.69 ppm [0–27 mg/m^3]; $n = 57$) for *n*-butyl acetate. Spotters were exposed to mean concentrations of 4.3 ppm [29 mg/m^3] (range, 0–38.7 ppm [0–262 mg/m^3]; $n = 115$) tetrachloroethylene and 1.4 ppm [7.5 mg/m^3] (range, 0–9.16 ppm [0–49 mg/m^3]; $n = 115$) trichloroethylene in air (Earnest, 1994).

In Japan, dry cleaned test samples of fabric were found to contain up to 13.6 mg/g tetrachloroethylene. In 53 air samples from 11 establishments where dry cleaned clothes were handled, the mean concentration was [434 µg/m^3] (range, 1–4813 µg/m^3). The mean level in alveolar air from 31 dry cleaners was [37.8 µg/m^3] (average, 36.9; range, 0–168 µg/m^3) (Kawauchi & Nishiyama, 1989).

The average concentration of tetrachloroethylene to which 22 workers were exposed in six dry cleaning shops in Belgium was 21 ppm [142 mg/m^3], with a range of 9–38 ppm [61.0–258 mg/m^3]. The mean concentration of tetrachloroethylene in alveolar air was 1.9 ppm [13 mg/m^3] before work and 5.1 ppm [35 mg/m^3] 30 min after exposure. The mean concentrations in the blood at the same time intervals were 0.4 and 1.2 mg/L, respectively. Trichloroacetic acid was not detected in any urine samples. The authors concluded that a blood concentration of < 1 mg/L tetrachloroethylene 16 h after exposure probably corresponded to an 8-h time-weighted average exposure of < 50 ppm [339 mg/m^3] (Lauwerys *et al.*, 1983).

In a study of 195 individuals in one industrial and 12 commercial dry cleaning facilities in Detroit, MI, United States, in 1986–87, the concentrations of tetrachloroethylene in 181 samples of exhaled breath and 54 personal samples were measured during use of dry-to-dry and transfer machinery. In facilities where transfer machines were used, the mean concentration in breath samples was 7.72 ppm [52.3 mg/m^3] (range, 2–14 ppm [12–95 mg/m^3]), which was over five times higher than that measured in facilities where dry-to-dry machinery was used (1.47 ppm [9.97 mg/m^3]). The mean concentration in air samples from facilities using transfer machines (29.5 ppm [200 mg/m^3]; range, 1–58 ppm [7–400 mg/m^3]) exceeded that for dry-to-dry machines (7.09 ppm [48.1 mg/m^3]) by more than four times. The mean concentrations of tetrachloroethylene in breath and air samples (ppm [mg/m^3]) for various job titles were, respectively: operator, 12.5 [84.8] and 46.5 [315]; presser, 5.88 [39.9] and 14.8 [100]; clerk, 4.52 [30.6] and 11.7 [79.3]; seamstress, 6.50 [44.1] and 19.0 [129]; managerial, 5.50 [37.3] and not reported;

other, 2.55 [17.3] and 5.00 [33.9]. Of 12 air samples taken from the breathing zone of operators, eight contained > 25 ppm [170 mg/m^3], 6 contained > 50 ppm [339 mg/m^3] and one contained > 100 ppm [678 mg/m^3]. None of the air samples from the four shops where only dry-to-dry machinery was used had concentrations > 25 ppm [170 mg/m^3] (Solet *et al.*, 1990). In the same population of dry cleaners, the ratios of total urinary protein, albumin and *n*-acetylglucosaminidase to creatinine in 192 urine samples were studied in relation to tetrachloroethylene levels in breath and estimates of chronic exposure; no consistent relationship was seen (Solet & Robins, 1991).

In a study in Croatia of 19 dry cleaners exposed to tetrachloroethylene for a mean duration of 16.6 years (range, 2–30), the concentrations of tetrachloroethylene and trichloroacetic acid in blood on a Monday before work were 2.96 (1.63–32) and 6.5 (1.65–17.1) µmol/L [0.5 (0.3–53) and 1.1 (0.3–2.8) mg/L], and those on a Thursday after work were 8.44 (4.8–80) and 8.57 (1.7–20.9) µmol/L [1.4 (0.8–13.3) and 1.4 (0.3–3.4) mg/L], respectively. The concentrations of trichloroethanol and trichloroacetic acid in urine were 0.03 (0–1.7) and 2.3 (0.5–15.8) µmol/mmol [0.04 (0–2.3) and 3 (0.7–23) mg/g] creatinine on a Monday before work, and 0.13 (0–0.95) and 2.3 (0.8–10.8) µmol/mmol [0.17 (0–1.3) and 3.3 (1.2–15.6) mg/g] creatinine on a Thursday after work (Skender *et al.*, 1986).

In a study in the United States (Petreas *et al.*, 1992), the mixed exhaled air of dry cleaning machine operators contained a mean concentration of 3.4 ppm [23.1 mg/m^3] tetrachloroethylene at the end of the work shift and 2.2 ppm [14.9 mg/m^3] the next morning; the mean concentrations for a presser were 3.3 and 1.4 ppm [22.4 and 9.49 mg/m^3], respectively. Exhaled air of operators using dry-to-dry machines contained 2.4 ppm [16.3 mg/m^3] at the end of the work shift and 1.5 ppm [10.2 mg/m^3] the next morning; the exhaled air of transfer machine operators contained 5.4 and 3.1 ppm [36.6 and 21.0 mg/m^3], respectively. The levels of tetrachloroethylene in 12 blood samples were 17–1154 µg/L, with a mean of [283 µg/L]. The mean blood:breath concentration ratio was thus 21.6, similar to that of 23 derived by Monster *et al.* (1979). Petreas *et al.* (1992) found strong correlations between tetrachloroethylene concentrations in mixed exhaled air and in blood ($r = 0.94$) and in mixed exhaled air and personal air samples collected over the entire shift ($r = 0.89$).

2. Studies of Cancer in Humans

2.1 Case reports

A man and four of his five children, all of whom worked in dry cleaning in the United States, developed chronic lymphocytic leukaemia (Blattner *et al.*, 1976).

Jalihal and Barlow (1984) reported a case of myeloid leukaemia diagnosed in a 60-year-old dry cleaner in the United Kingdom. He had had heavy exposure for many years, first to trichloroethylene and later to tetrachloroethylene.

2.2 Descriptive studies

Guralnick (1963) used death certificates to evaluate the cause-specific mortality by occupation for all men aged between 20 and 64 who had died in the United States during 1950. There were 1193 deaths among laundry or dry cleaning operatives. Nonsignificant excesses of cancer mortality were seen, both overall (standardized mortality ratio [SMR], 1.4) and for cancers at specific sites, including the stomach (1.5), intestine and rectum (1.6) and trachea, bronchus and lung (1.5).

Gallagher et al. (1989) used death certificates to study cause-specific mortality by occupation for all men aged 20–65 in British Columbia (Canada) for the years 1950–84. There were 165 deaths among launderers and dry cleaners, and significantly elevated proportionate mortality ratios (PMRs) were observed for cancer of the lip (PMR, 53; 95% confidence interval [CI], 6.4–100, based on two deaths) and for all tumours of the digestive organs (2.0; 1.3–3.0; 23 deaths), especially the stomach (2.5; 1.1–4.9; eight deaths).

Milham (1992) analysed information on occupation and cause of death from the death certificates of 1652 male laundry and dry cleaning operatives who died in Washington State, United States, between 1950 and 1989. Elevated PMRs were seen for cancers of the buccal cavity and pharynx and connective tissues.

Data from death certificates and censuses for Great Britain for the years 1979–80 and 1982–83 showed that mortality among men aged 20–64 and single women aged 20–59 [data for married women were not reported] who were launderers, dry cleaners or pressers was higher than that in the general population (males: SMR, 1.3 [95% CI, 1.2–1.5]; single females: 1.4 [1.1–1.7]) (Office of Population Censuses and Surveys, 1986). There was a significant overall excess of malignant neoplasms in males (1.4 [1.1–1.7]), and excesses were reported for cancers of the oesophagus (1.7 [0.5–4.5], four deaths), stomach (2.0 [0.9–3.6], 10 deaths), trachea, bronchus and lung (1.7 [1.2–2.3], 41 deaths) and urinary bladder (2.8 [0.9–6.5], five deaths). In single women, there was no overall excess of malignant neoplasms (SMR, 1.0 [0.7–1.5]), but some excess was reported for cancers of the digestive organs (1.4 [0.8–2.9], seven deaths), rectum (3.3 [0.7–9.7], three deaths) and bladder (7.3 [1.2–36], two deaths).

A proportionate mortality study in Wisconsin, United States, was based on the death certificates of 671 female laundry and dry cleaning workers in the period 1963–77 (Katz & Jowett, 1981). Comparisons were made with all occupations as well as with lower-wage occupations obtained from a data file of 66 230 deaths in the State among white women who had ever been employed. The PMRs given below are those obtained from comparisons with lower-wage occupations. Elevated risks were found for some cancers and for diabetes mellitus. For cancer of the kidney, the PMR was 2.5 [95% CI, 1.0–5.2], based on seven deaths; types of cancer associated with nonsignificantly increased PMRs included cancer of the skin (2.6 [0.73–6.8], four deaths), lymphosarcoma (1.8 [0.65–3.8], six deaths), urinary bladder cancer (1.9 [0.62–4.5], five deaths), cervical cancer (1.4 [0.68–2.6], 10 deaths) and rectal cancer (1.3 [0.45–2.7], six deaths). There was no overall excess of neoplasms.

The causes of death of 440 laundry and dry cleaning workers in Oklahoma, United States, were investigated in a study which covered the years 1975–81 (Duh & Asal, 1984). Petroleum solvents accounted for > 50% of the solvents used in the State. The distribution of deaths in the

United States in 1978 by age, sex, race and cause was used as the standard of comparison. Some SMRs were decreased, most notably that for ischaemic heart disease (0.8; 95% CI, 0.7–1.0), based on 134 deaths. For cancers at all sites, the standardized mortality odds ratio was 0.9 (0.7–1.2), based on 97 deaths. Cancer of the respiratory system was the cause of death for 39 persons (standardized mortality odds ratio, 1.8; 95% CI, 1.3–2.5), of whom 37 died of lung cancer (1.7; 1.2–2.5). For renal cancer, the standardized mortality odds ratio was 3.8 (1.9–7.6), based on seven deaths. There were two deaths from cervical cancer (1.3; 0.3–5.3), five from cancer of the ovary (1.5; 0.6–3.6), one from urinary bladder cancer (0.4) and one from breast cancer (0.1).

A further report on the Oklahoma dry cleaners was published as an abstract (Petrone, 1988). PMRs were based on the mortality of 474 white male dry cleaners whose exposure histories were obtained from plant records. The men had been exposed to petroleum solvents for an average of 10.5 years and had had no known exposure to 'synthetic solvents'. For all malignant neoplasms, the PMR was 1.0 (95% CI, 0.83–1.3), based on 84 deaths. For respiratory cancer, the PMR was 1.5 (1.0–2.1), based on 35 deaths; no relationship was seen with latency or degree of exposure. A PMR of 2.0 (0.89–3.7) was obtained for pancreatic cancer. The PMR for renal cancer was 2.0 (0.66–4.7), based on five deaths; when only men exposed to petroleum solvents were analysed, the PMR was 1.1 (0.13–3.9), based on two deaths. [The degree of overlap between this study and that of Duh & Asal (1984) is unknown.]

Nakamura (1985) conducted a proportionate mortality study of members of the All-Japan Laundry and Dry-cleaning Association. Trichloroethylene was reported to be the solvent used most commonly for dry cleaning in Japan between 1903 and 1940; more recently, 30% of dry cleaning was stated to be done using tetrachloroethylene and 65% with petroleum solvents. Death certificates were obtained for 1711 members who had died between 1971 and 1980, and expected numbers of deaths were calculated from mortality rates for Japanese males and females in 1975. When the results for men and women were combined, there was an increased risk for cancer of the bone ([PMR, 2.9; 95% CI, 0.9–6.9]; five deaths).

2.3 Cohort studies

A union of dry cleaning workers in Missouri, United States, had 11 062 members between 1945 and 1978, of whom 5790 had held membership for one year or more. After exclusion of 425 members for whom information on race, sex or date of birth was unavailable, the analysis was restricted to 5365 members (Blair *et al.*, 1986, abstract; Blair *et al.*, 1990). The members were followed-up from entry to the Union or 1 January 1948 (whichever came later) until 1 January 1979: follow-up was 88% successful, with slight variations for men and women and for race. The cohort contributed 98 818 person-years, and the expected numbers of deaths were calculated from national rates. The mortality rate was slightly lower than expected for all causes combined (SMR, 0.9; 95% CI, 0.9–1.0; 1129 deaths) but slightly raised for cancer (1.2; 1.0–1.3; 294 deaths). Excesses were found for cancers of the oesophagus (SMR, 2.1; 1.1–3.6; 13 deaths), larynx (1.6; 0.3–4.7; three deaths), lung (1.3; 0.9–1.7; 47 deaths), cervix uteri (1.7; 1.0–2.0; 21 deaths), bladder (1.7; 0.7–3.3; eight deaths) and thyroid (3.3; 0.7–9.8; three deaths) and for lymphosarcoma and reticulosarcoma (1.7; 0.7–3.4; seven deaths) and Hodgkin's disease (2.1; 0.6–5.3; four deaths). There was no excess of cancer of the liver or of the kidney. The excess risk for cancer of the oesophagus was restricted to black men (SMR, 3.5; 11 deaths). The relative risk for cancer at this site for all cohort members was related to estimated cumulative exposure to dry cleaning solvents and was 2.8 for those in the highest category of exposure. For deaths from causes other than cancer, an SMR of 2.0 (95% CI, 1.1–3.4) for emphysema was the most relevant finding. With the highest level of exposure to dry cleaning solvents, there were five deaths from lymphatic and haematopoietic malignancies, whereas 1.3 were expected (SMR, 4.0 [1.2–90]). The authors stated that the relative risks were similar for workers first employed before 1960 and those employed after that date, when use of tetrachloroethylene became predominant, but no details were provided.

Proportionate mortality among the same workers, reported in a previous study (Blair *et al.*, 1979), showed excess numbers of deaths from cancers of the liver, cervix and skin.

An updating of an earlier cohort mortality study in the United States (Kaplan, 1980; Brown & Kaplan, 1987) from California, Illinois, Michigan and New York, included 1701 dry cleaning workers in four labour unions (Ruder *et al.*, 1994). The inclusion criteria were employment for at least one year before 1960 in a shop where tetrachloroethylene was the primary solvent used and there was no known exposure to carbon tetrachloride. A survey in 1977–79 (Brown & Kaplan, 1987) showed geometric mean, time-weighted average concentrations of tetrachloroethylene in the range of 3–22 ppm [20.3–149 mg/m^3]; other solvents used for spot cleaning were not detected in the samples. In the analyses, two subcohorts were defined: people employed only in shops where tetrachloroethylene was the primary solvent, and people whose work also involved exposure to other solvents. The mean duration of employment in dry cleaning through 31 December 1990 was 6.4 years for those exposed only to tetrachloroethylene and 11.4 years for those also exposed to other solvents; the latter group had a mean duration of 6.0 years' exposure to tetrachloroethylene. Expected numbers of deaths were calculated from the national death rates. The follow-up reached 94% of the women and 96% of the men, contributing 47 273 person-years. There were 769 deaths (SMR, 1.0; 95% CI, 0.94–1.1) and 209 cancer deaths (1.2; 1.1–1.4). The SMR for urinary bladder cancer was significantly increased (2.5; 1.2–4.8; nine deaths), as were those for oesophageal cancer (2.1; 1.0–3.9; 10 deaths) and cancers of the colon and small intestine (1.6; 1.0–2.3; 26 deaths). Some excess of pancreatic cancer was noted (1.7; 0.93–2.8; 15 deaths). When the analysis was restricted to workers with a 20-year latency since first employment and with a length of employment ≥ 5 years, the SMRs were increased for cancers at all sites combined (1.5; 1.2–1.9; 83 deaths) and, notably, for cancers of the oesophagus (5.4; 2.3–11; eight deaths) and the urinary bladder (6.5; 2.8–13; eight deaths). Some overall excess was also seen for cervical cancer (1.8; 0.86–3.3; 10 deaths) and for renal cancer (1.5; 0.40–3.7; four deaths), but there was no association with time since first employment or length of employment. There were also three cancers of the tongue (3.5; 0.73–10). Only one death from cancer of the liver and biliary tract was found [with 4.8 expected].

A Danish cohort of 8567 women and 2033 men aged 20–64 years and employed in laundry and dry cleaning in 1970 was followed-up for 10 years (Lynge & Thygesen, 1990). One-fourth of the cohort was estimated to work in dry cleaning, but at the individual level there was no way of distinguishing those who worked in dry cleaning from those who worked in laundries. The dry cleaners in the cohort were mainly exposed to tetrachloroethylene after the late 1950s, but had also been exposed to trichloroethylene and chlorofluorocarbons CFC-11 and CFC-113. In measurements taken in 1979–80, the concentrations of tetrachloroethylene in clothes being taken out of the machines were 1000–7000 ppm [mg/kg] in some samples but were usually less than 100 ppm. The expected numbers of cases were calculated on the basis of the rates for the economically active proportion of the general population. A total of 510 cancer cases were observed [standardized incidence ratio (SIR), 1.0; 95% CI, 0.9–1.1]. For colon cancer, the SIR was [1.0] ([0.70–1.4]; 35 observed); for lung cancer, the SIR was [1.2] ([0.9–1.6]; 60 observed). Seven cases of liver cancer were observed [SIR, 2.2; 95% CI, 0.9–4.5], and there were 22 cases of pancreatic cancer (1.7; 1.1–2.6).

When the study was updated for the period 1981–87, a total of 10 additional cases of liver cancer were observed (4.5 expected) (Lynge, 1994).

The Swedish population at the time of the 1960 census was followed-up for cancer incidence, first for 1961–73 (Malker & Weiner, 1984) and later for 1961–79 (McLaughlin et al., 1987). In the first follow-up, data on the following cancers were reported for laundry and dry cleaning workers: all cancers (SIR, 1.0 [95% CI, 0.9–1.1]; 646 observed) and cancers of the buccal cavity and nasopharynx ([2.9; 1.7–4.7]; 16 observed), liver and biliary passages (1.2 [0.7–1.8]; 17 observed), lung (1.2 [0.9–1.7]; 34 observed) and breast (0.8 [0.7–1.0]; 113 observed). In the second follow-up, the only site for which data were reported for this occupational group was the kidney: the SIR was 0.99 ([0.6–1.6] 18 observed) in men and 0.86 ([0.6–1.3] 25 observed) in women.

Table 3 gives the details of these cohort studies. Results are shown for all cancer sites for which at least one of the cohort studies showed a significantly increased risk. In addition, results are shown for cancer sites on which a relevant case–control study was conducted (see Table 4).

2.4 Case–control studies

Dry cleaning (both in association with and independently of work in laundries) has been evaluated for possible associations with several types of cancer in case–control studies. Risks were investigated in relation to a variety of exposures and occupations and were not specifically concerned with dry cleaners. The studies addressed cancers of the colon (Fredriksson et al., 1989), liver (Stemhagen et al., 1983; Austin et al., 1987; Suarez et al., 1989), lung (Brownson et al., 1993), kidney (Asal et al., 1988; McCredie & Stewart, 1993; Mellemgaard et al., 1994), pancreas (Mack et al., 1985), oral cavity and pharynx (Huebner et al., 1992) and non-Hodgkin's lymphoma (Blair et al., 1992, 1993). Data from the National Bladder Cancer study of the United States National Cancer Institute conducted in 1978, in relation to dry cleaners, were used in four studies (Silverman et al., 1983; Schoenberg et al., 1984; Smith et al., 1985; Silverman et al., 1989). A study of multiple sites was reported by Siemiatycki (1991).

A record-linkage study from Denmark on parental employment at the time of conception and the risk of cancer in offspring mentions an increased risk for cancers at all sites combined among children of mothers who owned a laundry or dry cleaning establishment (odds ratio, 3.7; $p < 0.01$; 6 observed) (Olsen et al., 1991).

Table 4 summarizes the results of the case–control studies. [The Working Group noted that positive results may have been reported preferentially.]

Table 3. Results of cohort studies of dry cleaners

Site of cancer	Blair et al. (1990) 5365 US dry cleaners followed for mortality during 1948–78			Ruder et al. (1994) 1701 US dry cleaners followed for mortality during 1940–90			Lynge & Thygesen (1990) 10 600 Danes employed in laundry and dry cleaning in 1970, followed for cancer incidence during 1970–80			Malker & Weiner (1984) Swedes employed in laundry and dry cleaning in 1960, followed for cancer incidence during 1961–73		
	Obs.	SMR	95% CI	Obs.	SMR	95% CI	Obs.	SIR	95% CI	Obs.	SIR	95% CI
All	294	1.2	1.0–1.3	209	1.2	1.1–1.4	510	[1.0]	[0.9–1.1]	646	1.0	[0.9–1.1]
Oesophagus	13	2.1	1.1–3.6	10	2.1	1.0–3.9	–	–	–	–	–	–
Colon	25	1.0	0.6–1.4	26	1.6	1.0–2.3	35	[1.0]	[0.7–1.4]	–	–	–
Liver and biliary passages	5	0.7	0.2–1.7	1	0.21	0.0–1.7	15	[1.9]	[1.0–3.1]	17	1.2	[0.7–1.8]
Liver	–	–	–	–	–	–	7	[2.2]	[0.9–4.5]	–	–	–
Gall-bladder	–	–	–	–	–	–	8	[1.6]	[0.7–3.2]	–	–	–
Pancreas	15	1.2	0.7–1.9	15	1.7	0.93–2.8	22	1.7	1.1–2.6	–	–	–
Lung	47	1.3	0.9–1.7	43	1.2	0.85–1.6	60	[1.2]	[0.9–1.6]	34	1.2	[0.9–1.7]
Cervix	21	1.7	1.0–2.0	10	1.8	0.86–3.3	34	[0.8]	[0.6–1.2]	–	–	–
Kidney	2	0.5	0.1–1.8	4	1.5	0.40–3.7	11	[0.9]	[0.4–1.6]	43[a]	[0.9][a]	[0.7–1.2]
Urinary bladder	8	1.7	0.7–3.3	9	2.5	1.2–4.8	14	[0.7]	[0.4–1.2]	–	–	–
Non-Hodgkin's lymphoma	7[b]	1.7	0.7–3.4	2[b]	0.99	0.12–3.6	8[c]	[1.0]	[0.4–2.0]	–	–	–

Obs, observed; SMR, standardized mortality ratio; CI, confidence interval
[a] Follow-up for 1961–79 (McLaughlin et al., 1987)
[b] ICD 200
[c] ICD 200 and 202

Table 4. Risks for cancer associated with dry cleaning work in case–control studies of people with a variety of exposures and occupations

Reference (country)	Study design	Exposure	Sex	Numbers of cases/controls Total	Numbers of cases/controls Exposed	Odds ratio	95% CI	Comments
Colon cancer								
Fredriksson et al. (1989) (Sweden)	Population-based; response rates: 95% of cases and controls	Employment as dry cleaner	F	156/317	5/5	2.0	0.5–7.1	Adjusted for age and physical activity
Siemiatycki (1991) (Canada)	Population-based (case–case)[a]; response rate: 82%	Employment as launderer or dry cleaner: Any[b] 'Substantial'[c]	M	497/2056	5/NR 1/NR	0.6 0.2	[0.3–1.4] [0.0–1.5]	Adjusted for age, race, income, smoking and beer consumption
Oesophageal cancer								
Siemiatycki (1991) (Canada)	Population-based (case–case)[a]; response rate: 82%	Employment as launderer or dry cleaner: Any[b] 'Substantial'[c]	M	99/2546	0/NR 0/NR	0.0 0.0		
Liver cancer								
Sternhagen et al. (1983) (USA)	Population-based; response rates, 79% for cases and 77% for controls	Employed for ≥ 6 months in laundry, dry cleaning and garment services	M	178/356	10/8 8/7	2.5 2.3	1.0–6.1 0.85–6.1	All primary liver cancer Hepatocellular carcinoma Matched on age, race, county and vital status
Austin et al. (1987) (USA)	Hospital-based; response rates: 93% for cases and 91% for controls	Employment for ≥ 6 months in laundry and cleaning	M/F	80/146	0/4	—		Matched on sex, age, race and study centre
Suarez et al. (1989) (USA)	Based on death certificates	Usual occupation: in dry cleaning services as dry cleaning operator	M	1742/1742	11/12 4/8	0.98 0.55	0.44–2.2 0.17–1.8	Adjusted for age and race

Table 4 (contd)

Reference (country)	Study design	Exposure	Sex	Numbers of cases/controls		Odds ratio	95% CI	Comments
				Total	Exposed			
Lung cancer								
Siemiatycki (1991) (Canada)	Population-based (case–case)[a]; response rate: 82%	Employment as launderer or dry cleaner: Any[b] 'Substantial'[c]	M	857/1360	12/NR 5/NR	0.8 0.6	[0.4–1.5] [0.2–1.9]	Adjusted for age, race, income, smoking and alcohol consumption
Brownson et al. (1993) (USA)	Population-based; response rates: 66% for cases, 67% for controls	Any employment in dry cleaning	F	429/1021	30/39 23/31	1.8 2.9 2.1	1.1–3.0 1.5–5.4 1.2–3.7	Never and ex-smokers Employed for > 1.125 years Never smokers Adjusted for age
Prostate cancer								
Siemiatycki (1991) (Canada)	Population-based (case–case)[a]; response rate: 82%	Employment as launderer or dry cleaner: Any[b] 'Substantial'[c]	M	449/1550	9/NR 6/NR	1.5 2.1	[0.7–3.3] [0.7–6.0]	Adjusted for age, race, income, smoking and body-mass index
Kidney cancer								
Asal et al. (1988) (USA)	Population-based	Employed predominantly in dry cleaning	M F	315/336	3/6 8/1	0.7 2.8 8.7	0.2–2.3 0.8–9.8 0.9–81	Adjusted for age, weight and smoking From regression analysis
Siemiatycki (1991) (Canada)	Population-based (case–case)[a]; response rate: 82%	Employment in laundry or dry cleaning: Any[b] 'Substantial'[c]	M	177/2481	5/NR 2/NR	2.0 2.1	[0.8–5.1] [0.5–9.2]	Adjusted for age, race, income and smoking

Table 4 (contd)

Reference (country)	Study design	Exposure	Sex	Numbers of cases/controls Total	Numbers of cases/controls Exposed	Odds ratio	95% CI	Comments
Kidney cancer (contd)								
McCredie & Stewart (1993) (USA)	Population-based; response rates: 68% for renal cases, 75% for pelvic cases, 74% for controls	Any employment in dry cleaning	F M F M	179/292 310/231 89/292 58/231	MF 16/7 MF 8/7	2.7 2.5 6.1 4.7	1.1–6.7 0.97–6.4 2.0–19 1.3–17	Renal-cell cancer: adjusted for age and sex; also adjusted for smoking Renal pelvic cancer: adjusted for age, sex and education; also adjusted for smoking
Mellemgaard et al. (1994) (Denmark)	Population-based; response rates: 76% for cases, 88% for controls	Any employment ≥ 10 years before interview	M F	226/237 142/159	2/1 2/1	2.3 2.9	0.2–27 0.3–33	Adjusted for age, body-mass index and smoking
Bladder cancer[a]								
Silverman et al. (1983) (USA)	Response rates: 81% for cases, 84% for controls	Employed ≥ 6 months in laundry and dry cleaning	M	303/296	12/5	2.4	0.8–6.9	Detroit area, whites only
Schoenberg et al. (1984) (USA)	Response rates: 90% for cases, 87% for controls	Employed ≥ 6 months in dry cleaning	M	658/1258	7/10	1.3	0.50–3.6	New Jersey area, whites only; adjusted for age and smoking
Smith et al. (1985) (USA)		Employed ≥ 6 months in laundry and dry cleaning	M	NR	NR	1.3	0.85–2.0	All study areas, nonsmokers only
Silverman et al. (1989) (USA)		Employed ≥ 6 months as dry cleaner, ironer or presser	M	126/383	11/12	2.8	1.1–7.4	All study areas, non-whites only; adjusted for smoking

Table 4 (contd)

Reference (country)	Study design	Exposure	Sex	Numbers of cases/controls		Odds ratio	95% CI	Comments
				Total	Exposed			
Bladder cancer (contd)								
Siemiatycki (1991) (Canada)	Population-based (case–case)[a]; response rate: 82%	Employment as launderer or dry cleaner: Any[b] 'Substantial'[c]	M	484/1879	10/NR 7/NR	1.6 1.9	[0.8–3.3] [0.8–4.8]	Adjusted for age, income, smoking and coffee consumption
Pancreas cancer								
Mack et al. (1985) (USA)	Population-based; response rate: 67% for cases	Employment ≥ 6 months in laundry and dry cleaning > 10 years before diagnosis	M/F	490/490	23/36	[0.6]	[0.4–1.1]	Crude odds ratio
Siemiatycki (1991) (Canada)	Population-based (case–case)[a]; response rate: 82%	Employment as launderer or dry cleaner: Any[b] 'Substantial'[c]	M	116/2454	0/NR 0/NR	0.0 0.0		
Oral and pharyngeal cancer								
Huebner et al. (1992) (USA)	Population-based; response rates: 75% for cases, 76% for controls	Employed for ≥ 6 months: as laundry or dry cleaning worker in laundry or dry cleaning industry	M	762/837	14/22 18/26	0.39 0.64	0.17–0.88 0.32–1.3	Adjusted for age, race, smoking, alcohol and study location
Non-Hodgkin's lymphoma								
Siemiatycki (1991) (Canada)	Population-based (case–case)[a]; response rate, 82%	Employment as launderer or dry cleaner: Any[b] 'Substantial'[c]	M	215/2357	3/NR 0/NR	0.9 0.0	[0.3–3.1]	Adjusted for age, income and smoking

Table 4 (contd)

Reference (country)	Study design	Exposure	Sex	Numbers of cases/controls		Odds ratio	95% CI	Comments
				Total	Exposed			
Non-Hodgkin's lymphoma (contd)								
Blair et al. (1992, 1993) (USA)	Population-based; response rates: 87% for cases, 77–79% for controls	Employed for ≥ 1 year in laundry and dry cleaning	M	622/1245	16/14	2.0	0.97–4.3	Adjusted for smoking
Childhood cancer								
Olsen et al. (1991) (Denmark)	Population-based	Laundry and dry cleaner owners	M/F	1721/8340	6/8	3.7	[1.1–12]	Data relate to mothers' exposure

CI, confidence interval; NR, not reported; M, male; F, female
^aSiemiatycki (1991) studied 21 cancer sites; for each site, patients with other cancers served as controls
^b≥ 5 years before disease onset
^c≥ 10 years employment, ≥ 5 years before disease onset
^dThe first four studies involved cases and population controls from the National Bladder Cancer study of the United States National Cancer Institute. There was some overlap between the studies, which had different inclusion criteria, as noted.

3. Other Data Relevant to an Evaluation of Carcinogenicity and its Mechanisms

3.1 Absorption, distribution, metabolism and excretion in humans

Studies of the toxicokinetics of tetrachloroethylene are described in the relevant monograph (this volume).

3.2 Toxic effects in humans

Slight renal changes have been reported in workers exposed to tetrachloroethylene in dry cleaning shops in several studies. In a multi-centre cross-sectional investigation in several European countries, of nine men and 41 women (mean age, 41 years; range, 17–65) who had an average of 10 years of exposure to tetrachloroethylene (median air concentration, 15 ppm [102 mg/m^3]), high-relative-molecular-mass proteinuria and increased release of laminin fragments, fibronectin and glycosaminoglycans indicated slight nephrotoxicity. In addition, shedding of epithelial membrane components from tubular cells into the urine was observed. No such renal changes were observed in 50 blood donors matched by sex, age, smoking habits, alcohol consumption and use of medication. The presence of several other possible markers of renal damage, including urinary retinol-binding protein and β_2-microglobulin, was not increased (Mutti et al., 1992). [The Working Group noted that the period of employment was not taken into consideration.]

Biochemical markers of kidney damage were studied in 16 female dry cleaning workers in the former Czechoslovakia who were chronically exposed to tetrachloroethylene and compared with those in 13 women with only administrative functions and no known exposure to organic solvents. The concentration of tetrachloroethylene in the breathing zone of the dry cleaning workers was 157 mg/m^3 (time-weighted average exposure; range, 9–799 mg/m^3). Urinary excretion of lysozyme was 0.15 mg/g creatinine (range, 0–1.49) in the controls and 0.56 mg/g creatinine (range, 0–4.16) in the 16 exposed subjects, whereas there were no differences in the urinary excretion of albumin, β_2-microglobulin, lactate dehydrogenase, total protein or glucose. Moreover, no correlation was observed between the level of tetrachloroethylene in air and biochemical parameters of renal function (Vyskocil et al., 1990).

In contrast, several other studies found no relationship between exposure to tetrachloroethylene in dry cleaning and renal damage. Solet and Robins (1991) obtained histories of exposure and medical conditions and samples of urine and breath from 192 dry cleaning workers in the United States. No association was found between exposure and parameters indicative of renal damage. In a study of 22 subjects exposed to tetrachloroethylene in six dry cleaning shops in Belgium (Lauwerys et al., 1983), exposure was assessed by monitoring air in the breathing zone, by urinary analysis for trichloroacetic acid and by analysis of expired air and venous blood for tetrachloroethylene. The results were compared with those for 33 subjects who were not occupationally exposed to organic solvents. The time-weighted average exposure to tetrachloroethylene was 21 ppm [142 mg/m^3] (range, 9–38 ppm [61.0–258 mg/m^3]). The urinary concentrations of albumin, β_2-microglobulin and retinol-binding protein were similar in the exposed and control workers.

3.3 Reproductive and prenatal effects in humans

3.3.1 Endocrine and gonadal effects

In a small study on the association between dry cleaning work and menstrual disorders, 592 women involved in dry cleaning or laundry work in the Netherlands received a postal questionnaire; 471 (80%) responded, of whom 72 were excluded because of current pregnancy or lactation, chronic illness or gynaecological surgery. Of the remaining 399, 193 were potentially exposed to tetrachloroethylene and 206 were unexposed. Only the 68 exposed and 76 unexposed women who did not use oral contraceptives were included in the analysis. The mean length of the cycle was the same in the two groups, but virtually all of the menstrual disorders seen in exposed women occurred more frequently than in the reference group, the increase being significant for premenstrual syndrome (odds ratio, 3.6 [95% CI, 1.3–10]), menorrhagia (3.0 [1.4–6.3]) and dysmenorrhea (1.9 [1.0–3.8]) (Zielhuis *et al.*, 1989). [The main drawbacks of this study are its small sample size, the use of a postal questionnaire and the lack of exposure measurements; however, the two groups were investigated in the same way and had similar working conditions, except for potential exposure to tetrachloroethylene.]

The semen quality of 34 dry cleaners was compared with that of 48 laundry workers who were members of the Laundry and Dry Cleaners Union in the San Francisco Bay area and Greater Los Angeles, CA, United States (Eskenazi *et al.*, 1991a). Sperm concentrations and the overall percentage of abnormal forms were similar in the two groups. Sperm heads of dry cleaners had significantly more round forms, fewer narrow forms and showed greater amplitude of lateral head displacement and less linearity in the sperm swimming paths. These subtle effects on sperm quality were related to measures of exposure to tetrachloroethylene; it was not established whether these changes affected fertility.

3.3.2 Fertility

In a study of the effects of occupational exposures on fertility, 1069 infertile couples treated in hospital and resident on the island of Funen, Denmark, were compared with 4305 fertile couples. Occupational exposures, sociodemographic data and medical histories were obtained by postal questionnaires from 927 case and 3728 control couples (about 87% of the original sample), and associations with exposure to 18 groups of agents were tested. A significant association was reported between exposure to 'dry cleaning chemicals' and the risk for idiopathic infertility among women (odds ratio, 2.7; 95% CI, 1.0–7.1; adjusted for woman's age, education, residence and parity). Significant associations were also reported (but not quantified) between work as a dry cleaner and sperm abnormalities, hormonal disturbance in women and delay of conception for more than one year (Rachootin & Olsen, 1983).

The reproductive outcomes of the wives of 17 men working in dry cleaning were compared with those of the wives of 32 laundry workers who were not exposed to dry cleaning fluids. The mean number of pregnancies was 2.1 in both groups, and the rates of spontaneous abortion were not significantly different (11.1% for dry cleaners' wives and 15.2% for laundry workers' wives). The wives of dry cleaners were more than twice as likely to have a history of delayed conception for more than 12 months or to have sought care for an infertility problem. The ratio of the pregnancy rate per cycle for wives of dry cleaning workers to that for the comparison group was 0.54 (95% CI, 0.23–1.27) (Eskenazi *et al.*, 1991a,b). [The power of this study is obviously limited.]

3.3.3 Pregnancy

(a) Spontaneous abortion

In a survey in Finland, hospital discharge records were used to analyse the effects of parental occupations and exposures, obtained from census data, on the prevalence of spontaneous abortion. A total of 294 309 pregnancies were considered between 1973 and 1976; 416 occurred among dry cleaning and laundry workers. This group was found to have a higher rate of spontaneous abortions than other service workers (odds ratio, 1.5; 95% CI, 1.1–2.0; adjusted for age, place of residence and parity) (Lindbohm *et al.*, 1984).

A nested case–control study was conducted in Finland within a cohort of 5700 women who had been identified from union files and employers as dry cleaning or laundry workers between 1973 and 1983. They had experienced a total of 3279 pregnancies. One pregnancy per woman was randomly selected for analysis, and 243 cases of spontaneous abortion were identified from hospital discharge data; 680 age-matched controls were selected. A questionnaire was mailed to all subjects; the response rate was 68.3% for cases and 80.4% for controls. A total of 130 women who had had a spontaneous abortion and 289 controls reported the pregnancy under study, and the corresponding exposures were included in the analysis. Dry cleaning was associated with an increased risk for spontaneous abortion (odds ratio, 4.9; 95% CI, 1.3–20) and appeared to be closely related to exposure to tetrachloroethylene (Kyyrönen *et al.*, 1989).

In the Nordic countries (Denmark, Finland, Norway and Sweden), a common protocol was used in order to study reproductive outcomes among dry cleaning workers (Olsen *et al.* 1990). More than 18 000 women who had worked for at least one month between 1973 and 1983 in a laundry or dry cleaning establishment were enrolled in the cohort. All data related to their pregnancies were obtained from hospital registries, and exposure histories were collected by interview or postal questionnaire or from employers, depending on the country. The analysis of spontaneous abortions did not include data from Norway. Data on 159 women who had had a spontaneous abortion were compared with those on 436 controls within the cohort, matched for mother's age, year of pregnancy and parity (except in the Finnish study). The combined odds ratio for spontaneous abortion, adjusted for parity, smoking and drinking habits, was 1.2 (95% CI, 0.74–1.9) in the group considered to have had low exposure and 2.9 (95% CI, 0.98–8.4) in the group with high exposure (dry cleaning or spot removal for at least 1 h per day during the first trimester of pregnancy). [The Working Group noted that 118 of the 159 spontaneous abortions included in the combined study occurred in Finland. The results of the Finnish study are those reported by Kyyrönen *et al.* (1989).]

In Montréal (Québec, Canada), 56 012 women were interviewed in 11 obstetrical units over two years, after delivery (51 885) or after a spontaneous abortion (4127). Associations between spontaneous abortion (at the time of interview or previously) and occupations during pregnancy were studied in relation to 100 current and 123 previous pregnancies among women who had worked in laundries and dry cleaning shops. After adjustment for maternal age, parity, history of previous abortion, smoking habits and highest educational level reached, the odds ratios for spontaneous abortion were 1.2 (eight observed) for current pregnancies and 1.0 (31 observed) for previous pregnancies (McDonald et al., 1986).

All 67 women working in 53 dry cleaning shops in two large neighbourhoods in Rome, Italy, were interviewed about their work and obstetrical history. Exposure to dry cleaning solvents (mostly tetrachloroethylene) was evaluated by the presence of trichloroacetic acid in 24-h urine samples. Of the 56 pregnancies that occurred during employment in dry cleaning, five ended in spontaneous abortion, whereas of the 46 pregnancies of women who were not working outside the home, one ended in spontaneous abortion (χ^2, 3.05; $p < 0.10$) (Bosco et al., 1987).

(b) Stillbirths, congenital malformations, low birth weight

In the study in Montréal described above (McDonald et al., 1987), three stillbirths (observed/expected, 1.86), nine cases of congenital defects (observed/expected, 1.41) and 15 infants with low birth weight (≤ 2500 g) (observed/expected, 1.7) were seen. None of the results was significant.

In the study of Kyyrönen et al. (1989), described above, 24 cases of malformations in infants of dry cleaning and laundry workers were compared with 93 cases in controls. Dry cleaning was not associated with an excess of congenital malformations.

In the Nordic study described above (Olsen et al., 1990), there were 13 stillbirths, 38 cases of congenital malformations and 13 infants with low birth weights (< 1500 g). The odds ratio for these three outcomes combined was 1.7 (95% CI, 0.40–7.1) in association with low exposure and 0.87 (95% CI, 0.20–3.7) for high exposure.

[Most of the studies on outcomes other than spontaneous abortions are obviously limited by small numbers.]

3.4 Genetic and related effects in humans

3.4.1 Alkaline-labile sites/DNA single-strand breaks

No differences in DNA alkaline elution rates (indicating alkali-labile sites/single-strand breaks in DNA) were seen in peripheral lymphocytes from 16 female dry cleaners and from 18 control women in Germany; both groups were smokers. The dry cleaners were reported to have been exposed mainly to tetrachloroethylene, and in some of the facilities the workers were exposed to 'high levels of tetrachloroethylene as evidenced by the smell in the rooms where they worked with no visible signs of protection'. No data on exposure were available (Doerjer et al., 1988).

3.4.2 Urinary mutagenicity

The microfluctuation assay with *Salmonella typhimurium* TA98 and exogenous metabolic activation was used to study the mutagenicity of the urine of 35 dry cleaners (number of nonsmokers not reported) and eight controls. Concentrations of 1–1518 μg/L tetrachloroethylene and 181–320 μg/L fluorohydrocarbons were measured in blood (Kouros *et al.*, 1989). [No evaluation was possible, owing to the limited reporting of the study.]

A significant correlation ($r = 0.84$) was found between the tetrachloroethylene content of air and urinary mutagenicity (assayed in *Salmonella typhimurium* TA100 and TA98) in 24 Estonian dry cleaners (Pruul, 1992). [No evaluation was possible, owing to the limited reporting of the study, including the absence of information on smoking.]

3.4.3 Cytogenetic damage in lymphocytes

Sister chromatid exchange was studied in peripheral lymphocytes from 14 male and 13 female dry cleaners, 12 male and 14 female clerks and an expanded control group of 21 men and 23 women in Japan. The workers had been exposed to a geometric mean, 8-h time-weighted average concentration of tetrachloroethylene of 10 ppm [67.8 mg/m^3] (75th percentile, 27 ppm [183 mg/m^3]; maximum, 179 ppm [1214 mg/m^3]). An effect of exposure to tetrachloroethylene was reported in smoking men but not in nonsmoking women. This conclusion was based on a significant ($p < 0.05$) difference in the mean number of sister chromatid exchanges per cell in 12 exposed male smokers (5.91) from that in three control male nonsmokers (5.01). No difference was seen in comparison with nine nonsmoking men in the expanded control group (5.48) or with men in each control group (5.60; 5.70). No effect was seen in nonsmoking women (Seiji *et al.*, 1990). [The possible confounding effect of smoking was not accounted for.]

4. Summary and Evaluation

4.1 Exposure data

The process of cleaning fabrics with nonaqueous liquids is believed to have begun in France in 1825. The process has evolved into an industry called 'dry cleaning'. 'Camphene' (turpentine) was used initially; in the late 1800s, benzene, benzene soap, naphtha and gasoline began to be used. In the 1920s, Stoddard solvent (mineral spirits or white spirits) was introduced in the United States in order to minimize the fire hazards associated with use of the more volatile hydrocarbon-based solvents. Carbon tetrachloride, the first chlorinated solvent used for dry cleaning, was introduced because of the high cost of petroleum solvents and was widely used until the 1950s. Its use was discontinued because of its toxicity and corrosiveness. Trichloroethylene was introduced in the 1930s. It is still used to a limited extent in Europe and in industrial cleaning plants throughout the world, but it has had a limited market in dry cleaning in the United States because of its incompatibility with acetate dyes.

Use of tetrachloroethylene began to increase in the 1940s, and by the late 1950s it had virtually replaced carbon tetrachloride and trichloroethylene in commercial dry cleaning. Tetrachloroethylene is currently the solvent of choice in most of the world, except in regions, such as Japan, where petroleum-based solvents have remained important in the dry cleaning industry. In 1990, about 53% of the world demand for tetrachloroethylene was for dry cleaning, and about 75% of all dry cleaners used it to clean garments. Chlorofluorocarbon solvents (especially CFC-113) were introduced for use in dry cleaning in the 1970s; however, because of environmental concerns, their use is declining rapidly.

It is estimated that several million people are employed in dry cleaning worldwide. The predominant route of exposure to the solvents used in dry cleaning is by inhalation, although skin absorption and ingestion may also occur. In addition, a wide range of chemicals are used in 'spotting' (treatment of spots); they include chlorinated solvents, amyl acetate, bleaching agents, acetic acid, aqueous ammonia, oxalic acid, hydrogen peroxide and dilute hydrogen fluoride solutions.

Improvements in equipment, solvent reclamation and engineering controls in the dry cleaning industry are resulting in decreasing occupational exposures to chlorinated solvents. The trend to use of 'dry-to-dry' machines, as opposed to the transfer process, has also resulted in reduced emissions and exposures. Typical average exposures to tetrachloroethylene in dry cleaning declined from about 350–700 mg/m^3 in the 1970s to 70–350 mg/m^3 in the late 1980s. The differences in airborne concentrations between dry cleaning shops are often many times greater than the differences in the exposures of machine operators and other staff within a shop.

4.2 Human carcinogenicity data

The relationship between employment in dry cleaning and the occurrence of cancer has been assessed in proportionate mortality studies, case–control studies and four cohort studies. Two cohort studies restricted to dry-cleaning workers in the United States were given greater weight in the evaluation than were the results of cohort studies of laundry and dry-cleaning workers (from Denmark and Sweden).

The relative risks for mortality from urinary bladder cancer were elevated in both United States cohorts (relative risks of 1.7 and 2.5, total of 17 deaths), with evidence in one of the studies of an increasing risk with increasing duration of employment. These results are consistent with those from case–control studies in the United States and in Canada and with the findings of a proportionate mortality study in the United States (although the Danish cohort study found no elevated incidence of bladder cancer) and do not appear to be due to confounding by cigarette smoking.

The relative risk for mortality from oesophageal cancer was elevated by a factor of two in both United States cohorts (23 observed deaths in the two studies combined) and increased with increasing duration and/or intensity of employment. This cancer also occurred in slight excess in a proportionate mortality study in the United Kingdom with respect to launderers, dry cleaners and pressers. Risk estimates for oesophageal cancer were not provided in either of the two Nordic studies of laundry and dry cleaning workers. While in a case–control study of oesophageal cancer in Montréal, Canada, none of the case subjects had worked in dry cleaning, the study was relatively small. The relative incidence of oesophageal cancer is increased by consumption of alcohol drinking and cigarette smoking, but potential confounding by these exposures could not be explored directly in these studies.

The relative risk for mortality from cancer of the pancreas was modestly increased in both United States cohort studies; however, this result was not confirmed in two North American case–control studies.

The occurrence of lung cancer was increased slightly in each of the four cohort studies. The mortality rate from lung cancer in the subgroup with long duration of employment and a long interval since first employment (in the one study that evaluated these characteristics) was not elevated. Two case–control studies in North America gave conflicting results.

The relative risk for mortality from cervical cancer was increased by 70–80% in the two United States cohort studies but not at all among Danish dry cleaning and laundry workers. Socioeconomic characteristics were not adjusted for in these studies.

While in a Swedish case–control study and in one of the United States cohort studies moderate increases were found in the relative risk for cancer of the colon in association with employment in dry cleaning, there was no suggestion of an increase in the risk for this form of cancer in the other relevant studies. Furthermore, in the United States cohort study, there was no particular accentuation of the increased risk in relation to increasing duration of employment.

The Danish cohort study of dry cleaning and laundry workers showed some increase in the incidence of cancers of both the liver and the gall-bladder; however, this result was not confirmed in the two United States cohort studies. The results of the case–control studies of liver cancer in the United States are conflicting.

There was a suggestion of an increased risk for non-Hodgkin's lymphoma in one of the two United States cohort studies (relative risk, 1.7, based on seven deaths) and in a large case–control study in the United States; however, no increase in risk for non-Hodgkin's lymphoma was observed in the other United States cohort study, in the Danish cohort study or in a case–control study from Montréal, Canada.

The results of the four cohort studies do not suggest an increase in the risk for cancer of the kidney, while the results of proportionate mortality studies in Wisconsin and Oklahoma and of the case–control studies from Canada, Denmark and the United States indicate an increase in risk associated with a history of work as a dry cleaner. It may be noteworthy that the petroleum solvents used for dry cleaning in Oklahoma are not typical of those used in much of the rest of the world.

Variation within individual studies of dry cleaners may depend on the nature and level of exposure, which varies from shop to shop and across studies of dry cleaning workers. There is also variation in the types of solvents used over time and across geographic regions. These limitations notwithstanding, the epidemiological studies on dry cleaning indicate that the risks for cancers at two sites, urinary bladder and oesophagus, may be increased by employment in dry cleaning.

4.3 Other relevant data

Inconsistent evidence of slight renal damage among workers exposed to tetrachloroethylene in dry cleaning shops was found in two studies, whereas two other studies in which exposure to tetrachloroethylene was at least as high did not find such an association.

Disturbances of sperm quality and fertility have been observed among dry cleaning workers in a few studies of limited size. Several studies performed in Nordic countries have shown a consistent increase in the risk for spontaneous abortion among dry cleaners, but the studies are not entirely independent of each other. No effect has been observed on other reproductive outcomes, such as stillbirth, congenital malformation or low birth weight, but the power of the studies was limited.

In single studies, lymphocytes from dry cleaning workers showed no increase in the frequency of alkaline-labile sites/DNA single-strand breaks. There was inadequate information to evaluate the genetic effects in humans of exposures in dry cleaning.

4.4 Evaluation[1]

There is *limited evidence* in humans for the carcinogenicity of occupational exposures in dry cleaning.

Overall evaluation

Dry cleaning entails exposures that *are possibly carcinogenic to humans (Group 2B)*.

6. References

Aggazzotti, G., Fantuzzi, G., Righi, E., Predieri, G., Gobba, F.M., Paltrinieri, M. & Cavalleri, A. (1994) Occupational and environmental exposure to perchloroethylene (PCE) in dry cleaners and their family members. *Arch. environ. Health*, **49**, 487–493

Anon. (1991) A close look at dry cleaning shops. *Manuel tech. Nettoyeur Sec*, **111**, 4–11 (in French)

Asal, N.R., Geyer, J.R., Risser, D.R., Lee, E.T., Kadamani, S. & Cherng, N. (1988) Risk factors in renal cell carcinoma. II. Medical history, occupation, multivariate analysis, and conclusions. *Cancer Detect. Prevent.*, **13**, 263–279

Austin, H., Delzell, E., Grufferman, S., Levine, R., Morrison, A.S., Stolley, P.D. & Cole, P. (1987) Case–control study of hepatocellular carcinoma, occupation, and chemical exposures. *J. occup. Med.*, **29**, 665–669

Blair, A., Decoufle, P. & Grauman, D. (1979) Causes of death among laundry and dry cleaning workers. *Am. J. public Health*, **69**, 508–511

Blair, A., Tolbert, P., Thomas, T. & Grauman, D. (1986) Mortality among dry cleaners (Abstract). *Med. Lav.*, **77**, 82–83

Blair, A., Stewart, P.A., Tolbert, P.E., Grauman, D., Moran, F.X., Vaught, J. & Rayner, J. (1990) Cancer and other causes of death among a cohort of dry cleaners. *Br. J. ind. Med.*, **47**, 162–168

---

[1]For definitions of the italicized terms, see pp. 22–26. For separate evaluations of the carcinogenic risk to humans of some of the solvents used in dry cleaning, see this volume, pp. 75 and 159 and IARC (1987a–e, 1989).

Blair, A., Linos, A., Stewart, P.A., Burmeister, L.F., Gibson, R., Everett, G., Schuman, L. & Cantor, K.P. (1992) Comments on occupational and environmental factors in the origin of non-Hodgkin's lymphoma. *Cancer Res.*, **52** (Suppl.), 5501s–5502s

Blair, A., Linos, A., Stewart, P.A., Burmeister, L.F., Gibson, R., Everett, G., Schuman, L. & Cantor, K.P. (1993) Evaluation of risks for non-Hodgkin's lymphoma by occupation and industry exposures from a case–control study. *Am. J. ind. Med.*, **23**, 301–312

Blattner, W.A., Strober, W., Muchmore, A.V., Blaese, R.M., Broder, S. & Fraumeni, J.F., Jr (1976) Familial chronic lymphocytic leukemia. Immunologic and cellular characterization. *Ann. intern. Med.*, **84**, 554–557

Boillat, M.-A., Berode, M. & Droz, P.-O. (1986) Surveillance of workers exposed to tetrachloroethylene or styrene. *Soz. Präventivmed.*, **31**, 260–262 (in French)

Bosco, M.G., Figà-Talamanca, I. & Salerno, S. (1987) Health and reproductive status of female workers in dry-cleaning shops. *Int. Arch. occup. environ. Health*, **59**, 295–301

Brown, D.P. & Kaplan, S.D. (1987) Retrospective cohort mortality study of dry cleaner workers using perchloroethylene. *J. occup. Med.*, **29**, 535–541

Brownson, R.C., Alavanja, M.C.R. & Chang, J.C. (1993) Occupational risk factors for lung cancer among nonsmoking women: a case–control study in Missouri (United States). *Cancer Causes Control*, **4**, 449–454

Burr, G.A. & Todd, W. (1986) *Dutch Girl Cleaners, Springdale, OH* (HETA 86-005-1679), Cincinnati, OH, United States National Institute for Occupational Safety and Health

Cai, S.-X., Huang, M.-Y., Chen, Z., Liu, Y.-T., Jin, C., Watanabe, T., Nakatsuka, H., Seiji, K., Inoue, O. & Ikeda, M. (1991) Subjective symptom increase among dry-cleaning workers exposed to tetrachloroethylene vapor. *Ind. Health*, **29**, 111–121

Cantin, J. (1992) Overview of exposure pathways. In: *Proceedings, International Roundtable on Pollution Prevention and Control in the Drycleaning Industry, May 27–28, 1992, Falls Church, VA* (EPA/774R-92/002), Springfield, VA, National Technical Information Service, pp. 5–11

Center for Chemical Hazard Assessment (1985) *Monographs on Human Exposure to Chemicals in the Workplace: Tetrachloroethylene* (NIOSH Report No. 00167179); SCR TR 84-629), Bethesda, MD, United States National Cancer Institute

Davezies, P., Briant, V., Bergeret, A., Eglizeau, M. & Prost, G. (1983) Study of working conditions in the dry-cleaning business. *Arch. Mal. prof.*, **44**, 159–165 (in French)

Doerjer, G., Buchholz, U., Kreuzer, K. & Oesch, F. (1988) Biomonitoring of DNA damage by alkaline filter elution. *Int. Arch. occup. environ. Health*, **60**, 169–174

Douglas, V., Eli, A., Kemena, R., Marvin, C., McLaughlin, C., Vincent, R. & Woodhouse, L. (1993) *Proposed Airborne Toxic Control Measure and Proposed Environmental Training Program for Perchloroethylene Dry Cleaning Operations*, Sacramento, CA, California Environment Protection Agency

Duh, R.-W. & Asal, N.R. (1984) Mortality among laundry and dry cleaning workers in Oklahoma. *Am. J. public Health*, **74**, 1278–1280

Earnest, G.S. (1994) *Widmers Dry Cleaners, Cincinnati, OH (Health Hazard Evaluation Report)*, Cincinnati, OH, United States National Institute for Occupational Safety and Health

Earnest, G.S., Smith, S.S. & Song, R. (1994) *Control of Chemical Exposures and Ergonomic Risk Factors in Commercial Dry Cleaners* (Report No. ECTB 201-03), Cincinnati, OH, United States National Institute for Occupational Safety and Health

Eddleston, M.T. & Polakoff, P.L. (1974) *Swiss Cleansing Co., Providence, RI* (HHE Report No. 73-86-114; NTIS PB, 232-742), Cincinnati, OH, United States National Institute for Occupational Safety and Health

Edmondson, G.K. & Palin, M.J. (1993) *Occupational Exposure to Perchloroethylene in the UK Drycleaning Industry*, Harrogate, North Yorkshire, Fabric Care Research Association

Eskenazi, B., Wyrobek, A.J., Fenster, L., Katz, D.F., Sadler, M., Lee, J., Hudes, M. & Rempel, D.M. (1991a) A study of the effect of perchloroethylene exposure on semen quality in dry-cleaning workers. *Am. J. ind. Med.*, **20**, 575–591

Eskenazi, B., Fenster, L., Hudes, M., Wyrobek, A.J., Katz, D.F., Gerson, J. & Rempel, D.M. (1991b) A study of the effect of perchloroethylene exposure on the reproductive outcomes of wives of dry-cleaning workers. *Am. J. ind. Med.*, **20**, 593–600

Fredriksson, M., Bengtsson, N.-O., Hardell, L. & Axelson, O. (1989) Colon cancer, physical activity, and occupational exposures. A case–control study. *Cancer*, **63**, 1838–1842

Gallagher, R.P., Threlfall, W.J., Band, P.R. & Spinelli, J.J. (1989) *Occupational Mortality in British Columbia 1950–1984*, Vancouver, Cancer Control Agency of British Columbia

Grob, K., Artho, A. & Egli, J. (1991) Food contamination by perchloroethylene from the air: limits for the air rather than for foods! *Mitt. Geb. Lebensm. Hyg.*, **82**, 56–65 (in German)

Gulyas, H. & Hemmerling, L. (1990) Tetrachloroethylene air pollution originating from coin-operated dry cleaning establishments. *Environ. Res.*, **53**, 90–99

Guralnick, L. (1963) *Mortality by Occupation and Cause of Death among Men 20 to 64 Years of Age: United States 1950. Vital Statistics* (Special Reports 53, No. 3), Washington DC, United States Department of Health, Education, and Welfare

Hayashi, M., Kitazume, M., Yazawa, A., Sato, Y. & Kawamura, T. (1990) Investigation of the health effects associated with solvents used in dry cleaning workplace. Report 2: Personal exposure to tetrachloroethylene (TCE) and TCE levels in man. *Nippon Koshu Eisei Zasshi*, **37**, 177–185 (in Japanese)

Huebner, W.W., Schoenberg, J.B., Kelsey, J.L., Wilcox, H.B., McLaughlin, J.K., Greenberg, R.S., Preston-Martin, S., Austin, D.F., Stemhagen, A., Blot, W.J., Winn, D.M. & Fraumeni, J.F., Jr (1992) Oral and pharyngeal cancer and occupation: a case–control study. *Epidemiology*, **3**, 300–309

IARC (1987a) *IARC Monographs on the Evaluation of Carcinogenic Risks to Humans*, Suppl. 7, *Overall Evaluations of Carcinogenicity: An Updating of* IARC Monographs *Volumes 1–42*, Lyon, pp. 120–122

IARC (1987b) *IARC Monographs on the Evaluation of Carcinogenic Risks to Humans*, Suppl. 7, *Overall Evaluations of Carcinogenicity: An Updating of* IARC Monographs *Volumes 1–42*, Lyon, pp. 152–154

IARC (1987c) *IARC Monographs on the Evaluation of Carcinogenic Risks to Humans*, Suppl. 7, *Overall Evaluations of Carcinogenicity: An Updating of* IARC Monographs *Volumes 1–42*, Lyon, pp. 143–144

IARC (1987d) *IARC Monographs on the Evaluation of Carcinogenic Risks to Humans*, Suppl. 7, *Overall Evaluations of Carcinogenicity: An Updating of* IARC Monographs *Volumes 1–42*, Lyon, p. 64

IARC (1987e) *IARC Monographs on the Evaluation of Carcinogenic Risks to Humans*, Suppl. 7, *Overall Evaluations of Carcinogenicity: An Updating of* IARC Monographs *Volumes 1-42*, Lyon, p. 73

IARC (1989a) *IARC Monographs on the Evaluation of Carcinogenic Risks to Humans*, Vol. 45, *Occupational Exposures in Petroleum Refining; Crude Oil and Major Petroleum Fuels*, Lyon, pp. 39–117

IARC (1989b) *IARC Monographs on the Evaluation of Carcinogenic Risks to Humans*, Vol. 45, *Occupational Exposures in Petroleum Refining; Crude Oil and Major Petroleum Fuels*, Lyon, pp. 159–201

IARC (1989c) *IARC Monographs on the Evaluation of Carcinogenic Risks to Humans*, Vol. 47, *Some Organic Solvents, Resin Monomers and Related Compounds, Pigments and Occupational Exposures in Paint Manufacture and Painting*, Lyon, pp. 43–77

IARC (1989d) *IARC Monographs on the Evaluation of Carcinogenic Risks to Humans*, Vol. 47, *Some Organic Solvents, Resin Monomers and Related Compounds, Pigments and Occupational Exposures in Paint Manufacture and Painting*, Lyon, pp. 79–123

IARC (1989e) *IARC Monographs on the Evaluation of Carcinogenic Risks to Humans*, Vol. 47, *Some Organic Solvents, Resin Monomers and Related Compounds, Pigments and Occupational Exposures in Paint Manufacture and Painting*, Lyon, pp. 125–156

International Fabricare Institute (1987) Perchloroethylene vapor in dry cleaning plants. *Focus Dry Cleaning*, **11**, 1–17

Jalihal, S.S. & Barlow, A.M. (1984) Leukaemia in drycleaners (Letter to the Editor). *J. R. Soc. Health*, **104**, 42

Kaplan, S.D. (1980) *Dry Cleaners Workers Exposed to Perchloroethylene—A Retrospective Cohort Mortality Study* (PB81-231367), Springfield, VA, National Technical Information Service

Katz, R.M. & Jowett, D. (1981) Female laundry and dry cleaning workers in Wisconsin: a mortality analysis. *Am. J. public Health*, **71**, 305–307

Kawauchi, T. & Nishiyama, K. (1989) Residual tetrachloroethylene in dry-cleaned clothes. *Environ. Res.*, **48**, 296–301

Kerr, R.E. (1972) *Environmental Exposures to Tetrachloroethylene in the Dry-cleaning Industry*, M.S. Thesis, Cincinnati, OH, University of Cincinnati

Kirschner, E.M. (1994) Environment, health concerns force shift in use of organic solvents. *C&EN*, **20 June**, 13–20

Klein, P. & Kurz, J. (1994) *Untersuchung von Massnahmen zur Reduzierung der Lösemittelkonzentration in der Nachbarschaft von Textilreinigungen* [Study on preventive measures to reduce the concentration of solvents in the vicinity of textile cleaning shops], Bönnigheim, Hohenstein e.V. Physiological Institute on Clothing

Kouros, B.M., Böttger, A. & Dehnen, W. (1989) Investigations on correlation between thioether excretion and urinary mutagenicity in industrially exposed workers. *Zbl. Hyg.*, **189**, 125–134 (in German)

Kubota, J. (1992) Dry cleaning in Japan: current conditions and regulations. In: *Proceedings, International Roundtable on Pollution Prevention and Control in the Dry Cleaning Industry, May 27–28, 1992, Falls Church, VA* (EPA/774 R-92/002), Springfield, VA, United States National Technical Information Service, pp. 40–43

Kyyrönen, P., Taskinen, H., Lindbohm, M.-L., Hemminki, K. & Heinonen, O.P. (1989) Spontaneous abortions and congenital malformations among women exposed to tetrachloroethylene in drycleaning. *J. Epidemiol. Community Health*, **43**, 346–351

Lauwerys, R., Herbrand, J., Buchet, J.P., Bernard, A. & Gaussin, J. (1983) Health surveillance of workers exposed to tetrachloroethylene in dry-cleaning shops. *Int. Arch. occup. environ. Health*, **52**, 69–77

Linak, E., Leder, A. & Yoshida, Y. (1992) C_2 Chlorinated solvents. In: *Chemical Economics Handbook*, Menlo Park, CA, SRI International, pp. 632.3000–632.3002

Lindbohm, M.-L., Hemminki, K. & Kyyrönen, P. (1984) Parental occupational exposure and spontaneous abortions in Finland. *Am. J. Epidemiol.*, **120**, 370–378

Ludwig, H.R. (1981) *Occupational Exposure to Perchloroethylene in the Dry Cleaning Industry*, Cincinnati, OH, United States National Institute for Occupational Safety and Health

Ludwig, H.R., Meister, M.V., Roberts, D.R. & Cox, C. (1983) Worker exposure to perchloroethylene in the commercial dry cleaning industry. *Am. ind. Hyg. Assoc. J.*, **44**, 600–605

Lynge, E. (1994) Danish Cancer Registry as a resource for occupational research. *J. occup. Med.*, **36**, 1169–1173

Lynge, E. & Thygesen, L. (1990) Primary liver cancer among women in laundry and dry-cleaning work in Denmark. *Scand. J. Work Environ. Health*, **16**, 108–112

Mack, T.M., Peters, J.M., Yu, M.C., Hanisch, R., Wright, W.E. & Henderson, B.E. (1985) Pancreas cancer is unrelated to the workplace in Los Angeles. *Am. J. ind. Med.*, **7**, 253–266

Malker, H. & Weiner, J. (1984) Cancer environment register. Examples of use of register-based epidemiology in occupational health. *Arbete Hälsa*, **9**, 1–107 (in Swedish)

Materna, B.L. (1985) Occupational exposure to perchloroethylene in the dry cleaning industry. *Am. ind. Hyg. Assoc. J.*, **46**, 268–273

McCredie, M. & Stewart, J.H. (1993) Risk factors for kidney cancer in New South Wales. IV. Occupation. *Br. J. ind. Med.*, **50**, 349–354

McDonald, A.D., Armstrong, B., Cherry, N.M., Delorme, C., Diodati-Nolin, A., McDonald, J.C. & Robert, D. (1986) Spontaneous abortion and occupation. *J. occup. Med.*, **28**, 1232–1238

McDonald, A.D., McDonald, J.C., Armstrong, B., Cherry, N., Delorme, C., Nolin, A.D. & Robert, D. (1987) Occupation and pregnancy outcome. *Br. J. ind. Med.*, **44**, 521–526

McLaughlin, J.K., Malker, H.S.R., Stone, B.J., Weiner, J.A., Malker, B.K., Ericsson, J.L.E., Blot, W.J. & Fraumeni, J.F., Jr (1987) Occupational risks for renal cancer in Sweden. *Br. J. ind. Med.*, **44**, 119–123

Mellemgaard, A., Engholm, G., McLaughlin, J.K. & Olsen, J.H. (1994) Occupational risk factors for renal-cell carcinoma in Denmark. *Scand. J. Work Environ. Health*, **20**, 160–165

Michelsen, E.M. (1957) *Remembering the Years 1907–1957*, Silver Spring, MD, National Institute of Drycleaning

Milham, S., Jr (1992) *Occupational Mortality in Washington State 1950–1989*, Seattle, Washington State Department of Health

Missere, M., Giacomini, C., Gennari, P., Violante, F., Dallolio, L., Failla, G., Lodi, V. & Raffi, G. (1988) Evaluation of exposure to tetrachloroethylene in dry cleaning shops. *Med. Lav.*, **10**, 261–267 (in Italian)

Monster, A.C., Boersma, G. & Steenweg, H. (1979) Kinetics of tetrachloroethylene in volunteers; influence of exposure concentration and work load. *Int. Arch. occup. environ. Health*, **42**, 303–309

Mutti, A., Alinovi, R., Bergamaschi, E., Biagini, C., Cavazzini, S., Franchini, I., Lauwerys, R.R., Bernard, A.M., Roels, H., Gelpi, E., Rosello, J., Ramis, I., Price, R.G., Taylor, S.A., De Broe, M., Nuyts, G.D., Stolte, H., Fels, L.M. & Herbort, C. (1992) Nephropathies and exposure to perchloroethylene in dry-cleaners. *Lancet*, **340**, 189–193

Nakamura, K. (1985) Mortality patterns among cleaning workers. *Jpn. J. ind. Health*, **27**, 24–37

Office of Population Censuses and Surveys (1986) *Occupational Mortality—Decennial Supplement. 1979–80, 1982–83* (Series DS No.6), London, Her Majesty's Stationery Office

Olsen, J., Hemminki, K., Ahlborg, G., Bjerkedal, T., Kyyrönen, P., Taskinen, H., Lindbohm, M.-L., Heinonen, O.P., Brandt, L., Kolstad, H., Halvorsen, B.A. & Egenaes, J. (1990) Low birthweight, congenital malformations, and spontaneous abortions among dry-cleaning workers in Scandinavia. *Scand. J. Work Environ. Health*, **16**, 163–168

Olsen, J.H., de Nully Brown, P., Schulgen, G. & Møller Jensen, O. (1991) Parental employment at time of conception and risk of cancer in offspring. *Eur. J. Cancer*, **27**, 958–965

Pannier, R., Hübner, G. & Schulz, G. (1986) Results of biological exposure tests in connection with passive dosimetric measurements of ambient air concentrations in occupational exposure to tetrachloroethylene in dry-cleaning shops. *Z. ges. Hyg.*, **32**, 297–300 (in German)

Petreas, M.X., Rappaport, S.M., Materna, B.L. & Rempel, D.M. (1992) Mixed-exhaled air measurements to assess exposure to tetrachloroethylene in dry cleaners. *J. Exposure Anal. environ. Epidemiol.*, **Suppl. 1**, 25–39

Petrone, R.L. (1988) Cancer mortality among petroleum solvent exposed Oklahoma dry cleaners (Abstract). *Diss. Abstr. int.*, **48**, 1955-B

Pruul, R.K. (1992) The role of tetrachloroethylene on genetic system of dry-cleaning workers. *Gig. Sanit.*, **31**, 23–24 (in Russian)

Pryor, P. (1985) *Denver Laundry and Dry Cleaning, Denver, CO* (Health Hazard Evaluation Report No. 84-340-1606), Cincinnati, OH, United States National Institute for Occupational Safety and Health

Rachootin, P. & Olsen, J. (1983) The risk of infertility and delayed conception associated with exposures in the Danish workplace. *J. occup. Med.*, **25**, 394–402

Rantala, K., Riipinen, H. & Anttila, A. (1992) *Altisteet Työssä. 22. Halogeeni-hiilivedyt (Exposure at work. 22. Halogenated hydrocarbons)*, Helsinki, Finnish Institute of Occupational Health, Finnish Work Environment Fund

Reich, D.A. & Cormany, C.L. (1979) Dry cleaning. In: Mark, H.F., Othmer, D.F., Overberger, C.G., Seaborg, G.T. & Grayson, M., eds, *Kirk-Othmer Encyclopedia of Chemical Technology*, 3rd Ed., Vol. 8, New York, John Wiley & Sons, pp. 50–68

Rice, B. & Weinberg, J. (1994) *Dressed to Kill. The Dangers of Dry Cleaning and the Case for Chlorine-free Alternatives*, Washington DC, Greenpeace Pollution Probe

Ruder, A.M., Ward, E.M. & Brown, D.P. (1994) Cancer mortality in female and male dry-cleaning workers. *J. occup. Med.*, **36**, 867–874

Schoenberg, J.B., Stemhagen, A., Mogielnicki, A.P., Altman, R., Abe, T. & Mason, T.J. (1984) Case–control study of bladder cancer in New Jersey. I. Occupational exposures in white males. *J. natl Cancer Inst.*, **72**, 973–981

Seeber, A. (1989) Neurobehavioral toxicity of long-term exposure to tetrachloroethylene. *Neurotoxicol. Teratol.*, **11**, 579–583

Seiji, K., Jin, C., Watanabe, T., Nakatsuka, H. & Ikeda, M. (1990) Sister chromatid exchanges in peripheral lymphocytes of workers exposed to benzene, trichloroethylene, or tetrachloroethylene, with reference to smoking habits. *Int. Arch. occup. environ. Health*, **62**, 171–176

Shipman, A.J. & Whim, B.P. (1980) Occupational exposure to trichloroethylene in metal cleaning processes and to tetrachloroethylene in the dry cleaning industry in the UK. *Ann. occup. Hyg.*, **23**, 197–204

Siemiatycki, J. (1991) *Risk Factors for Cancer in the Workplace*, Boca Raton, FL, CRC Press

Silverman, D.T., Hoover, R.N., Albert, S. & Graff, K.M. (1983) Occupation and cancer of the lower urinary tract in Detroit. *J. natl Cancer Inst.*, **70**, 237–245

Silverman, D.T., Levin, L.I. & Hoover, R.N. (1989) Occupational risks of bladder cancer in the United States: II. Nonwhite men. *J. natl Cancer Inst.*, **81**, 1480–1483

Skender, L., Karacic, V., Prpic-Majic, D. & Kezic, S. (1986) Occupational exposure indicators to tetrachloroethylene in a dry-cleaning shop. In: Foa, V., Emmet, E.A., Maroni, M. & Colombi, A., eds, *Occupational and Environmental Chemical Hazards*, Chichester, Ellis Horwood, pp. 192–196

Smith, E.M., Miller, E.R., Woolson, R.F. & Brown, C.K. (1985) Bladder cancer risk among laundry workers, dry cleaners, and others in chemically-related occupations. *J. occup. Med.*, **27**, 295–297

Solet, D. & Robins, T.G. (1991) Renal function in dry cleaning workers exposed to perchloroethylene. *Am. J. ind. Med.*, **20**, 601–614

Solet, D., Robins, T.G. & Sampaio, C. (1990) Perchloroethylene exposure assessment among dry cleaning workers. *Am. ind. Hyg. Assoc. J.*, **51**, 566–574

Stemhagen, A., Slade, J., Altman, R. & Bill, J. (1983) Occupational risk factors and liver cancer. A retrospective case–control study of primary liver cancer in New Jersey. *Am. J. Epidemiol.*, **117**, 443–454

Stricoff, R.S. (1983) *Control of Occupational Health Hazards in the Dry Cleaning Industry. Instructor's Guide* (DHHS (NIOSH) Publ. No. 82-1876), Cincinnati, OH, United States National Institute for Occupational Safety and Health

Suarez, L., Weiss, N.S. & Martin, J. (1989) Primary liver cancer death and occupation in Texas. *Am. J. ind. Med.*, **15**, 167–175

Takeuchi, Y., Hisanaga, N., Ono, Y., Iwata, M., Oguri, S., Tauchi, T. & Tanaka, T. (1981) An occupational health survey on dry cleaning workers. *Jpn. J. ind. Health*, **23**, 407–418 (in Japanese)

van der Tuin, J. (1979) *Perchloroethylene in a Dry Cleaning Service* (Report F1647), Rijswyk, Institute of Environmental Hygiene and Health Engineering/Dutch Organization for Applied Research (in Dutch)

van der Tuin, J. & Hoevers, W. (1977a) *Perchloroethylene in a Dry Cleaning Service* (Report F1560), Rijswyk, Institute of Environmental Hygiene and Health Engineering/Dutch Organization for Applied Research (in Dutch)

van der Tuin, J. & Hoevers, W. (1977b) *Perchloroethylene in a Dry Cleaning Service* (Report F1590), Rijswyk, Netherlands, Institute of Environmental Hygiene and Health Engineering/Dutch Organization for Applied Research (in Dutch)

Tuttle, T.C., Wood, G.D. & Grether, C.B. (1977) *A Behavioral and Neurological Evaluation of Dry Cleaners Exposed to Perchloroethylene* (DHEW (NIOSH) Publ. No. 77-214), Cincinnati, OH, United States National Institute for Occupational Safety and Health

United States Environmental Protection Agency (1991a) *Economic Impact Analysis of Regulatory Controls in the Dry Cleaning Industry* (EPA 450/3-91-021; PB 92-126770), Research Triangle Park, NC, Office of Air Quality Planning and Standards

United States Environmental Protection Agency (1991b) *Dry Cleaning Facilities—Background Information for Proposed Standards* (EPA 450/3-91-020a), Research Triangle Park, NC, Office of Air Quality Planning and Standards

United States Environmental Protection Agency (1991c) National emission standards for hazardous air pollutants for source categories: perchloroethylene emissions from dry cleaning facilities. Proposed rule and notice of public hearing. *Fed. Reg.*, **56**, 64382–64402

United States National Institute for Occupational Safety and Health (1981) *NIOSH National Occupational Health Survey: Written Communication* (IWSB-071.25-0.71.54), Cincinnati, OH, US Department of Health and Human Services, Public Health Service, Centers for Disease Control

Vieths, S., Blaas, W., Fischer, M., Krause, C., Matissek, R., Mehlitz, I. & Weber, R. (1988) Contamination of foodstuffs by emissions from dry cleaning units. *Z. Lebensm. Unters.-Forsch.*, **186**, 393–397 (in German)

Vyskocil, A., Emminger, S., Tejral, J., Fiala, Z., Ettlerova, E. & Cermanová, A. (1990) Study on kidney function in female workers exposed to perchloroethylene. *Hum. exp. Toxicol.*, **9**, 377–380

Zielhuis, G.A., Gijsen, R. & van der Gulden, J.W.J. (1989) Menstrual disorders among dry-cleaning workers (Letter to the Editor). *Scand. J. Work Environ. Health*, **15**, 238

SOME CHLORINATED SOLVENTS AND RELATED CHEMICALS

TRICHLOROETHYLENE

This substance was considered by previous working groups, in June 1978 and March 1987 (IARC, 1979, 1987a). Since that time, new data have become available, and these have been incorporated into the monograph and taken into consideration in the present evaluation.

1. Exposure Data

1.1 Chemical and physical data

1.1.1 Nomenclature

Chem. Abstr. Serv. Reg. No.: 79-01-6
Deleted CAS Reg. No.: 52037-46-4
Chem. Abstr. Name: Trichloroethene
IUPAC Systematic Name: Trichloroethylene
Synonyms: Ethinyl trichloride; ethylene trichloride; TCE; 1,1,2-trichlorethylene

1.1.2 Structural and molecular formulae and relative molecular mass

$$\underset{Cl}{\overset{H}{}}C=C\underset{Cl}{\overset{Cl}{}}$$

C_2HCl_3 Relative molecular mass: 131.39

1.1.3 Chemical and physical properties of the pure substance

(a) *Description*: Mobile liquid with chloroform-like odour (Budavari, 1989)

(b) *Boiling-point*: 87 °C (Lide, 1993)

(c) *Melting-point*: –73 °C (Lide, 1993)

(d) *Density*: 1.4642 at 20 °C/4 °C (Lide, 1993)

(e) *Spectroscopy data*: Infrared (prism [185]; grating [62]), nuclear magnetic resonance (proton [9266]; C-13 [410]) and mass [583] spectral data have been reported (Sadtler Research Laboratories, 1980; Weast & Astle, 1985).

(f) *Solubility*: Slightly soluble in water (1.1 g/L at 25 °C); soluble in ethanol, diethyl ether, acetone and chloroform (Lide, 1993; PPG Industries, Inc., 1994)

(g) *Volatility*: Vapour pressure, 100 mm Hg [13.3 kPa] at 31.4 °C (Lide, 1993); relative vapour density (air = 1.0), 4.53 (Budavari, 1989)

(h) *Stability*: Photo-oxidized in air by sunlight (half-time, five days) giving phosgene and dichloroacetyl chloride (United States Environmental Protection Agency, 1985)

(i) *Reactivity*: Incompatible with strong caustics and alkalis and with chemically active metals such as barium, lithium, sodium, magnesium, titanium and beryllium (United States National Institute for Occupational Safety and Health, 1994a)

(j) *Octanol/water partition coefficient (P)*: log P, 2.61 (Hansch et al., 1995)

(k) *Conversion factor*: mg/m^3 = 5.37 × ppm[1]

1.1.4 Technical products and impurities

Commercial grades of trichloroethylene, formulated to meet use requirements, differ in the amount and type of added inhibitor. Typical grades contain > 99% trichloroethylene; they include a neutrally inhibited vapour-degreasing grade and a technical grade for use in formulations. Stabilizers that have been used in formulations of trichloroethylene include neutral inhibitors and free-radical scavengers, amyl alcohol, *n*-propanol, isobutanol, 2-pentanol, diethylamine, triethylamine, dipropylamine, diisopropylamine, diethanolamine, triethanolamine, morpholine (see IARC, 1989a), *N*-methylmorpholine, aniline (see IARC, 1987b), acetone, ethyl acetate, borate esters, ethylene oxide (see IARC, 1994a), propylene oxide (see IARC, 1994b), 1,2-epoxybutane (see IARC, 1989b), cyclohexene oxide, butadiene dioxide, styrene oxide (see IARC, 1994c), pentene oxide, 2,3-epoxy-1-propenol, 3-methoxy-1,2-epoxypropane, stearates, 2,2,4-trimethyl-1-pentene, 2-methyl-1,2-epoxypropanol, epoxycyclopentanol, epichlorohydrin (see IARC, 1987c), tetrahydrofuran, tetrahydropyran, 1,4-dioxane (see IARC, 1987d), dioxalane, trioxane, alkoxyaldehyde hydrazones, methyl ethyl ketone, nitromethanes, nitropropanes, phenol (see IARC, 1989c), *ortho*-cresol, thymol, *para-tert*-butylphenol, *para-tert*-amylphenol, isoeugenol, pyrrole, *N*-methylpyrrole, *N*-ethylpyrrole, (2-pyrryl)trimethylsilane, glycidyl acetate, isocyanates and thiazoles (United States Environmental Protection Agency, 1985; WHO, 1985).

Apart from added stabilizers, commercial grades of trichloroethylene should not contain more than the following amounts of impurities: water, 100 ppm [mg/L]; acidity (as HCl), 5 ppm; insoluble residue, 10 ppm (Mertens, 1993). Free chlorine should not be detectable (PPG Industries, Inc., 1994). Impurities that have been found in commercial trichloroethylene products include: carbon tetrachloride (see IARC, 1987e), chloroform (see IARC, 1987f), 1,2-dichloroethane (see IARC, 1987g), *trans*-1,2-dichloroethylene, *cis*-1,2-dichloroethylene, pentachloroethane (see IARC, 1987h), 1,1,1,2-tetrachloroethane (see IARC, 1987i), 1,1,2,2-tetrachloroethane (see IARC, 1987j), 1,1,1-trichloroethane (see IARC, 1987k), 1,1,2-trichloroethane (see IARC, 1991), 1,1-dichloroethylene, tetrachloroethylene (see monograph, this volume), bromodichloromethane, bromodichloroethylene and benzene (see IARC, 1987l) (WHO, 1985; Mertens, 1993).

Trade names for trichloroethylene include: Algylen, Anamenth, Chlorilen, Chlorylen, Densinfluat, Fluate, Germalgene, Narcogen, Narkosoid, Threthylen, Threthylene, Trethylene,

---

[1] Calculated from: mg/m^3 = (relative molecular mass/24.45) × ppm, assuming normal temperature (25 °C) and pressure (101 kPa)

Tri, Trichloran, Trichloren, Triclene, Trielene, Trielin, Trieline, Trilen, Trilene, Trimar and Westrosol.

1.1.5 Analysis

Selected methods for the analysis of trichloroethylene in various matrices are identified in Table 1.

Table 1. Methods for the analysis of trichloroethylene

Sample matrix	Sample preparation	Assay procedure	Limit of detection	Reference
Air	Adsorb on charcoal; desorb with carbon disulfide	GC/FID	0.01 mg/sample	Eller (1994); US Occupational Safety and Health Administration (1990)
	Draw air into sample bag; inject aliquot into gas chromatograph	GC/PID	0.25 ng/sample	Eller (1994)
	Draw air through Tenax sample tube; heat; desorb on cold trap	GC/MS	20 ng	US Environmental Protection Agency (1988a)
	Draw air into cryogenically cooled trap; heat	GC/FID and/or GC/EC	1–5 ng	US Environmental Protection Agency (1988a)
	Draw air into SUMMA® passivated stainless-steel canister; desorb on cold trap	GC/MS or GC/EC-FID-PID	NR	US Environmental Protection Agency (1988a)
Coffee	Isolate sample by closed-system vacuum distillation with toluene	GC/EC or GC/ECD	NR	US Food and Drug Administration (1983)
Grain	Add sample to acetone; store 48 h in the dark; add sodium chloride; add calcium chloride	GC/ECD	NR	Sawyer *et al.* (1990)
Spice oleoresins	Add sample to absolute alcohol/ 1,2-dichloropropane mixture; dilute with absolute alcohol and shake	GC	NR	Fazio (1990)
	Isolate sample by closed-system vacuum distillation with toluene	GC/EC	NR	US Food and Drug Administration (1983)
Water	Purge (inert gas); trap on suitable sorbent material; desorb as vapour onto packed gas chromatographic column	GC/ECD or GC/MCD	0.001 and 0.12 µg/L	US Environmental Protection Agency (1988b, 1994)
		GC/MS	0.4 and 1.9 µg/L	US Environmental Protection Agency (1988b, 1994)

Table 1 (contd)

Sample matrix	Sample preparation	Assay procedure	Limit of detection	Reference
Water (contd)	Purge and trap as above; desorb as vapour onto capillary gas chromatographic column	GC/PID-ECD GC/PID	0.01–0.06 μg/L 0.02–0.19 μg/L	US Environmental Protection Agency (1988b, 1994)
	Purge (inert gas); trap on suitable sorbent material; desorb as vapour onto gas chromatographic column	GC/PID	0.01 μg/L	US Environmental Protection Agency (1988b, 1994)
	Add internal standard (isotope-labelled trichloroethylene); purge, trap and desorb as above	GC/MS	10 μg/L	US Environmental Protection Agency (1994)
Liquid and solid wastes	Purge (inert gas); trap on suitable sorbent material; desorb as vapour onto packed gas chromatographic column	GC/ECD GC/MS	0.12 μg/L PQL	US Environmental Protection Agency (1986a) US Environmental Protection Agency (1986b)

GC, gas chromatography; FID, flame ionization detection; PID, photoionization detection; MS, mass spectrometry; NR, not reported; EC, electron capture detection; ECD, electrolytic conductivity detection; MCD, microcoulometric detection; PQL, practical quantification limit: 5 μg/L for groundwater; 5 μg/kg for soil and sediment samples; 250–2500 μg/kg for liquid wastes

Three gas chromatography/mass spectrometry (GC/MS) and three purge-and-trap GC methods for purgeable organic compounds, including trichloroethylene, are usually used for analysing aqueous samples (see also Table 1). The first method (EPA Method 624 and APHA/AWWA/WEF Method 6210B) is a packed-column method useful for the determination of trichloroethylene in municipal and industrial wastes. A similar purge-and-trap method (EPA Method 503.1 and APHA/AWWA/WEF Method 6220C), which includes photoionization detection, is applicable for the determination of trichloroethylene in drinking-water and raw source water. The second GC/MS method (EPA Method 524.1 and APHA/ AWWA/WEF Method 6210C), also involving a packed column, is also applicable for the determination of trichloroethylene in drinking-water and raw source water. Similar purge-and-trap methods (EPA Methods 601 and 502.1 and APHA/AWWA/WEF Methods 6230B and 6230C), including electrolytic conductivity and microcoulometric detection, are applicable for the determination of trichloroethylene in municipal and industrial discharges (6230B) and in drinking-water and raw source water (6230C). The third group of GC/MS and purge-and-trap methods (EPA Method 524.2 and APHA/AWWA/WEF Method 6210D; EPA Method 502.2 and APHA/AWWA/WEF Method 6230D) are identical to the previous ones except that a capillary column is used. The second and third methods are intended primarily for the detection of large numbers of contaminants at very low concentrations, which are not detectable with the first method (Greenberg et al., 1992).

Trichloroethylene can also be determined by colorimetry in the Fujiwara test, in which it is treated with pyridine in an alkaline environment. Solution absorbency is then determined at 535 or 470 nm, with a sensitivity of about 1 mg/kg. Trichloroethylene can also be determined by infrared spectroscopy. Gaseous compound is measured from the optical density of the mixture at a wavelength of 11.8 µm (detection sensitivity, ≥ 0.5 µg/L). High-resolution GC with electron capture detection has been used for determining trichloroethylene in soil. High-resolution GC with MS have been used for confirmation, with a detection threshold of about 10 mg/kg. Similar methods can be used to determine trichloroethylene and its major metabolites, trichloroacetic acid and trichloroethanol, in human tissues and fluids (WHO, 1985).

1.2 Production and use

1.2.1 Production

Trichloroethylene was first prepared in 1864 by Fischer in experiments on the reduction of hexachloroethane with hydrogen (Hardie, 1964). Commercial production of trichloroethylene began in Germany in 1920 and in the United States of America in 1925 (Mertens, 1993).

Until 1968, about 85% of United States production capacity of trichloroethylene was based on acetylene. The acetylene-based process consists of two steps: acetylene is first chlorinated to 1,1,2,2-tetrachloroethane, with a ferric chloride, phosphorus chloride or antimony chloride catalyst, and the product is then dehydrohalogenated to trichloroethylene (Mertens, 1993). The current method of manufacture is from ethylene or 1,2-dichloroethane. In a process used by one plant in the United States, trichloroethylene is produced by noncatalytic chlorination of ethylene dichloride or other C_2 chlorinated hydrocarbons. Another method is to react ethylene dichloride and other C_2 hydrocarbons with a mixture of oxygen and chlorine or hydrogen chloride (Linak et al., 1992).

Trichloroethylene can also be produced by direct chlorination of ethylene in the absence of oxygen, giving a mixture of tetrachloroethane and pentachloroethane. The products are thermally cracked to produce a mixture of trichloroethylene, tetrachloroethylene and hydrochloric acid. This process was developed in Japan and is used there (Linak et al., 1992).

Table 2 shows the production of trichloroethylene in selected countries between 1941 and 1990. Production has declined in recent years. Trichloroethylene is manufactured by one company each in Austria (with an annual capacity of 6000 tonnes), Germany (10 000 tonnes), Italy (15 000 tonnes) and Spain (29 000 tonnes). Two companies manufacture trichloroethylene in France (90 000 tonnes) and the United States (145 000 tonnes). Three companies in Japan produce trichloroethylene, with an estimated annual capacity of 85 000 tonnes (Linak et al., 1992). Two companies in Canada were the only domestic manufacturers of trichloroethylene. In 1976, the total capacity of these plants was 38 000 tonnes, and 22 500 tonnes were produced. One plant closed in 1985, and imports have increased as a result (Moore et al., 1991).

Trichloroethylene is also produced in Argentina, Australia, Belgium, China, India, Macedonia, Poland, Romania, the Russian Federation, Slovakia, South Africa and the United Kingdom (Chemical Information Services Ltd, 1994).

Table 2. Production of trichloroethylene in selected countries (thousand tonnes)

Year	Western Europe	Japan	USA[a]
1941			25
1945			84
1955			143[b]
1960			160
1965			197
1970			277
1975		85	133
1980	210	82	121
1981	205	74	177
1982	210	67	86
1983	200	67	91
1984	215	74	91
1985	205	73	79
1986	183	71	77
1987	166	64	88
1988	169	70	82
1989	154	65	79
1990	131	57	79

From Linak et al. (1992)
[a] The US International Trade Commission stopped reporting trichloroethylene production and sales in 1982. The data for 1983–90 are estimates from the *Chemical Economics Handbook* (Linak et al., 1992).
[b] From Su & Goldberg (1976)

1.2.2 Use

Trichloroethylene was used earlier as an extraction solvent for natural fats and oils, such as palm, coconut and soya bean oils. It was also an extraction solvent for spices, hops and the decaffeination of coffee (Linak et al., 1992). The United States Food and Drug Administration (1977) banned these uses of trichloroethylene because of its toxicity; its use in cosmetic and drug products was also discontinued (Mertens, 1993).

Demand for trichloroethylene was generated mainly by the development of vapour degreasing after the 1920s and by the growth of the dry cleaning industry in the 1930s, but trichloroethylene was replaced in dry cleaning by tetrachloroethylene in the mid-1950s. By 1989, about 85% of the trichloroethylene produced in the United States was used in metal cleaning; the remaining 15% was equally divided between exports and miscellaneous applications. The pattern in Japan was similar to that in the United States, at 83 and 17%, respectively. In western Europe, 95% was used in vapour degreasing and 5% in other uses (Mertens, 1993). Similar use patterns have been reported for Canada (Moore et al., 1991) and

Finland (Mroueh, 1993). Tables 3–5 present the uses of trichloroethylene in western Europe, Japan and the United States. Because of environmental and occupational health concerns, industry has attempted to restrict solvent emissions and maximize recovery and recycling. Trichloroethylene is, however, replacing 1,1,1-trichoroethane in some applications (Linak *et al.*, 1992).

Table 3. Use of trichloroethylene in western Europe (thousand tonnes)

Year	Metal cleaning (vapour degreasing)	Metal cleaning (cold cleaning)	Other
1980	164	25	26
1984	137	10	23
1987	124	10	16
1990	120	10	5

From Linak *et al.* (1992), estimates

Table 4. Use of trichloroethylene in Japan (thousand tonnes)

Year	Metal cleaning	Other
1980	49	16
1983	52	11
1987	49	12
1990	30	8

From Linak *et al.* (1992), estimates

Trichloroethylene has also been used, in limited quantities, to control relative molecular mass (by chain transfer) in the manufacture of polyvinyl chloride. An estimated 5500 tonnes are used annually for this application in the United States. It has also been used as a solvent in the rubber industry, some adhesive formulations and in research laboratories. In the textile industry, it is used as a carrier solvent for spotting fluids and as a solvent in dyeing and finishing (Fishbein, 1976; Linak *et al.*, 1992; Mertens, 1993). It is also used as a solvent in printing inks, paint, lacquers, varnishes, adhesives and paint strippers. It was used as both an anaesthetic and an analgesic in obstetrics (Smith, 1966). Trichloroethylene has been used in the aerospace industry for flushing liquid oxygen (Sax & Lewis, 1987). In a study of potential sources of indoor air pollution in the United States, 25 of 1159 (2.2%) common household products were found to contain trichloroethylene (Sack *et al.*, 1992).

The major use of trichloroethylene is in metal cleaning or degreasing. Degreasing is important in all metalworking and maintenance operations to remove oils, greases, waxes, tars and moisture before final surface treatments, such as galvanizing, electroplating, painting, anodizing and application of conversion coatings. Trichloroethylene is used in degreasing operations in five main industrial groups: furniture and fixtures, fabricated metal products, electric and electronic equipment, transport equipment and miscellaneous manufacturing industries. It is also used in plastics, appliances, jewellery, automobile, plumbing fixtures, textiles, paper, glass and printing (Papdullo *et al.*, 1985; Linak *et al.*, 1992).

Table 5. Use of trichloroethylene in the United States (thousand tonnes)

Year	Metal cleaning	Other
1971	200	15
1974	153	4
1977	102	20
1980	84	13
1984	72	14
1987	57	9
1990	46	5

From Linak *et al.* (1992), estimates

Metal cleaning operations are of two types: cold cleaning and vapour cleaning. In cold cleaning, trichloroethylene is applied at room temperature; in vapour degreasing, the solvent vapours are condensed on the part to be cleaned. In cold cleaning, the metal parts are either dipped into the solvent solution or the solution is sprayed and wiped onto the object. The cold process is frequently used in maintenance operations and on small parts. Vapour degreasing requires a tank with heating coils on the bottom and a condensing zone near the top. The solvent is heated to boiling, and the hot vapour fills the condensing zone near the top of the tank. Soiled objects are lowered into this zone, where the vapour condenses into a pure liquid solvent on the piece and dissolves and carries off dirt as it drains back into the tank. The part dries immediately (Papdullo *et al.*, 1985; Linak *et al.*, 1992).

1.3 Occurrence

1.3.1 Natural occurrence

Natural production of trichloroethylene has been reported in temperate, subtropical and tropical algae and in one red microalga (Abrahamsson *et al.*, 1995).

1.3.2 Occupational exposure

The United States National Institute for Occupational Safety and Health (1994b) indicated that about 401 000 employees in 23 225 plants in the United States are potentially exposed to trichloroethylene. This estimate is based on a survey of companies and did not involve actual measurements. Table 6 summarizes the results of studies of occupational exposure.

Table 6. Occupational exposures to trichloroethylene

Country	No. of plants	Job, task or industry	No. of samples[a]	Air concentration (mg/m³) Mean	Air concentration (mg/m³) Range	Reference
Finland 1982–85	11	Vapour degreasing	24 (A) 13 (P) TWA	[43.0] [37.6]	< [5.4–20.9] < [5.4–161]	Rantala et al. (1992)
	1	Rubber bonding	1 (A) TWA		[32.2]	
	1	Museum textile restoration	2 (P) 1-h		[3303]	
Netherlands	9	Rubber degreasing, cementing	137	4		Kromhout et al. (1994)
Sweden	14	Degreasing	336 (A)	[328]	[0–2230]	Ahlmark
	570	Degreasing	35 000–40 000 (A)	[86]	3% [> 161]	et al. (1963)
	19	Degreasing	29 (P)	27	3–144	Ulander et al. (1992)
Switzerland	10	Degreasing	96 (P)	[304]	[5.4–1799]	Grandjean et al. (1955)
United Kingdom	32	Degreasing	212 (P)	91% < [161] 97% < [269] 99% < [537]		Shipman & Whim (1980)
USA	60	Degreasing Condenser, nonvented Condenser, vented	433 (P) 187 149	 [725] [515]	 [16–4833] [27–2110]	Morse & Goldberg (1943)
	NR	Degreasing	146 (A)[b]	86% < [537] 96% < [1074]		Hargarten et al. (1961)
	1	Degreasing	11 (P)	[302]	[199–419]	Vandervort & Polakoff (1973)
	1	Degreasing ignition coils	(P)		0–[537]	Bloom et al. (1974)
	1	Electronic cleaning	3 (P)	[446]	[408–483]	Gilles & Philbin (1976)
	1	Semi-conductor degreasing	10 (P)	16.1	2–57	Gunter (1977)
	1	Degreasing operator Degreasing operator Degreasing operator Lathe operator next to degreaser	20 (P) 7 (P) 6 (P) 7 (P)	[736] [88.1] [67.7] [52.1]	[140–2024] [37.6–456] [37.6–199] [37.6–129]	Kominsky (1978)
	1	Aircraft degreasing	4 (P)	[21.5]	[5.4–37.6]	Okawa et al. (1978)
	1	Tank relining	8 (P)	[1.3]	ND–[5.4]	Burroughs (1980)

Table 6 (contd)

Country	No. of plants	Job, task or industry	No. of samples[a]	Air concentration (mg/m³)		Reference
				Mean	Range	
USA (contd)	1	Degreasing sheet metal	2 (P) 2 (A)	11 11	10–12 4–18	Johnson (1980)
	1	Degreasing, custom finishing	23 (P) 2 (A)	8.3 6	1–38 4–8	Ruhe & Donohue (1980)
	1	Vapour degreasing	14 (P)	[333]	[26.9–1670]	Burgess (1981)
	1	Degreasing, bus maintenance	3 (A)	3.0	ND–8.9	Love & Kern (1981)
	1	Degreasing	24 (STEL) 9 (TWA)	742 145	56–2000 37–357	Ruhe et al. (1981)
	1	Degreasing, plastics	2 (P)	[4.8]	[2.7–7.0]	Burroughs & Moody (1982)
	1	Degreasing, electronics	79 (P)	10.2	ND–209	Lee & Parkinson (1982)
	1	Degreasing, medical	5 (P) 2 (A)	5.4 6.5	1–16 4–9	Ruhe (1982)
	1	Degreasing, energy conservation products	2 (P) 10 (A)	[36.5] [1.1]	[22–51] [0.54–3.2]	Almaguer et al. (1984)
	1	Degreasing Silk screening	9 (P) 2 (A) 5 (P)	[716] [184] [23.6]	[39–2288] [0.54–367] [1.6–81.1]	Belanger & Coye (1984)
	1	Degreasing aircraft	29 (TWA, P) 11 (TWA, A) 22 (STEL)	[30.7] [28.5] [320]	[ND–208] [2–121] [ND–1256]	Gorman et al. (1984)
	1	Taxidermy	2 (A) 2 (P)	[8.9] [8.9]	[1.1–16.6] [1.7–16]	Kronoveter & Boiano (1984)
	1	Degreasing	(TWA) (STEL)	205 1084	117–357 413–2000	Landrigan et al. (1987)

ND, not detected; NR, not reported. Most measurements were taken after observation of operating deficiencies of degreasers between 1952 and 1957.

[a] P, personal air samples (breathing zone); A, area samples; STEL, short-term exposure limit; TWA, time-weighted average

1.3.3 Environmental occurrence

Trichloroethylene has been reported in the air, rainwater, surface waters, drinking-water, seawater, marine sediments, marine invertebrates, marine mammals, foods and human tissues (McConnell *et al.*, 1975).

(a) Air

The levels of trichloroethylene in air have been measured throughout the world (Table 7). In a compilation of the results of surveys of ambient air in the United States before 1981 (Brodzinsky & Singh, 1983; United States Agency for Toxic Substances and Disease Registry, 1989), representing 2353 monitoring points, the mean concentrations were 30 ppt [0.2 µg/m^3] in rural areas, 460 ppt [2.5 µg/m^3] in urban and suburban areas and 1200 ppt [64 µg/m^3] in industrialized areas near sources of trichloroethylene emissions. Industrial releases of trichloroethylene to the environment in the United States were 24 430 tonnes in 1988, 22 400 tonnes in 1989, 17 680 tonnes in 1990 and 15 950 tonnes in 1991 (United States Environmental Protection Agency, 1993).

Air emissions in western Europe in 1980 are reported in Table 8. In the Netherlands, emissions of trichloroethylene to the air were 6.5 tonnes in 1970, 5.4 tonnes in 1975, 4.2 tonnes in 1979, 3.7 tonnes in 1980, 2.6 tonnes in 1981 and 2.2 tonnes in 1982 (Besemer *et al.*, 1984).

Indoor air concentrations of trichloroethylene can increase when trichloroethylene-contaminated water is used domestically. A community water supply that contained 40 mg/L of trichloroethylene was estimated to contribute about 40 mg/m^3 to the air of a bathroom during showering, and the weekly dose through inhalation was estimated to be 48 mg trichloroethylene (assuming 1-h showering), due to off-gassing of trichloroethylene from the water. About 42 mg of trichloroethylene were ingested from the water per week (Andelman, 1985). Similar conclusions were reached by Bogen *et al.* (1988).

(b) Water

Trichloroethylene occurs at low levels in all water supplies and frequently in groundwater, owing to its widespread use and physical characteristics. Table 9 summarizes the concentrations of trichloroethylene found in surface waters, groundwater and drinking-water worldwide.

Trichloroethylene was detected in an estimated 3% of surface water samples and 19% of groundwater samples analysed, at geometric mean concentrations of 27.3 ppb [µg/L] in groundwater and 40.2 ppb in surface water (United States Environmental Protection Agency, 1989). In a computerized database on water quality, the reported median concentrations of trichloroethylene in 1983–84 were 5.0 µg/L in industrial effluents (19.6% detectable, 1480 samples), 0.1 µg/L (28% detectable, 9295 samples) in ambient water, < 50 µg/kg dry weight (6% detectable, 338 samples) in sediment and < 50 µg/kg (none detectable, 93 samples) in biota (Staples *et al.*, 1985).

The concentrations of trichloroethylene in sediment and animal tissue collected near the discharge zone of the Los Angeles County, CA, waste-treatment plant in 1980–81, were 17 µg/L in the effluent, < 0.5 µg/kg dry weight in sediment and 0.3–7 µg/kg wet weight in various marine animal tissues (Gossett *et al.*, 1983).

Table 7. Concentrations of trichloroethylene in ambient air

Area	Year	Concentration [µg/m³] Mean	Concentration [µg/m³] Range	Reference
Remote				
Pacific Ocean (latitude 37°N)	1977	[0.07]		US Environmental Protection Agency (1985)
Panama Canal Zone (latitude 9°N)	1977	[0.08]		US Environmental Protection Agency (1985)
Northern hemisphere	1985		[0.06–0.09]	US Environmental Protection Agency (1985)
Southern hemisphere	1981		[< 0.02]	Singh et al. (1983)
Rural				
Badger Pass, CA, USA	1977	[0.06]	[0.005–0.09]	US Environmental Protection Agency (1985)
Whiteface Mountains, NY, USA	1974	[0.5]	[< 0.3–1.9]	Lillian et al. (1975)
Reese River, NV, USA	1977	[0.06]	[0.005–0.09]	US Environmental Protection Agency (1985)
Jetmar, KS, USA	1978	[0.07]	[0.04–0.11]	US Environmental Protection Agency (1985)
Western Ireland	1974	[0.08]		Lovelock (1974)
Urban and suburban				
Phoenix, AZ, USA	1979	[2.6]	[0.06–16.7]	Singh et al. (1981)
Los Angeles, CA, USA	1976	[1.7]	[0.14–9.5]	US Environmental Protection Agency (1985)
Lake Charles, LA, USA	1976–78	[8.6]	[0.4–11.3]	US Environmental Protection Agency (1985)
New Jersey, USA	1973–79	[9.1]	[ND–97]	Lillian et al. (1975); US Environmental Protection Agency (1985)
New York City, NY, USA	1974	[3.8]	[0.6–5.9]	Lillian et al. (1975)
Denver, CO, USA	1980	[1.07]	[0.15–2.2]	US Environmental Protection Agency (1985)
St Louis, MO, US	1980	[0.6]	[0.1–1.3]	US Environmental Protection Agency (1985)
Portland, OR, USA	1984	[1.5]	[0.6–3.9]	Ligocki et al. (1985)
Philadelphia, PA, USA	1983–84	[1.9]	[1.6–2.1]	Sullivan et al. (1985)
Brussels, Belgium	1974–75	[21.5]	[5.9–31.2]	Su & Goldberg (1976)
Geneva, Switzerland	1974	[31.2]		Su & Goldberg (1976)
Moscow, Russian Federation	1974	[19.3]	[14.0–28.5]	Su & Goldberg (1976)
Paris, France	1975	[4.0]		Su & Goldberg (1976)
Grenoble, France	1975	[19.3]	[6.4–28.5]	Su & Goldberg (1976)
Kyoto, Japan	1975	[5.1]		Su & Goldberg (1976)
Tokyo, Japan	1975	[1.8]		Su & Goldberg (1976)
Yokohama, Nagoya and Kawasaki, Japan	1985–86	[5.4]	[3.4–7.5]	Urano et al. (1988)

Table 8. Estimated emissions of trichloroethylene to the air in western Europe, 1981

Country or region	Air emission (tonnes/year)
Netherlands	2.7
Belgium/Grand Duchy of Luxembourg	2.9
Western Germany	46.0
France	45.0
Italy	27.0
Spain	15.0
Austria	5.5
United Kingdom	50.0
Norway	0.9
Sweden	12.0
Finland	2
Portugal	1
Switzerland	7
Denmark	2

From Besemer *et al.* (1984); figures include secondary emissions from water and solid waste

Table 9. Concentrations of trichloroethylene in water

Area	Concentration (µg/L) Mean	Concentration (µg/L) Range	Reference
Surface waters			
Seawater			
Eastern Pacific Ocean	0.0003	0.0001–0.0007	Singh *et al.* (1983)
Coastal waters			
Sea coast, industrial area, United Kingdom		0.1–1	Herbert *et al.* (1986)
West coast, Sweden	0.015		Herbert *et al.* (1986)
Northern coast, Greece		0.06–2.8	Fytianos *et al.* (1985)
Rivers			
Tributaries of the Rhine		0.06–7.0	Herbert *et al.* (1986); Bauer (1981a); Hellman (1984)
Elbe, Germany		0.7–52.3	Hellman (1984)
Weser, Germany		0.5–1.5	Herbert *et al.* (1986)
Rhine		0.1–2.4	Herbert *et al.* (1986)
United Kingdom		0.01–1.0	Herbert *et al.* (1986)
Danube, Vienna, Austria	0.6		Herbert *et al.* (1986)
Netherlands		0.1–1.5	Herbert *et al.* (1986)

Table 9 (contd)

Area	Concentration (µg/L)		Reference
	Mean	Range	
Jackfish Bay, Canada		4.1–120	Comba et al. (1994)
Canada		< 0.001–42	Moore et al. (1991)
Rainwater			
Portland, OR, USA	0.006	0.002–0.02	Ligocki et al. (1985)
Groundwater			
Gloucester, Ontario, Canada		< 1–583	Lesage et al. (1990)
Zurich, Switzerland		1.1–1.9	Herbert et al. (1986)
Dubendorg, Germany	85		Herbert et al. (1986)
Northern Switzerland	0.92		Herbert et al. (1986)
Frankfurt, Germany		0.4–159	Herbert et al. (1986)
Mannheim, Germany		< 0.16–120	Herbert et al. (1986)
Italy		0.1–158	Ziglio et al. (1984a,b)
United Kingdom		< 0.1–70	Fielding (1981)
Netherlands		< 0.1–1100	Zoeteman et al. (1980); Trouwborst (1981)
Minnesota, USA, near landfill		0.7–125	Sabel & Clark (1984)
New Jersey, USA, near landfill		≤ 1530	Burmaster (1982)
Pennsylvania, near landfill		≤ 27 300	Burmaster (1982)
Japan, near electronics factory		≤ 10 000	Hirata et al. (1992)
Phoenix, Arizona, USA		8.9–29	Flood et al. (1990)
Drinking-water			
Southern Philippines		0.03	Trussell et al. (1980)
Northern Philippins		0.01	Trussell et al. (1980)
Egypt		1.2	Trussell et al. (1980)
United Kingdom		0.4	Trussell et al. (1980)
Nicaragua		0.05	Trussell et al. (1980)
USA 1976–77		0.2–49	Thomas (1989)
1977–81		Trace–53	
1978		0.5–210	
		Trace–35 000 (with local contamination)	
New Jersey	23.4	Max. 67	Cohn et al. (1994)
Woburn, Massachusetts		Max. 267	Lagakos et al. (1986)

1.3.4 Food

The concentrations of trichloroethylene in food in the United Kingdom were: 0.3–10 ppb [µg/kg] in dairy products, 12–22 ppb in meat, none detected (ND)–19 ppb in oils and fats, ND–60 ppb in beverages, ND–7 ppb in fruits and vegetables and 7 ppb in cereals. In marine organisms, the concentrations varied from ≤ 1 ppb in invertebrates to 10 ppb in the flesh of fish to a maximum of 50 ppb in the eggs of sea birds and the blubber of seals (McConnell et al., 1975). Molluscs from Liverpool Bay, United Kingdom, contained a mean of 85 µg/kg on a dry-weight basis (range, 2–250 µg/kg). Various fish had a mean concentration of 106.5 µg/kg (range, 7–479 µg/kg) (Dickson & Riley, 1976).

The average concentrations of trichloroethylene in food in the United States were 0.9 (0–2.7) µg/kg in grain-based foods, 1.8 (0–12) µg/kg in 'table-ready' foods, 73.6 (1.6–980) µg/kg in butter and margarine, 3.8 (0–9.5) µg/kg in cheese products, 0.5 (0–1.7) µg/kg in peanut butter, 3.0 (0–9.2) µg/kg in ready-to-eat cereal products and 1.3 (0–4) µg/kg in highly processed foods (Heikes & Hopper, 1986; Heikes, 1987). In an evaluation of process waters and food commodities collected at 15 food processing plants, trichloroethylene was found at 3–7.8 ppb [µg/L] in three process waters but in none of the food products (Uhler & Diachenko, 1987). It was detected in five of 372 fatty and non-fatty food samples at concentrations of 2–94 µg/kg, with a mean of 49 µg/kg (Daft, 1989).

Trichloroethylene was found at a concentration of 100–500 ppb [µg/kg] in one of 70 samples of margarine taken from shops in the United States in 1980–82 and 1984 but at < 50 ppb in 20 samples. In 1984, the levels were all < 50 ppb (Entz & Diachenko, 1988). The mean daily intake of trichloroethylene from food, water and air in Germany was estimated to be 32–51 µg/day (Bauer, 1981b; von Düszeln et al., 1982).

1.3.5 Biological monitoring

Individual exposure to trichloroethylene in Germany was determined in non-occupational and a number of occupational environments by biological monitoring. Trichloroethylene was detected in 31% of all blood samples from persons not occupationally exposed to volatile halogenated hydrocarbons (median, < 0.1 µg/L; range, < 0.1–1.3 µg/L). The median levels of trichloroacetic acid, a metabolite of trichloroethylene, were 21.4 µg/L (range, 4.8–221 µg/L) in 43 blood samples and 6.0 µg (range, 0.6–261 µg) in 94 samples of 24-h urine from these unexposed persons. The blood levels of trichloroethylene were < 0.1–0.2 µg/L in nine motor vehicle mechanics, < 0.1 µg/L in three painters, 0.1–15.5 µg/L in three precision instrument makers and 0.2–7.1 µg/L in six dry cleaners (Hajimiragha et al., 1986).

In a plant in the United States where trichloroethylene was used in five degreasing operations in the manufacture of steel tubing, the concentrations of trichloroethylene in air were 117–357 mg/m^3, with short-term exposures as high as 2000 mg/m^3. Urine samples collected from exposed workers before the shift contained, on average, 298 mg/L (range, 4–690 mg/L) of total trichloroethylene metabolites, while the mean concentration after the shift was 480 mg/L (range, 63–1050 mg/L) (Ruhe et al., 1981).

The average blood plasma levels of trichloroethylene of 157 employees at two metal-working plants in the United States were 2.5 ppb [µg/L] (range, 0–22 ppb) and undetectable; in the second plant, the major exposure was to a solvent that contained chloroform. A control population living several miles from the first plant also had undetectable levels of trichloro-ethylene (Pfaffenberger et al., 1984).

The concentration of total trichloro compounds in the urine of workers in a degreasing operation at a United States aircraft factory were 0.5–83.4 mg/g creatinine. These concentrations correlated well with the air concentrations, which averaged 5.7 ppm [30.6 mg/m^3] (Gorman et al., 1984).

The levels of trichloroacetic acid in the urine of 73 workers in 24 workshops in Switzerland where degreasing was performed were 8–444 mg/L, with a mean of 86.7 mg/L. The levels in the 96 air samples were 1–335 ppm [5.37–1800 mg/m^3] with a mean of 56.7 ppm [304 mg/m^3] (Grandjean et al., 1955).

The relationship between concentrations of trichloroethylene in the air near degreasing operations and urinary excretion of total trichloro compounds was reported in Japan. Eight workers had an average urinary concentration of 243.9 mg/L (range, 95–787 mg/L) total trichloro compounds after exposure to 40.7 ppm [217 mg/m^3] trichloroethylene in air. The calculated estimated air levels corresponding to the urine levels found were 41.7 (range, 22.3–67.4) ppm [224 (120–362) mg/m^3] (Nomiyama, 1971).

A total of 31 employees in 19 vapour degreasing plants in central Sweden were exposed to trichloroethylene at a mean level in ambient air of 27 mg/m^3; 86% of the air samples contained < 50 mg/m^3. A weak correlation was found between the concentrations of N-acetyl-β-D-glucosaminidase and trichloroacetic acid in urine ($r = 0.48$; $p < 0.01$), but no correlation was seen with ambient air levels ($r = 0.08$; $p = 0.66$) (Seldén et al., 1993).

In China, the relationship between the time-weighted average exposure to trichloroethylene at the end of a work week and the concentrations of metabolites in urine was investigated in 140 exposed and 114 control workers. In a plant where trichloroethylene was manufactured by chlorination of acetylene followed by dehydrochlorination, 61 men who were exposed to trichloroethylene in air at a concentration of 3–94 ppm [16.1–505 mg/m^3] and 17 women exposed to 2–47 ppm [11–253 mg/m^3] had ≤ 127 mg/L (men) and ≤ 111 mg/L (women) total trichloro compounds in their urine. In a metal-plating plant where trichloroethylene was used for degreasing, 52 men were exposed to concentrations of 1–63 ppm [5.37–338 mg/m^3] and 10 women were exposed to 2–13 ppm [10.7–69.8 mg/m^3]; the urinary levels were ≤ 89 mg/L for the men and ≤ 98 mg/L for the women (Inoue et al., 1989).

The Danish Labour Inspection Service conducted biological monitoring of workers exposed to trichloroethylene in various factories between 1947 and 1987. The concentrations of trichloroacetic acid in 2272 urine samples from workers in 330 factories were similar from the mid-1950s to the mid-1970s and then began to decrease. The average urinary concentrations were 82 mg/L (range, 0–750 mg/L) in 1947–51, 40 mg/L (0–1975 mg/L) in 1950–56, 32 mg/L (0–680 mg/L) in 1957–61, 55 mg/L (0–730 mg/L) in 1962–66, 53 mg/L (0–850 mg/L) in 1967–71, 35 mg/L (0–370 mg/L) in 1972–76, 30 mg/L (0–365 mg/L) in 1977–81 and 18 mg/L (0–130 mg/L) in 1982–86 (Christensen & Rasmussen, 1990).

Blood and urine samples were collected in 1990 from 10 people working in four dry cleaning shops in Croatia, where trichloroethylene was used as the cleaning solvent. The concentration of trichloroethylene in the air was 25–40 ppm [134–215 mg/m^3]. The mean blood levels of trichloroethylene were 0.38 µmol/L [50 µg/L] on Monday morning (range, 0.15–3.58 µmol/L) [20–470 µg/L] and 3.39 µmol/L [445 µg/L] on Wednesday afternoon (range, 0.46–12.71 µmol/L [60–1670 µg/L]). The mean trichloroethanol levels in blood were 3.02 µmol/L (0–10.7 µmol/L) [451 (0–1600 µg/L)] and 7.70 µmol/L (0–26.1 µmol/L) [1150 (0–3894 µg/l)] for the same period, respectively, and the results for trichloroacetic acid were 165 µmol/L (6.12–302 µmol/L) [27 (1–49 mg/L)] and 194 µmol/L (13.5–394 µmol/L) [31 (2–64 mg/L)]. The mean trichloroacetic acid level in urine was 32.5 mmol/mol creatinine (1.3–61.2) [47 (2–89) mg/g] on Monday morning and 37.2 mmol/mol creatinine (1.9–77.4) [54 (3–112) mg/g] on Wednesday afternoon. The mean trichloroethanol levels in urine were 9.7 mmol/mol creatinine (0.4–35.7) [13 (0.5–47 mg/g)] in the Monday morning sample and 54.9 mmol/mol creatinine (5.3–177.7) [73 (7–235) mg/g] in the Wednesday afternoon sample (Skender *et al.*, 1991).

A number of researchers have studied the influence of hourly and daily variations in exposure concentrations on the alveolar concentrations of trichloroethylene and on the urinary excretion of trichloroethanol and trichloroacetic acid (Ogata *et al.*, 1971; Droz & Fernández, 1978). The estimated concentrations of trichloroacetic acid in urine at the end of a workday in which workers were exposed to 270 mg/m^3 trichloroethylene for 8 h per day, five days a week, were 100 mg/g creatinine 0.5 h after exposure, 80 mg/g creatinine after 16 h and 50 mg/g creatinine after 64 h (Monster, 1984).

People exposed to 50 ppm (270 mg/m^3) trichloroethylene for 8 h per day on five days a week were estimated to have alveolar air concentrations of 10–15 ppm [53.7–80.6 mg/m^3] at the end of exposure and 0.1 ppm [0.5 mg/m^3] 64 h after exposure. The blood concentrations were estimated to range from 0.9 to 0.006 mg/L (Monster, 1984).

The airborne concentrations of trichloroethylene at a liquid–vapour degreasing operation in the United States in 1980 were 117–357 mg/m^3, with short-term sampling peaks of 413–2000 mg/m^3. Nine exposed workers had a mean pre-shift urinary concentration of total trichloroethylene metabolites of 298 µg/L; the mean post-shift concentration was 480 µg/L (Landrigan *et al.*, 1987).

Swedish producers of trichloroethylene offered an exposure control programme to customers using trichloroethylene in which free analysis of trichloroacetic acid in urine was conducted annually. On this basis, Axelson *et al.* (1994) categorized the average exposure of 1670 workers as 0–49 mg/L, 50–99 mg/L and ≥ 100 mg/L; 81% were placed in the lowest group. The analytical method used to determine trichloroacetic acid in urine indicated that 50 mg/L was approximately equivalent to an 8-h time-weighted average exposure to 20 ppm [107 mg/m^3] trichloroethylene.

In an ongoing biological monitoring study of workers in various occupations who are exposed to trichloroethylene, tetrachloroethylene or 1,1,1-trichloroethane, conducted by the Finnish Institute of Occupational Health, 11 534 samples representing 3976 workers in 600 workplaces were obtained for the three compounds between 1965 and 1983. Of these workers, 94.4% were monitored for one solvent, 5.2% for two solvents and 0.4% for three solvents. The overall median concentrations of trichloroethylene, reported as trichloroacetic acid in urine, were 63 µmol/L [10.3 mg/L] for women and 48 µmol/L [7.8 mg/L] for men. Before 1970, the mean urinary levels were 80–90 µmol/L [13.1–14.7 mg/L] for men and 60–80 µmol/L [9.8–13.1 mg/L] for women (Anttila *et al.*, 1995).

Trichloroethylene was detected in the blood of 22 of 39 subjects in Zagreb, Croatia, who had no known exposure to solvents, and trichloroacetic acid was found in all plasma and urine samples. The geometric mean concentrations of trichloroethylene were 0.023 µg/L (range, < 0.020–0.090 µg/L) in blood; those of trichloroacetic acid were 45.4 µg/L (13.5–160 µg/L) in plasma and 24.2 µg/L (1.67–292 µg/L) in urine. The concentration of trichloroethylene in the drinking-water was 4.20 µg/L (0.69–35.9 µg/L) (Skender et al., 1993).

The mean concentration of trichloroacetic acid in sera from 94 subjects who were not exposed to organic solvents in Germany was 23.8 µg/L (range, 4.8–221 µg/L), and the average level of trichloroacetic acid in 24-h urine samples was 7.6 µg (range, 0.6–261.4 µg) (Hajimiragha et al., 1986).

Of the 14 million inhabitants of the Netherlands in the 1980s, 14 000 were estimated to be exposed by all routes to an average trichloroethylene concentration of 10 µg/m^3, resulting in a daily intake of 200 µg; 350 000 were exposed to 4 µg/m^3 with a daily intake of 80 µg; and 13.6 million inhabitants were exposed to 0.8 µg/m^3 for a daily intake of 16 µg (Besemer et al., 1984).

The serum levels of trichloroacetic acid in inhabitants of Milan, Italy, who drank water containing > 2000 µg/L of trichloroethylene was 36.5 µg/L; that in an unexposed group was 8 µg/L (Ziglio et al., 1984c). The ambient air level of trichloroethylene in Milan in 1979 was 7.6 µg/m^3 (Ziglio et al., 1983).

Analysis of human tissue taken *post mortem* showed trichloroethylene concentrations of 2–32 µg/kg wet weight in body fat, 2–5.8 µg/kg in liver, < 1–3 µg/kg in kidney and ≤ 1 µg/kg in brain (McConnell et al., 1975).

1.4 Regulations and guidelines

Occupational exposure limits and guidelines for trichloroethylene in a number of countries are presented in Table 10.

WHO (1993) has established a provisional guideline of 70 µg/L trichloroethylene in drinking-water.

The American Conference of Governmental Industrial Hygienists (1994) has recommended several biological exposure indices for trichloroethylene. That for trichloroacetic acid in urine at the end of the work week is 100 mg/g creatinine; that for trichloroacetic acid and trichloroethanol in urine at the end of the shift at the end of the work week is 300 mg/g creatinine; and that for free trichloroethanol in blood at the end of the shift at the end of the work week is 4 mg/L. It is noted that these indices are nonspecific, i.e. other exposures can affect the measurement, and that trichloroethylene in exhaled air and in blood can be used as an indicator of exposure but interpretation of the measurement is only semiquantitative.

Biological indices for exposure to trichloroethylene have been reported. In Finland, the action level for trichloroacetic acid in urine is 360 µmol/L [47.3 mg/L] (Aitio et al., 1995); in Germany, the biological tolerance values are 5 mg/L trichloroethanol in blood and 100 mg/L trichloroacetic acid in urine (Deutsche Forschungsgemeinschaft, 1993); and in Switzerland, the biological tolerance values are 5 mg/L trichloroethanol in blood and 100 mg/g creatinine trichloroacetic acid in urine (Schweizerische Unfallversicherungsanstalt, 1994).

Table 10. Occupational exposure limits and guidelines for trichloroethylene

Country	Year	Concentration (mg/m³)	Interpretation
Australia	1993	270	TWA
		1080	STEL
Austria	1987	260	TWA
Belgium	1993	269	TWA
		1070	STEL
Brazil	1987	420	TWA
Bulgaria	1993	269	TWA
		537	STEL
Canada	1987	75	TWA
		402	STEL (15 min)
Chile	1987	428	TWA
China	1987	535	TWA
Colombia	1993	269	TWA
		537	STEL
Czech Republic	1993	250	TWA
		1250	STEL
Denmark	1993	160	TWA
Egypt	1987	269	TWA
Finland	1993	160	TWA
		240	STEL
France	1993	405	TWA
		1080	STEL
Germany	1993	270	TWA; suspected carcinogen
Hungary	1987	10	TWA
		40	STEL
India	1987	535	TWA
		800	STEL
Indonesia	1987	535	TWA
Italy	1987	400	TWA
Japan	1993	270	TWA
Jordan	1993	269	TWA
		537	STEL
Mexico	1987	535	TWA
Netherlands	1994	190	TWA
		538	STEL (15 min)
New Zealand	1993	269	TWA
		537	STEL
Norway	1984	105	TWA; carcinogen
Philippines	1993	535	TWA
Republic of Korea	1993	269	TWA
		537	STEL
Poland	1993	50	TWA

Table 10 (contd)

Country	Year	Concentration (mg/m³)	Interpretation
Romania	1987	200	TWA
		300	STEL
Russian Federation	1993	269	TWA
Singapore	1993	269	TWA
		537	STEL
Sweden	1993	50	TWA
		140	STEL
Switzerland	1994	260	TWA
		1300	STEL
Thailand	1993	537	TWA
		1074	STEL
Turkey	1993	535	TWA
United Kingdom	1993	535	TWA
		805	STEL
USA			
ACGIH	1994	269	TWA
		537	STEL
NIOSH	1994	134	TWA; carcinogen
		11	Ceiling (60 min[a])
OSHA	1994	537	TWA
		1074	Ceiling
		1611	Peak
Venezuela	1987	535	TWA
		800	STEL
Viet Nam	1993	269	TWA
		537	STEL

From Cook (1987); ILO (1991); Deutsche Forschungsgemeinschaft (1993); Työministeriö (1993); American Conference of Governmental Industrial Hygienists (ACGIH) (1994); Arbeidsinspectie (1994); Schweizerische Unfallversicherungsanstalt (1994); United Kingdom Health and Safety Executive (1994); United States National Institute for Occupational Safety and Health (NIOSH) (1994c); United States Occupational Safety and Health Administration (OSHA) (1994)

TWA, time-weighted average; STEL, short-term exposure limit; ceiling, level not to be exceeded during any part of the workday; peak, acceptable maximum peak above acceptable ceiling concentration for an 8-h shift (maximum duration, 5 min in any 2 h)

[a] During use as an anaesthetic

2. Studies of Cancer in Humans

2.1 Case reports

Málek *et al.* (1979) followed-up 57 men who had worked for at least one year in dry cleaning in Prague, Czech Republic, since the 1950s. Nearly 60% of those tested had a urinary trichloroacetic acid concentration in excess of 100 mg/L, with sporadic values in the region of 1000 mg/L. The follow-up period was 5–50 years. Six men were found to have cancer: three had lung cancer, one had cancer of the tongue, one had rectal cancer and one had a bladder cancer and two rectal tumours.

Novotná *et al.* (1979) reviewed the occupational histories of all 63 subjects diagnosed with histologically confirmed carcinoma of the liver in 1972 and 1974 in Prague, Czech Republic. None of them had been employed in workshops where trichloroethylene was used. Paraf *et al.* (1990) reported a case of gall-bladder cancer in a woman aged 64 who had worked as a technician in a laboratory in France where trichloroethylene was used for degreasing metal.

Jalihal and Barlow (1984) reported a case of acute myeloid leukaemia in a 60-year-old dry cleaner in the United Kingdom. He had had heavy exposure for many years first to trichloroethylene and later to tetrachloroethylene.

2.2 Descriptive studies

Risks for cancer among workers in industries where there is potential exposure to trichloroethylene have been addressed in a number studies but in which exposure to this compound was not specified (e.g. Krain, 1972; Blair, 1980; Blair & Mason, 1980; Brandt-Rauf *et al.*, 1982, 1986; Brandt-Rauf & Hathaway, 1986; Malker *et al.*, 1986; Dubrow & Gute, 1987). These descriptive studies were not considered relevant in view of the availability of cohort and case–control studies.

Paddle (1983) retrieved records from the Mersey Regional Cancer Registry (United Kingdom) for 1951–77 for all 95 subjects with a diagnosis of primary liver cancer and an address near Runcorn, where there is a plant in which trichloroethylene has been manufactured since 1909. Two members of the personnel department of the company compared the records of tens of thousands of people who had worked at the Runcorn site during 1934–76 with the registry list, and the records of two potential matched persons were subsequently checked at the Department of Health and Social Security. It was concluded that none of the subjects had ever worked at the Runcorn site. [The Working Group noted that the interpretation of this result was hindered by the lack of expected numbers.]

2.3 Cohort studies

The cohort studies available to the Working Group addressed three occupational groups: dry cleaners, workers who had undergone biological monitoring for exposure to trichloroethylene and workers employed in miscellaneous manufacturing industries. The Working Group did not consider that the first group of studies (see the monograph on dry cleaning) was relevant to an evaluation of trichloroethylene *per se*, given the extensive exposure of these people to other solvents. Workers who were biologically monitored were considered likely to have been exposed to trichloroethylene, but the proportion of workers in the third group of studies who were actually exposed to trichloroethylene varied.

2.3.1 Exposure evaluated by biological monitoring

Axelson *et al.* (1978, 1984 [abstract], 1994) studied a cohort of workers in Sweden who had been exposed to trichloroethylene. Between 1930 and 1986, only one plant in central Sweden produced trichloroethylene for the domestic market, and this producer offered its customers free surveillance of their exposed workers by analysis for trichloroacetic acid in the urine. Files containing data from such monitoring constitute the basis of the study, but some of the files had been destroyed. Axelson *et al.* (1978) originally retrieved records for 518 men, later expanded the cohort to 1424 men (Axelson *et al.*, 1984, abstract) and finally included 1727 persons drawn from 115 companies that had used the surveillance service at least once between 1955 and 1975 (Axelson *et al.*, 1994). Records were incomplete for 23 persons, four people could not be found in the population register, and 30 had emigrated. The final analysis was thus based on 1670 persons, 1421 men and 249 women, who were followed up for mortality from 1955 through 1986 and for cancer incidence from 1958 through 1987. Swedish national rates were used for the calculation of expected numbers. Exposure was assessed as the mean concentration of trichloroacetic acid in all urinary samples available for a given person: 78% of the person-years for men were accumulated in the category 0–49 mg/L, 14% in the category 50–99 mg/L and 8% in the > 100 mg/L category. A total of 253 deaths were observed [giving an overall standardized mortality ratio (SMR) of 1.0; 95% confidence interval (CI), 0.89–1.1]; and 129 incident cancer cases occurred [giving an overall standardized incidence ratio (SIR) of 1.0; 95% CI, 0.84–1.2]. Among men, a significant excess risk was found for skin cancer (SIR, 2.4; 95% CI, 1.0–4.7; eight observed). There were five cases of non-Hodgkin's lymphoma (1.6; 0.51–3.6) and four cases of liver and biliary tract cancer (1.4; 0.38–3.6). Of the incident cancer cases in men, 77 occurred in men in the lowest exposure category [SIR, 0.92], 18 in the medium category [SIR, 0.93] and 12 [SIR, 1.4] in the highest exposure category.

Anttila *et al.* (1995) studied a cohort of 3974 persons in Finland who were biologically monitored for occupational exposure to three halogenated hydrocarbons (3089 for trichloroethylene, 849 for tetrachloroethylene and 271 for 1,1,1-trichloroethane) during 1965–83. The cohort consisted of those people for whom 10 743 measurements were taken; the persons for whom a further 791 measurements were taken could not be identified. The overall median urinary concentration of trichloroacetic acid was higher for women (63 μmol/L [10.3 mg/L]) than for men (48 μmol/L [7.8 mg/L]). The cohort was followed up for incident cancer cases through 1992, and the expected numbers were calculated on the basis of Finnish national rates. There were 208 cancer cases among people monitored for exposure to trichloroethylene (SIR, 1.1; 95% CI, 0.92–1.2). A significant excess risk was seen for cervical cancer (2.4; 1.1–4.8; eight observed), and the risk was further increased for women with a mean level of exposure ≥ 100 μmol/L [≥ 16.3 mg/L] (4.4; 1.4–10; five observed); no further increase in risk was seen with increasing latency since the time the first measurement was made. The SIR for liver cancer among people with high exposure was 2.7 (0.33–9.9; two observed); a significantly increased SIR was seen with a 20-year latency since first measurement (6.1; 1.3–18; three observed). The

SIR for cancers of the lymphohaematopoietic tissues was increased among people with high exposure (2.1; 0.95–4.0; nine observed) and was further increased with the 20-year latency (3.0; 1.2–6.1; seven observed). The SIRs for stomach cancer were 0.91 (0.25–2.3; four cases) for high exposure and 3.0 (1.2–6.1; seven cases) with a 20-year latency. The SIR for prostatic cancer was 0.68 (0.08–2.4; two cases) with high exposure and 3.6 (1.5–7.0; eight cases) with a 20-year latency.

The population studied by Anttila *et al.* (1995) included most of the workers investigated in a previous study that comprised 2117 Finnish workers in whom urinary trichloroacetic acid was measured or were reported as having been exposed to trichloroethylene during 1963–76 (Tola *et al.*, 1980). A total of 11 cancer deaths (14.3 expected) was reported.

2.3.2 Exposure in miscellaneous manufacturing industries

Barret *et al.* (1984) reported in an abstract a study of the death certificates of 235 workers who had been exposed to trichloroethylene and cutting oils; a total of 14 500 had been so employed in 1983. In a comparison of SMRs [method not described] for each site of cancer, the authors found a high risk for cancer of the naso- and oropharynx (SMR, 2.5 [95% CI, 1.4–4.1]; 15 deaths).

Shindell and Ulrich (1985) studied a plant in northern Illinois, United States, where trichloroethylene had been used extensively as a degreasing agent and where the workers drank water containing traces (43 ppb [µg/L]) of trichloroethylene. The plant began operation in 1957. The study included all office employees at this plant and all production employees who had worked for three months or more in this or a nearby facility between 1 January 1957 and 31 July 1983. The cohort consisted of 2646 individuals, of whom 2140 were white men, 76 were non-white men and 430 were women. The cohort was followed up until 31 July 1983; vital status was determined for all but 52 persons. National mortality rates were used to calculate the expected numbers of deaths. A total of 141 persons had died, whereas 181.6 deaths were expected [SMR, 0.78; 95% CI, 0.65–0.92]. There were nine deaths from respiratory cancer [0.74; 0.34–1.4] and 12 deaths from non-respiratory cancer [0.49; 0.25–0.85]. The employees who had the greatest opportunity for occupational exposure to trichloroethylene were assemblers, but their mortality rate generally conformed to the expected value for all types of diseases.

Garabrant *et al.* (1988) followed a cohort of 14 067 persons who had worked for at least four years for a large aircraft manufacturing company in the United States and for at least one day at the company facility in San Diego County between January 1958 and 31 December 1982. The cohort was followed up through 1982. Persons lost to follow-up were included up to the last date at which they were known to be alive. United States national rates and rates from San Diego County were used to calculate the expected numbers of deaths. Data from a relatively small case–control study nested in the cohort indicated that 37% of the jobs held in the plant entailed exposure to trichloroethylene. A total of 1804 deaths was observed (SMR, 0.75; 95% CI, 0.72–0.79), and there were 453 deaths from cancer (0.84; 0.77–0.93). None of the SMRs for individual cancer sites was significantly elevated. There were eight deaths from cancer of the biliary passages and liver (0.94; 0.40–1.9).

Spirtas *et al.* (1991) analysed a cohort of 14 457 civilian employees who had worked for at least one year at an air force base in Utah, United States, between 1 January 1952 and 31

December 1956, where they maintained and overhauled aircraft and missiles, cleaning and repairing small parts. The analysis included 12 538 white workers and 1528 workers of unknown race, who were followed up until 31 December 1982; 97% were successfully traced. At the end of follow-up, 3832 persons had died, and their death certificates were obtained from the State vital statistics office and coded by a nosologist. The expected number of deaths was based on rates for the Utah population. In the early years of operation of the base, 1939–54, cold solvents were used to clean metal parts, and these were primarily Stoddard solvent, carbon tetrachloride, trichloroethylene and alcohols. Of these, Stoddard solvent was used most frequently; however, in 1955, trichloroethylene replaced Stoddard solvent, and in 1968 1,1,1-trichloroethane replaced trichloroethylene. Trichloroethylene was the primary solvent used in vapour degreasing in the base shops from 1939 to 1979, when it was replaced by 1,1,1-trichloroethane. Of the 14 467 cohort members, 10 256 were classified as having been exposed to mixed solvents, 7282 to trichloroethylene, 6977 to Stoddard solvent and 6737 to carbon tetrachloride (Stewart et al., 1991). Actual exposure levels could not be quantified, but for each combination of job and organization an index of exposure to trichloroethylene was calculated on the basis of the frequency of exposure, the frequency of peak exposure and duration of use. Cumulative exposure categories were derived by multiplying the exposure index assigned to each combination of job and organization by the time spent in this job and by adding these products. The 3832 deaths in the total cohort resulted in an overall SMR of 0.92 (95% CI, 0.90–0.95). Among white men exposed to trichloroethylene, there were 1508 deaths (0.92; 0.87–0.96), 248 of which were from cancer (0.92; 0.81–1.1). When the data for men and women exposed to trichloroethylene were combined, there were 1694 deaths from all causes [0.90; 0.86–0.95] and 281 deaths from cancer [0.88; 0.78–0.99]; there was an elevated risk for cancer of the biliary passages [2.2; 0.96–4.4]. Nonsignificantly excess risks were also seen for cancer of the bone in men (2.6; 0.54–7.7; three deaths) and for cancer of the cervix (2.2; 0.61–5.7; four deaths) and for non-Hodgkin's lymphoma (2.9; 0.78–7.3; four deaths) in women. There were two deaths from primary liver cancer [1.1; 0.12–4.0]. No evidence of a dose–response relationship was seen when the data were analysed by cumulative exposure to trichloroethylene (scored as categories of < 5, 5–25, > 25) for cancer at any site, including cancer of biliary passages, for which the SMRs were [2.5] (three deaths) for exposure to < 5, [4.3] (three deaths) for exposure to 5–25 and [1.3] (two deaths) for exposure to > 25. Both deaths from liver cancer occurred among men in the lowest category of cumulative exposure.

A retrospective cohort study of renal cancer among workers exposed to trichloroethylene in a cardboard manufacturing factory in Germany was reported by Henschler et al. (1995). Measurements of exposure were not available, and workers were classified as exposed or not exposed on the basis of categories of job held in the factory. The exposed group consisted of 169 men who had worked for at least one year during 1956–75; a control group consisting of 190 unexposed workers from the same factory was included for comparison. The average observation period was 34 years. Assessment of cancer occurrence was based on abdominal sonography, records of the medical, personnel and pension departments and interviews with relatives. Causes of death were obtained from hospital records or from the treating physician. During the period of follow-up, four histologically verified cases of renal-cell carcinoma and one case of urothelial cancer of the renal pelvis were seen in the exposed group, and no case was observed in the controls ($p = 0.03$). The five cancers occurred 18–34 years after first exposure; four of the five men had been exposed for more than 13 years. The excess was confirmed in comparisons with population rates for Denmark (SIR, 8.0; 95% CI, 2.6–19) and for the former German Democratic Republic (9.7; 3.1–23). The incidences of cancers at other sites were not reported. There were 50 deaths from all causes among exposed workers and 52 among controls; 16 cases of cancer of any organ were seen in both exposed and control workers; and two deaths from renal cancer occurred in exposed workers and none in controls. In a comparison with the local population, the SMR for renal cancer in the exposed group was 3.3 (95% CI, 0.40–12). [The Working Group noted that the use of sonography suggested that the study originated from the observation of a cluster of cases of renal cancer.]

The main cohort studies are summarized in Table 11.

2.4 Case–control studies

2.4.1 Primary liver cancer

Hernberg et al. (1984) identified 374 cases of primary liver cancer (ICD 155.0) that had been reported to the Finnish Cancer Registry in 1979–80. The notifying hospital could not be identified in nine cases, the hospital refused contact with 38 patients, and the diagnosis was incorrect in 83 cases. For the remaining 244 cases, a questionnaire was sent to either the patient or the next-of-kin. Three deceased patients had no relatives, and in 79 instances no reply was obtained. A further check of the diagnoses revealed that only 126 of the 162 cases for which a reply was obtained were primary liver cancers. For each of the 162 cases, two controls with coronary infarct and without cancer were selected, from the hospital register for living cases and from autopsy records for dead cases. Complete replies were obtained from only 174 controls or their next-of-kin. An industrial hygienist evaluated exposure to solvents on the basis of the reported occupational histories. Eight patients had been exposed to solvents for at least one year (odds ratio, 2.3; 95% CI, 0.8–7.0). Six of the exposed patients were women, one of whom had possibly been exposed to trichloroethylene; none of the female controls had been exposed.

Hernberg et al. (1988) subsequently identified 618 persons reported as having primary liver cancer to the Finnish Cancer Registry in 1976–78 and 1981. Five patients alive at the start of the study were excluded, and no relative was found for 87 patients. Questionnaires were sent to relatives of the remaining 526 cases, and a response was obtained from 377. Thirty-three cases were omitted on the basis of an incorrect or unconfirmed diagnosis, leaving 344 cases in the study. Two control groups were selected: one, as in the previous study, comprised 674 patients who had died with a coronary infarct, of whom 116 had no relatives and for 385 of whom the questionnaire was returned; the second control group consisted of 720 deceased stomach cancer patients, 66 of whom had no relatives and for 476 of whom a questionnaire was returned from next-of-kin. Two industrial hygienists coded occupational histories for potential exposure to solvents. In comparison with the two control groups combined, the odds ratios for exposure to solvents were 0.6 [95% CI, 0.3–1.4] for men and 3.4 [1.1–10] for women. None of the exposed women had been a heavy or moderate alcohol drinker. One of the seven solvent-exposed female patients and none of the solvent-exposed controls had been exposed to trichloroethylene.

Table 11. Summary of data from four cohort studies of trichloroethylene

Cancer site	Axelson et al. (1994) 1421 men using trichloroethylene and monitored for exposure (Sweden, 1958–87)			Anttila et al. (1995) 3089 men and women using trichloroethylene and monitored for exposure (Finland, 1967–92)			Spirtas et al. (1991) 7282 men and women employed in aircraft maintenance and exposed to trichloroethylene (USA, 1953–82)			Garabrant et al. (1988) 14 067 men and women employed in aircraft manufacture (USA, 1958–82)		
	SIR	95% CI	Obs	SIR	95% CI	Obs	SMR	95% CI	Obs	SMR	95% CI	Obs
All cancers	0.96	0.80–1.2	107	1.1	0.92–1.2	208	[0.88]	[0.78–0.99]	281	0.84	0.77–0.93	453
Oesophagus	NR			NR			[1.0]	[0.37–2.2]	6	1.1	0.62–1.9	14
Stomach	0.70	0.23–1.6	5	1.3	0.75–2.0	17	[0.78]	[0.43–1.3]	14	0.40	0.18–0.76	9
Colon	1.0	0.44–2.0	8	0.84	0.36–1.7	8	[1.0]	[0.67–1.4]	29	0.96	0.71–1.3	47
Liver and biliary tract	1.4	0.38–3.6	4	[1.9]	[0.86–3.6]	9	[1.9]	[0.91–3.5]	10	0.94	0.40–1.9	8
Primary liver cancer				2.3	0.74–5.3	5	[1.1]	[0.14–4.0]	2			
Biliary tract				1.6	0.43–4.0	4	[2.2]	[0.96–4.4]	8			
Cervix	NR			2.4	1.1–4.8	8	2.2	0.61–5.7	4	0.61[a]	0.25–1.3	7
Prostate	1.3	0.84–1.8	26	1.4	0.73–2.4	13	0.80	0.50–1.2	22	0.93	0.60–1.4	25
Kidney	1.2	0.42–2.5	6	0.87	0.32–1.9	6	[1.1]	[0.46–2.1]	8	0.93	0.48–1.6	12
Urinary bladder	1.0	0.44–2.0	8	0.82	0.27–1.9	5	[1.4]	[0.70–2.5]	11	1.3	0.74–2.0	17
Skin	2.4	1.0–4.7	8	NR			[1.0][b]	[0.38–2.3]	6	0.7[c]	0.29–1.5	7
Brain and nervous system	NR			1.1	0.50–2.1	9	[0.78]	[0.36–1.5]	9	0.78	0.42–1.3	13
Lymphohaemato-poietic system				1.5	0.92–2.3	20	[0.94]	[0.66–1.3]	37	0.78	0.56–1.1	38
Non-Hodgkin's lymphoma	[1.5][d]	0.5–3.6	5	1.8[d]	0.78–3.6	8	[1.3][d]	[0.68–2.1]	14	0.82	0.44–1.4	13
Hodgkin's disease	1.1	0.03–6.0	1	1.7	0.35–5.0	3	[0.87]	[0.24–2.2]	4	0.73	0.20–1.9	4
Leukaemia	NR			1.1	0.35–2.5	5	[0.73][e]	[0.37–1.3]	11	0.82[d]	0.47–1.3	16

SIR, standardized incidence ratio; CI, confidence interval; Obs, observed; SMR, standardized mortality ratio; NR, not reported
[a] Female genital organs
[b] Malignant melanoma
[c] Includes five cases of malignant melanoma
[d] Including ICD 202
[e] Including aleukaemia

Hardell *et al.* (1984) studied all cases of liver cancer reported to the Swedish Cancer Registry in 1974–81 in men aged 25–80 living in the Umeå region of Sweden. Six patients who were alive at the start of the study in 1981 were excluded, leaving 166 cases. The diagnosis of 114 cases was confirmed on review. Six patients had been used as controls in a previous study, relatives could not be identified for five patients, and the relatives of one patient refused participation, leaving 102 patients (78 with hepatocellular, 15 with cholangiocellular, five with mixed and four with other types of liver cancer) for whom completed questionnaires were obtained. Two deceased controls matched for age, sex, year of death and municipality were selected from the National Population Register for each case, excluding people who had died from suicide or cancer. Exposure to solvents was assessed on the basis of responses to a questionnaire. The risk ratio for all primary liver cancers (hepatocellular and/or cholangiocellular) was 1.8 (95% CI, 0.99–3.4); that for hepatocellular carcinoma was 2.1 (1.1–4.0). Two of the 22 solvent-exposed patients and one of the 27 solvent-exposed controls had been exposed to trichloroethylene.

2.4.2 Malignant lymphoma

Hardell *et al.* (1981) studied 169 men aged 25–85 with histologically confirmed malignant lymphoma (60 with Hodgkin's disease, 105 with non-Hodgkin's lymphoma and four with unclassified lymphomas) in the Umeå region of Sweden between 1974 and 1978. For each of the 107 living patients, two controls matched for sex, age and residence were selected from the National Population Registry. For each of the 62 deceased patients, two controls matched for sex, age, year of death and municipality were selected from the National Registry for Causes of Death, excluding people who had died from suicide or cancer. Exposure to solvents was assessed on the basis of responses to a questionnaire. Three of the 338 controls did not return the questionnaire but were considered not to have been exposed in matched analyses. The relative risk associated with exposure to styrene, trichloroethylene, tetrachloroethylene or benzene was 4.6 (95% CI, 1.9–11). Seven cases and three controls reported exposure to trichloroethylene.

2.4.3 Hodgkin's disease

Olsson and Brandt (1980) studied 25 men aged 20–65 who were admitted consecutively to the Department of Oncology at the University Hospital of Lund, Sweden, in 1978–79 with Hodgkin's disease. For each case, two male controls, matched for age and residence, were selected from the population register. Twelve of the patients had been exposed to organic solvents, giving a relative risk of 6.6 (95% CI, 1.8–24). Three cases and no control reported exposure to trichloroethylene.

2.4.4 Renal-cell carcinoma

Sharpe *et al.* (1989) identified 403 patients who had been diagnosed with renal-cell carcinoma in nine hospitals in Montréal, Canada, in 1982–87. Of these, 168 were still alive in 1987 and agreed to complete a questionnaire. For each case, one control originally suspected to have renal-cell carcinoma but for whom a non-neoplastic diagnosis was given was matched for sex, age and urologist. Ultimately, 164 patients and 161 controls provided information. Ten patients and three controls had been exposed to degreasing solvents (odds ratio, 3.4; 95% CI, 0.92–13). Tetrachloroethylene, 1,1,1-trichloroethane, trichloroethylene and dichloromethane were reported to be the agents most widely used.

2.4.5 *Cancer of the colon*

Fredriksson *et al.* (1989) carried out a case–control study of patients aged 30–75 in whom adenocarcinoma of the large bowel had been diagnosed in 1980–83 in the Umeå region of Sweden. A total of 402 incident cases were identified, but only patients alive in 1984–86 were included, leaving 344 patients, of whom 312 participated. Two population controls, matched by age, sex and county, were included for each case. Data on exposure were collected by a postal questionnaire. The odds ratio for exposure to trichloroethylene was 1.5 (95% CI, 0.4–5.7) and that for exposure to trichloroethylene among dry cleaners was 7.4 (1.1–47).

2.4.6 *Brain tumours*

Heineman *et al.* (1994) undertook a case–control study of 741 white men who had died from astrocytic brain tumours in two states of the United States between 1978 and 1981. Next-of-kin were identified for 654 patients; 483 of these were interviewed, and a hospital diagnosis of astrocytic brain tumour was confirmed in 300 cases. Of 741 selected deceased controls, 320 were included in the study. Exposure to solvents was assessed on the basis of a job–exposure matrix; 128 case patients had been employed in jobs with potential exposure to trichloroethylene (odds ratio, 1.1; 95% CI, 0.8–1.6). None of the risk estimates for subgroups reached significance.

2.4.7 *Childhood leukaemia*

Lowengart *et al.* (1987) identified 216 children aged 10 years or less from the Los Angeles County (United States) Cancer Surveillance Program in whom acute leukaemia had been diagnosed in 1980–84. Permission for contact with families was obtained for 202 patients; 159 mothers were interviewed, and information about the fathers was obtained for 154 cases. The mothers of the patients were asked to name a control child from among their child's friends. A total of 136 control mothers were interviewed; information about the fathers was obtained for 130 controls. Data on occupational exposure were obtained by telephone interview. The odds ratios associated with father's exposure to trichloroethylene were 2.0 ($p = 0.16$) for exposure one year before pregnancy, 2.0 ($p = 0.16$) for exposure during pregnancy and 2.7 ($p = 0.07$; 95% CI, 0.64–16) for exposure after delivery. The results of this study were also reported in an abstract (Peters *et al.*, 1984).

2.4.8 *Childhood brain tumours*

Peters *et al.* (1981) studied the occupations of the parents of 92 children under the age of 10 with brain tumours and of 92 matched controls in Los Angeles County, United States. Interviews with the fathers showed that those of 12 children with brain tumours and those of two controls had worked in the aircraft industry; the fathers of only two children with brain tumours reported exposure to trichloroethylene. The results of this study were also reported in an abstract (Peters *et al.*, 1984).

2.4.9 Multiple sites

Siemiatycki (1991) studied men aged 35–70 in Montréal, Canada, during 1979–85. A total of 3730 people with cancers at 21 sites and 533 population controls were interviewed about their occupations in detail, and their exposure to 293 agents or mixtures was then estimated by a group of chemists. The estimated prevalence of exposure to trichloroethylene was 2%. Both case–case and case–control comparisons were conducted. After control for confounding, increased odds ratios were found in the case–case comparison for cancer of the rectum (1.9 [95% CI, 0.9–3.9] and for skin melanoma (2.6 [1.2–5.8]) in relation to presumed exposure to trichloroethylene; for 'substantial' exposure (at least five years of exposure at a presumably medium or high concentration and frequency), elevated odds ratios were reported for prostatic cancer (1.8 [0.7–4.7]) and for skin melanoma (2.3 [0.8–7.0]), while the risk for rectal cancer was no longer elevated (0.8 [0.2–2.8]). The increased risk for skin melanoma was restricted to French Canadians; in the latter group, the risk for lung adenocarcinoma was also elevated (odds ratio for any exposure, 2.6 [0.8–8.4]; odds ratio for substantial exposure, 4.5 [1.1–18]). The risk was not increased for cancers of the bladder (0.6 [0.3–1.4]) or kidney (0.8 [0.3–2.1]) or for non-Hodgkin's lymphoma (1.1 [0.5–2.4]).

2.5 Studies of drinking-water

Cancer occurrence in populations exposed to drinking-water contaminated with various concentrations of trichloroethylene has been compared in a number of studies. The interpretation of some of these studies is complicated by several methodological problems:

(i) information on the concentration of trichloroethylene in water was obtained subsequently to or contemporaneously with the period over which cancer occurrence was measured, although cancer rates should be correlated with exposure before occurrence of the disease;

(ii) exposure was generally measured at the community level and does not necessarily reflect the exposure of individuals;

(iii) the problem of migration in and out of the populations under study was not addressed; and

(iv) the possible confounding effects of other characteristics of the populations being compared (socioeconomic, industrial and cultural factors) were not taken into account.

Isacson *et al.* (1985) tabulated the average annual age-adjusted incidence rates of cancers of the bladder, breast, colon, lung, prostate or rectum per 100 000 population in towns in Iowa, United States, in 1969–81 by the level of detectable volatile organic compounds in finished groundwater supplies. The levels of trichloroethylene were < 0.15 µg/L in one group of areas and ≥ 0.15 µg/L in another. There were virtually no differences in the incidences between these two groups.

Lagakos *et al.* (1986) studied childhood leukaemia in a community in Massachusetts, United States, where water from two wells was contaminated with trichloroethylene. Measurements made in 1979 showed a concentration of 267 ppb [µg/L] trichloroethylene in the well water. Twenty cases of childhood leukaemia were diagnosed in the community in 1964–83, and these were associated with a significantly higher estimated cumulative exposure to water from the two contaminated wells than a random sample of children from the community (observed cumulative exposure, 21.1; expected cumulative exposure, 10.6; $p = 0.03$).

A study conducted in New Jersey, United States, during 1979–87 included 75 towns (Cohn et al., 1994), of which 27 were included in a study reported by Fagliano et al. (1990). Trichloroethylene concentrations were measured during 1984–85, and an average level was assigned to each town. The highest level assigned was 67 µg/L. The water supply of six towns contained > 5 µg/L trichloroethylene (average, 23.4 µg/L). Women in these towns had a significantly higher total incidence of leukaemia than the inhabitants of towns where the concentration of trichloroethylene in drinking-water was < 0.1 µg/L (relative risk, 1.4; 95% CI, 1.1–1.9); no such effect was seen for men (1.1, 0.84–1.4). The risk among women was particularly elevated for acute lymphocytic leukaemia, chronic lymphocytic leukaemia and chronic myelogenous leukaemia. The risk for acute lymphocytic leukaemia in childhood was also significantly increased, in girls but not in boys. Increased risks for non-Hodgkin's lymphoma were apparent in towns in the highest category of trichloroethylene contamination (0.2; 0.94–1.5 for men and 1.4; 1.1–1.7 for women) and was particularly elevated for high-grade lymphomas.

Studies were conducted in two counties in Arizona, United States, to address the possible association between consumption of drinking-water from trichloroethylene-contaminated wells and childhood leukaemia (Maricopa County, Flood et al., 1990) or all childhood neoplasms and testicular cancer (Pima County, Arizona Department of Health Services, 1990). In Maricopa County, two wells that were occasionally used to supplement the water supply were found to contain 8.9 and 29.0 ppb [µg/L] trichloroethylene in 1982; they were then taken out of service. The concentrations of trichloroethylene in contaminated wells in Pima County were 1–239 µg/L, with levels as high as 4600 µg/L in wells at an Air Force facility in the area. No association was found between cancer at any of the sites examined and residence in the counties with contaminated wells, as opposed to residence in other areas of the county. The incidence rates in both Maricopa and Pima counties were comparable to those in other areas included in the United States SEER programme.

Vartiainen et al. (1993) collected 24-h urine samples from 95 and 21 inhabitants of two Finnish villages where the groundwater was contaminated with trichloroethylene (≤ 212 µg/L) and tetrachloroethylene (≤ 180 µg/L). The average excretion of trichloroethylene by inhabitants of the two villages was 0.55 and 0.45 µg/day, and that of two control groups was 0.36 and 0.32 µg/day; the corresponding figures for excretion of dichloroacetic acid were 0.78 and 1.3 µg/day versus 1.3 and 1.3 µg/day, and those for the excretion of trichloroacetic acid were 19 and 7.9 µg/day versus 2.0 and 4.0 µg/day. With the possible exception of non-Hodgkin's lymphoma, which occurred in a marginal excess in one of the villages (SIR, 1.4; 95% CI, 1.0–2.0; 31 cases) but not in the other (0.6; 0.3–1.1; 14 cases), neither overall cancer incidence nor the incidence of liver cancer or lymphohaematopoietic cancers was increased in the two villages.

3. Studies of Cancer in Experimental Animals

3.1 Oral administration

3.1.1 Mouse

Groups of 50 male and 50 female B6C3F1 mice, five weeks of age, were administered trichloroethylene (purity, > 99%; containing 0.19% epoxybutane and 0.09% epichlorohydrin [see IARC, 1987c] as stabilizers) in corn oil by gavage on five days a week for 78 weeks. The time-weighted average doses of trichloroethylene were 1169 and 2339 mg/kg bw per day for males and 869 and 1739 mg/kg bw per day for females. All surviving animals were killed 90 weeks after the start of treatment and submitted to complete necropsy and histopathological evaluation. Groups of 20 male and 20 female vehicle controls were included. The numbers of survivors at the end of the study were 8/20 male vehicle controls, 36/50 males at the low dose and 22/48 males at the high dose; and 20/20 female vehicle controls, 42/50 females at the low dose and 39/47 females at the high dose. The survival-adjusted (Cox and Tarone test) incidences of hepatocellular carcinomas were increased in animals of each sex in relation to dose; males: 1/20 in vehicle controls, 26/50 ($p = 0.004$) at the low dose, 31/48 ($p < 0.001$) at the high dose; females: 0/20 in vehicle controls, 4/50 at the low dose, 11/47 ($p = 0.008$) at the high dose. One male at the high dose developed a forestomach papilloma (United States National Cancer Institute, 1976).

In a subsequent study, groups of 50 male and 50 female B6C3F1 mice, eight weeks of age, were administered 1000 mg/kg bw trichloroethylene (purity, > 99.9%; containing no epichlorohydrin) in corn oil by gavage on five days a week for up to 103 weeks. Groups of 50 mice of each sex served as vehicle controls. Survival of treated males was significantly reduced ($p = 0.004$) in comparison with controls; at the end of the experiment, 33 control and 16 treated males and 32 control and 23 treated females were still alive. Histopathological evaluation revealed increased incidences (incidental tumour test) of hepatocellular tumours in treated animals. In males, hepatocellular adenomas occurred in 7/48 controls and 14/50 ($p = 0.048$) treated animals; hepatocellular carcinomas were found in 8/48 controls and 31/50 ($p < 0.001$) treated animals; and the combined numbers of animals bearing hepatocellular adenomas and/or carcinomas were 14/48 controls and 39/50 ($p < 0.001$) treated animals. In females, hepatocellular adenomas were seen in 4/48 control and 16/49 ($p = 0.001$) treated animals; hepatocellular carcinomas occurred in 2/48 control and 13/49 ($p = 0.002$) treated animals; and the combined numbers of animals bearing hepatocellular adenomas and/or carcinomas were 6/48 controls and 22/49 ($p < 0.001$) treated animals. There was no significant treatment-related increase in the incidence of tumours at other sites. Toxic nephrosis (cytomegaly) was seen in 90% of treated males and in 98% of treated females (United States National Toxicology Program, 1990).

Two groups of 30 male and 30 female ICR:Ha Swiss mice, six to eight weeks of age, were each administered 0 or 0.5 mg trichloroethylene [purity unspecified] by gavage in 0.1 ml trioctanoin once a week for at least 74 weeks. Only sections of lung, liver and stomach were taken for histopathological examination. The incidence of forestomach tumours was reported not to be increased; findings were not given for other sites (Van Duuren et al., 1979). [The Working Group noted the low dose used and the inadequate conduct and reporting of the study.]

3.1.2 *Rat*

Groups of 50 male and 50 female Osborne-Mendel rats, six weeks of age, were administered trichloroethylene (purity, > 99%; containing 0.19% epoxybutane and 0.09% epichlorohydrin as stabilizers) in corn oil by gavage on five days a week for 78 weeks. The time-weighted average doses of trichloroethylene were 549 (low dose) and 1097 mg/kg bw per day (high dose) for animals of each sex. All surviving animals were killed 110 weeks after the start of treatment and were submitted to complete necropsy. Groups of 20 male and 20 female vehicle controls were included. Large proportions of treated and control rats died during the experiment; the numbers of animals alive at the end of the study were 3/20 male vehicle controls, 8/50 males at the low dose and 3/50 males at the high dose; of the females, there were 8/20 vehicle controls, 13/48 at the low dose and 13/50 at the high dose. There was no significant difference in tumour incidence at any site between treated and control rats (United States National Cancer Institute, 1976). [The Working Group noted the high rates of early mortality in both control and treated rats and the limited duration of treatment.]

In a subsequent study, groups of 50 male and 50 female Fischer 344/N rats, eight weeks of age, were administered 0, 500 or 1000 mg/kg bw trichloroethylene (purity, > 99.9%; containing no epichlorohydrin) in corn oil by gavage on five days a week for up to 103 weeks. A group of 50 male and 50 female rats were used as untreated controls. Survival of low-dose and high-dose males was significantly reduced ($p < 0.005$) in comparison with vehicle controls; the numbers of survivors at the end of the experiment were 35 male vehicle controls, 20 at the low dose and 16 at the high dose; and 37 female vehicle controls, 33 at the low dose and 26 at the high dose. An increased incidence of renal tubular-cell adenocarcinomas was seen in males: 0/49 untreated controls, 0/48 vehicle controls, 0/49 at the low dose and 3/49 at the high dose ($p = 0.028$; incidental tumour test). Two males at the low dose had renal tubular-cell adenomas. The incidence of tumours in female rats was not increased at any site. Toxic nephrosis of the kidney occurred in 96/98 treated males and in all of the treated females but not in vehicle control rats of either sex (United States National Toxicology Program, 1990). [The Working Group noted the uncommon occurrence of renal tubular-cell tumours in untreated Fischer 344/N rats.]

Groups of 50 males and 50 females of four strains (ACI, August, Marshall and Osborne-Mendel), 6.5–8 weeks of age, were administered 0, 500 or 1000 mg/kg bw trichloroethylene (purity, > 99.9%) in corn oil by gavage on five days a week for 103 weeks. Additional groups of 50 rats of each sex and strain served as untreated controls. Survival was reduced significantly in low-dose and high-dose males and high-dose females of the ACI strain, in both treated groups of males and females of the Marshall strain, and in high-dose female Osborne-Mendel rats. The numbers of survivors at the end of the study were: ACI males – 36 untreated controls, 37 vehicle controls, 19 at the low dose, 11 at the high dose; ACI females – 36 untreated controls, 33 vehicle controls, 20 at the low dose, 17 at the high dose; August males – 24 untreated controls, 21 vehicle controls, 13 at the low dose, 15 at the high dose; August females – 26 untreated controls, 23 vehicle controls, 26 at the low dose, 24 at the high dose; Marshall males – 32 untreated controls, 26 vehicle controls, 12 at the low dose, 6 at the high dose; Marshall females – 31 untreated controls, 30 vehicle controls, 12 at the low dose, 10 at the high dose; Osborne-Mendel males – 18 untreated controls, 22 vehicle controls, 17 at the low dose, 14 at the high dose; Osborne-Mendel females – 19 untreated controls, 18 vehicle controls, 10 at the low dose, 7 at the high dose. Many early deaths occurred accidentally. The incidence of renal cytomegaly was > 80% in all treated males and females, and toxic nephropathy (described as dilated tubules lined by elongated and flattened epithelial cells) occurred at rates of 17–80% in the treated groups; however, there was no difference in kidney toxicity between males and females of any strain. Neither of these two renal lesions was seen in untreated or vehicle controls. The incidences of renal tubular-cell hyperplasia and tubular-cell adenoma were increased in male Osborne-Mendel rats at the low dose: hyperplasia – 0/50 untreated controls, 0/50 vehicle controls, 5/50 at the low dose, 3/50 at the high dose; adenoma – 0/50 untreated controls, 0/50 vehicle controls, 6/50 ($p = 0.007$; survival-adjusted incidental tumour test) at the low dose, 1/50 at the high dose. One renal tubular-cell adenocarcinoma occurred in a male at the high dose. The incidences of interstitial-cell tumours of the testis were increased in Marshall rats exposed to trichloroethylene: 16/46 untreated controls, 17/46 vehicle controls, 21/48 ($p < 0.001$; survival-adjusted incidental tumour test) at the low dose, 32/48 ($p < 0.001$) at the high dose. No significant increase in tumour incidence was reported for ACI or August rats (United States National Toxicology Program, 1988). [The Working Group noted the poor survival among all strains and the fact that five of the six renal adenomas in male Osborne-Mendel rats at the low dose occurred among the 17 rats alive at the end of the study.]

Groups of 30 male and 30 female Sprague-Dawley rats, 12–13 weeks of age, were administered 0, 50 or 250 mg/kg bw trichloroethylene (purity, 99.9%; containing no epoxide) in olive oil by gavage on four to five days per week for 52 weeks and observed for life. Data on survival were not provided, but the authors reported a nonsignificant increase in mortality among treated females. Renal tubular-cell cytokaryomegaly was observed only in male rats at the high dose (46.7% [14/30]; $p < 0.01$). A nonsignificant increase in the incidence of leukaemias was observed in males: none in controls, 6.7% [2/30] at the low dose and 10.0% [3/30] at the high dose (Maltoni *et al.*, 1986). [The Working Group noted the short period of exposure.]

3.2 Inhalation

3.2.1 *Mouse*

Groups of 30 male and 30 female NMRI mice [age unspecified] were exposed to air containing trichloroethylene (purity, > 99.9%; stabilized with 0.0015% triethanolamine) at a concentration of 0, 100 or 500 ppm (0, 540 or 2700 mg/m^3) for 6 h per day on five days per week for 18 months. The experiment was terminated after 30 months. At the end of exposure (75 weeks), there was no difference in the probability of survival among the females; in males, the probability of survival was reduced from 83% in controls to 63% in low-dose and 56% in high-dose groups. Histopathological examination of spleen, liver, kidney, lung, heart, stomach, central nervous system and all tumours indicated increased age-adjusted incidences of lymphomas in treated female mice: 9/29 controls, 17/30 at the low dose ($p < 0.001$) and 18/28 ($p = 0.01$) at the high dose (Henschler *et al.*, 1980).

Groups of 49–50 female ICR mice, seven weeks of age, were exposed to air containing trichloroethylene (purity, 99.8%; containing 0.13% carbon tetrachloride and > 0.02% benzene and epichlorohydrin) at concentrations of 0, 50, 150 or 450 ppm (0, 270, 810 or 2430 mg/m^3) for 7 h per day on five days per week for up to 104 weeks. There were no significant differences in survival between the control and exposed groups. Complete necropsy was carried out on all animals. Histopathological evaluation revealed a significant increase (Fisher's exact test) in the incidence of lung adenocarcinomas: 1/49 controls, 3/50 at the low dose, 8/50 ($p < 0.05$) at the middle dose and 7/46 ($p < 0.05$) at the high dose. [The Working Group found a significant dose–response trend: $p = 0.034$, Cochran-Mantel-Haenszel test.] The incidences of adenomas and adenocarcinomas of the lung combined in the groups at the middle (13/50) and high doses (11/46) were not significantly increased in comparison with controls (6/49). The average number of lung tumours was, however, increased in mice at the middle and high doses in comparison with controls: 0.12 in controls, 0.10 at the low dose, 0.46 at the middle dose and 0.39 at the high dose (Fukuda et al., 1983).

Groups of 90 male and 90 female Swiss mice, 11 weeks of age, and groups of 90 male and 90 female B6C3F1 mice, 12 weeks of age, were exposed to air containing trichloroethylene (purity, 99.9%; containing no epoxide) at concentrations of 0, 100, 300 or 600 ppm (0, 540, 1620 or 3240 mg/m^3) for 7 h per day on five days a week for 78 weeks and were then observed for life. Data on survival were not provided, but the authors reported that mortality was higher ($p < 0.05$) in treated male B6C3F1 mice than in controls. Dose-related increases in the incidences of lung and liver tumours were observed in male Swiss mice [Fisher's exact test or Cochran-Armitage linear trend test]. The percentages of male Swiss mice bearing a malignant pulmonary tumour were: control, 11.1% [10/90]; low-dose, 12.2% [11/90]; mid-dose, 25.5% [23/90] ($p < 0.05$); and high-dose, 30.0% [27/90] ($p < 0.01$); the percentages of male mice bearing a hepatoma were: control, 4.4% [4/90]; low-dose, 2.2% [2/90]; mid-dose, 8.9% [8/90]; and high-dose, 14.4% [13/90] ($p < 0.05$). In B6C3F1 mice, a dose-related increase in the incidence of lung tumours was observed in females: control, 4.4% [4/90]; low-dose, 6.7% [6/90]; mid-dose, 7.8% [7/90]; and high-dose, 16.7% [15/90] ($p < 0.05$) (Maltoni et al., 1986, 1988).

3.2.2 Rat

Groups of 30 male and 30 female Wistar rats [age unspecified] were exposed to air containing trichloroethylene (purity, > 99.9%; stabilized with 0.0015% triethanolamine) at concentrations of 0, 100 or 500 ppm (0, 540 or 2700 mg/m^3) for 6 h per day on five days per week for 18 months. The experiment was terminated after 36 months. No differences in survival were reported; the probability of survival in each group at the end of the experiment was: 46.7% of male controls, 23.3% of males at the low dose, 36.7% of males at the high dose, 16.7% of female controls, 13.3% of females at the low dose and 16.7% of females at the high dose. Histopathological and gross examination of spleen, liver, kidney, lung, heart, stomach, central nervous system and all tumours revealed no increase in tumour incidence (Henschler et al., 1980).

Groups of 49–51 female Sprague-Dawley rats, seven weeks of age, were exposed to air containing trichloroethylene (purity, 99.8%) at concentrations of 0, 50, 150 or 450 ppm (0, 270, 810 or 2430 mg/m^3) for 7 h per day on five days per week for 104 weeks. Survival was significantly higher in the exposed groups than in controls: about 75% of the rats in the three treated groups and 50% of controls were alive at 100 weeks. Gross and histopathological examination revealed no difference in the incidence of tumours between the control and exposed groups (Fukuda et al., 1983).

Groups of 130–145 male and female Sprague-Dawley rats, 12 weeks of age, were exposed to air containing trichloroethylene (purity, 99.9%; containing no epoxide) at a concentration of 0, 100, 300 or 600 ppm (0, 540, 1620 or 3240 mg/m^3) for 7 h per day on five days per week for 104 weeks. All animals were observed for their lifetime. Data on survival were not provided, but the authors reported no excess mortality in any of the exposed groups. A significant, dose-related increase in the incidence of Leydig cell (interstitial) tumours of the testis was observed [$p < 0.001$; Cochran-Mantel-Haenszel test]; the percentages of male rats bearing these tumours were 4.4% [6/135] of controls, 12.3% [16/130] at the low dose [$p < 0.05$; Fisher's exact test], 23.1% [30/130] at the middle dose [$p < 0.01$; Fisher's exact test] and 23.8% [31/130] at the high dose [$p < 0.01$; Fisher's exact test]. Four renal tubular adenocarcinomas (3.1%) were observed in the high-dose male rats; no such tumours were observed in the lower dose groups, in controls or in the historical control database for Sprague-Dawley rats at the study laboratory. Cytokaryomegaly of renal tubular cells was also observed: in none of the control or low-dose rats, in 16.9% at the middle dose and in 77.7% at the high dose (Maltoni et al., 1986, 1988).

3.2.3 Hamster

Groups of 30 male and 30 female Syrian hamsters [age unspecified] were exposed to air containing trichloroethylene (purity, > 99.9%; stabilized with 0.0015% triethanolamine) at concentrations of 0, 100 or 500 ppm (0, 540 or 2700 mg/m^3) for 6 h per day on five days per week for 18 months. The experiment was terminated after 30 months. The probability of survival was similar in exposed and control groups. Histopathological examination of spleen, liver, kidney, lung, heart, stomach, central nervous system and all tumours revealed no significant increase in tumour incidence (Henschler et al., 1980).

3.3 Topical application

Mouse: In a study of two-stage carcinogenesis on mouse skin, single doses of 1.0 mg trichloroethylene [purity unspecified] in 0.1 ml of acetone were applied to the shaven dorsal skin of 30 female ICR:Ha Swiss mice aged six to eight weeks; 14 days later, topical applications of 12-*O*-tetradecanoylphorbol 13-acetate (TPA; 2.5 µg in 0.1 ml of acetone, three times per week) were begun, for at least 49 weeks. Nine skin papillomas were found in 4/30 treated mice, and 10 papillomas were found in 9/120 TPA-treated controls. Trichloroethylene was also administered by repeated topical application (three times per week) to groups of 30 female ICR:Ha Swiss mice, six to eight weeks of age, for 83 weeks at a dose of 1.0 mg per mouse. No tumours were observed at the site of application (Van Duuren et al., 1979).

3.4 Subcutaneous injection

Mouse: Groups of 30 female ICR:Ha Swiss mice, six to eight weeks of age, were given subcutaneous injections of 0.5 mg trichloroethylene [purity unspecified] in 0.05 ml trioctanoin once a week for at least 74 weeks, or received the vehicle alone. No tumours were observed at the injection site in either group (Van Duuren et al., 1979).

3.5 Administration with known carcinogens

Mouse: Five groups of 50 male and 50 female ICR:Ha Swiss mice, five weeks of age, were administered either industrial-grade trichloroethylene (purity, 99.4%; containing 0.11% epichlorohydrin and 0.20% 1,2-epoxybutane) in corn oil by gavage, purified trichloroethylene (purity, > 99.9%) in corn oil by gavage, purified trichloroethylene with added epichlorohydrin (0.8%), purified trichloroethylene with added 1,2-epoxybutane (0.8 %) or purified trichloroethylene with 0.25% epichlorohydrin plus 0.25% 1,2-epoxybutane, on five days per week for 18 months. The doses of trichloroethylene that were administered were 2.4 g/kg bw for males and 1.8 g/kg bw for females. Groups of 50 mice of each sex given corn oil served as vehicle controls. The treatment period was followed by a six-month observation period. The probabilities of survival were significantly reduced ($p < 0.001$) in all groups of treated males in comparison with controls; in females, the probabilities of survival were reduced ($p < 0.05$) in the group receiving purified trichloroethylene and in that receiving purified trichloroethylene plus epichlorohydrin ($p < 0.001$). At the end of the study, there were no more than two survivors in any treatment group. Complete necropsies were performed on all animals. The incidence of squamous-cell carcinomas of the forestomach was increased in several of the treatment groups over that in controls (0/50 for males and females); males: purified trichloroethylene, 0/50; industrial-grade trichloroethylene, 0/49; purified trichloroethylene plus epichlorohydrin, 5/49 ($p < 0.001$); purified trichloroethylene plus 1,2-epoxybutane, 3/49 ($p = 0.029$); and purified trichloroethylene plus epichlorohydrin and 1,2-epoxybutane, 2/49 ($p = 0.036$); females: controls, 0/50; purified trichloroethylene, 0/50; industrial-grade trichloroethylene, 3/50; purified trichloroethylene plus epichlorohydrin, 9/50 ($p < 0.001$); purified trichloroethylene plus 1,2-epoxybutane, 1/48; and purified trichloroethylene plus epichlorohydrin and 1,2-epoxybutane, 9/50 ($p < 0.001$). No significant increase in the incidences of tumours at other sites was reported. The authors attributed the increased incidence of forestomach cancers to the direct alkylating effects of epichlorohydrin and 1,2-epoxybutane (Henschler *et al.*, 1984). [The Working Group noted that the incidences of hepatocellular tumours (adenomas and carcinomas combined) in male mice were: controls, 3/50; purified trichloroethylene, 6/50; and industrial-grade trichloroethylene, 9/50; and that no survival-adjusted analysis of tumour incidence was performed.]

Groups of 23–33 male B6C3F1 mice, 15 days of age, were given a single intraperitoneal injection of *N*-ethylnitrosourea in 0.1 mol/L sodium acetate at doses of 0, 2.5 or 10 mg/kg bw. When the mice were four weeks of age, a 61-week treatment period was begun with 0, 3 or 40 mg/L trichloroethylene (purity, > 99%) in the drinking-water. The highest concentration of trichloroethylene was equivalent to a daily dose of 6 mg/kg bw. The incidences of hepatocellular adenomas and carcinomas were not increased in mice that received trichloroethylene alone in comparison with vehicle controls, and trichloroethylene did not promote liver tumours in mice initiated with *N*-ethylnitrosourea (Herren-Freund *et al.*, 1987). [The Working Group noted the low dose of trichloroethylene used.]

3.6 Carcinogenicity of metabolites

Studies of the carcinogenicity of the known metabolites of trichloroethylene, dichloroacetic acid, trichloroacetic acid and chloral hydrate, are summarized in separate monographs in this volume.

3.6.1 Mouse

A single dose of 1.0 mg of trichloroethylene oxide, a putative metabolite [purity unspecified], in 0.1 ml of acetone was applied to the dorsal skin of 30 female ICR:Ha Swiss mice, six to eight weeks of age; 14 days later, topical applications of TPA (2.5 µg in 0.1 ml of acetone, three times per week) were begun and continued for more than 61 weeks. The incidence of tumours at the site of application was not increased in the group treated with trichloroethylene plus TPA (three mice each had a single papilloma) in comparison with mice receiving TPA alone (10 papillomas in 9/120 mice) (Van Duuren *et al.*, 1979).

Trichloroethylene oxide was administered to a group of 30 female ICR:Ha Swiss mice, six to eight weeks of age, by repeated skin application for 82 weeks (2.5 mg/mouse in 0.1 ml acetone three times weekly); 30 mice served as vehicle controls. No tumour was observed at the site of application in either group. Further groups of 30 female ICR:Ha Swiss mice, six to eight weeks of age, were given 0 or 500 µg/mouse trichloroethylene oxide in 0.05 ml tricaprylin once a week for up to 80 weeks. One fibrosarcoma occurred at the injection site in treated animals (Van Duuren *et al.*, 1983).

1,2-Dichlorovinyl cysteine, a minor metabolite [purity unspecified], was administered at a concentration of 0, 10 or 50 mg/L in drinking-water to three groups of 30 Swiss-Webster mice [age and sex unspecified] for 14 weeks, beginning one day after administration of *N*-nitrosodimethylamine (NDMA) (six intraperitoneal injections of 5.0 mg/kg bw administered every other day). The average daily doses of 1,2-dichlorovinyl cysteine were 2.4 and 12.6 mg/kg bw, respectively. Renal tumours occurred after 50 weeks in 2/16 mice receiving NDMA alone, 2/15 receiving NDMA plus the low dose of 1,2-dichlorovinyl cysteine and 3/16 receiving NDMA plus the high dose of 1,2-dichlorovinyl cysteine [not significant]. Multiple renal tumours were found in 7/40 mice treated with NDMA plus 1,2-dichlorovinyl cysteine, whereas none were found in 21 mice treated with NDMA alone [$p = 0.043$; Fisher's exact test] (Meadows *et al.*, 1988).

4. Other Data Relevant to an Evaluation of Carcinogenicity and its Mechanisms

4.1 Absorption, distribution, metabolism and excretion

4.1.1 Humans

The biotransformation and the kinetics of trichloroethylene have been described in many studies of workers and of volunteers. Pulmonary uptake of trichloroethylene is rapid, the rate of uptake being dependent on the rate of respiration, and uptake increases about twofold with exercise (Monster et al., 1976). Distribution to the tissues has not been described, but the concentrations of trichloroethylene should be proportional to the duration and concentration of exposure, and the distribution is probably similar to that in animals. The blood:air partition coefficient for trichloroethylene in human volunteers was about 15 (Monster et al., 1979), and the fat:air partition coefficient was about 700 (Sherwood, 1976; Steward et al., 1973); there is therefore a tendency for deposition in fat from blood, the fat:blood partition coefficient being about 50 (700/15).

After inhalation, 40–70% of an administered dose of trichloroethylene is metabolized, the unmetabolized fraction being cleared by exhalation. Metabolism was proportional to the concentration of trichloroethylene in air up to 315 mg/m^3 for 3 h (Ikeda & Imamura, 1973; Monster et al., 1976; Ikeda, 1977; Nomiyama & Nomiyama, 1977). No saturation of biotransformation has been detected with concentrations up to 380 ppm [1976 mg/m^3].

Trichloroethanol, its glucuronide and trichloroacetic acid are major metabolites in urine, and chloral hydrate is a transient metabolite in blood (Cole et al., 1975). After controlled exposure of males to 200 ppm (1040 mg/m^3) trichloroethylene for 6 h, oxalic acid and N-(hydroxyacetyl)aminoethanol were detected as minor metabolites (Dekant et al., 1984). Traces of N-acetyl-S-(1,2-dichlorovinyl)-L-cysteine and N-acetyl-S-(2,2-dichlorovinyl)-L-cysteine were present in the urine of workers exposed to unknown concentrations of trichloroethylene in air (Birner et al., 1993). Trichloroethanol and its glucuronide are rapidly eliminated in urine, with half-lives of about 10 h, and trichloroacetic acid is eliminated slowly, with a half-life of about 52 h (range, 35–70 h) (Müller et al., 1972, 1974). Repeated exposure of volunteers to 50 ppm (260 mg/m^3) trichloroethylene for 4 h per day on five consecutive days resulted in slightly higher concentrations of trichloroethylene and trichloroethanol in blood than after a single exposure to 40 ppm (208 mg/m^3) for 4 h (Ertle et al., 1972). Urinary excretion of trichloroethanol by five male volunteers exposed to 70 ppm [364 mg/m^3] for 4 h per day for five days stabilized rapidly and remained constant until the end of the exposure, whereas urinary excretion of trichloroacetic acid continued to rise (Monster et al., 1979).

4.1.2 Experimental systems

The biotransformation of trichloroethylene has been reviewed (Bonse & Henschler, 1976; Kimbrough et al., 1985; Dekant, 1986; Bruckner et al., 1989; Davidson & Beliles, 1991).

The absorption, distribution, metabolism and excretion of trichloroethylene at doses outside the range of those tested experimentally have been predicted from a number of physiologically based pharmacokinetic models constructed from the existing experimental data (Dallas et al., 1991; Fisher et al., 1991; Allen & Fischer, 1993).

The absorption and excretion of trichloroethylene have been studied in rats and mice. The compound is rapidly absorbed from the gastrointestinal tract and through the lungs; skin absorption after exposure to the vapour is negligible. In male Sprague-Dawley rats exposed to 50 ppm [260 mg/m^3] or 500 ppm [2600 mg/m^3] trichloroethylene for 2 h through a miniaturized one-way breathing valve (Dallas et al., 1991), the uptake decreased from > 95% at the beginning of exposure to a relatively constant, almost steady-state level of 70%. The concentrations of trichloroethylene in exhaled breath towards the end of the exposure period were 34.6 ± 1.1 ppm

[185 ± 6 mg/m^3] after exposure to 50 ppm and 340.8 ± 10.6 ppm [1830 ± 60 mg/m^3] after exposure to 500 ppm. This direct proportionality was not reflected in the arterial blood concentrations, where the 10-fold increase in dose resulted in a 25- to 30-fold increase in blood levels and only an 8.7-fold increase in total absorbed dose.

The blood:air partition coefficient is about 14 in mice (Fisher et al., 1991) and about 18 in rats (Andersen et al., 1987; Fisher et al., 1989). The corresponding fat:blood values are about 36 and 27, and the liver:blood partition coefficients are about 1.8 and 1.3. At the end of 4-h exposures of Fischer 344 rats to 529 ppm [2751 mg/m^3] (males) and 600 ppm [3120 mg/m^3] (females), the concentrations of trichloroethylene in blood were about 35.5 µg/ml (males) and 25.8 µg/ml (females). The concentrations of trichloroethylene in the blood of B6C3F1 mice were much lower: the highest mean blood concentrations seen during exposure of males to 110–748 ppm [572–3890 mg/m^3] and females to 42–889 ppm [218–4623 mg/m^3] were 7.3 µg/ml after exposure to 748 ppm [3890 mg/m^3] (males) and 6.3 µg/ml after exposure to 368 ppm [1914 mg/m^3] (females) (Fisher et al., 1991).

The distribution of trichloroethylene in mice after a 10-min inhalation (approximate dose, 280 mg/kg bw) was studied by whole-body autoradiography of animals killed at intervals over 8 h. Trichloroethylene was distributed throughout the body into well-perfused organs; after 30 min, redistribution to adipose tissues had occurred (Bergman, 1983a).

The urinary excretion of trichloroacetic acid by rats exposed to 55 ppm [286 mg/m^3] trichloroethylene for 8 h per day for 14 weeks reached a maximum after two days and remained constant until the end of the exposure, whereas urinary excretion of trichloroethanol increased steadily over the first 10 weeks of the study (Kimmerle & Eben, 1973).

Mice have consistently higher rates of biotransformation than rats (Fisher et al., 1991). The metabolism of trichloroethylene in rats can be described by Michaelis-Menten kinetics and is saturated after exposure by inhalation to more than 500–600 ppm (2600–3120 mg/m^3). Saturation of metabolism in rats at 500 ppm was also seen in the experiments of Dallas et al. (1991), described above. The atmospheric concentration at which elimination shifts from first-order to zero-order kinetics was found to be 65 ppm [338 mg/m^3] in rats in a closed exposure system (Filser & Bolt, 1979). Metabolic saturation occurs after oral administration of > 200–500 mg/kg bw trichloroethylene to rats; in mice, the rate of biotransformation is linear up to a dose of 2000 ppm (10 400 mg/m^3) by inhalation and up to 2000 mg/kg bw by oral administration (Stott et al., 1982; Buben & O'Flaherty, 1985; Green & Prout, 1985; Prout et al., 1985).

Mice have been shown to biotransform 2.6 times more trichloroethylene on a body weight basis than rats after exposure by inhalation to 600 ppm (3120 mg/m^3) (Dekant et al., 1986a). Trichloroacetic acid concentrations in blood reached significantly higher values in B6C3F1 mice than in Fischer 344 rats at the end of a 4-h exposure by inhalation. The peak concentrations were 23.3 µg/ml in male rats and 39.6 µg/ml in female rats exposed to 505 ppm [2626 mg/m^3] and 600 ppm [3120 mg/m^3], respectively, while the values for mice were 129.6 µg/ml in males exposed to 748 ppm [3890 mg/m^3] and 94.3 µg/ml in females exposed to 889 ppm [4623 mg/m^3] (Fisher et al., 1991). After exposure to low doses, the rate of metabolism in mice and rats is similar, and about 90% of an oral dose of 2 or 10 mg/kg bw trichloroethylene was eliminated as metabolites within 72 h by female Wistar and NMRI mice (Dekant et al., 1986b). After an oral dose of 2000 mg/kg bw, 78% of the dose was exhaled as unchanged trichloroethylene by rats but only 14% by mice (Prout et al., 1985).

As a result of the higher biotransformation rate in mice, their blood levels of trichloroethanol and trichloroacetic acid were four- and sixfold higher than those in rats, and peak concentrations were reached within 2 h in mice and up to 10 h in rats. In mice, the high levels of trichloroacetate in blood persisted for over 30 h (Prout et al., 1985). After dosing by gavage with 1.5 mmol/kg bw (200 mg/kg bw) trichloroethylene, the peak blood concentrations of trichloroacetic acid and the area under the integrated time–concentration curve were higher in mice (216 nmol/ml [35 µg/ml] and 2.5 µmol-h/ml) [408 µg-h/ml] than in rats (81 nmol/ml [13 µg/ml] and 1.5 µmol-h/ml [245 µg-h/ml]) (Larson & Bull, 1992a). The highest concentration of trichloroacetic acid that was found in the blood of rats after oral administration of trichloroethylene in corn oil was equivalent to about 50 mg/kg bw of trichloroacetic acid (Elcombe, 1985). Blood concentrations of the chloroacetic acids resulting from their administration to mice and rats are described in the relevant monographs in this volume.

Several excretory metabolites have been identified in mice and rats (see Figure 1). Most of the metabolites in urine can be accounted for by cytochrome P450-catalysed oxidation reactions of trichloroethylene to chloral hydrate. Trichlorethanol and its glucuronide are formed by reduction of chloral hydrate; trichloroacetic acid is formed by oxidation of this intermediate (Butler, 1949; Daniel, 1963; Kimmerle & Eben, 1973). The glucuronide of trichloroacetic acid has been identified in the urine of non-human primates treated by intramuscular injection with trichloroethylene (Müller et al., 1982). The mechanism of formation of dichloroacetic acid has been postulated as a rearrangement of 1,1,2-trichlorooxirane and subsequent hydrolysis (Hathway, 1980), but it may also be formed by biotransformation of chloral hydrate or trichloroacetic acid (Larson & Bull, 1992b). Oxalic acid may be formed as a urinary metabolite of trichloroethylene as an end-product of 1,1,2-trichlorooxirane, by enzymatic or non-enzymatic cleavage of the epoxide followed by spontaneous elimination of two equivalents of hydrochloric acid, reaction with water and oxidation (Dekant et al., 1984). Oxalic acid may also be formed by oxidation of dichloroacetic acid (Larson & Bull, 1992a,b). The formation of N-(hydroxyacetyl)-aminoethanol is proposed to proceed by the reaction of trichloroethylene-derived oxidative intermediates with ethanol amine or with phosphatidylethanol amine and enzymic breakdown of the acylated lipids (Dekant et al., 1984).

Traces of metabolites indicative of conjugation of trichloroethylene with glutathione are also excreted in urine after high oral doses of trichloroethylene. The presence of N-acetyl-S-(1,2-dichlorovinyl)-L-cysteine and N-acetyl-S-(2,2-dichlorovinyl)-L-cysteine indicates trichloroethylene conjugation with glutathione followed by catabolism and acetylation by the enzymes of the mercapturic acid pathway (Dekant et al., 1986a; Commandeur & Vermeulen, 1990; Dekant et al., 1990). Chloroacetic acid is another trace metabolite of trichloroethylene in rats (Green & Prout, 1985); it may be formed by hydrolysis of the intermediate electrophile, chlorothioketene, which is a cysteine conjugate β-lyase-catalysed cleavage product of S-(1,2-dichlorovinyl)-L-cysteine (Dekant et al., 1986c, 1988). Monochloroacetate may be formed by reduction of dichloroacetic acid (Larson & Bull, 1992b).

Species and strain differences in the biotransformation of trichloroethylene have been reported. Higher peak blood levels of dichloroacetic acid were reported in B6C3F1 mice

Figure 1. Proposed biotransformation of trichloroethylene to urinary metabolites in rats

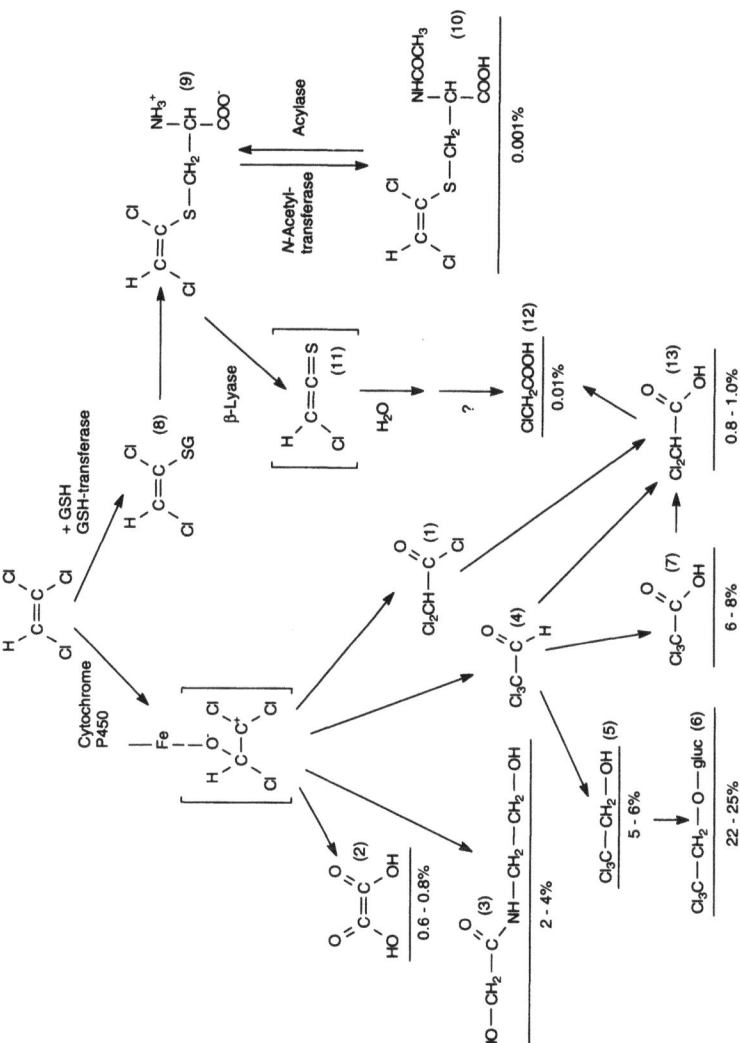

Modified from Dekant et al. (1984); Dekant (1986)
Identified urinary metabolites are underlined; percentages are those of an oral dose of 200 mg/kg bw excreted as individual metabolites
1, Dichloroacetyl chloride; 2, oxalic acid; 3, N-(hydroxyacetyl)aminoethanol; 4, chloral; 5, trichloroethanol; 6, trichloroethanol glucuronide; 7, trichloroacetic acid; 8, dichlorovinylglutathione; 9, S-1,2-dichlorovinylcysteine; 10, S-1,2-dichlorovinyl-N-acetylcysteine; 12, monochloroacetic acid; 13, dichloroacetic acid

(35 nmol/ml [4.5 µg/ml]) dosed with 1.5 mmol/kg bw (200 mg/kg bw) trichloroethylene orally than in rats (< 4 nmol/ml [< 0.5 µg/ml]) receiving 23 mmol/kg bw (3000 mg/kg bw) (Larson & Bull, 1992a). These differences are not, however, reflected in the urinary excretion of dichloroacetic acid (Green & Prout, 1985; Dekant et al., 1986b): In both mice and rats, the blood levels of dichloroacetic acid are at least one order of magnitude lower than those of trichloroacetic acid (Larson & Bull, 1992a). Strain differences among mice in the metabolism of trichloroethylene to trichloroacetic acid are also apparent: In Swiss and B6C3F1 mice, trichloroacetic acid in urine accounts for 7–12% of an oral dose of trichloroethylene; in NMRI mice, trichloroacetic acid is only a trace metabolite of trichloroethylene (Dekant et al., 1986b).

The elimination rates of the major trichloroethylene metabolites differ markedly. Trichloroethanol and chloral hydrate are cleared from the blood with a half-life of 1–2 h, whereas high concentrations of trichloroacetic acid are present for up to 30 h and are cleared only slowly (Kimmerle & Eben, 1973). The amounts of trichloroethylene that are cleared by exhalation depend on the administered dose.

No changes in metabolite profiles were observed after exposure of rats to 55 ppm (286 mg/m^3) trichloroethylene by inhalation for 14 weeks (Kimmerle & Eben, 1973). Daily administration by gavage of 1000 mg/kg bw trichloroethylene to male B6C3F1 mice for 180 days did not induce the overall metabolism of trichloroethylene (Green & Prout, 1985).

Cytochrome P450 activity in mouse lung Clara cells was reduced following exposure to 100 ppm [537 mg/m^3] trichloroethylene for 6 h; the activities of glutathione S-transferases were unaffected. Studies with isolated mouse lung Clara cells showed oxidative metabolism of trichloroethylene, leading to accumulation of chloral in the cells, which were presumably unable to metabolize chloral further to trichloroethanol, as occurs in the liver (Odum et al., 1992).

The metabolism of trichloroethylene in liver microsomes from mice and rats has been studied by determining changes in trichloroethylene concentrations in the headspace of incubation vials containing liver subfractions. The apparent Michaelis-Menten constant (K_m) and the maximal metabolic velocity (V_{max}) in microsomal fractions were 4.2 µmol/L and 8.0 mmol/mg protein per 10 min, respectively, for substrate concentrations of 0.3–34 µmol/L (Kim et al., 1994). Chloral hydrate was found consistently as an end-product of trichloroethylene biotransformation. The formation of chloral hydrate and the cofactor requirements suggest that a cytochrome P450 (probably 2E1) catalyses the formation of chloral hydrate from trichloroethylene (Byington & Leibman, 1965; Leibman & McAllister, 1967; Leibman, 1968; Costa et al., 1980; Guengerich et al., 1991). Other cytochrome P450 enzymes may also catalyse the oxidation of trichloroethylene but have a lower affinity (Nakajima et al., 1990, 1992).

An epoxide (1,2,2-trichlorooxirane) was postulated as an intermediate during the oxidation of trichloroethylene to chloral hydrate (Bonse et al., 1975; Greim et al., 1975; Bonse & Henschler, 1976; Henschler, 1977; Henschler & Bonse, 1977; Hathway, 1980); however, later studies on the biotransformation of trichloroethylene and other chlorinated olefins and knowledge of the mechanisms of oxidation by cytochrome P450 enzymes suggest a stepwise oxidation of trichloroethylene to chloral hydrate, in which the epoxide is not an obligatory intermediate (Miller & Guengerich, 1982; Liebler & Guengerich, 1983; Miller & Guengerich, 1983). Mouse liver microsomes had a threefold higher capacity for the oxidative biotransformation of trichloroethylene than rat liver microsomes (Miller & Guengerich, 1982).

Incubation of trichloroethylene with liver microsomes and liver cytosol from rats in the absence of cofactors for oxidative biotransformation by cytochrome P450 and in the presence of glutathione resulted in the formation of *S*-(1,2-dichlorovinyl)glutathione at low rates (Dekant *et al.*, 1990).

4.1.3 Comparison of humans and animals

A quantitative comparison of the metabolism of trichloroethylene in humans and rats and mice by application of physiologically based pharmacokinetic models suggests that humans have a lower rate of metabolism (14.9 mg/kg bw per h) than B6C3F1 mice (23.2 mg/kg bw per h in females and 32.7 mg/kg bw per h in males) but a slightly higher rate than Fischer 344 rats (11 mg/kg bw per h) (Allen & Fisher, 1993). In the absence of comparative studies, the role of saturable metabolism in humans cannot be assessed; however, in the occupationally and environmentally relevant range of exposures, the metabolism of trichloroethylene after exposure by inhalation seems to be similar in humans and rats. Qualitatively, the pathways of biotransformation in humans and animals are identical, and most metabolites identified in experimental animals have also been found in humans; however, whereas the urinary excretion of trichloroacetic acid remains constant in rats exposed repeatedly to trichloroethylene, the quantity increases steadily in humans over five days. The opposite trend is observed for trichloroethanol, the urinary excretion increasing in rats and remaining constant in humans. The kinetics of the biotransformation of trichloroethylene to trichloroacetic acid in isolated hepatocytes was markedly species dependent: The V_{max}/K_m values ('intrinsic clearance') in mouse, rat and human hepatocytes were 3.8×10^{-6}, 1.2×10^{-7} and 3.25×10^{-8} L/min per 10^6 cells, respectively (Elcombe, 1985).

4.2 Toxic effects

4.2.1 Humans

The acute toxicity of trichloroethylene in humans is characterized mainly by depression of the central nervous system: In 288 cases of acute intoxication with trichloroethylene, effects on the central nervous system were the major toxic manifestations. Liver toxicity was seen in only five individuals, and there was no renal damage (McCarthy & Jones, 1983).

Chronic exposure to trichloroethylene has been reported to be hepatotoxic, and trichloroethylene has also been implicated in the so-called 'psycho-organic syndrome' (McCarthy & Jones, 1983). There was no direct evidence for renal toxicity in humans exposed chronically to low levels of trichloroethylene (50 mg/m^3) (Seldén *et al.*, 1993).

4.2.2 Experimental systems

The oral LD_{50} values for trichloroethylene are 7183 mg/kg bw in rats (Smyth *et al.*, 1969) and 2400–2850 mg/kg bw in mice (Aviado *et al.*, 1976; Tucker *et al.*, 1982). The LC_{50} in rats was 26 300 ppm [136 760 mg/m^3] for a 1-h exposure (Vernot *et al.*, 1977) and 12 500 ppm [65 000 mg/m^3] for a 4 h-exposure (Siegel *et al.*, 1971).

The major toxic effects in animals are depression of central nervous function and sensitization of cardiac function to adrenalin. After acute exposure of Fischer 344 rats to high doses of trichloroethylene, liver damage was observed, characterized by increased activities of serum glutamic–oxaloacetic acid and glutamic–pyruvic transaminases. Administration of high doses of trichloroethylene after pretreatment with phenobarbital also induced renal damage (Chakrabarti & Tuchweber, 1988). High oral doses of trichloroethylene (> 2000 mg/kg bw) damaged Clara cells in mouse lung (Scott *et al.*, 1988; Forkert & Birch, 1989), and dose-dependent damage to mouse Clara cells was observed after single exposures to 200–1000 ppm [1040–5200 mg/m^3] by inhalation for 6 h; no effect was seen at 20 ppm [104 mg/m^3]. The effect seems to be species-specific, since inhalation of 1000 ppm [5200 mg/m^3] trichloroethylene for 6 h had no toxic effects on the rat lung (Odum *et al.*, 1992).

In male Sprague-Dawley rats injected once intraperitoneally with trichloroethylene at 1 mmol/kg bw [131 mg/kg bw], the activities of serum bile acids, particularly cholic and taurocholic acids, were increased 4 and 8 h after dosing. These times reflect those at which high levels of trichloroethylene and trichloroethanol appear in serum and liver. The selected dose did not induce hepatotoxic effects, and it was suggested that the changes in bile acid activity were due to perturbation of a physiological process (Bai & Stacey, 1993; Hamdan & Stacey, 1993).

Studies on the longer-term toxicity of trichloroethylene in rats and mice exposed orally and by inhalation showed consistent increases in relative liver weight and associated histopathological and biochemical changes. The effects described in kidney included increased relative weights in mice exposed continuously to > 75 ppm (> 390 mg/m^3) trichloroethylene for 30 days and renal dysfunction in the absence of marked histopathological changes in rats exposed to > 50 ppm [> 260 mg/m^3] for 12 weeks (Kjellstrand *et al.*, 1981a,b; Stott *et al.*, 1982; Tucker *et al.*, 1982; Kjellstrand *et al.*, 1983a,b; Elcombe *et al.*, 1985; Nomiyama *et al.*, 1986).

Oral administration of 500–1500 mg/kg bw trichloroethylene for 10 consecutive days increased the weight of the liver and the synthesis of DNA and decreased hepatic DNA concentrations in B6C3F1 and Alderley Park mice (Elcombe *et al.*, 1985). Increased hepatic DNA synthesis and mitosis, but no unscheduled DNA synthesis (see section 4.4.2), have been reported in mice dosed with trichloroethylene by gavage or inhalation (Stott *et al.*, 1982; Dees & Travis, 1993).

Trichloroethylene has been shown to induce hepatic peroxisome proliferation in mice, causing substantial increases in cyanide-insensitive palmitoyl coenzyme-A oxidase activity and peroxisomal volume density. The minimal daily dose of trichloroethylene reported to induce this effect in mice is 100 mg/kg bw over 10 days (Elcombe, 1985). Increased hepatic cyanide-insensitive palmitoyl coenzyme A oxidase activity has been reported in Fischer 344 rats treated by gavage with much higher doses of trichloroethylene (1200 mg/kg bw for 14 days, 130%; 1000 mg/kg bw for 10 days, 180% increase) (Goldsworthy & Popp, 1987; Melnick *et al.*, 1987). Increases of 786% and 625% in the activity of this enzyme were reported in B6C3F1 mice treated with 1000 mg/kg bw per day for 10 days (Elcombe *et al.*, 1985; Goldsworthy & Popp, 1987).

Trichloroethylene has been shown to induce a small increase in cyanide-insensitive palmitoyl coenzyme A oxidation activity in the kidneys of both mice and rats after oral dosing with 1000 mg/kg bw per day for 10 days. Greater effects were observed in mice than in rats (Goldsworthy & Popp, 1987).

Two metabolites of trichloroethylene, dichloroacetic acid and trichloroacetic acid (see monographs, this volume), have also been shown to induce peroxisome proliferation in mice and rats (Elcombe, 1985; Goldsworthy & Popp, 1987; DeAngelo *et al.*, 1989). Trichloroacetic acid induced peroxisome proliferation in the kidney of mice, but not rats (Goldsworthy & Popp, 1987).

Trichloroethylene has been reported to inhibit the activity of the natural immune system (natural killer, natural cytotoxic and natural P815 killer cells) in Sprague-Dawley rats and B6C3F1 mice (Wright *et al.*, 1991). The inhibition was particularly evident in the liver after administration *in vivo* and in both liver and spleen after exposure *in vitro*. The background activities of natural immune activities had previously been reported to be higher in species and strains with lower background incidences of liver tumours (Wright & Stacey, 1991). More recently, trichloroethylene has been shown to inhibit aspects of the natural immune system in cells isolated from human liver (Wright *et al.*, 1994). Inhibition of natural immunity may therefore enhance the likelihood of tumour development.

Nuclear magnetic resonance was used to show that trichloroethylene interacts non-specifically with lipid molecules and that, in phosphatidylcholine bilayers, interaction occurs predominantly with the interfacial region rather than the hydrocarbon interior (Bhakuni & Roy, 1994).

4.3 Reproductive and prenatal effects

4.3.1 Humans

(a) Endocrine and gonadal effects

Out of a group of 99 metal workers in Aarhus (Denmark), 15 men who degreased parts with trichloroethylene for more than 20 h per week were asked to deliver a semen specimen (Rasmussen *et al.*, 1988). Twelve were included in the analysis and compared with 14 unexposed physicians. There was no difference between the two groups in terms of sperm count or morphology, but the exposed group had a small, non-significant increase in the prevalence of mature spermatozoa containing two fluorescent Y bodies, which may indicate Y-chromosomal nondisjunction.

(b) Fertility

Taskinen *et al.* (1989) conducted a nested case–control study of 120 cases of spontaneous abortion and 251 controls on the basis of a file of 6000 Finnish workers who had been biologically monitored for exposure to solvents. Information about their marriages and their wives' pregnancies and spontaneous abortions were obtained from national registries; data on paternal occupational exposure to solvents were collected by means of a questionnaire sent to workers and covered the period of spermatogenesis. The likelihood of exposure was defined in three categories: unexposed, potentially exposed (i.e. use of solvents was possible but no exposure was reported or measured) and probably exposed (i.e. exposure was measured or reported). No association was found between paternal occupational exposure to trichloroethylene and spontaneous abortion (crude odds ratio, 1.0; 95% CI, 0.6–2.0).

(c) Pregnancy

Pregnancies occurring among 3265 women biologically monitored for exposure to solvents in 1965–83 were identified from a Finnish database (Lindbohm *et al.*, 1990). Only one pregnancy per woman was included, resulting in a total of 120 cases of spontaneous abortion; 336 age-matched controls were randomly selected among women who had only normal births during the study period. Data on workplace, occupational exposure, medical history, alcohol and smoking habits were obtained from a postal questionnaire, to which 85.5% of subjects responded. For each potential exposure, women were classified, without knowledge of their case or control status, into one of three categories: unexposed, potentially exposed (i.e. work tasks might have involved use of solvents, but exposure was not reported or measured) or exposed (i.e. exposure was measured or reported). The analysis addressed 73 women who had had a spontaneous abortion and 167 controls who reported a pregnancy of interest and detailed information on occupational exposures during pregnancy. The odds ratio for spontaneous abortion, adjusted for previous spontaneous abortions, parity, smoking, use of alcohol and exposure to other solvents, was 0.6 (95% CI, 0.2–2.3) for exposure to trichloroethylene.

The 852 women for whom a spontaneous abortion was certified in one of the 11 hospital laboratories in Santa Clara County, CA (United States) were compared with 1618 controls randomly selected among County residents who had had a live birth and frequency matched by date of last menstrual period and hospital (Windham *et al.*, 1991). All participants were contacted by telephone and asked about occupational use of 18 solvents or products during the first 20 weeks of pregnancy. An excess risk for spontaneous abortion was observed for those women who reported exposure to trichloroethylene (crude odds ratio, 3.1; 95% CI, 0.92–10.4) [adjusted odds ratio not calculated]; four of the seven women who reported exposure to trichloroethylene had also used tetrachloroethylene. The odds ratio increased for women who reported more 'intense' exposure, primarily on the basis of detection of odour (odds ratio, 3.9; $p = 0.04$). Odds ratios adjusted for maternal age, race, education, prior fetal loss, smoking, average number of hours worked and quality of response were nonsignificant when the whole group of halogenated solvents was considered (odds ratio for any use, 1.0; 95% CI, 0.65–1.6; odds ratio for use > 10 h per week, 1.5, 95%CI, 0.73–3.0).

Information on 7316 pregnancies was obtained from the hospital discharge register for 9186 women identified as working in Finnish laboratories (Taskinen *et al.*, 1994).The pregnancies resulted in 5663 births, 687 spontaneous abortions and 966 induced abortions, and a case–referent study was conducted within the cohort. Questionnaires were posted requesting confirmation of the study pregnancy and data on exposures; the response rate was 78%. The 206 women with only one registered spontaneous abortion and 329 controls randomly selected among women who had given birth to a normal infant were included in the analysis of spontaneous abortion. The analysis of congenital malformations involved 36 cases and 105 referents. Seven women who had had a spontaneous abortion and nine controls reported exposure to trichloroethylene, giving an odds ratio of 1.6 (95% CI, 0.5–4.8), adjusted for employment, smoking, alcohol consumption, parity, previous miscarriages, failed birth control and febrile disease during pregnancy. The odds ratios associated with exposure to halogenated solvents as a group were 0.6 (0.4–1.1) for exposure on one to two days per week and 1.8 (0.9–3.7) for exposure on three to five days per week. The odds ratio for congenital malformations associated with exposure to halogenated solvents was 0.8 (0.2–2.5), adjusted for alcohol consumption, parity, previous miscarriages and failed birth control.

In 1981, the groundwater in a small area in the southwestern part of the city of Tucson, Arizona (United States), was found to be contaminated with trichloroethylene and, to a lesser extent, with dichloroethylene and chromium (Goldberg et al., 1990). The parents of 707 children with congenital heart disease who had conceived their child and spent the beginning of the pregnancy (one month before and the first trimester) in the Tucson valley between 1969 and 1987 were interviewed. The prevalence of congenital heart disease among children born to mothers who had been exposed (0.68%) was higher than that of mothers who lived outside the area (0.26%; $p < 0.001$). The ratio decreased to near unity for new arrivals in the contaminated area after closure of the well.

4.3.2 Experimental systems

Trichloroethylene and its metabolites appear to cross the placenta readily in many species (Helliwell & Hutton, 1949, 1950; Lanham, 1970; Withey & Karpinski, 1985; Ghantous et al., 1986). In mice, inhalation of trichloroethylene resulted in accumulation of its metabolite, trichloroacetic acid (see also Land et al., 1981), in amniotic fluid (Ghantous et al., 1986).

A significant increase in the percentage of abnormal spermatozoa was observed in mice exposed to 0.2% trichloroethylene for 4 h per day for five days over that in controls and in mice exposed to 0.02% trichloroethylene (Land et al., 1981). No sperm toxicity was induced in male Long-Evans rats exposed by gavage to up to 1000 mg/kg bw, trichloroethylene on five days per week for six weeks (Zenick et al., 1984). Mating of untreated female NMRI mice with male mice that had been exposed to up to 450 ppm [2417 mg/m^3] trichloroethylene by inhalation for 24 h did not influence fertilization or pre- or post-implantation rates and did not induce dominant lethal mutation (Slacik-Erben et al., 1980). No modification of mating performance or female fertility was observed in groups of female Long-Evans rats exposed to trichloroethylene by gavage for two weeks before mating at doses up to 1000 mg/kg bw, which was a toxic dose (Manson et al., 1984). Administration of trichloroethylene in the diet of mice and rats at concentrations equivalent to doses of up to 300 mg/kg bw per day for two generations resulted in marginal effects on testicular weight and on survival of pups of both the F_1 and F_2 generations at the highest dose. No other signs of reproductive toxicity were observed (United States National Toxicology Program, 1985, 1986).

Female Long-Evans rats were exposed by inhalation to 1800 ± 200 ppm [9666 ± 1074 mg/m^3] trichloroethylene for two weeks before and/or during gestation. Post-natal body weight was decreased in the offspring of mothers that had been exposed before gestation. Significant increases in the incidence of skeletal and soft-tissue anomalies, indicative of developmental delay in maturation rather than teratogenesis, were observed in the group exposed during pregnancy alone (Dorfmueller et al., 1979). A significant increase in the incidence of cardiac malformations was reported in newborn Sprague-Dawley rats after maternal exposure to trichloroethylene in drinking-water (1.5 or 1100 ppm [mg/L]) for seven days before and throughout gestation. [The actual dose could not be calculated from the available data.] No signs of maternal toxicity or other signs of fetal toxicity were observed (Dawson et al., 1993). No increase in the frequency of birth defects has been reported in most other studies of rat or mouse dams exposed by various routes to various concentrations of trichloroethylene, except for a predictable impairment of fetal growth associated with maternally toxic doses (Schwetz et al., 1975; Leong et al., 1975; Healy & Wilcox, 1978; Hardin et al., 1981; Cosby & Dukelow, 1992).

The male offspring of female rats exposed to trichloroethylene in the drinking-water at up to 1250 mg/L before and during gestation and postpartum up to day 21 had enhanced locomotor activity and exploratory behaviour (Taylor et al., 1985). Impairment of myelinization of the central nervous system and decreased glucose uptake by whole brain and cerebellum were observed in the offspring of rats exposed to 312 or 625 mg/L trichloroethylene in the drinking-water before and during gestation and postpartum (Noland-Gerbee et al., 1986; Isaacson & Taylor, 1989). The specific gravity of brain tissue was reduced in the offspring of mice exposed to 150 ppm [806 mg/m^3] trichloroethylene by inhalation four weeks before and during gestation (Westergren et al., 1984).

4.4 Genetic and related effects

4.4.1 Humans

Cytogenetic damage in lymphocytes: In a study of 28 male degreasers exposed to trichloroethylene, nine were reported to have > 13% hypodiploid cells in cultured peripheral lymphocytes (Konietzko et al., 1978). These men had been exposed to a higher mean maximal concentration of trichloroethylene (206 ppm [1106 mg/m^3]) than those considered to have normal rates of hypodiploidy (116 ppm [623 mg/m^3]). A correlation ($r = 0.46$; $p < 0.05$) was also seen between the hypodiploidy rate and the average daily or average maximal exposure to trichloroethylene. The mean rate of hypodiploid cells was 10.9% (SD, 4.5; $n = 27$, excluding one man with karyotype 47, XY, +mar), in comparison with 6.5% (SD, 3.2) among 10 male controls. The exposed workers also had a fivefold higher mean rate of chromosomal breaks per 100 mitoses (3.1; SD, 3.7; $n = 27$) than the controls (0.6; SD, 0.7; $n = 10$), but these data were not commented upon. The effects of age and cigarette smoking could not be judged from the report. [The Working Group noted that the hypodiploidy rate among controls was very high.]

In a study of 22 workers who had constantly used trichloroethylene in their [unspecified] jobs for an average of 9.7 years (range, 0.7–34) and 22 controls matched for age, sex and smoking habits, no increase in the frequency of sister chromatid exchange was seen in peripheral lymphocyte (Nagaya et al., 1989). Spot urine samples collected at the same time as the blood samples from the exposed workers showed a concentration of 19.1–1066.4 mg/L (mean, 183.6 mg/L) total trichloro compounds. Smoking increased the frequency of sister chromatid exchange.

A group of 15 workers involved in metal degreasing with trichloroethylene for more than 20 h per week in a half-open vapour plant had a significantly greater frequency of chromosomal aberrations, excluding gaps and hyperdiploid cells, in cultured lymphocytes than 669 controls; seven of the degreasers were also painters. The mean urinary concentration of trichloroacetate was fairly low: 3.7 mg/L (range, 0.02–26.9), and the mean number of cumulative working years was 4.6 (range, 0.8–22.0) (Rasmussen et al., 1988). The effects of smoking and age could not be judged from the paper. The authors considered the reference group 'not ideal' but reported that the distribution of confounding factors was no different from that in the average population.

Sperm counts and the frequencies of abnormal sperm heads and of sperm with two fluorescent Y bodies were not significantly different in the 12 workers and 14 controls from whom semen samples containing sperm were taken.

Sister chromatid exchange was analysed in 22 male and 16 female workers in trichloroethylene synthesis and degreasing and in 26 control male and 25 female subjects who worked filling tanks with hydrogen, nitrogen and oxygen or as lathe operators (Seiji et al., 1990). No effect of the occupational exposure was seen among nonsmokers, but the eight exposed smokers (all males) had a significantly higher mean frequency of sister chromatid exchange per cell (7.06) than seven male smoking controls (5.10). Sister chromatid exchange was also studied in nine male and 10 female tetrachloroethylene synthesis workers who had been exposed to an 8-h time-weighted geometric mean concentration of 8 ppm [43.0 mg/m^3] trichloroethylene (75th percentile, 49 ppm [263 mg/m^3]; maximum, 521 ppm [2798 mg/m^3]) and 17 ppm [115 mg/m^3] tetrachloroethylene (75th percentile, 28 ppm [190 mg/m^3]; maximum, 567 ppm [3844 mg/m^3]). They were compared with a control group of nine men and nine women and an extended control group consisting of 21 men and 23 women. Occupational exposure was reported to have affected the frequency of sister chromatid exchange in exposed male smokers, on the basis of a comparison of the frequency in these five men (7.33) with that in six nonsmoking male controls in the small (5.72; $p < 0.05$) and nine controls in the extended (5.48; $p < 0.01$) groups; the mean frequency of sister chromatid exchange in exposed male smokers was also higher than that in the 12 male smokers in the extended control group (5.7). No significant differences were reported between exposed and unexposed smokers. [Comparison of exposed smokers and unexposed nonsmokers may not be justified, especially as smoking usually induces sister chromatid exchange, although in this study such an effect could not be shown.]

4.4.2 *Experimental systems* (see also Tables 12 and 13 and Appendices 1 and 2)

The genetic toxicology of trichloroethylene has been reviewed (Baden & Simmon, 1980; Fabricant & Chalmers, 1980; Vainio et al., 1985; Crebelli & Carere, 1989; Candura & Faustman, 1991; Jackson et al., 1993; European Centre for Ecotoxicology and Toxicology of Chemicals, 1994). The mechanisms of the possible genotoxicity of trichloroethylene were discussed by Henschler (1987).

(a) DNA binding

Trichloroethylene was reported to bind to DNA *in vitro* after metabolic activation; the binding was enhanced by the addition of glutathione and reduced by addition of SKF-525-A, an inhibitor of mixed-function oxidases. High-performance liquid chromatography indicated a possible DNA adduct, which could not be identified (Mazzullo et al., 1992). DNA binding could not be demonstrated *in vivo* in several tissues of mice in one study (Bergman, 1983b) or in the liver of rats in another study (Parchman & Magee, 1982); however, the latter authors noted incorporation of label into normal nucleosides. A low level of covalent interaction was reported with the DNA of rat and mouse liver, kidney, lungs and stomach (estimated at 0.15 adducts per 10^6 nucleotides; Mazzullo et al., 1992) and of mouse liver (maximum, 0.62 alkylations per 10^6 nucleotides; Stott et al., 1982).

Table 12. Genetic and related effects of trichloroethylene without mutagenic stabilizers

Test system	Result[a] Without exogenous metabolic system	Result[a] With exogenous metabolic system	Dose[b] (LED/HID)	Reference
PRB, SOS chromotest, *Escherichia coli* PQ37	–	–	7325[c]	Mersch-Sundermann et al. (1989)
SAF, *Salmonella typhimurium* BAL13, forward mutation (*ara* test)	–	–	190	Roldán-Arjona et al. (1991)
SA0, *Salmonella typhimurium* TA100, reverse mutation	–	(+)	160 vapour[c]	Simmon et al. (1977)
SA0, *Salmonella typhimurium* TA100, reverse mutation	–	–	160 vapour[c]	Baden et al. (1979)
SA0, *Salmonella typhimurium* TA100, reverse mutation	–	–	420 (8% vapour) 16h	Bartsch et al. (1979)
SA0, *Salmonella typhimurium* TA100, reverse mutation	–	+	18 vapour	Crebelli et al. (1982)
SA0, *Salmonella typhimurium* TA100, reverse mutation	–	–	260 vapour[d]	Shimada et al. (1985)
SA0, *Salmonella typhimurium* TA100, reverse mutation	–	–	167[c]	Mortelmans et al. (1986)
SA0, *Salmonella typhimurium* TA100, reverse mutation	0	–	1050 vapour	McGregor et al. (1989)
SA5, *Salmonella typhimurium* TA1535, reverse mutation	–	–	526 vapour[c]	Baden et al. (1979)
SA5, *Salmonella typhimurium* TA1535, reverse mutation	(+)	0	50[c]	Kringstad et al. (1981)
SA5, *Salmonella typhimurium* TA1535, reverse mutation	(+)	–	50 vapour[d]	Shimada et al. (1985)
SA5, *Salmonella typhimurium* TA1535, reverse mutation	–	–	167[c]	Mortelmans et al. (1986)
SA7, *Salmonella typhimurium* TA1537, reverse mutation	–	–	167[c]	Mortelmans et al. (1986)
SA9, *Salmonella typhimurium* TA98, reverse mutation	–	–	167[c]	Mortelmans et al. (1986)
SA9, *Salmonella typhimurium* TA98, reverse mutation	0	–	1050 vapour	McGregor et al. (1989)
SCG, *Saccharomyces cerevisiae* D7, gene conversion	–	+	2600	Bronzetti et al. (1978)
SCR, *Saccharomyces cerevisiae* D7, reverse mutation	–	+	1300	Bronzetti et al. (1978)
ANG, *Aspergillus nidulans*, diploid yA2/+ strain 35x17, quiescent conidia, mitotic crossing-over	–	0	3660	Crebelli et al. (1985)
ANG, *Aspergillus nidulans*, diploid yA2/+ strain 35x17, growth-mediated assay, mitotic crossing-over	–	0	90 vapour	Crebelli et al. (1985)
SZF, *Schizosaccharomyces pombe* P1, stationary phase, forward mutation	–	–	3280	Rossi et al. (1983)
SZF, *Schizosaccharomyces pombe* P1, growing cells, forward mutation	–	–	13 140	Rossi et al. (1983)
ANF, *Aspergillus nidulans*, haploid strain 35, quiescent conidia, forward mutation (methionine suppressor)	–	0	100 vapour	Crebelli et al. (1985)
ANF, *Aspergillus nidulans*, haploid strain 35, 'growth-mediated assay', forward mutation (methionine suppressor)	+	0	13 vapour	Crebelli et al. (1985)

Table 12 (contd)

Test system	Result[a] Without exogenous metabolic system	Result[a] With exogenous metabolic system	Dose[b] (LED/HID)	Reference
ANN, *Aspergillus nidulans*, diploid yA2/+ strain 35x17, quiescent conidia, nondisjunctional diploids	–	0	3660	Crebelli *et al.* (1985)
ANN, *Aspergillus nidulans*, diploid yA2/+ strain 35x17, quiescent conidia, haploids	–	0	3660	Crebelli *et al.* (1985)
ANN, *Aspergillus nidulans*, diploid yA2/+ strain 35x17, 'growth-mediated assay', nondisjunctional diploids	+	0	40 vapour	Crebelli *et al.* (1985)
ANN, *Aspergillus nidulans*, diploid yA2/+ strain 35x17, 'growth-mediated assay', haploids	+	0	90 vapour	Crebelli *et al.* (1985)
DMX, *Drosophila melanogaster*, sex-linked recessive lethal mutation	–		2500[c] injection	Foureman *et al.* (1994)
DMX, *Drosophila melanogaster*, sex-linked recessive lethal mutation	?		5000 feeding[c]	Foureman *et al.* (1994)
URP, Unscheduled DNA synthesis, rat primary hepatocytes *in vitro*	–	0	130 vapour[d]	Shimada *et al.* (1985)
G5T, Gene mutation, mouse lymphoma L5178Y cells, *tk* locus *in vitro*	–	+	146[c]	Caspary *et al.* (1988)
SIC, Sister chromatid exchange, Chinese hamster ovary (CHO) cells *in vitro*	(+)	(+)	401[c]	Galloway *et al.* (1987)
CIC, Chromosomal aberrations, Chinese hamster ovary (CHO) cells *in vitro*	–	–	14 900[c]	Galloway *et al.* (1987)
TRR, Cell transformation, RLV/Fischer rat F1706 embryo cells *in vitro*	+	0	144	Price *et al.* (1978)
GIH, Gene mutation, human lymphoblastoid TK6 cells *in vitro*	–	–	600	Caspary *et al.* (1988)
ICR, Inhibition of intercellular communication, B6C3F1 mouse hepatocytes *in vitro*	+	0	1.3	Klaunig *et al.* (1989)
ICR, Inhibition of intercellular communication, F344 rat hepatocytes *in vitro*	–	0	13	Klaunig *et al.* (1989)
HMM, Host-mediated assay, gene conversion in *Saccharomyces cerevisiae* D4 recovered from CD-1 mouse liver, lungs and kidneys	+		400 po × 1[c]	Bronzetti *et al.* (1978)
HMM, Host-mediated assay, gene conversion in *Saccharomyces cerevisiae* D7 recovered from CD-1 mouse liver and kidneys	+		400 po × 1	Bronzetti *et al.* (1978)
HMM, Host-mediated assay, gene conversion in *Saccharomyces cerevisiae* D7 recovered from CD-1 mouse lungs	–		400 po × 1	Bronzetti *et al.* (1978)
HMM, Host-mediated assay, reverse mutation in *Saccharomyces cerevisiae* D7 from CD-1 mouse liver, lungs and kidneys	+		400 po × 1	Bronzetti *et al.* (1978)
HMM, Host-mediated assay, forward mutation in *Schizosaccharomyces pombe* P1, CD-1 × C57Bl hybrid mouse	–		2000 iv or ip × 1	Rossi *et al.* (1983)
DVA, DNA single-strand breaks, mouse liver *in vivo*	–[f]		2000 ip × 1	Parchman & Magee (1982)
DVA, DNA single-strand breaks (alkaline unwinding) in liver and kidney of male NMRI mice *in vivo*	+[f]		790 ip × 1	Walles (1986)

Table 12 (contd)

Test system	Result[a] Without exogenous metabolic system	With exogenous metabolic system	Dose[b] (LED/HID)	Reference
DVA, DNA single-strand breaks (alkaline unwinding), mouse liver in vivo	+		1500 po × 1[c]	Nelson & Bull (1988)
DVA, DNA single-strand breaks (alkaline unwinding), rat liver in vivo	+		3000 po × 1[c]	Nelson & Bull (1988)
MST, Mouse spot test in vivo	–		350 ip × 1	Fahrig (1977)
UVM, Unscheduled DNA synthesis, CD-1 mouse primary hepatocytes in vivo	–		1000 po × 1	Doolittle et al. (1987)
MVM, Micronucleus induction, mouse bone-marrow erythrocytes in vivo	+		750 po × 2	Duprat & Gradiski (1980)
MVM, Micronucleus induction, B6C3F1 mouse bone-marrow erythrocytes in vivo	–		2500 ip × 3[c]	Shelby et al. (1993)
MVM, Micronucleus induction, mouse spermatocytes in vivo (spermatids examined) in vivo	–		565 inh 6 h/d × 5	Allen et al. (1994)
MVM, Micronucleus induction, mouse splenocytes in vivo	–		9800 inh 6 h	Kligerman et al. (1994)
MVR, Micronucleus induction, rat bone-marrow erythrocytes in vivo	+		5 inh 6 h	Kligerman et al. (1994)
MVR, Micronucleus induction, rat bone-marrow erythrocytes in vivo	–		960 inh 6 h × 4	Kligerman et al. (1994)
MVR, Micronucleus induction, rat peripheral lymphocytes in vivo	–		8800 inh 6 h	Kligerman et al. (1994)
MVR, Micronucleus induction, rat peripheral lymphocytes in vivo	–		960 inh 6 h × 4	Kligerman et al. (1994)
SVA, Sister chromatid exchange, rat peripheral lymphocytes in vivo	–		8800 inh 6 h	Kligerman et al. (1994)
SVA, Sister chromatid exchange, rat peripheral lymphocytes in vivo	–		960 inh 6 h × 4	Kligerman et al. (1994)
SVA, Sister chromatid exchange, mouse splenocytes in vivo	–		9800 inh 6 h	Kligerman et al. (1994)
CLA, Chromosomal aberrations, rat peripheral lymphcytes in vivo	–		8800 inh 6 h	Kligerman et al. (1994)
CLA, Chromosomal aberrations, rat peripheral lymphcytes in vivo	–		960 inh 6 h × 4	Kligerman et al. (1994)
CVA, Chromosomal aberrations, mouse splenocytes in vivo	–		9800 inh 6 h	Kligerman et al. (1994)
DLM, Dominant lethal mutation, male NMRI-Han/BGA mice in vivo	–		3400 inh 24 h[c]	Slacik-Erben et al. (1980)
BID, Binding (covalent) to salmon sperm DNA in vitro		+	270	Banerjee & Van Duuren (1978)
BID, Binding (covalent) to calf thymus DNA in vitro	–	+	340[c]	Bergman (1983b)
BID, Binding (covalent) to calf thymus DNA in vitro	0	+	13	Miller & Guengerich (1983)
BID, Binding (covalent) to DNA of isolated rat hepatocytes in vitro	+	0	13	Miller & Guengerich (1983)
BID, Binding (covalent) to DNA of isolated mouse hepatocytes in vitro	+	0	13	Miller & Guengerich (1983)
BID, Binding (covalent) to calf thymus DNA in vitro	0	+	131	DiRenzo et al. (1982)
BVP, Binding (covalent) to RNA of NMRI mouse spleen, lung, liver, kidney, pancreas, testis and brain in vivo	–[g]		67 ip × 5[c]	Bergman (1983b)

Table 12 (contd)

Test system	Result[a] Without exogenous metabolic system	Result[a] With exogenous metabolic system	Dose[b] (LED/HID)	Reference
BVD, Binding (covalent) to DNA of NMRI mouse spleen, pancreas, lung, testis, kidney and brain *in vivo*	–[x]		67 ip × 5	Bergman (1983b)
BVD, Binding (covalent) to DNA of NMRI mouse liver *in vivo*	?		67 ip × 5	Bergman (1983b)
BVD, Binding (covalent) to DNA of B6C3F1 mouse liver *in vivo*	?		1200 po × 1	Stott *et al.* (1982)
BVD, Binding (covalent) to DNA of B6C3F1 mouse liver *in vivo*	?		250 ip × 1	Parchman & Magee (1982)
BVD, Binding (covalent) to DNA of rat liver *in vivo*	?		1000 ip × 1	Parchman & Magee (1982)
Dichloroacetyl chloride				
PRB, λ Prophage induction, *Escherichia coli* WP2	–	–	10 000	DeMarini *et al.* (1994)
SA0, *Salmonella typhimurium* TA100, reverse mutation	–	(+)	3	DeMarini *et al.* (1994)

[a] +, considered to be positive; (+), considered to be weakly positive in an inadequate study; –, considered to be negative; ?, considered to be inconclusive (variable responses in several experiments within an inadequate study); 0, not tested
[b] LED, lowest effective dose; HID, highest effective dose. In-vitro tests, μg/ml; in-vivo tests, mg/kg bw; ip, intraperitoneally; po, orally
[c] 99% purity or greater
[d] 0.001% stabilizers
[e] Also positive by gavage at 150 mg/kg for 5 days a week, 22 times with 400 mg/kg on the last day
[f] No DNA strand breaks in lungs of mice treated with 1300 mg/kg ip × 1
[x] Metabolic incoporation of ¹⁴C into nucleotides was observed.

Table 13. Genetic and related effects of trichloroethylene containing mutagenic stabilizers or for which information on purity was not sufficiently clear

Test system	Result[a] Without exogenous metabolic system	Result[a] With exogenous metabolic system	Dose[b] (LED/HID)	Reference
PRB, SOS chromotest, *Escherichia coli* PQ37	–	–	0.00	von der Hude et al. (1988)
***, Mutatox assay, derepression of luminescence operon, *Photobacterium phosphorium*	–	0	0.00	Elmore & Fitzgerald (1990)
SA0, *Salmonella typhimurium* TA100, reverse mutation	–	–	14 650	Henschler et al. (1977)
SA0, *Salmonella typhimurium* TA100, reverse mutation	–	–	525 vapour	Waskell (1978)
SA0, *Salmonella typhimurium* TA100, reverse mutation	+	+	260 vapour[c]	Shimada et al. (1985)
SA0, *Salmonella typhimurium* TA100, reverse mutation	+	–	0.00	Milman et al. (1988)
SA0, *Salmonella typhimurium* TA100, reverse mutation	(+)	(+)	130 vapour	McGregor et al. (1989)
SA5, *Salmonella typhimurium* TA1535, reverse mutation	+	+	50 vapour[c]	Shimada et al. (1985)
SA5, *Salmonella typhimurium* TA1535, reverse mutation	+	–	0.00	Milman et al. (1988)
SA5, *Salmonella typhimurium* TA1535, reverse mutation	+	+	33 vapour	McGregor et al. (1989)
SA7, *Salmonella typhimurium* TA1537, reverse mutation	–	–	0.00	Milman et al. (1988)
SA9, *Salmonella typhimurium* TA98, reverse mutation	–	–	525 vapour	Waskell (1978)
SA9, *Salmonella typhimurium* TA98, reverse mutation	–	–	0.00	Milman et al. (1988)
SA9, *Salmonella typhimurium* TA98, reverse mutation	–	–	65 vapour	McGregor et al. (1989)
ECK, *Escherichia coli* K12, forward mutation	–	–	434	Greim et al. (1975)
ECK, *Escherichia coli* K12, reverse mutation (*arg*⁺)	–	+	434	Greim et al. (1975)
ECK, *Escherichia coli* K12, reverse mutation (*gal*⁻)	–	–	434	Greim et al. (1975)
ECK, *Escherichia coli* K12, reverse mutation (*nad*⁻)	–	–	434	Greim et al. (1975)
SCG, *Saccharomyces cerevisiae* D7, log-phase cultures, gene conversion	0	+	1970	Callen et al. (1980)
SCG, *Saccharomyces cerevisiae* D7, log-phase and stationary cultures, gene conversion	–	–	2900	Koch et al. (1988)
SCG, *Saccharomyces cerevisiae* XV185-14C, reverse mutation (*lys*1-1, *his*1-7, *hom*3-10)	0	+	1460	Shahin & Von Borstel (1977)
SCR, *Saccharomyces cerevisiae* D7, log-phase cultures, reverse mutation	0	+	1970	Callen et al. (1980)
SCH, *Saccharomyces cerevisiae* D7, log-phase cultures, mitotic recombinants or otherwise genetically altered colonies (*ade*2)	0	+	1970	Callen et al. (1980)

Table 13 (contd)

Test system	Result[a] Without exogenous metabolic system	Result[a] With exogenous metabolic system	Dose[b] (LED/HID)	Reference
SCR, *Saccharomyces cerevisiae* D7, log-phase and stationary cultures, reverse mutation	–	(+)	2900	Koch et al. (1988)
SZF, *Schizosaccharomyces pombe* P1, stationary phase, forward mutation	–	–	3280	Rossi et al. (1983)
SZF, *Schizosaccharomyces pombe* P1, growing cells, forward mutation	–	–	13 140	Rossi et al. (1983)
SCN, *Saccharomyces cerevisiae* D61.M, growing cells, aneuploidy	0	+	725	Koch et al. (1988)
TSM, *Tradescantia* species, mutation	+	0	0.0003	Schairer & Sautkulis (1982)
URP, Unscheduled DNA synthesis, rat primary hepatocytes *in vitro*	–	0	130 vapour	Shimada et al. (1985)
URP, Unscheduled DNA synthesis, rat primary hepatocytes *in vitro*	–	0	0.00	Milman et al. (1988)
URP, Unscheduled DNA synthesis, rat primary hepatocytes *in vitro*	+	0	1445	Williams et al. (1989)
UIA, Unscheduled DNA synthesis, B6C3F1 mouse primary hepatocytes *in vitro*	+	0	0.00	Milman et al. (1988)
SIC, Sister chromatid exchange, Chinese hamster ovary (CHO) cells *in vitro*	0	–	9	White et al. (1979)
CIC, Chromosomal aberrations, Chinese hamster lung (CHL) cells *in vitro*	–	0	1000	Sofuni et al. (1985)
TBM, BALB/c–3T3 mouse cells, cell transformation *in vitro*	(+)	0	250	Tu et al. (1985)
TFS, Syrian hamster embryo cells, morphological transformation *in vitro*	(+)	0	25	Amacher & Zelljadt (1983)
SHL, Sister chromatid exchange, human lymphocytes *in vitro*	(+)	0	178	Gu et al. (1981)
HMM, Host-mediated assay, forward mutation in *Schizosaccharomyces pombe* P1 recovered from CD-1 mouse kidneys and lungs	–	0	2000 po × 1	Loprieno & Abbondandolo (1980)
HMM, Host-mediated assay, forward mutation in *Schizosaccharomyces pombe* P1 recovered from CD-1 mouse liver	(+)	0	2000 po × 1	Loprieno & Abbondandolo (1980)
HMM, Host-mediated assay, *Schizosaccharomyces pombe* P1, forward mutation, in CD-1 mouse peritoneum	(+)	0	1000 po × 1	Loprieno & Abbondandolo (1980)
HMM, Host-mediated assay, *Schizosaccharomyces pombe* P1, forward mutation, in Sprague-Dawley rat peritoneum	–	0	1000 po × 1	Loprieno & Abbondandolo (1980)
HMM, Host-mediated assay, forward mutation in *Schizosaccharomyces pombe* P1, CD-1 × 7BL hybrid mouse	–	0	2000 iv or ip × 1	Rossi et al. (1983)
UPR, Unscheduled DNA synthesis, Fischer-344 male rat hepatocytes *in vivo*	–		1000 po × 1	Mirsalis et al. (1989)
UVM, Unscheduled DNA synthesis, male and female B6C3F1 mouse hepatocytes *in vivo*	–		1000 po × 1	Mirsalis et al. (1989)

Table 13 (contd)

Test system	Result[a]		Dose[b] (LED/HID)	Reference
	Without exogenous metabolic system	With exogenous metabolic system		
CBA, Chromosomal aberrations, CD-1 mouse bone marrow cells *in vivo*	–		1000 po × 1	Loprieno & Abbondandolo (1980)
CBA, Chromosomal aberrations, mouse bone marrow cells *in vivo*	–		1200 po × 1	Sbrana *et al.* (1985) (abstract)
CBA, Chromosomal aberrations, mouse bone marrow cells *in vivo*	–		795 inh 7 h × 50[c]	Sbrana *et al.* (1985) (abstract)
MVM, Micronucleus induction, mouse bone marrow erythrocytes *in vivo*	+		1200 po × 1	Sbrana *et al.* (1985) (abstract)
MVM, Micronucleus induction, mouse bone marrow erythrocytes *in vivo*	+		460 ip × 1	Hrelia *et al.* (1995)
SLH, Sister chromatid exchange, human lymphocytes *in vivo*	(+)		0.00	Gu *et al.* (1981)
BID, Binding (covalent) to calf thymus DNA *in vitro*	0	+	3.2	Mazzullo *et al.* (1992)
BVD, Binding (covalent to DNA of BALB/c mouse liver, kidney, lung and stomach *in vivo*	(+)		0.76 ip × 1	Mazzullo *et al.* (1992)
BVD, Binding (covalent) to DNA of Wistar rat liver, kidney, lung and stomach *in vivo*	(+)		0.76 ip × 1	Mazzullo *et al.* (1992)
***, Enzyme-altered foci in male Osborne-Mendel rat liver *in vivo*, promotion protocol, with and without NDEA as an initiator	–		1300 mg/kg, 5 d/week, 7 weeks	Milman *et al.* (1988)
***, Enzyme-altered foci in male Osborne-Mendel rat liver *in vivo*, initiation protocol, phenobarbital as promoter	–		1300 mg/kg	Milman *et al.* (1988)
***, S-Phase induction, male and female B6C3F1 mouse hepatocytes *in vivo*	+		200 mg/kg	Mirsalis *et al.* (1989)

NDEA, *N*-nitrosodiethylamine

[a] +, considered to be positive; (+), considered to be weakly positive in an inadequate study; –, considered to be negative; ?, considered to be inconclusive (variable responses in several experiments within an inadequate study); 0, not tested

[b] LED, lowest effective dose; HID, highest effective dose. In-vitro tests, mg/ml; in-vivo tests, mg/kg bw; 0.00, dose not reported; ip, intraperitoneally; po, orally

[c] 5 days/week, 10 weeks

***, Not included on profile

(b) Mutation and allied effects

The stabilizers often used in commercial preparations of trichloroethylene, such as epichlorohydrin and 1,2-epoxybutane, are mutagenic, rendering problematic the interpretation of positive results in assays for the mutagenicity of trichloroethylene *per se* (McGregor et al., 1989). Humans are exposed mostly, if not exclusively, to preparations containing stabilizers.

Apart from two reports in which trichloroethylene weakly induced mutation in *Salmonella typhimurium* TA1535, purified trichloroethylene did not induce gene mutation in various strains of *Salmonella* in the absence of metabolic activation; however, trichloroethylene containing directly mutagenic epoxide stabilizers did. Purified trichloroethylene also did not usually induce mutation in *Salmonella* in the presence of exogenous metabolic activation systems, except in two tests with *S. typhimurium* TA100.

Trichloroethylene (pure or of unspecified purity) gave negative results in the SOS chromotest in *Escherichia coli* with and without metabolic activation and in the Mutatox assay in the absence of metabolic activation. In the presence of metabolic activation, analytical-grade trichloroethylene induced arg^+ reverse mutations, but not forward mutations or gal^+ or nad^+ reversions, in *E. coli*.

Trichloroethylene (pure or of unspecified purity) induced gene conversion in *Saccharomyces cerevisiae* in two of three studies and induced reverse mutation in all four studies available in the presence of a metabolic activation system. In a single study, pure trichloroethylene or trichloroethylene containing stabilizers did not induce forward mutation in *Schizosaccharomyces pombe*. Pure trichloroethylene induced forward mutation in one study of growing cultures of *Aspergillus nidulans*, which are capable of some metabolic activation reactions, whereas no such effect was seen in quiescent conidia. Trichloroethylene (of unspecified purity) induced aneuploidy in *S. cerevisiae* in the presence of growth-mediated metabolic activation, and the pure compound induced aneuploidy in *A. nidulans*. In a single study, trichloroethylene (of unspecified purity) induced gene mutation in *Tradescantia*. Pure trichloroethylene did not cause recessive lethal mutations in *Drosophila melanogaster* after injection, and equivocal results were obtained after feeding.

Unscheduled DNA synthesis *in vitro* was reported in four studies, one with mouse and three with rat hepatocytes. Positive results were obtained with trichloroethylene (of unspecified purity) in mouse cells and in one study of rat cells, while negative results were obtained in the other two studies of rat primary hepatocytes, in one of which trichloroethylene of high and of unspecified purity were compared. Pure trichloroethylene induced gene mutation in mouse lymphoma L5178Y cells in the presence of exogenous metabolic activation. In a single study, pure trichloroethylene weakly induced sister chromatid exchange in Chinese hamster cells *in vitro* with and without metabolic activation. Pure trichloroethylene did not increase the frequency of chromosomal aberrations in Chinese hamster cells *in vitro*. In three different assays, trichloroethylene (of unspecified purity) weakly induced cell transformation in mouse, Syrian hamster and (pure trichloroethylene) rat cells *in vitro*, without exogenous metabolic activation. Pure trichloroethylene inhibited intercellular communication in mouse hepatocytes but not in rat hepatocytes *in vitro*.

A 95% pure formulation weakly induced sister chromatid exchange in the absence of metabolic activation in one study. No induction of gene mutation was seen in human lymphoblastoid cells exposed to pure trichloroethylene.

In a host-mediated assay, gene conversion and reverse mutation were induced in *S. cerevisiae* recovered from the liver, lungs and kidneys of mice treated orally with pure trichloroethylene. Forward mutation was weakly induced by trichloroethylene of unspecified purity in *Schizosaccharomyces pombe* cells injected into the peritoneum of mice in one of two studies; no effect was seen in the only study available in rats. *S. pombe* cells recovered from mice after intravenous injection showed no forward mutation in one study; a positive result was seen in another study in mouse liver, but not in kidneys or lungs, after treatment with trichloroethylene of unspecified purity.

Pure trichloroethylene induced DNA single-strand breaks/alkaline-labile sites *in vivo* in mouse liver and kidney and in rat liver. Unscheduled DNA synthesis was not augmented in mouse or rat hepatocytes after treatment with trichloroethylene (pure or of unspecified purity) *in vivo*, and pure trichloroethylene did not induce a significant response in a mouse spot test. Trichloroethylene did not induce chromosomal aberrations in mouse bone marrow *in vivo*. Micronuclei were reported to be induced by trichloroethylene (pure or of unknown purity) in mouse bone-marrow polychromatic erythrocytes in two studies (one was reported in an abstract), while two other studies showed no such effect. A significant increase ($p = 0.028$) observed in one of the latter studies was considered to be due to an exceptionally low control value. Micronuclei were not induced in mouse spermatocytes. In a study in which mice and rats were exposed by inhalation to reagent-grade trichloroethylene (purity, > 99%), micronuclei were induced in the bone-marrow cells of rats but not of mice; neither micronuclei, chromosomal aberrations nor sister chromatid exchange were induced in the peripheral lymphocytes of rats or the splenocytes of mice (Kligerman *et al.*, 1994). In a single study, pure trichloroethylene did not induce dominant lethal mutation in mice. Trichloroethylene increased the frequency of S-phase in mouse hepatocytes *in vivo* but did not produce enzyme-altered foci in rat liver.

(c) Genetic effects of trichloroethylene metabolites

The genetic toxicology of dichloroacetic and trichloroacetic acids is reviewed in the relevant monographs in this volume. Dichloroacetyl chloride, a presumed metabolite of trichloroethylene, did not induce prophage in *E. coli*, but was weakly mutagenic in *S. typhimurium* TA100 in the absence of metabolic activation in one study.

The minor urinary metabolite, *N*-acetyl-*S*-(1,2-dichlorovinyl)-L-cysteine, was mutagenic in *S. typhimurium* TA2638 in the presence of kidney cytosol, which allows deacetylation to the corresponding cysteine conjugate (Vamvakas *et al.*, 1987). The presumed intermediate metabolite, *S*-(1,2-dichlorovinyl)-L-cysteine, was mutagenic to *S. typhimurium* TA100 and TA2638 in the presence and absence of metabolic activation (Green & Odum, 1985; Dekant *et al.*, 1986c). *S*-(1,2-Dichlorovinyl)glutathione, the precursor of the cysteine conjugate, was also mutagenic to *S. typhimurium* TA2638 in the presence of rat kidney microsomes, which allow degradation to the cysteine conjugate (Vamvakas *et al.*, 1988). Both the cysteine and the glutathione conjugate induced a low rate of unscheduled DNA synthesis in a cultured pig kidney cell line (Vamvakas *et al.*, 1989). In the same cell line, *S*-(1,2-dichlorovinyl)-L-cysteine induced

DNA double-strand breaks and expression of the proto-oncogenes c-*fos* and c-*myc* (Vamvakas *et al.*, 1992; Vamvakas & Köster, 1993; Vamvakas *et al.*, 1993). DNA single-strand breaks were observed in mouse kidney and double-strand breaks in rat kidney after intraperitoneal injection of *S*-(1,2-dichlorovinyl)-L-cysteine (Jaffe *et al.*, 1985; McLaren *et al.*, 1994).

^{35}S-(1,2-Dichlorovinyl)-L-cysteine metabolites bound to isolated DNA *in vitro* (Bhattacharya & Schulze, 1972).

(d) Mutations in proto-oncogenes in tumours from trichloroethylene-treated animals

A group of 110 male B6C3F1 mice, eight weeks of age, were given trichloroethylene in corn oil orally by gavage at a dose of 1700 mg/kg bw per day on five days per week for up to 76 weeks. There were two concurrent control groups, each consisting of 50 male mice: one was untreated and the other received corn oil at a dose of 10 ml/kg bw. Ten control mice in each group were killed at 76 weeks, and the remainder were killed at 96, 103 and 134 weeks [numbers not stated]. At death, liver tumours 0.5 cm in diameter were taken for histological examination and for analysis of oncogenes. At the time of the terminal kill, there were 24 untreated controls, 32 vehicle controls and 75 animals treated with trichloroethylene. The numbers of hepatocellular adenomas per mouse in animals in these three groups were 0.9 ± 0.06 (8%), 0.13 ± 0.06 (13%) and 1.27 ± 0.14 (67%); the corresponding numbers of hepatocellular carcinomas were 0.09 ± 0.06 (8%), 0.12 ± 0.06 (12%) and 0.57 ± 0.10 (39%), respectively. The authors noted numerous foci of cellular alteration (presumed preneoplastic lesions) in the livers of treated mice but only rare foci in the livers of controls. No neoplasms related to treatment were found at other sites. The frequency of mutations in codon 61 of H-*ras* was not significantly different in 76 hepatocellular tumours from trichloroethylene-treated mice and in those from the 74 combined historical and concurrent controls (51% *versus* 69%). The spectra of these mutations, however, showed a significant decrease in AAA and an increase in CTA in the tumours from treated mice in comparison with those from controls. Other H-*ras* and K-*ras* mutations each contributed 4% to the total in the treated mice, whereas their frequency appeared to be very low in the concurrent controls and none were seen in the historical controls. The authors interpreted these findings as suggesting that exposure to trichloroethylene provides the environment for a selective growth advantage for spontaneous CTA mutations in codon 61 of H-*ras* (Anna *et al.*, 1994).

5. Summary and Evaluation

5.1 Exposure data

Trichloroethylene, a chlorinated solvent, has been produced commercially since the 1920s in many countries by chlorination of ethylene or acetylene. Its use in vapour degreasing began in the 1920s. In the 1930s, it was introduced for use in dry cleaning, but it has had limited use in that way since the 1950s. Currently, 80–90% of trichloroethylene worldwide is used for degreasing metals. Use for all applications in western Europe, Japan and the United States in 1990 was about 225 thousand tonnes.

Trichloroethylene has been detected in air, water, soil, food and animal tissues. The most heavily exposed people are those working in the degreasing of metals, who are exposed by inhalation.

5.2 Human carcinogenicity data

Three cohort studies were considered to be particularly relevant for the evaluation of trichloroethylene. Two of these studies, conducted in Sweden and Finland, involved people who had been monitored for exposure to trichloroethylene by measurement of trichloroacetic acid in urine. The levels in samples from most of the people in the two cohorts indicated relatively low levels of exposure. The third study, from the United States, covered workers exposed to trichloroethylene during maintenance of military aircraft and missiles, some of whom were also exposed to other solvents.

A fourth cohort study included all workers in an aircraft manufacturing company in the United States. This study was considered less relevant, as only one-third of the jobs in the plant entailed exposure to trichloroethylene and the exposures of the workers could not be classified.

In none of the available cohort studies was it possible to control for potential confounding factors, such as those associated with social class with regard to cervical cancer and smoking in respect of urinary bladder cancer.

Case–control studies have been conducted to investigate a number of cancer sites, including a multisite study from Montréal, Canada, in which other cancer cases were used as controls. Most of these studies do not provide risk estimates for exposure to trichloroethylene separately but only for groups of chemicals.

The results of the three most informative cohort studies consistently indicate an excess relative risk for cancer of the liver and biliary tract, with a total of 23 observed cases, whereas 12.87 were expected. The risk for these cancers was not elevated in the fourth, less informative cohort study. Results for liver cancer were given separately in the study from Finland and for the maintenance workers in the study in the United States. A total of seven cases were observed, whereas 4.00 were expected. Three case–control studies of primary liver cancer indicated elevated relative risks for people exposed to solvents, but only a few of the subjects in each study reported exposure to trichloroethylene.

With regard to non-Hodgkin's lymphoma, the results of the three most informative cohort studies were consistent; the data indicated a modest excess relative risk, with 27 cases observed and 18.9 expected. The risk for non-Hodgkin's lymphoma was not increased in the fourth, less informative study. In a case–control study covering all malignant lymphomas, an elevated odds ratio for exposure to trichloroethylene was indicated on the basis of seven exposed cases. The risk for non-Hodgkin's lymphoma was not increased among people assumed to have been exposed to trichloroethylene in the study in Montréal.

A twofold risk for cervical cancer was observed in two cohort studies.

The occurrence of cancer of the kidney was not elevated in the cohort studies; however, a study of German workers exposed to trichloroethylene revealed five cases of renal cancer whereas no case was found in an unexposed comparison group. The study may, however, have been initiated after the observation of a cluster. A case–control study and the multisite cancer study, both from Montréal, Canada, provided discordant results with regard to cancer of the kidney.

The incidence of urinary bladder cancer was not increased in the two cohort studies from Sweden and Finland, whereas slightly increased numbers of deaths were seen in the two United States cohorts. The incidence of urinary bladder cancer was not increased in people assumed to be exposed to trichloroethylene in the Montréal study.

Data on cancer incidence or mortality have been reported from five areas in which groundwater was contaminated with trichloroethylene. A weak association between contamination and the incidence of leukaemia was indicated in two of these studies, from Massachusetts and New Jersey, United States. The cohort studies of trichloroethylene-exposed workers did not indicate an association with the occurrence of leukaemia. Two studies, from Finland and New Jersey, suggested a marginal increase in the occurrence of non-Hodgkin's lymphoma in areas with contaminated groundwater.

Overall, the most important observations are the elevated risk for cancer of the liver and biliary tract and the modestly elevated risk for non-Hodgkin's lymphoma in all three of the most informative cohort studies. Two of these studies reported data for primary liver cancer separately. Finally, the suggested marginally increased risk for non-Hodgkin's lymphoma in areas with trichloroethylene-contaminated groundwater is noted.

5.3 Animal carcinogenicity data

Trichloroethylene, with and without stabilizers, was tested for carcinogenicity by oral administration in two adequate experiments in mice. The studies showed significant increases in the incidences of benign and malignant liver tumours. Of seven studies in which trichloroethylene was given orally to rats, most were inconclusive because of reduced survival or a too short treatment. In two of the studies, the incidence of uncommonly occurring renal-cell tumours was significantly increased in male rats, and in one study an increased incidence of interstitial-cell testicular tumours was seen.

Trichloroethylene was tested for carcinogenicity by inhalation in four experiments in mice. One study showed an increased incidence of lymphomas, one study showed increased incidences of liver tumours, and three studies showed increased incidences of lung tumours. One of three experiments in which rats were exposed by inhalation showed an increased incidence of interstitial testicular tumours and a marginal increase in that of renal-cell tumours in males. No increase in tumour incidence was observed in one study in hamsters exposed by inhalation.

In limited studies, trichloroethylene and its proposed metabolite trichloroethylene oxide did not increase the incidence of skin tumours or local sarcomas in mice when administered by topical application or subcutaneous injection.

5.4 Other relevant data

In rodents, trichloroethylene is rapidly absorbed from the gastrointestinal tract and through the lungs, whereas absorption of the vapour through the skin is negligible. The major pathway is oxidative metabolism leading to the formation of chloroacetic acids. Mice showed consistently higher rates of oxidative biotransformation than rats. A minor pathway in rodents and humans involves the formation of mercapturic acids.

The acute toxicity of trichloroethylene in rodents and humans is low. After high doses of trichloroethylene are administered repeatedly to rodents, damage is seen in liver and kidney (in mice and rats) and in lung (in mice only). Repeated exposure of humans in the workplace appears to have no marked toxic effects on the kidney or liver. Trichloroethylene is a more potent peroxisome proliferator in the livers of mice than of rats.

The available studies show no consistent effect of trichloroethylene on the human reproductive system. Trichloroethylene is metabolized to trichloroacetic acid in the placenta or fetus of many species. There is little evidence of toxic effects in developing rats or mice.

Studies of structural chromosomal aberrations, aneuploidy and sister chromatid exchange in peripheral lymphocytes of workers exposed to trichloroethylene were inconclusive.

Pure trichloroethylene did not induce chromosomal aberrations, dominant lethal mutations, sister chromatid exchange or unscheduled DNA synthesis in rodents, whereas an increased induction of micronuclei and DNA single-strand breaks/alkaline labile sites was observed.

In single studies with human cells *in vitro*, trichloroethylene of low purity slightly increased the frequencies of sister chromatid exchange and unscheduled DNA synthesis. Pure trichloroethylene did not induce gene mutation in human cells. In mammalian cells *in vitro*, pure trichloroethylene induced cell transformation, sister chromatid exchange and gene mutation, but not chromosomal aberrations. In fungi, trichloroethylene (pure or of unspecified purity) induced aneuploidy, gene mutation and mitotic recombination and induced gene conversion in the presence of metabolic activation.

Gene mutation or DNA damage was usually not induced in prokaryotes by pure trichloroethylene, while preparations containing epoxide stabilizers were mutagenic. Sulfur-containing metabolites formed by a minor trichloroethylene biotransformation pathway were genotoxic in bacteria and cultured renal cells.

5.5 Evaluation[1]

There is *limited evidence* in humans for the carcinogenicity of trichloroethylene.

There is *sufficient evidence* in experimental animals for the carcinogenicity of trichloroethylene.

Overall evaluation[2]

Trichloroethylene *is probably carcinogenic to humans (Group 2A)*.

In making the overall evaluation, the Working Group considered the following evidence:

(i) Although the hypothesis linking the formation of mouse liver tumours with peroxisome proliferation is plausible, trichloroethylene also induced tumours at other sites in mice and rats.

---

[1] For definition of the italicized terms, see Preamble, pp. 22–26.
[2] Dr N.H. Stacey disassociated himself from the overall evaluation.

(ii) Several epidemiological studies showed elevated risks for cancer of the liver and biliary tract and for non-Hodgkin's lymphoma.

6. References

Abrahamsson, K., Ekdahl, A., Collén, J., Fahlström, E., Sporrong, N. & Pedersén, M. (1995) Marine algae— a source of trichloroethylene and perchloroethylene. *Limnol. Oceanogr.* (in press)

Ahlmark, A., Gerhardsson, G. & Holm, A. (1963) Trichloroethylene exposure in Swedish engineering workshops. In: *Proceedings of the 14th International Congress on Occupational Health*, London, Permanent Commission and International Association on Occupational Health, pp. 448–450

Aitio, A., Luotamo, M. & Kiilunen, M., eds (1995) *Kemikaalialtistumisen Biomonitorointi* [Biological Monitoring of Exposure to Chemicals], Helsinki, Finnish Institute of Occupational Health, pp. 292–294 (in Finnish)

Allen, B.C. & Fisher, J.W. (1993) Pharmacokinetic modeling of trichloroethylene and trichloroacetic acid in humans. *Risk Anal.*, 13, 71–86

Allen, J.W., Collins, B.W. & Evansky, P.A. (1994) Spermatid micronucleus analyses of trichloroethylene and chloral hydrate effects in mice. *Mutat. Res.*, 323, 81–88

Almaguer, D., Kramkowski, R.S. & Orris, P. (1984) *Johnson Controls, Inc., Watertown, WI* (Health Hazard Evaluation Report No. 83-296-1491), Cincinnati, OH, United States National Institute for Occupational Safety and Health

Amacher, D.E. & Zelljadt, I. (1983) The morphological transformation of Syrian hamster embryo cells by chemicals reportedly nonmutagenic to *Salmonella typhimurium*. *Carcinogenesis*, 4, 291–295

American Conference of Governmental Industrial Hygienists (1994) *1994–1995 Threshold Limit Values for Chemical Substances and Physical Agents and Biological Exposure Indices*, Cincinnati, OH, pp. 34, 62

Andelman, J.B. (1985) Inhalation exposure in the home to volatile organic contaminants of drinking water. *Sci. total Environ.*, 47, 443–460

Andersen, M.E., Gargas, M.L., Clewell, H.J., III & Severyn, K.M. (1987) Quantitative evaluation of the metabolic interactions between trichloroethylene and 1,1-dichloroethylene *in vivo* using gas uptake methods. *Toxicol. appl. Pharmacol.*, 89, 149–157

Anna, C.H., Maronpot, R.R., Pereira, M.A., Foley, J.F., Malarkey, D.E. & Anderson, M.W. (1994) *ras* Proto-oncogene activation in dichloroacetic-, trichloroethylene- and tetrachloroethylene-induced liver tumors in B6C3F1 mice. *Carcinogenesis*, 15, 2255–2261

Anttila, A., Pukkala, E., Sallmén, M., Hernberg, S. & Hemminki, K. (1995) Cancer incidence among Finnish workers exposed to halogenated hydrocarbons. *J. occup. Med.* (in press)

Arbeidsinspectie (1994) *De Nationale MAC-lijst 1994* [The national MAC list 1994], The Hague, p. 40

Arizona Department of Health Services (1990) *The Incidence of Childhood Leukemia and Testicular Cancer in Pima County: 1970–1986*, Tucson

Aviado, D.M., Zakhari, S., Simaan, J. & Ulsamer, A.G. (1976) *Methyl Chloroform and Trichloroethylene in the Environment*, Cleveland, OH, CRC Press

Axelson, O., Andersson, K., Hogstedt, C., Holmberg, B., Molina, G. & de Verdier, A. (1978) A cohort study on trichloroethylene exposure and cancer mortality. *J. occup. Med.*, 20, 194–196

Axelson, O., Andersson, K., Seldén, A. & Hogstedt, C. (1984) Cancer morbidity and exposure to trichloroethylene (Abstract). *Arb. Hälsa*, **29**, 126

Axelson, O., Seldén, A., Andersson, K. & Hogstedt, C. (1994) Updated and expanded Swedish cohort study on trichloroethylene and cancer risk. *J. occup. Med.*, **36**, 556–562

Baden, J.M. & Simmon, V.F. (1980) Mutagenic effects of inhalational anesthetics. *Mutat. Res.*, **75**, 169–189

Baden, J.M., Kelley, M., Mazze, R.I. & Simmon, V.F. (1979) Mutagenicity of inhalation anaesthetics: trichloroethylene, divinyl ether, nitrous oxide and cyclopropane. *Br. J. Anaesthesiol.*, **51**, 417–421

Bai, C.-L. & Stacey, N.H. (1993) Mechanism of trichloroethylene-induced elevation of individual serum bile acids. II. In vitro and in vivo interference by trichloroethylene with bile acid transport in isolated rat hepatocytes. *Toxicol. appl. Pharmacol.*, **121**, 296–302

Banerjee, S. & Van Duuren, B.L. (1978) Covalent binding of the carcinogen trichloroethylene to hepatic microsomal proteins and to exogenous DNA *in vitro*. *Cancer Res.*, **38**, 776–780

Barret, L., Faure, J. & Danel, V. (1984) Epidemiological study of cancer in a community of workers occupationally exposed to trichloroethylene and cutting oils (Abstract no. 34). Stockholm, Association Européenne des Centres Anti-poison

Bartsch, H., Malaveille, C., Barbin, A. & Planche, G. (1979) Mutagenic and alkylating metabolites of halo-ethylenes, chlorobutadienes and dichlorobutenes produced by rodent or human liver tissues. *Arch. Toxicol.*, **41**, 249–277

Bauer, U. (1981a) Human exposure to environmental chemicals. Investigations on volatile organic halogenated compounds in water, air, food, and human tissues. I. Properties, distribution and effects of volatile organic halogenated compounds. Analytical methods. *Zbl. Bakt. Hyg. I. Abt. B*, **174**, 15–56 (in German)

Bauer, U. (1981b) Human exposure to environmental chemicals. Investigations on volatile organic halogenated compounds in water, air, food, and human tissues. IV. Calculation of human exposure to organic halogenated compounds from the environment. *Zbl. Bakt. Hyg. I. Abt. B*, **174**, 556–583 (in German)

Belanger, P.L. & Coye, M.J. (1984) *Basic Tool and Supply, Oakland, CA* (Health Hazard Evaluation Report No. 84-196-1527), Cincinnati, OH, United States National Institute for Occupational Safety and Health

Bergman, K. (1983a) Application and results of whole-body autoradiography in distribution studies of organic solvents. *Arch. Toxicol.*, **12**, 59–118

Bergman, K. (1983b) Interactions of trichloroethylene with DNA *in vitro* and with RNA and DNA of various mouse tissues *in vivo*. *Arch. Toxicol.*, **54**, 181–193

Besemer, A.C., Eggels, P.G., van Esch, G.J., Hollander, J.C.T., Huldy, H.J. & Maas, R.J.M. (1984) *Criteria Document over Trichlooetheen* [Criteria for Trichloroethylene] (Publication No. 33), The Hague, Ministerie van Volkshuisvesting, Ruimtelijke Ordening en Milieubeheer

Bhakuni, V. & Roy, R. (1994) Interaction of trichloroethylene with phosphatidylcholine and its localization in the phosphatidylcholine vesicles: a ^{1}H-NMR study. *Biochem. Pharmacol.*, **47**, 1461–1464

Bhattacharya, R.K. & Schulze, M.O. (1972) Properties of DNA treated with S-(1,2-dichlorovinyl)-L-cysteine and a lyase. *Arch. Biochem. Biophys.*, **153**, 105–115

Birner, G., Vamvakas, S., Dekant, W. & Henschler, D. (1993) Nephrotoxic and genotoxic *N*-acetyl-S-dichlorovinyl-L-cysteine is a urinary metabolite after occupational 1,1,2-trichloroethene exposure in humans: implications for the risk of trichloroethene exposure. *Environ. Health Perspectives*, **99**, 281–284

Blair, A. (1980) Mortality among workers in the metal polishing and plating industry, 1951–1969. *J. occup. Med.*, **22**, 158–162

Blair, A. & Mason, T.J. (1980) Cancer mortality in United States counties with metal electroplating industries. *Arch. environ. Health*, **35**, 92–94

Bloom, T.F., Kramkowski, R.S. & Cromer, J.W. (1974) *Essex Wire Corp., Kenton, OH* (Health Hazard Evaluation Determination Report No. 73-151-141), Cincinnati, OH, United States National Institute for Occupational Safety and Health

Bogen, K.T., Hall, L.C., Perry, L., Fish, R., McKone, T.E., Dowd, P., Patton, S.E. & Mallon, B. (1988) *Health Risk Assessment of Trichloroethylene (TCE) in California Drinking Water* (DE88-005364), Livermore, CA, Lawrence Livermore National Laboratory, Environmental Sciences Division

Bonse, G. & Henschler, D. (1976) Chemical reactivity, biotransformation, and toxicity of polychlorinated aliphatic compounds. *Crit. Rev. Toxicol.*, **4**, 395–409

Bonse, G., Urban, T., Reichert, D. & Henschler, D. (1975) Chemical reactivity, metabolic oxirane formation and biological reactivity of chlorinated ethylenes in the isolated perfused rat liver preparation. *Biochem. Pharmacol.*, **24**, 1829–1834

Brandt-Rauf, P.W. & Hathaway, J.A. (1986) Biliary tract cancer in the chemical industry: a proportional mortality study. *Br. J. ind. Med.*, **43**, 716–717

Brandt-Rauf, P.W., Pincus, M.R. & Adelson, S. (1982) Cancer of the gallbladder: a review of forty-three cases. *Hum. Pathol.*, **13**, 48–53

Brandt-Rauf, P.W., Pincus, M.R. & Adelson, S. (1986) Carcinoma of the ampulla of Vater. *Dig. Dis.*, **4**, 43–48

Brodzinsky, R. & Singh, H.B. (1983) *Volatile Organic Chemicals in the Atmosphere: An Assessment of Available Data (Final Report)*, Menlo Park, CA, SRI International, pp. 107–108

Bronzetti, G., Zeiger, E. & Frezza, D. (1978) Genetic activity of trichloroethylene in yeast. *J. environ. Pathol. Toxicol.*, **1**, 411–418

Bruckner, J.V., Davis, B.D. & Blancato, J.N. (1989) Metabolism, toxicity, and carcinogenicity of trichloroethylene. *Crit. Rev. Toxicol.*, **20**, 31–50

Buben, J.A. & O'Flaherty, E.J. (1985) Delineation of the role of metabolism in the hepatotoxicity of trichloroethylene and perchloroethylene: a dose–effect study. *Toxicol. appl. Pharmacol.*, **78**, 105–122

Budavari, S., ed. (1989) *The Merck Index*, 11th Ed., Rahway, NJ, Merck & Co., pp. 1516–1517

Burgess, W.A. (1981) *Recognition of Health Hazards in Industry. A Review of Materials and Processes*, New York, John Wiley & Sons, pp. 20–32

Burmaster, D.E. (1982) The new pollution—groundwater contamination. *Environment*, **24**, 6–36

Burroughs, G.E. (1980) *Protective Coatings Corporation, Fort Wayne, IN* (Health Hazard Evaluation Report No. 79-96-729), Cincinnati, OH, United States National Institute for Occupational Safety and Health

Burroughs, G.E. & Moody, P.L. (1982) *Industrial Plastics, Valley City, OH* (Health Hazard Evaluation Report No. 81-029-1088), Cincinnati, OH, United States National Institute for Occupational Safety and Health

Butler, T.C. (1949) Metabolic transformations of trichloroethylene. *J. Pharmacol. exp. Ther.*, **97**, 84–92

Byington, K.H. & Leibman, K.C. (1965) Metabolism of trichloroethylene in liver microsomes. II. Identification of the reaction product as chloral hydrate. *Mol. Pharmacol.*, **1**, 247–254

Callen, D.F., Wolf, C.R. & Philpot, R.M. (1980) Cytochrome P-450 mediated genetic activity and cytotoxicity of seven halogenated aliphatic hydrocarbons in *Saccharomyces cerevisiae*. *Mutat. Res.*, **77**, 55–63

Candura, S.M. & Faustman, E.M. (1991) Trichloroethylene: toxicology and health hazards. *Med. Lav.*, **13**, 17–25

Caspary, W.J., Langenbach, R., Penman, B.W., Crespi, C., Myhr, B.C. & Mitchell, A.D.(1988) The mutagenic activity of selected compounds at the TK locus: rodent vs. human cells. *Mutat. Res.*, **196**, 61–81

Chakrabarti, S.K. & Tuchweber, B. (1988) Studies of acute nephrotoxic potential of trichloroethylene in Fischer 344 rats. *J. Toxicol. environ. Health*, **23**, 147–158

Chemical Information Services, Ltd (1994) *Directory of World Chemical Producers 1995/96 Standard Edition*, Dallas, TX, p. 678

Christensen, J.M. & Rasmussen, K. (1990) Exposure of Danish workers to trichloroethylene 1947–1987. *Ugeskr. Laeg.*, **152**, 464–466 (in Danish)

Cohn, P., Klotz, J., Bove, F., Berkowitz, M. & Fagliano, J. (1994) Drinking water contamination and the incidence of leukemia and non-Hodgkin's lymphoma. *Environ. Health Perspectives*, **102**, 556–561

Cole, W.J., Mitchell, R.G. & Salamonsen, R.F. (1975) Isolation, characterization and quantitation of chloral hydrate as a transient metabolite of trichloroethylene in man using electron capture gas chromatography and mass fragmentography. *J. Pharm. Pharmacol.*, **27**, 167–171

Comba, M.E., Palabrica, V.S. & Kaiser, K.L.E. (1994) Volatile halocarbons as tracers of pulp mill effluent plumes. *Environ. Toxicol. Chem.*, **13**, 1065–1074

Commandeur, J.N.M. & Vermeulen, N.P.E. (1990) Identification of *N*-acetyl(2,2-dichlorovinyl)- and *N*-acetyl(1,2-dichlorovinyl)-L-cysteine as two regioisomeric mercapturic acids of trichloroethylene in the rat. *Chem. Res. Toxicol.*, **3**, 212–218

Cook, W.A. (1987) *Occupational Exposure Limits—Worldwide*, Akron, OH, American Industrial Hygiene Association, pp. 89, 111, 126, 155, 220

Cosby, N.C. & Dukelow, W.R. (1992) Toxicology of maternally ingested trichloroethylene (TCE) on embryonal and fetal development in mice and of TCE metabolites on in vitro fertilization. *Fundam. appl. Toxicol.*, **19**, 268–274

Costa, A.K., Katz, I.D. & Ivanetich, K.M. (1980) Trichloroethylene: its interaction with hepatic microsomal cytochrome P-450 *in vitro*. *Biochem. Pharmacol.*, **29**, 433–439

Crebelli, R. & Carere, A. (1989) Genetic toxicology of 1,1,2-trichloroethylene. *Mutat. Res.*, **221**, 11–37

Crebelli, R., Bignami, M., Conti, L. & Carere, A. (1982) Mutagenicity of trichloroethylene in *Salmonella typhimurium* TA100. *Ann. Ist. super. Sanità*, **18**, 117–122

Crebelli, R., Conti, G., Conti, L. & Carere, A. (1985) Mutagenicity of trichloroethylene, trichloroethanol and chloral hydrate in *Aspergillus nidulans*. *Mutat. Res.*, **155**, 105–111

Daft, J.L. (1989) Determination of fumigants and related chemicals in fatty and non-fatty foods. *J. agric. Food Chem.*, **37**, 560–564

Dallas, C.E., Gallo, J.M., Ramanathan, R., Muralidhara, S. & Bruckner, J.V. (1991) Physiological pharmacokinetic modeling of inhaled trichloroethylene in rats. *Toxicol. appl. Pharmacol.*, **110**, 303–314

Daniel, J.W. (1963) The metabolism of ^{36}Cl-labelled trichloroethylene and tetrachloroethylene in the rat. *Biochem. Pharmacol.*, **12**, 795–802

Davidson, I.W.F. & Beliles, R.P. (1991) Consideration of the target organ toxicity of trichloroethylene in terms of metabolite toxicity and pharmacokinetics. *Drug Metab. Rev.*, **23**, 493–599

Dawson, B.V., Johnson, P.D., Goldberg, S.J. & Ulreich, J.B. (1993) Cardiac teratogenesis of halogenated hydrocarbon-contaminated drinking water. *J. Am. Coll. Cardiol.*, **21**, 1466–1472

DeAngelo, A.B., Daniel, F.B., McMillan, L., Wernsing, P. & Savage, R.E., Jr (1989) Species and strain sensitivity to the induction of peroxisome proliferation by chloroacetic acids. *Toxicol. appl. Pharmacol.*, **101**, 285–298

Dees, C. & Travis, C. (1993) The mitogenic potential of trichloroethylene in B6C3F1 mice. *Toxicol. Lett.*, **69**, 129–137

Dekant, W. (1986) Metabolic conversions of tri- and tetrachloroethylene: formation and deactivation of genotoxic intermediates. In: Chambers, P.L., Gehring, P. & Sakai, F., eds, *New Concepts and Developments in Toxicology*, Amsterdam, Elsevier Science Publishers, pp. 211–221

Dekant, W., Metzler, M. & Henschler, D. (1984) Novel metabolites of trichloroethylene through dechlorination reactions in rats, mice and humans. *Biochem. Pharmacol.*, **33**, 2021–2027

Dekant, W., Metzler, M. & Henschler, D. (1986a) Identification of S-1,2-dichlorovinyl-N-acetylcysteine as a urinary metabolite of trichloroethylene: a possible explanation for its nephrocarcinogenicity in male rats. *Biochem. Pharmacol.*, **35**, 2455–2458

Dekant, W., Schulz, A., Metzler, M. & Henschler, D. (1986b) Absorption, elimination and metabolism of trichloroethylene: a quantitative comparison between rats and mice. *Xenobiotica*, **16**, 143–152

Dekant, W., Vamvakas, S., Berthold, K., Schmidt, S., Wild, D. & Henschler, D. (1986c) Bacterial β-lyase mediated cleavage and mutagenicity of cysteine conjugates derived from the nephrocarcinogenic alkenes trichloroethylene, tetrachloroethylene and hexachlorobutadiene. *Chem.-biol. Interactions*, **60**, 31–45

Dekant, W., Berthold, K., Vamvakas, S., Henschler, D. & Anders, M.W. (1988) Thioacylating intermediates as metabolites of S-(1,2-dichlorovinyl)-L-cysteine and S-(1,2,2-trichlorovinyl)-L-cysteine formed by cysteine conjugate β-lyase. *Chem. Res. Toxicol.*, **1**, 175–178

Dekant, W., Koob, M. & Henschler, D. (1990) Metabolism of trichloroethene—in vivo and in vitro evidence for activation by glutathione conjugation. *Chem.-Biol. Interactions*, **73**, 89–101

DeMarini, D.M., Perry, E. & Shelton, M.L. (1994) Dichloroacetic acid and related compounds: induction of prophage in *E. coli* and mutagenicity and mutation spectra in *Salmonella* TA100. *Mutagenesis*, **9**, 429–437

Deutsche Forschungsgemeinschaft (1993) *MAK- und BAT Werte Liste 1993* [MAK and BAT value list 1993] (Report No. 29), Weinheim, VCH Verlagsgesellschaft, pp. 75, 129

Dickson, A.G. & Riley, J.P. (1976) The distribution of short-chain halogenated aliphatic hydrocarbons in some marine organisms. *Mar. Pollut. Bull.*, **7**, 167–169

DiRenzo, A.B., Gandolfi, A.J. & Sipes, I.G. (1982) Microsomal bioactivation and covalent binding of aliphatic halides to DNA. *Toxicol. Lett.*, **11**, 243–252

Doolittle, D.J., Muller, G. & Scribner, H.E. (1987) The *in vivo–in vitro* hepatocyte assay for assessing DNA repair and DNA replication: studies in the CD-1 mouse. *Food chem. Toxicol.*, **25**, 399–405

Dorfmueller, M.A., Henne, S.P., York, R.G., Bornschein, R.L. & Manson, J.M. (1979) Evaluation of teratogenicity and behavioral toxicity with inhalation exposure of maternal rats to trichloroethylene. *Toxicology*, **14**, 153–166

Droz, P.O. & Fernández, J.G. (1978) Trichloroethylene exposure. Biological monitoring by breath and urine analysis. *Br. J. ind. Med*, **35**, 35–42

Dubrow, R. & Gute, D.M. (1987) Cause-specific mortality among Rhode Island jewelry workers. *Am. J. ind. Med.*, **12**, 579–593

Duprat, P. & Gradiski, D. (1980) Cytogenetic effect of trichloroethylene in the mouse as evaluated by the micronucleus test. *IRCS med. Sci.*, **8**, 182

von Düszeln, J., Lahl, U., Bätjer, K., Cetinkaya, M., Stachel, B. & Thiemann, W. (1982) Occurrence of volatile halogenated hydrocarbons in air, water and food in East Germany. *Dtsch. Lebensm. Rundsch.*, **78**, 253–356 (in German)

Elcombe, C.R. (1985) Species differences in carcinogenicity and peroxisome proliferation due to trichloroethylene: a biochemical human hazard assessment. *Arch. Toxicol.*, **Suppl. 8**, 6–17

Elcombe, C.R., Rose, M.S. & Pratt, I.S. (1985) Biochemical, histological, and ultrastructural changes in rat and mouse liver following the administration of trichloroethylene: possible relevance to species differences in hepatocarcinogenicity. *Toxicol. appl. Pharmacol.*, **79**, 365–376

Eller, P.M., ed. (1994) *NIOSH Manual of Analytical Methods*, 4th Ed., Vol. 3 (DHHS (NIOSH) Publ. No. 94-113), Washington DC, United States Government Printing Office, Methods 1022 and 3701

Elmore, E. & Fitzgerald, M.P. (1990) Evaluation of the bioluminescence assays as screens for genotoxic chemicals. *Progr. clin. biol. Res.*, **340D**, 379–387

Entz, R.C. & Diachenko, G.W. (1988) Residues of volatile halocarbons in margarines. *Food Addit. Contam.*, **5**, 267–276

Ertle, T., Henschler, D., Müller, G. & Spassowski, M. (1972) Metabolism of trichloroethylene in man. I. The significance of trichloroethanol in long-term exposure conditions. *Arch. Toxicol.*, **29**, 171–188

European Centre for Ecotoxicology and Toxicology of Chemicals (1994) *Trichloroethylene: Assessment of Human Carcinogenic Hazard* (ECETOC Technical Report No. 60), Brussels

Fabricant, J.D. & Chalmers, J.H., Jr (1980) Evidence of the mutagenicity of ethylene dichloride and structurally related compounds. In: Ames, B., Infante, P. & Reitz, R., eds, *Ethylene Dichloride: A Potential Health Risk* (Banbury Report 5), Cold Spring Harbor, NY, CSH Press, pp. 309–329

Fagliano, J., Berry, M., Bove, F. & Burke, T. (1990) Drinking water contamination and the incidence of leukemia: an ecologic study. *Am. J. public Health*, **80**, 1209–1212

Fahrig, R. (1977) The mammalian spot test (Fallfleckentest) with mice. *Arch. Toxicol.*, **38**, 87–98

Fazio, T. (1990) Food additives: direct. In: Helrich, K., ed., *Official Methods of Analysis of the Association of Official Analytical Chemists*, 15th Ed., Vol. 2, Arlington, VA, Association of Official Analytical Chemists, p. 1175

Fielding, M., Gibson, T.M. & James, H.A. (1981) Levels of trichloroethylene, tetrachloroethylene and p-dichlorobenzene in groundwaters. *Environ. Technol. Lett.*, **2**, 545–550

Filser, J.G. & Bolt, H.M. (1979) Pharmacokinetics of halogenated ethylenes in rats. *Arch. Toxicol.*, **42**, 123–136

Fishbein, L. (1976) Industrial mutagens and potential mutagens. I. Halogenated aliphatic derivatives. *Mutat. Res.*, **32**, 267–308

Fisher, J.W., Whittaker, T.A., Taylor, D.H., Clewell, H.J., III & Andersen, M.E. (1989) Physiologically based pharmacokinetic modeling of the pregnant rat: a multiroute exposure model for trichloroethylene and its metabolite, trichloroacetic acid. *Toxicol. appl. Pharmacol.*, **99**, 395–414

Fisher, J.W., Gargas, M.L., Allen, B.C. & Andersen, M.E. (1991) Physiologically based pharmacokinetic modeling with trichloroethylene and its metabolite, trichloroacetic acid, in the rat and mouse. *Toxicol. appl. Pharmacol.*, **109**, 183–195

Flood, T.J., Aickin, M., Lucier, J.L. & Petersen, N.J. (1990) *Incidence Study of Childhood Cancer in Maricopa County: 1965–1986* (Final Report), Phoenix, Arizona Department of Health Services

Forkert, P.G. & Birch, D.W. (1989) Pulmonary toxicity of trichloroethylene in mice. Covalent binding and morphological manifestations. *Drug Metab. Disposition*, **17**, 106–113

Foureman, P., Mason, J.M., Valencia, R. & Zimmering, S. (1994) Chemical mutagenesis testing in *Drosophila*: IX. Results of 50 coded compounds tested for the National Toxicology Program. *Environ. mol. Mutag.*, **23**, 51–63

Fredriksson, M., Bengtsson, N.-O., Hardell, L. & Axelson, O. (1989) Colon cancer, physical activity, and occupational exposures. A case–control study. *Cancer*, **63**, 1838–1842

Fukuda, K., Takemoto, K. & Tsuruta, H. (1983) Inhalation carcinogenicity of trichloroethylene in mice and rats. *Ind. Health*, **21**, 243–254

Fytianos, K., Vasilikiotis, G. & Weil, L. (1985) Identification and determination of some trace organic compounds in coastal seawater of northern Greece. *Bull. environ. Contam. Toxicol.*, **34**, 390–395

Galloway, S.M., Armstrong, M.J., Reuben, C., Colman, S., Brown, B., Cannon, C., Bloom, A.D., Nakamura, F., Ahmed, M., Duk, S., Rimpo, J., Margolin, B.H., Resnick, M.A., Anderson, B. & Zeiger, E. (1987) Chromosome aberrations and sister chromatid exchanges in Chinese hamster ovary cells: evaluations of 108 chemicals. *Environ. mol. Mutag.*, **10** (Suppl. 10), 1–175

Garabrant, D.H., Held, J., Langholz, B. & Bernstein, L. (1988) Mortality of aircraft manufacturing workers in southern California. *Am. J. ind. Med.*, **13**, 683–693

Ghantous, H., Danielsson, B.R.G., Dencker, L., Gorczak, J. & Vesterberg, O. (1986) Trichloroacetic acid accumulates in murine amniotic fluid after tri- and tetrachloroethylene inhalation. *Acta pharmacol. toxicol.*, **58**, 105–114

Gilles, D. & Philbin, E. (1976) *TRW Inc., Philadelphia, PA* (Health Hazard Evaluation Determination Report No. 76-61-337), Cincinnati, OH, United States National Institute for Occupational Safety and Health

Goldberg, S.J., Lebowitz, M.D., Graver, E.J. & Hicks, S. (1990) An association of human congenital cardiac malformations and drinking water contaminants. *J. Am. Coll. Cardiol.*, **16**, 155–164

Goldsworthy, T.L. & Popp, J.A. (1987) Chlorinated hydrocarbon-induced peroxisomal enzyme activity in relation to species and organ carcinogenicity. *Toxicol. appl. Pharmacol.*, **88**, 225–233

Gorman, R., Rinsky, R., Stein, G. & Anderson, K. (1984) *Pratt & Whitney Aircraft, West Palm Beach, FL* (HETA Report No. 82-075-1545), Cincinnati, OH, United States National Institute for Occupational Safety and Health

Gossett, R.W., Brown, D.A. & Young, D.R. (1983) Predicting the bioaccumulation of organic compounds in marine organisms using octanol/water partition coefficients. *Mar. Pollut. Bull.*, **14**, 387–392

Grandjean, E., Münchinger, R., Turrian, V., Haas, P.A., Knoepfel, H.-K. & Rosenmund, H. (1955) Investigations into the effects of exposure to trichloroethylene in mechanical engineering. *Br. J. ind. Med.*, **12**, 131–142

Green, T. & Odum, J. (1985) Structure/activity studies of the nephrotoxic and mutagenic action of cysteine conjugates of chloro- and fluoroalkenes. *Chem.-biol. Interactions*, **54**, 15–31

Green, T. & Prout, M.S. (1985) Species differences in response to trichloroethylene. II. Biotransformation in rats and mice. *Toxicol. appl. Pharmacol.*, **79**, 401–411

Greenberg, A.E., Clesceri, L.S. & Eaton, A.D., eds (1992) *Standard Methods for the Examination of Water and Wastewater*, 18th Ed., Washington DC, American Public Health Association/American Water Works Association/Water Environment Federation, pp. 6-17–6-36, 6-42–6-57

Greim, H., Bonse, G., Radwan, Z., Reichert, D. & Henschler, D. (1975) Mutagenicity *in vitro* and potential carcinogenicity of chlorinated ethylenes as a function of metabolic oxirane formation. *Biochem. Pharmacol.*, **24**, 2013–2017

Gu, Z.W., Sele, B., Chmara, D., Jalbert, P., Vincent, M., Vincent, F., Marka, C. & Faure, J. (1981) Effects of trichlorethylene and some of its metabolites on the rate of sister chromatid exchange. Study *in vivo* and *in vitro* on human lymphocytes. *Ann. Génét.*, **24**, 105–106 (in French)

Guengerich, F.R., Kim, D.-H. & Iwasaki, M. (1991) Role of human cytochrome P-450 II E1 in the oxidation of many low molecular weight cancer suspects. *Chem. Res. Toxicol.*, **4**, 168–179

Gunter, B.J. (1977) *FMC Corp., Bloomfield, CO* (Health Hazard Evaluation Determination Report No. 76-101-376), Cincinnati, OH, United States National Institute for Occupational Safety and Health

Hajimiragha, H., Ewers. U., Jansen-Rosseck, R. & Brockhaus, A. (1986) Human exposure to volatile halogenated hydrocarbons from the general environment. *Int. Arch. occup. environ. Health*, **58**, 141–150

Hamdan, H. & Stacey, N.H. (1993) Mechanism of trichloroethylene-induced elevation of individual serum bile acids. 1. Correlation of trichloroethylene concentrations to bile acids in rat serum. *Toxicol. appl. Pharmacol.*, **121**, 291–295

Hansch, C., Leo, A. & Hoekman, D.H. (1995) *Exploring QSAR*, Washington, DC, American Chemical Society

Hardell, L., Eriksson, M., Lenner, P. & Lundgren, E. (1981) Malignant lymphoma and exposure to chemicals, especially organic solvents, chlorophenols and phenoxy acids: a case–control study. *Br. J. Cancer*, **43**, 169–176

Hardell, L., Bengtsson, N.O., Jonsson, U., Eriksson, S. & Larsson, L.G. (1984) Aetiological aspects on primary liver cancer with special regard to alcohol, organic solvents and acute intermittent porphyria—an epidemiological investigation. *Br. J. Cancer*, **50**, 389–397

Hardie, D.W.F. (1964) Chlorocarbons and chlorohydrocarbons. Trichloroethylene. In: Kirk, R.E. & Othmer, D.F., eds, *Encyclopedia of Chemical Technology*, 2nd Ed., Vol. 5, New York, John Wiley & Sons, pp. 183–195

Hardin, B.D., Bond, G.P., Sikov, M.R., Andrew, F.D., Beliles, R.P. & Niemeier, R.W. (1981) Testing of selected workplace chemicals for teratogenic potential. *Scand. J. Work Environ. Health*, **7** (Suppl. 4), 66–75

Hargarten, J.J., Hetrick, G.H. & Fleming, A.J. (1961) Industrial safety experience with trichloroethylene: its use as a vapor degreasing solvent, 1948–1957. *Arch. environ. Health*, **3**, 461–467

Hathway, D.E. (1980) Consideration of the evidence for mechanisms of 1,1,2-trichloroethylene metabolism, including new identification of its dichloroacetic acid and trichloroacetic acid metabolites in mice. *Cancer Lett.*, **8**, 263–269

Healy, T.E.J. & Wilcox, A. (1978) Chronic exposures of rats to inhalational anaesthetic agents. *J. Physiol.*, **276**, 24P–25P

Heikes, D.L. (1987) Purge and trap method for determination of volatile halocarbons and carbon disulfide in table-ready foods. *J. Assoc. off. anal. Chem.*, **70**, 215–226

Heikes, D.L. & Hopper, M.L. (1986) Purge and trap method for determination of fumigants in whole grains, milled grain products, and intermediate grain-based foods. *J. Assoc. off. anal. Chem.*, **69**, 990–998

Heineman, E.F., Cocco, P., Gómez, M.R., Dosemeci, M., Stewart, P.A., Hayes, R.B., Hoar Zahm, S., Thomas, T.L. & Blair, A. (1994) Occupational exposure to chlorinated aliphatic hydrocarbons and risk of astrocytic brain cancer. *Am. J. ind. Med.*, **26**, 155–169

Helliwell, P.J. & Hutton, A.M. (1949) Analgesia in obstetrics. *Anaesthesia*, **4**, 18–21

Helliwell, P.J. & Hutton, A.M. (1950) Trichloroethylene anaesthesia. 1. Distribution in the foetal and maternal circulation of pregnant sheep and goats. *Anaesthesia*, **5**, 4–13

Hellmann, H. (1984) Volatile organochlorine hydrocarbons in water in East Germany. *Haustech. Bauphys. Umwelttech. Gesund. Ing.*, **105**, 269–278 (in German)

Henschler, D. (1977) Metabolism and mutagenicity of halogenated olefins—a comparison of structure and activity. *Environ. Health Perspectives*, **21**, 61–64

Henschler, D. (1987) Mechanisms of genotoxicity of chlorinated aliphatic hydrocarbons. In: De Matteis, F. & Lock, E.A., eds, *Selectivity and Molecular Mechanisms of Toxicity*, New York, MacMillan, pp.153–181

Henschler, D. & Bonse, G. (1977) Metabolic activation of chlorinated ethylenes: dependence of mutagenic effect on electrophilic reactivity of the metabolically formed epoxides. *Arch. Toxicol.*, **39**, 7–12

Henschler, D., Eder, E., Neudecker, T. & Metzler, M. (1977) Carcinogenicity of trichloroethylene: fact or artifact. *Arch. Toxicol.*, **37**, 233–236

Henschler, D., Romen, W., Elsässer, H.M., Reichert, D., Eder, E. & Radwan, Z. (1980) Carcinogenicity study of trichloroethylene by longterm inhalation in three animal species. *Arch. Toxicol.*, **43**, 237–248

Henschler, D., Elsässer, H., Romen, W. & Eder, E. (1984) Carcinogenicity study of trichloroethylene, with and without epoxide stabilizers, in mice. *J. Cancer Res. clin. Oncol.*, **107**, 149–156

Henschler, D., Vamvakas, S., Lammert, M., Dekant, W., Kraus, B., Thomas, B. & Ulm, K. (1995) Increased incidence of renal cell tumors in a cohort of cardboard workers exposed to trichloroethylene. *Arch. Toxicol.* (in press)

Herbert, P., Charbonnier, P., Rivolta, L., Servais, M., Van Mensch, F. & Campbell, I. (1986) The occurrence of chlorinated solvents in the environment. *Chem. Ind.*, **December**, 861–869

Hernberg, S., Korkala, M.-L., Asikainen, U. & Riala, R. (1984) Primary liver cancer and exposure to solvents. *Int. Arch. occup. environ. Health*, **54**, 147–153

Hernberg, S., Kauppinen, T., Riala, R., Korkala, M.-L. & Asikainen, U. (1988) Increased risk for primary liver cancer among women exposed to solvents. *Scand. J. Work Environ. Health*, **14**, 356–365

Herren-Freund, S.L., Pereira, M.A., Khoury, M.D. & Olson, G. (1987) The carcinogenicity of trichloroethylene and its metabolites, trichloroacetic acid and dichloroacetic acid, in mouse liver. *Toxicol. appl. Pharmacol.*, **90**, 183–189

Hirata, T., Nakasugi, O., Yoshioka, M. & Sumi, K. (1992) Groundwater pollution by volatile organochlorines in Japan and related phenomena in the subsurface environment. *Water Sci. Technol.*, **2**, 9–16

Hrelia, P., Maffei, F., Vivagni, F., Flori, P., Stanzani, R. & Cantelli Forti, G. (1995) Interactive effects between trichloroethylene and pesticides at metabolic and genetic level in mice. *Environ. Health Perspectives* (in press)

von der Hude, W., Behm, C., Gürtler, R. & Basler, A. (1988) Evaluation of SOS chromotest. *Mutat. Res.*, **203**, 81–94

IARC (1979) *IARC Monographs on the Evaluation of the Carcinogenic Risk of Chemicals to Humans*, Vol. 20, *Some Halogenated Hydrocarbons*, Lyon, pp. 545–572

IARC (1987a) *IARC Monographs on the Evaluation of Carcinogenic Risks to Humans*, Suppl. 7, *Overall Evaluations of Carcinogenicity: An Updating of* IARC Monographs *Volumes 1–42*, Lyon, pp. 364–366

IARC (1987b) *IARC Monographs on the Evaluation of Carcinogenic Risks to Humans*, Suppl. 7, *Overall Evaluations of Carcinogenicity: An Updating of* IARC Monographs *Volumes 1–42*, Lyon, pp. 99–100

IARC (1987c) *IARC Monographs on the Evaluation of Carcinogenic Risks to Humans*, Suppl. 7, *Overall Evaluations of Carcinogenicity: An Updating of* IARC Monographs *Volumes 1–42*, Lyon, pp. 202–203

IARC (1987d) *IARC Monographs on the Evaluation of Carcinogenic Risks to Humans*, Suppl. 7, *Overall Evaluations of Carcinogenicity: An Updating of* IARC Monographs *Volumes 1–42*, Lyon, p. 201

IARC (1987e) *IARC Monographs on the Evaluation of Carcinogenic Risks to Humans*, Suppl. 7, *Overall Evaluations of Carcinogenicity: An Updating of* IARC Monographs *Volumes 1–42*, Lyon, pp. 143–144

IARC (1987f) *IARC Monographs on the Evaluation of Carcinogenic Risks to Humans*, Suppl. 7, *Overall Evaluations of Carcinogenicity: An Updating of* IARC Monographs *Volumes 1–42*, Lyon, pp. 152–154

IARC (1987g) *IARC Monographs on the Evaluation of Carcinogenic Risks to Humans*, Suppl. 7, *Overall Evaluations of Carcinogenicity: An Updating of* IARC Monographs *Volumes 1–42*, Lyon, p. 62

IARC (1987h) *IARC Monographs on the Evaluation of Carcinogenic Risks to Humans*, Suppl. 7, *Overall Evaluations of Carcinogenicity: An Updating of* IARC Monographs *Volumes 1–42*, Lyon, p. 69

IARC (1987i) *IARC Monographs on the Evaluation of Carcinogenic Risks to Humans*, Suppl. 7, *Overall Evaluations of Carcinogenicity: An Updating of* IARC Monographs *Volumes 1–42*, Lyon, p. 72

IARC (1987j) *IARC Monographs on the Evaluation of Carcinogenic Risks to Humans*, Suppl. 7, *Overall Evaluations of Carcinogenicity: An Updating of* IARC Monographs *Volumes 1–42*, Lyon, p. 72

IARC (1987k) *IARC Monographs on the Evaluation of Carcinogenic Risks to Humans*, Suppl. 7, *Overall Evaluations of Carcinogenicity: An Updating of* IARC Monographs *Volumes 1–42*, Lyon, p. 73

IARC (1987l) *IARC Monographs on the Evaluation of Carcinogenic Risks to Humans*, Suppl. 7, *Overall Evaluations of Carcinogenicity: An Updating of* IARC Monographs *Volumes 1–42*, Lyon, pp. 120–122

IARC (1989a) *IARC Monographs on the Evaluation of Carcinogenic Risks to Humans*, Vol. 47, *Some Organic Solvents, Resin Monomers and Related Compounds, Pigments and Occupational Exposures in Paint Manufacture and Painting*, Lyon, pp. 199–213

IARC (1989b) *IARC Monographs on the Evaluation of Carcinogenic Risks to Humans*, Vol. 47, *Some Organic Solvents, Resin Monomers and Related Compounds, Pigments and Occupational Exposures in Paint Manufacture and Painting*, Lyon, pp. 217–228

IARC (1989c) *IARC Monographs on the Evaluation of Carcinogenic Risks to Humans*, Vol. 47, *Some Organic Solvents, Resin Monomers and Related Compounds, Pigments and Occupational Exposures in Paint Manufacture and Painting*, Lyon, pp. 263–287

IARC (1991) *IARC Monographs on the Evaluation of Carcinogenic Risks to Humans*, Vol. 52, *Chlorinated Drinking-water; Chlorination By-products; Some Other Halogenated Compounds; Cobalt and Cobalt Compounds*, Lyon, pp. 337–359

IARC (1994a) *IARC Monographs on the Evaluation of Carcinogenic Risks to Humans*, Vol. 60, *Some Industrial Chemicals*, Lyon, pp. 73–159

IARC (1994b) *IARC Monographs on the Evaluation of Carcinogenic Risks to Humans*, Vol. 60, *Some Industrial Chemicals*, Lyon, pp. 181–213

IARC (1994c) *IARC Monographs on the Evaluation of Carcinogenic Risks to Humans*, Vol. 60, *Some Industrial Chemicals*, Lyon, pp. 321–346

Ikeda, M. (1977) Metabolism of trichlorethylene and tetrachlorethylene in human subjects. *Environ. Health Perspectives*, **21**, 239–245

Ikeda, M. & Imamura, T. (1973) Biological half-life of trichloroethylene and tetrachloroethylene in human subjects. *Int. Arch. Arbeitsmed.*, **31**, 209–224

ILO (1991) *Occupational Exposure Limits for Airborne Toxic Substances: Values of Selected Countries* (Occupational Safety and Health Series No. 37), 3rd Ed., Geneva, pp. 396–397

Inoue, O., Seiji, K., Kawai, T., Jin, C., Liu, Y.-T., Chen, Z., Cai, S.-X., Yin, S.-N., Li, G.-L., Nakatsuka, H., Watanabe, T. & Ikeda, M. (1989) Relationship between vapor exposure and urinary metabolite excretion among workers exposed to trichloroethylene. *Am. J. ind. Med.*, **15**, 103–110

Isaacson, L.G. & Taylor, D.H. (1989) Maternal exposure to 1,1,2-trichloroethylene affects myelin in the hippocampal formation of the developing rat. *Brain Res.*, **488**, 403–407

Isacson, P., Bean, J.A., Splinter, R., Olson, D.B. & Kohler, J. (1985) Drinking water and cancer incidence in Iowa. III. Association of cancer with indices of contamination. *Am. J. Epidemiol.*, **121**, 856–869

Jackson, M.A., Stack, H.F. & Waters, M.D. (1993) The genetic toxicology of putative nongenotoxic carcinogens. *Mutat. Res.*, **296**, 241–277

Jaffe, D.R., Hassall, C.D., Gandolfi, A.J. & Brendel, K. (1985) Production of DNA single strand breaks in rabbit renal tissue after exposure to 1,2-dichlorovinylcysteine. *Toxicology*, **35**, 25–33

Jalihal, S.S. & Barlow, A.M. (1984) Leukaemia in dry cleaners (Letter to the Editor). *J. R. Soc. Health*, **104**, 42

Johnson, P. (1980) *Miami Carey, Inc., Monroe, OH* (Health Hazard Evaluation Report No. 80-48-689), Cincinnati, OH, United States National Institute for Occupational Safety and Health

Kim, C., Manning, R.O., Brown, R.P. & Bruckner, J.V. (1994) A comprehensive evaluation of the vial equilibration method for quantification of the metabolism of volatile organic chemicals. Trichloroethylene. *Drug Metab. Disposition*, **22**, 858–865

Kimbrough, R.D., Mitchell, F.L. & Houk, V.N. (1985) Trichloroethylene: an update. *J. Toxicol. environ. Health*, **15**, 369–383

Kimmerle, G. & Eben, A. (1973) Metabolism, excretion and toxicology of trichloroethylene after inhalation. 1. Experimental exposure on rats. *Arch. Toxicol.*, **30**, 115–126

Kjellstrand, P., Bjerkemo, M., Mortensen, I., Månsson, L., Lanke, J. & Holmquist, B. (1981a) Effects of long-term exposure to trichloroethylene on the behavior of Mongolian gerbils (*Meriones unguiculatus*). *J. Toxicol. environ. Health*, **8**, 787–793

Kjellstrand, P., Kanje, M., Månsson, L., Bjerkemo, M., Mortensen, I., Lanke, J. & Holmquist, B. (1981b) Trichloroethylene: effects on body and organ weights in mice, rats and gerbils. *Toxicology*, **21**, 105–115

Kjellstrand, P., Holmquist, B., Alm, P., Kanje, M., Romare, S., Jonsson, I., Månsson, L. & Bjerkemo, M. (1983a) Trichloroethylene: further studies on the effects on body and organ weights and plasma butyrylcholinesterase activity in mice. *Acta pharmacol. toxicol.*, **53**, 375–384

Kjellstrand, P., Holmquist, B., Mandahl, N. & Bjerkemo, M. (1983b) Effects of continous trichloroethylene inhalation on different strains of mice. *Acta pharmacol. toxicol.*, **53**, 369–374

Klaunig, J.E., Ruch, R. & Lin, E.L.C. (1989) Effects of trichloroethylene and its metabolites on rodent hepatocyte intercellular communication. *Toxicol. appl. Pharmacol.*, **99**, 454–465

Kligerman, A.D., Bryant, M.F., Doerr, C.L., Erexson, G.L., Evansky, P.A., Kwanyuen, P. & McGee, J.K. (1994) Inhalation studies of the genotoxicity of trichloroethylene to rodents. *Mutat. Res.*, **322**, 87–96

Koch, R., Schlegelmilch, R. & Wolf, H.U. (1988) Genetic effects of chlorinated ethylenes in the yeast *Saccharomyces cerevisiae*. *Mutat. Res.*, **206**, 209–216

Kominsky, J.R. (1978) *Dana Corp., Tipton, IN* (Health Hazard Evaluation Report No. 76-24-350; US NTIS PB-273716), Cincinnati, OH, United States National Institute for Occupational Safety and Health

Konietzko, H., Haberlandt, W., Heilbronner, H., Reill, G. & Weichardt, H. (1978) Chromosome studies on trichloroethylene workers. *Arch. Toxicol.*, **40**, 201–206 (in German)

Krain, L.S. (1972) Gallbladder and extrahepatic bile duct carcinoma. Analysis of 1,808 cases. *Geriatrics*, **27**, 111–117

Kringstad, K.P., Ljungquist, P.O., de Sousa, F. & Strömberg, L.M. (1981) Identification and mutagenic properties of some chlorinated aliphatic compounds in the spent liquor from kraft pulp chlorination. *Environ. Sci. Technol.*, **15**, 562–566

Kromhout, H., Swuste, P. & Boleij, J.S.M. (1994) Empirical modelling of chemical exposure in the rubber-manufacturing industry. *Ann. occup. Hyg.*, **38**, 3–22

Kronoveter, K.J. & Boiano, J.L. (1984) *Charlies' Taxidermy and Gifts, Fleetwqod, PA* (Health Hazard Evaluation Report No. 83-276-1499), Cincinnati, OH, United States National Institute for Occupational Safety and Health

Lagakos, S.W., Wessen, B.J. & Zelen, M. (1986) An analysis of contaminated well water and health effects in Woburn, Massachusetts. *J. Am. stat. Assoc.*, **81**, 583–596

Land, P.C., Owen, E.L. & Linde, H.W. (1981) Morphologic changes in mouse spermatozoa after exposure to inhalational anesthetics during early spermatogenesis. *Anesthesiology*, **54**, 53–56

Landrigan, P.J., Stein, G.F., Kominsky, J.R., Ruhe, R.L. & Watanabe, A.S. (1987) Common-source community and industrial exposure to trichloroethylene. *Arch. environ. Health*, **42**, 327–332

Lanham, S. (1970) Studies on placental transfer. *Ind. Med.*, **39**, 46–49

Larson, J.L. & Bull, R.J. (1992a) Species differences in the metabolism of trichloroethylene to the carcinogenic metabolites trichloroacetate and dichloroacetate. *Toxicol. appl. Pharmacol.*, **115**, 278–285

Larson, J.L. & Bull, R.J. (1992b) Metabolism and lipoperoxidative activity of trichloroacetate and dichloroacetate in rats and mice. *Toxicol. appl. Pharmacol.*, **115**, 268–277

Lee, S. & Parkinson, D. (1982) *Rola-Esmark Co., Dubois, PA* (Health Hazard Evaluation Report No. 80-168-1204), Cincinnati, OH, United States National Institute for Occupational Safety and Health

Leibman, K.C. (1968) On the metabolism of trichloroethylene (Letter to the Editor). *Anesthesiology*, **29**, 1066–1067

Leibman, K.C. & McAllister, W.J., Jr (1967) Metabolism of trichloroethylene in liver microsomes. III. Induction of the enzymic activity and its effect on excretion of metabolites. *J. Pharmacol. exp. Ther.*, **157**, 574–580

Leong, B.K.J., Schwetz, B.A. & Gehring, P.J. (1975) Embryo- and fetotoxicity of inhaled trichloroethylene, perchloroethylene, methylchloroform, and methylene chloride in mice and rats (Abstract No. 34). *Toxicol. appl. Pharmacol.*, **33**, 136

Lesage, S., Jackson, R.E., Priddle, M.W. & Reiemann, P. (1990) Occurrence and fate of organic solvent residues in anoxic ground-water at the Gloucester landfill, Canada. *Environ. Sci. Technol.*, **24**, 559–566

Lide, D.R., ed. (1993) *CRC Handbook of Chemistry and Physics*, 74th Ed., Boca Raton, FL, CRC Press, pp. 3–242, D-197

Liebler, D.C. & Guengerich, F.P. (1983) Olefin oxidation by cytochrome P-450: evidence for group migration in catalytic intermediates formed with vinylidene chloride and *trans*-1-phenyl-1-butene. *Biochemistry*, **22**, 5482–5489

Ligocki, M.P., Leuenberger, C. & Pankow, J.F. (1985) Trace organic compounds in rain—II. Gas scavenging of neutral organic compounds. *Atmos. Environ.*, **19**, 1609–1617

Lillian, D., Singh, H.B., Appleby, A., Lobban, L., Arnts, R., Gumpert, R., Hague, R., Toomey, J., Kazazis, J., Antell, M., Hansen, D. & Scott, B. (1975) Atmosphere fates of halogenated compounds. *Environ. Sci. Technol.*, **9**, 1042–1048

Linak, E., Leder, A. & Yoshida, Y. (1992) C_2 Chlorinated solvents. In: *Chemical Economics Handbook*, Menlo Park, CA, SRI International, pp. 632.3000–632.3001

Lindbohm, M.-L., Taskinen, H., Sallmén, M. & Hemminki, K. (1990) Spontaneous abortions among women exposed to organic solvents. *Am. J. ind. Med.*, **17**, 449–463

Loprieno, N. & Abbondandolo, A. (1980) Comparative mutagenic evaluation of some industrial compounds. In: Norpoth, K.H. & Garner, R.C., eds, *Short-term Test Systems for Detecting Carcinogens*, Berlin, Springer-Verlag, pp. 333–356

Love, J.R. & Kern, M. (1981) *Metro Bus Maintenance Shop, Washington, DC* (Health Hazard Evaluation Report No. 81-065-938), Cincinnati, OH, United States National Institute for Occupational Safety and Health

Lovelock, J.E. (1974) Atmospheric halocarbons and stratospheric ozone. *Nature*, **252**, 292–294

Lowengart, R.A., Peters, J.M., Cicioni, C., Buckley, J., Bernstein, L., Preston-Martin, S. & Rappaport, E. (1987) Childhood leukemia and parents' occupational and home exposures. *J. natl Cancer Inst.*, **79**, 39–46

Málek, B., Krcmárová, B. & Rodová, O. (1979) An epidemiological study of hepatic tumour incidence in subjects working with trichloroethylene. II. Negative result of retrospective investigations in dry cleaners. *Prac. Lék*, **31**, 124–126 (in Czech)

Malker, H.S.R., McLaughlin, J.K., Malker, B.K., Stone, B.J., Weiner, J.A., Ericsson, J.L.E. & Blot, W.J. (1986) Biliary tract cancer and occupation in Sweden. *Br. J. ind. Med.*, **43**, 257–262

Maltoni, C., Lefemine, G. & Cotti, G. (1986) Experimental research on trichloroethylene carcinogenesis. In: Maltoni, C. & Mehlman, M.A., eds, *Archives of Research on Industrial Carcinogenesis*, Vol. V, Princeton, NJ, Princeton Scientific Publishing Co., pp. 1–393

Maltoni, C., Lefemine, G., Cotti, G. & Perino, G. (1988) Long-term carcinogenic bioassays on trichloroethylene administered by inhalation to Sprague-Dawley rats and Swiss and B6C3F1 mice. *Ann. N.Y. Acad. Sci.*, **534**, 316–351

Manson, J.M., Murphy, M., Richdale, N. & Smith, M.K. (1984) Effects of oral exposure to trichloroethylene on female reproductive function. *Toxicology*, **32**, 229–242

Mazzullo, M., Bartoli, S., Bonora, B., Colacci, A., Lattanzi, G., Niero, A., Silingardi, P. & Grilli, S. (1992) In vivo and in vitro interaction of trichloroethylene with macromolecules from various organs of rat and mouse. *Res. Commun. chem. Pathol. Pharmacol.*, **76**, 192–208

McCarthy, T.B. & Jones, R.D. (1983) Industrial gassing poisonings due to trichloroethylene, perchloroethylene, and 1,1,1-trichloroethane, 1961–80. *Br. J. ind. Med.*, **40**, 450–455

McConnell, G., Ferguson, D.M. & Pearson, C.R. (1975) Chlorinated hydrocarbons and the environment. *Endeavor*, **34**, 13–18

McGregor, D.B., Reynolds, D.M. & Zeiger, E. (1989) Conditions affecting the mutagenicity of trichloroethylene in *Salmonella*. *Environ. mol. Mutag.*, **13**, 197–202

McLaren, J., Boulikas, T. & Vamvakas, S. (1994) Induction of poly(ADP-ribosyl)ation in the kidney after in vivo application of renal carcinogens. *Toxicology*, **88**, 101–112

Meadows, S.D., Gandolfi, A.J., Nagle, R.B. & Shively, J.W. (1988) Enhancement of DMN-induced kidney tumors by 1,2-dichlorovinylcysteine in Swiss-Webster mice. *Drug chem. Toxicol.*, **11**, 307–318

Melnick, R.L., Jameson, C.W., Goehl, T.J., Maronpot, R.R., Collins, B.J., Greenwell, A., Harrington, F.W., Wilson, R.E., Tomaszewski, K.E. & Agarwal, D.K. (1987) Application of microencapsulation for toxicology studies. II. Toxicity of microencapsulated trichloroethylene in Fischer 344 rats. *Fundam. appl. Toxicol.*, **8**, 432–442

Mersch-Sundermann, V., Müller, G. & Hofmeister, A. (1989) Examination of mutagenicity of organic microcontaminations of the environment. IV. Communication: the mutagenicity of halogenated aliphatic hydrocarbons with the SOS-chromotest. *Zbl. Hyg.*, **189**, 266–271 (in German)

Mertens, J.A. (1993) Chlorocarbons and chlorohydrocarbons. In: Kroschwitz, J.I. & Howe-Grant, M., eds, *Kirk-Othmer Encyclopedia of Chemical Technology*, 4th Ed., Vol. 6, New York, John Wiley & Sons, pp. 40–50

Miller, R.E. & Guengerich, F.P. (1982) Oxidation of trichloroethylene by liver microsomal cytochrome P-450: evidence for chlorine migration in a transition state not involving trichloroethylene oxide. *Biochemistry*, **21**, 1090–1097

Miller, R.E. & Guengerich, F.P. (1983) Metabolism of trichloroethylene in isolated hepatocytes, microsomes and reconstituted enzyme systems containing cytochrome P-450. *Cancer Res.*, **43**, 1145–1152

Milman, H.A., Story, D.L., Riccio, E.S., Sivak, A., Tu, A.S., Williams, G.M., Tong, C. & Tyson, C.A. (1988) Rat liver foci and in vitro assays to detect initiating and promoting effects of chlorinated ethanes and ethylenes. *Ann. N.Y. Acad. Sci.*, **534**, 521–530

Mirsalis, J.C., Tyson, C.K., Steinmetz, K.L., Loh, E.K., Hamilton, C.M., Bakke, J.P.& Spalding, J.W. (1989) Measurement of unscheduled DNA synthesis and S-phase synthesis in rodent hepatocytes following in vivo treatment: testing of 24 compounds. *Environ. mol. Mutag.*, **14**, 155–164

Monster, A.C. (1984) Trichloroethylene. In: Aitio, A., Riihimaki, V. & Vainio, H., eds, *Biological Monitoring and Surveillance of Workers Exposed to Chemicals*, Washington, DC, Hemisphere Publishing Corp., pp. 111–130

Monster, A.C., Boersma, G. & Duba, W.C. (1976) Pharmacokinetics of trichloroethylene in volunteers, influence of workload and exposure concentration. *Int. Arch. occup. environ. Health*, **38**, 87–102

Monster, A.C., Boersma, G. & Duba, W.C. (1979) Kinetics of trichloroethylene in repeated exposure of volunteers. *Int. Arch. occup. environ. Health*, **42**, 283–292

Moore, D.R.J., Walker, S.L. & Ansari, R. (1991) *Canadian Water Quality Guidelines for Trichloroethylene* (Scientific Series No. 183), Ottawa, Inland Waters Directorate, Water Quality Branch

Morse, K.M. & Goldberg, L. (1943) Chlorinated solvent exposures at degreasing operations. *Ind. Med.*, **12**, 706–713

Mortelmans, K., Haworth, S., Lawlor, T., Speck, W., Tainer, B. & Zeiger, E. (1986) *Salmonella* mutagenicity tests: II. Results from the testing of 270 chemicals. *Environ. Mutag.*, **8** (Suppl. 7), 1–119

Mroueh, U.-M. (1993) *Orgaanisten Liuotteiden Käyttö Suomessa* [Solvent use in Finland], Helsinki, National Board of Waters and the Environment

Müller, G., Spassovski, M. & Henschler, D. (1972) Trichloroethylene exposure and trichloroethylene metabolites in urine and blood. *Arch. Toxicol.*, **29**, 335–340

Müller, G., Spassovski, M. & Henschler, D. (1974) Metabolism of trichloroethylene in man. II. Pharmacokinetics of metabolites. *Arch. Toxicol.*, **32**, 283–295

Müller, W.F., Coulston, F. & Korte, F. (1982) Comparative metabolism of [^{14}C]trichloroethylene in chimpanzees, baboons, and rhesus monkeys. *Chemosphere*, **11**, 215–218

Nagaya, T., Ishikawa, N. & Hata, H. (1989) Sister-chromatid exchanges in lymphocytes of workers exposed to trichloroethylene. *Mutat. Res.*, **222**, 279–282

Nakajima, T., Wang, R.-S., Murayama, N. & Sato, A. (1990) Three forms of trichloroethylene-metabolizing enzymes in rat liver induced by ethanol, phenobarbital, and 3-methylcholanthrene. *Toxicol. appl. Pharmacol.*, **102**, 546–552

Nakajima, T., Wang, R.-S., Elovaara, E., Park, S.S., Gelboin, H.V. & Vainio, H. (1992) A comparative study on the contribution of cytochrome P450 isoenzymes to metabolism of benzene, toluene, and trichloroethylene in rat liver. *Biochem. Pharmacol.*, **43**, 251–257

Nelson, M.A. & Bull, R.J. (1988) Induction of strand breaks in DNA by trichloroethylene and metabolites in rat and mouse liver *in vivo*. *Toxicol. appl. Pharmacol.*, **94**, 45–54

Noland-Gerbee, E.A., Pfohl, R.J., Taylor, D.H. & Bull, R.J. (1986) 2-Deoxyglucose uptake in developing rat brain upon pre- and postnatal exposure to trichloroethylene. *Neurotoxicology*, **7**, 157–164

Nomiyama, K. (1971) Estimation of trichloroethylene exposure by biological material. *Int. Arch. Arbeitsmed.*, **27**, 281–292

Nomiyama, K. & Nomiyama, H. (1977) Dose–response relationship for trichlorethylene in man. *Int. Arch. occup. environ. Health*, **39**, 237–248

Nomiyama, K., Nomiyama, H. & Arai, H. (1986) Reevaluation of subchronic toxicity of trichloroethylene (Abstract No. P16-4). *Toxicol. Lett.*, **31**, 225

Novotná, E., David, A. & Málek, B. (1979) An epidemiological study on hepatic tumour incidence in subjects working with trichloroethylene: I. Negative result of retrospective investigations in subjects with primary liver carcinoma. *Prac. Lék.*, **31**, 121–123 (in Czech)

Odum, J., Foster, J.R. & Green, T. (1992) A mechanism for the development of Clara cell lesions in the mouse lung after exposure to trichloroethylene. *Chem.-biol. Interactions*, **83**, 135–153

Ogata, M., Takatsuka, Y. & Tomokuni, K. (1971) Excretion of organic chlorine compounds in the urine of persons exposed to vapours of trichloroethylene and tetrachloroethylene. *Br. J. ind. Med.*, **28**, 386–391

Okawa, M.T., Blejer, H.P. & Solomon, R.M. (1978) *Trans World Airlines Corp., Los Angeles, CA* (Health Hazard Evaluation Determination Report No. 78-38-312), Cincinnati, OH, United States National Institute for Occupational Safety and Health

Olsson, H. & Brandt, L. (1980) Occupational exposure to organic solvents and Hodgkin's disease in men. A case–referent study. *Scand. J. Work Environ. Health*, 6, 302–305

Paddle, G.M. (1983) Incidence of liver cancer and trichloroethylene manufacture: joint study by industry and a cancer registry. *Br. med. J.*, 286, 846

Papdullo, R.F., Shareef, S.A., Kincaid, L.E. & Murphy, P.V. (1985) *Survey of Trichloroethylene Emission Sources* (EPA-450/3-85-021; PB86-107943), Research Triangle Park, NC, Office of Air Quality Planning and Standards, Environmental Protection Agency

Paraf, F., Paraf, A. & Barge, J. (1990) Is industrial exposure a risk factor in gallbladder cancer? *Gastroenterol. clin. Biol.*, 14, 877-880 (in French)

Parchman, L.G. & Magee, P.N. (1982) Metabolism of [^{14}C]trichloroethylene to $^{14}CO_2$ and interaction of a metabolite with liver DNA in rats and mice. *J. Toxicol. environ. Health*, 9, 797–813

Peters, J.M., Preston-Martin, S. & Yu, M.C. (1981) Brain tumors in children and occupational exposure of parents. *Science*, 213, 235–237

Peters, J.M., Garabrant, D.H., Preston-Martin, S. & Yu, M.C. (1984) Is trichloroethylene a human carcinogen? (Abstract). *Scand. J. Work Environ. Health*, 13, 180

Pfaffenberger, C.D., Peoples, A.J. & Briggle, T.V. (1984) Blood plasma levels of volatile chlorinated solvents and metabolites in occupationally exposed workers. In: Veziroglu, T.N., ed., *The Biosphere: Problems and Solutions*, Amsterdam, Elsevier Science Publishers, pp. 559–569

PPG Industries, Inc. (1994) *Product Information Sheet, Trichloroethylene*, Pittsburgh, PA

Price, P.J., Hassett, C.M. & Mansfield, J.I. (1978) Transforming activities of trichloroethylene and proposed industrial alternatives. *In Vitro*, 14, 290–293

Prout, M.S., Provan, W.M. & Green, T. (1985) Species differences in response to trichloroethylene. I. Pharmacokinetics in rats and mice. *Toxicol. appl. Pharmacol.*, 79, 389–400

Rantala, K., Riipinen, H. & Anttila, A. (1992) *Altisteettyössä, 22—Halogeenihiitivedyt* [Exposure at work—halogenated hydrocarbons], Helsinki, Finnish Institute of Occupational Health—Finnish Work Environment Fund

Rasmussen, K., Sabroe, S., Wohlert, M., Ingerslev, H.J., Kappel, B. & Nielsen, J. (1988) A genotoxic study of metal workers exposed to trichloroethylene. Sperm parameters and chromosome aberrations in lymphocytes. *Int. Arch. occup. environ. Health*, 60, 419–423

Roldán-Arjona, T., García-Pedrajas, M.D., Luque-Romero, F.L., Hera, C. & Pueyo, C. (1991) An association between mutagenicity of the Ara test of *Salmonella typhimurium* and carcinogenicity in rodents for 16 halogenated aliphatic hydrocarbons. *Mutagenesis*, 6, 199–205

Rossi, A.M., Migliore, L., Barale, R. & Loprieno, N. (1983) In vivo and in vitro mutagenicity studies of a possible carcinogen, trichloroethylene, and its two stabilizers, epichlorohydrin and 1,2-epoxybutane. *Teratog. Carcinog. Mutag.*, 3, 75–87

Ruhe, R.L. (1982) *Synthes Ltd, Monument, CO* (Health Hazard Evaluation Report No. 82-040-1119), Cincinnati, OH, United States National Institute for Occupational Safety and Health

Ruhe, R.L. & Donohue, M. (1980) *Duralectra, Inc., Natick, MA* (Health Hazard Evaluation Report No. 79-147-702), Cincinnati, OH, United States National Institute for Occupational Safety and Health

Ruhe, R.L., Watanabe, A. & Stein, G. (1981) *Superior Tube Company, Collegeville, PA* (Health Hazard Evaluation Report No. 80-49-808), Cincinnati, OH, United States National Institute for Occupational Safety and Health

Sabel, G.V. & Clark, T.P. (1984) Volatile organic compounds as indicators of municipal solid waste leachate contamination. *Waste Manage. Res.*, **2**, 119–130

Sack, T.M., Steele, D.H., Hammerstrom, K. & Remmers, J. (1992) A survey of household products for volatile organic compounds. *Atmos. Environ.*, **26A**, 1063–1070

Sadtler Research Laboratories (1980) *1980 Cumulative Index*, Philadelphia, PA

Sawyer, L.D., McMahon, B.M., Newsome, W.H. & Parker, G.A., eds (1990) Pesticide and industrial chemical residues. In: Helrich, K., ed., *Official Methods of Analysis of the Association of Official Analytical Chemists*, 15th Ed., Vol. 1, Arlington, VA, Association of Official Analytical Chemists, pp. 290–291

Sax, N.I. & Lewis, R.J. (1987) *Hawley's Condensed Chemical Dictionary*, 11th Ed., New York, Van Nostrand Reinhold Company, p. 1176

Sbrana, I., Lascialfari, D. & Loprieno, N. (1985) Trichloroethylene induces micronuclei but not chromosomal aberrations in mouse bone marrow cells (Abstract). In: Ramel, C., ed., *Fourth International Conference on Environmental Mutagens, Stockholm, June 24–28, 1985*, New York, Alan R. Liss, p. 163

Schairer, L.A. & Sautkulis, R.C. (1982) Detection of ambient levels of mutagenic atmospheric pollutants with the higher plant *Tradescantia*. *Environ. Mutag. Carcinog. Plant Biol.*, **2**, 154–194

Schweizerische Unfallversicherungsanstalt [Swiss Accident Insurance Company] (1994) *Grenzwerte am Arbeitsplatz* [Limit values at the workplace], Lucerne, pp. 101, 131

Schwetz, B.A., Leong, B.K.J. & Gehring, P.J. (1975) The effect of maternally inhaled trichloroethylene, perchloroethylene, methyl chloroform, and methylene chloride on embryonal and fetal development in mice and rats. *Toxicol. appl. Pharmacol.*, **32**, 84–96

Scott, J.E., Forkert, P.G., Oulton, M., Rasmusson, M.-G., Temple, S., Fraser, M.O. & Whitefield, S. (1988) Pulmonary toxicity of trichloroethylene: induction of changes in surfactant phospholipids and phospholipase A2 activity in the mouse lung. *Exp. mol. Pharmacol.*, **49**, 141–150

Seiji, K., Jin, C., Watanabe, T., Nakatsuka, H. & Ikeda, M. (1990) Sister chromatid exchanges in peripheral lymphocytes of workers exposed to benzene, trichloroethylene, or tetrachloroethylene, with reference to smoking habits. *Int. Arch. occup. environ. Health*, **62**, 171–176

Seldén, A., Hultberg, B., Ulander, A. & Ahlborg, G., Jr (1993) Trichloroethylene exposure in vapour degreasing and the urinary excretion of *N*-acetyl-β-D-glucosaminidase. *Arch. Toxicol.*, **67**, 224–226

Shahin, M.M. & Von Borstel, R.C. (1977) Mutagenic and lethal effects of α-benzene hexachloride, dibutyl phthalate and trichloroethylene in *Saccharomyces cerevisiae*. *Mutat. Res.*, **48**, 173–180

Sharpe, C.R., Rochon, J.E., Adam, J.M. & Suissa, S. (1989) Case–control study of hydrocarbon exposures in patients with renal cell carcinoma. *Can. med. Assoc. J.*, **140**, 1309–1318

Shelby, M.D., Erexson, G.L., Hook, G.J. & Tice, R.R. (1993) Evaluation of a three-exposure mouse bone marrow micronucleus protocol: results with 49 chemicals. *Environ. mol. Mutag.*, **21**, 160–179

Sherwood, R.J. (1976) Ostwald solubility coefficients of some industrially important substances. *Br. J. ind. Med.*, **33**, 106–107

Shimada, T., Swanson, A.F., Leber, P. & Williams, G.M. (1985) Activities of chlorinated ethane and ethylene compounds in the *Salmonella*/rat microsome mutagenesis and rat hepatocyte/DNA repair assays under vapor phase exposure conditions. *Cell Biol. Toxicol.*, **1**, 159–179

Shindell, S. & Ulrich, S. (1985) A cohort study of employees of a manufacturing plant using trichloroethylene. *J. occup. Med.*, **27**, 577–579

Shipman, A.J. & Whim, B.P. (1980) Occupational exposure to trichloroethylene in metal cleaning processes and to tetrachloroethylene in the drycleaning industry in the UK. *Ann. occup. Hyg.*, **23**, 197–204

Siegel, J., Jones, R.A., Coon, R.A. & Lyon, J.P. (1971) Effects on experimental animals of acute, repeated and continuous inhalation exposures to dichloroacetylene mixtures. *Toxicol. appl. Pharmacol.*, **18**, 168–174

Siemiatycki, J. (1991) *Risk Factors for Cancer in the Workplace*, Boca Raton, FL, CRC Press

Simmon, V.F., Kauhanen, K. & Tardiff, R.G. (1977) Mutagenic activity of chemicals identified in drinking water. In: Scott, D., Bridges, B.A. & Sobels, F.H., eds, *Progress in Genetic Toxicology*, Amsterdam, Elsevier/North-Holland, pp. 249–258

Singh, H.B., Salas, L.J., Smith, A.J. & Shigeishi, H. (1981) Measurement of some potentially hazardous organic chemicals in urban environments. *Atmos. Environ.*, **15**, 601–612

Singh, H.B., Salas, L.J. & Stiles, R.E. (1983) Selected man-made halogenated chemicals in the air and oceanic environment. *J. geophys. Res.*, **88**, 3675–3683

Skender, L.J., Karacic, V. & Prpic-Majic, D. (1991) A comparative study of human levels of trichloroethylene and tetrachloroethylene after occupational exposure. *Arch. environ. Health*, **46**, 174–178

Skender, L., Karacic, V., Bosner, B. & Prpic-Majic, D. (1993) Assessment of exposure to trichloroethylene and tetrachloroethylene in the population of Zagreb, Croatia. *Int. Arch. occup. environ. Health*, **65**, S163–S165

Slacik-Erben, R., Roll, R., Franke, G. & Uehleke, H. (1980) Trichloroethylene vapours do not produce dominant lethal mutations in male mice. *Arch. Toxicol.*, **45**, 37–44

Smith, G.F. (1966) Trichloroethylene: a review. *Br. J. ind. Med.*, **23**, 249–262

Smyth, H.F., Jr, Carpenter, C.P., Weil, C.S., Pozzani, U.C., Striegel, J.A. & Nycum, J.S. (1969) Range-finding toxicity data. List VII. *Am. ind. Hyg. Assoc. J.*, **30**, 470–476

Sofuni, T., Hayashi, M., Matsuoka, A., Sawada, M., Hatanaka, M. & Ishidate, M., Jr (1985) Mutagenicity tests on organic chemical contaminants in city water and related compounds. II. Chromosome aberration tests in cultured mammalian cells. *Bull. natl Inst. Hyg. Sci. (Tokyo)*, **103**, 64–75

Spirtas, R., Stewart, P.A., Lee, J.S., Marano, D.E., Forbes, C.D., Grauman, D.J., Pettigrew, H.M., Blair, A., Hoover, R.N. & Cohen, J.L. (1991) Retrospective cohort mortality study of workers at an aircraft maintenance facility. I. Epidemiological results. *Br. J. ind. Med.*, **48**, 515–530

Staples, C.A., Werner, A.F. & Hoogheem, T.J. (1985) Assessment of priority pollutant concentrations in the United States using STORET database. *Environ. Toxicol. Chem.*, **4**, 131–142

Steward, A., Allott, P.R., Cowles, A.L. & Mapleson, W.W. (1973) Solubility coefficients for inhaled anaesthetics for water, oil and biological media. *Br. J. Anaesthesiol.*, **45**, 282–293

Stewart, P.A., Lee, J.S., Marano, D.E., Spirtas, R., Forbes, C.D. & Blair, A. (1991) Retrospective cohort mortality study of workers at an aircraft maintenance facility. II. Exposures and their assessment. *Br. J. ind. Med.*, **48**, 531–537

Stott, W.T., Quast, J.F. & Watanabe, P.G. (1982) The pharmacokinetics and macromolecular interactions of trichloroethylene in mice and rats. *Toxicol. appl. Pharmacol.*, **62**, 137–151

Su, C. & Goldberg, E.D. (1976) Environmental concentrations and fluxes of some halocarbons. In: Windom, H.L. & Duce, R.A., eds, *Marine Pollutant Transfer*, Lexington, MA, Lexington Books, pp. 353–374

Sullivan, D.A., Jones, A.D. & Williams, J.G. (1985) Results of the United States Environmental Protection Agency's air toxics analysis in Philadelphia. In: *Proceedings of the 78th Annual Meeting of the Air Pollution Control Association*, Vol. 2, Detroit, MI, Air Pollution Control Association, Paper 85-17.5

Taskinen, H., Anttila, A., Lindbohm, M.-L., Sallmén, M. & Hemminki, K. (1989) Spontaneous abortions and congenital malformations among the wives of men occupationally exposed to organic solvents. *Scand. J. Work Environ. Health*, **15**, 345–352

Taskinen, H., Kyyrönen, P., Hemminki, K., Hoikkala, M., Lajunen, K. & Lindbohm, M.-L. (1994) Laboratory work and pregnancy outcome. *J. occup. Med.*, **36**, 311–319

Taylor, D.H., Lagory, K.E., Zaccaro, D.J., Pfohl, R.J. & Laurie, R.D. (1985) Effect of trichloroethylene on the exploratory and locomotor activity of rats exposed during development. *Sci. total Environ.*, **47**, 415–420

Thomas, R.D. (1989) Epidemiology and toxicology of volatile organic chemical contaminants in water absorbed through the skin. *J. Am. Coll. Toxicol.*, **8**, 779–795

Tola, S., Vilhunen, R., Järvinen, E. & Korkala, M.-L. (1980) A cohort study on workers exposed to trichloroethylene. *J. occup. Med.*, **22**, 737–740

Trouwborst, T. (1981) Groundwater pollution by volatile halogenated hydrocarbons: source of pollution and methods to estimate their relevance. *Sci. total Environ.*, **21**, 41–46

Trussell, A.R., Cromer, J.L., Umphres, M.D., Kelly, P.E. & Moncur, J.G. (1980) Monitoring of volatile halogenated organics: a survey of twelve drinking waters from various parts of the world. In: Jolley, R.L., Brungs, W.A., Cumming, R.B. & Jacobs, V.A., eds, *Water Chlorination. Environment Impact and Health Effects*, Vol. 3, Ann Arbor, MI, Ann Arbor Science, pp. 39–53

Tu, A.S., Murray, T.A., Hatch, K.M., Sivak, A. & Milman, H.A. (1985) In vitro transformation of BALB/c-3T3 cells by chlorinated ethanes and ethylenes. *Cancer Lett.*, **28**, 85–92

Tucker, A.N., Sanders, V.M., Barnes, D.W., Bradshaw, T.J., White, K.L., Jr, Sain, L.E., Borzelleca, J.F. & Munson, A.E. (1982) Toxicology of trichloroethylene in the mouse. *Toxicol. appl. Pharmacol.*, **62**, 351–357

Työministeriö [Ministry of Labour] (1993) *HTP-Arvot 1993* [Limit values 1993], Tampere, p. 19

Uhler, A.D. & Diachenko, G.W. (1987) Volatile halocarbon compounds in process water and processed foods. *Bull. environ. Contam. Toxicol.*, **39**, 601–607

Ulander, A., Seldén, A. & Ahlborg, G., Jr (1992) Assessment of intermittent trichloroethylene exposure in vapor degreasing. *Am. ind. Hyg. Assoc. J.*, **53**, 742–743

United Kingdom Health and Safety Executive (1994) *Occupational Exposure Limits 1994* (Guidance Note EH 40/94), London, Her Majesty's Stationary Office, p. 26

United States Agency for Toxic Substances and Disease Registry (1989) *Toxicology Profile for Trichloroethylene* (ATSDR/TP-88-24), Oak Ridge, TN, Oak Ridge National Laboratory

United States Environmental Protection Agency (1985) *Health Assessment Document for Trichloroethylene (Final Report)* (EPA-600/8-82006F; US NTIS PB85-249696), Washington DC

United States Environmental Protection Agency (1986a) Method 8010. Halogenated volatile organics. In: *Test Methods for Evaluating Solid Waste—Physical/Chemical Methods* (US EPA No. SW-846), 3rd Ed., Vol. 1A, Washington DC, Office of Solid Waste and Emergency Response, pp. 1–13

United States Environmental Protection Agency (1986b) Method 8240. Gas chromatography/mass spectrometry for volatile organics. In: *Test Methods for Evaluating Solid Waste—Physical/Chemical Methods* (US EPA No. SW-846), 3rd Ed., Vol. 1A, Washington DC, Office of Solid Waste and Emergency Response, pp. 1–43

United States Environmental Protection Agency (1988a) *Compendium of Methods for the Determination of Toxic Organic Compounds in Ambient Air* (EPA-600/4-89-017; US NTIS PB90-116989), Research Triangle Park, NC, Office of Research and Development, pp. TO1-1–TO1-38; TO3-1–TO3-22; TO14-1–TO14-94

United States Environmental Protection Agency (1988b) *Methods for the Determination of Organic Compounds in Drinking Water* (EPA-600/4-88-039; US NTIS PB89-220461), Cincinnati, OH, Environmental Monitoring Systems Laboratory, pp. 5–87, 255–323 (Methods 502.1, 502.2, 503.1, 524.1, 524.2)

United States Environmental Protection Agency (1989) *Contract Laboratory Program Statistical Database*, Washington DC

United States Environmental Protection Agency (1993) *1991 Toxic Release Inventory* (EPA 745-R-93-003), Washington DC, Office of Pollution Prevention and Toxics, p. 250

United States Environmental Protection Agency (1994) Protection of the environment. *US Code fed. Regul.*, **Title 40**, Part 136, Appendix A, pp. 400–413, 559–573, 602–614 (Methods 601, 624, 1624)

United States Food and Drug Administration (1977) Trichloroethylene. Removal from food additive use. *Fed. Regist.*, 42, 49465–49471

United States Food and Drug Administration (1983) Trichloroethylene. In: Warner, C., Modderman, J., Fazio, T., Beroza, M., Schwartzman, G., Fominaya, K. & Sherma, J., eds, *Food Additives Analytical Manual*, Vol. I, *A Collection of Analytical Methods for Selected Food Additives*, Washington DC/Arlington VA, Bureau of Foods/Association of Official Analytical Chemists, pp. 176–185, 224–232, 317

United States National Cancer Institute (1976) *Carcinogenesis Bioassay of Trichloroethylene (CAS No. 79-01-6)* (Tech. Rep. Ser. No. 2), Bethesda, MD

United States National Institute for Occupational Safety and Health (1994a) *NIOSH Pocket Guide to Chemical Hazards* (DHHS (NIOSH) Publ. No. 94-116), Cincinnati, OH, pp. 316–317, 350

United States National Institute for Occupational Safety and Health (1994b) *National Occupational Exposure Survey, 1981–1983*, Cincinnati, OH

United States National Institute for Occupational Safety and Health (1994c) *RTECS Chem-Bank*, Cincinnati, OH

United States National Toxicology Program (1985) *Trichloroethylene: Reproduction and Fertility Assessment in CD-1 Mice When Administered in the Feed* (NTP-86-068), Bethesda, MD, Department of Health and Human Services, National Institutes of Health

United States National Toxicology Program (1986) *Trichloroethylene: Reproduction and Fertility Assessment in F344 Rats When Administered in the Feed, Final Report* (NTP-86-085), Bethesda, MD, Department of Health and Human Services, National Institutes of Health

United States National Toxicology Program (1988) *Toxicology and Carcinogenesis Studies of Trichloroethylene (CAS No. 79-01-6) in Four Strains of Rats (ACI, August, Marshall, Osborne-Mendel) (Gavage Studies)* (Tech. Rep. Ser. No. 273), Research Triangle Park, NC

United States National Toxicology Program (1990) *Carcinogenesis Studies of Trichloroethylene (Without Epichlorohydrin) (CAS No. 79-01-6) in F344/N Rats and B6C3F$_1$ Mice (Gavage Studies)* (Tech. Rep. Ser. No. 243), Research Triangle Park, NC

United States Occupational Safety and Health Administration (1990) *OSHA Analytical Methods Manual*, 2nd Ed., Part 1, Vol. 1, Salt Lake City, UT, Method 7

United States Occupational Safety and Health Administration (1994) Air contaminants. *US Code fed. Regul.*, **Title 29**, Part 1910.1000, p. 18

Urano, K., Kawamoto, K., Abe, Y. & Otake, M. (1988) Chlorinated organic compounds in urban air in Japan. *Sci. total Environ.*, **74**, 121–131

Vainio, H., Waters, M.D. & Norppa, H. (1985) Mutagenicity of selected organic solvents. *Scand. J. Work Environ. Health*, **11** (Suppl. 1), 75–82

Vamvakas, S. & Köster, U. (1993) The nephrotoxin dichlorovinylcysteine induces expression of the protooncogenes *c-fos* and *c-myc* in LLC-PK$_1$ cells. A comparative investigation with growth factors and 12-*O*-tetradecanoylphorbolacetate. *Cell Biol. Toxicol.*, **9**, 1–13

Vamvakas, S., Dekant, W., Berthold, K., Schmidt, S., Wild, D. & Henschler, D. (1987) Enzymatic transformation of mercapturic acids derived from halogenated alkenes to reactive and mutagenic intermediates. *Biochem. Pharmacol.*, **36**, 2741–2748

Vamvakas, S., Elfarra, A.A., Dekant, W., Henschler, D. & Anders, M.W. (1988) Mutagenicity of amino acid and glutathione *S*-conjugates in the Ames test. *Mutat. Res.*, **206**, 83–90

Vamvakas, S., Dekant. W. & Henschler, D. (1989) Assessment of unscheduled DNA synthesis in a cultured line of renal epithelial cells exposed to cysteine S-conjugates of haloalkenes and haloalkanes. *Mutat. Res.*, **222**, 329–335

Vamvakas, S., Bittner, D., Dekant, W. & Anders, M.W. (1992) Events that precede and that follow S-(1,2-dichlorovinyl)-L-cysteine-induced release of mitochondrial Ca^{2+} and their association with cytotoxicity to renal cells. *Biochem. Pharmacol.*, **44**, 1131–1138

Vamvakas, S., Bittner, D. & Köster, U. (1993) Enhanced expression of the protooncongenes c-*myc* and c-*fos* in normal and malignant renal growth. *Toxicol. Lett.*, **67**, 161–172

Vandervort, R. & Polakoff, A. (1973) *Dunham-Bush, Inc., West Hartford, CN* (Health Hazard Evaluation/Toxicity Determination Report No. 72-84-31; US NTIS PB-229-627), Cincinnati, OH, United States National Institute for Occupational Safety and Health

Van Duuren, B.L., Goldschmidt, B.M., Loewengart, G., Smith, A.C., Melchionne, S., Seidman, I. & Roth, D. (1979) Carcinogenicity of halogenated olefinic and aliphatic hydrocarbons in mice. *J. natl Cancer Inst.*, **63**, 1433–1439

Van Duuren, B.L., Kline, S.A., Melchionne, S. & Seidman, I. (1983) Chemical structure and carcinogenicity relationships of some chloroalkene oxides and their parent olefins. *Cancer Res.*, **43**, 159–162

Vartiainen, T., Pukkala, E., Rienoja, T., Strandman, T. & Kaksonen, K. (1993) Population exposure to tri- and tetrachloroethene and cancer risk: two cases of drinking water pollution. *Chemosphere*, **27**, 1171–1181

Vernot, E.H., MacEwen, J.D., Haun, C.C. & Kinkead, E.R. (1977) Acute toxicity and skin corrosion data for some organic and inorganic compounds and aqueous solutions. *Toxicol. appl. Pharmacol.*, **42**, 417–423

Walles, S.A.S. (1986) Induction of single-strand breaks in DNA of mice by trichloroethylene and tetrachloroethylene. *Toxicol. Lett.*, **31**, 31–35

Waskell, L. (1978) A study of the mutagenicity of anesthetics and their metabolites. *Mutat. Res.*, **57**, 141–153

Weast, R.C. & Astle, M.J. (1985) *CRC Handbook of Data on Organic Compounds*, Vols I & II, Boca Raton, FL, CRC Press Inc., pp. 627(I), 592(II)

Westergren, I., Kjellstrand, P., Linder, L.E. & Johansson, B.B. (1984) Reduction of brain specific gravity in mice prenatally exposed to trichloroethylene. *Toxicol. Lett.*, **23**, 223–226

White, A.E., Takehisa, S., Eger, E.I., II, Wolff, S. & Sterens, W.C. (1979) Sister chromatid exchanges induced by inhaled anesthetics. *Anesthesiology*, **50**, 426–430

WHO (1985) *Trichloroethylene* (Environmental Health Criteria 50), Geneva

WHO (1993) *Guidelines for Drinking-water Quality*, Vol. 1, *Recommendations*, 2nd Ed., Geneva, pp. 62–63, 175

Williams, G.M., Mori, H. & McQueen, C.A. (1989) Structure-activity relationships in the rat hepatocyte DNA-repair test for 300 chemicals. *Mutat. Res.*, **221**, 263–286

Windham, G.C., Shusterman, D., Swan, S.H., Fenster, L. & Eskenazi, B. (1991) Exposure to organic solvents and adverse pregnancy outcome. *Am. J. ind. Med.*, **20**, 241–259

Withey, J.R. & Karpinski, K. (1985) The fetal distribution of some aliphatic chlorinated hydrocarbons in the rat after vapor phase exposure. *Biol. Res. Pregnancy Perinatol.*, **6**, 79–88

Wright, P.F.A & Stacey, N.H. (1991) A species/strain comparison of hepatic natural lymphocytotoxic activities in rats and mice. *Carcinogenesis*, **12**, 1365–1370

Wright, P.F.A., Thomas, W.D. & Stacey, N.H. (1991) Effects of trichloroethylene on hepatic and splenic lymphocytotoxic activities in rodents. *Toxicology*, **70**, 231–242

Wright, P.F.A., Schlichting, L.M. & Stacey, N.H. (1994) Effects of chlorinated solvents on the natural lymphocytotoxic activities of human liver immune cells. *Toxicol. in vitro*, **8**, 1037–1039

Zenick, H., Blackburn, K., Hope, E., Richdale, N. & Smith, M.K. (1984) Effects of trichloroethylene exposure on male reproductive function in rats. *Toxicology*, **31**, 237–250

Ziglio, G., Fara, G.M., Beltramelli, G. & Pregliasco, F. (1983) Human environmental exposure to trichloro- and tetrachloroethylene from water and air in Milan, Italy. *Arch. environ. Contam. Toxicol.*, **12**, 57–64

Ziglio, G., Beltramelli, G., Pregliasco, F., Arosio, D. & De Donato, S. (1984a) Ambient exposure to chlorinated solvents of the student population of a commune in northern Italy. *Ig. Mod.*, **82**, 133–161 (in Italian)

Ziglio, G., Beltramelli, G. & Giovanardi, A. (1984b) Occurrence of halogenated organic compounds in drinking-water of some cities of northern Italy. *Ig. Mod.*, **82**, 419–435 (in Italian)

Ziglio, G., Beltramelli, G., Pregliasco, F. & Mazzocchi, M.A. (1984c) Exposure to trichloroethylene through drinking-water of some families of Porto Mantovano (Mn). *Ig. Mod.*, **82**, 591–605 (in Italian)

Zoeteman, B.C.J., Harmsen, K., Linders, J.B.H.J., Morra, C.F.H. & Slooff, W. (1980) Persistent organic pollutants in river water and ground water of the Netherlands. *Chemosphere*, **9**, 231–249

TETRACHLOROETHYLENE

This substance was considered by previous working groups, in June 1978 and March 1987 (IARC, 1979, 1987). Since that time, new data have become available, and these have been incorporated into the monograph and taken into consideration in the present evaluation.

1. Exposure Data

1.1 Chemical and physical data

1.1.1 Nomenclature

Chem. Abstr. Serv. Reg. No.: 127-18-4
Chem. Abstr. Name: Tetrachloroethene
IUPAC Systematic Name: Tetrachloroethylene
Synonyms: Ethylene tetrachloride; PCE; 'per'; PER; perchlorethylene; perchloroethene; perchloroethylene; tetrachlorethylene; 1,1,2,2-tetrachloroethene; 1,1,2,2-tetrachloroethylene

1.1.2 Structural and molecular formulae and relative molecular mass

$$\begin{array}{c}Cl\\ \diagdown\\ \diagup\end{array}C=C\begin{array}{c}Cl\\ \diagup\\ \diagdown\end{array}$$

C_2Cl_4 Relative molecular mass: 165.83

1.1.3 Chemical and physical properties of the pure substance

(a) *Description*: Colourless liquid with an ether-like odour (Budavari, 1989)
(b) *Boiling-point*: 121 °C (Lide, 1993)
(c) *Melting-point*: −19 °C (Lide, 1993)
(d) *Density*: 1.6227 at 20 °C/4 °C (Lide, 1993)
(e) *Spectroscopy data*: Infrared (prism [5422]; grating [469]), ultraviolet [1485] and mass [1053] spectral data have been reported (Sadtler Research Laboratories, 1980; Weast & Astle, 1985).
(f) *Solubility*: Slightly soluble in water (0.15 g/L at 25 °C) (PPG Industries, Inc., 1994); soluble in ethanol, diethyl ether and benzene (Lide, 1993)
(g) *Volatility*: Vapour pressure, 9.975 mm Hg [1.33 kPa] at 13.8 °C (Hickman, 1993)

(h) *Stability*: Photooxidized in air with sunlight (half-time, about 12 h), giving phosgene and trichloroacetyl chloride (Gilbert *et al.*, 1982; Hickman, 1993)

(i) *Reactivity:* Incompatible with chemically active metals such as barium, lithium and beryllium and with caustic soda, sodium hydroxide, potash and strong oxidizers (United States National Institute for Occupational Safety and Health, 1994a)

(j) *Octanol/water partition coefficient (P)*: log P, 3.40 (Hansch *et al.*, 1995)

(k) *Conversion factor*: $mg/m^3 = 6.78 \times ppm$[1]

1.1.4 Technical products and impurities

Commercial grades of tetrachloroethylene available in the United States include a vapour degreasing grade, a dry cleaning grade, a technical or industrial grade for use in formulations, a high-purity, low-residue grade and a grade specifically formulated for use as a transformer fluid (Hickman, 1993). Isomerization and fluorocarbon grades have purities of 99.995% (Vulcan Chemicals, 1994). The various grades differ in the amount and type of added stabilizers. Stabilizers that have been used include amines, phenols and epoxides [not specified] in various combinations at levels of 0.01–0.35% (Gilbert *et al.*, 1982). Commercial grades should not contain more than 50 ppm [mg/L] water, 0.0005 wt% acidity (as hydrochloric acid) or 0.001 wt% insoluble residue (Hickman, 1993; Vulcan Chemicals, 1994; PPG Industries, Inc., 1994).

Trade names for tetrachloroethylene include Ankilostin, Antisol 1, Didakene, Dilatin PT, Fedal-Un, Nema, Perchlor, Perclene, PerSec, Tetlen, Tetracap, Tetraleno, Tetroguer and Tetropil.

1.1.5 Analysis

Selected methods for the analysis of tetrachloroethylene in various matrices are identified in Table 1. Several methods for the analysis of tetrachloroethylene in air, solids, liquids, water, food, blood and breath were reviewed by WHO (1984) and the United States Environmental Protection Agency (1985a).

Three gas chromatography/mass spectrometry (GC/MS) and three purge-and-trap GC methods for purgeable organics, including tetrachloroethylene, are typically used for aqueous samples (see also Table 1). The first (EPA Method 624 and APHA/AWWA/WEF Method 6210B) is a packed-column method useful for the determination of tetrachloroethylene in municipal and industrial wastes. A similar purge-and-trap method (EPA Method 503.1 and APHA/AWWA/WEF Method 6220C), which involves photoionization detection, is applicable for the determination of tetrachloroethylene in drinking-water and raw source water. The second method (EPA Method 524.1 and APHA/ AWWA/WEF Method 6210C), which is also a packed-column method, is applicable for the determination of tetrachloroethylene in drinking-water and raw source water. Similar purge-and-trap methods (EPA Methods 601 and 502.1 and APHA/-AWWA/WEF Methods 6230B and 6230C), which involve electrolytic conductivity or microcoulometric detection, are applicable for the determination of tetrachloroethylene in municipal

---

[1]Calculated from: mg/m^3 = (relative molecular mass/24.45) × ppm, assuming normal temperature (25 °C) and pressure (101 kPa)

and industrial discharges (6230B) and in drinking-water and raw source water (6230C). The third group of GC/MS and purge-and-trap methods (EPA Method 524.2 and APHA/AWWA/WEF Method 6210D; EPA Method 502.2 and APHA/AWWA/WEF Method 6230D) are identical to the previous ones, except that a capillary column is used. The second and third methods are intended primarily for the detection of large numbers of contaminants at low concentrations, which are not detectable with the first method (Greenberg et al., 1992).

Table 1. Methods for the analysis of tetrachloroethylene

Sample matrix	Sample preparation	Assay procedure	Limit of detection	Reference
Air	Adsorb on charcoal; desorb with carbon disulfide	GC/FID	0.01 mg	Eller (1994)
	Draw air through Tenax sample tube; heat; desorb on cold trap	GC/MS	20 ng	US Environmental Protection Agency (1988a)
	Draw air into cryogenically cooled trap; heat	GC/FID or GC/EC	1–5 ng	US Environmental Protection Agency (1988a)
	Draw air into SUMMA® passivated stainless-steel canister; desorb on cold trap	GC/MS or GC/EC-FID-PID	NR	US Environmental Protection Agency (1988a)
Water	Purge (inert gas); trap on suitable sorbent material; desorb as vapour onto packed gas chromatographic column	GC/ECD or GC/MCD	0.001–0.03 µg/L	US Environmental Protection Agency (1988b, 1994)
		GC/MS	0.3–1.9 µg/L	
	Purge and trap as above; desorb as vapour into capillary gas chromatographic column	GC/PID-ECD	0.02–0.05 µg/L	US Environmental Protection Agency (1988b, 1994)
		GC/MS	0.05–0.14 µg/L	
	Purge (inert gas); trap on suitable sorbent material; desorb as vapour onto gas chromatographic column	GC/PID	0.01 µg/L	US Environmental Protection Agency (1988b, 1994)
	Add internal standard (isotope-labelled tetrachloroethylene); purge, trap and desorb as above	GC/MS	10 µg/L	US Environmental Protection Agency (1994)
Liquid and solid wastes	Purge (inert gas); trap on suitable sorbent material; desorb as vapour onto packed gas chromatographic column	GC/ECD	0.03 µg/L	US Environmental Protection Agency (1986a)
		GC/MS	5 µg/L (groundwater)[a] 5 µg/kg (soil/sediment)[a] 250–2500 µg/kg (liquid wastes)[a]	US Environmental Protection Agency (1986b)

Table 1 (contd)

Sample matrix	Sample preparation	Assay procedure	Limit of detection	Reference
Food	Purge (inert gas) in water at 100 °C; trap on Tenax TA and XAD-4; elute with hexane	GC/ECD	0.4 ppb [µg/kg]	Heikes & Hopper, 1986; Heikes, 1987

GC, gas chromatography; FID, flame ionization detection; MS, mass spectrometry; EC, electron capture detection; PID, photoionization detection; NR, not reported; ECD, electrolytic conductivity detection; MCD, microcoulometric detection
^a Practical quantification limit

The DuPont Pro-Tek® Organic Vapor Monitoring Badge has been used extensively in the dry cleaning industry to monitor time-weighted average exposures to tetrachloroethylene. The badge adsorbs vapours onto charcoal for subsequent gas chromatographic determination (International Fabricare Institute, 1987; Solet et al., 1990). Good correlations have been reported between time-weighted average concentrations evaluated with passive dosimeters and conventional sampling pumps and concentrations in blood (Schaller & Triebig, 1987).

1.2 Production and use

1.2.1 Production

Tetrachloroethylene was first prepared in 1821 by Faraday by thermal decomposition of hexachloroethane (Hickman, 1993). Many commercial processes have since been developed for the production of tetrachloroethylene. Either single or multiple products result.

In Europe, tetrachloroethylene is produced by oxychlorination of C_2 chlorinated hydrocarbons and by chlorination of C_1–C_3 hydrocarbons or their partially chlorinated derivatives. One company in Japan manufactures tetrachloroethylene by the chlorination of ethylene followed by thermal dehydrogenation (Linak et al., 1992).

Between 1925 and the 1970s, the main commercial process for producing tetrachloroethylene in the United States involved the chlorination of acetylene to form tetrachloroethane, followed by dehydrochlorination to trichloroethylene (see monograph, this volume); the trichloroethylene was further chlorinated to pentachloroethane and dehydrochlorinated to form tetrachloroethylene. Owing partly to the high cost of acetylene, most tetrachloroethylene is now produced by chlorination and oxychlorination of other hydrocarbons (C_1–C_3) or chlorinated compounds. The raw materials include ethylene dichloride, methane, ethane, propane, propylene (see IARC, 1994a), propylene dichloride and various other chlorinated hydrocarbons (Linak et al., 1992; Hickman, 1993).

Production of tetrachloroethylene has been declining. Table 2 shows the production of tetrachloroethylene in selected countries between 1941 and 1991. In 1992, annual capacity was 10 thousand tonnes in Austria, 30 thousand tonnes in Belgium, 62 thousand tonnes in France, 100 thousand tonnes in Germany and in Italy, 21 thousand tonnes in Spain and 130 thousand tonnes in the United Kingdom. Tetrachloroethylene has been produced commercially in Japan since 1972; in 1992, seven plants in Japan had a capacity of 96 thousand tonnes. The only producer of tetrachloroethylene in Canada ceased production in 1992. Four companies in the United States had an annual capacity of 223 thousand tonnes of tetrachloroethylene (Linak *et al.*, 1992).

Table 2. Production of tetrachloroethylene in selected countries (thousand tonnes)

Year	Western Europe	Japan	USA
1941			5
1945			27
1955			81[a]
1960			95
1965			195
1970		NR	320
1975		48	308
1980	295	64	347
1981	290	57	313
1982	295	60	265
1983	300	61	248
1984	315	63	260
1985	315	71	225
1986	341	70	188
1987	323	84	215
1988	343	97	226
1989	317	91	218
1990	280	83	169
1991	NR	NR	109[b]

From Linak *et al.* (1992); NR, not reported
[a] From Su & Goldberg (1976)
[b] Preliminary

Tetrachloroethylene is also produced in Australia, Brazil, the Czech Republic, China, India, Mexico, Poland, Romania, the Russian Federation and South Africa (Chemical Information Services, Inc, 1994).

1.2.2 Use

Tetrachloroethylene is commercially important as a chlorinated solvent and as a chemical intermediate. In 1990, about 53% of world demand for tetrachloroethylene was for dry cleaning, and it was the cleaning fluid used by about 75% of all dry cleaners. About 23% was used as a chemical intermediate, principally for making 1,1,2-trichloro-1,2,2-trifluoroethane (CFC-113); 13% was used for metal cleaning and 11% for other uses (Linak *et al.*, 1992).

Tables 3–5 present the uses of tetrachloroethylene in western Europe, Japan and the United States (Linak et al., 1992). Countries in the southern hemisphere are reported to receive only 1–2% of worldwide sales of tetrachloroethylene (Coopers & Lybrand, Ltd, 1993).

Table 3. Use of tetrachloroethylene in western Europe (thousand tonnes)

Year	Metal cleaning (vapour degreasing)	Metal cleaning (cold cleaning)	Dry cleaning	Precursor[a]	Other
1980	71	10	150	34	20
1984	61	5	133	36	15
1987	50	5	122	65	15
1990	45	5	115	60	10

From Linak et al. (1992); estimates
[a] Almost exclusively for production of 1,1,2-trichloro-1,2,2-trifluoroethane (CFC-113), dichlorotetrafluoroethane (CFC-114) and chloropentafluoroethane (CFC-115)

Table 4. Use of tetrachloroethylene in Japan (thousand tonnes)

Year	Metal cleaning	Dry cleaning	Other[a]
1980	10	26	20
1983	12	23	29
1987	11	25	63
1990	13	20	69

From Linak et al. (1992); estimates
[a] Almost exclusively for production of CFC-113, CFC-114 and CFC-115

Table 5. Use of tetrachloroethylene in the United States (thousand tonnes)

Year	Metal cleaning	Dry cleaning and textile processing	Precursor[a]	Other
1971	50	163	32	38
1974	54	193	39	45
1977	59	181	39	28
1980	45	172	50	59
1984	34	136	70	39
1987	27	127	84	14
1990	16	111	45	6

From Linak et al. (1992); estimates
[a] Almost exclusively for production of CFC-113, CFC-114 and CFC-115

In addition to its widespread use in dry cleaning, tetrachloroethylene has been used in the textile industry as a scouring solvent, for removing lubricants, oil and grime from fabrics, and as a carrier solvent for fabric dyes and finishes, as a water repellant and for sizing and desizing textiles. Tetrachloroethylene dissolves fats, greases, waxes and oils in fabric without harming the fibres (United States Agency for Toxic Substances and Disease Registry, 1990).

Another major use of tetrachloroethylene was as a chemical intermediate in the manufacture of two-carbon chlorofluorocarbons (CFCs), principally CFC-113. As a result of international agreements to protect against depletion of the ozone layer, this use is rapidly being phased out (Linak *et al.*, 1992).

Tetrachloroethylene is also used for vapour and liquid degreasing and for cold cleaning of metals (Keil, 1979). In vapour degreasing, the part to be cleaned is placed in a zone of warm solvent vapour. The vapour condenses on the metal and subjects the surface to a solvent-flushing action. The liquid drops are collected in a reservoir and revaporized (Schneberger, 1981).

Tetrachloroethylene is also used in paint removers, printing inks, adhesive formulations, paper coatings, leather treatments and as a carrier solvent for silicones. It is used in aerosol formulations, e.g. water repellants, automotive cleaners, solvent soaps, silicone lubricants and spot removers. Tetrachloroethylene has also reportedly been used to remove soot from industrial boilers. Other reported uses are as an insulating fluid for power transformers, as a heat transfer medium, as an extractant in the pharmaceutical industry and as a pesticide and chemical intermediate. It has been reported to be used as an anthelminthic in the treatment of hookworm and some trematode infestations (Budavari, 1989; United States Agency for Toxic Substances and Disease Registry, 1990).

In a study of potential sources of indoor air pollution in the United States, 63 of 1159 (5.4%) common household products were found to contain tetrachloroethylene (Sack *et al.*, 1992).

1.3 Occurrence

1.3.1 Natural occurrence

Natural production of tetrachloroethylene has been reported in temperate, subtropical and tropical algae and in one red microalga (Abrahamsson *et al.*, 1995).

1.3.2 Occupational exposure

The National Occupational Exposure Survey conducted between 1981 and 1983 by the United States National Institute for Occupational Safety and Health (1994b) indicated that about 566 000 employees in 42 700 plants were potentially exposed to tetrachloroethylene. The estimate was based on a survey of United States companies and did not involve actual measurements of exposures. An independent estimate prepared by industry in 1994 indicated that about 450 000 workers in the United States may be exposed (Center for Emissions Control, 1994).

There is considerable potential exposure to tetrachloroethylene, by both skin contact and inhalation, during its use in dry cleaning and degreasing (Hake & Stewart, 1977). Acute cases of intoxication due to inhalation of tetrachloroethylene have been reported in the literature (Foot *et al.*, 1943; Coler & Rossmiller, 1953; Lob, 1957; Stewart *et al.*, 1961a; Dumortier *et al.*, 1964; Meckler & Phelps, 1966; Gold, 1969; Morgan, 1969; Stewart, 1969; Lackore & Perkins, 1970; Patel *et al.*, 1973; see also section 4.2.1).

Table 6 presents a range of exposures to tetrachloroethylene by occupation in the United States. Dry cleaning is specifically excluded, as occupational exposures in dry cleaning are covered in a separate monograph in this volume. Occupational exposure in dry cleaning was generally to 350–700 mg/m^3 in the 1970s and to 70–350 mg/m^3 in the late 1980s.

Table 6. Occupational exposures to tetrachloroethylene in the United States

No. of plants	Job, task or industry	No. of samples	Air concentration [mg/m^3]		Reference
			Mean	Range	
1	Degreasing, auto industry	Short-term	NR	[1573–2610]	Coler & Rossmiller (1953)
1	Urethane foam	3 (A)	0.6	0.344–0.714	Costello (1979)
1	Protective coatings	11	[2.7]	[ND–27]	Burroughs (1980)
1	Polyether urethane foam, car industry	3 (A) 9 (P)	2.1 4.2	0.4–4.2 < 0.03–8.0	White & Wegman (1978)
1	Degreasing, medical equipment	3 (A)	[106]	[48–197]	Tharr & Donahue (1980)
1	Degreasing	6 (P)	[271]	[34–1180]	Burgess (1981)
1	Degreasing, printing plates	4 (A) 2 (P)	78 57	25–99.3 28–86.4	Pryor (1977)
1	Cutlery manufacture, blade degreasing	2 (P)	[115]	[104–126]	Center for Chemical Hazard Assessment (1985)
1	Filling aerosol cans with carburettor cleaner	30 (A) 30 (P)	[311] [403]	[31–1248] [104–1010]	Hervin *et al.* (1972)
1	Coal testing laboratory	1 (P) Several (A)	[1010]	[746–1315]	Jankovic (1980)
1	Automotive brake manufacture	11 (P) 11 (A)	103 145	10–350 10–350	Hervin & Lucas (1973)
8	Degreasing	14 24	[1220] (work in) [3282] (work out)	[170–2102] [542–12204]	Crowley *et al.* (1945)
60	Degreasing (11 non-condensing machines	68	[1498]	[163–5966]	Morse & Goldberg (1943)
1	Specialty packaging	4 (P) 5 (A)	[4] [15]	[1.4–5.4] [1.8–41]	Hanley (1993)
1	Rubber moulding	15 (P) 1 (A)	[17.6] [8]	[ND–36]	Cook & Parker (1994)
1	Plating, degreasing	1 (P) 1 (A)	11 2		Abundo *et al.* (1994)
14	Motion picture film processing	119 (P) 51 (A)	189 111	2.7–1606 2.2–965	Mosely (1980)

Table 6 (contd)

No. of plants	Job, task or industry	No. of samples	Air concentration [mg/m³]		Reference
			Mean	Range	
1	Electroplating	5 (P)	[753]	[557–1253]	Daniels & Kramkowski (1986)
1	Degreasing, foundry	1 (P)	86.1		Hartle & Aw
		3 (A)	130.3	40.9–250	(1984)
1	Spray painting	9 (P)	21.4	4.4–50	Hartle & Aw (1983)
1	Automative parts, fasteners	2 (A)	1.3	1.1–1.5	Ahrenholz & Anderson (1982)
1	Motion picture film processing	4 (A)	9.5	6.5–11.3	Okawa & Coye
		7 (P)	16.4	7.8–54.5	(1982)
1	Graphic arts	4 (P)	13	0.01–30	Love (1982)
1	Painters, power plant	6 (P)	0.13	< 0.01–0.46	Chrostek & Levine
		2 (A)	0.29	< 0.01–0.88	(1981)
1	Taxidermy	9 (P)	403	0.01–1546	Gunter & Lybarger (1979)

NR, not reported; ND, not detected; A, area air sample; P, personal air sample (breathing zone)

In measurements made in Finland in 1982–85, 13 area samples taken in five plants for the manufacture of printing plates contained a mean of 9.6 ppm [65.1 mg/m³] tetrachloroethylene, with a range of 3–19 ppm [20.3–129 mg/m³]. The mean concentration in four samples taken in three plants where workers degreased metal parts was 3 ppm [20 mg/m³], with a range of 1.8–5 ppm [12.2–33.9 mg/m³] (Rantala et al., 1992).

1.3.3 Environmental occurrence

Tetrachloroethylene has been reported in air, rainwater, surface water, drinking-water, seawater, marine sediments, marine invertebrates, fish, waterbirds, marine mammals, foods and human tissues (McConnell et al., 1975). Human exposure to tetrachloroethylene has been estimated by modelling multiple exposure pathways (McKone & Daniels, 1991).

(a) Air

Concentrations of tetrachloroethylene in air have been reported in numerous studies. Table 7 presents levels measured in remote, rural, urban and suburban locations worldwide. In a compilation of the results of surveys of ambient air in the United States before 1981 (Brodzinsky & Singh, 1983; United States Agency for Toxic Substances and Disease Registry, 1990), representing 2553 monitoring points, the mean concentrations were 160 ppt [1.1 µg/m³] in rural areas, 790 ppt [5.4 µg/m³] in urban communities and 1300 ppt [8.8 µg/m³] in areas near sources of tetrachloroethylene emissions. A similar study, conducted by the United States Environmental Protection Agency (1985a), estimated the average ambient levels to be 40 ppt [0.3 µg/m³] in the northern hemisphere and 12 ppt [0.08 µg/m³] in the southern hemisphere. The average urban level was reported to be 800 ppt [5.4 µg/m^3]. According to the United States Environmental Protection Agency (1993) Toxic Chemical Release Inventory, industrial releases of tetrachloroethylene to the environment were 16 335 tonnes in 1988, 12 527 tonnes in 1989, 10 183 tonnes in 1990 and 7596 tonnes in 1991; however, total emissions to the atmosphere, including those from dry cleaning, were estimated to be 87 000 tonnes in 1991 (United States Environmental Protection Agency, 1991)

Table 7. Concentrations of tetrachloroethylene in air

Area	Concentration [ng/m^3]		Reference
	Mean	Range	
Remote			
Atlantic Ocean			
Northern hemisphere	[230]		Penkett (1982)
Northern hemisphere	[197]		Singh et al. (1983)
Southern hemisphere	[34]		
Northern hemisphere	[380]		Rasmussen & Khalil (1982)
Southern hemisphere	[95]		
Northern hemisphere		[102–203]	Class & Ballschmiter (1986)
Southern hemisphere		[14–68]	
Northern hemisphere	[102]		Class & Ballschmiter (1987)
Southern hemisphere	[14]		
Northern hemisphere	[88]		European Centre for Ecotoxicology and Toxicology of Chemicals (1995)
Southern hemisphere	[20]		
Pacific Ocean			European Centre for Ecotoxicology and Toxicology of Chemicals (1995)
Northern hemisphere	[54]		
Southern hemisphere	[20]		
Rural			
Southern Washington, USA	[136]		Grimsrud & Rasmussen (1975)
Western Ireland	[190]		Lovelock (1974)
Rural California, USA	[210]		Singh et al. (1977)
Central Michigan, USA		200–300	Russell & Shadoff (1977)
Southern Germany		[197–1763]	European Centre for Ecotoxicology and Toxicology of Chemicals (1995)
Netherlands		[136–407]	European Centre for Ecotoxicology and Toxicology of Chemicals (1995)
Brittany, France		[136–183]	European Centre for Ecotoxicology and Toxicology of Chemicals (1995)
USA, 577 sites	[1085]		European Centre for Ecotoxicology and Toxicology of Chemicals (1995)

Table 7 (contd)

Area	Concentration [ng/m³]		Reference
	Mean	Range	
Urban and suburban			
Germany, 92 sites	6600		Bauer (1991)
Tokyo, Japan	[8136]		Ohta et al. (1976)
Brussels, Belgium		[6441–21 700]	Su & Goldberg (1976)
Geneva, Switzerland	[46 104]		Su & Goldberg (1976)
Russian Federation		[203–1356]	Su & Goldberg (1976)
Paris, France	[2102]		Su & Goldberg (1976)
Grenoble, France	[9153]	[6507–12 204]	Su & Goldberg (1976)
Kyoto, Japan	[9492]		Su & Goldberg (1976)
Tokyo, Japan	[15 594]		Su & Goldberg (1976)
New York City, NY, USA	[9017]	[1085–71 936]	Evans et al. (1979)
Houston, TX, USA	[2644]	[< 678–30 510]	Evans et al. (1979)
Detroit, MI, USA	[3119]	[678–14 916]	Evans et al. (1979)
Los Angeles, CA, USA	[10 034]	[1180–14 001]	Singh et al. (1981)
Phoenix, AZ, USA	[6739]	[875–25 066]	Singh et al. (1981)
Oakland, CA, USA	[2054]	[359–9831]	Singh et al. (1981)
Bochum, Germany	6100	1100–67 000	Bauer (1981)
Frankfurt, Germany		700–1800	Bauer (1981)
Munich, Germany		< 1000–25 000	Bauer (1981)
Milan, Italy (polluted)		9500–14 800	Ziglio et al. (1983)
Yokohama and Kawasaki, Japan		[2712–4543]	Urano et al. (1988)
Turin, Italy		[4746–13 357]	European Centre for Ecotoxicology and Toxicology of Chemicals (1995)
Fonte, Portugal		[197–949]	European Centre for Ecotoxicology and Toxicology of Chemicals (1995)
San Diego, CA, USA	[1831]		Douglas et al. (1993)
San Francisco, CA, USA	[1559]		Douglas et al. (1993)
Sacramento, CA, USA	[475]		Douglas et al. (1993)

The concentrations of tetrachloroethylene in New Jersey (United States) were 6 µg/m³ in the autumn of 1981, 6.2 µg/m³ in the summer of 1982 and 4.2 µg/m³ in the winter of 1983. The mean indoor air concentrations were 45, 11 and 28 µg/m³, respectively. The mean outdoor air level in Los Angeles, CA, in February 1984 was 10 µg/m³ (Wallace, 1986), and the geometric mean concentrations of tetrachloroethylene in three cities in New Jersey were 0.45 ppb [3.05 µg/m³] (n = 38), 0.31 ppb [2.10 µg/m³] (n = 37) and 0.24 ppb [1.63 µg/m³] (n = 35) in summer 1981 (Harkov et al., 1984).

(b) Occurrence in air near dry cleaning shops

In 1990 in New York State, the levels of tetrachloroethylene in 12-h samples of air inside apartments located above dry cleaners were 300–55 000 µg/m³ (mean, 13 000 µg/m³) for morning samples, compared with < 6.7–103 µg/m³ (mean, 28 µg/m³) in control apartments. Afternoon 12-h samples contained 100–36 500 µg/m³ (mean, 10 000 µg/m³), while those from control apartments contained < 6.7–77 µg/m³. The concentrations of tetrachloroethylene in apartments above dry cleaners using transfer machines were 1730–17 000 µg/m³ (mean, 7500 µg/m³) in the morning and 1350–14 000 (7900) in the afternoon; those in apartments above dry-to-dry machines were 300–440 µg/m³ (mean, 370 µg/m³) in the morning and 100–160 (130) in the afternoon. The ambient morning levels outside the apartments were 195–2600 µg/m³ (mean, 1000 µg/m³), and those in the afternoon were 66–1400 µg/m³ (mean, 580 µg/m³). The ambient levels outside the control apartments were < 6.7–21 µg/m³ (mean, 8.4 µg/m³) in the morning and < 6.7–6.9 µg/m³ (mean, 3.9 µg/m³) in the afternoon (Schreiber *et al.*, 1993).

Levels as high as 197 mg/m³ were detected in an apartment next to a dry cleaning facility that operated transfer machines in New York State in 1990. In an apartment 30 feet [9.1 m] from a dry cleaners, the level was 1.5 mg/m³. Shops in a mall where there was a dry cleaning establishment had levels as high as 50.4 mg/m³. The air levels were 1.5 mg/m³ in a fish market in the mall, 34.5 mg/m³ in a pizza parlour and 15.7 mg/m³ in a delicatessen (New York Department of Health, 1994).

Tetrachloroethylene was determined in the alveolar air of 136 people living near 12 dry cleaning shops in the Netherlands. The geometric mean concentration in the breath of residents was 5 mg/m³ in apartments above the dry cleaning shops, 1 mg/m³ in houses next door to the shops, 0.2 mg/m³ in the next house and < 0.1 mg/m³ in a house across the street from the shop. The mean alveolar air concentration of 18 employees in the dry cleaning shops was 73 mg/m³ (Verberk & Scheffers, 1980).

In an apartment above a dry cleaning shop in Switzerland, 0.1 mg/m³ tetrachloroethylene and 1 mg/m³ CFC-113 were measured. At the counter of the shop, the concentrations were 10 mg/m³ tetrachloroethylene and 150 mg/m³ CFC-113, while in the room where the dry cleaning took place, 25 mg/m³ tetrachloroethylene and 1000 mg/m³ CFC-113 were measured (Grob *et al.*, 1991).

The concentrations of tetrachloroethylene in the air of five apartments in the vicinity of dry cleaners in Germany in 1987 were 17.1–2296 µg/m³; the concentrations outside the apartments were 3.07–138 µg/m³ (Reinhard *et al.*, 1989).

Elevated levels of tetrachloroethylene have been measured in the homes of dry cleaning workers that were not near dry cleaning shops. For example, the levels in six homes in New Jersey were 21–560 µg/m³ (European Centre for Ecotoxicology and Toxicology of Chemicals, 1995), and those in 25 homes in Italy were 25–9600 µg/m³ (Aggazzotti *et al.*, 1994). The concentrations were suggested to result from the exhaled breath and clothing of the workers (Thompson & Evans, 1993).

Off-gassing of tetrachloroethylene from dry cleaned clothes has also been studied. A concentration of 9.3 mg/m³ was measured 2 min after deposition into a private car of a cleaned down-filled jacket, which increased to 24.8 mg/m³ 108 min after deposition (Gulyas &

Hemmerling, 1990). Levels of 70–2100 mg/m^3 were reported inside a car containing dry cleaned clothes in another study (Jensen & Ingvordsen, 1977).

The concentrations of tetrachloroethylene in the air of four homes in Japan where there were freshly dry cleaned clothes were 1.3–7.4 µg/m^3 (mean, 2.6 µg/m^3). The levels outside the homes were 0.3–1.6 µg/m^3 (mean, 1.2 µg/m^3) (Kawauchi & Nishiyama, 1989). Levels up to 300 µg/m^3 were measured at seven of nine houses in New Jersey when freshly dry cleaned clothes were brought home (Thomas *et al.*, 1991).

(c) Water

Tetrachloroethylene occurs ubiquitously at low levels in water supplies and frequently occurs as a contaminant of groundwater, owing to its widespread use and physical characteristics. Table 8 presents measurements of tetrachloroethylene in surface waters, groundwater and drinking-water.

Table 8. Concentrations of tetrachloroethylene in water

Location (year)	Concentration (µg/L)		Reference
	Mean	Range	
Surface waters			
Seawater			
Eastern Pacific		0.0001–0.0021	Singh *et al.* (1983)
North Atlantic		0.00012–0.0008	Pearson & McConnell (1975); Murray & Riley (1973)
Coastal waters			
Sweden, coast	0.007		European Centre for Ecotoxicology and Toxicology of Chemicals (1995)
Greece, northern coast		0.27–3	Fytianos *et al.* (1985)
Jackfish Bay, Canada		2.1–190	Comba *et al.* (1994)
Birmingham, United Kingdom		0.02–500	Rivett *et al.* (1990); Burston *et al.* (1993)
Rainwater			
Los Angeles, CA, 1982	0.021		Kawamura & Kaplan (1983)
La Jolla, CA	0.006		Su & Goldberg (1976)
Portland, OR		0.0008–0.009	Ligocki *et al.* (1985)
Germany	0.08		European Centre for Ecotoxicology and Toxicology of Chemicals (1995)
Switzerland		< 0.010–0.115	European Centre for Ecotoxicology and Toxicology of Chemicals (1995)
Kobe, Japan	0.050		European Centre for Ecotoxicology and Toxicology of Chemicals (1995)

Table 8 (contd)

Location (year)	Concentration (µg/L)		Reference
	Mean	Range	
Rivers			
USA, five states, surface water (14% of samples positive)		max, 21	Dyksen & Hess (1982)
Rhine, Germany		0.2–78	Dietz & Traud (1973); Bauer (1978); Bauer (1981); Hellmann (1984)
Elbe		0.2–9.9	Hellmann (1984)
Mosel		0.3–1.3	Hellmann (1984)
Rhine, Switzerland	~1		Zürcher & Giger (1976); Zoeteman et al. (1980)
Lake Zurich		0.07–1.4	Zürcher & Giger (1976); Schwarzenbach et al. (1979)
Danube	0.6		Bolzer (1982)
Netherlands		0.1–1.5	Herbert et al. (1986)
Drinking-water			
Zagreb, Croatia		0.36–7.8	Skender et al. (1993)
China, southern	0.2		Trussell et al. (1980)
Egypt	0.3		Trussell et al. (1980)
Australia, southern	0.1		Trussell et al. (1980)
New Jersey, USA, 1981–83	0.4		Wallace et al. (1987)
New Jersey, USA	7.7	max, 14	Cohn et al. (1994)
Woburn, MA, USA		66–212	Lagakos et al. (1986)
Germany, 1977	0.6	< 0.1–35.3	Bauer (1981)
Finland, south, 1992, two industrialized cities		≤ 200	Vartiainen et al. (1993)
Germany, 1985–86, > 90% of drinking-water supplies		51% < 0.001 40% 0.001–0.5 9% > 5	Bauer (1991)
Milan, Italy		2–68	Ziglio et al. (1985)
Groundwater			
Japan, 1983–84, 41 wells	23 samples, GM of 1.27 33 samples, GM of 0.94 35 samples, GM of 0.92		Kido et al. (1989)
USA, CA, 945 water supplies		max, 0.58–69	Westrick et al. (1984)
USA, five sites, 28% of samples positive		max, 1500	Dyksen & Hess (1982)
Netherlands, 1976		max, 22	Zoeteman et al. (1980)

GM, geometric mean

Tetrachloroethylene was reported in drinking-water in the Cape Cod, MA, region of the United States in the late 1970s (Webler & Brown, 1993). About 650 miles [1046 km] of vinyl-lined asbestos–cement water pipes were found to have contaminated the municipal water supply with tetrachloroethylene. The concentrations ranged from 1600–7750 µg/L at sites of low use to 1.5–80 µg/L in areas of medium and high use (Aschengrau et al., 1993). A peak of 18 mg/L was found in a vinyl-coated asbestos–cement water pipe in Falmouth, MA (Wakeham et al., 1980).

In a computerized database on water quality in the United States, the reported median concentrations of tetrachloroethylene in 1983–84 were 5.0 µg/L in industrial effluents (10.1% detectable, 1390 samples), 0.1 µg/L (38% detectable, 9323 samples) in ambient water, < 5.0 µg/kg dry weight (5% detectable, 359 samples) in sediment and < 50 µg/kg (7.0% detectable, 101 samples) in biota (Staples et al., 1985).

The United States Environmental Protection Agency (1985b) estimated in 1985 that 11 430 000 individuals (5.3% of the United States population using municipal water supplies) were exposed to tetrachloroethylene at concentrations ≥ 0.5 µg/L. In addition, 874 000 individuals (0.4% of the United States population) were exposed to levels > 5 µg/L.

In the National Organic Monitoring Survey in the United States in 1976–77, 113 drinking-water samples were found to contain tetrachloroethylene at concentrations of 0.2–3.1 µg/L. In the National Screening Program in 1977–81, 142 drinking-water samples were reported to contain levels from traces to 3.2 µg/L. In the Community Water Supply Survey, 452 drinking-water samples contained 0.5–30 µg/L. An aggregate of various state reports on 1652 drinking-water samples showed a range of trace levels to 3000 µg/L (Thomas, 1989).

1.3.4 Food

The concentrations in milk and meat products in Switzerland ranged from 3 to 3490 µg/kg. The total daily intake was calculated to be 160 µg/day (Zimmerli et al., 1982).

In the United Kingdom, tetrachloroethylene was found at concentrations of 0.3–13 µg/kg in dairy products, 0.9–5.0 µg/kg in meat, 7 µg/kg in margarine, 0.01–7 µg/kg in oils, 3 µg/kg in instant coffee, 3 µg/kg in tea and 0.7–2.0 µg/kg in fruits and vegetables (McConnell et al., 1975). In the United Kingdom, 81 of 98 samples of olive oil contained < 10 µg/kg and the other 17 contained 1–70 µg/kg tetrachloroethylene (Norman, 1991).

In the United States, tetrachloroethylene was found in 93 of 231 food samples (40%) at a mean concentration of 13 µg/kg (1–124 µg/kg). Residues were determined in a variety of cereals (mean, 22 µg/kg; range, 1–108 µg/kg), in corn oil (21 µg/kg), pork and beans (2 µg/kg), peas (2 µg/kg), onion rings (5 µg/kg), fried potatoes (9 µg/kg), a wide variety of baked goods (mean, 12 µg/kg; range, 3–48 µg/kg), peanut butter (3 µg/kg), pecan nuts (120 µg/kg), dairy products (mean, 9 µg/kg; range, 2–30 µg/kg), milk chocolate (20 µg/kg), meat products (mean, 13 µg/kg; range, 1–124 µg/kg), baby foods (mean, 2.5 µg/kg; range, 1–5 µg/kg), bananas (2 µg/kg), grapes (1 µg/kg) and avocados (14 µg/kg) (Daft, 1988).

The concentrations of tetrachloroethylene in fish in the United Kingdom were 0.3–11 µg/kg in tissue and 1–41 µg/kg in liver, with an average concentration in water of 0.12 µg/L. Molluscs from Liverpool Bay contained a mean of 4 µg/kg on a dry weight basis, with a range of 1–15 µg/kg (Pearson & McConnell, 1975). In Lake Pontchartrain, LA (United States), tetrachloroethylene was found at a concentration of 3.3 ppb [µg/kg] in clams and 10 ppb in oysters; the concentration in the sediment was reported to be 0.3 ppb wet weight (Ferrario et al., 1985).

The concentration of tetrachloroethylene in sediment and marine animal tissue collected near the discharge zone of the waste treatment plant of Los Angeles County, CA, was 2.9 µg/L in the effluent and < 0.5 µg/kg in sediment and in various marine animals (range, < 0.3–29 µg/kg wet tissue) (Gossett et al., 1983).

Tetrachloroethylene has frequently been found at higher levels in fatty foods in residences and markets near dry cleaning establishments (Vieths et al., 1987, 1988; Reinhard et al., 1989). In Germany, the concentrations of tetrachloroethylene in foods in a supermarket near a dry cleaning shop were 110 µg/kg in margarine, 7 µg/kg in herb butter, 36 µg/kg in a cheese spread, 21 µg/kg in butter, 25 µg/kg in flour and 36 µg/kg in cornstarch. The concentrations found in foods taken into a dry cleaning shop were 2 µg/kg in a fruit sherbert, 1330 µg/kg in chocolate-coated ice cream, 4450 µg/kg in chocolate- and nut-coated ice cream and 18 750 µg/kg in an ice-cream confection (Vieths et al., 1988). Butter obtained from a supermarket located near a dry cleaning establishment contained concentrations of 100–> 1000 ppb [µg/kg], whereas butter obtained from other shops generally contained < 50 ppb (Miller & Uhler, 1988). The concentrations in butter in apartments above a coin-operated dry cleaners in Hamburg, Germany, were as high as 58 000 µg/kg (Gulyas et al., 1988). Concentrations of 500–5000 ppb were found in four of 56 margarine samples bought from supermarkets in the Washington DC metropolitan area; the concentrations in the remaining samples were 100–500 µg/kg in one, 50–100 µg/kg in one, < 50 µg/kg in nine, < 4 µg/kg in 18 and undetectable in 23 (Entz & Diachenko, 1988).

1.3.5 Other

Tetrachloroethylene was reported as a contaminant of cosmetic products (range, 0.3–400 µg/L) and of cough mixtures (range, 0.2–97.1 µg/kg) (Bauer, 1981). In Germany, the mean daily intake of tetrachloroethylene from air, food, and water has been estimated to range from 113 to 144 µg/day (Bauer, 1981; von Düszeln et al., 1982).

1.3.6 Biological monitoring

The results of the United States Third National Health and Nutrition Survey showed a mean blood concentration of tetrachloroethylene in 590 non-occupationally exposed volunteers of 0.19 µg/L (Ashley et al., 1994).

The median blood level of tetrachloroethylene in people not occupationally exposed to volatile halogenated hydrocarbons in Germany was 0.4 µg/L (range, < 0.1–3.7 µg/L), whereas the ranges were 1.7–29.2 µg/L in the blood of nine motor vehicle mechanics, 0.4–0.7 µg/L in three painters, 0.5–1.1 µg/L in three precision instrument makers and 17.6–2497.9 µg/L (mean, 748.8 µg/L) in six dry cleaners and pressers (Hajimiragha et al., 1986).

In a Japanese factory where tetrachloroethylene was used for degreasing, the air concentrations were 30–220 ppm [203–1492 mg/m^3] (geometric mean, 92 ppm [624 mg/m^3]); the geometric mean concentration of total trichloro compounds in urine was 50.7 mg/L (Ikeda et al., 1980).

In a metal degreasing process in the Republic of Korea, the mean concentration of tetrachloroethylene in the air was 22.4 ppm [152 mg/m^3] (range, 0–61 ppm [0–414 mg/m^3]). The mean blood concentration was 0.85 mg/L (range, 0.2–2.5 mg/L), and the mean concentration of trichloroacetic acid in urine was 1.76 mg/L (range, 0.6–3.5 mg/L) (Jang et al., 1993).

In an ongoing biological monitoring study of workers in various occupations exposed to trichloroethylene, tetrachloroethylene or 1,1,1-trichloroethane conducted by the Finnish Institute of Occupational Health, 11 534 samples representing 3976 workers in 600 workplaces were obtained for the three compounds between 1965 and 1983. Of these workers, 94.4% were monitored for one solvent, 5.2% for two solvents and 0.4% for three solvents. The mean concentrations of tetrachloroethylene measured in blood samples collected between 1974 and 1983 were 0.7 µmol/L [116 µg/L] for men and 0.4 µmol/L [66 µg/L] for women (Anttila et al., 1995). A maximal concentration of 0.5 µmol/L [83 µg/L] tetrachloroethylene was measured in the blood of Finnish workers manufacturing printing plates (Rantala et al., 1992).

Tetrachloroethylene was detected at a geometric mean concentration of 0.069 µg/L (range, < 0.015–2.54 µg/L) in the blood of 31 of 39 subjects in Zagreb, Croatia, with no known exposure to solvents. The geometric mean concentration in drinking-water was 1.98 µg/L (range, 0.36–7.80 µg/L) (Skender et al., 1993). Similar observations were made in 79 subjects (Skender et al., 1994).

Tetrachloroethylene was detected in seven of 42 samples of breast milk from the general population in four urban areas of the United States (Pellizzari et al., 1982). In a case of jaundice and hepatomegaly reported in a six-week-old breast-fed infant in Halifax, Canada, the mother's blood was found to contain 3000 µg/L and her milk contained 10 000 µg/L. The father, who worked as a leather and suede cleaner at a dry cleaning plant, had a blood level of 30 000 µg/L. The mother regularly visited her husband at the plant during lunch (Bagnell & Ellenberger, 1977; see also section 4.3.1, p. 195).

In a study of non-occupationally exposed nursing mothers in the United States in 1985, the concentrations of tetrachloroethylene in milk ranged from undetectable to 43 µg/L (mean, 6.2 µg/L). Those in personal air samples were 1.1–210 µg/m^3 (mean, 27 µg/m^3), and their blood concentrations were 2.9–8.7 µg/L (Schreiber, 1992, 1993).

In a study in the Netherlands, the mean concentrations of tetrachloroethylene in alveolar air were 24 µg/m^3 for six children in a school located near a factory and 2.8 µg/m^3 in 11 control children; the indoor ambient air levels were 13 µg/m^3 in the school near the factory and 1–3 µg/m^3 in the control school. The mean alveolar air concentrations of 10 residents on the first floor of a retirement home located near a chemical waste dump was 7.8 µg/m^3, whereas that of 19 residents on higher floors was 1.8 µg/m^3 (Monster & Smolders, 1984).

Tetrachloroethylene was found at concentrations of 0.4–29.2 µg/kg in the body fat of eight people *post mortem* in the United Kingdom, with an average of 7.9 µg/kg wet weight (McConnell et al., 1975). The maximal concentration of tetrachloroethylene in fat from the cadavers of 15 people who had lived in industrialized areas of Germany was 36.9 µg/kg wet weight, with an average of 14 µg/kg (Bauer, 1981).

1.4 Regulations and guidelines

Occupational exposure limits and guidelines for tetrachloroethylene in a number of countries are presented in Table 9. In 1947, the American Conference of Governmental Industrial Hygienists (ACGIH) reduced its recommended threshold limit value from 200 ppm (1370 mg/m^3) to 100 ppm (670 mg/m^3). It was reduced again in 1984 to 50 ppm (335 mg/m^3) and was further reduced in 1993 to 25 ppm (170 mg/m^3) (Coler & Rossmiller, 1953; American Conference of Governmental Industrial Hygienists, 1984, 1994).

WHO (1993) has established a guideline for tetrachloroethylene in drinking-water of 40 µg/L. In Switzerland, the tolerance limit for tetrachloroethylene in food is 0.05 mg/kg, while that in the fat of meat and milk is 0.2 mg/kg (Grob *et al.*, 1991).

The ACGIH (1994) has recommended several biological exposure indices for tetrachloroethylene. That for exhaled air before the last shift of a work week is 10 ppm [67.8 mg/m^3]; that for blood before the last shift of the work week is 1 mg/L; and that for trichloroacetic acid in urine at the end of the work week is 7 mg/L. It is noted that measurement of trichloroacetic acid in urine is a nonspecific, semiquantitative determinant of exposure to tetrachloroethylene.

Biological indices for tetrachloroethylene have been reported; in Finland: tetrachloroethylene in blood, 6 µmol/L [995 µg/L] (Aitio *et al.*, 1995); in Germany: tetrachloroethylene in blood, 1 mg/L; in alveolar air, 9.5 ml/m^3 (Deutsche Forschungsgemeinschaft, 1993) and in Switzerland: tetrachloroethylene in blood, 1 mg/L; trichloroacetic acid in urine, 7 mg/L; tetrachloroethylene in alveolar air, 9.5 ml/m^3 (Schweizerische Unfallversicherungsanstalt, 1994).

In some countries, emissions of tetrachloroethylene from dry cleaning establishments are specifically regulated. In Germany, for example, the concentration of tetrachloroethylene in the air leaving a dry cleaning machine may be no higher than 2000 mg/m^3. Tetrachloroethylene concentrations in air emissions from dry cleaning plants must not exceed 20 mg/m^3, and the concentration in the outlet air of the workroom cannot exceed 35 mg/m^3. The threshold value for tetrachloroethylene in a neighbouring apartment is 0.1 mg/m^3. The concentration in contaminated water from distillation and in the water from the drying cycle must not exceed 0.5 mg/m^3 (Kurz, 1992).

2. Studies of Cancer in Humans

2.1 Case reports

Ratnoff and Gress (1980) in the United States reported a case of polycythemia vera in a 44-year-old salesman who distributed tetrachloroethylene to dry cleaning plants and checked for leaks. He had been exposed transiently to concentrations of 50–1000 ppm [339–6780 mg/m^3]. Polycythemia vera had been diagnosed in his father, who was the director of the chemical distribution company, at the age of 53 years.

Jalihal and Barlow (1984) reported a case of myeloid leukaemia diagnosed in a 60-year-old dry cleaner in the United Kingdom. He had had heavy exposure for many years first to trichloroethylene and later to tetrachloroethylene.

Table 9. Occupational exposure limits and guidelines for tetrachloroethylene

Country	Year	Concentration (mg/m^3)	Interpretation
Australia	1993	335	TWA, suspected carcinogen
		1005	STEL
Austria	1987	345	TWA
Belgium	1993	339	TWA
		1368	STEL
Brazil	1987	525	TWA; skin notation
Bulgaria	1993	170	TWA
		685	STEL
Canada (Saskatchewan)	1987	335	TWA
		420	STEL; 5 min
Chile	1987	536	TWA; skin notation
China	1987	670	TWA
Czech Republic	1993	250	TWA
		1250	STEL
Denmark	1993	200	TWA; skin notation
Egypt	1993	35	TWA
Finland	1993	335	TWA
		520	STEL
France	1993	335	TWA
Germany	1993	345	TWA; suspected carcinogen
Indonesia	1987	670	TWA
Italy	1987	600	TWA; skin notation
Japan	1993	335	TWA
Mexico	1987	670	TWA
Netherlands	1994	240	TWA; skin notation
Poland	1993	60	TWA; skin notation
Republic of Korea	1993	170	TWA
		685	STEL
Romania	1987	400	TWA
		500	STEL
Russian Federation	1993	339	TWA
Sweden	1993	70	TWA
		170	STEL
Switzerland	1993	345	TWA; skin notation
		678	STEL
United Kingdom	1994	335	TWA
		1000	STEL
USA			
ACGIH	1994	170	TWA; animal carcinogen
		685	STEL
NIOSH	1994	None	TWA; carcinogen
OSHA	1994	678	TWA
		1356	Ceiling
		2034	PEAK; 5 min in any 3 h

Table 9 (contd)

Country	Year	Concentration (mg/m^3)	Interpretation
Venezuela	1987	670	TWA; skin notation
		1000	STEL; skin notation

From Cook (1987); Deutsche Forschungsgemeinschaft (1993); Työministeriö (1993); American Conference of Governmental Industrial Hygienists (ACGIH) (1994); Arbeidsinspectie (1994); United Kingdom Health and Safety Executive (1994); US National Institute for Occupational Safety and Health (NIOSH) (1994a,c); US Occupational Safety and Health Administration (OSHA) (1994)

TWA, time-weighted average; STEL, short-term exposure limit; Ceiling, level not to be exceeded during any part of the workday

2.2 Cohort studies

Olsen *et al.* (1989) studied a cohort of 2610 white men employed for one or more years between 1956 and 1980 in a chemical company in Louisiana, United States. Tetrachloroethylene was one of many chemicals produced in the plant. The cohort was followed until 1 January 1981, and vital status was ascertained for 98.9% of the men. The expected numbers of deaths were calculated on the basis of both national and local mortality rates. There were 48 deaths, giving a standardized mortality ratio (SMR) of 0.56 (95% confidence interval [CI], 0.42–0.75), and 11 cancer deaths, giving an SMR of 0.76 (0.38–1.4), both in comparison with local rates. Three deaths from leukaemia and aleukaemia gave a significant SMR of 4.9 (1.0–14.4); however, the leukaemias were all of different types and occurred in men with different employment histories.

In the study of Blair *et al.* (1990), described in the monograph on dry cleaning on p. 48, there were increased SMRs for cancers at all sites (SMR, 1.2; 95% CI, 1.0–1.3; 294 deaths) and for cancers of the oesophagus (2.1; 1.1–3.6; 13 deaths), cervix (1.7; 1.0–2.0; 21 deaths) and urinary bladder (1.7; 0.7–3.3; 8 deaths). Specific exposures could not be accounted for, but the mortality rates were similar for those entering before and after 1960, when use of tetrachloroethylene became predominant.

In a study described in the monograph on trichloroethylene (p. 97), Spirtas *et al.* (1991) also evaluated exposure to other specified chemicals, including tetrachloroethylene. These analyses showed two deaths from multiple myeloma ([0.12 expected] SMR, 17) in women and four deaths from non-Hodgkin's lymphoma [SMR, 3.2; 95% CI, 0.87–8.1] in men and women exposed to tetrachloroethylene. Data for other cancer sites were not provided.

In a study described in the monograph on trichloroethylene (p. 96), Anttila *et al.* (1995) also included 849 persons in Finland who had been biologically monitored for occupational exposure to tetrachloroethylene. The median measured level of tetrachloroethylene in blood during 1974–83 was 0.7 µmol/L [116 µg/L] in men and 0.4 µmol/L [66 µg/L] in women. A total of 31 cancer cases were observed (standardized incidence ratio [SIR], 0.90; 95% CI, 0.61–1.3), but significant excess risks were not seen for cancer at any site. There were two cases of cervical cancer (3.2;

0.39–12) and three cases of non-Hodgkin's lymphoma (3.8; 0.77–11). The observed numbers were not provided for several sites of potential interest, such as the oesophagus and urinary bladder.

In the study of Ruder *et al.* (1994), described in the monograph on dry cleaning (p. 49), in the subcohort of 625 workers employed only in shops where tetrachloroethylene was the primary solvent used, an excess was seen for cancer of the oesophagus (SMR, 2.6; 95% CI, 0.72–6.8; four deaths); no excess was seen for cancers of the intestine (1.0; 0.32–2.3; five deaths), urinary bladder (no case [1.0 expected]), pancreas (0.73; 0.09–2.6; two deaths) or female genital organs (1.6; 0.68–3.1; eight deaths).

These studies are summarized in Table 10.

2.3 Case–control studies

2.3.1 Cancer of the liver and bile duct

Bond *et al.* (1990) conducted a case–control study nested in a cohort of 21 437 hourly workers employed at a chemical company in Midland City and Bay City, MI, United States. Among the 6259 men who had died in 1940–82, 44 had cancer of the liver or bile duct mentioned on their death certificates (11 primary liver cancer, 14 cancer of the gall-bladder or bile ducts, 19 cancer of the liver not specified as primary or secondary). A random sample of 1888 men was chosen to serve as controls. Company records were searched for potential exposure to 11 chemical agents. Exposure to tetrachloroethylene was recorded for six cases, giving an odds ratio of 1.8 (95% CI, 0.8–4.3).

Hardell *et al.* (1984) (in a study described in detail in the monograph on trichloroethylene, p. 101) studied 98 patients with liver cancer and 200 matched controls in the Umeå region of Sweden. One patient and no control reported exposure to tetrachloroethylene.

2.3.2 Malignant lymphoma

Hardell *et al.* (1981) (in a study described in detail in the monograph on trichloroethylene, p. 101) studied 169 patients with malignant lymphoma and 338 controls in the Umeå region of Sweden. One patient and no control reported exposure to tetrachloroethylene.

2.3.3 Brain tumours

In a study described in detail in the monograph on trichloroethylene (p. 102), Heineman *et al.* (1994) studied white men in the United States with astrocytic brain tumours. A total of 111 of these 300 men had job titles that were compatible with exposure to tetrachloroethylene (odds ratio, 1.2; 95% CI, 0.8–1.6). None of the risk estimates for subgroups reached statistical significance.

2.3.4 Renal-cell carcinoma

In the study of Sharpe *et al.* (1989) (described in the monograph on trichloroethylene, p. 101) in Montréal, Canada, 10 of 164 patients and three of 161 controls had been exposed to

Table 10. Summary of data from cohort studies of exposure to tetrachloroethylene

Cancer site	Studies in which subjects were exposed predominantly to tetrachloroethylene							Studies in which subjects had mixed exposures, including tetrachloroethylene								
	Anttila et al. (1995) 292 men and 557 women monitored for exposure (Finland, 1974–92)			Ruder et al. (1994) 625 men and women employed in dry cleaning (USA, 1960–90)				Blair et al. (1990) 5365 men and women employed in dry cleaning (USA, 1948–78)				Olsen et al. (1989) 2610 white men employed at a chemical company (USA, 1956–80)				Spirtas et al. (1991) 14 457 men and women employed in aircraft maintenance (subcohort exposed to tetrachloroethylene) (USA, 1953–82)			
	SIR	95% CI	Obs	SMR	95% CI	Obs	SMR	95% CI	Obs	SMR[a]	95% CI	Obs	SMR	95% CI	Obs
All cancers	0.90	0.61–1.3	31	1.0	0.76–1.3	54	1.2	1.0–1.3	294	0.76	0.38–1.4	11	NR		
Oesophagus	NR			2.6	0.72–6.8	4	2.1	1.1–3.6	13	NR			NR		
Stomach	NR			—		0	0.8	0.4–1.4	11	NR			NR		
Colon	NR			1.0	0.32–2.3	5	1.0	0.6–1.4	25	NR			NR		
Cervix	3.2	0.39–12	2	1.6[b]	0.68–3.1	8	1.7	1.0–2.0	21	NR			NR		
Kidney	1.8	0.22–6.6	2	1.2	0.03–6.5	1	0.5	0.1–1.8	2	NR			NR		
Urinary bladder	NR			—	[1.0 expected]	0	1.7	0.7–3.3	8	NR			NR		
Brain and nervous system	1.2	0.14–4.2	2	NR			0.2	0.0–1.2	1	3.2	0.67–9.4	3	NR		
Lymphohaemato-poietic system	1.4	0.28–4.0	3	0.49	0.06–1.8	2	1.2	0.8–1.8	24	NR			NR		
Non-Hodgkin's lymphoma	3.8[c]	0.77–11	3	NR			1.7	0.7–3.4	7	NR			[3.2][c]	[0.87–8.1]	4
Leukaemias	NR			NR			0.9	0.4–1.8	7	4.9	1.0–14	3	NR		

SIR, standardized incidence ratio; CI, confidence interval; Obs, number of cases or deaths observed; SMR, standardized mortality ratio; NR, not reported
[a] In comparison with local mortality rates
[b] Female genital organs
[c] Includes ICD 202

degreasing solvents (odds ratio, 3.4; 95% CI, 0.92–13). The agents most widely used were reported to be tetrachloroethylene, 1,1,1-trichloroethane, trichloroethylene and dichloromethane.

2.3.5 Multiple sites

In the study of Siemiatycki (1991), described in the monograph on trichloroethylene (p. 103), the estimated prevalence of exposure to tetrachloroethylene was 1%. The odds ratio for prostatic cancer was 1.8 ([95% CI, 0.8–4.1]; nine cases) for any exposure and 3.2 ([1.1–9.3]; eight cases) for 'substantial' exposure.

2.4 Studies of drinking-water

The reservations expressed in the monograph on trichloroethylene (p. 103) with regard to the relevance of some of the studies on drinking-water for evaluating carcinogenicity apply equally to tetrachloroethylene, especially in view of the fact that some people may have been exposed to high concentrations of both of these compounds and perhaps others.

Aschengrau *et al.* (1993) studied residents of the five Upper Cape towns in Massachusetts, United States, in whom cancers of the urinary bladder and kidney and leukaemia had been diagnosed in 1983–86. During the late 1970s, tetrachloroethylene had leached into drinking-water from the inner vinyl lining of water distribution pipes (see p. 173); the concentrations at sites of low use (with slow water flow, such as in dead-end sites in the distribution system) had been 1600–7750 µg/L, and that at sites of medium and high use was 1.5–80 µg/L. Population controls were selected for living persons under the age of 65 by random-digit dialling, for living persons over the age of 65 from the Medicare files and for deceased persons from the death lists. Residential history was collected by personal interview, and relative exposure to tetrachloroethylene in drinking-water was estimated. The final study groups included 61 people with cancer of the urinary bladder and 852 controls, 35 patients with cancer of the kidney and 777 controls and 34 people with leukaemia and 737 controls. Odds ratios of 4.0 (95% CI, 0.65–25; based on four exposed cases) and 8.3 (1.5–45; two exposed cases) were found for urinary bladder cancer and leukaemia, respectively, in people with estimated exposure to tetrachloroethylene-contaminated drinking-water at above the 90th percentile. The estimates were controlled for potential confounding factors. None of the patients with cancer of the kidney had been exposed to tetrachloroethylene at this level.

Isacson *et al.* (1985) tabulated cancer incidence data in Iowa, United States, in 1969–81 by two groups of areas, with levels of < 0.3 µg/L and ≥ 0.3 µg/L tetrachloroethylene in groundwater. There were virtually no differences in the incidence rates of cancers of the urinary bladder, breast, colon, lung, prostate or rectum between the two groups of areas.

In a study described in detail in the monograph on trichloroethylene (p. 103; Lagakos *et al.*, 1986), the occurrence of childhood leukaemia in a community in Massachusetts, United States, was significantly related to consumption of water from two wells that had been contaminated with chlorinated organic chemicals, including tetrachloroethylene. The level of tetrachloroethylene measured in 1979 was 21 ppb [µg/L].

In a study described in detail in the monograph on trichloroethylene (p. 104; Cohn *et al.*, 1994), the highest level of tetrachloroethylene in groundwater was 14 µg/L; 11 towns had levels

> 5 μg/L. When these towns were compared with towns where no tetrachloroethylene was detected in drinking-water, a slight increase in the incidence of leukaemia was noted among women (odds ratio, 1.2; 95% CI, 0.9–1.5) but not among men (0.8; 0.7–1.1). The odds ratio for the incidence of non-Hodgkin's lymphoma in women was 1.1 (0.9–1.3) and that in men was 1.1 (0.9–1.4).

Vartiainen et al. (1993) (in a study described in detail in the monograph on trichloroethylene, p. 104) studied cancer incidence in two Finnish villages where the groundwater was contaminated with trichloroethylene and tetrachloroethylene. The average urinary excretion of tetrachloroethylene was 0.19 and 0.10 μg per day for inhabitants of the two exposed villages and 0.11 and 0.09 μg per day for two groups of controls. With the possible exception of non-Hodgkin's lymphoma, which occurred in marginal excess of one of the villages (SIR, 1.4, 95% CI, 1.0–2.0; 31 cases) but not in the other (0.6; 0.3–1.1; 14 cases), neither overall cancer incidence nor the incidence of liver cancer or lymphohaematopoietic cancers was increased for inhabitants of the two villages.

3. Studies of Cancer in Experimental Animals[1]

3.1 Oral administration

3.1.1 Mouse

Groups of 50 male and 50 female B6C3F1 mice, five weeks of age, were administered tetrachloroethylene (USP grade; purity, > 99%) in corn oil by gavage on five days per week for 78 weeks. The time-weighted average doses of tetrachloroethylene were 536 and 1072 mg/kg bw per day for males and 386 and 772 mg/kg bw per day for females. The treatment period was followed by a 12-week observation period. Groups of 20 vehicle controls and 20 untreated controls of each sex were included. Mortality was significantly increased in animals treated with tetrachloroethylene in comparison with controls. The numbers of survivors at the end of the study were 11/20 untreated control males, 10/20 vehicle control males, 19/50 low-dose males and 10/48 high-dose males; and 11/20 untreated control females, 18/20 vehicle control females, 11/50 low-dose females and 7/49 high-dose females. All animals were submitted to a complete necropsy and histopathological evaluation. Significantly increased incidences (Fisher exact test) of hepatocellular carcinomas were seen in all treated groups: in males, the rates were 2/17 untreated controls, 2/20 vehicle controls, 32/49 ($p < 0.001$) animals at the low dose and 27/48 ($p < 0.001$) at the high dose; in females, the rates were 2/20 untreated controls, 0/20 vehicle controls, 19/48 ($p < 0.001$) at the low dose and 19/48 ($p < 0.001$) at the high dose (United States National Cancer Institute, 1977).

---

[1] The Working Group was aware of a study in progress in which rats were exposed to tetrachloroethylene by inhalation (IARC, 1994b).

3.1.2 Rat

Groups of 50 male and 50 female Osborne-Mendel rats, seven weeks of age, were administered two doses of tetrachloroethylene (purity, > 99%) in corn oil by gavage on five days per week for 78 weeks. The time-weighted average doses of tetrachloroethylene were 471 and 941 mg/kg bw per day for males and 474 and 949 mg/kg bw per day for females. The treatment period was followed by a 32-week observation period. Groups of 20 vehicle controls and 20 untreated controls of each sex were included. Mortality was significantly increased in treated animals in comparison with controls. The numbers of survivors at the end of the study were 5/20 untreated control males, 2/20 vehicle control males, 6/50 low-dose males and 2/50 high-dose males; and 12/20 untreated control females, 8/20 vehicle control females, 17/50 at the low dose and 14/50 at the high dose. All animals were submitted to a complete necropsy. There was no difference in tumour incidence between control and treated animals. Toxic nephropathy occurred in 88 and 94% of treated males and in 50 and 80% of treated females but not in controls (United States National Cancer Institute, 1977). [The Working Group noted that the high mortality precluded an evaluation of carcinogenicity.]

3.2 Inhalation

3.2.1 Mouse

Groups of 49 or 50 male and 49 or 50 female B6C3F1 mice, eight to nine weeks of age, were exposed to air containing tetrachloroethylene (purity, 99.9%) at concentrations of 0, 100 or 200 ppm (0, 680 or 1360 mg/m^3) for 6 h per day on five days per week for 103 weeks. Survival was significantly reduced ($p < 0.05$) among exposed male mice and among females at the high dose: the numbers of survivors at the end of the study were 46/49 control males, 25/50 at the low and 32/50 at the high dose; and 36/49 control females, 31/50 at the low and 19/50 at the high dose. All animals were submitted to a complete necropsy and histopathological evaluation. Exposure-related increases in the incidences of liver neoplasms were seen in all treated animals. In males, the incidences of hepatocellular adenomas were 12/49 controls, 8/49 at the low dose and 19/50 ($p = 0.012$, incidental tumour test) at the high dose; the incidences of hepatocellular carcinomas were 7/49 controls, 25/49 ($p = 0.016$) at the low dose and 26/50 ($p = 0.001$) at the high dose; and the combined incidences of hepatocellular adenomas or carcinomas were 17/49 controls, 31/49 ($p = 0.026$) at the low dose and 41/50 ($p < 0.001$) at the high dose. In females, the incidences of hepatocellular carcinomas were 1/48 controls, 13/50 ($p < 0.001$) at the low dose and 36/50 ($p < 0.001$) at the high dose; and the combined incidences of hepatocellular adenomas or carcinomas were 4/48 controls, 17/50 ($p < 0.001$) at the low dose and 38/50 ($p < 0.001$) at the high dose. Degeneration and necrosis of the liver in treated mice were also related to exposure. In males, degeneration was seen in 2/49 controls, 8/49 animals at the low dose and 14/50 at the high dose; necrosis was seen in 1/49 controls, 6/49 at the low dose and 15/50 at the high dose. In females, degeneration was seen in 1/49 controls, 2/50 at the low dose and 13/50 at the high dose; necrosis was seen in 3/48 controls, 5/40 at the low dose and 9/50 at the high dose (Mennear *et al.*, 1986; United States National Toxicology Program, 1986).

3.2.2 Rat

In a study reported as an abstract, groups of 96 male and 96 female weanling Sprague-Dawley rats were exposed to air containing 300 or 600 ppm (2030 or 4070 mg/m^3) tetrachloroethylene (purity, 99.9%) for 6 h per day on five days per week for 12 months. The exposure period was followed by an observation period that extended through the rats' lifetime (up to 19 additional months). The control groups consisted of 192 male and 192 female rats. Slightly higher mortality was reported among males exposed to the dose of 600 ppm. All rats were submitted to necropsy and histopathological examination. No dose-related increase in tumour incidence was seen in exposed rats in comparison with controls (Rampy et al., 1977). [The Working Group noted that the duration of exposure was too short to adequately evaluate the carcinogenicity of tetrachloroethylene.]

Groups of 50 male and 50 female Fischer 344/N rats, eight to nine weeks of age, were exposed to air containing tetrachloroethylene (purity, 99.9%) at concentrations of 0, 200 or 400 ppm (0, 1360 or 2720 mg/m^3) for 6 h per day on five days per week for up to 103 weeks. Survival of male rats exposed to the high dose was significantly lower ($p < 0.05$) than that of controls; the numbers of survivors at the end of the study were 23 male controls, 19 males at the low dose and 11 at the high dose; and 23 female controls, 21 at the low dose and 24 at the high dose. All animals were submitted to a complete necropsy and histopathological evaluation. Dose-related increases (life-table test) in the incidence of mononuclear-cell leukaemia were seen in animals of each sex; the incidences were 28/50 control males, 37/50 ($p = 0.046$) at the low dose and 37/50 ($p = 0.004$) at the high dose; 18/50 control females, 30/50 ($p = 0.023$) at the low dose and 29/50 ($p = 0.053$) at the high dose. On the basis of life-table analysis (with adjustment for survival), the incidences of advanced (stage 3) mononuclear-cell leukaemia were increased in animals of each sex: 20/50 control males, 24/50 at the low dose and 27/50 ($p = 0.022$) at the high dose; 10/50 control females, 18/50 at the low dose and 21/50 ($p = 0.029$) at the high dose. The historical incidence of mononuclear-cell leukaemia in rats at the same laboratory was 47% in males and 29% in females. Uncommonly occurring renal tubular-cell adenomas or adenocarcinomas were found in male rats; adenomas were seen in 1/49 controls, 3/49 at the low dose and 2/50 at the high dose; adenocarcinomas occurred in 0/49 controls, 0/49 at the low dose and 2/50 at the high dose (not statistically significant even when combined). The incidence of renal tubular-cell hyperplasia was also increased in treated males, with 0/49 controls, 3/49 at the low dose and 5/50 at the high dose; karyomegaly was found in 1/49 control males, 37/49 at the low dose and 45/50 at the high dose and in 0/50 control females, 8/49 at the low dose and 20/50 at the high dose (Mennear et al., 1986; United States National Toxicology Program, 1986).

3.3 Intraperitoneal injection

Mouse: In a screening assay based on the increased multiplicity and incidence of lung tumours in a strain of mice highly susceptible to development of this neoplasm, groups of 20 strain A/St male mice, six to eight week of age, were given intraperitoneal injections of tetrachloroethylene [purity unspecified] in tricaprylin three times a week at doses of 80 mg/kg bw (14 injections) or 200 or 400 mg/kg bw (24 injections). Twenty-four weeks after the first injection, the mice were killed and their lungs were examined under a dissecting microscope.

Tetrachloroethylene did not increase the incidence of pulmonary adenomas in treated mice in comparison with vehicle control mice (Theiss et al., 1977).

3.4 Topical application

Mouse: In a study of two-stage carcinogenesis on mouse skin, single doses of 163 mg tetrachloroethylene [purity unspecified] in 0.2 ml of acetone were applied to the dorsal skin of 30 female ICR:Ha Swiss mice aged six to eight weeks; 14 days later, topical applications of 12-*O*-tetradecanoylphorbol 13-acetate (TPA; 5 µg in 0.2 ml of acetone, three times per week) were begun, for at least 61 weeks. Seven skin papillomas were found in 4/30 treated mice, and six papillomas were found in 6/90 TPA-treated controls. Trichloroethylene was also administered in 0.2 ml of acetone by repeated topical application (three times per week) to groups of 30 female ICR:Ha Swiss mice, six to eight weeks of age, for at least 63 weeks at doses of 18 or 54 mg per mouse. One papilloma occurred in a mouse treated with the lower dose; no skin tumours were observed among controls or mice given the higher dose (Van Duuren et al., 1979).

3.5 Carcinogenicity of metabolites

Carcinogenicity studies on a known metabolite, trichloroacetic acid, are summarized in a separate monograph in this volume.

Mouse: Tetrachloroethylene oxide, a presumed metabolite of tetrachloroethylene, was administered to groups of 30 female ICR/Ha Swiss mice, six to eight weeks of age, by repeated skin application for 66 weeks (7.5 mg/mouse three times weekly) or by a subcutaneous injection of 500 µg/mouse in 0.05 ml of trioctanoin once a week for up to 80 weeks. The incidence of tumours at the site of application was not increased (Van Duuren et al., 1983).

4. Other Data Relevant to an Evaluation of Carcinogenicity and its Mechanisms

4.1 Absorption, distribution, metabolism and excretion

4.1.1 Humans

The toxicokinetics of tetrachloroethylene in humans has been reviewed (Hake & Stewart, 1977; Reichert, 1983).

The mean blood concentration of tetrachloroethylene in a group of 590 nonoccupationally exposed people in the United States was 0.19 ppb [µg/L] (Ashley et al., 1994).

Exposure of nine men and 41 women employed in dry cleaning shops to concentrations of tetrachloroethylene ranging from traces to 85 ppm [576 mg/m^3] (4-h samples randomly collected over the working week) led to blood concentrations of 9–900 µg/L (samples collected during the working day). The median values were 14.8 ppm [100 mg/m^3] in air and 143 µg/L in blood (Mutti et al., 1992).

The kinetics and metabolism of tetrachloroethylene have been reported in a large number of studies of volunteers. Pulmonary uptake of tetrachloroethylene is rapid, but complete tissue equilibrium is achieved only after several hours. In six male volunteers (mean body weight, 77 kg; range, 67–86), the estimated uptake of tetrachloroethylene after exposure to 72 ppm [488 mg/m^3] for 4 h at rest was 455 mg (range, 370–530). Uptake from an atmosphere containing 144 ppm [976 mg/m^3] was 945 mg (range, 670–1210), i.e. 2.08-fold higher, at rest and 1318 mg (range 1060–1510) when the 4 h exposure was combined with a work load of 100 W for two half-hour periods [3.6 kJ]. The alveolar retention at rest at the end of the exposure was calculated to be about 60% (Monster et al., 1979).

The human blood:air partition coefficients range from 10.3 to 14.0 (Hattis et al., 1990; Gearhart et al., 1993).

Excretion of tetrachloroethylene in breath is proportional to the level of exposure (Fernandez et al., 1976; Solet & Robins, 1991). Several studies have indicated slow elimination of tetrachloroethylene by exhalation; for example, tetrachloroethylene was still detectable at a concentration of about 1 ppm [~ 6.78 mg/m^3] 162 h after exposure to 496 and 992 mg/m^3 for 4 h. More than 80% of the estimated uptake was recovered in breath (Monster et al., 1979). Terminal half-lives of 34–55 h were reported for exhalation of tetrachloroethylene (Stewart et al., 1961b, 1970). Unchanged compound is excreted by exhalation at three different rate constants, with half-lives of 12–16 h, 30–40 h and about 55 h (Monster et al., 1979). Only a minor amount (< 2 %) of the retained tetrachloroethylene is recovered as identified metabolites, which include trichloroacetic acid and trichloroethanol (Ikeda et al., 1972; Ikeda & Imamura, 1973; Ikeda, 1977; Monster et al., 1979; Ohtsuki et al., 1983). In these studies, 20–40% of the tetrachloroethylene was not recovered. None of the other metabolites identified in animals has yet been confirmed in humans. As is seen with trichloroethylene, trichloroacetic acid is slowly cleared over the course of several days, with an estimated half-life of 144 h (Ikeda & Imamura, 1973). Urinary recoveries of trichloroacetic acid indicate that biotransformation of tetrachloroethylene in humans is no longer linear after exposure by inhalation to > 100 ppm [> 678 mg/m^3] (Ohtsuki et al., 1983). Conjugation of tetrachloroethylene with glutathione was not detectable in either cytosolic or microsomal fractions of human liver from seven donors (Green et al., 1990).

4.1.2 Experimental systems

The absorption of tetrachloroethylene has been studied in rodents after inhalation, oral administration and skin contact.

The level of tetrachloroethylene in the blood of adult male Sprague-Dawley rats reached a maximum 1 h after oral ingestion and was maximal immediately after the end of a 6-h inhalation period. The peak blood levels were 10 μg/ml immediately after exposure to 573 ppm [3885 mg/m^3] for 6 h and 40 μg/ml 1 h after oral administration of 500 mg/kg bw. Blood levels declined following first-order kinetics, with a single half-life of about 7 h and an elimination rate constant (K_e) of about 0.10/h (Pegg et al., 1979). The mean blood:air partition coefficients in a number of mouse strains range from 16.9 to 24.4 (Hattis et al., 1990; Gearhart et al., 1993).

Percutaneous absorption of ^{14}C-tetrachloroethylene from dilute aqueous solutions (10–100 ppb [μg/L]) was studied in hairless guinea-pigs exposed in an air-tight glass chamber containing no headspace for 70 min (the head of each sedated animal protruded through a latex diaphragm). Uptake of tetrachloroethylene, as determined by the reduction in radioactivity in the aqueous phase and the amount of radiolabel in the urine and faeces of the animals, was fairly rapid and amounted to 25–30% of the tetrachloroethylene present (Bogen et al., 1992).

Unanaesthetized male Sprague-Dawley rats were exposed to 50 or 500 ppm [339 or 3390 mg/m^3] tetrachloroethylene for 2 h through a one-way breathing valve. The tetrachloroethylene concentrations in exhaled breath increased to a steady-state level within 20 min and were directly proportional to the inhaled concentrations. Blood levels increased throughout the exposure period at both concentrations. The maximal blood concentrations were about 1.2 μg/ml in the group exposed to the lower concentration and 20 μg/ml at the higher level (Dallas et al., 1994a).

Toxicokinetic parameters were estimated in male Sprague-Dawley rats after either inhalation of 500 ppm [3390 mg/m^3] tetrachloroethylene for 2 h or intravenous injection of 10 mg/kg bw. The tissue concentrations (area under the concentration–time curve, based upon measurements from 1 min to 72 h) and maximal tissue concentrations were higher after the inhalation regime, and the tissue concentrations were at least five times higher in fat than in any other tissue examined. Tetrachloroethylene was found throughout the body. The half-lives in all tissues varied around 430 min after both exposure regimes (Dallas et al., 1994b).

Toxicokinetic parameters were compared in male Sprague-Dawley rats and male beagle dogs after oral administration of 10 mg/kg bw tetrachloroethylene. Neither the tissue concentrations (area under the concentration–time curve, based on measurements from 1 min to 72 h) nor the maximal tissue concentrations were appreciably different in the two species, although maximal concentrations occurred earlier in rats. Tetrachloroethylene had much longer half-lives in tissue and blood of dogs than rats; the values in blood were 384 ± 145 min in four rats and 865 ± 385 min in three dogs, while the values for other tissues were about 400 min in rats and 2000 min in dogs (Dallas et al., 1994c).

The metabolism of tetrachloroethylene was studied in male B6C3F1 mice individually housed in 0.7-L closed exposure chambers for 8 h. The starting concentrations were 200, 1000 and 3500 ppm [1356, 6780 and 23 730 mg/m^3]. The maximal metabolic velocity (V_{max}) was 0.2 mg/h per kg, the Michaelis-Menten constant (K_m) was 2.0 mg/L and the first-order metabolic rate was 2.0 g/kg. Dermal absorption was also rapid, and peak concentrations of tetrachloroethylene were reached in blood 30 min after application (Gearhart et al., 1993). Less than 5% of the administered radioactivity was recovered in the animals 72 h after oral administration of 500 mg/kg bw tetrachloro[1,2-C^{14}]ethylene by gavage or after exposure by inhalation to 600 ppm [4068 mg/m^3] for 6 h (Pegg et al., 1979).

Elimination of tetrachloroethylene and its metabolites by the main routes is dependent on dose. In general, mice have a greater capacity than rats to biotransform tetrachloroethylene. After single oral doses of 500 and 800 mg/kg bw ^{14}C-tetrachloroethylene, male B6C3F1 and female NMRI mice excreted 10.3 and 7.1% of the recovered radiolabel in urine within 72 h; 94.6 and 96.0% of the administered radiolabel was recovered (Schumann et al., 1980; Dekant et al., 1986). After male Sprague-Dawley rats were given oral doses of 1 and 500 mg/kg bw, 16.5 and 4.6% of the radiolabel was recovered in urine within 72 h; 71.5 and 89.9% of the recovered radiolabel was found in expired tetrachloroethylene, and 2.5 and 0.5%, respectively, was expired as carbon dioxide; 103 and 91% of the administered radiolabel was recovered (Pegg et al., 1979). After female Wistar rats were given 800 mg/kg bw orally, 2.3% of the recovered radiolabel was found in urine; 98.9% of the administered dose was recovered (Dekant et al., 1986). After male and female Fischer 344 rats and B6C3F1 mice were exposed once for 6 h to 400 ppm [2712 mg/m^3] tetrachloroethylene, the peak blood levels of the major metabolite, trichloroacetic acid, were 13 times higher in mice than in rats. Comparisons of tissue concentrations, measured as the area under the concentration–time curve, showed that the mice had been exposed to 6.7 times more trichloroacetic acid than the rats (Odum et al., 1988).

After exposure to 10 ppm [67.8 mg/m^3] tetrachloroethylene for 6 h, male B6C3F1 mice excreted 62.5% of the recovered radiolabel in urine and 7.9% as ^{14}C-carbon dioxide over 72 h; only 12% was excreted as unchanged tetrachloroethylene in expired air (Schumann et al., 1980). Male Sprague-Dawley rats exposed to 9.1 ppm [61.7 mg/m^3] or 573 ppm [3885 mg/m^3] tetrachloro[1,2-^{14}C]ethylene for 6 h expired 68.1 and 88.0% as unchanged tetrachloroethylene and 3.6 and 0.7% as ^{14}C-carbon dioxide over 72 h; 18.7 and 6.0%, respectively, were recovered in urine over the same time (Pegg et al., 1979).

The biotransformation of tetrachloroethylene in male Sprague-Dawley rats and B6C3F1 mice is a saturable process, and saturation occurs at higher doses in mice than rats. Furthermore, mice metabolized 8.5 and 1.6 times more tetrachloroethylene per kilogram of body weight after inhalation of 10 ppm [67.8 mg/m^3] or a single oral dose of 500 mg/kg bw. The higher rate of metabolism in mice is paralleled by greater irreversible binding of tetrachloroethylene metabolites to hepatic macromolecules than in rats after inhalation of 10 or 600 ppm [67.8 or 4068 mg/m^3] or a single oral dose of 500 mg/kg bw (Schumann et al., 1980).

Other studies also showed that mice have a greater capacity for the biotransformation of tetrachloroethylene than rats (standardized for body weight), the magnitude of which depends on the dose and route of administration. Male and female DD mice excreted 3.9 times more trichloroacetic acid after inhalation of 200 ppm (1380 mg/m^3) tetrachloroethylene for 8 h than male Wistar rats; and after intraperitoneal injection of 2.78 mmol/kg [461 mg/kg] bw tetrachloroethylene to the same strains of animals, mice excreted 4.3 times more trichloroacetic acid than rats (Ikeda & Ohtsuji, 1972). After exposure of Wistar rats, Sprague-Dawley rats and B6C3F1 mice to 60 ppm [407 mg/m^3] tetrachloroethylene by inhalation, biotransformation to urinary trichloroacetic acid was 2.7 times higher in mice than in rats (Bolt & Link, 1980).

Several excretory metabolites have been identified in rodents (see Figure 1). In most studies, trichloroacetic acid was reported to be the major metabolite, representing about 50% of the radiolabel in urine (Yllner, 1961; Daniel, 1963; Dekant et al., 1986). Oxalic acid, trichloroethanol and N-trichloroacetyl-2-aminoethanol are derived from the oxidative biotransformation of tetrachloroethylene and further reactions of trichloroacetyl chloride with water and phospholipids, respectively (Pegg et al., 1979; Dekant et al., 1986). N-Acetyl-S-(1,2,2-trichlorovinyl)-L-cysteine is the end-product of the conjugation of tetrachloroethylene with glutathione and represents a minor metabolite in urine (< 2% of the radioactivity in urine pooled over 72 h after administration of 800 mg/kg bw tetrachloroethylene). The concentrations of N-acetyl-S-(1,2,2-trichlorovinyl)-L-cysteine in pooled urine samples from five Fischer 344 rats and five B6C3F1 mice, one, seven and 14 days after exposure to atmospheres of 400 ppm [2760 mg/m^3] tetrachloroethylene for 6 h per day for up to 14 days, were 0.75, 2.04 and

0.55 μg/ml in male rats; 1.30, 0.90 and 1.04 μg/ml in female rats; 0, 0.07 and 0.20 μg/ml in male mice; and 0, 0 and 0.15 μg/ml in female mice. The concentrations of this metabolite in the urine of male Fischer 344 rats given 1500 mg/kg bw per day by gavage were 23.0 μg/ml after one day, 41.1 μg/ml after 17 days and 32.7 μg/ml after 42 days (Green et al., 1990). Dichloroacetic acid, which was found to represent no more than 5.1% of the administered radioactivity in urine pooled over 72 h after administration of 800 mg/kg bw tetrachloroethylene, has been suggested to be the end-product of renal processing of S-(1,2,2-trichlorovinyl)glutathione by the mercapturic pathway and cleavage by cysteine conjugate β-lyase, followed by hydrolysis of the thioketene intermediate (Dekant et al., 1988, 1991).

N^ϵ-(Trichloroacetyl)-L-lysine residues are formed in liver proteins of rats treated with tetrachloroethylene. Western blot analysis of renal fractions from rats treated with either ^{14}C-tetrachloroethylene or S-(1,2,2-trichlorovinyl)-L-cysteine also suggested the presence of modified proteins in mitochondria and cytosol, but not in microsomes. The modified proteins that were separated after each of these treatments had identical relative molecular masses. Incubation of rat renal mitochondria with S-(1,2,2-trichlorovinyl)-L-cysteine in vitro resulted in the formation of N^ϵ-(dichloroacetyl)-L-lysine. This residue was quantified in kidney mitochondria and liver microsomes (which are the fractions in each organ with the largest quantities of rabbit anti-trifluoroacetyl serum cross-reactive material), 24 h after treatment of male rats with 1000 mg/kg bw tetrachloroethylene: 12 nmol/mg protein were found in kidney and only 46 pmol/mg in liver. N^ϵ-(Dichloroacetyl)-L-lysine is probably formed by the reaction of dithioketene with lysine in proteins after cleavage of S-(1,2,2-trichlorovinyl)-L-cysteine by β-lyase (Birner et al., 1994).

In studies of the biotransformation of tetrachloroethylene in rat and mouse liver microsomes, trichloroacetic acid was identified as an end-product (Costa & Ivanetich, 1980). As with trichloroethylene, an epoxide has been proposed as an intermediate, and its rearrangement is postulated to result in trichloroacetyl chloride (Bonse & Henschler, 1976; Henschler & Bonse, 1977). Consideration of the mechanism of cytochrome P450-mediated olefin oxidation, however, suggests that trichloroacetyl chloride may also be formed in the absence of the epoxide (Guengerich & Macdonald, 1984). Tetrachlorooxirane may react with water to give oxalic acid; trichloroacetyl chloride is hydrolysed to trichloroacetic acid or may react in small amounts with proteins and phospholipids (Costa & Ivanetich, 1980; Dekant et al., 1987). The rates of oxidation of tetrachloroethylene seen in mouse liver microsomes were significantly higher than those seen in rat liver microsomes. Both cytosolic and microsomal glutathione S-transferases catalysed the formation of S-(1,2,2-trichlorovinyl)glutathione from tetrachloroethylene (Dekant et al., 1987; Green et al., 1990). Conjugation in rat liver fractions occurred at a higher rate in the cytosol (18.2 pmol/min per mg protein) than in microsomes (6.4 pmol/min per mg protein), and the conjugation rate in mouse liver cytosol was only 3.4 pmol/min mg protein. The kinetic constants for kidney cytosolic β-lyase in mice, rats and humans were measured using S-(1,2,2-trichlorovinyl)-L-cysteine (which cleaves cysteine conjugates to give pyruvate and ammonia) as the substrate: the K_m values were about 5.1 mmol/L for mice, 1.0 mmol/L for rats and 2.6 mmol/L for humans; and the V_{max} values were about 1.4 nmol/min per mg for mice, 3.8 nmol/min per mg for rats and 0.6 nmol/min per mg for humans (Green et al., 1990).

Figure 1. Metabolism of tetrachloroethylene in rats

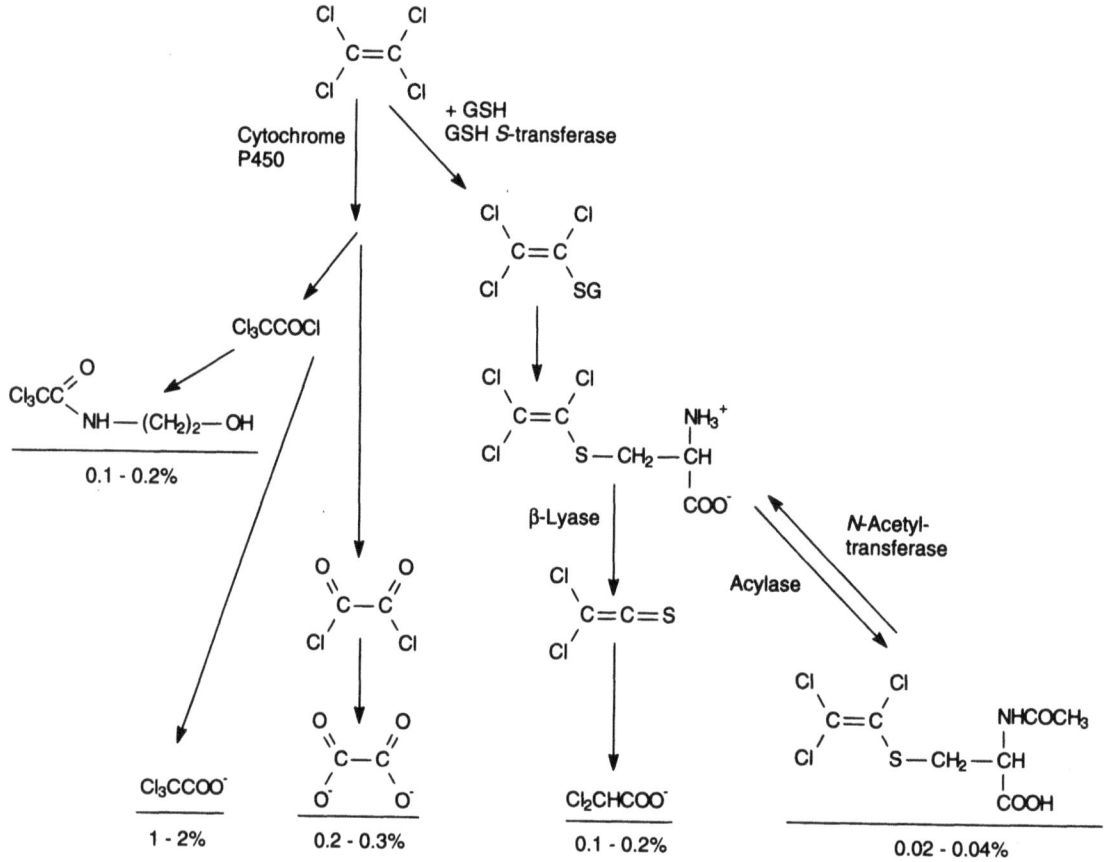

From Dekant et al. (1986)
Identified urinary metabolites are underlined; percentages are those of an oral dose of 800 mg/kg bw excreted as individual metabolites.

4.1.3 Comparison of humans and animals

In one study, the blood:air partition coefficient for mice (21.5) was about twice the human value (11.6). The reason for this significant species difference is not known (Gearhart et al., 1993).

Toxicokinetic modelling of the uptake and elimination of tetrachloroethylene showed that human metabolic parameters could be predicted by scaling rat metabolic parameters for tetrachloroethylene as a function of body weight (Ward et al., 1988). Trichloroacetic acid and trichloroethanol have been reported as urinary metabolites of tetrachloroethylene in both humans and experimental animals. A comparison of the enzyme kinetics of hepatic conjugation of tetrachloroethylene with glutathione and cleavage of the cysteine S-conjugate by β-lyase suggested that this pathway may not be relevant for humans. Human kidney contains β-lyase activity, but biosynthesis of glutathione conjugates, a prerequisite for β-lyase activation, could not be demonstrated (Green *et al.*, 1990).

4.2 Toxic effects

4.2.1 Humans

Acute exposure to tetrachloroethylene by inhalation results in central nervous system depression. Liver and kidney toxicity have been reported as effects of acute exposures to very high doses (Reichert, 1983). In dry cleaners chronically exposed to tetrachloroethylene, increased levels of markers of early renal damage and/or dysfunction were attributed to the exposure (Mutti *et al.*, 1992).

4.2.2 Experimental systems

The acute toxicity of tetrachloroethylene is low. LD_{50} values of 6000–8571 mg/kg bw have been reported in mice after oral administration, and values of 2400–13 000 mg/kg bw have been reported in rats (Dybing & Dybing, 1946; Witey & Hall, 1975). In rats, the 6-h LC_{50} was 4100 ppm [27 798 mg/m^3] and the 8-h LC_{50} was 5000 ppm [33 900 mg/m^3]; in mice, the 4-h LC_{50} was 5200 ppm [35 256 mg/m^3] and the 6-h LC_{50} was 2978 ppm [20 191 mg/m^3]. Histopathological examination of mice 24 h after intraperitoneal injection of 4600 mg/kg bw showed minimal hepatic damage (Klaassen & Plaa, 1966). After injection of 3800 mg/kg bw, the kidneys of mice showed swelling of the proximal convoluted tubules on histological examination (Plaa & Larson, 1965).

After subchronic exposure, the major target for the toxicity of tetrachloroethylene is the liver in mice and the kidney in rats. Male and female Sprague Dawley rats that received theoretical doses of 14, 400 or 1400 mg/kg bw tetrachloroethylene per day for 90 consecutive days in their drinking-water had higher liver: and kidney:body weight ratios at the higher doses (Hayes *et al.*, 1986). In male and female Fischer 344 rats dosed orally with 1000 mg/kg bw tetrachloroethylene daily for 10 days, protein droplet accumulation in the P2 segment of the proximal tubule and increased cell replication rates were observed in males but not in females. The tritiated thymidine-labelling index was increased by two- to threefold in male rats. The cells of the P2 segment have short microvilli, relatively few, small apical vacuoles and varying amounts of protein droplets. In untreated female Fischer 344 rats, these protein droplets were rare and very small, whereas they were clearly visible in about 25% of control males. Immunohistochemical staining for $\alpha_{2\mu}$-globulin showed it to be present in these droplets, and its quantity increased when the droplets were larger and more frequent (Goldsworthy *et al.*, 1988); however, no data are available on the binding of tetrachloroethylene or its metabolites to $\alpha_{2\mu}$-globulin.

Male Swiss mice dosed by gavage with 20–2000 mg/kg bw per day of tetrachloroethylene for six weeks showed dose-related increases in their liver:body weight ratio at 100 mg/kg bw per day or more and histological effects and elevated alanine transaminase activity at 200 mg/kg bw per day or more (Buben & O'Flaherty, 1985). Mice also developed histopathological hepatic changes after an 11-day treatment with daily oral doses as low as 100 mg/kg bw tetrachloroethylene (Schumann *et al.*, 1980). In male and female B6C3F1 mice exposed to 200 or 400 ppm [1356 or 2712 mg/m^3] tetrachloroethylene for 6 h for 14, 21 or 28 consecutive days, a two- to threefold increase in the number of peroxisomes and in the activity of peroxisomal enzymes was seen in liver. Peroxisome proliferation was more marked in male than in female mice but was not seen in Fischer 344 rats of either sex after identical treatment (Odum *et al.*, 1988).

Administration by gavage of 1000 mg/kg bw tetrachloroethylene in corn oil daily for 10 days to male Fischer 344 rats and male B6C3F1 mice resulted in peroxisome proliferation in the livers of animals of both species but in the kidneys only of mice, although the latter response was marginal (Goldsworthy & Popp, 1987).

In studies to determine the doses that could be administered chronically by inhalation to male and female Fischer 344 rats, it was found that a dose of 1750 ppm [11 865 mg/m^3] for 6 h per day for 14 days killed about one-half of the animals, whereas 875 ppm [5933 mg/m^3] was not lethal (United States National Toxicology Program, 1986). Four of 10 male and 7/20 female rats exposed to 1600 ppm [10 848 mg/m^3] for 13 weeks died before the end of the study. The incidence and severity of hepatic congestion was dose-related, this effect clearly persisting at 400 ppm [2712 mg/m^3]. In the subsequent two-year study, no remarkable treatment-related pathological effects were seen in the liver at either of the doses used (200 and 400 ppm [1356 and 2712 mg/m^3]); however, the frequency of renal enlargement and cytomegaly was increased in both males and females, particularly (but not only) in the proximal convoluted tubules and the inner half of the cortex. The incidences of karyomegaly and tubular hyperplasia are given in section 3.2.2 (p. 184).

In parallel studies with male and female B6C3F1 mice, the highest dose used during 14 days of exposure to 1750 ppm [11 865 mg/m^3], 6 h per day, was not lethal, and there was only a slight deficit in body weight gain (United States National Toxicology Program, 1986). In the 13-week study, 2/10 male and 4/10 female mice exposed to 1600 ppm [10 848 mg/m^3] died before the end of the observation period. Liver lesions (leukocyte infiltration, centrilobular necrosis and biliary stasis) were seen in males and females at 400, 800 and 1600 ppm [2712, 5424 and 10 848 mg/m^3], and karyomegaly was seen in renal tubular epithelial cells of all animals exposed to > 200 ppm [> 1356 mg/m^3]. In the subsequent two-year study, dose-related increases were observed in the prevalence of liver degeneration (both sexes), hepatocellular necrosis (both sexes) and nuclear inclusions (males only). Dose-related increases were also observed in the prevalence of renal tubular epithelial cell karyomegaly (both sexes), casts (both sexes) and nephrosis (females only).

Tetrachloroethylene has been reported to inhibit the activity of natural cytotoxic cells from liver and spleen of mice and rats after exposure *in vitro*, while only natural killer and natural P-815 killer cell activities were inhibited in rat cells (Schlichting *et al.*, 1992). Administration *in vivo* did not affect subsequently isolated immune cells. Tetrachloroethylene also inhibited the activity of natural immune cells isolated from human liver (Wright *et al.*, 1994). Direct inhibition of immune function may increase the possibility of tumour development in affected organs.

4.3 Reproductive and prenatal effects

4.3.1 Humans

(a) Endocrine and gonadal effects

The semen quality of 34 dry cleaners was compared with that of 48 laundry workers who were members of the Laundry and Dry Cleaners Union in the San Francisco Bay area and Greater Los Angeles, CA, United States (Eskenazi *et al.*, 1991a). The sperm concentrations and the overall percentage of abnormal forms were similar in the two groups. The sperm heads of dry cleaners had significantly more round forms, fewer narrow forms and showed greater amplitude of lateral head displacement and less linearity in the sperm swimming paths. These subtle effects on sperm quality were related to the concentration of tetrachloroethylene in expired air; it was not established whether these changes affected fertility.

(b) Fertility

Taskinen *et al.* (1989) conducted a nested case–control study of 120 cases of spontaneous abortion and 251 controls on the basis of a file of 6000 Finnish workers who had been biologically monitored for exposure to solvents. Information about their marriages and their wives' pregnancies and spontaneous abortions were obtained from national registries; data on paternal occupational exposure to solvents were collected by means of a questionnaire sent to workers and covered the period of spermatogenesis. The likelihood of exposure was defined in three categories: unexposed, potentially exposed (i.e. use of solvents was possible but no exposure was reported or measured) and probably exposed (i.e. exposure was measured or reported). No association was found between paternal occupational exposure to tetrachloroethylene and spontaneous abortion (crude odds ratio, 0.5; 95% CI, 0.2–1.5).

The reproductive outcomes of the wives of 17 men exposed to tetrachloroethylene in dry cleaning were compared with those of the wives of 32 laundry workers who were not exposed to dry cleaning fluids. The mean number of pregnancies was 2.1 in both groups, and the rates of spontaneous abortion were not significantly different (11.1% for dry cleaners' wives and 15.2% for laundry workers' wives; odds ratio, 2.5; 95% CI, 0.6–10.9). A small, nonsignificant, trend was seen for an association between the length of time to conception and the concentration of tetrachloroethylene in the husband's expired air (Eskenazi *et al.*, 1991a,b). [Interpretation of these findings is necessarily limited by the small sample size.]

(c) Pregnancy

(i) *Spontaneous abortion*

Several studies of the reproductive outcomes of women involved in dry cleaning reported specific results in relation to exposure to tetrachloroethylene.

In the nested case–control study conducted in Finland, described in detail in the monograph on dry cleaning (p.), exposure to tetrachloroethylene was assessed as high when tasks included dry cleaning for an average of at least 1 h daily or when the woman reported handling tetrachloroethylene at least once a week. Exposure was considered to be low when tasks included pressing at a dry cleaners or removing spots or when the woman reported handling tetrachloroethylene less than once a week. High exposure to tetrachloroethylene was associated with an excess risk for spontaneous abortion (odds ratio, 3.4; 95% CI, 1.0–11.2), adjusted for use of other solvents, heavy lifting at work and frequent use of alcohol (Kyyrönen et al., 1989).

In the study of reproductive outcomes among dry cleaning workers in the Nordic countries (Olsen *et al.*, 1990), described in detail in the monograph on dry cleaning (p. 59), the authors noted that CFC-113 was increasingly used in the later years of the study. Women were grouped into three categories of potential exposure to tetrachloroethylene: no exposure, low exposure (women working in dry cleaning but who were not in the high exposure group) and high exposure (women who actually did dry cleaning or spot removal for at least 1 h per day). The combined odds ratios for spontaneous abortion, adjusted for parity, smoking and drinking habits, were 1.17 (95% CI, 0.74–1.85) for the women with low exposure and 2.88 (95% CI, 0.98–8.44) for those with high exposure. [The Working Group noted that 118 of the 159 spontaneous abortions included in the combined study occurred in Finland. The results of the Finnish study are those reported by Kyyrönen *et al.* (1989).]

The study of Bosco *et al.* (1987), described in detail in the monograph on dry cleaning (p. 60), is relevant to tetrachloroethylene. Of the 56 pregnancies occurring during employment as a dry cleaner, five ended in a spontaneous abortion, whereas one of the 46 pregnancies that occurred in women who were not employed ended in a spontaneous abortion ($\chi_2 = 3.05$, $p < 0.10$). The mean concentration of trichloroacetic acid in urine (a marker of exposure to dry cleaning solvents) was higher among cleaners (5.01 µg/L) than among women who only did ironing (1.35 µg/L) and among controls (1.56 µg/L) [standard deviations not reported].

In the study of Lindbohm *et al.* (1990), described in detail in the monograph on trichloroethylene (p. 120), the odds ratios for spontaneous abortion were 0.5 (95% CI, 0.1–2.9) for low exposure to tetrachloroethylene and 2.5 (0.6–10.5) for high exposure. The mean concentration of tetrachloroethylene in the blood samples taken nearest to the pregnancy was higher among six dry cleaning workers (2.11 µmol/L [350 µg/L]) than among seven other workers monitored for exposure to tetrachloroethylene (0.43 µmol/L [71 µg/L]).

In the study in Santa Clara County, CA (United States) (Windham *et al.*, 1991), described in detail in the monograph on trichloroethylene (p. 120), a significant association was found between exposure to tetrachloroethylene and the risk for spontaneous abortion (crude odds ratio, 4.7; 95% CI, 1.1–21.1). The odds ratio increased for women reporting more intense exposure involving skin contact, odour or symptoms (odds ratio, 6.3; $p = 0.04$). Four of the seven women who reported exposure to tetrachloroethylene had also used trichloroethylene. Odds ratios adjusted individually for maternal age, race, education, prior fetal loss, smoking, average number of hours worked and quality of response ranged from 4.2 (adjusted for number of hours worked) to 6.0 (adjusted for age).

(ii) Stillbirths, congenital malformations, low birth weight

Most early studies involved samples that were too small for investigating rare outcomes.

In the study of Kyyrönen *et al.* (1989), 24 cases of malformations in infants of dry cleaning and laundry workers were compared with 93 cases in controls. The odds ratio in relation to exposure to tetrachloroethylene during the first trimester was 0.8 (95% CI, 0.2–3.5), adjusted for exposure to other solvents, alcohol use, smoking habits and the prevalence of febrile disease.

In the Nordic study described above (Olsen et al., 1990), there were 13 stillbirths, 38 cases of congenital malformations and 13 infants with low birth weights (< 1500 g). No association was found with high exposure to tetrachloroethylene (odds ratio, 0.87; 95% CI; 0.20–3.69).

(iii) Lactation and posnatal effects

Bagnell and Ellenberger (1977) reported a case of obstructive jaundice and hepatomegaly in a six-week-old breast-fed infant. Tetrachloroethylene was detected in the mother's milk and blood (see section 1.3.6, p. 175). [The Working Group noted that the levels reported are very unlikely.] After breast-feeding was discontinued, clinical improvement followed rapidly, and serum bilirubin and transaminase and alkaline phosphatase activities returned to normal values.

4.3.2 Experimental systems

Placental transfer of tetrachloroethylene has been demonstrated in rats, suggesting that tetrachloroethylene may be metabolized by the uterus, placenta or fetus, causing accumulation of trichloroacetic acid in the amniotic fluid after maternal inhalation of tetrachloroethylene (Ghantous et al., 1986).

Swiss-Webster mice and Sprague-Dawley rats were exposed by inhalation to tetrachloroethylene (purity > 99%) at 300 ppm [2034 mg/m^3] for 7 h per day on days 6–15 of gestation. The mean body weight of mouse fetuses was significantly reduced in comparison with controls, and delayed ossification of the skull bones was observed. In rats, the resorption rate was significantly increased, from 4% in controls to 9% in the exposed group (Schwetz et al., 1975).

Groups of about 30 Wistar and Sprague-Dawley rats and about 20 New Zealand white rabbits were exposed by inhalation to 500 ppm [3390 mg/m^3] tetrachloroethylene for 6–7 h per day during gestation. No sign of maternal toxicity, fetal toxicity or malformations was observed (Hardin et al., 1981).

Neuromotor function was impaired in newborn Sprague-Dawley rats exposed *in utero* by maternal inhalation of 900 ppm [6102 mg/m^3] tetrachloroethylene for 7 h per day. No significant difference in the performance of behavioural tests was observed between the offspring of dams exposed to 100 ppm [678 mg/m^3] and those of controls (Nelson et al., 1980).

4.4 Genetic and related effects

4.4.1 Humans

Cytogenetic damage in lymphocytes

Of seven male and three female factory workers occupationally exposed to tetrachloroethylene, six were employed in degreasing, with exposure to concentrations of 30–220 ppm [203–1492 mg/m^3] (geometric mean, 92 ppm [624 mg/m^3]) and work histories of 10–18 years (except for one person who had been exposed for only two years); the four others were employed in a support department, had lower exposure (range, 10–40 ppm [67.8–271 mg/m^3], with rare peaks up to 80 ppm [542 mg/m^3]), and had worked for periods of three months to three years. A control group consisted of six men and five women. No increase in the frequency of chromosomal aberrations or sister chromatid exchange and no decrease in mitotic indices or in the frequencies of second- or third-cycle metaphases were seen in cultured peripheral lymphocytes from the workers. There was, however, a slightly higher frequency than in the controls of peripheral lymphocytes with numerical chromosomal abnormalities (2.5 versus 1.6%; $p < 0.036$). Cultured lymphocytes from the workers responded similarly to those from controls to treatment with mitomycin C with regard to the induction of sister chromatid exchange and chromosomal aberrations (Ikeda *et al.*, 1980). [The Working Group noted that no information was available on the smoking habits of the subjects.]

In the study of Seiji *et al.* (1990), described in detail the monograph on trichloroethylene (p. 123), sister chromatid exchange frequency was studied in dry cleaning workers who had been exposed to tetrachloroethylene and in tetrachloroethylene synthesis workers who were exposed to both tetrachloroethylene and trichloroethylene. An effect was reported in all smoking workers in comparison with nonsmoking controls. [The Working Group noted that the confounding effect of smoking could not be ruled out.]

4.4.2 Experimental systems (see Table 11 and Appendices 1 and 2)

The genetic toxicology of tetrachloroethylene has been reviewed (Fabricant & Chalmers, 1980; Reichert, 1983; WHO, 1984; Vainio *et al.*, 1985; Illing *et al.*, 1987; European Centre for Ecotoxicology and Toxicology of Chemicals, 1990; Jackson *et al.*, 1993). The mechanisms of the genotoxicity of tetrachloroethylene were discussed by Henschler (1987).

(a) DNA binding

Binding of radioactively labelled tetrachloroethylene to calf thymus DNA *in vitro* was reported in one study in the presence of cytosol from mouse and rat kidney, lungs and stomach, microsomes plus cytosol and liver microsomes.

No radioactivity was found in one study in purified hepatic DNA of mice treated *in vivo* with ^{14}C-labelled tetrachloroethylene (detection limit calculated as 10–14.5 alkylations per 10^6 nucleotides). In another study, radioactively labelled tetrachloroethylene bound to DNA and proteins in mouse and rat liver, kidney, lung and stomach *in vivo* (Mazzullo *et al.*, 1987), but the DNA was not purified to constant radioactivity.

(b) Mutation and allied effects

Tetrachloroethylene was not active in the SOS chromotest with *Escherichia coli* and was not mutagenic to bacteria in the absence of metabolic activation. Purified tetrachloroethylene was not mutagenic to *Salmonella typhimurium* or *E. coli* when tested in the presence of a metabolizing system prepared from rat liver microsomes; however, a doubling of revertant frequencies was seen in *S. typhimurium* TA1535 in one study at both doses tested. [The authors considered the finding to be negative.] In another study, purified tetrachloroethylene clearly increased the number of revertants in *S. typhimurium* TA100 in the presence of rat liver glutathione *S*-transferase, glutathione and kidney microsomes. This study was intended to simulate the multistep bioactivation pathway by glutathione conjugation (Vamvakas *et al.*, 1989). The mutagenicity was accompanied by time-dependent formation of *S*-(1,2,2-trichlorovinyl)glutathione, a promutagen activated by kidney microsomes, and did not occur in the absence of glutathione, glutathione *S*-transferase or kidney microsomes. Bile collected from an isolated rat liver system perfused with tetrachloroethylene was clearly mutagenic in the presence

Table 11. Genetic and related effects of tetrachloroethylene

Test system	Result[a] Without exogenous metabolic system	Result[a] With exogenous metabolic system	Dose[b] (LED/HID)	Reference
PRB, SOS chromotest, *Escherichia coli* PQ37	–	–	8150	Mersch-Sundermann et al. (1989)
PRB, SOS chromotest, *Escherichia coli* PQ37	–	–	0.00	von der Hude et al. (1988)
PRB, λ Prophage induction, *Escherichia coli* WP2	–	–	10 000	DeMarini et al. (1994)
SAF, *Salmonella typhimurium* BAL13, forward mutation (*ara* test)	–	–	76	Roldán-Arjona et al. (1991)
SA0, *Salmonella typhimurium* TA100, reverse mutation	–	–	660	Bartsch et al. (1979)
SA0, *Salmonella typhimurium* TA100, reverse mutation	–	–	167	Haworth et al. (1983)
SA0, *Salmonella typhimurium* TA100, reverse mutation	–	–	1000	Connor et al. (1985)
SA0, *Salmonella typhimurium* TA100, reverse mutation	–[c]	–	166 vapour	Shimada et al. (1985)
SA0, *Salmonella typhimurium* TA100, reverse mutation	–	–	0.00	Milman et al. (1988)
SA0, *Salmonella typhimurium* TA100, reverse mutation	–	–[d]	332	Vamvakas et al. (1989)
SA0, *Salmonella typhimurium* TA100, reverse mutation	–	–	1.3 vapour	DeMarini et al. (1994)
SA5, *Salmonella typhimurium* TA1535, reverse mutation	–	0	50	Kringstad et al. (1981)
SA5, *Salmonella typhimurium* TA1535, reverse mutation	–	–	167	Haworth et al. (1983)
SA5, *Salmonella typhimurium* TA1535, reverse mutation	–[c]	(+)	66 vapour	Shimada et al. (1985)
SA5, *Salmonella typhimurium* TA1535, reverse mutation	–	–	0.00	Milman et al. (1988)
SA7, *Salmonella typhimurium* TA1537, reverse mutation	–	–	167	Haworth et al. (1983)
SA7, *Salmonella typhimurium* TA1537, reverse mutation	–	–	0.00	Milman et al. (1988)
SA9, *Salmonella typhimurium* TA98, reverse mutation	–	–	167	Haworth et al. (1983)
SA9, *Salmonella typhimurium* TA98, reverse mutation	–	–	1000	Connor et al. (1985)
SA9, *Salmonella typhimurium* TA98, reverse mutation	–	–	0.00	Milman et al. (1988)
SAS, *Salmonella typhimurium* UTH8413, reverse mutation	–	–	1000	Connor et al. (1985)
SAS, *Salmonella typhimurium* UTH8414, reverse mutation	–	–	1000	Connor et al. (1985)
ECK, *Escherichia coli* K12, forward mutation	–	–	150	Greim et al. (1975)
ECK, *Escherichia coli* K12, reverse mutation (*arg*⁺)	–	–	150	Greim et al. (1975)
ECK, *Escherichia coli* K12, reverse mutation (*gal*⁻)	–	–	150	Greim et al. (1975)
ECK, *Escherichia coli* K12, reverse mutation (*nad*⁻)	–	–	150	Greim et al. (1975)
SCG, *Saccharomyces cerevisiae* D7, log-phase cultures, gene conversion	+	0	1100	Callen et al. (1980)

Table 11 (contd)

Test system	Result[a] Without exogenous metabolic system	Result[a] With exogenous metabolic system	Dose[b] (LED/HID)	Reference
SCG, *Saccharomyces cerevisiae* D7, gene conversion	–	–	9960	Bronzetti et al. (1983)
SCG, *Saccharomyces cerevisiae* D7, log-phase and stationary cultures, gene conversion	–	–	2440	Koch et al. (1988)
SCH, *Saccharomyces cerevisiae* D7, log-phase cultures, mitotic recombination or other genetic alterations (*ade2*)	+	0	1100	Callen et al. (1980)
SCH, *Saccharomyces cerevisiae* D7, mitotic recombination	–	–	9960	Bronzetti et al. (1983)
SCR, *Saccharomyces cerevisiae* D7, log-phase cultures, reverse mutation	(+)	0	810	Callen et al. (1980)
SCR, *Saccharomyces cerevisiae* D7, reverse mutation	–	–	9960	Bronzetti et al. (1983)
SCR, *Saccharomyces cerevisiae* D7, log-phase and stationary cultures, reverse mutation	–	–	2440	Koch et al. (1988)
SCN, *Saccharomyces cerevisiae* D61.M, growing cells, aneuploidy	(+)	(+)	810	Koch et al. (1988)
TSM, *Tradescantia* species, mutation	+	0	7 vapour	Schairer & Sautkulis (1982)
TSI, *Tradescantia* species, micronucleus induction	–	0	600	Sandhu et al. (1989)
TSI, *Tradescantia* species, micronucleus induction	(+)	0	2 vapour	Sandhu et al. (1989)
DMX, *Drosophila melanogaster*, sex-linked recessive mutation	–		1000 injection	Valencia et al. (1985)
DMX, *Drosophila melanogaster*, sex-linked recessive mutation	–		4000 feeding	Valencia et al. (1985)
URP, Unscheduled DNA synthesis, rat primary hepatocytes *in vitro*	–[c]	0	166 vapour	Shimada et al. (1985)
URP, Unscheduled DNA synthesis, Osborne Mendel rat primary hepatocytes *in vitro*	–	0	0.00	Milman et al. (1988)
UIA, Unscheduled DNA synthesis, B6C3F1 mouse primary hepatocytes *in vitro*	–	0	0.00	Milman et al. (1988)
G5T, Gene mutation, mouse lymphoma L5178Y cells, *tk* locus	–	–	245	US National Toxicology Program (1986)
SIC, Sister chromatid exchange, Chinese hamster ovary (CHO) cells *in vitro*	–	–	164	Galloway et al. (1987)
CIC, Chromosomal aberrations, Chinese hamster lung (CHL) cells *in vitro*	–		500	Sofuni et al. (1985)
CIC, Chromosomal aberrations, Chinese hamster ovary (CHO) cells *in vitro*	–	–	136	Galloway et al. (1987)
TRR, Cell transformation, RLV/Fischer rat embryo F1706 cells *in vitro*	+	0	16	Price et al. (1978)
TBM, BALB/c-3T3 mouse cells, cell transformation *in vitro*	–	0	250	Tu et al. (1985)
HMM, Gene conversion and reverse mutation in *Saccharomyces cerevisiae* D7 recovered from liver, lungs and kidneys of CD-1 mice	–	0	11 000 po × 1	Bronzetti et al. (1983)

Table 11 (contd)

Test system	Result[a]		Dose[b] (LED/HID)	Reference
	Without exogenous metabolic system	With exogenous metabolic system		
HMM, Gene conversion and reverse mutation in *Saccharomyces cerevisiae* D7 recovered from liver, lungs and kidneys of CD-1 mice	0	–	2000 po × 12	Bronzetti *et al.* (1983)
DVA, DNA single-strand breaks (alkaline unwinding) in liver and kidney of male NMRI mice *in vivo*	+[c]		660 ip × 1	Walles (1986)
SLH, Sister chromatid exchange, human lymphocytes *in vivo*	–		88 inh	Ikeda *et al.* (1980)
CLH, Chromosomal aberrations, human lymphocytes *in vivo*	–		88 inh	Ikeda *et al.* (1980)
BID, Binding (covalent) to calf thymus DNA *in vitro*	0	+	9	Mazzullo *et al.* (1987)
BVD, Binding (covalent) to DNA in male B6C3F1 mouse liver *in vivo*	–		1400 inh 6 h	Schumann *et al.* (1980)
BVD, Binding (covalent) to DNA in male B6C3F1 mouse liver *in vivo*	–		500 po × 1	Schumann *et al.* (1980)
BVD, Binding (covalent) to DNA in male BALB/c mouse and Wistar rat liver, kidney, lung and stomach *in vivo*	+		1.3 ip × 1	Mazzullo *et al.* (1987)
BVP, Binding (covalent) to RNA and protein in male BALB/c mouse and Wistar rat liver, kidney, lung and stomach *in vivo*	+		1.3 ip × 1	Mazzullo *et al.* (1987)
***, Enzyme-altered foci in male Osborne Mendel rat liver *in vivo*, promotion protocol, with or without NDEA as an initiator	+		1000, 5 d/week, for 7 weeks	Milman *et al.* (1988)
***, Enzyme-altered foci in male Osborne Mendel rat liver *in vivo*, initiation protocol, phenobarbital as a promoter	–		1000	Milman *et al.* (1988)
Trichloroacetyl chloride				
PRB, λ Prophage induction, *Escherichia coli* WP2	–	–	10 000	DeMarini *et al.* (1994)
SA0, *Salmonella typhimurium* TA100, reverse mutation	+	+	2.6	DeMarini *et al.* (1994)

NDEA, N-nitrosodiethylamine

[a] +, considered to be positive; (+), considered to be weakly positive in an inadequate study; –, considered to be negative; ?, considered to be inconclusive (variable responses in several experiments within an inadequate study); 0, not tested

[b] LED, lowest effective dose; HID, highest effective dose. In-vitro tests, µg/ml; in-vivo tests, mg/kg bw; 0.00, dose not reported; ip, intraperitoneally; po, orally

[c] Tetrachloroethylene with stabilizers was positive with and without metabolic activation

[d] Weak increase in activity with rat liver S9, rat kidney microsomes and glutathione (GSH); fourfold increase with rat kidney microsomes, GSH and GSH S-transferase

[e] Negative in lung

***, Not included on profile

of rat kidney particulate fractions. In both assay systems, the mutagenicity was reduced by the presence of the γ-glutamyl transpeptidase inhibitor, serine borate, and the β-lyase inhibitor, aminooxoacetic acid, suggesting that the important steps in the metabolic activation of tetrachloroethylene in renal fractions are glutathione S-transferase-mediated glutathione conjugation, with subsequent γ-glutamyl-transpeptidase-mediated formation of an S-cysteine conjugate and bioactivation of the S-cysteine conjugate by β-lyase.

Tetrachloroethylene did not induce gene conversion, mitotic recombination or reverse mutations in yeast in stationary cultures. Both negative and positive findings were reported from studies of cultures in logarithmic growth phase, which stimulates xenobiotic metabolism; and positive results were obtained with tetrachloroethylene containing 0.01% thymol as a stabilizer. In a single study in yeast, tetrachloroethylene (analytical grade) weakly induced aneuploidy in growing cells capable of xenobiotic metabolism.

In single studies with *Tradescantia*, tetrachloroethylene [purity not given] induced mutations, and a 99% pure compound induced a slight increase in the frequency of micronuclei.

In one study with *Drosophila*, tetrachloroethylene did not induce sex-linked recessive lethal mutations.

Highly pure tetrachloroethylene did not induce unscheduled DNA synthesis in rat primary hepatocytes, and tetrachloroethylene did not induce chromosomal aberrations or (in a single study) sister chromatid exchange in Chinese hamster cells *in vitro*. It did not induce gene mutation in mouse lymphoma cells in one study, in the presence or absence of exogenous metabolic activation. In single studies, tetrachloroethylene induced cell transformation in Fischer rat embryo cells but not in mouse BALB/c-3T3 cells in the absence of exogenous metabolic activation.

In a single study, the frequency of gene conversion and reverse mutation was not increased in yeast recovered from the liver, lungs and kidneys of mice after treatment with tetrachloroethylene *in vivo*. In another study, tetrachloroethylene (purity, 99.8%) induced DNA single-strand breaks/alkaline-labile sites in mouse liver and kidney, but not lung, after treatment *in vivo*.

(c) Genetic effects of tetrachloroethylene metabolites

The genetic toxicology of dichloroacetate and trichloroacetate is reviewed in the relevant monographs in this volume.

Trichloroacetyl chloride, a presumed intermediate metabolite of tetrachloroethylene, did not induce λ prophage in *E. coli*, but was mutagenic to *S. typhimurium* TA100.

The minor urinary metabolite N-acetyl-S-(1,2,2-trichlorovinyl)-L-cysteine was mutagenic to *S. typhimurium* TA100 in the presence of kidney cytosol, which allows deacylation to the corresponding cysteine conjugate (Vamvakas *et al.*, 1987). The presumed intermediate metabolite, S-(1,2,2-trichlorovinyl)-L-cysteine, was mutagenic to *S. typhimurium* TA100 in the presence and absence of metabolic activation (Green & Odum, 1985; Dekant *et al.*, 1986). S-(1,2,2-Trichlorovinyl)glutathione, the precursor of the cysteine conjugate, was also mutagenic to *S. typhimurium* TA100 in the presence of rat kidney microsomes, which catalyse its degradation to the cysteine conjugate. Both the cysteine and glutathione conjugates induced a low rate of unscheduled DNA synthesis in a cultured porcine kidney cell line (Vamvakas *et al.*, 1989).

(d) Mutations in proto-oncogenes in tumours from tetrachloroethylene-treated animals

A group of 160 male B6C3F1 mice, eight weeks of age, was administered tetrachloroethylene in corn oil by gavage at a dose of 800 mg/kg bw daily on five days per week for up to 76 weeks. There were two concurrent control groups, each consisting of 50 male mice: one was untreated, while the other received corn oil at a dose of 10 ml/kg bw. Ten control mice in each group were killed at 76 weeks, and the remainder were killed at 96, 103 and 134 weeks [numbers not stated]. At death, liver tumours ≥ 0.5 cm diameter were taken for histological examination and for oncogene analysis. At the time of the terminal kill, there were 24 untreated controls, 32 vehicle controls and 112 mice treated with tetrachloroethylene. The numbers of hepatocellular adenomas per mouse in these three groups were 0.9 ± 0.06 (8%), 0.13 ± 0.06 (13%) and 1.43 ± 0.11 (80%). The corresponding numbers of hepatocellular carcinomas were 0.09 ± 0.06 (8%), 0.12 ± 0.06 (12%) and 0.29 ± 0.05 (23%). The authors noted numerous foci of cellular alteration (presumed preneoplastic lesions) in the livers of treated mice but only rare foci in the livers of controls. No neoplasms related to treatment were found at other sites. The frequency of mutations in codon 61 of H-*ras* was significantly *reduced* in 53 hepatocellular tumours from tetrachloroethylene-treated mice in comparison with the 74 combined historical and concurrent controls (24% versus 69%). The spectra of these mutations showed a smaller proportion of AAA and a higher proportion of CTA in the tumours from treated mice in comparison with those of controls, but these variations were not statistically significant. Other H-*ras* mutations contributed 4% and K-*ras* mutations contributed 13% to the total in the treated mice, whereas their frequency appeared to be very low in the concurrent controls, and none were seen in the historical controls. The authors interpreted these findings as suggesting that exposure to tetrachloroethylene provides the environment for a selective growth advantage for spontaneous CTA mutations in codon 61 of H-*ras* (Anna *et al.*, 1994).

5. Summary and Evaluation

5.1 Exposure data

Tetrachloroethylene is one of the most important chlorinated solvents worldwide and has been produced commercially since the early 1900s. Most of the tetrachloroethylene produced is used for dry cleaning garments; smaller amounts are used in the production of chlorofluorocarbons and for degreasing metals. About 513 thousand tonnes were used in all applications in western Europe, Japan and the United States in 1990.

Tetrachloroethylene has been detected in air, water, food and animal and human tissues. The greatest exposure occurs via inhalation, and workers in dry cleaning and degreasing are the most heavily exposed. Individuals living or working in the vicinity of such operations have been shown to be exposed to lower concentrations.

5.2 Human carcinogenicity data

Results relevant to assessing the relationship between exposure to tetrachloroethylene and cancer risk are available from five cohort studies. In one study in Finland and one in four states of the United States, exposure was specifically to tetrachloroethylene; biological monitoring was conducted in the Finnish study. In a cohort study in Missouri, United States, in which follow-up was from 1948 to 1978, tetrachloroethylene was the chemical to which predominant exposure had occurred since about 1960. Data for a few cancer sites were reported in two other cohort studies, one in Louisiana and one in Utah, United States, in which exposure was to both tetrachloroethylene and other chemicals. Although data on different levels or duration of exposure were available in some of the cohort studies, the number of observed cases in each category was generally too small to allow adequate statistical power for testing for a dose–response relationship. Data from six relevant case–control studies have also been reported.

In the two cohort studies in which results for oesophageal cancer were reported, namely the four-state United States and Missouri studies, the relative risks were 2.6 and 2.1. Lack of data on smoking or alcohol consumption, both strong risk factors for this cancer, indicates caution in interpreting this observation.

The relative risks for cervical cancer were increased in three cohort studies in which such results were reported; however, potential confounding factors associated with socioeconomic status could not be adjusted for.

Elevated relative risks for non-Hodgkin's lymphoma were observed in all three cohort studies in which such results were reported.

With respect to cancer of the kidney, no consistent pattern of elevated risk was seen in the three cohort studies in which such results were reported. Although a case–control study conducted in Montréal, Canada, showed an odds ratio of 3.4, this was not statistically significant, and the exposure in question was to degreasing solvents and not specifically to tetrachloroethylene. In the cohort study in Missouri, the relative risk for urinary bladder cancer was elevated but not statistically significant; little or no information was available from other studies.

Five studies of people exposed to drinking-water contaminated with tetrachloroethylene have been reported. In four of these, no consistent pattern of risk for any specific cancers was observed. In the fifth study, in Massachusetts, United States, although the increase in the relative risk for leukaemia was significant, the result was based on only two cases. No consistent evidence for an elevated risk for leukaemia was seen in the cohort studies.

In summary, there is evidence for consistently positive associations between exposure to tetrachloroethylene and the risks for oesophageal and cervical cancer and non-Hodgkin's lymphoma. These associations appear unlikely to be due to chance, although confounding cannot be excluded and the total numbers in the cohort studies combined are relatively small.

5.3 Animal carcinogenicity data

Tetrachloroethylene was tested for carcinogenicity by oral administration in one experiment in mice, and a significant increase in the incidence of hepatocellular carcinomas was observed in animals of each sex. A study in rats treated orally was inadequate for an evaluation of carcinogenicity. Tetrachloroethylene was tested for carcinogenicity by inhalation in one experiment in mice and in one experiment in rats. The incidence of hepatocellular adenomas and carcinomas was significantly increased in mice of each sex, and the incidence of mononuclear-cell leukaemia was significantly increased in rats of each sex. A nonsignificant increase in the incidence of uncommonly occurring renal-cell adenomas and adenocarcinomas was also observed in male rats. Tetrachloroethylene did not induce skin tumours in mice after administration by topical application in one study.

A presumed metabolite of tetrachloroethylene, tetrachloroethylene oxide, did not increase the incidence of local tumours in mice when given by topical application or subcutaneous injection.

5.4 Other relevant data

Tetrachloroethylene is rapidly absorbed after inhalation and from the gastrointestinal tract, but dermal absorption from the gaseous phase is negligible. The biotransformation of tetrachloroethylene is species- and dose-dependent; mice consistently had a greater capacity to biotransform tetrachloroethylene than rats. Two metabolic pathways have been demonstrated in rodents: cytochrome P450-catalysed oxidation and, as a minor route, glutathione conjugation.

Tetrachloroethylene shows only low acute toxicity in humans and in experimental animals. After repeated administration, the major target organ is the liver in mice and the kidney in rats. Tetrachloroethylene induced peroxisome proliferation in mouse liver after oral administration; a marginal response was observed in mouse kidney and rat liver.

Disturbances of sperm quality and fertility have been observed among dry cleaners exposed to tetrachloroethylene in a few studies of limited size. The results of studies of women exposed to tetrachloroethylene in dry cleaning shops and other settings are generally consistent in showing an increase in the rate of spontaneous abortions; however, other solvents were also present in most of these workplaces. Effects on other reproductive outcomes such as stillbirths, congenital malformations and low birth weight could not be evaluated in these studies.

Tetrachloroethylene can cross the placenta of rats and is metabolized in the placenta or fetus to trichloroacetic acid. Tetrachloroethylene appears to have little toxicity in developing rats and rabbits; high atmospheric concentrations produced delayed fetal development in mice in one study.

The frequencies of gene conversion and gene mutation were not increased in yeast recovered from mice treated with tetrachloroethylene *in vivo*. Tetrachloroethylene increased the frequency of DNA single-strand breakage/alkaline-labile sites in the liver and kidney of mice *in vivo* in one study, but binding to DNA was not demonstrated in mouse liver.

It did not induce gene mutation (in a single study), chromosomal aberrations, sister chromatid exchange (in a single study) or DNA damage in mammalian cells *in vitro*. In single studies, it induced morphological transformation in virus-infected rat embryo cells but not in BALBc/3T3 cells. The only study available showed no induction of gene mutation by tetrachloroethylene in insects. Tetrachloroethylene did not usually induce gene conversion in yeasts; the results with regard to induction of aneuploidy in one study were inconclusive. Tetrachloroethylene did not increase the frequency of mutations in bacteria, except in one study in which a metabolic activation system consisting of liver and kidney fractions, which favours glutathione conjugation and further activation, was used. The metabolites formed from tetrachloroethylene in rats by minor biotransformation pathways, S-1,2,2-trichloroglutathione and derived sulfur conjugates, were genotoxic in bacteria and cultured renal cells.

The frequency of H-*ras* mutations was lower in hepatocellular tumours from tetrachloroethylene-treated mice than in tumours from control animals, whereas the frequency in hepatocellular tumours from trichloroethylene-treated mice was not significantly different from that in controls. The frequency of K-*ras* mutations was higher in liver tumours from tetrachloroethylene-treated mice than in liver tumours from control animals.

5.5 Evaluation[1]

There is *limited evidence* in humans for the carcinogenicity of tetrachloroethylene.

There is *sufficient evidence* in experimental animals for the carcinogenicity of tetrachloroethylene.

Overall evaluation[2]

Tetrachloroethylene *is probably carcinogenic to humans (Group 2A)*.

In making the overall evaluation, the Working Group considered the following evidence:

(i) Although tetrachloroethylene is known to induce peroxisome proliferation in mouse liver, a poor quantitative correlation was seen between peroxisome proliferation and tumour formation in the liver after administration of tetrachloroethylene by inhalation. The spectrum of mutations in proto-oncogenes in liver tumours from mice treated with tetrachloroethylene is different from that in liver tumours from mice treated with trichloroethylene.

(ii) The compound induced leukaemia in rats.

(iii) Several epidemiological studies showed elevated risks for oesophageal cancer, non-Hodgkin's lymphoma and cervical cancer.

6. References

Abrahamsson, K., Ekdahl, A., Collén, J., Fahlström, E., Sporrong, N. & Pedersén, M. (1995) Marine algae—a source of trichloroethylene and perchloroethylene. *Limnol. Oceanogr.* (in press)

Abundo, M.-L., Almaguer, D. & Driscoll, R. (1994) *Electrode Corp., Chadon, OH* (HETA 93-1133-2425), Cincinnati, OH, United States National Institute for Occupational Safety and Health

---

[1] For definitions of the italicized terms, see Preamble, pp. 22–26.
[2] Dr N.H. Stacey disassociated himself from the overall evaluation.

Aggazzotti, G., Fantuzzi, G., Righi, E., Predieri, G., Gobba, F.M., Paltrinieri, M. & Cavalleri, A. (1994) Occupational and environmental exposure to perchloroethylene (PCE) in dry cleaners and their family members. *Arch. environ. Health*, **49**, 487–493

Ahrenholz, S.H. & Anderson, K.E. (1982) *Industrial and Automotive Fasteners, Inc., Royal Oak, MI* (NIOSH Hazard Evaluation Report No. 82-037-1120), Cincinnati, OH, United States National Institute for Occupational Safety and Health

Aitio, A., Luotamo, M. & Kiilunen, M., eds (1995) *Kemikaalialtistumisen Biomonitorointi* [Biological monitoring of exposure to chemicals], Helsinki, Finnish Institute of Occupational Health, pp. 263–266 (in Finnish)

American Conference of Governmental Industrial Hygienists (1984) *1984–1985 Threshold Limit Values for Chemical Substances and Physical Agents and Biological Exposure Indices*, Cincinnati, OH, p. 26

American Conference of Governmental Industrial Hygienists (1994) *1994–1995 Threshold Limit Values for Chemical Substances and Physical Agents and Biological Exposure Indices*, Cincinnati, OH, pp. 29, 61, 64

Anna, C.H., Maronpot, R.R., Pereira, M.A., Foley, J.F., Malarkey, D.E. & Anderson, M.W. (1994) *ras* Proto-oncogene activation in dichloroacetic-, trichloroethylene- and tetrachloroethylene-induced liver tumors in B6C3F1 mice. *Carcinogenesis*, **15**, 2255–2261

Anttila, A., Pukkala, E., Sallmén, M., Hernberg, S. & Hemminki, K. (1995) Cancer incidence among Finnish workers exposed to halogenated hydrocarbons. *J. occup. Med.* (in press)

Arbeidsinspectie (1994) *De Nationale MAC-Lijst 1994* [The national MAC list 1994], The Hague, p. 39

Aschengrau, A., Ozonoff, D., Paulu, C., Coogan, P., Vezina, R., Heeren, T. & Zhang, Y. (1993) Cancer risk and tetrachloroethylene-contaminated drinking water in Massachusetts. *Arch. environ. Health*, **48**, 284–292

Ashley, D.L., Bonin, M.A., Cardinali, F.L., McCraw, J.M. & Wooten, J.V. (1994) Blood concentrations of volatile organic compounds in a non-occupationally exposed United States population and in groups with suspected exposure. *Clin. Chem.*, **40**, 1401–1404

Bagnell, P.C. & Ellenberger, H.A. (1977) Obstructive jaundice due to chlorinated hydrocarbon in breast milk. *Can. med. Assoc. J.*, **117**, 1047–1048

Bartsch, H., Malaveille, C., Barbin, A. & Planche, G. (1979) Mutagenic and alkylating metabolites of halo-ethylenes, chlorobutadienes and dichlorobutenes produced by rodent or human liver tissues. *Arch. Toxicol.*, **41**, 249–277

Bauer, U. (1978) Halogenated hydrocarbons in drinking and surface waters. Results of the 1976/77 measurements in the Federal Republic of Germany (Drinking water from 100 cities, surface water from the Ruhr, Lippe, Main, Rhein). In: Sonneborn, M., ed., *Gesundheitliche Probleme der Wasserchlorung und Bewertung der dabei gebildeten halogenierten organischen Verbindungen* [Hygienic problems of water chlorination and estimation of halogenated organic compounds formed therefrom], Berlin, Dietrich Reimer Verlag, pp. 64–74 (in German)

Bauer, U. (1981) Human exposure to environmental chemicals. Investigations on volatile organic halogenated compounds in water, air, food and human tissues. I, III, IV. *Zbl. Bakt. Hyg. I Abt. Orig. B.*, **174**, 15–56, 200–237, 556–583 (in German)

Bauer, U. (1991) Occurrence of tetrachloroethylene in the Federal Republic of Germany. *Chemosphere*, **23**, 1777–1781

Birner, G., Richling, C., Henschler, D., Anders, M.W. & Dekant, W. (1994) Metabolism of tetrachloroethene in rats: identification of N^ε-(dichloroacetyl)-L-lysine and N^ε-(trichloroacetyl)-L-lysine as protein adducts. *Chem. Res. Toxicol.*, **7**, 726–732

Blair, A., Stewart, P.A., Tolbert, P.E., Grauman, D., Moran, F.X., Vaught, J. & Rayner, J. (1990) Cancer and other causes of death among a cohort of dry cleaners. *Br. J. ind. Med.*, **47**, 162–168

Bogen, K.T., Colston, B.W., Jr & Machicao, L.K. (1992) Dermal absorption of dilute aqueous chloroform, trichloroethylene and tetrachloroethylene in hairless guinea pigs. *Fundam. appl. Toxicol.*, **18**, 30–39

Bolt, H.M. & Link, B. (1980) Toxicology of tetrachloroethylene. *Verh. Dtsch. ges. Arbeitsmed.*, **20**, 463–470 (in German)

Bolzer, W.E. (1982) Volatile halogenated hydrocarbons: a new risk to our drinking water. *Osterreich. Wasserrwirtsch.*, **34**, 198–204 (in German)

Bond, G.G., McLaren, E.A., Sabel, F.L., Bodner, K.M., Lipps, T.E. & Cook, R.R. (1990) Liver and biliary tract cancer among chemical workers. *Am. J. ind. Med.*, **18**, 19–24

Bonse, G. & Henschler, D. (1976) Chemical reactivity, biotransformation, and toxicity of polychlorinated aliphatic compounds. *Crit. Rev. Toxicol.*, **4**, 395–409

Bosco, M.G., Figà-Talamanca, I. & Salerno, S. (1987) Health and reproductive status of female workers in dry-cleaning shops. *Int. Arch. occup. environ. Health*, **59**, 295–301

Brodzinsky, R. & Singh, H.B. (1983) *Volatile Organic Chemicals in the Atmosphere: An Assessment of Available Data* (EPA-60018-83-027a), Menlo Park, CA, SRI International, pp. 109–110

Bronzetti, G., Bauer, C., Corsi, C., Del Carratore, R., Galli, A., Nieri, R. & Paolini, M. (1983) Genetic and biochemical studies on perchloroethylene '*in vitro*' and '*in vivo*'. *Mutat. Res.*, **116**, 323–331

Buben, J.A. & O'Flaherty, E.J. (1985) Delineation of the role of metabolism in the hepatoxicity of trichloroethylene and perchloroethylene: a dose–effect study. *Toxicol. appl. Pharmacol.*, **78**, 105–122

Budavari, S., ed. (1989) *The Merck Index*, 11th Ed., Rahway, NJ, Merck & Co., p. 1449

Burgess, W.A. (1981) *Recognition of Health Hazards in Industry. A Review of Materials & Processes*, New York, John Wiley & Sons, pp. 20–33

Burroughs, G.E. (1980) *Protective Coatings Corporation, Fort Wayne, IN* (Health Hazard Evaluation Report No. 79-96-729), Cincinnati, OH, United States National Institute for Occupational Safety and Health

Burston, M.W., Nazari, M.M., Bishop, P.K. & Lerner, D.N. (1993) Pollution of groundwater in the Coventry region (UK) by chlorinated hydrocabon solvlents. *J. Hydrol.*, **149**, 137–161

Callen, D.F., Wolf, C.R. & Philpot, R.M. (1980) Cytochrome P-450 mediated genetic activity and cytotoxicity of seven halogenated aliphatic hydrocarbons in *Saccharomyces cerevisiae*. *Mutat. Res.*, **77**, 55–63

Center for Chemical Hazard Assessment (1985) *Monographs on Human Exposure to Chemicals in the Worplace: Tetrachloroethylene*, Syracuse, NY

Center for Emissions Control (1994) *Worker Exposure to Perchloroethylene and Trichloroethylene*, Washington DC

Chemical Information Services, Inc. (1994) *Directory of World Chemical Producers 1995/96 Standard Edition*, Dallas, TX, p. 556

Chrostek, W.J. & Levine, M.J. (1981) *Bechtel Power Corp., Berwick, PA* (Health Hazard Evaluation Report No. 80-154-1027), Cincinnati, OH, United States National Institute for Occupational Safety and Health

Class, T. & Ballschmiter, K. (1986) Chemistry of organic traces in air. VI: Distribution of chlorinated C_1-C_4 hydrocarbons in air over the northern and southern Atlantic Ocean. *Chemosphere*, **15**, 413–427

Class, T. & Ballschmiter, K. (1987) Atmospheric halocarbons: global budget estimation for tetrachloroethylene, 1,2-dichloroethane, 1,1,1,2-tetrachloroethane, hexachloroethane, and hexachlorobutadiene. Estimation of the hydroxyl radical concentrations in the troposphere of the northern and southern hemisphere. *Fres. Z. anal. Chem.*, **327**, 198–204

Cohn, P., Klotz, J., Bove, F., Berkowitz, M. & Fagliano, J. (1994) Drinking water contamination and the incidence of leukemia and non-Hodgkin's lymphoma. *Environ. Health Perspectives*, **102**, 556–561

Coler, H.R. & Rossmiller, H.R. (1953) Tetrachloroethylene exposure in a small industry. *Arch. ind. Hyg.*, **8**, 227–233

Comba, M.E., Palabrica, V.S. & Kaiser, K.L.E. (1994) Volatile halocarbons as tracers of pulp mill effluent plumes. *Environ. Toxicol. Chem.*, **13**, 1065–1074

Connor, T.H., Theiss, J.C., Hanna, H.A., Monteith, D.K. & Matney, T.S. (1985) Genotoxicity of organic chemicals frequently found in the air of mobile homes. *Toxicol. Lett.*, **25**, 33–40

Cook, W.A. (1987) *Occupational Exposure Limits—Worldwide*, Akron, OH, American Industrial Hygiene Association, pp. 124, 150, 207

Cook, C.K. & Parker, M. (1994) *Geneva Rubber Co., Geneva, OH* (HETA 91-0377-2383), Cincinnati, OH, United States National Institute for Occupational Safety and Health

Coopers & Lybrand, Ltd (1993) *Methylene Chloride, Perchloroethylene, and Trichloroethylene. Worldwide Statistics for Years from 1988–1992*, Brussels

Costa, A.K. & Ivanetich, K.M. (1980) Tetrachloroethylene metabolism by the hepatic microsomal cytochrome P-450 system. *Biochem. Pharmacol.*, **29**, 2863–2869

Costello, R.J. (1979) *Western Electric Co., Denver, CO* (Health Hazard Evaluation Determination Report No. 78-111-588), Cincinnati, OH, United States National Institute for Occupational Safety and Health

Crowley, R.C., Ford, C.B. & Stern, A.C. (1945) A study of perchloroethylene degreasers. *J. ind. Hyg. Toxicol.*, **27**, 140–144

Daft, J.L. (1988) Rapid determination of fumigant and industrial chemical residues in food. *J. Assoc. off. anal. Chem.*, **71**, 748–760

Dallas, C.E., Muralidhara, S., Chen, X.-M., Ramanathan, R., Varkonyi, P., Gallo, J.M. & Bruckner, J.V. (1994a) Use of a physiologically based model to predict systemic uptake and respiratory elimination of perchloroethylene. *Toxicol. appl. Pharmacol.*, **128**, 60–68

Dallas, C.E., Chen, X.-M., O'Barr, K., Muralidhara, S., Varkonyi, P. & Bruckner, J.V. (1994b) Development of a physiologically based pharmacokinetic model for perchloroethylene using tissue concentration–time data. *Toxicol. appl. Pharmacol.*, **128**, 50–59

Dallas, C.E., Chen, X.-M., Muralidhara, S., Varkonyi, P., Tackett, R.L. & Bruckner, J.V. (1994c) Use of tissue disposition data from rats and dogs to determine species differences in input parameters for a physiological model for perchloroethylene. *Environ. Res.*, **67**, 54–67

Daniel, J.W. (1963) The metabolism of ^{36}Cl-labelled trichloroethylene and tetrachloroethylene in the rat. *Biochem. Pharmacol.*, **12**, 795–802

Daniels, W. & Kramkowski, R. (1986) *Summit Finishing Co., Inc., Mooresville, IN* (Health Hazard Evaluation Report No. 85-083-1705), Cincinnati, OH, United States National Institute for Occupational Safety and Health

Dekant, W., Metzler, M. & Henschler, D. (1986) Identification of S-1,2,2-trichlorovinyl-N-acetylcysteine as a urinary metabolite of tetrachloroethylene: bioactivation through glutathione conjugation as a possible explanation of its nephrocarcinogenicity. *J. biochem. Toxicol.*, **1**, 57–72

Dekant, W., Martens, G., Vamvakas, S., Metzler, M. & Henschler, D. (1987) Bioactivation of tetrachloroethylene—role of glutathione S-transferase-catalyzed conjugation versus cytochrome P-450-dependent phospholipid alkylation. *Drug Metab. Disposition*, **15**, 702–709

Dekant, W., Berthold, K., Vamvakas, S., Henschler, D. & Anders, M.W. (1988) Thioacylating intermediates as metabolites of S-(1,2-dichlorovinyl)-L-cysteine and S-(1,2,2-trichlorovinyl)-L-cysteine formed by cysteine conjugate β-lyase. *Chem. Res. Toxicol.*, **1**, 175–178

Dekant, W., Urban, G., Görsmann, C. & Anders, M.W. (1991) Thioketene formation from α-haloalkenyl 2-nitrophenyl disulfides: models for biological reactive intermediates of cytotoxic S-conjugates. *J. Am. chem. Soc.*, **113**, 5120–5122

DeMarini, D.M., Perry, E. & Shelton, M.L. (1994) Dichloroacetic acid and related compounds: induction of prophage in *E. coli* and mutagenicity and mutation spectra in Salmonella TA100. *Mutagenesis*, **9**, 429–437

Deutsche Forschungsgemeinschaft (1993) *MAK-und BAT-werte-liste 1993* [MAK and BAT list 1993], Weinheim, VCH Verlagsgesellschaft, pp. 72, 129

Dietz, F. & Traud, J. (1973) Determination of low molecular hydrocarbons in water and sludge with GC. *Wasser*, **41**, 137–155 (in German)

Douglas, V., Eli, A., Kemena, R., Marvin, C., McLaughlin, C., Vincent, R. & Woodhouse, L. (1993) *Proposed Airborne Toxic Control Measure and Proposed Environmental Training Program For Perchloroethylene Dry Cleaning Operations*, Sacramento, CA, California Environmental Protection Agency, Air Resources Board, Stationary Source Division

Dumortier, L., Nicolas, G. & Nicolas, F. (1964) A case of hepato-nephritis due to perchloroethylene. Artificial kidney cure. *Arch. mal. Prof.*, **25**, 519

von Düszeln, J., Lahl, U., Bätjer, K., Cetinkaya, M., Stachel, B. & Thiemann, W. (1982) Volatile halogenated carbohydrates in air, water, and food in the Federal Republic of Germany. *Dtsch. Lebensmittel. Rundsch.*, **78**, 352–356 (in German)

Dybing, F. & Dybing, O. (1946) The toxic effect of tetrachloromethane and tetrachloroethylene in oily solution. *Acta pharmacol. toxicol.*, **2**, 223–226

Dyksen, J.E. & Hess, A.F., III (1982) Alternatives for controlling organics in groundwater supplies. *J. Am. Water Works Assoc.*, 74, 394–403

Eller, P.M., ed. (1994) *NIOSH Manual of Analytical Methods*, 4th Ed., Vol. 2 (DHHS (NIOSH) Publ. No. 94-113), Washington DC, United States Government Printing Office, Method 1003

Entz, R.C. & Diachenko, G.W. (1988) Residues of volatile halocarbons in margarines. *Food Addit. Contam.*, **5**, 267–276

Eskenazi, B., Wyrobek, A.J., Fenster, L., Katz, D.F., Sadler, M., Lee, J., Hudes, M. & Rempel, D.M. (1991a) A study of the effect of perchloroethylene exposure on semen quality in dry-cleaning workers. *Am. J. ind. Med.*, **20**, 575–591

Eskenazi, B., Fenster, L., Hudes, M., Wyrobek, A.J., Katz, D.F., Gerson, J. & Rempel, D.M. (1991b) A study of the effect of perchloroethylene exposure on the reproductive outcomes of wives of dry-cleaning workers. *Am. J. ind. Med.*, **20**, 593–600

European Centre for Ecotoxicology and Toxicology of Chemicals (1990) *Tetrachloroethylene: Assessment of Human Carcinogenic Hazard* (ECETOC Technical Report No. 37), Brussels

European Centre for Ecotoxicology and Toxicology of Chemicals (1995) *Tetrachloroethylene*, Brussels (in press)

Evans, G., Baumgardner, R., Bumgarner, J., Finkelstein, P., Knoll, J., Martin, B., Sykes, A., Wagoner, D. & Decker, L. (1979) *Measurement of Perchloroethylene in Ambient Air* (EPA 600/4-79-047), Research Triangle Park, NC, United States Environmental Protection Agency

Fabricant, J.D. & Chalmers, J.H., Jr (1980) Evidence of the mutagenicity of ethylene dichloride and structurally related compounds. In: Ames, B., Infante, P. & Reitz, R., eds, *Ethylene Dichloride: A Potential Health Risk* (Banbury Report 5), Cold Spring Harbor, NY, CSH Press, pp. 309–329

Fernandez, J., Guberan, E. & Caperos, J. (1976) Experimental human exposures to tetrachloroethylene vapor and elimination in breath after inhalation. *Am. ind. Hyg. Assoc. J.*, **37**, 143–150

Ferrario, J.B., Lawler, G.C., DeLeon, I.R. & Laseter, J.L. (1985) Volatile organic pollutants in biota and sediment of Lake Pontchartrain. *Bull. environ. Contam. Toxicol.*, **34**, 246–255

Foot, E., Bishop, K. & Apgar, V. (1943) Tetrachloroethylene as an anesthetic agent. *Anesthesiology*, **4**, 283–292

Fytianos, K., Vasilikiotis, G. & Well, L. (1985) Identification and determination of some trace organic compounds in coastal seawater of northern Greece. *Bull. Environ. Contam. Toxicol.*, **34**, 390–395

Galloway, S.M., Armstrong, M.J., Reuben, C., Colman, S., Brown, B., Cannon, C., Bloom, A.D., Nakamura, F., Ahmed, M., Duk, S., Rimpo, J., Margolin, B.H., Resnick, M.A., Anderson, B. & Zeiger, E. (1987) Chromosome aberrations and sister chromatid exchanges in Chinese hamster ovary cells: evaluations of 108 chemicals. *Environ. mol. Mutag.*, **10** (Suppl. 10), 1–175

Gearhart, J.M., Mahle, D.A., Greene, R.J., Seckel, C.S., Flemming, C.D., Fisher, J.W. & Clewell, H.J., III (1993) Variability of physiologically based pharmacokinetic (PBPK) model parameters and their effects on PBPK model predictions in a risk assessment for perchloroethylene (PCE). *Toxicol. Lett.*, **68**, 131–144

Ghantous, H., Danielsson, B.R.G., Dencker, L., Gorczak, J. & Vesterberg, O. (1986) Trichloroacetic acid accumulates in murine amniotic fluid after tri- and tetrachloroethylene inhalation. *Acta pharmacol. toxicol.*, **58**, 105–114

Gilbert, D., Goyer, M., Lyman, W., Magil, G., Walker, P., Wallace, D., Wechsler, A. & Yee, J. (1982) *An Exposure and Risk Assessment for Tetrachloroethylene* (EPA-440/4-85-015), Washington DC, United States Environmental Protection Agency, Office of Water Regulations and Standards

Gold, J.H. (1969) Chronic perchloroethylene poisoning. *Can. Psychiatr. Assoc. J.*, **14**, 627

Goldsworthy, T.L. & Popp, J.A. (1987) Chlorinated hydrocarbon-induced peroxisomal enzyme activity in relation to species and organ carcinogenicity. *Toxicol. appl. Pharmacol.*, **88**, 225–233

Goldsworthy, T.L., Lyght, O., Burnett, V.L. & Popp, J.A. (1988) Potential role of α-2μ-globulin, protein droplet accumulation, and cell replication in the renal carcinogenicity of rats exposed to trichloroethylene, perchloroethylene, and pentachloroethane. *Toxicol. appl. Pharmacol.*, **96**, 367–379

Gossett, R.W., Brown, D.A. & Young, D.R. (1983) Predicting the bioaccumulation of organic compounds in marine organisms using octanol/water partition coefficients. *Mar. Pollut. Bull.*, **14**, 387–392

Green, T. & Odum, J. (1985) Structure/activity studies of the nephrotoxic and mutagenic action of cysteine conjugates of chloro- and fluoroalkenes. *Chem.-biol. Interactions*, **54**, 15–31

Green, T., Odum, J., Nash, J.A. & Foster, J.R. (1990) Perchloroethylene-induced rat kidney tumors: an investigation of the mechanisms involved and their relevance to humans. *Toxicol. appl. Pharmacol.*, **103**, 77–89

Greenberg, A.E., Clesceri, L.S. & Eaton, A.D., eds (1992) *Standard Methods for the Examination of Water and Wastewater*, 18th Ed., Washington DC, American Public Health Association/American Water Works Association/Water Environment Federation, pp. 6-17–6-36, 6-42–6-57

Greim, H., Bonse, G., Radwan, Z., Reichert, D. & Henschler, D. (1975) Mutagenicity *in vitro* and potential carcinogenicity of chlorinated ethylenes as a function of metabolic oxirane formation. *Biochem. Pharmacol.*, **24**, 2013–2017

Grimsrud, E.P. & Rasmussen, R.A. (1975) Survey and analysis of halocarbons in the atmosphere by gas chromatography–mass spectroscopy. *Atmos. Environ.*, **9**, 1014–1017

Grob, K., Artho, A. & Egli, J. (1991) Food contamination by perchloroethylene from the air: limits for the air rather than for foods! *Mitt. Geb. Lebensm. Hyg.*, **82**, 56–65 (in German)

Guengerich, F.P. & Macdonald, T.L. (1984) Chemical mechanisms of catalysis by cytochrome P-450: a unified view. *Acc. chem. Res.*, **17**, 9–16

Gulyas, H. & Hemmerling, L. (1990) Tetrachloroethylene air pollution originating from coin-operated dry cleaning establishments. *Environ. Res.*, **53**, 90–99

Gulyas, H., Billing, E. & Hemmerling, L. (1988) Sample measurements of tetrachloroethylene emissions in dry cleaning establishments in Hamburg. *Forum Städte-Hyg.*, **39**, 97–100 (in German)

Gunter, B.J. & Lybarger, J.A. (1979) *Jonas Brothers Taxidermy, Co., Denver, CO* (Health Hazard Evaluation Determination Report No. 78-95-596), Cincinnati, OH, United States National Institute for Occupational Safety and Health

Hajimiragha, H., Ewers, U., Jansen-Rosseck, R. & Brockhaus, A. (1986) Human exposure to volatile halogenated hydrocarbons from the general environment. *Int. Arch. occup. environ. Health*, **58**, 141–150

Hake, C.L. & Stewart, R.D. (1977) Human exposure to tetrachloroethylene: inhalation and skin contact. *Environ. Health Perspectives*, **21**, 231–238

Hanley, K.W. (1993) *Daubert Coated Products, Inc., Dixon, IL* (HETA 91-0004-2316), Cincinnati, OH, United States National Institute for Occupational Safety and Health

Hansch, C., Leo, A. & Hoekman, D.H. (1995) *Exploring QSAR*, Washington DC, American Chemical Society

Hardell, L., Eriksson, M., Lenner, P. & Lundgren, E. (1981) Malignant lymphoma and exposure to chemicals, especially organic solvents, chlorophenols and phenoxy acids: a case–control study. *Br. J. Cancer*, **43**, 169–176

Hardell, L., Bengtsson, N.O., Jonsson, U., Eriksson, S. & Larsson, L.G. (1984) Aetiological aspects on primary liver cancer with special regard to alcohol, organic solvents and acute intermittent porphyria—an epidemiological investigation. *Br. J. Cancer*, **50**, 389–397

Hardin, B.D., Bond, G.P., Sikov, M.R., Andrew, F.D., Beliles, R.P. & Niemeier, R.W. (1981) Testing of selected workplace chemicals for teratogenic potential. *Scand. J. Work Environ. Health*, **7** (Suppl. 4), 66–75

Harkov, R., Kebbekus, B., Bozzelli, J., Lioy, P.J. & Daisey, J. (1984) Comparison of selected volatile organic compounds during the summer and winter at urban sites in New Jersey. *Sci. total Environ.*, **38**, 259–274

Hartle, R. & Aw, T.-C. (1983) *Hoover Co., North Canton, OH* (Health Hazard Evaluation Report No. 82-127-1370), Cincinnati, OH, United States National Institute for Occupational Safety and Health

Hartle, R. & Aw, T.-C. (1984) *Hoover Co., North Canton, OH* (Health Hazard Evaluation Report No. 82-280-1407), Cincinnati, OH, United States National Institute for Occupational Safety and Health

Hattis, D., White, P., Marmorstein, L. & Koch, P. (1990) Uncertainties in pharmacokinetic modeling for perchloroethylene. I. Comparison of model structure, parameters, and predictions for low-dose metabolism rates for models derived by different authors. *Risk Anal.*, **10**, 449–458

Haworth, S., Lawlor, T., Mortelmans, K., Speck, W. & Zeiger, E. (1983) *Salmonella* mutagenicity test results for 250 chemicals. *Environ. Mutag.*, **Suppl. 1**, 3–142

Hayes, J.R., Condie, L.W., Jr & Brozelleca, J.F. (1986) The subchronic toxicity of tetrachloroethylene (perchloroethylene) administered in the drinking water of rats. *Fundam. appl. Toxicol.*, **7**, 119–125

Heikes, D.L. (1987) Purge and trap method for determination of volatile halocarbons and carbon disulfide in table-ready foods. *J. Assoc. off. anal. Chem.*, **70**, 215–226

Heikes, D.L. & Hopper, M.L. (1986) Purge and trap method for determination of fumigants in whole grains, milled grain products, and intermediate grain-based foods. *J. Assoc. off. anal. Chem.*, **69**, 990–998

Heineman, E.F., Cocco, P., Gómez, M.R., Dosemeci, M., Stewart, P.A., Hayes, R.B., Hoar Zahm, S., Thomas, T.L. & Blair, A. (1994) Occupational exposure to chlorinated aliphatic hydrocarbons and risk of astrocytic brain cancer. *Am. J. ind. Med.*, **26**, 155–169

Hellmann, H. (1984) Volatile chlorohydrocarbons in waters of East Germany. *Haustechn. Bauphys. Umwelttechn. Gesundheit*, **105**, 269–278 (in German)

Henschler, D. (1987) Mechanisms of genotoxicity of chlorinated aliphatic hydrocarbons. In: De Matteis, F. & Lock, E.A., eds, *Selectivity and Molecular Mechanisms of Toxicity*, New York, MacMillan, pp. 153–181

Henschler, D. & Bonse, G. (1977) Metabolic activation of chlorinated ethylenes: dependence of mutagenic effect on electrophilic reactivity of the metabolically formed epoxides. *Arch. Toxicol.*, **39**, 7–12

Herbert, P., Charbonnier, P., Rivolta, L., Servais, M., Van Mensch, F. & Campbell, I. (1986) The occurrence of chlorinated solvents in the environment. *Chem. Ind.*, **December**, 861–869

Hervin, R.L. & Lucas, J. (1973) *The Budd Company–Automotive Division, Clinton, MI* (Health Hazard Evaluation Report No. 72-35-34), Cincinnati, OH, United States National Institute for Occupational Safety and Health

Hervin, R., Shama, S. & Ruhe, R.L. (1972) *Aerosol Techniques, Inc., Danville, IL* (Health Hazard Evaluation. Report No. 71-25-20), Cincinnati, OH, United States National Institute for Occupational Safety and Health

Hickman, J.C. (1993) Chlorocarbons and chlorohydrocarbons. In: Kroschwitz, J.I. & Howe-Grant, M., eds, *Kirk-Othmer Encyclopedia of Chemical Technology*, 4th Ed., Vol. 6, New York, John Wiley & Sons, pp. 50–59 von der Hude, W., Behm, C., Gürtler, R. & Basler, A. (1988) Evaluation of SOS chromotest. *Mutat. Res.*, **203**, 81–94

IARC (1979) *IARC Monographs on the Evaluation of the Carcinogenic Risk of Chemicals to Humans*, Vol. 20, *Some Halogenated Hydrocarbons*, Lyon, pp. 491–514

IARC (1987) *IARC Monographs on the Evaluation of Carcinogenic Risks to Humans*, Suppl. 7, *Overall Evaluations of Carcinogenicity: An Updating of* IARC Monographs *Volumes 1–42*, Lyon, pp. 355–357

IARC (1994a) *IARC Monographs on the Evaluation of Carcinogenic Risks to Humans*, Vol. 60, *Some Industrial Chemicals*, Lyon, pp. 161–180

IARC (1994b) *Directory of Agents Being Tested for Carcinogenicity*, No. 16, Lyon, p. 170

Ikeda, M. (1977) Metabolism of trichlorethylene and tetrachlorethylene in human subjects. *Environ. Health Perspectives*, **21**, 239–245

Ikeda, M. & Imamura, T. (1973) Biological half-life of trichloroethylene and tetrachloroethylene in human subjects. *Int. Arch. Arbeitsmed.*, **31**, 209–224

Ikeda, M. & Ohtsuji, H. (1972) A comparative study of the excretion of Fujiwara reaction-positive substances in urine of humans and rodents given trichloro- or tetrachloro-derivates of ethane and ethylene. *Br. J. ind. Med.*, **29**, 99–104

Ikeda, M., Ohtsuji, H., Imamura, T. & Komoike, Y. (1972) Urinary excretion of total trichloro-compounds, trichloroethanol, and trichloroacetic acid as a measure of exposure to trichloroethylene and tetrachloroethylene. *Br. J. ind. Med.*, **29**, 328–333

Ikeda, M., Koizumi, A., Watanabe, T., Endo, A. & Sato, K. (1980) Cytogenetic and cytokinetic investigations on lymphocytes from workers occupationally exposed to tetrachloroethylene. *Toxicol. Lett.*, **5**, 251–256

Illing, H.P.A., Mariscotti, S.P. & Smith, A.M. (1987) *Tetrachloroethylene (Tetrachloroethene, Perchloroethylene)* (Health and Safety Executive Toxicity Review, 7), London, Her Majesty's Stationery Office

International Fabricare Institute (1987) Perchloroethylene vapor in drycleaning plants. In: *Focus on Drycleaning*, Vol. 11(1), Silver Spring, MD

Isacson, P., Bean, J.A., Splinter, R., Olson, D.B. & Kohler, J. (1985) Drinking water and cancer incidence in Iowa. III. Association of cancer with indices of contamination. *Am. J. Epidemiol.*, **121**, 856–869

Jackson, M.A., Stack, H.F. & Waters, M.D. (1993) The genetic toxicology of putative nongenotoxic carcinogens. *Mutat. Res.*, **296**, 241–277

Jalihal, S.S. & Barlow, A.M. (1984) Leukaemia in dry cleaners (Letter to the Editor). *J. R. Soc. Health*, **104**, 42

Jang, J.-Y., Kang, S.-K. & Chung, H.-K. (1993) Biological exposure indices of organic solvents for Korean workers. *Int. Arch. occup. environ. Health*, **65**, S219–S222

Jankovic, J. (1980) *Patriot Coal Company Laboratory, Kingwood, WV* (Health Hazard Evaluation Report No. M-TA-80-109-110), Cincinnati, OH, United States National Institute for Occupational Safety and Health

Jensen, S.A. & Ingvordsen, C. (1977) Tetrachloroethylene in garments. *Ugeskr. Laeg.*, **139**, 239–297 (in Danish)

Kawamura, K. & Kaplan, I.R. (1983) Organic compounds in the rainwater of Los Angeles. *Environ. Sci. Technol.*, **17**, 497–501

Kawauchi, T. & Nishiyama, K. (1989) Residual tetrachloroethylene in dry-cleaned clothes. *Environ. Res.*, **48**, 296–301

Keil, S.L. (1979) Tetrachloroethylene. In: Mark, H.F., Othmer, D.F., Overberger, C.G., Seaborg, G.T. & Grayson, N., eds, *Kirk-Othmer Encyclopedia of Chemical Technology*, 3rd Ed., Vol. 5, New York, John Wiley & Sons, pp. 754–762

Kido, K.M., Nagara, Y., Furuichi, T. & Ikeda, M. (1989) Statistical approach towards point sources of groundwater pollution with tetrachloroethylene: a field study. *Tohoku J. exp. Med.*, **157**, 229–239

Klaassen, C.D. & Plaa, G.L. (1966) Relative effects of various chlorinated hydrocarbons on liver and kidney function in mice. *Toxicol. appl. Pharmacol.*, **9**, 139–151

Koch, R., Schlegelmilch, R. & Wolf, H.U. (1988) Genetic effects of chlorinated ethylenes in the yeast *Saccharomyces cerevisiae*. *Mutat. Res.*, **206**, 209–216

Kringstad, K.P., Ljungquist, P.O., de Sousa, F. & Strömberg, L.M. (1981) Identification and mutagenic properties of some chlorinated aliphatic compounds in the spent liquor from kraft pulp chlorination. *Environ. Sci. Technol.*, **15**, 562–566

Kurz, J. (1992) German drycleaning regulations and technology. In: *Proceedings. International Roundtable on Pollution Prevention and Control in the Drycleaning Industry, 27–28 May 1992, Falls Church, VA* (EPA/774/R-92/002), Springfield, VA, United States National Technical Information Service

Kyyrönen, P., Taskinen, H., Lindbohm, M.-L., Hemminki, K.& Heinonen, O.P. (1989) Spontaneous abortions and congenital malformations among women exposed to tetrachloroethylene in drycleaning. *J. Epidemiol. Community Health*, **43**, 346–351

Lackore, L.K. & Perkins, H.M. (1970) Accidental narcosis. Contamination of compressed air system. *J. Am. med. Assoc.*, **211**, 1846

Lagakos, S.W., Wessen, B.J. & Zelen, M. (1986) An analysis of contaminated well water and health effects in Woburn, Massachusetts. *J. Am. stat. Assoc.*, **81**, 583–596

Lide, D.R., ed. (1993) *CRC Handbook of Chemistry and Physics*, 74th Ed., Boca Raton, FL, CRC Press, pp. 3–242

Ligocki, M.P., Leuenberger, C. & Pankow, J.F. (1985) Trace organic compounds in rain—II. Gas scavenging of neutral organic compounds. *Atmos. Environ.*, **19**, 1609–1617

Linak, E., Leder, A. & Yoshida, Y. (1992) C_2 Chlorinated solvents. In: *Chemical Economics Handbook*, Menlo Park, CA, SRI International, pp. 632.3000–632.3001

Lindbohm, M.-L., Taskinen, H., Sallmén, M. & Hemminki, K. (1990) Spontaneous abortions among women exposed to organic solvents. *Am. J. ind. Med.*, **17**, 449–463

Lob, M. (1957) The dangers of perchloroethylene. *Arch. Gewerbepath.*, **16**, 45 (in French)

Lochner, F. (1976) Tetrachloroethylene. An inventory. *Umwelt*, **6**, 434–436 (in German)

Love, J.R. (1982) *King-Smith Printing Co., Detroit, MI* (Health Hazard Evaluation Report No. 81-310-1039), Cincinnati, OH, United States National Institute for Occupational Safety and Health

Lovelock, J.E. (1974) Atmospheric halocarbons and statospheric ozone. *Nature*, **252**, 292–294

Mazzullo, M., Grilli, S., Lattanzi, G., Prodi, G., Turina, M.P. & Colacci, A. (1987) Evidence of DNA binding activity of perchloroethylene. *Res. Commun. chem. Pathol. Pharmacol.*, **58**, 215–235

McConnell, G., Ferguson, D.M. & Pearson, C.R. (1975) Chlorinated hydrocarbons and the environment. *Endeavor*, **34**, 13–18

McKone, T.E. & Daniels, J.I. (1991) Estimating human exposure through multiple pathways from air, water, and soil. *Regul. Toxicol. Pharmacol.*, **13**, 36–61

Meckler, L.C. & Phelps, D.K. (1966) Liver disease secondary to tetrachloroethylene exposure, a case report. *J. Am. med. Assoc.*, **197**, 662

Mennear, J., Maronpot, R., Boorman, G., Eustis, S., Huff, J., Haseman, J., McConnell, E., Ragan, H. & Miller, R. (1986) Toxicologic and carcinogenic effects of inhaled tetrachloroethyhlene in rats and mice. In: Chambers, P.L., Gehring, P. & Sakai, F., eds, *New Concepts and Developments in Toxicology*, Amsterdam, Elsevier, pp. 201–210

Mersch-Sundermann, V., Müller, G. & Hofmeister, A. (1989) Examination of mutagenicity of organic microcontaminations of the environment. IV. Communication: the mutagenicity of halogenated aliphatic hydrocarbons with the SOS chromotest. *Zbl. Hyg.*, **189**, 266–271 (in German)

Miller, L.J. & Uhler, A.D. (1988) Volatile halocarbons in butter: elevated tetrachloroethylene levels in samples obtained in close proximity to dry-cleaning establishments. *Bull. environ. Contam. Toxicol.*, **41**, 469–474

Milman, H.A., Story, D.L., Riccio, E.S., Sivak, A., Tu, A.S., Williams, G.M., Tong, C. & Tyson, C.A. (1988) Rat liver foci and in vitro assays to detect initiating and promoting effects of chlorinated ethanes and ethylenes. *Ann. N.Y. Acad. Sci.*, **534**, 521–530

Monster, A.C. & Smolders, J.F.J. (1984) Tetrachloroethylene in exhaled air of persons living near pollution sources. *Int. Arch. occup. environ. Health*, **53**, 331–336

Monster, A.C., Boersma, G. & Steenweg, H. (1979) Kinetics of tetrachloroethylene in volunteers; influence of exposure concentration and work load. *Int. Arch. occup. environ. Health*, **42**, 303–309

Morgan, B. (1969) Dangers of perchloroethylene. *Br. med. J.*, **ii**, 513

Morse, K.M. & Goldberg, L. (1943) Chlorinated solvent exposures at degreasing operations. *Ind. Med.*, **12**, 706–713

Mosely, C.L. (1980) *Motion Pictures Screen Cartoonist, Local 841, New York, NY* (Health Hazard Evaluation Determination Report No. HE 79-42-685), Cincinnati, OH, United States National Institute for Occupational Safety and Health

Murray, A.J. & Riley, J.P. (1973) Occurrence of some chlorinated aliphatic hydrocarbons in the environment. *Nature*, **242**, 37–38

Mutti, A., Alinovi, R., Bergamaschi, E., Biagini, C., Cavazzini, S., Franchini, I., Lauwerys, R.R., Bernard, A.M., Roels, H., Gelpi, E., Rosello, J., Ramis, I., Price, R.G., Taylor, S.A., De Broe, M., Nuyts, G.D., Stolte, H., Fels, L.M. & Herbort, C. (1992) Nephropathies and exposure to perchloroethylene in dry-cleaners. *Lancet*, **340**, 189–193

Nelson, B.K., Taylor, B.J., Setzer, J.V. & Hornung, R.W. (1980) Behavioral teratology of perchloroethylene in rats. *J. environ. Pathol. Toxicol.*, **3**, 233–250

New York Department of Health (1994) *Residences Tested for Perchloroethylene*, New York City

Norman, K.N.T. (1991) Determination of tetrachloroethylene in olive oil by automated headspace gas chromatography. *Food Addit. Contaminants*, **8**, 513–516

Odum, J., Green, T., Foster, J.R. & Hext, P.M (1988) The role of trichloroacetic acid and peroxisome proliferation in the differences in carcinogenicity of perchloroethylene in the mouse and rat. *Toxicol. appl. Pharmacol.*, **92**, 103–112

Ohta, T., Morita, M. & Mizoguchi, I. (1976) Local distribution of chlorinated hydrocarbons in the ambient air in Tokyo. *Atmos. Environ.*, **10**, 557–560

Ohtsuki, T., Sato, K., Koizumi, A., Kumai, M. & Ikeda, M. (1983) Limited capacity of humans to metabolize tetrachloroethylene. *Int. Arch. occup. environ. Health*, **51**, 381–390

Okawa, M.T. & Coye, M.J. (1982) *Film Processing Industry, Hollywood, CA* (Health Hazard Evaluation Report No. 80-144-1109), Cincinnati, OH, United States National Institute for Occupational Safety and Health

Olsen, G.W., Hearn, S., Cook, R.R., Currier, M.F. & Allen, S. (1989) Mortality experience of a cohort of Louisiana chemical workers. *J. occup. Med.*, **31**, 32–34

Olsen, J., Hemminki, K., Ahlborg, G., Bjerkedal, T., Kyyrönen, P., Taskinen, H., Lindbohm, M.-L., Heinonen, O.P., Brandt, L., Kolstad, H., Halvorsen, B.A. & Egenaes, J. (1990) Low birthweight, congenital malformations, and spontaneous abortions among dry-cleaning workers in Scandinavia. *Scand. J. Work Environ. Health*, **16**, 163–168

Pàtel, R., Janakiraman, N., Johnson, R. & Elman, J.B. (1973) Pulmonary edema and coma from perchloroethylene (Letter to the Editor). *J. Am. med. Assoc.*, **223**, 1510

Pearson, C.R. & McConnell, G. (1975) Chlorinated C_1 and C_2 hydrocarbons in the marine environment. *Proc. R. Soc. Lond. B.*, **189**, 305–332

Pegg, D.G., Zempel, J.A., Braun, W.H. & Watanabe, P.G. (1979) Disposition of tetrachloro(^{14}C)ethylene following oral and inhalation exposure in rats. *Toxicol. appl. Pharmacol.*, **51**, 465–474

Pellizzari, E.D., Hartwell, T.D., Harris, B.S.H., III, Waddell, R.D., Whitaker, D.A. & Erickson, M.D. (1982) Purgeable organic compounds in mother's milk. *Bull. Environ. Contam. Toxicol.*, **28**, 322–328

Penkett, S.A. (1982) Non-methane organics in the remote troposphere. In: Goldberg, E.D., ed., *Atmospheric Chemistry*, New York, Springer, pp. 329–355

Plaa, G.L. & Larson, R.E. (1965) Relative nephrotoxic properties of chlorinated methane, ethane and ethylene derivatives in mice. *Toxicol. appl. Pharmacol.*, **7**, 37–44

PPG Industries, Inc. (1994) *Product Information Sheet: Perchloroethylene*, Pittsburgh, PA

Price, P. J., Hassett, C.M. & Mansfield, J.I. (1978) Transforming activities of trichloroethylene and proposed industrial alternatives. *In Vitro*, **14**, 290–293

Pryor, P. (1977) *McCourt Label Company, Bradford, PA* (Health Hazard Evaluation Determination Report No. 77-84-450), Cincinnati, OH, United States National Institute for Occupational Safety and Health

Rampy, L.W., Quast, J.F., Leong, B.K.J. & Gehring, P.J. (1977) Results of long-term inhalation toxicity studies on rats of 1,1,1-trichloroethane and perchloroethylene formulations (Abstract). In: Plaa, G.L. & Duncan, W.A.M., eds, *Proceedings of the First International Congress on Toxicology*, New York, Academic Press, p. 27

Rantala, K., Riipinen, H. & Anttila, A. (1992) *Altisteettyössä, 22—Halogeenihiitivedyt* [Exposure at work—halogenated hydrocarbons], Helsinki, Finnish Institute of Occupational Health–Finnish Work Environment Fund (in Finnish)

Rasmussen, R.A. & Khalil, M.A.K. (1982) Latitudinal distribution of trace gases in and above the boundary layer. *Chemosphere*, **11**, 227–235

Ratnoff, W.D. & Gress, R.E. (1980) The familial occurrence of polycythemia vera: report of a father and son, with consideration of the possible etiologic role of exposure to organic solvents, including tetrachloroethylene. *Blood*, **56**, 233–236

Reichert, D. (1983) Biological actions and interactions of tetrachloroethylene. *Mutat. Res.*, **123**, 411–429

Reinhard, K., Dulson, W. & Exner, M. (1989) Concentrations of tetrachloroethylene in indoor air and food in apartments in the vicinity of dry cleaning shops. *Zbl. Hyg.*, **189**, 111–116 (in German)

Rivett, M.O., Lerner, D.N., Lloyd, J.W. & Clark, L. (1990) Organic contamination of the Birmingham aquifer, UK. *J. Hydrol.*, **113**, 307–323

Roldán-Arjona, T., García-Pedrajas, M.D., Luque-Romero, F.L., Hera, C. & Pueyo, C. (1991) An association between mutagenicity of the Ara test of *Salmonella typhimurium* and carcinogenicity in rodents for 16 halogenated aliphatic hydrocarbons. *Mutagenęsis*, **6**, 199–205

Ruder, A.M., Ward, E.M. & Brown, D.P. (1994) Cancer mortality in female and male dry-cleaning workers. *J. occup. Med.*, **36**, 867–874

Russell, J.W. & Shadoff, L.A. (1977) The sampling and determination of halocarbons in ambient air using concentration on porous polymer. *J. Chromatogr.*, **134**, 375–384

Sack, T., Steele, D.H., Hammerstrom, K. & Remmers, J. (1992) A survey of household products for volatile organic compounds. *Atmos. Environ.*, **26A**, 1063–1070

Sadtler Research Laboratories (1980) *1980 Cumulative Index*, Philadelphia, PA

Sandhu, S.S., Ma, T.-H., Peng, Y. & Zhou, X. (1989) Clastogenic evaluation of seven chemicals commonly found at hazardous industrial waste sites. *Mutat.Res.*, **224**, 437–445

Schairer, L.A. & Sautkulis, R.C. (1982) Detection of ambient levels of mutagenic atmospheric pollutants with the higher plant *Tradescantia*. *Environ. Mutag. Carcinog. Plant Biol.*, **2**, 154–194

Schaller, K.H. & Triebig, G. (1987) Active and passive air monitoring in comparison to biological monitoring in tetrachloroethylene and xylene exposed workers. In: Berlin, A., Brown, R.H. & Saunders, K.J., eds, *Diffusive Sampling: An Alternative Approach to Workplace Air Monitoring*, Middlesex, BP Research Centre, pp. 111–114

Schlichting, L.M., Wright, P.F.A. & Stacey, N.H. (1992) Effects of tetrachloroethylene on hepatic and splenic lymphocytotoxic activities in rodents. *Toxicol. ind. Health*, **8**, 255–266

Schneberger, G.L. (1981) Metal surface treatments (cleaning). In: Mark, H.F., Othmer, D.F., Overberger, C.G., Seaborg, G.T. & Grayson, N., eds, *Kirk-Othmer Encyclopedia of Chemical Technology*, 3rd Ed., Vol. 15, New York, John Wiley & Sons, pp. 296–299

Schreiber, J.S. (1992) An assessment of tetrachloroethylene in human breast milk. *J. exp. anal. environ. Epidemiol.*, **2** (Suppl. 2), 15–26

Schreiber, J.S. (1993) Predicted infant exposure to tetrachloroethylene in human breast milk. *Risk Anal.*, **13**, 515–524

Schreiber, J.S., House, S., Prohonic, E., Smead, G., Hudson, C., Styk, M. & Lauber, J. (1993) An investigation of indoor air contamination in residences above dry cleaners. *Risk Anal.*, **13**, 335–344

Schumann, A.M., Quast, J.F. & Watanabe, P.G. (1980) The pharmacokinetic and macromolecular interactions of perchloroethylene in mice and rats as related to oncogenicity. *Toxicol. appl. Pharmacol.*, **55**, 207–219

Schwarzenbach, R.P., Molnar-Kubica, E., Giger, W. & Wakeham, S.G. (1979) Distribution, residence time, and fluxes of tetrachloroethylene and 1,4-dichlorobenzene in Lake Zurich, Switzerland. *Environ. Sci. Technol.*, **13**, 1367–1373

Schweizerische Unfallversicherungsanstalt [Swiss Accident Insurance Company] (1994) *Grenzwerte am Arbeitsplatz* [Limit values at the workplace], Lucerne, p. 131

Schwetz, B.A., Leong, B.K.J. & Gehring, P.J. (1975) The effect of maternally inhaled trichloroethylene, perchloroethylene, methyl chloroform, and methylene chloride on embryonal and fetal development in mice and rats. *Toxicol. appl. Pharmacol.*, **32**, 84–96

Seiji, K., Jin, C., Watanabe, T., Nakatsuka, H. & Ikeda, M. (1990) Sister chromatid exchanges in peripheral lymphocytes of workers exposed to benzene, trichloroethylene, or tetrachloroethylene, with reference to smoking habits. *Int. Arch. occup. environ. Health*, **62**, 171–176

Sharpe, C.R., Rochon, J.E., Adam, J.M. & Suissa, S. (1989) Case–control study of hydrocarbon exposures in patients with renal cell carcinoma. *Can. med. Assoc. J.*, **140**, 1309–1318

Shimada, T., Swanson, A.F., Leber, P. & Williams, G.M. (1985) Activities of chlorinated ethane and ethylene compounds in the *Salmonella*/rat microsome mutagenesis and rat hepatocyte/DNA repair assays under vapor phase exposure conditions. *Cell Biol. Toxicol.*, **1**, 159–179

Siemiatycki, J. (1991) *Risk Factors for Cancer in the Workplace*, Boca Raton, FL, CRC Press

Singh, H.B., Salas, L., Shigeishi, H. & Crawford, A. (1977) Urban–nonurban relationships of halocarbons, SF_6, N_2O, and other atmospheric trace constituents. *Atmos. Environ.*, **11**, 819–828

Singh, H.B., Salas, L.J., Smith, A.J. & Shigeishi, H. (1981) Measurement of some potentially hazardous organic chemicals in urban environments. *Atmos. Environ.*, **15**, 601–612

Singh, H.B., Salas, L.J. & Stiles, R.E. (1983) Selected man-made halogenated chemicals in the air and oceanic environment. *J. geophys. Res.*, **88**, 3675–3683

Skender, L., Karacic, V., Bosner, B. & Prpic-Majic, D. (1993) Assessment of exposure to trichloroethylene and tetrachloroethylene in the population of Zagreb, Croatia. *Int. Arch. occup. environ. Health*, **65**, S163–S165

Skender, L., Karacic, V., Bosner, B. & Prpic-Majic (1994) Assessment of urban population exposure to trichloroethylene and tetrachloroethylene by means of biological monitoring. *Arch. environ. Health*, **49**, 445–451

Sofuni, T., Hayashi, M., Matsuoka, A., Sawada, M., Hatanaka, M. & Ishidate, M., Jr (1985) Mutagenicity tests on organic chemical contaminants in city water and related compounds. II. Chromosome aberration tests in cultured mammalian cells. *Bull. natl Inst. Hyg. Sci. (Tokyo)*, **103**, 64–75

Solet, D. & Robins, T.G. (1991) Renal function in dry cleaning workers exposed to perchloroethylene. *Am. J. ind. Med.*, **20**, 601–614

Solet, D., Robins, T. & Sampaio, C. (1990) Perchloroethylene exposure assessment among dry cleaning workers. *Am. ind. Hyg. Assoc. J.*, **51**, 566–574

Spirtas, R., Stewart, P.A., Lee, J.S., Marano, D.E., Forbes, C.D., Grauman, D.J., Pettigrew, H.M., Blair, A., Hoover, R.N. & Cohen, J.L. (1991) Retrospective cohort mortality study of workers at an aircraft maintenance facility. I. Epidemiological results. *Br. J. ind. Med.*, **48**, 515–530

Staples, C.A., Werner, A.F. & Hoogheem, T.J. (1985) Assessment of priority pollutant concentrations in the United States using STORET database. *Environ. Toxicol. Chem.*, **4**, 131–142

Stewart, R.D. (1969) Acute tetrachloroethylene intoxication. *J. Am. med. Assoc.*, **208**, 1490–1492

Stewart, R.D., Erley, D.S., Schaffer, A.W. & Gay, H.H. (1961a) Accidental vapor exposure to anesthetic concentrations of a solvent containing tetrachloroethylene. *Ind. Med. Surg.*, **30**, 327–330

Stewart, R.D., Gay, H.H., Erley, D.S., Hake, C.L. & Schaffer, A.W. (1961b) Human exposure to tetrachloroethylene vapor. *Arch. environ. Health*, **2**, 516–522

Stewart, R.D., Baretta, E.D., Dodd, H.C. & Torkelson, T.R. (1970) Experimental human exposure to tetrachloroethylene. *Arch. environ. Health*, **20**, 224–229

Su, C. & Goldberg, E.D. (1976) Environmental concentrations and fluxes of some halocarbons. In: Windom, H.L. & Duce, R.A., eds, *Marine Pollutant Transfer*, Lexington, MA, Lexington Books, pp. 353–374

Taskinen, H., Anttila, A., Lindbohm, M.-L., Sallmén, M. & Hemminki, K. (1989) Spontaneous abortions and congenital malformations among the wives of men occupationally exposed to organic solvents. *Scand. J. Work Environ. Health*, **15**, 345–352

Tharr, D. & Donahue, M. (1980) *Cobo Laboratories, Inc., Lakewood & Arvada, CO* (Health Hazard Evaluation 79-80-746), Cincinnati, OH, United States National Institute for Occupational Safety and Health

Theiss, J.C., Stoner, G.D., Shimkin, M.B. & Weisburger, E.K. (1977) Test for carcinogenicity of organic contaminants of United States drinking waters for pulmonary tumor response in strain A mice. *Cancer Res.*, **37**, 2717–2720

Thomas, R.D. (1989) Epidemiology and toxicology of volatile organic chemical contaminants in water absorbed through the skin. *J. Am. Coll. Toxicol.*, **8**, 779–795

Thomas, K.W., Pellizzari, E.D., Perritt, R.L. & Nelson, W.C. (1991) Effect of dry-cleaned clothes on tetrachloroethylene levels in indoor air, personal air, and breath for residents of several New Jersey homes. *J. Exposure Anal. environ. Epidemiol.*, **1**, 475–490

Thompson, K.M. & Evans, J.S. (1993) Workers' breath as a source of perchloroethylene (perc) in the home. *J. Exposure Anal. environ. Epidemiol.*, **3**, 417–450

Trussell, A.R., Cromer, J.L., Umphres, M.D., Kelley, P.E. & Moncur, J.G. (1980) Monitoring of volatile halogenated organics: a survey of twelve drinking waters from various parts of the world. In: Jolley, R.L., Brungs, W.A. & Cumming, R.B., eds, *Water Chlorination–Environmental Impact and Health Effects*, Vol. 3, Ann Arbor, MI, Ann Arbor Science, pp. 39–53

Tu, A., Murray, T.A., Hatch, K.M., Sivak, A. & Milman, H.A. (1985) In vitro transformation of BALB/c-3T3 cells by chlorinated ethanes and ethylenes. *Cancer Lett.*, **28**, 85–92

Työministeriö [Ministry of Labour] (1993) *HTP-Arvot 1993* [Limit Values 1993], Tampere, p. 16

United Kingdom Health and Safety Executive (1994) *Occupational Exposure Limits 1994* (Guidance Note EH 40/94), London, Her Majesty's Stationery Office, p. 35

United States Agency for Toxic Substances and Disease Registry (1990) *Toxicological Profile for Tetrachloroethylene*, Oak Ridge, TN, Oak Ridge National Laboratory

United States Environmental Protection Agency (1985a) *Health Assessment Document for Tetrachloroethylene (Perchloroethylene)—Final Report* (EPA-600/8-82-005F; US NTIS PB85-249704), Washington DC

United States Environmental Protection Agency (1985b) *Drinking Water Criteria Document for Tetrachloroethylene—Final Report* (PB86-118114), Washington DC

United States Environmental Protection Agency (1986a) Method 8010. Halogenated Volatile Organics. In: *Test Methods for Evaluating Solid Waste—Physical/Chemical Methods* (US EPA No. SW-846), 3rd Ed., Vol. 1A, Washington DC, Office of Solid Waste and Emergency Response

United States Environmental Protection Agency (1986b) Method 8240. Gas chromatography/mass spectrometry for volatile organics. In: *Test Methods for Evaluating Solid Waste—Physical/Chemical Methods* (US EPA No. SW-846), 3rd Ed., Vol. 1A, Washington DC, Office of Solid Waste and Emergency Response

United States Environmental Protection Agency (1988a) *Compendium of Methods for the Determination of Toxic Organic Compounds in Ambient Air* (EPA-600/4-89-017; US NTIS PB90-116989), Research Triangle Park, NC, Office of Research and Development, pp. TO1-1–TO1-38; TO3-1–TO3-22; TO14-1–TO14-94

United States Environmental Protection Agency (1988b) *Methods for the Determination of Organic Compounds in Drinking Water* (EPA-600/4-88-039; US NTIS PB89-220461), Cincinnati, OH, Environmental Monitoring Systems Laboratory, pp. 5–87, 255–323 (Methods 502.1, 502.2, 503.1, 524.1, 524.2)

United States Environmental Protection Agency (1991) National emission standards for hazardous air pollutants for source categories: perchloroethylene emissions from dry cleaning facilities. Proposed rule and notice of public hearing. *Fed. Reg.*, **56**, 64382–64402

United States Environmental Protection Agency (1993) *1991 Toxic Release Inventory* (EPA 745-R-93-003), Washington DC, Office of Pollution Prevention and Toxics, p. 250

United States Environmental Protection Agency (1994) Protection of the environment. *US Code fed. Regul.*, **Title 40**, Part 136, Appendix A, pp. 400–413, 559–573, 602–614 (Methods 601, 624, 1624)

United States National Cancer Institute (1977) *Bioassay of Tetrachloroethylene for Possible Carcinogenicity (CAS No. 127-18-4)* (Tech. Rep. Ser. No. 13), Bethesda, MD

United States National Institute for Occupational Safety and Health (1994a) *Pocket Guide to Chemical Hazards* (DHHS (NIOSH) Publ. No. 94-116), Cincinnati, OH, pp. 300–301

United States National Institute for Occupational Safety and Health (1994b) *National Occupational Exposure Survey (1981–1983)*, Cincinnati, OH

United States National Institute for Occupational Safety and Health (1994c) *RTECs Chem.-Bank*, Cincinnati, OH

United States National Toxicology Program (1986) *Toxicology and Carcinogenesis Studies of Tetrachloroethylene (Perchloroethylene) (CAS No. 127-18-4) in F344/N Rats and B6C3F1 Mice (Inhalation Studies)* (Tech. Rep. Ser. No. 311), Research Triangle Park, NC

United States Occupational Safety and Health Administration (1994) Air contaminants. *US Code fed. Regul.*, **Title 29**, Part 1910.1000, p. 18

Urano, K., Kawamoto, K., Abe, Y. & Otake, M. (1988) Chlorinated organic compounds in urban air in Japan. *Sci. total Environ.*, **74**, 121–131

Vainio, H., Waters, M.D. & Norppa, H. (1985) Mutagenicity of selected organic solvents. *Scand. J. Work Environ. Health*, **11** (Suppl. 1), 75–82

Valencia, R., Mason, J.M., Woodruff, R.C. & Zimmering, S. (1985) Chemical mutagenesis testing in *Drosophila*. III. Results of 48 coded compounds tested for the National Toxicology Program. *Environ. Mutag.*, **7**, 325–348

Vamvakas, S., Dekant, W., Berthold, K., Schmidt, S., Wild, D. & Henschler, D. (1987) Enzymatic transformation of mercapturic acids derived from halogenated alkenes to reactive and mutagenic intermediates. *Biochem. Pharmacol.*, **36**, 2741–2748

Vamvakas, S., Herkenhoff, M., Dekant, W. & Henschler, D. (1989) Mutagenicity of tetrachloroethylene in the Ames test—metabolic activation by conjugation with glutathione. *J. Biochem. Toxicol.*, **4**, 21–27

Van Duuren, B.L., Goldschmidt, B.M., Loewengart, G., Smith, A.C., Melchionne, S., Seidman, I. & Roth, D. (1979) Carcinogenicity of halogenated olefinic and aliphatic hydrocarbons in mice. *J. natl Cancer Inst.*, **63**, 1433–1439

Van Duuren, B.L., Kline, S.A., Melchionne, S. & Seidman, I. (1983) Chemical structure and carcinogenicity relationships of some chloroalkene oxides and their parent olefins. *Cancer Res.*, **43**, 159–162

Vartiainen, T., Pukkala, E., Rienoja, T., Strandman, T. & Kaksonen, K. (1993) Population exposure to tri- and tetrachloroethene and cancer risk: two cases of drinking water pollution. *Chemosphere*, **27**, 1171–1181

Verberk, M. & Scheffers, T.M.L. (1980) Tetrachloroethylene in exhaled air of residents near dry cleaning shops. *Environ. Res.*, **21**, 432–437

Vieths, S., Blaas, W., Fischer, M., Krause, C., Mehlitz, I. & Weber, R. (1987) Contamination of foodstuffs by emission of tetrachloroethene from a dry-cleaning unit. *Z. Lebensm. Unters. Forsch.*, **185**, 267–270 (in German)

Vieths, S., Blaas, W., Fischer, M., Krause, C., Matissek, R., Mehlitz, I. & Weber, R. (1988) Contamination of food stuffs by emissions from dry cleaning units. *Z. Lebensm. Unters. Forsch.*, **186**, 393–397 (in German)

Vulcan Chemicals (1994) *Technical Data Sheets: Perchloroethylene*, Birmingham, AL

Wakeham, S.G., Davis, A.C., Witt, R.T., Tripp, B.W. & Frew, N.M. (1980) Tetrachloroethylene contamination of drinking water by vinyl-coated asbestos–cement pipe. *Bull. environ. Contam. Toxicol.*, **25**, 639–645

Wallace, L. (1986) Personal exposures, indoor and outdoor air concentrations and exhaled breath concentrations of selected volatile organic compounds measured for 600 residents of New Jersey, North Dakota, North Carolina, and California. *Toxicol. environ. Chem.*, **12**, 215–236

Wallace, L.A., Pellizzari, E.D., Hartwell, T.D., Sparacino, C., Whitmore, R., Sheldon, L., Zelon, H. & Perritt, R. (1987) The TEAM study: personal exposures to toxic substances in air, drinking water, and breath of 400 residents of New Jersey, North Carolina, and North Dakota. *Environ. Res.*, **43**, 290–307

Walles, S.A.S. (1986) Induction of single-strand breaks in DNA of mice by trichloroethylene and tetrachloroethylene. *Toxicol. Lett.*, **31**, 31–35

Ward, R.C., Travis, C.C., Hetrick, D.M., Andersen, M.E. & Gargas, M.L. (1988) Pharmacokinetics of tetrachloroethylene. *Toxicol. appl. Pharmacol.*, **93**, 108–117

Weast, R.C. & Astle, M.J. (1985) *CRC Handbook of Data on Organic Compounds*, Vols I & II, Boca Raton, FL, CRC Press Inc., pp. 627 (I), 591 (II)

Webler, T. & Brown, H.S. (1993) Exposure to tetrachloroethylene via contaminated drinking water in pipes in Massachusetts: a predictive model. *Arch. environ. Health*, **48**, 293–297

Westrick, J.J., Mello, J.W. & Thomas, R.F. (1984) The groundwater supply survey. *J. Am. Water Works Assoc.*, **76**, 52–59

White, G.L. & Wegman, D.H. (1978) *Lear Siegler, Inc., Marblehead, MA* (Health Hazard Evaluation Determination Report No. 78-68-546), Cincinnati, OH, United States National Institute of Occupational Safety and Health

WHO (1984) *Tetrachloroethylene* (Environmental Health Criteria 31), Geneva

WHO (1993) *Guidelines for Drinking-water Quality*, 2nd Ed., Vol. 1, *Recommendations*, Geneva, pp. 63–64, 175

Windham, G.C., Shusterman, D., Swan, S.H., Fenster, L. & Eskenazi, B. (1991) Exposure to organic solvents and adverse pregnancy outcome. *Am. J. ind. Med.*, **20**, 241–259

Witey, R.J. & Hall, J.W. (1975) The joint toxic action of perchloroethylene with benzene or toluene in rats. *Toxicology*, **4**, 5–15

Wright, P.F.A., Schlichting, L.M. & Stacey, N.H. (1994) Effects of chlorinated solvents on the natural lymphocytotoxic activities of human liver immune cells. *Toxicol. in vitro*, **8**, 1037–1039

Yllner, S. (1961) Urinary metabolites of ^{14}C-tetrachloroethylene in mice. *Nature*, **191**, 820

Ziglio, G., Fara, G.M., Beltramelli, G. & Pregliasco, F. (1983) Human environmental exposure to trichlorethylene and tetrachloroethylene from water and air in Milan, Italy. *Arch. environ. Contam. Toxicol.*, **12**, 57–64

Ziglio, G., Beltramelli, F., Pregliasco, F. & Ferrari, G. (1985) Metabolites of chlorinated solvents in blood and urine of subjects exposed at environmental level. *Sci. total Environ.*, **47**, 473–477

Zimmerli, B., Zimmermann, H. & Müller, F. (1982) Perchloroethylene in foodstuffs. *Mittel. Geb. Lebensmittel. Hyg.*, **73**, 71–81 (in German)

Zoeteman, B.C.J., Harmsen, K., Linders, J.B.H.J., Morra, C.F.H. & Slooff, W. (1980) Persistent organic pollutants in river water and ground water of the Netherlands. *Chemosphere*, **9**, 231–249

Zürcher, F. & Giger, W. (1976) Volatile organic components in the Glatt. *Wasser*, **47**, 37–55 (in German)

1,2,3-TRICHLOROPROPANE

1. Exposure Data

1.1 Chemical and physical data

1.1.1 Nomenclature

Chem. Abstr. Serv. Reg. No.: 96-18-4
Chem. Abstr. Name: 1,2,3-Trichloropropane
IUPAC Systematic Name: 1,2,3-Trichloropropane
Synonyms: Allyl trichloride; glycerol trichlorohydrin; glyceryl trichlorohydrin; trichlorohydrin; trichloropropane

1.1.2 Structural and molecular formulae and relative molecular mass

```
      H   Cl  H
      |   |   |
  Cl—C — C — C—Cl
      |   |   |
      H   H   H
```

$C_3H_5Cl_3$ Relative molecular mass: 147.43

1.1.3 Chemical and physical properties of the pure substance

(a) *Description*: Colourless to straw-coloured liquid with chloroform-like odour (Verschueren, 1983; United States National Institute for Occupational Safety and Health, 1994)

(b) *Boiling-point*: 156.8 °C (Lide, 1993)

(c) *Melting-point*: −14.7 °C (Lide, 1993)

(d) *Density*: 1.3889 at 20 °C/4 °C (Lide, 1993)

(e) *Spectroscopy data*: Infrared (prism [4653], grating [10777]), nuclear magnetic resonance (proton [6769], C-13 [624]) and mass [814] spectral data have been reported (Sadtler Research Laboratories, 1980; Weast & Astle, 1985).

(f) *Solubility*: Slightly soluble in water (1.75 g/L); soluble in chloroform, diethyl ether and ethanol (Riddick *et al.*, 1986; American Conference of Governmental Industrial Hygienists, 1991; Lide, 1993)

(g) *Volatility*: Vapour pressure, 2 mm Hg [0.266 kPa] at 20 °C (Verschueren, 1983)

(h) *Reactivity*: Reacts with active metals, strong caustics and strong oxidizers (Sittig, 1985; United States National Institute for Occupational Safety and Health, 1994)

(i) *Octanol:water partition coefficient (P)*: log P, 1.98 (United States Agency for Toxic Substances and Disease Registry, 1992)

(j) *Conversion factor*: mg/m^3 = 6.03 × ppm[1]

1.1.4 Technical products and impurities

1,2,3-Trichloropropane is available commercially at a purity of > 98–99.9% (Crescent Chemical Co., 1990; Fluka Chemical Corp., 1993; Aldrich Chemical Co., 1994; TCI America, 1994). The material tested by the United States National Toxicology Program (1993) contains the following impurities: 0.066% water, 0.14% unspecified chlorohexene, two unspecified chlorohexadienes (0.24 and 0.13%), several unidentified impurities (each < 0.1%) and 48 ppm [mg/L] total acidity (as hydrochloric acid) (Alessandri, 1993).

1.1.5 Analysis

Selected methods for the analysis of 1,2,3-trichloropropane are presented in Table 1.

Two gas chromatography/mass spectrometry (GC/MS) and two purge-and-trap GC methods for purgeable organic compounds, including 1,2,3-trichloropropane, are usually used for analysing aqueous samples (see also Table 1). The first method (EPA Method 524.1 and APHA/AWWA/WEF Method 6210C), involving use of a packed column, and similar purge-and-trap methods (EPA Method 502.1 and APHA/AWWA/WEF Method 6230C), involving detection by electrolytic conductivity or microcoulometric methods, are applicable for determining 1,2,3-trichloropropane in raw source water or in drinking-water at any stage of treatment. The second group of methods (EPA Method 524.2 and APHA/AWWA/WEF Method 6210D; EPA Method 502.2 and APHA/AWWA/WEF Method 6230D) is identical to the previous set, except that a capillary column is used. These methods are intended primarily for the detection of large numbers of contaminants at very low concentrations (Greenberg *et al.*, 1992).

Methods for the analysis of 1,2,3-trichloropropane in the exhaled air, urine, faeces, bile, major tissues and blood of rats have been described. The first method involved drawing dry air through a cage and into a trap filled with ethanol at –15 °C, followed by GC and electron capture detection. The second involved sample homogenization, extraction with hexane and analysis by GC with electron capture detection. Blood samples were added to water, and bile samples were added to ethanol before extraction (United States Agency for Toxic Substances and Disease Registry, 1992).

In a method for determining fumigants, including 1,2,3-trichloropropane, in citrus fruit (lemon, orange, grapefruit), samples were blended with water, distilled into cyclohexane in an apparatus for essential oils, cleaned-up on a Florisil column and analysed by GC with electron capture detection (Tonogai *et al.*, 1986).

---

[1] Calculated from: mg/m^3 = (relative molecular mass/24.45) × ppm, assuming normal temperature (25 °C) and pressure (101 kPa)

Table 1. Methods for the analysis of 1,2,3-trichloropropane

Sample matrix	Sample preparation	Assay procedure	Limit of detection	Reference
Air	Adsorb on charcoal; desorb with carbon disulfide	GC/FID	0.01 mg/ sample	Eller (1994); US Occupational Safety and Health Administration (1990)
Water	Purge (inert gas); trap on suitable adsorbent material; desorb as vapour onto packed gas chromatographic column	GC/ECD or GC/MCD GC/MS	ND NR	US Environmental Protection Agency (1988)
	Purge and trap as above; desorb as vapour onto capillary gas chromatographic column	GC/PID-ECD GC/MS	0.02–0.4 µg/L 0.03–0.32 µg/L	US Environmental Protection Agency (1988)
Liquid and solid wastes	Purge (inert gas); trap on suitable adsorbent material; desorb as vapour onto packed gas chromatographic column	GC/ECD GC/MS	NR NR	US Environmental Protection Agency (1986a,b)

GC/FID, gas chromatography/flame ionization detection; ECD, electrolytic conductivity detection; MCD, microcoulometric detection; PID-ECD, photoionization detector in series with an electrolytic conductivity detector; ND, not determined; NR, not reported

1.2 Production and use

1.2.1 Production

1,2,3-Trichloropropane can be produced by chlorination of propylene (Sax & Lewis, 1987; see IARC, 1994a). Other reported methods for producing 1,2,3-trichloropropane include the addition of chlorine to allyl chloride (see IARC, 1987a), the reaction of thionyl chloride with glycerol and the reaction of phosphorus pentachloride with either 1,3- or 2,3-dichloropropanol. 1,2,3-Trichloropropane may also be produced in potentially significant amounts as a by-product of the production of other chemicals, including dichloropropene (a soil fumigant and nematocide; see IARC, 1987b), propylene chlorohydrin, propylene oxide (see IARC, 1994b), dichlorohydrin and glycerol (United States Agency for Toxic Substances and Disease Registry, 1992).

No data were available regarding recent production of 1,2,3-trichloropropane. The volume estimated to have been produced in 1977 in the United States was 21–110 million pounds (9.5–50 thousand tonnes) (United States Agency for Toxic Substances and Disease Registry, 1992). It is currently produced by one company each in Germany, Japan, the Russian Federation, the United Kingdom and the United States (United States International Trade Commission, 1992; Chemical Information Services, Inc., 1994).

1.2.2 Use

1,2,3-Trichloropropane was used in the past mainly as a solvent and extractive agent, including as a paint and varnish remover and as a cleaning and degreasing agent (Sax & Lewis, 1987). It is now used mainly as a chemical intermediate, for example, in the production of polysulfone liquid polymers, dichloropropene and hexafluoropropylene, and as a cross-linking agent in the synthesis of polysulfides (United States Agency for Toxic Substances and Disease Registry, 1992). These manufacturing processes generally occur in closed systems. 1,2,3-Trichloropropane is also found as an impurity in mixtures used as soil fumigants and fungicides, for instance, during the manufacture of the nematocide DD (a dichloropropane–dichloropropene mixture), which was introduced in 1942. Estimates of the amount of trichloropropanes in the DD mixture vary from 0.4 to 6–7% by weight (Oki & Giambelluca, 1987).

1.3 Occurrence

1.3.1 Natural occurrence

1,2,3-Trichloropropane is not known to occur as a natural product.

1.3.2 Occupational exposure

The National Occupational Exposure Survey conducted in the United States in 1981–83 indicated that 492 workers were potentially exposed to 1,2,3-trichloropropane (United States National Institute for Occupational Safety and Health, 1990). Any occupational exposure that occurred would probably result from inhalation and dermal contact. No data on occupational exposures were available to the Working Group.

1.3.3 Air

1,2,3-Trichloropropane was not detected in more than 400 samples of urban and rural air in Germany (von Düszeln *et al.*, 1982).

1.3.4 Water

1,2,3-Trichloropropane was found in groundwater in California (0.1–5 µg/L) and Hawaii (United States) as a result of use of pesticides in agriculture. It was detected in groundwater at two of 10 sites in an agricultural community in Suffolk County, New York, at concentrations of 6 and 10 µg/L and was found in 39% of 941 samples of groundwater in the United States, at a median concentration of 0.69 µg/L, an average concentration of 1.0 µg/L and a range of trace (below the detection limit) to 2.5 µg/L. It was found in groundwater at 0.71% of hazardous waste sites at a geometric mean concentration of 57.3 µg/L (United States Agency for Toxic Substances and Disease Registry, 1992). Concentrations of 0.3–2.8 µg/L were measured in well water on Oahu, Hawaii, probably as a result of the use and handling of the nematicide DD on pineapple plantations (Oki & Giambelluca, 1987).

Surface water from the Delaware River basin (United States) contained trichloropropane (unspecified isomer) at concentrations > 1 µg/L in 3% of the samples. Trichloropropane was also found at unspecified concentrations in seawater of Narragansett Bay, RI (United States Environmental Protection Agency, 1989).

Drinking-water from a water plant in New Orleans, LA (United States), contained 1,2,3-trichloropropane at < 0.2 µg/L (Keith *et al.*, 1976). The compound was also detected in drinking-water in Ames, IA, but the concentrations were not reported (United States Environmental Protection Agency, 1989). It has also been detected in drinking-water in the Netherlands (Kool *et al.*, 1982).

1,2,3-Trichloropropane was found in 69 of 141 samples of sewage sludge from municipal sewage treatment plants in Michigan (United States) in 1980, at a median concentration of 0.35 mg/kg, an average of 1.07 mg/kg and a range of 0.005–19.5 mg/kg on a dry-weight basis (United States Agency for Toxic Substances and Disease Registry, 1992).

The half-life for evaporation of 1,2,3-trichloropropane from water was about 1 h (Dilling, 1977; United States Environmental Protection Agency, 1989). Experiments in showers showed that the fraction of 1,2,3-trichloropropane volatilized from tap water in the United States was around 20% (Tancrède *et al.*, 1992).

1.3.5 Soil

1,2,3-Trichloropropane was reported to have a half-life of 2.7 days in a silt loam and a sandy loam (Anderson *et al.*, 1991). An unspecified isomer of trichloropropane was reported to be relatively easily decomposed by microbes in activated sludge (United States Environmental Protection Agency, 1989).

In the study described above in California and Hawaii, 1,2,3-trichloropropane was found in soil samples at levels of 0.2–2 ppb [µg/kg]. It was found at a depth of at least 10 feet [3.05 m] in soil profiles in Hawaii (Cohen *et al.*, 1987).

1.4 Regulations and guidelines

Occupational exposure limits and guidelines in several countries are given in Table 2.

2. Studies of Cancer in Humans

No data were available to the Working Group.

Table 2. Occupational exposure limits and guidelines for 1,2,3-trichloropropane

Country	Year	Concentration (mg/m³)	Interpretation
Australia	1991	60	TWA; skin notation
Belgium	1991	60	TWA; skin notation
Denmark	1991	300	TWA
Finland	1993	300	TWA
		450	STEL
Germany	1993	None	Carcinogenic to animals
Netherlands	1994	60	TWA; skin notation
Russian Federation	1991	2	STEL
Switzerland	1991	300	TWA
		1500	STEL
United Kingdom	1993	300	TWA
		450	STEL
USA			
ACGIH	1994	60	TWA; skin notation
NIOSH	1994	None	Potential carcinogen
OSHA	1994	300	TWA

From ILO (1991); Deutsche Forschungsgemeinschaft (1993); Työministeriö (1993); United Kingdom Health and Safety Executive (1993); American Conference of Governmental Industrial Hygienists (ACGIH) (1994); Arbeidsinspectie (1994); United States Occupational Safety and Health Administration (1994). TWA, time-weighted average; STEL, short-term exposure limit

3. Studies of Cancer in Experimental Animals

3.1 Oral administration

3.1.1 Mouse

Groups of 60 male and 60 female B6C3F1 mice, six weeks of age, were administered 1,2,3-trichloropropane (purity, > 99%) in corn oil by gavage at doses of 0 (vehicle control), 6, 20 or 60 mg/kg bw on five days per week for 104 weeks. Four to 10 mice per group were removed for histopathological evaluation at 15 months. Survival of treated mice was significantly lower ($p < 0.001$) than that of vehicle controls; the numbers of survivors at the end of the experiment were: 42 control males and 41 control females and 18 males at the low dose and 13 females at the low dose; none of the animals at the middle or high doses survived. Histopathological evaluation revealed increased incidences (life-table test or logistic regression analysis) of neoplasms of the forestomach, liver and Harderian gland in males and females and neoplasms of the oral mucosa and uterus in females (see Table 3). The incidence of focal hyperplasia of the forestomach epithelium was increased in treated mice, occurring in 8/52 control males, 29/51 at

Table 3. Incidences of neoplastic lesions in B6C3F1 mice during gavage with 1,2,3-trichloropropane for two years

Site and tumour type	Males				Females			
	Vehicle control	6 mg/kg bw	20 mg/kg bw	60 mg/kg bw	Vehicle control	6 mg/kg bw	20 mg/kg bw	60 mg/kg bw
Forestomach								
Squamous-cell papilloma	3/52	28/51 $p < 0.001$	22/54 $p < 0.001$	33/56 $p < 0.001$	0/50	23/50 $p < 0.001$	18/51 $p < 0.001$	29/55 $p < 0.001$
Squamous-cell carcinoma	0/52	40/51 $p < 0.001$	50/54 $p < 0.001$	51/56 $p < 0.001$	0/50	46/50 $p < 0.001$	49/51 $p < 0.001$	49/55 $p < 0.001$
Papilloma or carcinoma	3/52	50/51 $p < 0.001$	53/54 $p < 0.001$	55/56 $p < 0.001$	0/50	48/50 $p < 0.001$	50/51 $p < 0.001$	54/55 $p < 0.001$
Liver								
Hepatocellular adenoma	11/52	18/51 $p = 0.073$	21/54 $p = 0.028$	29/56 $p < 0.001$	6/50	9/50	8/51	31/55 $p < 0.001$
Hepatocellular carcinoma	4/52	11/51 $p = 0.015$	5/54	3/56	1/50	3/50	0/51	2/55
Adenoma or carcinoma	13/52	24/51 $p = 0.008$	24/54 $p = 0.007$	31/56 $p < 0.001$	7/50	11/50	8/51	31/55 $p < 0.001$
Harderian gland, adenoma	1/52	2/51	10/54 $p = 0.002$	11/56 $p = 0.008$	2/50	6/50	7/51	10/55 $p = 0.06$
Oral mucosa								
Squamous-cell papilloma					1/50	0/50	1/51	0/55
Squamous-cell carcinoma					0/50	0/50	1/51	5/55 $p = 0.006$
Papilloma or carcinoma					1/40	0/50	2/51	5/55 $p = 0.006$
Uterus								
Stromal polyp					0/50	2/50	1/51	6/54
Endometrial adenoma or adenocarcinoma					0/50	5/50	3/51	9/54 $p = 0.030$

From United States National Toxicology Program (1993)
p values are given for incidences that are significantly greater than those of controls on the basis of life-table tests or logistic regression analysis (adjusted for survival)

the low dose, 27/54 at the middle dose and 34/56 at the high dose; and in 10/50 control females, 15/50 at the low dose, 14/51 at the middle dose and 31/55 at the high dose (United States National Toxicology Program, 1993).

3.1.2 Rat

Groups of 60 male and 60 female Fischer 344/N rats, six weeks of age, were administered 1,2,3-trichloropropane (purity, > 99%) in corn oil by gavage at doses of 0 (vehicle control), 3, 10 or 30 mg/kg bw on five days per week for up to 104 weeks. Eight to 10 rats per group were removed for histopathological evaluation at 15 months. Survival rates of rats that received 10 or 30 mg/kg were significantly lower ($p < 0.001$) than that of vehicle controls; the numbers of survivors at the end of the experiment were 34 control males, 32 at the low dose and 14 at the middle dose; and 31 control females, 30 at the low dose and 8 at the middle dose; none of the animals at the high dose survived. Histopathological evaluation revealed increased incidences (life-table test or logistic regression analysis) of neoplasms of the oral mucosa and forestomach in males and females, neoplasms of the pancreas, preputial gland and kidney in males, and neoplasms of the clitoral gland and mammary gland in females (Table 4). The incidence of focal hyperplasia of the forestomach epithelium was increased in treated rats, occurring in 3/50 control males, 28/50 at the low dose, 13/49 at the middle dose and 6/52 at the high dose; and in 1/50 female controls, 25/49 at the low dose, 11/51 at the middle dose and 15/52 at the high dose. The incidence of focal hyperplasia of the renal tubular epithelium was increased in male rats receiving 10 or 30 mg/kg bw, being seen in 0/50 controls, 1/50 at the low dose, 21/49 at the middle dose and 29/52 at the high dose (United States National Toxicology Program, 1993).

3.2 Carcinogenicity of metabolites

Mouse: 1,3-Dichloroacetone, a metabolite of 1,2,3-trichloropropane, was tested for initiation in a two-stage mouse skin tumour model. 1,3-Dichloroacetone (purity, > 99%) was applied topically in 0.2 ml ethanol to groups of 40 female SENCAR mice [age unspecified] at a dose of 0, 50, 75 or 100 mg/kg bw, three times a week for two weeks. Two weeks after the final application, 1.0 μg of 12-*O*-tetradecanoylphorbol 13-acetate (TPA) in 0.2 ml acetone was applied three times a week for 20 weeks. The numbers of animals with skin tumours at one year were: 23/199 vehicle controls, 12/25 at the low dose ($p < 0.02$; log rank test), 18/40 at the middle dose ($p < 0.02$) and 12/38 at the high dose ($p < 0.02$). In a second experiment, 1,3-dichloroacetone was given as a single application to groups of 30 female SENCAR mice at a dose of 0, 37.5, 75, 150 or 300 mg/kg bw, and TPA was applied as in the multiple-dose study. After 24 weeks, the numbers of animals with skin tumours were: 23/199 vehicle controls, 14/30 at 37.5 mg/kg bw ($p < 0.02$), 14/30 at 75 mg/kg bw ($p < 0.02$), 19/30 at 150 mg/kg bw ($p < 0.02$) and 4/20 at 300 mg/kg bw (Robinson *et al.*, 1989).

Table 4. Incidences of neoplastic lesions in Fischer 344/N rats during gavage with 1,2,3-trichloropropane for two years

Site and tumour type	Males				Females			
	Vehicle control	3 mg/kg bw	10 mg/kg bw	30 mg/kg bw	Vehicle control	3 mg/kg bw	10 mg/kg bw	30 mg/kg bw
Oral cavity								
Squamous-cell papilloma	0/50	4/50	9/49 $p < 0.001$	19/52 $p < 0.001$	1/50	5/49	10/52 $p = 0.003$	18/52 $p < 0.001$
Squamous-cell carcinoma	1/50	0/50	11/49 $p < 0.001$	25/52 $p < 0.001$	0/50	1/49	21/52 $p < 0.001$	21/52 $p < 0.001$
Papilloma or carcinoma	1/50	4/50	18/49 $p < 0.001$	40/52 $p < 0.001$	1/50	6/49	28/52 $p < 0.001$	32/52 $p < 0.001$
Forestomach								
Squamous-cell papilloma	0/50	29/50 $p < 0.001$	33/49 $p < 0.001$	38/52 $p < 0.001$	0/50	13/49 $p < 0.001$	32/51 $p < 0.001$	17/52 $p < 0.001$
Squamous-cell carcinoma	0/50	9/50 $p = 0.003$	27/49 $p < 0.001$	13/52 $p < 0.001$	0/50	3/49 $p < 0.001$	9/51 $p < 0.001$	4/52 $p = 0.001$
Papilloma or carcinoma	0/50	33/50 $p < 0.001$	42/49 $p < 0.001$	43/52 $p < 0.001$	0/50	16/49 $p < 0.001$	37/51 $p < 0.001$	19/52 $p < 0.001$
Preputial gland/clitoral gland								
Adenoma	5/49	3/47	5/49	11/50 $p = 0.023$	5/46	10/46 $p < 0.001$	13/50 $p = 0.001$	10/51 $p = 0.030$
Carcinoma	0/49	3/47	3/49	5/50	0/46	0/46	4/50	6/51
Adenoma or carcinoma	5/49	6/47	8/49	16/50 $p = 0.007$	5/46	10/46	17/50 $p < 0.001$	15/51 $p = 0.013$
Pancreas								
Acinar adenoma	5/50	21/50 $p < 0.001$	36/49 $p < 0.001$	29/52 $p < 0.001$				
Adenocarcinoma	0/50	0/50	2/49	1/52				
Adenoma or adenocarcinoma	5/50	21/50 $p < 0.001$	36/49 $p = 0.001$	29/52 $p < 0.001$				
Kidney, renal tubular adenoma	0/50	2/50	20/49 $p < 0.001$	21/52 $p < 0.001$				
Mammary gland, adeno-carcinoma					1/50	6/49	12/52 $p < 0.001$	21/52 $p < 0.001$

From United States National Toxicology Program (1993)
p values are given for incidences that are significantly greater than those of controls on the basis of life-table tests or logistic regression analysis (adjusted for survival)

4. Other Data Relevant to an Evaluation of Carcinogenicity and its Mechanisms

4.1 Absorption, distribution, metabolism and excretion

4.1.1 Humans

No data were available to the Working Group.

4.1.2 Experimental systems

[2-^{14}C]1,2,3-Trichloropropane (specific activity, 57 mCi/mmol; radiochemical purity, > 93%; about 5% assumed to be [2-^{14}C]2,3-dichloropropene) was administered by gavage to male and female Fischer 344 rats at a single dose of 30 mg/kg bw in corn oil. The compound was rapidly absorbed, metabolized and excreted. Six hours after treatment, the highest concentration of radiolabel was found in the tissue of the forestomach, followed by the glandular stomach, intestine, fat, liver and kidney. Tissue concentrations declined thereafter. After 60 h, most of the radiolabel derived from 1,2,3-trichloropropane was concentrated in the liver, kidney and forestomach. Extraction with organic solvents removed 17% (from liver) to 50% (from the forestomach) of the radiolabel. By 60 h, ≥ 90% of the compound had been cleared, with 50% (in females) to 57% (in males) of the dose excreted in urine. Exhalation of ^{14}C-carbon dioxide and excretion in the faeces each accounted for about 20% of the dose, with no marked difference between males and females; < 2% of the dose was exhaled as unchanged compound. More than half of the dose was excreted within 24 h. N-Acetyl-S-(3-chloro-2-hydroxypropyl)-L-cysteine was identified as the major urinary metabolite in male rats (representing 40% of the radiolabel in urine after 6 h); smaller amounts were detected in the urine of female rats. Another urinary metabolite was identified as S-(3-chloro-2-hydroxypropyl)-L-cysteine (Mahmood et al., 1991).

[1,3-^{14}C]1,2,3-Trichloropropane (specific activity, 8.8 mCi/mmol [radiochemical purity not given]) was injected intravenously into male Fischer 344 rats at a single dose of 3.6 mg/kg bw in Emulphor EL-620:ethanol:water (1:1:3 by volume). 1,2,3-Trichloropropane was distributed and eliminated rapidly: the initial half-life of unchanged compound in the blood was 0.29 h and the terminal half-life was 23 h. Adipose tissue accumulated 37% of the dose within 15 min. After 4 h, the liver contained the largest fraction of the dose, primarily as metabolites. Excretion was nearly complete (90% of the dose) within 24 h and occurred predominantly via the urine (47% of the dose); 25% of the dose was exhaled as carbon dioxide and 5% as unchanged 1,2,3-trichloropropane. None of the numerous other urinary and biliary metabolites accounted for more than 10% of the dose (Volp et al., 1984). [The Working Group noted that these metabolites were not identified.]

[2-^{14}C]1,2,3-Trichloropropane (specific activity, 57 mCi/mmol; radiochemical purity, > 93%; about 5% assumed to be [2-^{14}C]2,3-dichloropropene) was administered by gavage to male B6C3F1 mice at a single dose of 30 or 60 mg/kg bw in corn oil. The compound was extensively absorbed and rapidly metabolized and excreted. By 60 h after treatment, the highest concentrations of radiolabel were found in liver, kidney and forestomach, and in most tissues the concentrations were proportional to the dose. By 60 h, 65% of the dose had been excreted in the urine, 20% was exhaled as carbon dioxide and less than 1% as volatile compounds, whereas 16% was excreted in the faeces. Male mice exhaled ^{14}C-carbon dioxide significantly more rapidly than male rats, and the amounts and patterns of urinary metabolites in male mice were different from those in male rats. *N*-Acetyl-*S*-(3-chloro-2-hydroxypropyl)-L-cysteine accounted for about 3% of the urinary radiolabel; no other metabolites were identified (Mahmood *et al.*, 1991).

[2-^{14}C]1,2,3-Trichloropropane (specific activity, 5.7 mCi/mmol; radiochemical purity, > 96%) was injected intraperitoneally into male Fischer 344 rats at doses of 30 mg/kg bw daily for one to three days in soya bean oil. 1,2,3-Trichloropropane bound covalently to hepatic protein, DNA and RNA, with binding levels of radioequivalents 4 h after administration of 418, 244 and 432 pmol/mg, respectively. Binding to hepatic DNA did not change significantly over 1–48 h, whereas binding to hepatic protein was maximal after 4 h. The binding to hepatic DNA and protein was cumulative when two and three doses were given 24 h apart. Covalent binding to hepatic DNA and protein was increased in animals treated with SKF 525-A; binding was decreased after treatment with phenobarbital; whereas binding was unaffected by treatment with β-naphthoflavone. Pretreatment of rats with L-buthionine (*R,S*)-sulfoximine increased binding to protein by 342% and decreased binding to DNA by 56%. Intraperitoneal administration of 1,2,3-trichloropropane depleted hepatic glutathione by 41% 2 h after a dose of 30 mg/kg bw and by 61% after a dose of 100 mg/kg bw (Weber & Sipes, 1990a).

The metabolism of [2-^{14}C]1,2,3-trichloropropane (specific activity, 5.7 mCi/mmol; radiochemical purity, > 99.5%) was studied *in vitro* in subfractions of liver from male Fischer 344 rats and human organ donors. 1,3-Dichloroacetone was identified as a microsomal metabolite of 1,2,3-trichloropropane in the presence of NADPH; it was formed at a rate of 0.27 nmol/min per mg protein with microsomes from rat liver and at 0.03 nmol/min per mg protein with microsomes from one sample of human liver. Formation of 1,3-dichloroacetone was increased after treatment with phenobarbital and dexamethasone but was decreased by treatment with β-naphthoflavone or after addition of SKF 525-A or 1-aminobenzotriazol. Addition of alcohol dehydrogenase and NADH to the microsomal incubations resulted in the formation of 1,3-dichloro-2-propanol and 2,3-dichloropropanol by reduction of 1,3-dichloroacetone and 2,3-dichloropropanol, respectively. 1,2,3-Trichloropropane was found to bind covalently to rat liver microsomal protein. The rate of binding was increased eightfold by treatment with phenobarbital, while addition of glutathione and *N*-acetylcysteine completely inhibited binding. In the presence of *N*-acetylcysteine, 1,3-(2-propanone)bis-*S,N*-acetylcysteine was the only conjugate detected. No binding occurred in the presence of rat liver cytosol and glutathione, but water-soluble metabolites were formed (Weber & Sipes, 1992).

4.1.3 Comparison of humans and animals

No data were available on the toxicokinetics of 1,2,3-trichloropropane in humans. 1,3-Dichloracetone is generated in human liver microsomes at a rate one-tenth of that in rat liver microsomes.

4.2 Toxic effects

4.2.1 Humans

The limited information available indicates that brief exposure (15 min) to 100 ppm [about 600 mg/m^3] 1,2,3-trichloropropane (purity unknown) can irritate the eye and throat (Silverman et al., 1946).

4.2.2 Experimental systems

The toxicology of 1,2,3-trichloropropane in experimental animals and humans has been reviewed (United States Agency for Toxic Substances and Disease Registry, 1992; Anon., 1993).

The oral LD$_{50}$ of 1,2,3-trichloropropane in rats is reported to be 320 mg/kg bw, whereas its approximate lethal concentration after exposure by inhalation for 4 h is 1000 ppm [about 6000 mg/m^3] (Kennedy & Graepel, 1991). A similar LD$_{50}$ was observed in rats after dermal (Anon., 1993) and oral administration, indicating that it is absorbed by these routes.

Groups of 10 male and female Sprague-Dawley rats were administered 1,2,3-trichloropropane (purity, 99.3%; 0.7% tetrachloropropane) in corn oil by gavage at doses of 0, 0.01, 0.05, 0.20 or 0.80 mmol/kg bw [0, 1.5, 7, 29 or 117 mg/kg bw] daily for 10 days, or 0, 0.01, 0.05, 0.10 or 0.40 mmol/kg bw [0, 1.5, 7, 14 or 29 mg/kg bw] daily for 90 days [the doses are given incorrectly as mmol/kg bw in Table 1 and Figure 1 of the report]. The primary histological finding was cardiopathy associated with inflammation. Myocardial necrosis and degeneration occurred in a diffuse pattern, with marked eosinophilia at the highest dose after exposure for 10 days; cardiopathy was also noted at the lower doses in the 90-day study. A mild hepatotoxic response was noted in animals receiving the high dose in each study. Bile-duct hyperplasia was observed after exposure for 90 days (Merrick et al., 1991).

Groups of 20 male and female Fischer 344/N rats and B6C3F1 mice received 1,2,3-trichloropropane (purity, > 99%) in corn oil by gavage at doses of 8, 16, 32, 63, 125 or 250 mg/kg bw on five days per week for eight (interim sacrifice) or 17 weeks; groups of 30 animals of each sex served as controls. Rats receiving the highest dose that died during the first several weeks had severe multifocal, centrilobular hepatocellular necrosis. The necrosis was more extensive in female rats, and the necrosis seen in animals at 125 mg/kg bw was less extensive than that seen at 250 mg/kg bw. Rats that received the high dose and died had severe nephrotoxicity with diffuse acute tubular necrosis in the outer stripe of the outer medulla. The nephrotoxicity seen at the time of the eight-week interim sacrifice in rats given 125 mg/kg bw was characterized by regenerative hyperplasia with karyomegaly. At the end of the study, chronic renal inflammation was also seen. Rats given 250 mg/kg bw had extensive necrosis of the olfactory and respiratory epithelium in the nose; these lesions were also seen at 125 mg/kg bw later in the study. Mice at the two highest doses had focal hepatocellular necrosis; those that died early while receiving the high dose also had necrosis, regeneration and hyperplasia of the bronchiolar epithelium. Minimal pulmonary changes were noted at the end of the study in the group receiving 125 mg/kg bw. At the time of the eight-week interim evaluation and at the end of the study, a number of mice receiving 250 mg/kg bw had minimal acanthosis and hyperkeratosis of the forestomach (United States National Toxicology Program, 1993).

Groups of 15 male and female CD rats were exposed by inhalation to 0, 5, 15 or 50 ppm (0, 30, 90 or 302 mg/m^3) 1,2,3-trichloropropane (purity, 98.9%) for 6 h per day, on five days per week for 13 weeks. Hepatocellular hypertrophy was observed in male rats at all doses. Dose-related focal peribronchial lymphoid hyperplasia was observed primarily in males and splenic extramedullary haematopoiesis only in females (Johannsen et al., 1988).

Groups of 10 male and female Sprague-Dawley rats were administered 1,2,3-trichloropropane (stated purity, 99%), solubilized with 0.5% Emulphor, in their drinking-water at concentrations of 0, 1, 10, 100 or 1000 mg/L for 13 weeks. Animals of each sex receiving the highest concentration showed mild changes, consisting of anisokaryosis, accentuated zonation and occasional fatty vacuolation of the liver; female rats also had biliary hyperplasia. In addition, mild cellular changes were seen in the kidneys and thyroids of animals at the highest dose. The activities of hepatic aminopyrine demethylase and aniline hydroxylase were increased in animals receiving the highest concentration (Villeneuve et al., 1985).

4.3 Reproductive and prenatal effects

4.3.1 Humans

No data were available to the Working Group.

4.3.2 Experimental systems

Maternal toxicity was observed in female Sprague-Dawley rats administered 1,2,3-trichloropropane by intraperitoneal injection of 37 mg/kg bw on days 1–15 of gestation and killed on day 21, but there was no sign of fetal toxicity or malformations (Hardin et al., 1981).

Groups of 10 male and 20 female CD rats were exposed to 1,2,3-trichloropropane in air at concentrations of 0, 30 or 90 mg/m^3 for 6 h per day on five days per week during a 10-week premating period; females were further exposed during a mating period of up to 40 days. Histopathological examination of the testes, epididymides and ovaries revealed no treatment-related toxic effect. Except for a marginal decrease in mating rates among animals at the highest dose, 1,2,3-trichloropropane neither altered fertility nor induced embryotoxic or teratogenic effects (Johannsen et al., 1988).

1,2,3-Trichloropropane was administered by gavage at doses of 0, 30, 60 or 120 mg/kg bw to groups of 20 male and female CD-1 mice during a seven-day precohabitation and a 98-day cohabitation period. The controls and the group receiving 120 mg/kg bw were additionally used in a cross-over mating experiment, in which treated females were mated with control males and vice versa. 1,2,3-Trichloropropane induced a dose-related impairment of fertility in the absence of gross general toxicity, determined on the basis of body weight. Fewer pairs at the high dose delivered third, fourth or fifth litters, and the litters included fewer live pups. A similar impairment of survival rates was observed among pups of treated females that had been mated with untreated males. The weights of the livers of both male and female F_0 mice were increased, and the weights of the kidneys and ovaries of female mice were reduced; epididymal weight was slightly reduced in animals at the high dose, but sperm parameters were not influenced. Taken together, the data indicate impairment of the female reproductive system. Assessment of the fertility of the last second-generation litter born from the F_0 generation also revealed a decreased number of pups, although other indications of fertility were not altered. In summary, the study indicates that 1,2,3-trichloropropane is toxic to the reproductive system of Swiss CD-1 mice (Gulati et al., 1990).

4.4 Genetic and related effects

4.4.1 Humans

No data were available to the Working Group.

4.4.2 Experimental systems (see also Table 5 and Appendices 1 and 2)

1,2,3-Trichloropropane was mutagenic to *Salmonella typhimurium* in the presence of an exogenous metabolic activation system, but it did not induce SOS repair functions in *Escherichia coli* PQ37.

It did not induce unscheduled DNA synthesis in primary rat hepatocytes. It induced gene mutation at the thymidine kinase locus in L5178Y mouse lymphoma cells, sister chromatid exchange in Chinese hamster V79 and ovary cells and chromosomal aberrations in Chinese hamster ovary cells, only in the presence of metabolic activation. As reported in an abstract, 1,2,3-trichloropropane enhanced the transformation of Syrian hamster embryo cells by simian adenovirus SA7 [dose not given].

Binding to hepatic DNA, RNA and protein was observed after male Fischer 344 rats, aged 9–11 weeks, were treated intraperitoneally with single or two to three repeated injections of 30 mg/kg bw ^{14}C-1,2,3-trichloropropane. DNA binding was decreased by pretreating animals with phenobarbital or the glutathione depleting agent, L-buthionine(R,S)sulfoximine, was increased by pretreatment with SKF 525-A and was not altered by pretreatment with β-naphthoflavone (Weber & Sipes, 1990a). In rats of the same strain and sex treated intraperitoneally with 30, 100 or 300 mg/kg bw, DNA strand breaks, but not DNA–DNA or DNA–protein cross-links, were detected in the alkaline elution assay. It was reported in an abstract that 1,2,3-trichloropropane did not induce unscheduled DNA synthesis in the hepatocytes of male Fischer 344 rats treated orally [dose not given].

1,2,3-Trichoropropane did not induce dominant lethal effects in the offspring of Sprague-Dawley rats that were treated at the age of 11 weeks or more by gastric intubation once a day for five days with a dose of 80 mg/kg bw and then mated with virgin females for eight successive weeks.

4.4.3 Mutagenicity of metabolites

1,3-Dichloro-2-propanol induced SOS repair functions in *E. coli* GC 4798 and mutation in *S. typhimurium* strains in the absence of exogenous metabolic systems. The addition of aldehyde dehydrogenase did not alter the activity. The genetic activity *in* vitro may have been due to formation of epichlorohydrin in the buffer system (Hahn et al., 1991). Sister chromatid exchange was induced in cultured mammalian cells.

Table 5. Genetic and related effects of 1,2,3-trichloropropane

Test system	Result[a] Without exogenous metabolic system	Result[a] With exogenous metabolic system	Dose[b] (LED/HID)	Reference
PRB, SOS chromotest, *Escherichia coli* PQ37	–	–	0.00	von der Hude *et al.* (1988)
SA0, *Salmonella typhimurium* TA100, reverse mutation	–	+	5.5	Stolzenberg & Hine (1980)
SA0, *Salmonella typhimurium* TA100, reverse mutation	–	+	4	Haworth *et al.* (1983)
SA0, *Salmonella typhimurium* TA100, reverse mutation	–	+	2	Ratplan & Plaumann (1988)
SA0, *Salmonella typhimurium* TA100, reverse mutation	–	+	4.0	US National Toxicology Program (1993)
SA0, *Salmonella typhimurium* TA100, reverse mutation	–	(+)	7	Låg *et al.* (1994)
SA5, *Salmonella typhimurium* TA1535, reverse mutation	–	+	0.4	Haworth *et al.* (1983)
SA5, *Salmonella typhimurium* TA1535, reverse mutation	–	+	2	Ratplan & Plaumann (1988)
SA5, *Salmonella typhimurium* TA1535, reverse mutation	–	+	4.0	US National Toxicology Program (1993)
SA7, *Salmonella typhimurium* TA1537, reverse mutation	–	–	128	Haworth *et al.* (1983)
SA7, *Salmonella typhimurium* TA1537, reverse mutation	–	–	37.0	Ratplan & Plaumann (1988)
SA8, *Salmonella typhimurium* TA1538, reverse mutation	–	–	37.0	Ratplan & Plaumann (1988)
SA9, *Salmonella typhimurium* TA98, reverse mutation	–	+	39	Haworth *et al.* (1983)
SA9, *Salmonella typhimurium* TA98, reverse mutation	–	–	37.0	Ratplan & Plaumann (1988)
SA9, *Salmonella typhimurium* TA98, reverse mutation	–	+	40.0	US National Toxicology Program (1993)
SAS, *Salmonella typhimurium* TA97, reverse mutation	–	+	4	US National Toxicology Program (1993)
URP, Unscheduled DNA synthesis, Fischer 344 rat primary hepatocytes *in vitro*	–		10.0	Williams *et al.* (1989)
G5T, Gene mutation, mouse lymphoma L5178Y cells, *tk* locus *in vitro*	–	+	14	US National Toxicology Program (1993)
SIC, Sister chromatid exchange, Chinese hamster lung V79 cells *in vitro*	–	+	44	von der Hude *et al.* (1987)
SIC, Sister chromatid exchange, Chinese hamster ovary (CHO) cells *in vitro*	–	+	14	US National Toxicology Program (1993)
CIC, Chromosomal aberrations, Chinese hamster ovary (CHO) cells *in vitro*	–	+	60	US National Toxicology Program (1993)

Table 5 (contd)

Test system	Result[a] Without exogenous metabolic system	Result[a] With exogenous metabolic system	Dose[b] (LED/HID)	Reference
T7S, Cell transformation, SA7/Syrian hamster embryo cells in vitro	+	0	0.00	Hatch et al. (1983) (Abstract)
DVA, DNA strand breaks, rat hepatocytes in vivo	+		30.0 ip × 1	Weber & Sipes (1990b)
DVA, DNA–DNA and DNA–protein cross-links, rat hepatocytes in vivo	–		300.0 ip × 1	Weber & Sipes (1990b)
UPR, Unscheduled DNA synthesis, rat hepatocytes in vivo	–		0.00 po	Mirsalis et al. (1983) (Abstract)
DLR, Dominant lethal mutation, rats in vivo	–		80.0 po × 5	Saito-Suzuki et al. (1982)
BVD, Covalent DNA binding, rat hepatocytes in vivo	+		30.0 ip × 1	Weber & Sipes (1990a)
Metabolites of 1,2,3-trichloropropane				
1,3-Dichloro-2-propanol				
PRB, SOS chromotest, *Escherichia coli* GC4798	+	0	369.0	Hahn et al. (1991)
SA0, *Salmonella typhimurium* TA100, reverse mutation	+	+	582.0	Hahn et al. (1991)
SA5, *Salmonella typhimurium* TA1535, reverse mutation	+	+	192.0	Hahn et al. (1991)
SIC, Sister chromatid exchange, Chinese hamster lung V79 cells in vitro	+	+	147.4	von der Hude et al. (1987)
1,3-Dichloroacetone (1,3-Dichloropropanone)				
PRB, SOS chromotest, *Escherichia coli* PQ37	(+)	+	0.7	Le Curieux et al. (1994)
SA0, *Salmonella typhimurium* TA100, reverse mutation	+	0	0.6	Meier et al. (1985)
SA0, *Salmonella typhimurium* TA100, reverse mutation	+	+	0.8	Merrick et al. (1987)
SA0, *Salmonella typhimurium* TA100, reverse mutation, fluctuation test	+	+	0.03	Le Curieux et al. (1994)
SA5, *Salmonella typhimurium* TA1535, reverse mutation	+	+	0.8	Merrick et al. (1987)
SA9, *Salmonella typhimurium* TA98, reverse mutation	–	0	5.0	Meier et al. (1985)
***, *Pleurodeles waltl*, micronucleus induction in vivo	+		0.03, 12 d	Le Curieux et al. (1994)
SIC, sister chromatid exchange, Chinese hamster lung V79 cells in vitro	+	+	0.3	von der Hude et al. (1987)

[a] +, considered to be positive; (+), considered to be weakly positive in an inadequate study; –, considered to be negative; 0, not tested; for in-vivo tests, no entry of a result under 'with exogenous metabolic system'

[b] LED, lowest effective dose; HID, highest ineffective dose; in-vitro tests, µg/ml; in-vivo tests, mg/kg bw; 0.00, dose not reported; ip, intraperitoneally; po, orally

1,3-Dichloroacetone (1,3-dichloropropanone) induced SOS repair functions in *E. coli* PQ37 and mutations in *S. typhimurium* strains in the absence of exogenous metabolic systems.

Sister chromatid exchange was induced in cultured mammalian cells and micronuclei were induced in the peripheral erythrocytes of the newt, *Pleurodeles waltl, in vivo*.

5. Summary and Evaluation

5.1 Exposure data

1,2,3-Trichloropropane, a chlorinated solvent, has been produced commercially for use as a paint and varnish remover and as a cleaning and degreasing agent. Currently, it is used primarily as a chemical intermediate. It has been detected in water, including drinking-water, and in soil as a result of its presence as an impurity in a commercial nematocide.

5.2 Human carcinogenicity data

No data were available to the Working Group.

5.3 Animal carcinogenicity data

1,2,3-Trichloropropane was tested for carcinogenicity by oral administration in one experiment in mice and in one experiment in rats. It produced tumours of the oral mucosa and of the uterus in female mice and increased the incidences of tumours of the forestomach, liver and Harderian gland in mice of each sex. In rats, increased incidences of tumours were observed in the preputial gland, kidney and pancreas of males, in the clitoral gland and mammary gland of females and in the oral cavity and forestomach of both males and females.

The metabolite, 1,3-dichloroacetone, initiated skin tumour development in mice when applied topically.

5.4 Other relevant data

No data are available on the toxicokinetics of 1,2,3-trichloropropane in humans. It is rapidly absorbed and excreted after oral administration to rats and mice. Its metabolic products bind covalently to rat hepatic protein and DNA. The reactive and mutagenic metabolite, 1,3-dichloroacetone, was formed by hepatic metabolism in rat and human microsomes *in vitro*.

1,2,3-Trichloropropane causes tissue necrosis in a number of organs in rats and mice; the liver and kidney are the main target organs in the rat. In addition, myocardial and nasal epithelial damage is observed; in mice, hepatic and bronchiolar necrosis are seen.

There are no data on the effects of 1,2,3-trichloropropane on human reproduction. Studies performed in rats provided no evidence of alteration of fertility or of embryotoxic effects. In a two-generation study in mice, there was evidence of impairment of the female reproductive system.

In single studies, DNA binding and induction of DNA breaks, but not of dominant lethal mutations, were reported in rodents treated *in vivo*.

Gene mutation, sister chromatid exchange and chromosomal aberrations, but not DNA damage, were induced in rodent cells *in vitro* (all single studies, except for sister chromatid exchange). 1,2,3-Trichloropropane was mutagenic to bacteria.

5.5 Evaluation[1]

There is *inadequate evidence* in humans for the carcinogenicity of 1,2,3-trichloropropane.

There is *sufficient evidence* in experimental animals for the carcinogenicity of 1,2,3-trichloropropane.

Overall evaluation

1,2,3-Trichloropropane *is probably carcinogenic to humans (Group 2A)*.

In making the overall evaluation, the Working Group took into account the following evidence:

(i) 1,2,3-Trichloropropane causes tumours at multiple sites and at high incidence in mice and rats.

(ii) The metabolism of 1,2,3-trichloropropane is qualitatively similar in human and rodent microsomes.

(iii) 1,2,3-Trichloropropane is mutagenic to bacteria and to cultured mammalian cells and binds to DNA of animals treated *in vivo*.

6. References

Aldrich Chemical Co. (1994) *Specification Sheet: 1,2,3-Trichloropropane*, Milwaukee, WI

Alessandri, B. (1993) *Propane, 1,2,3-trichloro—CAS No. 96-18-4*, Switzerland, Dow Europe SA

American Conference of Governmental Industrial Hygienists (1991) *Documentation of the Threshold Limit Values and Biological Exposure Indices*, 6th Ed., Cincinnati, OH, pp. 1626–1630

American Conference of Governmental Industrial Hygienists (1994) *1994–1995 Threshold Limit Values for Chemical Substances and Physical Agents and Biological Exposure Indices*, Cincinnati, OH, p. 35

Anderson, T.A., Beauchamp, J.J. & Walton, B.T. (1991) Organic chemicals in the environment. Fate of volatile and semivolatile organic chemicals in soils: abiotic versus biotic losses. *J. environ. Qual.*, **20**, 420–424

Anon. (1993) *1,2,3-Trichloropropane, MAK-Werte-Liste 1993* [1,2,3-Trichloropropane, MAC values list, 1993], Weinheim, VCH Verlagsgesellsschaft

---

[1] For definition of the italicized terms, see Preamble, pp. 22–26.

Arbeidsinspectie [Labour Inspection] (1994) *De Nationale MAC-lijst* [National MAC list], The Hague, p. 40

Chemical Information Services, Inc. (1994) *Directory of World Chemical Producers 1995/96 Standard Edition*, Dallas, TX, p. 678

Crescent Chemical Co. (1990) *Laboratory Chemicals Catalog*, Hauppauge, NY, p. 1266

Deutsche Forschungsgemeinschaft (1993) *MAK-und-BAT-Werte-Liste 1993* [MAK and BAT values list 1993] (Report No. 29), Weinheim, VCH Verlagsgesellschaft, p. 76

Dilling, W.L. (1977) Interphase transfer processes. II. Evaporation rates of chloromethanes, ethanes, ethylenes, propanes, and propylenes from dilute aqueous solutions. Comparison with theoretical predictions. *Environ. Sci. Technol.*, **11**, 405–409

von Düszeln, J., Lahl, U., Bätjer, K., Cetinkaya, M., Stachel, B. & Thiemann, W. (1982) Volatile halogenohydrocarbons in air, water and food in east Germany. Analysis and intake estimation for men. *Dtsch. Lebensm. Rundsch.*, **78**, 352–356 (in German)

Eller, P.M., ed. (1994) *NIOSH Manual of Analytical Methods*, 4th Ed., Vol. 3 (DHHS (NIOSH) Publ. No. 94–113), Washington DC, United States Government Printing Office, Method 1003

Fluka Chemical Corp. (1993) *Fluka Chemika-BioChemika*, Buchs, p. 1297

Greenberg, A.E., Clesceri, L.S. & Eaton, A.D. (1992) *Standard Methods for the Examination of Water and Wastewater*, 18th Ed., Washington DC, American Public Health Association/American Water Works Association/Water Environment Federation, pp. 6–17 – 6–36, 6–42 – 6–57

Gulati, D.K., Mounce, R.C., Russell, S., Poonacha, K.B., Chapin, R.E. & Heindel, J. (1990) *Final Report on the Reproduction Toxicity of 1,2,3-Trichloropropane in CD-1 Swiss Mice* (NTP Report 90-209), Research Triangle Park, NC, United States National Toxicology Program

Hahn, H., Eder, E. & Deininger, C. (1991) Genotoxicity of 1,3-dichloro-2-propanol in the SOS chromotest and in the Ames test. Elucidation of the genotoxic mechanism. *Chem.-biol. Interactions*, **80**, 73–88

Hardin, B.D., Bond, G.P., Sikov, M.R., Andrew, F.D., Beliles, R.P. & Niemeier, R.W. (1981) Testing of selected workplace chemicals for teratogenic potential. *Scand. J. Work Environ. Health*, **7** (Suppl. 4), 66–75

Hatch, G., Anderson, T., Elmore, E. & Nesnow, S. (1983) Status of enhancement of DNA viral trasformation for determination of mutagenic and carcinogenic potential of gaseous and volatile compounds (Abstract Cd-26). *Environ. Mutag.*, **5**, 422

Haworth, S., Lawlor, T., Mortelmans, K., Speck, W. & Zeiger, E. (1983) *Salmonella* mutagenicity test results for 250 chemicals. *Environ. Mutag.*, **Suppl. 1**, 3–142

von der Hude, W., Scheutwinkel, M., Gramlich, U., Fissler, B. & Basler, A. (1987) Genotoxicity of three-carbon compounds evaluated in the SCE test *in vitro*. *Environ. Mutag.*, **9**, 401–410

von der Hude, W., Behm, C., Gürtler, R. & Basler, A. (1988) Evaluation of the SOS chromotest. *Mutat. Res.*, **203**, 81–94

IARC (1987a) *IARC Monographs on the Evaluation of Carcinogenic Risks to Humans*, Suppl. 7, *Overall Evaluations of Carcinogenicity: An Updating of* IARC Monographs *Volumes 1–42*, Lyon, p. 56

IARC (1987b) *IARC Monographs on the Evaluation of Carcinogenic Risks to Humans*, Suppl. 7, *Overall Evaluations of Carcinogenicity: An Updating of* IARC Monographs *Volumes 1–42*, Lyon, pp. 195–196

IARC (1994a) *IARC Monographs on the Evaluation of Carcinogenic Risks to Humans*, Vol. 60, *Some Industrial Chemicals*, Lyon, pp. 161–180

IARC (1994b) *IARC Monographs on the Evaluation of Carcinogenic Risks to Humans*, Vol. 60, *Some Industrial Chemicals*, Lyon, pp. 181–213

ILO (1991) *Occupational Exposure Limits for Airborne Toxic Substances: Values of Selected Countries* (Occupational Safety and Health Series No. 37), 3rd Ed., Geneva, pp. 398–399

Johannsen, F.R., Levinskas, G.J., Rusch, G.M., Terrill, J.B. & Schroeder, R.E. (1988) Evaluation of the subchronic and reproductive effects of a series of chlorinated propanes in the rat. 1. Toxicity of 1,2,3-trichloropropane. *J. Toxicol. environ. Health*, **25**, 299–315

Keith, L.H., Garrison, A.W., Allen, F.R., Carter, M.H., Floyd, T.L., Pope, J.D. & Thruston, A.D., Jr (1976) Identification of organic compounds in drinking water from thirteen US cities. In: Keith, L.H., ed., *Identification and Analysis of Organic Pollutants in Water*, Ann Arbor, MI, Ann Arbor Science, pp. 329–373

Kennedy, G.L., Jr & Graepel, G.J. (1991) Acute toxicity in the rat following either oral or inhalation exposure. *Toxicol. Lett.*, **56**, 317–326

Kool, H.J., van Kreijl, C.F. & Zoeteman, B.C.J. (1982) Toxicology assessment of organic compounds in drinking water. *Crit. Rev. environ. Control*, **12**, 307–357

Låg, M., Omichinski, J.G., Dybing, E., Nelson, S.D. & Søderlund, E.J. (1994) Mutagenic activity of halogenated propanes and propenes: effect of bromine and chlorine positioning. *Chem.-biol. Interactions*, **93**, 73–84

Le Curieux, F., Marzin, D. & Erb, F. (1994) Study of the genotoxic activity of five chlorinated propanones using the SOS chromotest, the Ames-fluctuation test and the newt micronucleus test. *Mutat. Res.*, **341**, 1–15

Lide, D.R., ed. (1993) *CRC Handbook of Chemistry and Physics*, 74th Ed., Boca Raton, FL, CRC Press, p. 3-414

Mahmood, N.A., Overstreet, D. & Burka, L.T. (1991) Comparative disposition and metabolism of 1,2,3-trichloropropane in rats and mice. *Drug Metab. Disposition*, **19**, 411–418

Meier, J.R., Ringhand, H.P., Coleman, W.E., Munch, J.W., Streicher, R.P., Kaylor, W.H. & Schenck, K.M. (1985) Identification of mutagenic compounds formed during chlorination of humic acid. *Mutat. Res.*, **157**, 111–122

Merrick, B.A., Smallwood, C.L., Meier, J.R., McKean, D.L., Kaylor, W.H. & Condie, L.W. (1987) Chemical reactivity, cytotoxicity, and mutagenicity of chloropropanones. *Toxicol. appl. Pharmacol.*, **91**, 46–54

Merrick, B.A., Robinson, M. & Condie, L.W. (1991) Cardiopathic effect of 1,2,3-trichloropropane after subacute and subchronic exposure in rats. *J. appl. Toxicol.*, **11**, 179–187

Mirsalis, J., Tyson, K., Beck, J., Loh, E., Steinmetz, K., Contreras, C., Austere, L., Martin, S. & Spalding, J. (1983) Induction of unscheduled DNA synthesis (UDS) in hepatocytes following in vitro and in vivo treatment (Abstract Ef-5). *Environ. Mutag.*, **5**, 482

Oki, D.S. & Giambelluca, T.W. (1987) DBCP, EDB, and TCP contamination of ground water in Hawaii. *Ground Water*, **25**, 693–702

Ratplan, F. & Plaumann, H. (1988) Mutagenicity of halogenated propanes and their methylated derivatives. *Environ. mol. Mutag.*, **12**, 253–259

Riddick, J.A., Bringer, W.B. & Sakano, T.K. (1986) *Organic Solvents, Physical Properties and Methods of Purification Techniques of Chemistry*, 4th Ed., New York, Wiley Interscience, pp. 1018–1019

Robinson, M., Bull, R.J., Olson, G.R. & Stober, J. (1989) Carcinogenic activity associated with halogenated acetones and acroleins in the mouse skin assay. *Cancer Lett.*, **48**, 197–203

Sadtler Research Laboratories (1980) *1980 Cumulative Index*, Philadelphia, PA

Saito-Suzuki, R., Teramoto, S. & Shirasu Y. (1982) Dominant lethal studies in rats with 1,2-dibromo-3-chloropropane and its structurally related compounds. *Mutat. Res.*, **101**, 321–327

Sax, N.I. & Lewis, R.J. (1987) *Hawley's Condensed Chemical Dictionary*, 11th Ed., New York, Van Nostrand Reinhold, p. 1178

Silverman, L., Schulte, H.F. & First, M.W. (1946) Further studies on sensory response to certain industrial solvent vapors. *J. ind. Hyg. Toxicol.*, **28**, 262–266

Sittig, M. (1985) *Handbook of Toxic and Hazardous Chemicals and Carcinogens*, 2nd Ed., Park Ridge, NJ, Noyes Publications, p. 887

Stolzenberg, S.J. & Hine, C.H. (1980) Mutagenicity of 2- and 3-carbon halogenated compounds in the *Salmonella*/mammalian microsome test. *Environ. Mutag.*, **2**, 59–66

Tancrède, M., Yanagisawa, Y. & Wilson, R. (1992) Volatilization of volatile organic compounds from showers. I. Analytical method and quantitative assessment. *Atmos. Environ.*, **26A**, 1103–1111

TCI America (1994) *Certificate of Analysis: 1,2,3-Trichloropropane*, Portland, OR

Tonogai, Y., Ito, Y., Ogawa, S. & Iwaida, M. (1986) Determination of dibromochloropropane and related fumigants in citrus fruit. *J. Food Prot.*, **49**, 909–913

Työministeriö [Ministry of Labour] (1993) *HTP-Arvot 1993* [Limit values 1993], Tampere, p. 19

United Kingdom Health and Safety Executive (1993) *Occupational Exposure Limits* (EH40/93), London, Her Majesty's Stationery Office

United States Agency for Toxic Substances and Disease Registry (1992) *Toxicological Profile for 1,2,3-Trichloropropane* (ATSDR/TP-91/28), Washington DC, United States Department of Health and Human Services

United States Environmental Protection Agency (1986a) Method 8010. Halogenated volatile organics. In: *Test Methods for Evaluating Solid Waste—Physical/Chemical Methods* (US EPA No. SW-846), 3rd Ed., Vol. 1A, Washington DC, Office of Solid Waste and Emergency Response, pp. 1–13

United States Environmental Protection Agency (1986b) Method 8240. Gas chromatography/mass spectrometry for volatile organics. In: *Test Methods for Evaluating Solid Waste—Physical/Chemical Methods* (US EPA No. SW-846), 3rd Ed., Vol. 1A, Washington DC, Office of Solid Waste and Emergency Response, pp. 1–43

United States Environmental Protection Agency (1988) *Methods for the Determination of Organic Compounds in Drinking Water*, Cincinnati, OH, Environmental Monitoring Systems Laboratory, Methods 502.1, 524.1

United States Environmental Protection Agency (1989) *1,2,3-Trichloropropane* (PB91-160697), Washington DC

United States International Trade Commission (1992) *Synthetic Organic Chemicals. United States Production And Sales, 1992* (USITC Publication 2720), Washington DC, United States Government Printing Office

United States National Institute for Occupational Safety and Health (1990) *National Occupational Exposure Survey (NOES)*, Cincinnati, OH

United States National Institute for Occupational Safety and Health (1994) *NIOSH Pocket Guide to Chemical Hazards* (DHHS (NIOSH) Publ. No. 94-116), Cincinnati, OH, pp. 316–317

United States National Toxicology Program (1993) *Toxicology and Carcinogenesis Studies of 1,2,3-Trichloropropane (CA No. 96-18-4) in F344/N Rats and B6C3F1 Mice (Gavage Studies)* (NTP Tech. Rep. No. 384; NIH Publ. No. 94-2829), Research Triangle Park, NC

United States Occupational Safety and Health Administration (1990) *OSHA Analytical Methods Manual*, 2nd Ed., Part 1, Vol. 1, Salt Lake City, UT, United States Department of Labor, Method 7

United States Occupational Safety and Health Administration (1994) Air contaminants. *US Code fed. Regul.*, **Title 29**, Part 1910.1000, p. 17

Verschueren, K. (1983) *Handbook of Environmental Data on Organic Chemicals*, 2nd Ed., New York, Van Nostrand Reinhold, p. 1146

Villeneuve, D.C., Chu, I., Secours, V.E., Coté, M.G., Plaa, G.L. & Valli, V.E. (1985) Results of a 90-day toxicity study on 1,2,3- and 1,1,2-trichloropropane administered via the drinking water. *Sci. total Environ.*, **47**, 421–426

Volp, R.F., Sipes, I.G., Falcoz, C., Carter, D.E. & Gross, J.F. (1984) Disposition of 1,2,3-trichloropropane in the Fischer 344 rat: conventional and physiological pharmacokinetics. *Toxicol. appl. Pharmacol.*, **75**, 8–17

Weast, R.C. & Astle, M.J. (1985) *CRC Handbook of Data on Organic Compounds*, Vol. II, Boca Raton, FL, CRC Press, pp. 149, 704

Weber, G.L. & Sipes, I.G. (1990a) Rat hepatic DNA damage induced by 1,2,3-trichloropropane. In: Witmer, C.M., Snyder, R.R., Jollow, D.L., Kalf, G.F., Kocsis, J.J. & Sipes, I.G., eds, *Advances in Experimental Medicine and Biology: Biological Reactive Intermediates IV*, New York, Plenum Press, pp. 853–855

Weber, G.L. & Sipes, I.G. (1990b) Covalent interaction of 1,2,3-trichloropropane with hepatic macromolecules: studies in the male F-344 rat. *Toxicol. appl. Pharmacol.*, **104**, 395–402

Weber, G.L. & Sipes, I.G. (1992) In vitro metabolism and bioactivation of 1,2,3-trichloropropane. *Toxicol. appl. Pharmacol.*, **113**, 152–158

Williams, G.M., Mori, H. & McQueen, C.A. (1989) Structure–activity relationships in the rat hepatocyte DNA-repair test for 300 chemicals. *Mutat. Res.*, **221**, 263–286

CHLORAL AND CHLORAL HYDRATE

1. Exposure Data

1.1 Chemical and physical data

1.1.1 Nomenclature

Chloral

Chem. Abstr. Serv. Reg. No.: 75-87-6
Chem. Abstr. Name: Trichloroacetaldehyde
IUPAC Systematic Name: Chloral
Synonyms: Anhydrous chloral; 2,2,2-trichloroacetaldehyde; trichloroethanal; 2,2,2 trichloroethanal

Chloral hydrate

Chem. Abstr. Serv. Reg. No.: 302-17-0
Chem. Abstr. Name: 2,2,2-Trichloro-1,1-ethanediol
IUPAC Systematic Name: Chloral hydrate
Synonyms: Chloral monohydrate; trichloroacetaldehyde hydrate; trichloroacetaldehyde monohydrate; 1,1,1-trichloro-2,2-dihydroxyethane

1.1.2 Structural and molecular formulae and relative molecular mass

C_2HCl_3O Chloral   Relative molecular mass: 147.39

$C_2H_3Cl_3O_2$ Chloral hydrate   Relative molecular mass: 165.42

1.1.3 Chemical and physical properties of the pure substance

Chloral

(a) *Description*: Colourless, oily hygroscopic liquid with pungent, irritating odour (Budavari, 1989; EniChem America Inc., 1994)

(b) *Boiling-point*: 97.8 °C (Lide, 1993)

(c) *Melting-point*: –57.5 °C (Lide, 1993)

(d) *Density*: 1.51214 at 20 °C/4 °C (Lide, 1993)

(e) *Spectroscopy data*: Infrared (prism [4626], grating [36780]), ultraviolet [5-3], nuclear magnetic resonance [8241] and mass [814] spectral data have been reported (Sadtler Research Laboratories, 1980; Weast & Astle, 1985).

(f) *Solubility*: Soluble in water, carbon tetrachloride, chloroform, diethyl ether and ethanol (Lide, 1993; EniChem America, Inc., 1994)

(g) *Volatility*: Vapour pressure, 35 mm Hg [4.67 kPa] at 20 °C; relative vapour density (air = 1), 5.1 (Verschueren, 1983; EniChem America Inc., 1994)

(h) *Stability*: Polymerizes under the influence of light and in the presence of sulfuric acid, forming a white solid trimer called metachloral (Budavari, 1989)

(i) *Reactivity*: Forms chloral hydrate when dissolved in water and forms chloral alcoholate with alcohol (Budavari, 1989)

(j) *Conversion factor*: $mg/m^3 = 6.03 \times ppm$[1]

Chloral hydrate

(a) *Description*: Monoclinic plates from water with aromatic, penetrating and slightly acrid odour and slightly bitter, caustic taste (Budavari, 1989; Lide, 1993)

(b) *Boiling-point*: 98 °C (Budavari, 1989)

(c) *Melting-point*: 57 °C (Lide, 1993)

(d) *Density*: 1.9081 at 20 °C/4 °C (Lide, 1993)

(e) *Spectroscopy data*: Infrared (prism [158], grating [41020P]), nuclear magnetic resonance (proton [10362], C-13 [4005]) and mass [1054] spectral data have been reported (Sadtler Research Laboratories, 1980; Weast & Astle, 1985).

(f) *Solubility*: Soluble in water, acetone, benzene, chloroform, diethyl ether, ethanol and methyl ethyl ketone (Budavari, 1989; Lide, 1993)

(g) *Stability*: Slowly volatilizes on exposure to air (Budavari, 1989)

(h) *Octanol/water partition coefficient (P)*: log P, 0.99 (Hansch et al., 1995)

(i) *Conversion* factor: $mg/m^3 = 6.76 \times ppm$[1]

---

[1] Calculated from: mg/m^3 = (relative molecular mass/24.45) × ppm, assuming normal temperature (25 °C) and pressure (101 kPa)

1.1.4 Technical products and impurities

Chloral is available commercially at a purity of 94–99.5% and containing the following typical impurities (max.): water, 0.06%; 2,2-dichloroethanal, 0.3%; 2,2,3-trichlorobutanal, 0.01%; hydrogen chloride, 0.06%; chloroform (see IARC, 1987a), dichloroacetaldehyde and phosgene (Jira *et al.*, 1986; EniChem America Inc., 1994). The *United States Pharmacopeia* specifies that chloral hydrate for pharmaceutical use must contain 99.5–102.5% $C_2H_3Cl_3O_2$ (United States Pharmacopeial Convention, 1989); the *British Pharmacopoiea* specifies values of 98.5–101.0% (Medicines Commission, 1988).

Trade names for chloral have included: Grasex and Sporotal 100. Trade names for chloral hydrate have included: Aquachloral, Bi 3411, Dormal, EPA Pesticide Chemical Code 268100, Felsules, Hydral, Kessodrate, Lorinal, Noctec, Nycoton, Nycton, Phaldrone, Rectules, Somnos, Sontec, Tosyl and Trawotox.

1.1.5 Analysis

Gas chromatography (GC) can be used for quantitative analysis of chloral and its hydrate, which releases chloral on vaporization (Jira *et al.*, 1986). High-performance liquid chromatography has been used for the determination of nanogram amounts of aldehydes, including chloral, in air, water and other environmental samples. Chloral was separated as the 2,4-dinitrophenylhydrazone derivative using isocratic solvent elution and ultraviolet detection (Fung & Grosjean, 1981).

Determination of trichloroethylene metabolites, including chloral hydrate, in rat liver homogenate has been reported on the basis of selective thermal conversion of chloral hydrate into chloroform, which is determined by headspace GC and electron capture detection (Køppen *et al.*, 1988).

A multi-channel, microwave-induced plasma atomic spectroscopic GC detector has been used to characterize the profiles of chlorinated humic acid on capillary columns and the content of carbon, chlorine and bromine in drinking-water. This technique makes it possible to estimate the empirical formulae of separated compounds with sufficient accuracy for useful peak identification. Chloral was among the compounds characterized by this method (Italia & Uden, 1988).

Headspace analysis and GC–mass spectrometry were used to identify volatile organic substances in the presence of aggressive oxidants, including chloral in drinking-, natural, demineralized and wastewater (Pilipenko *et al.*, 1988).

A spectrophotometric method for the determination of chloral hydrate in drugs is based on the reaction of quinaldine ethyl iodide with chloral hydrate to produce a stable blue cyanine dye, with an absorption maximum at about 605 nm (Helrich, 1990).

Chloral hydrate has been determined by GC in biological materials using four columns with different packings. Elution of the compound was monitored with two flame ionization detectors. The limit of detection was about 0.01 mg per sample (Mishchikhin & Felitsyn, 1988).

The iodide ion produced by oxidation of chloral hydrate with iodine in chloroform solution was measured using an iodide ion-selective electrode, by either direct measurement, addition of a standard or potentiometric titration with silver nitrate solution (Zaki, 1985).

1.2 Production and use

1.2.1 Production

Chloral was synthesized by J. von Liebig in 1832 and introduced (as the hydrate) as the first hypnotic drug in 1869. It is made by chlorination of ethanol (Jira *et al.*, 1986) but has also been prepared by chlorination of a mixture of ethanol and acetaldehyde (see IARC, 1987b) (French patent 612 396, 1929), from chloral hydrate by azeotropic distillation (United States patent 2584 036, 1952) or from hypochlorous acid and trichloroethylene (see monograph, this volume) (United States patent 2 759 978, 1956) (Budavari, 1989). Chloral is also formed as a by-product of the oxychlorination of ethylene (IARC, 1994a) to produce vinyl chloride (see IARC, 1987c; Cowfer & Magistro, 1983).

Chloral is an intermediate in the production of the insecticide DDT (1,1,1-trichloro-2,2-bis(4-chlorophenyl)ethane; see IARC, 1991a). After the discovery of DDT in 1939, the demand for chloral increased, and production reached a peak around 1963, when 40 000 tonnes were produced in the United States. When use of DDT was banned in the United States in 1972, and subsequently in many other countries, the demand rapidly declined. DDT is still produced in the United States for use in tropical countries. Use of chloral in the production of other pesticides was 1400 tonnes in 1972 (Jira *et al.*, 1986).

Chloral hydrate has been produced for use as a hypnotic drug in relatively low volume for many years. United States production for this purpose was about 135 tonnes in 1978 (Jira *et al.*, 1986).

Anhydrous chloral is produced by 11 companies in China and by one company each in Brazil, France, Germany, Italy, Japan, Mexico and the Russian Federation. Chloral hydrate is produced by two companies each in Brazil, Japan and Germany and by one company in Spain (Chemical Information Services, Inc., 1994).

Estimated production and use of chloral in the Member States of the European Union in 1984 was 2500 tonnes (Environmental Chemicals Data and Information Network, 1993).

1.2.2 Use

The principal use of chloral is in production of the insecticide DDT (Sax & Lewis, 1987). Much smaller amounts are used to make other insecticides, including methoxychlor (see IARC, 1987d), naled, trichlorfon (see IARC, 1987e) and dichlorvos (see IARC, 1991b). Chloral is also used as an intermediate in the production of the herbicide trichloroacetic acid (see monograph, this volume) and the hypnotic drugs chloral hydrate, chloral betaine, α-chloralose and triclofos sodium (Jira *et al.*, 1986). Chloroform was first prepared by treating chloral with alkali (DeShon, 1979). Chloral has also been used in the production of rigid polyurethane foam (see IARC, 1987f; Boitsov *et al.*, 1970) and to induce swelling of starch granules at room temperature (Whistler & Zysk, 1978).

Estimated use of chloral in the United States in 1975 was about 40% in the manufacture of DDT, about 10% in the manufacture of methoxychlor, dichlorvos and naled and about 50% in other applications (SRI International, 1975). Chloral hydrate is used as a sedative and hypnotic drug (Medicines Commission, 1988; Goodman Gilman *et al.*, 1991).

1.3 Occurrence

1.3.1 Natural occurrence

Chloral is not known to occur as a natural product.

1.3.2 Occupational exposures

The National Occupational Exposure Survey conducted between 1981 and 1983 indicated that 2757 employees in the United States of America were potentially exposed to chloral (United States National Institute for Occupational Safety and Health, 1994). The estimate is based on a survey of companies and did not involve measurements of actual exposures.

Chloral has been detected in the work environment during spraying and casting of polyurethane foam (Boitsov et al., 1970). It has also been identified as an autoxidation product of trichloroethylene during extraction of vegetable oil (McKinney et al., 1955). It has been identified at the output of etching chambers in semiconductor processing (Ohlson, 1986).

1.3.3 Air

No data were available to the Working Group.

1.3.4 Water

Chloral is formed during aqueous chlorination of humic substances and amino acids (Miller & Uden, 1983; Sato et al., 1985; Trehy et al., 1986; Italia & Uden, 1988). It may therefore occur in drinking-water as a result of chlorine disinfection of raw waters containing natural organic substances (see IARC, 1991c). The concentrations of chloral measured in drinking-water in the United States are summarized in Table 1.

Table 1. Concentrations of chloral (as chloral hydrate) in drinking-water in the United States

Water type (location)	Concentration ($\mu g/L$)	Reference
Tap water (reservoir)	7.2–18.2	Uden & Miller (1983)
Surface, reservoir, lake and groundwater	1.7–3.0	Krasner et al. (1989)
Tap water	0.01–5.0	US Environmental Protection Agency (1988)
Distribution system	0.14–6.7	Koch & Krasner (1989)
Surface water	6.3–28	Jacangelo et al. (1989)

Chloral has also been detected in the spent chlorination liquor from bleaching of sulfite pulp after oxygen treatment, at concentrations of < 0.1–0.5 g/tonne of pulp (Carlberg et al.,

1986). It has been found in trace amounts from photocatalytic degradation of trichloroethylene in water (Glaze et al., 1993).

1.3.5 Other

Chloral is a reactive intermediate metabolite of trichloroethylene (Cole et al., 1975; Davidson & Beliles, 1991).

1.4 Regulations and guidelines

In most countries, no exposure limits have been recommended. A guideline limit of 5 mg/m³ for short-term occupational exposure (ILO, 1991) and a tentative safe exposure limit of 0.01 mg/m³ in ambient air have been set for chloral in the Russian Federation (Environmental Chemicals Data and Information Network, 1993).

The United States Environmental Protection Agency (1994) has proposed that the maximum level of chloral hydrate in drinking-water be 0.04 mg/L. WHO (1993) recommends a provisional guideline value of 10 µg/L for chloral hydrate.

2. Studies of Cancer in Humans

No data were available to the Working Group.

3. Studies of Cancer in Experimental Animals[1]

Chloral

No data were available to the Working Group.

Chloral hydrate

Mouse: Groups of 25 and 20 male B6C3F1 mice, aged 15 days, were treated by gavage with a single dose of 5 or 10 mg/kg bw chloral hydrate (USP purity) in distilled water. A control group of 35 mice received distilled water only (0.01 ml/g bw). In order to study acute effects on the liver, 6–10 mice from each group were killed 24 h after treatment [numbers per group not specified]. Six mice at the high dose, seven at the low dose and 15 controls were killed when moribund and examined before termination. At week 92, all surviving mice were killed. From week 48 onwards, 19 controls, nine mice at the low dose and eight at the high dose were available for histological examination. Hepatic nodules described as 'hyperplastic', 'adenomatous' or 'trabecular' were found in two control mice, 3/9 at the low dose and 6/8 at the high

---

[1] The Working Group was aware of studies in progress by oral administration to rats (IARC, 1994b).

dose [$p < 0.002$, Fisher's exact test] (Rijhsinghani et al., 1986). [The Working Group noted the poor reporting of survival and the unusual histological terminology and that only single low doses were tested.]

A group of 40 male B6C3F1 mice, four weeks of age, received 1 g/L chloral hydrate (purity, > 95%; impurities unspecified) in distilled drinking-water (mean dose, 166 mg/kg bw per day) for 104 weeks; 33 controls received distilled water only. Five mice per group were killed after 30 weeks and another five after 60 weeks, for interim evaluation. Three control and six treated mice died before week 104. All mice were subjected to complete necropsy. Hepatocellular carcinomas were found in 2/5 mice killed at 60 weeks and in none of five controls. Of those killed at 104 weeks, 11/24 and 2/20 controls [$p = 0.01$ Fisher's exact test] had hepatocellular carcinomas, 7/24 treated mice and 1/20 controls [$p = 0.04$] had hepatocellular adenomas and 17/24 treated mice and 3/20 controls [$p = 0.001$] had carcinomas or adenomas. One treated mouse had a hyperplastic liver nodule. The authors reported several non-neoplastic hepatic changes: 10/24 treated mice and 1/20 controls had hepatocellular necrosis and 19/24 treated mice and 1/20 controls had cytomegaly (Daniel et al., 1992a).

4. Other Data Relevant to an Evaluation of Carcinogenicity and its Mechanisms

4.1 Absorption, distribution, metabolism and excretion

4.1.1 Humans

After oral administration, chloral hydrate is rapidly absorbed from the gastrointestinal tract. Its biotransformation to trichloroethanol must be rapid, since no parent compound could be detected in even the first samples taken 10 min after administration of 15 mg/kg bw to volunteers. Peak levels of trichloroethanol and trichloroethanol glucuronide were reached within 20–60 min after oral administration of aqueous solutions. The average half-life of trichloroethanol glucuronide was 6.7 h (Breimer et al., 1974). The average plasma half-life for chloral hydrate metabolites was 8.2 h; the half-life of the third chloral hydrate metabolite, trichloroacetic acid, was about four days (Breimer et al., 1974; Gorecki et al., 1990), as it binds extensively to plasma proteins (Sellers & Koch-Weser, 1971). As < 50% of an administered dose of chloral hydrate was recovered as metabolites in urine, yet unknown biotransformation reactions may exist for chloral hydrate in humans (Müller et al., 1974).

4.1.2 Experimental systems

In mammalian species, chloral hydrate is rapidly reduced to trichloroethanol, the metabolite that appears to be responsible for the hypnotic properties of the drug (Breimer, 1977). In dogs and horses, trichloroethanol is subsequently excreted with urine and bile as trichloroethanol glucuronide and in the urine after oxidation to trichloroacetic acid (Butler, 1948; Alexander, 1967). In rodents, a slightly different metabolic pattern is seen, as chloral hydrate is oxidized directly to trichloroacetic acid, and the oxidative pathway from trichloroethanol to trichloroacetate that is observed in humans seems to be absent (Daniel, 1963; Cabana & Gessner, 1970). A serum half-life of 0.2 h was reported for chloral hydrate in mice, but its rate of disappearance in dogs was two to seven times faster. Trichloroacetic acid formed by oxidation of chloral hydrate persisted in the serum of both mice and dogs (Butler, 1948; Cabana & Gessner, 1970; Breimer et al., 1974).

4.1.3 Comparison of humans and animals

Chloral hydrate is biotransformed along similar pathways in humans and all animal species tested. Trichloroethanol and its glucuronide are rapidly eliminated with urine, whereas trichloroacetate persists in both humans and animals. There appear to be no major quantitative differences in the kinetics of the metabolites.

4.2 Toxic effects

4.2.1 Humans

The lethal dose of chloral hydrate in humans is about 10 g; however, a fatal outcome was reported after ingestion of 4 g, and recovery has been seen after a dose of 30 g. The toxic effects that have been described after overdosing with chloral hydrate include irritation of the mucous membranes in the alimentary tract, depression of respiration and induction of cardiac arrhythmia. Habitual use of chloral hydrate is reported to cause unspecified hepatic and renal damage (Goodman Gilman et al., 1991).

4.2.2 Experimental systems

The oral LD_{50} of chloral hydrate in rats was 480 mg/kg bw (Goldenthal, 1971); those in mice were reported as 1442 mg/kg bw in males and 1265 mg/kg bw in females (Sanders et al., 1982). The cause of death after administration of lethal doses of chloral hydrate appeared to be inhibition of respiration.

The subchronic toxicity of chloral hydrate has been studied in CD1 mice and Sprague-Dawley rats. Administration of chloral hydrate to mice by gavage at daily doses of 14.4 and 144 mg/kg bw for 14 consecutive days resulted in an increase in relative liver weight and a decrease in spleen size. No other changes were seen. Administration of chloral hydrate to mice in drinking-water for 90 days at concentrations of 0.07 and 0.7 mg/ml resulted in dose-related hepatomegaly and significant changes in serum enzymes indicative of hepatic toxicity. In male mice, increased relative liver weights were also seen. After chloral hydrate was administered for 90 days in drinking-water to male and female Sprague-Dawley rats at a concentration of 0.3, 0.6, 1.2 or 2.4 mg/ml, the animals receiving the highest dose showed significant decreases in food and water consumption and weight gain. Males also had an apparent increase in the incidence of focal hepatocellular necrosis and increased activities of serum enzymes. No liver damage was seen in female rats (Daniel et al., 1992b).

Exposure of female CD1 mice to 100 ppm [603 mg/m^3] chloral for 6 h induced deep anaesthesia, which was fully reversible on cessation of exposure. Vacuolation of lung Clara cells, alveolar necrosis, desquamation of the bronchiolar epithelium and alveolar oedema were observed. Cytochrome P450 enzyme activity was reduced, although the activities of ethoxycoumarin *O*-diethylase and glutathione *S*-transferase were unaffected (Odum *et al.*, 1992).

Metabolism of chloral hydrate by male B6C3F1 mouse liver microsomes resulted in increased amounts of lipid peroxidation products (malonaldehyde and formaldehyde); the reactions could be inhibited by α-tocopherol or menadione (Ni *et al.*, 1994).

4.3 Reproductive and prenatal effects

4.3.1 Humans

Little information is available on the possible adverse effects of chloral on human pregnancy. Chloral hydrate is known to cross the human placenta at term (Bernstine *et al.*, 1954), but its use during relatively few pregnancies did not cause a detectable increase in abnormal outcomes (Heinonen *et al.*, 1977). Some data suggest that prolonged administration of sedative doses of chloral hydrate to newborns increases the likelihood of hyperbilirubinaemia (Lambert *et al.*, 1990).

Low levels of chloral hydrate have been found in breast milk. Although breast-feeding infants may be sedated by chloral hydrate in breast milk, the peak concentration measured (about 8 μg/ml) was considerably lower than the clinically active dose (Bernstine *et al.*, 1956; Wilson, 1981).

4.3.2 Experimental systems

Administration of one to five times the human therapeutic dose of chloral hydrate to pregnant mice (21.3 and 204.8 mg/kg per day in drinking-water during gestation) did not increase the incidence of gross external malformations in the offspring and did not impair normal development of pups (Kallman *et al.*, 1984).

4.4 Genetic and related effects (see also Table 2 and Appendices 1 and 2)

4.4.1 Humans

No data were available to the Working Group.

4.4.2 Experimental systems

The results obtained with chloral hydrate in a collaborative European Union project on aneuploidy have been summarized (Adler, 1993; Natarajan, 1993; Parry, 1993).

(a) DNA binding

In a single study in mice *in vivo*, radioactively labelled chloral hydrate did not bind to liver DNA.

(b) Mutation and allied effects

Chloral hydrate did not induce mutation in most strains of *Salmonella typhimurium*, but did in two of four studies with *S. typhimurium* TA100 and in a single study with *S. typhimurium*

Table 2. Genetic and related effects of chloral hydrate

Test system	Result[a]		Dose[b] (LED/HID)	Reference
	Without exogenous metabolic system	With exogenous metabolic system		
SA0, *Salmonella typhimurium* TA100, reverse mutation	–	–	2500	Waskell (1978)
SA0, *Salmonella typhimurium* TA100, reverse mutation	+	+	500	Haworth et al. (1983)
SA0, *Salmonella typhimurium* TA100, reverse mutation	–	–	1850	Leuschner & Leuschner (1991)
SA0, *Salmonella typhimurium* TA100, reverse mutation	+	+	2000	Ni et al. (1994)
SA4, *Salmonella typhimurium* TA104, reverse mutation	+	+	1000	Ni et al. (1994)
SA5, *Salmonella typhimurium* TA1535 reverse mutation	–	–	5000	Waskell (1978)
SA5, *Salmonella typhimurium* TA1535, reverse mutation	–	–	1850	Leuschner & Leuschner (1991)
SA5, *Salmonella typhimurium* TA1535, reverse mutation	–	–	5000	Haworth et al. (1983)
SA7, *Salmonella typhimurium* TA1537, reverse mutation	–	–	5000	Haworth et al. (1983)
SA7, *Salmonella typhimurium* TA1537, reverse mutation	–	–	1850	Leuschner & Leuschner (1991)
SA8, *Salmonella typhimurium* TA1538, reverse mutation	–	–	1850	Leuschner & Leuschner (1991)
SA9, *Salmonella typhimurium* TA98, reverse mutation	–	–	5000	Waskell (1978)
SA9, *Salmonella typhimurium* TA98, reverse mutation	–	–	5000	Haworth et al. (1983)
SA9, *Salmonella typhimurium* TA98, reverse mutation	–	–	1850	Leuschner & Leuschner (1991)
SCR, *Saccharomyces cerevisiae* D7, reverse mutation	–	–	3300	Bronzetti et al. (1984)
ANG, *Aspergillus nidulans*, diploid strain 35×17, mitotic crossing-over	–	0	1650	Crebelli et al. (1985)
ANG, *Aspergillus nidulans*, diploid strain 30, mitotic crossing-over	–	0	6600	Käfer (1986)
ANG, *Aspergillus nidulans*, diploid strain NH, mitotic crossing-over	–	0	1000	Kappas (1989)
ANG, *Aspergillus nidulans*, diploid strain P1, mitotic crossing-over	–	0	990	Crebelli et al. (1991)
SCG, *Saccharomyces cerevisiae* D7, gene conversion	–	(+)	2500	Bronzetti et al. (1984)
ANN, *Aspergillus nidulans*, diploid strain 35×17, haploids and nondisjunctional diploids	+	0	825	Crebelli et al. (1985)

Table 2 (contd)

Test system	Result[a] Without exogenous metabolic system	Result[a] With exogenous metabolic system	Dose[b] (LED/HID)	Reference
ANN, *Aspergillus nidulans*, diploid strain 30 conidia, aneuploidy	+	0	825	Käfer (1986)
ANN, *Aspergillus nidulans*, haploid conidia, aneuploidy and polyploidy	+	0	1650	Käfer (1986)
ANN, *Aspergillus nidulans*, diploid strain NH, nondisjunctional mitotic segregants	+	0	450	Kappas (1989)
ANN, *Aspergillus nidulans*, diploid strain P1, nondisjunctional diploids and haploids	+	0	660	Crebelli et al. (1991)
ANN, *Aspergillus nidulans*, haploid strain 35, hyperploidy	+	0	2640	Crebelli et al. (1991)
SCN, *Saccharomyces cerevisiae*, meiotic recombination	?	0	3300	Sora & Agostini Carbone (1987)
SCN, *Saccharomyces cerevisiae*, disomy in meiosis	+	0	2500	Sora & Agostini Carbone (1987)
SCN, *Saccharomyces cerevisiae*, diploids in meiosis	+	0	3300	Sora & Agostini Carbone (1987)
SCN, *Saccharomyces cerevisiae* D61.M, mitotic chromosomal malsegregation	+	0	1000	Albertini (1990)
SCN, *Saccharomyces cerevisiae* diploid strain D6, monosomy	+	0	1000	Parry et al. (1990)
***Seedlings of hexaploid Chinese spring wheat, Neatby's strain, chromosomal loss and gain	–	0	5000	Sandhu et al. (1991)
DMM, *Drosophila melanogaster*, somatic mutation wing spot test	+		825	Zordan et al. (1994)
DIA, DNA-protein cross-links, rat liver nuclei *in vitro*	–	0	41 250	Keller & Heck (1988)
DIA, DNA single-strand breaks (alkaline unwinding), rat primary hepatocytes *in vitro*	–	0	1650	Chang et al. (1992)
MIA, Kinetochore-positive micronuclei, Chinese hamster C1-1 cells *in vitro*, with antikinetochore antibodies	+	0	165	Degrassi & Tanzarella (1988)
MIA, Kinetochore-negative micronuclei, Chinese hamster C1-1 cells *in vitro*, with antikinetochore antibodies	–	0	250	Degrassi & Tanzarella (1988)
MIA, Kinetochore-positive micronuclei, Chinese hamster LUC2 cells *in vitro*	+	0	400	Parry et al. (1990)
MIA, Kinetochore-positive micronuclei, Chinese hamster LUC2 cells *in vitro*	+	0	400	Lynch & Parry (1993)
MIA, Micronuclei, Chinese hamster V79 cells *in vitro*	+	0	316	Seelbach et al. (1993)

Table 2 (contd)

Test system	Result[a] without exogenous metabolic system	Result[a] with exogenous metabolic system	Dose[b] (LED/HID)	Reference
ICR, Inhibition of intercellular communication, B6C3F1 mouse hepatocytes in vitro	−	0	83	Klaunig et al. (1989)
ICR, Inhibition of intercellular communication, F344 rat hepatocytes in vitro	−	0	83	Klaunig et al. (1989)
CIC, Chromosomal aberrations, Chinese hamster CHED cells in vitro	+	0	20	Furnus et al. (1990)
AIA, Aneuploidy, Chinese hamster CHED cells in vitro	+[c]	0	10	Furnus et al. (1990)
AIA, Aneuploidy, primary Chinese hamster embryonic cells in vitro	+[c]	0	250	Natarajan et al. (1993)
AIA, Aneuploidy (hypoploidy), Chinese hamster LUC2 p4 cells in vitro	+	0	250	Warr et al. (1993)
***, Tetraploidy and endoreduplication, Chinese hamster LUC2 p4 cells in vitro	+	0	500	Warr et al. (1993)
***, Apolar mitosis, Haemanthus katherinae endosperm in vitro	+	0	200	Molè-Bajer (1969)
***, Inhibition of spindle elongation, PtK2 rat kangaroo kidney epithelial cells in vitro	+	0	1000	Lee et al. (1987)
***, Inhibition of chromosome-to-pole movement, PtK2 rat kangaroo kidney epithelial cells in vitro	−	0	1000	Lee et al. (1987)
***, Breakdown of mitotic microtubuli, PtK2 rat kangaroo kidney epithelial cells in vitro	+	0	1000	Lee et al. (1987)
***, Multipolar mitotic spindles, Chinese hamster DON.Wg.3H cells in vitro	+	0	500	Parry et al. (1990)
***, Chromosomal dislocation from mitotic spindle, Chinese hamster DON.Wg.3H cells in vitro	+	0	500	Parry et al. (1990)
***, Lacking mitotic spindle, Chinese hamster DON.Wg.3H cells in vitro	+	0	250	Parry et al. (1990)
***, Metaphase defects, lacking mitotic spindle, Chinese hamster LUC1 cells in vitro	+	0	50	Parry et al. (1990)
***Multipolar mitotic spindles, Chinese hamster DON.Wg.3H cells in vitro	+	0	50	Warr et al. (1993)
***, Chromosomal dislocation from mitotic spindle, Chinese hamster DON.Wg.3H cells in vitro	+	0	500	Warr et al. (1993)
DIH, DNA single–strand breaks (alkaline unwinding), human lymphoblastoid CCRF-CEM cells in vitro	−	0	1650	Chang et al. (1992)
SHL, Sister chromatid exchange, human lymphocytes in vitro	(+)	0	54	Gu et al. (1981)
MIH, Micronucleus induction, isolated human lymphocytes in vitro	−	−	1500	Vian et al. (1995)
MIH, Micronucleus induction, human lymphocytes in whole blood in vitro	+	0	100	Migliore & Nieri (1991)

Table 2 (contd)

Test system	Result[a]		Dose[b] (LED/HID)	Reference
	Without exogenous metabolic system	With exogenous metabolic system		
MIH, Micronucleus induction, human lymphocytes in vitro	(+)	0	100	Ferguson et al. (1993)
MIH, Micronucleus induction, human lymphocytes in vitro	+	-	100	Van Hummelen & Kirsch-Volders (1992)
MIH, Kinetochore-positive micronuclei, human diploid LEO fibroblasts in vitro	+	0	120	Bonatti et al. (1992)
***, Aneuploidy (double Y induction), human lymphocytes in vitro	+	0	250	Vagnarelli et al. (1990)
AIH, Aneuploidy (hyperdiploidy and hypoploidy), human lymphocytes in vitro	+	0	50	Sbrana et al. (1993)
AIH, Polyploidy, human lymphocytes in vitro	+	0	137	Sbrana et al. (1993)
***, C-Mitosis, human lymphocytes in vitro	+	0	75	Sbrana et al. (1993)
HMM, Host-mediated assay, Saccharomyces cerevisiae D7 recovered from CD1 mouse lungs	(+)	0	500 po × 1	Bronzetti et al. (1984)
DVA, DNA single-strand breaks (alkaline unwinding), rat liver in vivo	+		300 po × 1	Nelson & Bull (1988)
DVA, DNA single-strand breaks (alkaline unwinding), mouse liver in vivo	+		100 po × 1	Nelson & Bull (1988)
DVA, DNA single-strand breaks (alkaline unwinding), male Fischer 344 rat liver in vivo	-		1650 po × 1	Chang et al. (1992)
DVA, DNA single-strand breaks (alkaline unwinding), male B6C3F₁ mouse liver in vivo	-		825 po × 1	Chang et al. (1992)
CGC, Chromosomal aberrations, (C57B1/Cne×C3H/Cne)F₁ mouse secondary spermatocytes (staminal gonia–pachytene treated)	+		82.7 ip × 1	Russo et al. (1984)
CGC, Chromosomal aberrations (translocations, breaks and fragments), (C57B1/Cne × C3H/Cne)F₁ mouse primary and secondary spermatocytes (from differentiating spermatogonia–pachytene stages treated)	-		413 ip × 1	Liang & Pacchierotti (1988)
CBA, Chromosomal aberrations, male and female (102/E1 × C3H/E1)F₁ mouse bone-marrow cells in vivo	-		600 ip × 1	Xu & Adler (1990)
CBA, Chromosomal aberrations, rat bone-marrow cells in vivo	-		1000 po × 1	Leuschner & Leuschner (1991)
CGG, Chromosomal aberrations, BALB/c mouse spermatogonia treated, spermatogonia observed in vivo	-		83 ip × 1	Russo & Levis (1992a)
COE, Chromosomal aberrations, ICR mouse oocytes treated in vivo	-		600 ip × 1	Mailhes et al. (1993)

Table 2 (contd)

Test system	Result[a] Without exogenous metabolic system	Result[a] With exogenous metabolic system	Dose[b] (LED/HID)	Reference
MVM, Micronuclei, male and female NMRI mice, bone-marrow erythrocytes in vivo	–		500 ip × 1	Leuschner & Leuschner (1991)
MVM, Micronuclei, mouse spermatids in vivo (preleptotene spermatocytes treated)	–		83 ip × 1	Russo & Levis (1992b)
MVM, Micronuclei, male BALB/c mouse bone marrow erythrocytes in vivo	+		83 ip × 1	Russo & Levis (1992a)
MVM, Micronuclei, BALB/c mouse early spermatids in vivo (diakinesis/metaphase I and metaphase II stages treated)	+		83 ip × 1	Russo & Levis (1992a)
MVM, Kinetochore-positive and -negative micronuclei, male BALB/c mouse bone-marrow erythrocytes in vivo	+		200 ip × 1	Russo et al. (1992)
MVM, Micronuclei, male (C57Bl/Nce × C3H/Cne)F₁ mouse bone–marrow erythrocytes in vivo	–		400 ip × 1	Leopardi et al. (1993)
MVM, Micronuclei, mouse spermatids in vivo (spermatogonial stem cells and preleptotene spermatocytes treated)	+		41 ip × 1	Allen et al. (1994)
AVA, Aneuploidy, (C57Bl/Nce × C3H/Cne)F₁ mouse secondary spermatocytes in vivo	+		82.7 ip × 1	Russo et al. (1984)
AVA, Aneuploidy (C57B1/Cne × C3H/Cne) F₁ mouse secondary spermatocytes (from differentiating spermatogonia–pachytene stages treated)	(+)		165 ip × 1	Liang & Pacchierotti (1988)
AVA, Aneuploidy (hypoploidy), ICR mouse oocytes in vivo	–[d]		200 ip × 1	Mailhes et al. (1988)
AVA, Polyploidy, male and female 102/E1 × C3H/E1)F₁ mouse bone-marrow cells in vivo	–		600 ip × 1	Xu & Adler (1990)
AVA, Aneuploidy, (102/E1 × C3H/E1)F₁ mouse secondary spermatocytes in vivo	+		200 ip × 1	Miller & Adler (1992)
AVA, Aneuploidy, male (C57Bl/Nce × C3H/Cne)F₁ mouse bone marrow in vivo	+[c]		400 ip × 1	Leopardi et al. (1993)
AVA, Aneuploidy, (C57Bl/Nce × C3H/Cne)F₁ mouse secondary spermatocytes in vivo	–		400 ip × 1	Leopardi et al. (1993)
AVA, Hypoploidy, ICR mouse oocytes in vivo	–[d]		600 ip × 1	Mailhes et al. (1993)
BVD, Binding to DNA, male B6C3F1 mouse liver in vivo	–		800 ip × 1	Keller & Heck (1988)

Table 2 (contd)

Test system	Result[a]		Dose[b] (LED/HID)	Reference
	Without exogenous metabolic system	With exogenous metabolic system		
***, Gonosomal and autosomal univalents (C57Bl/Cne × C3H/Cne)F$_1$ mouse primary spermatocytes (from differentiating spermatogonia–pachytene stages treated)	−		413 ip × 1	Liang & Pacchierotti (1988)
***, Porcine brain tubulin assembly inhibition *in vitro*	+	0	9900	Brunner *et al.* (1991)
***, Porcine brain tubulin disassembly inhibition *in vitro*	+	0	40	Brunner *et al.* (1991)
***, Bovine brain tubulin assembly inhibition *in vitro*	(+)	0	165	Wallin & Hartley-Asp (1993)
***, Centriole migration block, Chinese hamster cells clone 237 *in vitro*	+	0	1000	Alov & Lyubskii (1974)
Trichloroethanol				
***, λ Prophage induction, WP2 in *Escherichia coli*	−	−	155 000	DeMarini *et al.* (1994)
SA0, *Salmonella typhimurium* TA100, reverse mutation	−	−	0.5 vapour	DeMarini *et al.* (1994)

[a] +, considered to be positive; (+), considered to be weakly positive in an inadequate study; −, considered to be negative; ?, considered to be inconclusive (variable responses in several experiments within an inadequate study); 0, not tested
[b] LED, lowest effective dose; HID, highest effective dose. In-vitro tests, µg/ml; in-vivo tests, mg/kg bw; ip, intraperitoneally; po, orally
[c] Negative for induction of polyploidy
[d] Slight induction of hypoploid cells may have been due to technical artefacts.
***, Not included on profile

TA104. The latter response was inhibited by the free-radical scavengers α-tocopherol and menadione (Ni *et al.*, 1994).

Chloral hydrate did not induce mitotic crossing over in *Aspergillus nidulans* in the absence of metabolic activation, but weak induction of meiotic recombination in the presence of metabolic activation and of gene conversion in the absence of metabolic activation were seen in *Saccharomyces cerevisiae*. It did not induce reverse mutation in *S. cerevisiae* in one study. Chloral hydrate clearly induced aneuploidy in various fungi in the absence of metabolic activation. The results of a single study in Chinese spring wheat were inconclusive with respect to induction of chromosome loss and gain.

Chloral hydrate induced somatic mutations in *Drosophila melanogaster* in a wing-spot test.

In single studies, chloral hydrate did not produce DNA–protein cross-links in rat liver nuclei or DNA single-strand breaks/alkaline-labile sites in rat primary hepatocytes *in vitro*. It increased the frequency of micronuclei in Chinese hamster cell lines. Although a single study suggested that chloral hydrate induces chromosomal aberrations in Chinese hamster CHED cells *in vitro*, the micronuclei produced probably contained whole chromosomes and not chromosome fragments, as the micronuclei could all be labelled with antikinetochore antibodies. In a study of rat kangaroo kidney epithelial cells, chloral hydrate inhibited spindle elongation and broke down mitotic microtubuli, although it did not inhibit pole-to-pole movement of chromosomes. It produced multipolar spindles, chromosomal dislocation from the mitotic spindle and a total lack of mitotic spindles in Chinese hamster Don.Wg.3H cells. It did not inhibit cell-to-cell communication in mouse or rat hepatocytes *in vitro*.

In a single study, chloral hydrate weakly induced sister chromatid exchange in cultured human lymphocytes. It induced micronuclei, aneuploidy, C-mitosis and polyploidy in human cells *in vitro*. In human diploid fibroblasts, the micronuclei contained kinetochores. Micronuclei were induced in studies with human whole blood cultures but not in one study with isolated lymphocytes. The differences seen in the micronucleus test have been attributed to differences between whole blood and purified lymphocyte cultures (Vian *et al.*, 1995), but this hypothesis has not been tested.

Chloral hydrate increased the rate of mitotic gene conversion in a host-mediated assay with *S. cerevisiae* recovered from mouse lungs. One study showed induction of single-strand breaks in liver DNA of both rats and mice treated *in vivo*; another study in both species found no such effect. The frequency of chromosomal aberrations in mouse bone marrow, spermatogonia, primary and secondary spermatocytes and oocytes was not increased in single studies after treatment with chloral hydrate *in vivo*. In one study, it induced chromosomal aberrations in mouse secondary spermatocytes after treatment of animals *in vivo*. Micronuclei were induced in mouse bone-marrow erythrocytes in two of four studies after treatment with chloral hydrate *in vivo*; in one of these studies, the use of antikinetochore antibodies suggested induction of micronuclei containing both whole chromosomes and fragments. Chloral hydrate induced micronuclei in the spermatids of mice treated *in vivo* in two studies but not in a third in which the stage of spermatogenesis studied, the premeiotic S-phase (preleptotene), was concluded to be sensitive only to clastogenic agents. In one of the studies that showed an effect, only kinetochore-negative micronuclei were induced, but kinetochore-negative micronuclei were also produced by another established aneuploidogen, vincristine sulfate. The finding may therefore suggest not induction of fragments harbouring micronuclei but an inability of the antibody to label kinetochores in the micronuclei. Chloral hydrate induced aneuploidy in the bone marrow of mice treated *in vivo* in one study. It increased the rate of aneuploidy in mouse secondary spermatocytes in three of four studies, and one study also suggested increased hypodiploidy in mouse oocytes. It did not produce polyploidy in bone marrow or oocytes or gonosomal or autosomal univalents in primary spermatocytes of mice treated *in vivo*.

Trichloroethanol, a reduction product of chloral hydrate, did not induce λ prophage in *E. coli* or mutation in *S. typhimurium* TA100.

5. Summary and Evaluation

5.1 Exposure data

Chloral has been produced commercially since the 1940s by chlorination of ethanol. Until the early 1970s, its major use was in the production of the insecticide DDT. Chloral is also used as an intermediate in the production of the insecticides methoxychlor, naled, trichlorfon and dichlorvos, the herbicide trichloroacetic acid and the hypnotic drugs chloral hydrate, chloral betaine, α-chloralose and triclofos sodium.

Human exposure to chloral (or its hydrate) can occur during its production and use, from drinking chlorinated water and from pharmaceutical use.

Chloral is rapidly converted to its hydrate in contact with aqueous solutions.

5.2 Human carcinogenicity data

No data were available to the Working Group.

5.3 Animal carcinogenicity data

Chloral hydrate was tested for carcinogenicity in one adequate study in male mice by oral administration. It increased the incidence of hepatocellular adenomas and carcinomas.

5.4 Other relevant data

Chloral hydrate is metabolized rapidly in both humans and experimental animals to trichloroethanol and trichloroacetate. Its main acute toxic effects in humans are inhibition of respiration and induction of cardiac arrhythmia. Repeated administration of chloral hydrate damages the liver in mice and in male rats. Exposure of mice by inhalation results in damage to Clara cells in the lung.

Chloral hydrate crosses the human placenta, but there have been no reports of adverse results other than an increased likelihood of hyperbilirubinaemia in infants. No malformations and no effect on development were observed in the offspring of mice administered chloral throughout gestation.

Chloral hydrate is a well-established aneuploidogenic agent. It clearly induced aneuploidy and micronuclei in mammals treated *in vivo*, whereas chromosomal aberrations were not found in most studies. Conflicting results were obtained with regard to the induction of DNA damage in mammals treated with chloral hydrate *in vivo*.

Chloral hydrate induced aneuploidy and micronuclei in cultured human cells *in vitro*, but the results with regard to the induction of sister chromatid exchange were inconclusive. In rodent cells *in vitro*, chloral hydrate increased the induction of micronuclei but did not induce DNA damage; chromosomal aberrations were induced in a single study *in vitro*. In fungi, chloral hydrate clearly induced aneuploidy, while the results of studies on mitotic recombination and gene conversion were inconclusive. A single study showed induction of somatic mutation by chloral hydrate in insects. The results of assays for mutagenicity in bacteria were inconsistent.

5.5 Evaluation[1]

There is *inadequate evidence* in humans for the carcinogenicity of chloral and chloral hydrate.

There is *inadequate evidence* in experimental animals for the carcinogenicity of chloral.

There is *limited evidence* in experimental animals for the carcinogenicity of chloral hydrate.

Overall evaluation

Chloral and chloral hydrate are *not classifiable as to their carcinogenicity to humans (Group 3)*.

6. References

Adler, I.-D. (1993) Synopsis of the in vivo results obtained with the 10 known or suspected aneugens tested in the CEC collaborative study. *Mutat. Res.*, **287**, 131–137

Albertini, S. (1990) Analysis of nine known or suspected spindle poisons for mitotic chromosome malsegregation using *Saccharomyces cerevisiae* D61.M. *Mutagenesis*, **5**, 453–459

Alexander, F. (1967) The salivary secretion and clearance in the horse of chloral hydrate and its metabolites. *Biochem. Pharmacol.*, **16**, 1305–1311

Allen, J.W., Collins, B.W. & Evansky, P.A. (1994) Spermatid micronucleus analyses of trichloroethylene and chloral hydrate effects in mice. *Mutat. Res.*, **323**, 81–88

Alov, I.A. & Lyubskii, L. (1974) Experimental study of the functional morphology of the kinetochore in mitosis. *Byull. éksp. Biol. Med.*, **78**, 91–94

Bernstine, J.B., Meyer, A.E. & Hayman, H.B. (1954) Maternal and foetal blood estimation following the administration of chloral hydrate during labour. *J. Obstet. Gynaecol. Br. Emp.*, **61**, 683–685

---

[1]For definition of the italicized terms, see Preamble, pp. 22–26.

Bernstine, J.B., Meyer, A.E. & Bernstine, R.L. (1956) Maternal blood and breast milk estimation following the administration of chloral hydrate during the puerperium. *J. Obstet. Gynaecol. Br. Emp.*, **63**, 228–231

Boitsov, A.N., Rotenberg, Y.S. & Mulenkova, V.G. (1970) On the toxicological evaluation of chloral in the process of its liberation during filling and pouring of foam polyurethanes. *Gig. Tr. prof. Zabol.*, **14**, 26–29 (in Russian)

Bonatti, S., Cavalieri, Z., Viaggi, S. & Abbondandolo, A. (1992) The analysis of 10 potential spindle poisons for their ability to induce CREST-positive micronuclei in human diploid fibroblasts. *Mutagenesis*, **7**, 111–114

Breimer, D.D. (1977) Clinical pharmacokinetics of hypnotics. *Clin. Pharmacokin.*, **2**, 93–109

Breimer, D.D., Ketelaars, H.C.J. & van Rossum, J.M. (1974) Gas chromatographic determination of chloral hydrate, trichloroethanol and trichloroacetic acid in blood and in urine employing head-space analysis. *J. Chromatogr.*, **88**, 55–63

Bronzetti, G., Galli, A., Corsi, C., Cundari, E., Del Carratore, R., Nieri, R. & Paolini, M. (1984) Genetic and biochemical investigation on chloral hydrate *in vitro* and *in vivo*. *Mutat. Res.*, **141**, 19–22

Brunner, M., Albertini, S. & Würgler, F.E. (1991) Effects of 10 known or suspected spindle poisons in the in vitro porcine brain tubulin assay. *Mutagenesis*, **6**, 65–70

Budavari, S., ed. (1989) *The Merck Index*, 11th Ed., Rahway, NJ, Merck & Co., pp. 317, 1515

Butler, T.C. (1948) The metabolic fate of chloral hydrate. *J. Pharmacol. exp. Ther.*, **92**, 49–58

Cabana, B.E. & Gessner, P.K. (1970) The kinetics of chloral hydrate metabolism in mice and the effect thereon of ethanol. *J. Pharmacol. exp. Ther.*, **174**, 260–275

Carlberg, G.E., Drangsholt, H. & Gjøs, N. (1986) Identification of chlorinated compounds in the spent chlorination liquor from differently treated sulphite pulps with special emphasis on mutagenic compounds. *Sci. total Environ.*, **48**, 157–167

Chang, L.W., Daniel, F.B. & DeAngelo, A.B. (1992) Analysis of DNA strand breaks induced in rodent liver *in vivo*, hepatocytes in primary culture, and a human cell line by chlorinated acetic acids and chlorinated acetaldehydes. *Environ. mol. Mutag.*, **20**, 277–288

Chemical Information Services, Inc. (1994) *Directory of World Chemical Producers 1995/96 Standard Edition*, Dallas, TX, p. 162

Cole, W.J., Mitchell, R.G. & Salamonsen, R.F. (1975) Isolation, characterization and quantitation of chloral hydrate as a transient metabolite of trichloroethylene in man using electron capture gas chromatography and mass fragmentography. *J. Pharm. Pharmacol.*, **27**, 167–171

Cowfer, J.A. & Magistro, A.J. (1983) Vinyl polymers, vinyl chloride. In: Mark, H.F., Othmer, D.F., Overberger, C.G., Seaborg, G.T. & Grayson, N., eds, *Kirk-Othmer Encyclopaedia of Chemical Technology*, 3rd Ed., Vol. 23, New York, John Wiley & Sons, pp. 865–885

Crebelli, R., Conti, G., Conti, L. & Carere, A. (1985) Mutagenicity of trichloroethylene, trichloroethanol and chloral hydrate in *Aspergillus nidulans*. *Mutat. Res.*, **155**, 105–111

Crebelli, R., Conti, G., Conti, L. & Carere, A. (1991) In vitro studies with nine known or suspected spindle poisons: results in tests for chromosome malsegregation in *Aspergillus nidulans*. *Mutagenesis*, **6**, 131–136

Daniel, J.W. (1963) The metabolism of ^{36}Cl-labelled trichloroethylene and tetrachloroethylene in the rat. *Biochem. Pharmacol.*, **12**, 795–802

Daniel, F.B., DeAngelo, A.B., Stober, J.A., Olson, G.R. & Page, N.P. (1992a) Hepatocarcinogenicity of chloral hydrate, 2-chloroacetaldehyde, and dichloroacetic acid in the male B6C3F1 mouse. *Fundam. appl. Toxicol.*, **19**, 159–168

Daniel, F.B., Robinson, M., Stober, J.A., Page, N.P. & Olson, G.R. (1992b) Ninety-day toxicity study of chloral hydrate in the Sprague-Dawley rat. *Drug chem. Toxicol.*, **15**, 217–232

Davidson, I.W.F. & Beliles, R.P. (1991) Consideration on the target organ toxicity of trichloroethylene in terms of metabolite toxicity and pharmacokinetics. *Drug Metab. Rev.*, **23**, 493–599

Degrassi, F. & Tanzarella, C. (1988) Immunofluorescent staining of kinetochores in micronuclei: a new assay for the detection of aneuploidy. *Mutat. Res.*, **203**, 339–345

DeMarini, D.M., Perry, E. & Shelton, M.L. (1994) Dichloroacetic acid and related compounds: induction of prophage in *E. coli* and mutagenicity and mutation spectra in Salmonella TA100. *Mutagenesis*, **9**, 429–437

DeShon, H.D. (1979) Chlorocarbons and chlorohydrocarbons—chloroform. In: Mark, H.F., Othmer, D.F., Overberger, C.G., Seaborg, G.T. & Grayson, N., eds, *Kirk-Othmer Encyclopaedia of Chemical Technology*, 3rd Ed., Vol. 5, New York, John Wiley & Sons, pp, 693–703

EniChem America Inc. (1994) *Chemical Information Sheet: Chloral Anhydrous*, New York

Environmental Chemicals Data and Information Network (1993) *Chloral*, Ispra, JRC-CEC, last update: 02.09.1993

Ferguson, L.R., Morcombe, P. & Triggs, C.N. (1993) The size of cytokinesis-blocked micronuclei in human peripheral blood lymphocytes as a measure of aneuploidy induction by set A compounds in the EEC trial. *Mutat. Res.*, **287**, 101–112

Fung, K. & Grosjean, D. (1981) Determination of nanogram amounts of carbonyls as 2,4-dinitrophenylhydrazones by high-performance liquid chromatography. *Anal. Chem.*, **53**, 168–171

Furnus, C.C., Ulrich, M.A., Terreros, C. & Dulout, F.N. (1990) The induction of aneuploidy in cultured Chinese hamster cells by propionaldehyde and chloral hydrate. *Mutagenesis*, **5**, 323–326

Glaze, W.H., Kenneke, J.F. & Ferry, L.J. (1993) Chlorinated by-products from the TiO_2-mediated photodegradation of trichloroethylene and tetrachloroethylene in water. *Environ. Sci. Technol.*, **27**, 177–184

Goldenthal, E.I. (1971) A compilation of LD_{50} values in newborn and adult animals. *Toxicol. appl. Pharmacol.*, **18**, 185–207

Goodman Gilman, A., Rall, T.W., Nies, A.S. & Taylor, P., eds (1991) *Goodman and Gilman's. The Pharmacological Basis of Therapeutics*, New York, Pergamon Press, pp. 357, 364

Gorecki, D.K.J., Hindmarsh, K.W., Hall, C.A., Mayers, D.J. & Sankaran, K. (1990) Determination of chloral hydrate metabolism in adult and neonate biological fluids after single-dose administration. *J. Chromatogr.*, **528**, 333–341

Gu, Z.W., Sele, B., Jalbert, P., Vincent, M., Vincent, F., Marka, C., Chmara, D. & Faure, J. (1981) Induction of sister chromatid exchange by trichloroethylene and its metabolites. *Toxicol. Eur. Res.*, **3**, 63–67 (in French)

Hansch, C., Leo, A. & Hoekman, D.H. (1995) *Exploring QSAR*, Washington, DC, American Chemical Society

Haworth, S., Lawlor, T., Mortelmans, K., Speck, W. & Zeiger, E. (1983) Salmonella mutagenicity test results for 250 chemicals. *Environ. Mutag.*, **Suppl. 1**, 3–142

Heinonen, O.P., Slone, D. & Shapiro, S. (1977) *Birth Defects and Drugs in Pregnancy*, Littleton, MA, Littleton Publishing Sciences Group

Helrich, K., ed. (1990) *Official Methods of Analysis of the Association of Official Analytical Chemists*, 15th Ed., Vol. 1, Arlington, VA, Association of Official Analytical Chemists, p. 562

IARC (1987a) *IARC Monographs on the Evaluation of Carcinogenic Risks to Humans*, Suppl. 7, *Overall Evaluations of Carcinogenicity: An Updating of* IARC Monographs *Volumes 1–42*, Lyon, pp. 152–154

IARC (1987b) *IARC Monographs on the Evaluation of Carcinogenic Risks to Humans*, Suppl. 7, *Overall Evaluations of Carcinogenicity: An Updating of* IARC Monographs *Volumes 1–42*, Lyon, pp. 77–78

IARC (1987c) *IARC Monographs on the Evaluation of Carcinogenic Risks to Humans*, Suppl. 7, *Overall Evaluations of Carcinogenicity: An Updating of* IARC Monographs *Volumes 1–42*, Lyon, pp. 373–376

IARC (1987d) *IARC Monographs on the Evaluation of Carcinogenic Risks to Humans*, Suppl. 7, *Overall Evaluations of Carcinogenicity: An Updating of* IARC Monographs *Volumes 1–42*, Lyon, p. 66

IARC (1987e) *IARC Monographs on the Evaluation of Carcinogenic Risks to Humans*, Suppl. 7, *Overall Evaluations of Carcinogenicity: An Updating of* IARC Monographs *Volumes 1–42*, Lyon, p. 73

IARC (1987f) *IARC Monographs on the Evaluation of Carcinogenic Risks to Humans*, Suppl. 7, *Overall Evaluations of Carcinogenicity: An Updating of* IARC Monographs *Volumes 1–42*, Lyon, p. 70

IARC (1991a) *IARC Monographs on the Evaluation of the Carcinogenic Risks to Humans*, Vol. 53, *Occupational Exposures in Insecticide Application, and Some Pesticides*, Lyon, pp. 179–249

IARC (1991b) *IARC Monographs on the Evaluation of the Carcinogenic Risks to Humans*, Vol. 53, *Occupational Exposures in Insecticide Application, and Some Pesticides*, Lyon, pp. 267–307

IARC (1991c) *IARC Monographs on the Evaluation of the Carcinogenic Risks to Humans*, Vol. 52, *Chlorinated Drinking-water; Chlorination By-products; Some Other Halogenated Compounds; Cobalt and Cobalt Compounds*, Lyon, pp. 55–141

IARC (1994a) *IARC Monographs on the Evaluation of the Carcinogenic Risks to Humans*, Vol. 60, *Some Industrial Chemicals*, Lyon, pp. 45–71

IARC (1994b) *Directory of Agents Being Tested for Carcinogenicity*, No. 16, Lyon, p. 146

ILO (1991) *Occupational Exposure Limits for Airborne Toxic Substances: Values of Selected Countries* (Occupational Safety and Health Series No. 37), 3rd Ed., Geneva, pp. 80–81

Italia, M.P. & Uden, P.C. (1988) Multiple element emission spectral detection gas chromatographic profiles of halogenated products from chlorination of humic acid and drinking water. *J. Chromatogr.*, **438**, 35–43

Jacangelo, J.G., Patania, N.L., Reagan, K.M., Aieta, E.M., Krasner, S.W. & McGuire, M.J. (1989) Ozonation: assessing its role in the formation and control of disinfection by-products. *J. Am Water Works Assoc.*, **81**, 74–84

Jira, R., Kopp, E. & McKusick, B.C. (1986) Chloroacetaldehydes. In: Gerhartz, W., Yamamoto, Y.S., Campbell, F.T., Pfefferkorn, R. & Rounsaville, J.F., eds, *Ullmann's Encyclopedia of Industrial Chemistry*, 5th rev. Ed., Vol. A6, New York, VCH Publishers, pp. 533–536

Käfer, E. (1986) Tests which distinguish induced crossing-over and aneuploidy from secondary segregation in Aspergillus treated with chloral hydrate and γ-rays. *Mutat. Res.*, **164**, 145–166

Kallman, M.J., Kaempf, G.L. & Balster, R.L. (1984) Behavioral toxicity of chloral hydrate in mice: an approach to evaluation. *Neurobehav. Toxicol. Teratol.*, **6**, 137–146

Kappas, A. (1989) On the mechanisms of induced aneuploidy in *Aspergillus nidulans* and validation of tests for genomic mutations. *Prog. clin. Biol. Res.*, **318**, 377–384

Keller, D.A. & Heck, H.d'A. (1988) Mechanistic studies on chloral toxicity: relationship to trichloroethylene carcinogenesis. *Toxicol. Lett.*, **42**, 183–191

Klaunig, J.E., Ruch, R.J. & Lin, E.L. (1989) Effects of trichloroethylene and its metabolites on rodent hepatocyte intercellular communication. *Toxicol. appl. Pharmacol.*, **99**, 454–465

Koch, B. & Krasner, S.W. (1989) Occurrence of disinfection by-products in a distribution system. *J. Am. Water Works Assoc.*, **81**, 1203–1230

Køppen, B., Dalgaard, L. & Christensen, J.M. (1988) Determination of trichloroethylene metabolites in rat liver homogenate using headspace gas chromatography. *J. Chromatogr.*, **442**, 325–332

Krasner, S.W., McGuire, M.J., Jacangelo, J.G., Patania, N.L., Reagan, K.M. & Aieta, E.M. (1989) The occurrence of disinfection by-products in US drinking water. *J. Am. Water Works Assoc.*, **81**, 41–53

Lambert, G.H., Muraskas, J., Anderson, C.L. & Myers, T.F. (1990) Direct hyperbilirubinemia associated with chloral hydrate administration in the newborn. *Pediatrics*, **86**, 277–281

Lee, G.M., Diguiseppi, J., Gawdi, G.M. & Herman, B. (1987) Chloral hydrate disrupts mitosis by increasing intracellular free calcium. *J. Cell Sci.*, **88**, 603–612

Leopardi, P., Zijno, A., Bassani, B. & Pacchierotti, F. (1993) In vivo studies on chemically induced aneuploidy in mouse somatic and germinal cells. *Mutat. Res.*, **287**, 119–130

Leuschner, J. & Leuschner, F. (1991) Evaluation of the mutagenicity of chloral hydrate *in vitro* and *in vivo*. *Arzneimittel.-Forsch.*, **41**, 1101–1103

Liang, J.C. & Pacchierotti, F. (1988) Cytogenetic investigations of chemically-induced aneuploidy in mouse spermatocytes. *Mutat. Res.*, **201**, 325–335

Lide, D.R., ed. (1993) *CRC Handbook of Chemistry and Physics*, 74th Ed., Boca Raton, FL, CRC Press, p. 3–172

Lynch, A.M. & Parry, J.M. (1993) The cytochalasin-B micronucleus/kinetochore assay *in vitro*: studies with 10 suspected aneugens. *Mutat. Res.*, **287**, 71–86

Mailhes, J.B., Preston, R.J., Yan, Z.P. & Payne, H.S. (1988) Analysis of mouse metaphase II oocytes as an assay for chemically induced aneuploidy. *Mutat. Res.*, **198**, 145–152

Mailhes, J.B., Aardema, M.J. & Marchetti, F. (1993) Investigation of aneuploidy induction in mouse oocytes following exposure to vinblastine-sulfate, pyrimethamine, diethylstilbestrol diphosphate, or chloral hydrate. *Environ. mol. Mutag.*, **22**, 107–114

McKinney, L.L., Uhing, E.H., White, J.L. & Picken, J.C., Jr (1955) Vegetable oil extraction. Autoxidation products of trichloroethylene. *Agric. Food Chem.*, **3**, 413–419

Medicines Commission (1988) *British Pharmacopoeia*, London, Her Majesty's Stationery Office, p. 116

Migliore, L. & Nieri, M. (1991) Evaluation of twelve potential aneuploidogenic chemicals by the *in vitro* human lymphocyte micronucleus assay. *Toxicol. in vitro*, **5**, 325–336

Miller, B.M. & Adler, I.-D. (1992) Aneuploidy induction in mouse spermatocytes. *Mutagenesis*, **7**, 69–76

Miller, J.W. & Uden, P.C. (1983) Characterization of non-volatile aqueous chlorination products of humic substances. *Environ. Sci. Technol.*, **17**, 150–157

Mishchikhin, V.A. & Felitsyn, F.P. (1988) Gas-chromatographic determination of chloroform, carbon tetrachloride, dichloromethane, trichloroethylene and chloral hydrate in biological material. *Sud. Med. Ekspert.*, **31**, 30–33 (in Russian) [*Anal. Abstr.*, **51**, D111]

Molè-Bajer, J. (1969) Fine structural studies of apolar mitosis. *Chromosoma*, **26**, 427–448

Müller, G., Spassovski, M. & Henschler, D. (1974) Metabolism of trichloroethylene in man. II. Pharmacokinetics of metabolites. *Arch. Toxicol.*, **32**, 283–295

Natarajan, A.T. (1993) An overview of the results of testing of known or suspected aneugens using mammalian cells *in vitro*. *Mutat. Res.*, **287**, 113–118

Natarajan, A.T., Duivenvoorden, W.C.M., Meijers, M. & Zwanenburg, T.S.B. (1993) Induction of mitotic aneuploidy using Chinese hamster primary embryonic cells. Test results of 10 chemicals. *Mutat. Res.*, **287**, 47–56

Nelson, M.A. & Bull, R.J. (1988) Induction of strand breaks in DNA by trichloroethylene and metabolites in rat and mouse liver *in vivo*. *Toxicol. appl. Pharmacol.*, **94**, 45–54

Ni, Y.-C., Wong, T.-Y., Kadlubar, F.F. & Fu, P.P. (1994) Hepatic metabolism of chloral hydrate to free radical(s) and induction of lipid peroxidation. *Biochem. Biophys. Res. Comm.*, **204**, 937–943

Odum, J., Foster, J.R. & Green, T. (1992) A mechanism for the development of Clara cell lesions in the mouse lung after exposure to trichloroethylene. *Chem.-biol. Interactions*, **83**, 135–153

Ohlson, J. (1986) Dry etch chemical safety. *Solid State Technol.*, **July**, 69–73

Parry, J.W. (1993) An evaluation of the use of in vitro tubulin polymerisation, fungal and wheat assays to detect the activity of potential chemical aneugens. *Mutat. Res.*, **287**, 23–28

Parry, J.M., Parry, E.M., Warr, T., Lynch, A. & James, S. (1990) The detection of aneugens using yeasts and cultured mammalian cells. In: Mendelsohn, M.L. & Albertini, R.J., eds, *Mutation and the Environment*. Part B, *Metabolism, Testing Methods, and Chromosomes*, New York, Wiley-Liss, pp. 247–266

Pilipenko, A.T., Milyukin, M.V., Kuzema, A.S. & Tulyupa, F.M. (1988) Use of headspace analysis for determining volatile organic substances in liquid media by combined chromatography–mass spectrometry. *Zh. anal. Khim.*, **43**, 136–142 (in Russian)

Rijhsinghani, K.S., Abrahams, C., Swerdlow, M.A., Rao, K.V.N. & Ghose, T. (1986) Induction of neoplastic lesions in the livers of C57Bl × C3HF1 mice by chloral hydrate. *Cancer Detect. Prev.*, **9**, 279–288

Russo, A. & Levis, A.G. (1992a) Further evidence for the aneuploidogenic properties of chelating agents: induction of micronuclei in mouse male germ cells by EDTA. *Environ. mol. Mutag.*, **19**, 125–131

Russo, A. & Levis, A.G. (1992b) Detection of aneuploidy in male germ cells of mice by means of a meiotic micronucleus assay. *Mutat. Res.*, **281**, 187–191

Russo, A., Pacchierotti, F. & Metalli, P. (1984) Nondisjunction induced in mouse spermatogenesis by chloral hydrate, a metabolite of trichloroethylene. *Environ. Mutag.*, **6**, 695–703

Russo, A., Stocco, A. & Majone, F. (1992) Identification of kinetochore-containing (CREST$^+$) micronuclei in mouse bone marrow erythrocytes. *Mutagenesis*, **7**, 195–197

Sadtler Research Laboratories (1980) *1980 Cumulative Index*, Philadelphia, PA

Sanders, V.M., Kauffmann, B.M., White, K.L., Jr, Douglas, K.A., Barnes, D.W., Sain, L.E., Bradshaw, T.J., Borzelleca, J.F. & Munson, A.E. (1982) Toxicology of chloral hydrate in the mouse. *Environ. Health Perspectives*, **44**, 137–146

Sandhu, S.S., Dhesi, J.S., Gill, B.S. & Svendsgaard, D. (1991) Evaluation of 10 chemicals for aneuploidy induction in the hexaploid wheat assay. *Mutagenesis*, **6**, 369–373

Sato, T., Mukaida, M., Ose, Y., Nagase, H. & Ishikawa, T. (1985) Chlorinated products from structural compounds of soil humic substances. *Sci. total Environ.*, **43**, 127–140

Sax, N.I. & Lewis, R.J. (1987) *Hawley's Condensed Chemical Dictionary*, 11th Ed., New York, Van Nostrand Reinhold, p. 256

Sbrana, I., Di Sibio, A., Lomi, A. & Scarelli, V. (1993) C-Mitosis and numerical chromosome aberration analyses in human lymphocytes: 10 known or suspected spindle poisons. *Mutat. Res.*, **287**, 57–70

Seelbach, A., Fissler, B. & Madle, S. (1993) Further evaluation of a modified micronucleus assay with V79 cells for detection of aneugenic effects. *Mutat. Res.*, **303**, 163–169

Sellers, E.M. & Koch-Weser, J. (1971) Kinetics and clinical importance of displacement of warfarin from albumin by acidic drugs. *Ann. N.Y. Acad. Sci.*, **179**, 213–225

Sora, S. & Agostini Carbone, M.L. (1987) Chloral hydrate, methylmercury hydroxide and ethidium bromide affect chromosomal segregation during meiosis of *Saccharomyces cerevisiae*. *Mutat. Res.*, **190**, 13–17

SRI International (1975) *Chemical Economics Handbook*, Menlo Park, CA

Trehy, M.L., Yost, R.A. & Miles, C.J. (1986) Chlorination byproducts of amino acids in natural waters. *Environ. Sci. Technol.*, **20**, 1117–1122

Uden, P.C. & Miller, J.W. (1983) Chlorinated acids and chloral in drinking water. *J. Am. Water Works Assoc.*, **75**, 524–527

United States Environmental Protection Agency (1988) *Health and Environmental Effects Document for Chloral* (EPA-600/8-89/012; PB91-21 6481), Cincinnati, OH

United States Environmental Protection Agency (1994) Drinking water; national primary drinking water regulations: disinfectants and disinfection byproducts. *Fed. Regist.*, **59**, 38668–38710

United States National Institute for Occupational Safety and Health (1994) *National Occupational Exposure Survey (1981–1983)*, Cincinnati, OH, p. 10

United States Pharmacopeial Convention (1989) *The United States Pharmacopeia. The National Formulary 1990* (USP XXII, NF XVII), Rockville, MD, pp. 269–270

Vagnarelli, P., De Sario, A. & De Carli, L. (1990) Aneuploidy induced by chloral hydrate detected in human lymphocytes with the Y97 probe. *Mutagenesis*, **5**, 591–592

Van Hummelen, P. & Kirsch-Volders, M. (1992) Analysis of eight known or suspected aneugens by the in vitro human lymphocyte micronucleus test. *Mutagenesis*, **7**, 447–455

Verschueren, K. (1983) *Handbook of Environmental Data on Organic Chemicals*, 2nd Ed., New York, Van Nostrand Reinhold Co., p. 349

Vian, L., Van Hummelen, P., Bichet, N., Gouy, D. & Kirsch-Volders, M. (1995) Evaluation of hydroquinone and chloral hydrate on the in vitro micronucleus test on isolated lymphocytes. *Mutat. Res.*, **334**, 1–7

Wallin, M. & Hartley-Asp, B. (1993) Effects of potential aneuploidy inducing agents on microtubule assembly *in vitro*. *Mutat. Res.*, **287**, 17–22

Warr, T.J., Parry, E.M. & Parry, J.M. (1993) A comparison of two in vitro mammalian cell cytogenetic assays for the detection of mitotic aneuploidy using 10 known or suspected aneugens. *Mutat. Res.*, **287**, 29–46

Waskell, L. (1978) A study of the mutagenicity of anesthetics and their metabolites. *Mutat. Res.*, **57**, 141–153

Weast, R.C. & Astle, M.J. (1985) *CRC Handbook of Data on Organic Compounds*, Vols I & II, Boca Raton, FL, CRC Press, pp. 7, 414 (I), 448, 543 (II)

Whistler, R.L. & Zysk, J.R. (1978) Carbohydrates. In: Mark, H.F., Othmer, D.F., Overberger, C.G., Seaborg, G.T. & Grayson, N., eds, *Kirk-Othmer Encyclopedia of Chemical Technology*, 3rd Ed., Vol. 4, New York, John Wiley & Sons, pp. 535–555

WHO (1993) *Guidelines for Drinking-water Quality*, 2nd Ed., Vol. 1, *Recommendations*, Geneva, pp. 103, 177

Wilson, J.T. (1981) *Drugs in Breast Milk*, Lancaster, MTP

Xu, W. & Adler, I.-D. (1990) Clastogenic effects of known and suspect spindle poisons studied by chromosome analysis in mouse bone marrow cells. *Mutagenesis*, **5**, 371–374

Zaki, M.T.M. (1985) Application of the iodide ion-selective electrode in the potentiometric determination of chloral hydrate. *Anal. Lett.*, **18**, 1697–1702

Zordan, M., Osti, M., Pesce, M. & Costa, R. (1994) Chloral hydrate is recombinogenic in the wing spot test in *Drosophila melanogaster*. *Mutat. Res.*, **322**, 111–116

DICHLOROACETIC ACID

1. Exposure Data

1.1 Chemical and physical data

1.1.1 Nomenclature

Chem. Abstr. Serv. Reg. No.: 79-43-6
Deleted CAS Reg. No.: 42428-47-7
Chem. Abstr. Name: Dichloroacetic acid
IUPAC Systematic Name: Dichloroacetic acid
Synonyms: Bichloracetic acid; DCA; DCA (acid); DCAA; dichloracetic acid; dichlorethanoic acid; dichloroethanoic acid; 2,2-dichloroethanoic acid

1.1.2 Structural and molecular formulae and relative molecular mass

$$H-\underset{\underset{Cl}{|}}{\overset{\overset{Cl}{|}}{C}}-\overset{\overset{O}{\parallel}}{C}-OH$$

$C_2H_2Cl_2O_2$ Relative molecular mass: 128.94

1.1.3 Chemical and physical properties of the pure substance

From Lide (1993), unless otherwise noted

(a) *Description*: Colourless to slightly yellowish liquid with a pungent acid-like odour (Budavari, 1989; Hoechst Chemicals, 1990)

(b) *Boiling-point*: 194 °C

(c) *Melting-point*: 13.5 °C

(d) *Density*: 1.5634 at 20 °C/4 °C

(e) *Spectroscopy data*: Infrared (prism [2806]; grating [36771]), nuclear magnetic resonance (proton [166], C-13 [500]) and mass spectral data have been reported (Sadtler Research Laboratories, 1980; Weast & Astle, 1985).

(f) *Solubility*: Soluble in water, acetone, ethanol and diethyl ether; also soluble in ketones, hydrocarbons and chlorinated hydrocarbons (Hoechst Chemicals, 1990). In aqueous solution, dichloroacetic acid and dichloroacetate exist as an equilibrium mixture, the proportions of each depending primarily on the pH of the solution. The pK_a of dichloroacetic acid is 1.48 at 25 °C.

(g) *Volatility*: Vapour pressure: 0.19 mbar [19 Pa] at 20 °C (Hoechst Chemicals, 1990)

(h) *Reactivity*: Highly corrosive and attacks metals; releases hydrogen chloride gas (see IARC, 1992) when heated (Hoechst Chemicals, 1990)

(i) *Octanol:water partition coefficient (P)*: log P, 0.92 (Hansch *et al.*, 1995)

(j) *Conversion factor*: mg/m^3 = 5.27 × ppm[1]

1.1.4 Technical products and impurities

Dichloroacetic acid is available commercially at a purity of 99% with a maximum of 0.3% water (Hoechst Chemicals, 1990; Spectrum Chemical Mfg Corp., 1994).

1.1.5 Analysis

Two ion chromatography methods are available for determining haloacetic acids, including dichloroacetic acid. The first is based on anion-exchange separation with suppressed electrical conductivity detection. The second is based on anion-exclusion separation with ultraviolet detection. The detection limits for dichloroacetic acid were 16.0 µg/L with the ion-exchange method and 8.0 µg/L with the ion-exclusion method (Nair *et al.*, 1994).

A high-performance liquid chromatography method has been used to separate and quantify dichloroacetic acid. The samples were chromatographed with aqueous ammonium sulfate as the mobile phase, and the chromatograph was equipped with an ultraviolet-visible radiation detector set at 210 nm (Husain *et al.*, 1993).

A method for the microdetermination of chloroacetic acids, including dichloroacetic acid, in water involved conversion to the difluoroanilide derivative by reaction with difluoroaniline and dicyclohexylcarbodiimide. The derivative was extracted into ethyl acetate and determined by gas chromatography with electron capture detection. The detection limit for dichloroacetic acid was about 1 µg/L (Ozawa & Tsukioka, 1990).

A multi-channel, microwave-induced plasma atomic spectroscopic gas chromatographic detector has been used to characterize the profiles of chlorinated humic acid on capillary columns and the content of carbon, chlorine and bromine in drinking-water. This technique makes it possible to estimate the empirical formulae of separated compounds with sufficient accuracy for useful peak identification. Dichloroacetic acid was among the compounds characterized by this method (Italia & Uden, 1988).

The United States Environmental Protection Agency (1990) reported a method for the determination of haloacetic acids, including dichloroacetic acid, in drinking-water, groundwater, raw water and water in any intermediate treatment stage. The method involves adjusting the pH to 11.5 and extraction with methyl-*tert*-butyl ether to remove neutral and basic organic compounds. The aqueous sample is then acidified to a pH of 0.5, and the acids are extracted into methyl-*tert*-butyl ether. The acids are then converted to their methyl esters with diazomethane.

---

[1] Calculated from: mg/m^3 = (relative molecular mass/24.45) × ppm, assuming normal temperature (25 °C) and pressure (101 kPa)

The methyl esters are determined by capillary gas chromatography with electron capture detection. The detection limit for this method is 0.015 µg/L.

1.2 Production and use

1.2.1 Production

The most efficient method for producing dichloracetic acid is hydrolysis of dichloroacetyl chloride produced by oxidation of trichloroethylene (see monograph, this volume) (Koenig et al., 1986). Dichloroacetic acid is also prepared by hydrolysis of pentachloroethane (see IARC, 1987a) with concentrated sulfuric acid, by oxidation of 1,1-dichloroacetone with nitric acid and air or by catalytic dechlorination of trichloroacetic acid (see monograph, this volume) with hydrogen over a palladium catalyst. Dichloroacetic acid can be produced in the laboratory by reacting chloral hydrate (see monograph, this volume) with potassium or sodium cyanide (Koenig et al., 1986).

Figures for the production and use of dichloroacetic acid throughout the world are not available (Koenig et al., 1986). In the Member States of the European Union, 3000 tonnes were estimated to have been produced and used in 1984 (Environmental Chemicals Data and Information Network, 1993). Dichloroacetic acid is produced by two companies in Japan and by one company each in Germany and India (Chemical Information Services, Inc., 1994).

1.2.2 Use

Dichloroacetic acid is presently of little economic importance. Its acid chloride and methyl ester, however, are used as intermediates in the manufacture of agrochemicals and the pharmaceutical, chloramphenicol (see IARC, 1990) (Koenig et al., 1986). Dichloroacetic acid is also a starting material for the production of glyoxylic acid, dialkyloxy and diaryloxy acids and sulfonamides. The compound is used as a test reagent for analytical measurements during the manufacture of poly(ethylene terephthalate) and as a medical disinfectant, in particular as a substitute for formaldehyde (Koenig et al., 1986). It has been considered for use in the treatment of lactic acidosis, diabetes mellitus, hyperlipoproteinaemia and several other disorders; however, it has never been marketed for any of these purposes (Budavari, 1989; Stacpoole, 1989).

1.3 Occurrence

1.3.1 Natural occurrence

Dichloroacetic acid is not known to occur as a natural product.

1.3.2 Occupational exposure

The National Occupational Exposure Survey conducted between 1981 and 1983 indicated that 1592 employees in the United States were potentially exposed to dichloroacetic acid in 39 facilities (United States National Institute for Occupational Safety and Health, 1994).

1.3.3 Air

No data were available to the Working Group.

1.3.4 Water

Dichloroacetic acid is produced as a by-product during aqueous chlorination of humic substances (Christman *et al.*, 1983; Miller & Uden, 1983; Legube *et al.*, 1985; Reckhow *et al.*, 1990). Consequently, it may occur in drinking-water after chlorine disinfection of raw waters containing natural organic substances (Hargesheimer & Satchwill, 1989; see IARC, 1991a). The concentrations of dichloroacetic acid measured in various water sources are summarized in Table 1. It has been identified as a major chlorinated by-product of the photocatalytic degradation of tetrachloroethylene in water but a minor by-product of the degradation of trichloroethylene (Glaze *et al.*, 1993).

Table 1. Concentrations of dichloracetic acid in water

Water type (location)	Concentration range (µg/L)	Reference
Drinking-water (chlorinated tap water) (USA)	63.1–133	Uden & Miller (1983)
Chlorinated treated water (Australia)	200 max	Nicholson *et al.* (1984)
Drinking-water (chlorinated surface, reservoir, lake and groundwater) (USA)	5.0–7.3	Krasner *et al.* (1989)
Chlorinated surface water (USA)	9.4–23	Jacangelo *et al.* (1989)
Chlorinated drinking-water (USA)	8–79	Reckhow & Singer (1990)
Chlorinated drinking-water (Japan)	4.5	Ozawa (1993)
Rainwater (Germany)	1.35	Clemens & Schöler (1992a)
Swimming pool (Germany)	indoors: 0.2–10.6 open air: 83.5–181.0[a]	Clemens & Schöler (1992b)
Surface water (downstream from a paper mill) (Austria)	< 3–522	Geist *et al.* (1991)
Biologically treated kraft pulp mill effluent (Malaysia)	14–18	Mohamed *et al.* (1989)

[a]The higher levels found in open-air swimming pools may be due to the input of organic material by swimmers.

1.3.5 Other

In humans, dichloroacetic acid is a reactive intermediate metabolite of trichloroethylene and an end-metabolite of 1,1,2,2-tetrachloroethane (see IARC, 1987b). As dichloroacetic acid has also been reported as a biotransformation product of methoxyflurane (Mazze & Cousins, 1974) and dichlorvos (see IARC, 1991b; WHO, 1989), it may occur in the tissues and fluids of animals treated with dichlorvos for helminthic infections (Schultz *et al.*, 1971).

Dichloroacetic acid has been detected in spruce needles from the Black Forest in Germany and the Montafon region in Austria, both considered to be relatively unpolluted areas, in the range of ten to several hundreds of micrograms per kilogram (Frank *et al.*, 1989).

1.4 Regulations and guidelines

In most countries, no limits have been recommended for exposure to dichloroacetic acid. A guideline limit of 4 mg/m^3 for short-term occupational exposure has been set in the Russian Federation (ILO, 1991).

The United States Environmental Protection Agency (1994) proposed that the maximal level of haloacetic acids (the sum of the concentration of mono-, di- and trichloroacetic acids and mono- and dibromoacetic acids) in drinking-water be 0.06 mg/L.

WHO (1993) recommends a provisional guideline value for dichloroacetic acid in drinking-water of 50 µg/L.

2. Studies of Cancer in Humans

No studies were available on people exposed to dichloroacetic acid. In view of the fact that it is a metabolite of trichloroethylene and tetrachloroethylene, the results of studies on populations exposed to those compounds may be relevant (see pp. 95 and 176). In particular, urinary levels of dichloroacetic acid were measured in one descriptive study of a population exposed to trichloroethylene and tetrachloroethylene in drinking-water (Vartiainen *et al.*, 1993).

3. Studies of Cancer in Experimental Animals

Oral administration

Mouse: A group of 26 male B6C3F1 mice, four weeks of age, received drinking-water containing 5 g/L dichloroacetic acid (purity, > 99%) neutralized with sodium hydroxide to a pH of 6.5–7.5. A control group of 27 mice received drinking-water containing 2 g/L sodium chloride. Both groups were kept for 61 weeks, at which time they were killed and necropsied. Two of 22 control mice had hepatic adenomas and none had hepatic carcinomas, whereas 25/26 mice that received dichloroacetic acid had hepatic adenomas and 21/26 had hepatocellular carcinomas ($p < 0.01$; Fisher's exact test) (Herren-Freund *et al.*, 1987). [The Working Group noted that this study also included groups pretreated with intraperitoneal injections of *N*-nitrosoethylurea in order to test for promoting effects of dichloroacetic acid; however, because of the high incidence and early appearance of liver tumours induced by dichloroacetic acid alone, promoting effects could not be evaluated.]

Groups of male and female B6C3F1 mice, 37 days old, received dichloroacetic acid in drinking-water (neutralized to pH 6.8-7.2 with sodium hydroxide) for up to 52 weeks, at which time the experiment was terminated. A group of 11 male mice received a dose of 1 g/L for 52 weeks, 24 male mice received 2.0 g/L for 52 weeks, and a further group of 11 males received 2.0 g/L for 37 weeks and then water alone until week 52. Two groups of 35 and 11 male control mice were kept until the end of the experiment. Groups of 10 female mice received either 0 or 2.0 g/L dichloroacetic acid for 52 weeks. Livers and kidneys were weighed and examined macroscopically. Microscopic examination was undertaken only of lesions found in the livers of 35 male control mice, the 11 male mice treated with 2.0 g/L dichloroacetic acid for 37 weeks and other groups [numbers unspecified] chosen at random. The lesions were classified histologically as hyperplastic nodules, adenomas or hepatocellular carcinomas. The incidences of these lesions were increased in mice receiving 2 g/L dichloroacetic acid (see Table 2). Only hyperplastic nodules and adenomas were found in mice treated for 37 weeks, and only hyperplastic nodules were observed in 3/10 treated female mice [no further details reported]. Other pathological signs seen at 37 or 52 weeks in males and females treated with dichloroacetic acid included cytomegaly, massive accumulation of glycogen and focal necrotic lesions (Bull *et al.*, 1990). [The Working Group noted that histopathological examination was limited to selected macroscopic lesions.]

Table 2. Lesions in the livers of male B6C3F1 mice given dichloroacetic acid in the drinking-water

Dose (g/L) × no. of weeks	No. of mice	No. with macroscopic lesions	Total no. of lesions	No. of macroscopic lesions examined histologically (no. of mice)	No. of hyperplastic nodules (no. of mice)	No. of hepatic adenomas (no. of mice)	No. of hepatocellular carcinomas (no. of mice)
Control × 52 (water alone)	35	2	2	2 (2)	1 (1)	0	0
1 × 52	11	2	3	1 (1)	1 (1)	0	0
2 × 52	24	23	92	23 (10)	15 (9)	2 (2)	6 (5)
2 × 37	11	7	23	19 (7)	15 (6)	2 (2)	0

From Bull *et al.* (1990)

Groups of 50 male B6C3F1 mice, four weeks old, were given 0.05, 0.5 or 5 g/L dichloroacetic acid (purity, > 99%; adjusted to pH 6.8–7.2 by the addition of 10 N sodium hydroxide) in the drinking-water. A control group of 50 mice was given drinking-water containing 2 g/L sodium chloride. In a second experiment, groups of 50 male B6C3F1 mice were given drinking-water containing either 3.5 g/L dichloroacetic acid or 1.5 g/L acetic acid (control group) in order to examine the metabolic appropriateness of an alternative control group. Interim kills of five mice were made in all treatment groups at four, 15, 30 and 45 weeks, except in the group given 3.5 g/L dichloroacetic acid. After 60 weeks of treatment, nine mice treated with 2 g/L saline or with 0.05 or 0.5 g/L dichloroacetic acid and 30 mice given 5.0 g/L dichloroacetic acid were killed. The remaining animals were killed at 75 weeks. In the second experiment, 12 mice receiving 3.5 g/L dichloroacetic acid and 10 mice given acetic acid were killed at 60 weeks. [The fate of the remaining mice in these two groups is not described.] Drinking-water intake and final body weight were lower in the groups receiving 3.5 or 5.0 g/L dichloroacetic acid than among their respective controls; there was no difference in survival. Proliferative lesions of the liver were classified as hyperplastic nodules, hepatocellular adenomas or hepatocellular carcinomas; the prevalence of the two tumour types was reported only as percentages on the basis of the number of animals examined. Hyperplastic nodules occurred mainly among animals receiving dichloroacetic acid; the prevalence rates [presented graphically] at 60 weeks were 58% among those given 3.5 g/L and 83% for those given 5.0 g/L. Hepatocellular carcinomas were first observed at 30 weeks in mice at 3.5 g/L. At 60 weeks, the group given 5.0 g/L dichloroacetic acid had prevalences of 80% hepatic adenomas and 83% hepatocellular carcinomas ($p < 0.001$); the prevalences in the group given 3.5 g/L were 100% hepatic adenomas and 67% hepatocellular carcinomas [combined prevalence of tumours at 60 and 75 weeks] ($p < 0.001$). In contrast, the prevalences of hepatic adenomas and carcinomas (combined) were 11.1% in the group given 0.5 g/L dichloroacetic acid and 24.1% in that given 0.05 g/L; these values were not significantly different from that in the saline controls (7.1%). No liver tumours were found in 10 controls given acetic acid and killed at 60 weeks (DeAngelo et al., 1991). [The Working Group noted the unconventional design and reporting of the study.]

A group of 33 male B6C3F1 mice (initially two groups of 23 and 10 mice but analysed as one), four weeks of age, received 0.5 g/L dichloroacetic acid (purity, > 95%; impurities unspecified; pH adjusted to 6.8–7.2 with 10 N sodium hydroxide) in distilled drinking-water (pooled estimated mean dose, 88 mg/kg bw per day) for 104 weeks; 33 control mice received distilled water only. Five mice per group were killed at 30 weeks and a further five in the control group at 60 weeks, for interim evaluation. Three control mice and four treated mice died before week 104. Of the animals killed at week 104, 15/24 treated mice and 2/20 controls had hepatocellular carcinomas [$p = 0.001$, Fisher's exact test]; 10/24 treated mice and 1/20 control mice had hepatocellular adenomas [$p = 0.005$]; and 18/24 treated mice and 3/20 controls had carcinomas or adenomas [$p = 0.001$]. Two treated mice had hyperplastic liver nodules; 8/24 treated mice and 1/20 controls had hepatocellular necrosis, and 22/24 treated mice and 1/20 controls had cytomegaly (Daniel et al., 1992).

4. Other Data Relevant to an Evaluation of Carcinogenicity and its Mechanisms

4.1 Absorption, distribution, metabolism and excretion

4.1.1 Humans

Oral or intravenous administration of 50 mg/kg bw dichloroacetic acid to healthy volunteers resulted in the excretion of oxalate and dichloroacetic acid itself as the major urinary components; however, only 5% of the administered dose was found in the urine. The plasma half-life of dichloroacetic acid after a single oral or intravenous dose was 0.5–2 h. Negligible amounts of dichloroacetic acid appear to bind to plasma proteins (Curry *et al.*, 1985; Stacpoole, 1989). Identification of oxalate as a metabolite in urine indicates cytochrome P450-dependent dechlorination to glyoxylate, which can be converted to oxalate either directly or via glycolate. The plasma half-life of dichloroacetic acid is several times longer after repeated than after single oral or intravenous administration, possibly because of enzyme inhibition by dichloroacetic acid or its metabolites (Curry *et al.*, 1985).

4.1.2 Experimental systems

The toxicokinetics of dichloroacetic acid were investigated in male Fischer 344 rats during 48 h after oral administration of 28.2 or 282 mg/kg bw ^{14}C-dichloroacetic acid (Lin *et al.*, 1993). The percentage of radiolabel excreted in the urine increased from 12.7% at the lower dose to 35.2% at the high dose. Unmetabolized dichloroacetic acid comprised > 20% of the urinary radiolabel at the high dose and < 1% at the low dose. The percentage of the dose excreted as carbon dioxide decreased from 34.4% at the lower dose to 25% at the higher. Significant percentages of the administered dose were retained in the liver (4.9–7.9%), muscle (4.5–9.9%), skin (3.3–4.5%) and intestines (1.0–1.7%). [The Working Group noted that the tissue retention might be related to the formation of one-carbon fragments.]

After dichloroacetic acid was administered as a single oral dose in water at 5, 20 or 100 mg/kg bw to male Fischer 344 rats and male B6C3F1 mice, only 2% was found unchanged in urine, indicating that it is extensively metabolized in both species. The total radiolabel in the urine comprised about 20–24% of the administered dose in rats and 28% in mice; the mean plasma half-life of dichloroacetic acid in both species was 0.9–1.6 h. The blood concentration over time was much greater in rats than in mice. Since this difference was magnified by increasing dose, the clearance mechanisms for dichloroacetic acid are probably more susceptible to saturation in rats than in mice (Larson & Bull, 1992a,b).

Dichloroacetic acid is metabolized by both oxidative and reductive pathways (Figure 1). Both pathways lead ultimately to oxalate and carbon dioxide, glycolate and glyoxylate being probable intermediate metabolites in the reductive pathway. Reductive dechlorination of dichloroacetic acid to monochloroacetate, followed by glutathione conjugation to give thiodiacetic acid as the ultimate metabolite, has also been demonstrated. Oxalic, glycolic and thiodiacetic acids are the major urinary metabolites of dichloroacetic acid in both rats and mice. The metabolic reactions possibly involve free-radical intermediates (Crabb & Harris, 1979; Stacpoole *et al.*, 1990; Larson & Bull, 1992a). The ability of dichloroacetic acid to elicit a lipid peroxidative response in liver was therefore investigated in male Fischer 344 rats and male B6C3F1 mice after administration of a single oral dose of 100–2000 mg/kg bw in water. A dose-dependent response was induced up to 300 mg/kg bw dichloroacetic acid in both species. A further increase to 1000 mg/kg bw resulted in only minimal increases in the lipid peroxidative response in mice and in a decreased response in rats (Larson & Bull, 1992a).

Figure 1. Metabolism of dichloroacetic acid

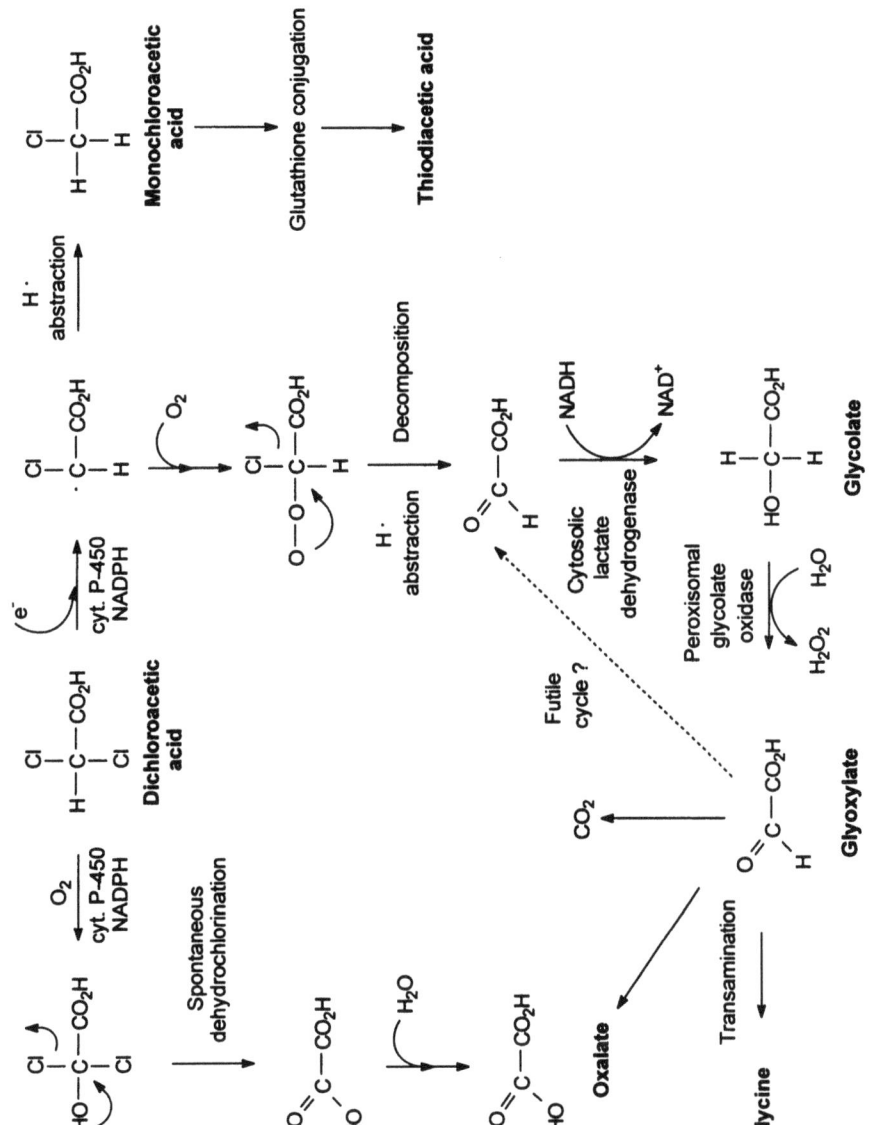

From Larson and Bull (1992a)

4.1.3 Comparison of humans and animals

Both humans and rodents metabolize dichloroacetic acid to glyoxylate by oxidative dechlorination; the plasma half-life of the parent compound is 0.5–2 h. A significantly greater percentage of the dose is excreted in the urine of rodents (about 20–30%), however, than by humans. Since only a negligible percentage of administered radiolabel is bound to plasma proteins or taken up by erythrocytes, dichloroacetic acid and its metabolites may be distributed to other tissues, although there is no direct evidence for this suggestion. Administration of repeated doses results in a marked decrease in the clearance of dichloroacetic acid from human plasma, but a similar inhibition by dichloroacetic acid of its own metabolism has not been reported in rodents.

4.2 Toxic effects

4.2.1 Humans

The pharmacological and toxic effects of dichloroacetic acid in humans have been studied extensively owing to the potential use of this chloroacetic acid for the treatment of various disorders (see section 1.2.2). Dichloroacetic acid lowers blood sugar levels in animals and humans with diabetes mellitus by stimulating peripheral glucose use and inhibiting gluconeogenesis. In addition, long-term administration of dichloroacetic acid reduces plasma triglyceride and cholesterol levels (a particularly important effect in patients with congenital hypercholesterolaemia, who have no cholesterol receptors), and it facilitates oxidation of lactate by activating pyruvate dehydrogenase in patients with acquired and congenital forms of lactic acidosis. Its capacity to activate pyruvate dehydrogenase also made dichloroacetic acid a candidate for use in the treatment of conditions involving myocardial ischaemia, because when oxygen delivery to heart muscle is limited, a shift from fatty acid to carbohydrate oxidation may increase the ratio of ATP production:oxygen consumption (Stacpoole, 1989).

The neurotoxic effects of dichloroacetic acid observed repeatedly in experimental animals have rarely been documented in clinical trials. Drowsiness is a fairly frequent side-effect of dichloroacetic acid and has been observed in healthy volunteers, adults with type II diabetes and patients with lactic acidosis. A patient with homozygous familial hypercholesterolaemia who received single doses of 50 mg/kg bw dichloroacetic acid daily for four months developed reversible peripheral neuropathy characterized by loss of reflexes and muscle weakness; the effect subsided several weeks after cessation of administration of dichloroacetic acid (Moore et al., 1979).

4.2.2 Experimental systems

More total radioactivity was associated with albumin and haemoglobin in male Fischer 344 rats than in male B6C3F1 mice 4–120 h after a single oral dose of 5 mg/kg bw ^{14}C-dichloroacetic acid. In contrast, incorporation of radiolabelled amino acids resulting from the metabolism of chloroacetate (glycine being derived from glyoxylic acid) was more extensive in mice; this result is consistent with the more extensive metabolism in this species (Stevens et al., 1992).

Exposure of male and female Sprague-Dawley rats to dichloroacetic acid at target doses of 10–600 mg/kg bw per day in the drinking-water for 14 days resulted in reduced weight gain only in the group given the highest dose. Treatment also increased urinary excretion of ammonia and changed the activities of enzymes of ammoniagenesis, indicating renal compensation for an acid load (Davis, 1986).

Male Sprague-Dawley rats administered dichloroacteic acid in the drinking-water for 90 days at concentrations providing daily doses of about 4, 35 or 350 mg/kg bw had decreased body weights. Animals given the high dose also showed histological and biochemical signs of liver and kidney damage and increased hepatic peroxisomal β-oxidation activity (Mather et al., 1990).

Male Sprague-Dawley rats were given dichloroacetic acid in the drinking-water at a concentration of 80.5 mmol/L [10 g/L] to provide an approximate intake of 1100 mg/kg bw per day. After 90 days, body weights were decreased, and there was an 11% increase in liver weight and a 34% decrease in testicular weight; histopathological changes were seen in the liver and lung (Bhat et al., 1991).

Ocular toxicity was observed in beagle dogs (which are susceptible to drug-induced cataract formation) that were treated for 13 weeks with an approximate dose of 1100 mg/kg bw dichloroacetic acid in the drinking-water. No similar organ-specific effect has been seen in other studies or in other species (Katz et al., 1981).

Administration of the sodium salt of dichloroacetic acid at a target dose of 50 or 1100 mg/kg bw to male Sprague-Dawley rats in the drinking-water for seven weeks resulted in severe hind limb weakness, demyelinization of cerebral and cerebellar parenchyma and thiamine depletion in the group at the high dose. The neurotoxic effects could be partially prevented by providing thiamine supplementation during the treatment period (Stacpoole et al., 1990). These results confirmed the observed association between neurotoxicity induced by dichloroacetic acid and the histopathological changes in the brain seen in thiamine deficiency. The underlying mechanism may involve stimulation of thiamine-dependent enzymes by dichloroacetic acid, resulting in increased turnover of this vitamin (Katz et al., 1981; Yount et al., 1982). Oxalate, a metabolite of dichloroacetic acid in humans and rodents, has been shown to cause both peripheral neuropathy and cataracts; however, the renal and testicular oxalate crystals seen commonly in such cases have not been observed after administration of high doses of dichloroacetic acid (Yount et al., 1982; Stacpoole et al., 1990).

Dichloroacetic acid, trichloroacetic acid and chloroform are formed during the chlorination of drinking-water (see IARC, 1991a). After concomitant administration of dichloroacetic acid (at 0.92 and 2.45 mmol [118 and 316 mg]/kg bw by gavage, three times over 24 h) and chloroform (one intraperitoneal injection of 0.75 mg/kg bw once after the last dose of dichloroacetic acid) to male and female Sprague-Dawley rats, the toxicity of chloroform to the liver and kidney was increased (Davis, 1992).

Exposure of male and female B6C3F1 mice to dichloroacetic acid at 1000 and 2000 mg/L in drinking-water for up to 52 weeks induced severe cytomegaly associated with extensive accumulation of glycogen, the effects progressing to multiple focal areas of necrosis, regenerative cell division and hepatomegaly (Bull et al., 1990; Sanchez & Bull, 1990; Bull et al., 1993).

Induction of peroxisome proliferation has been repeatedly associated with the chronic toxicity and carcinogenicity of dichloroacetic acid to the liver (DeAngelo et al., 1989). It can induce peroxisome proliferation in the livers of both mice and rats, as indicated by increased activities of palmitoyl-coenzyme A oxidase and carnitine acetyl transferase, the appearance of a peroxisome proliferation-associated protein and increased volume-density of peroxisomes after exposure to dichloroacetic acid for 14 days.

4.3 Reproductive and prenatal effects

4.3.1 Humans

No data were available to the Working Group.

4.3.2 Experimental systems

Dichloroacetic acid and its metabolites accumulate in rat fetuses after treatment of the dam (Roth et al., 1991). The main effect of maternal doses of 140–2400 mg/kg bw per day on days 6–15 of gestation was altered development of the heart and major vessels and, less frequently, the kidneys and the orbits of the eyes (Randall et al., 1991; Smith et al., 1991; Epstein et al., 1992; Smith et al., 1992).

Long-term administration of dichloroacetic acid orally at up to 72 mg/kg bw per day to dogs and 80.5 mmol/L [10 g/L] in drinking-water to rats (calculated dose, 1100 mg/kg bw per day) for 90 days induced testicular toxicity in both species, with degeneration of the seminiferous epithelium (Bhat et al., 1991; Cicmanec et al., 1991). Earlier studies in rats, including one with a similar dose (1100 mg/kg bw per day for seven weeks), showed normal testicular histopathology and sperm production (Yount et al., 1982; Stacpoole et al., 1990). In male Long-Evans rats given 0, 31.3, 62.5 or 125 mg/kg bw per day by gavage for 10 weeks, toxic effects were seen on the male reproductive accessory organs (preputial glands and epididymes) and sperm at 31.3 or 62.5 mg/kg bw, whereas toxic effects on the testis and a reduction in late-step spermatid head count were observed only in the group given the highest dose. The number of viable implants on day 14 of gestation after an overnight mating with unexposed controls was decreased only in group at the high dose (Toth et al., 1992).

4.4 Genetic and related effects

4.4.1 Humans

No data were available to the Working Group.

4.4.2 Experimental systems (see also Table 3 and Appendices 1 and 2)

Dichloroacetic acid did not induce differential toxicity in DNA repair-deficient strains of *Salmonella typhimurium* but did induce prophage in *Escherichia coli* in one study. It was mutagenic to *S. typhimurium* TA100 and TA98 in single studies. Most of the mutations in 400 revertants of dichloroacetic acid-treated *S. typhimurium* TA100 cultures were GC→AT transitions (DeMarini et al., 1994).

Table 3. Genetic and related effects of dichloroacetic acid

Test system	Result[a] Without exogenous metabolic activation	Result[a] With exogenous metabolic activation	Dose[b]	Reference
PRB, λ Prophage induction, *Escherichia coli* WP2s	–	+	2500	DeMarini et al. (1994)
SAD, *Salmonella typhimurium*, DNA repair-deficient TS24 strain	–	–	31 000	Waskell (1978)
SAD, *Salmonella typhimurium*, DNA repair-deficient TA2322 strain	–	–	31 000	Waskell (1978)
SAD, *Salmonella typhimurium*, DNA repair-deficient TA1950 strain	–	–	31 000	Waskell (1978)
SA0, *Salmonella typhimurium* TA100, reverse mutation	–	–	0.00	Herbert et al. (1980)
SA0, *Salmonella typhimurium* TA100, reverse mutation	(+)	(+)	1.0	DeMarini et al. (1994)
SA5, *Salmonella typhimurium* TA1535, reverse mutation	–	–	0.00	Herbert et al. (1980)
SA7, *Salmonella typhimurium* TA1537, reverse mutation	–	–	0.00	Herbert et al. (1980)
SA8, *Salmonella typhimurium* TA1538, reverse mutation	–	–	0.00	Herbert et al. (1980)
SA9, *Salmonella typhimurium* TA98, reverse mutation	(+)	(+)	5	Herbert et al. (1980)
DIA, DNA strand breaks, B6C3F1 mouse hepatocytes *in vitro*	–	0	2580	Chang et al. (1992)
DIA, DNA strand breaks, Fischer 344 rat hepatocytes *in vitro*	–	0	1290	Chang et al. (1992)
DIH, DNA strand breaks, human CCRF-CEM cells *in vitro*	–	0	1290	Chang et al. (1992)
DVA, DNA strand breaks, B6C3F1 mouse hepatic cells *in vivo*	+		13 po × 1	Nelson & Bull (1988)
DVA, DNA strand breaks, B6C3F1 mouse hepatic cells *in vivo*	+		10 po × 1	Nelson et al. (1989)
DVA, DNA strand breaks, B6C3F1 mouse hepatic cells *in vivo*	–		1290 po × 1	Chang et al. (1992)
DVA, DNA strand breaks, B6C3F1 mouse splenocytes *in vivo*	–		1290 po × 1	Chang et al. (1992)
DVA, DNA strand breaks, B6C3F1 mouse epithelial cells from stomach and duodenum *in vivo*	–		1290 po × 1	Chang et al. (1992)
DVA, DNA strand breaks, B6C3F1 mouse hepatic cells *in vivo*	–		830 dw × 7–14 d	Chang et al. (1992)
DVA, DNA strand breaks, Sprague-Dawley rat hepatic cells *in vivo*	+		30 po × 1	Nelson & Bull (1988)
DVA, DNA strand breaks, Fischer 344 rat hepatic cells *in vivo*	–		645 po × 1	Chang et al. (1992)
DVA, DNA strand breaks, Fischer 344 rat hepatic cells *in vivo*	–		250 dw × 30 weeks	Chang et al. (1992)

[a] +, considered to be positive; (+), considered to be weakly positive in an inadequate study; –, considered to be negative; 0, not tested
[b] LED, lowest effective dose; HID, highest effective dose. In-vitro tests, μg/ml; in-vivo tests, mg/kg bw; 0.00, dose not reported; po, orally; dw, drinking water

DNA strand breaks were not induced in mammalian cells *in vitro* in the absence of an exogenous metabolic activation system, but contradictory results were obtained *in vivo*. No effect was seen in either mouse or rat hepatic cells after single or repeated dosing, and no effects were observed in epithelial cells from spleen, stomach or duodenum after a single dose (Chang *et al.*, 1992).

DNA strand breaks were reported in one laboratory in the livers of mice and rats treated 4 h previously with dichloroacetate. None were observed 24 h after a single dose of 500 mg/kg bw or after repeated daily dosing (Nelson & Bull, 1988; Nelson *et al.*, 1989). Peroxisome proliferation, as indicated by β-oxidation of palmitoyl-coenzyme A, was observed only after induction of DNA damage (Nelson *et al.*, 1989). In another laboratory, DNA strand breakage was not observed in the livers of either mice or rats, while there was increased peroxisomal enzyme activity (Chang *et al.*, 1992). [The reasons for the contrasting results obtained using similar techniques are unclear.]

Mutations of proto-oncogenes in tumours induced by dichloroacetic acid

A group of 110 male B6C3F1 mice, eight weeks of age, were administered dichloroacetic acid, neutralized with sodium hydroxide, at a concentration of 0.5% in their drinking water for up to 76 weeks. Of two concurrent control groups, each consisting of 50 male mice, one was untreated and the other received corn oil at a dose of 10 ml/kg bw; 10 mice in each group were killed at 76 weeks and the remainder at 96, 103 and 134 weeks [numbers not stated]. At death, liver tumours measuring ≥ 0.5 cm in diameter were taken for histological examination and for oncogene analysis. At the time of the terminal kill, there were 24 untreated controls, 32 corn oil controls and 89 treated animals. The numbers of hepatocellular adenomas per mouse in these three groups were 0.9 ± 0.06 (8%), 0.13 ± 0.06 (13%) and 4.98 ± 0.38 (93%). The corresponding numbers of hepatocellular carcinomas were 0.09 ± 0.06 (8%), 0.12 ± 0.06 (12%) and 1.73 ± 0.17 (74%). The authors noted numerous foci of cellular alteration (presumed preneoplastic lesions) in the livers of treated mice but only rare foci in the livers of controls. No neoplasms related to treatment were found at other sites. The frequency of mutations in codon 61 of H-*ras* was not significantly different in the hepatocelluar tumours from 64 treated mice (62%) and in those from 74 combined historical and concurrent controls (69%); however, the spectra of these mutations showed a significant decrease in AAA and an increase in CTA in the treated mice in comparison with the controls. No other H-*ras* mutations were found, and only one K-*ras* mutation was detected in tumours from the treated and concurrent control groups. The authors interpreted these findings as suggesting that exposure to dichloroacetic acid provides the environment for a selective growth advantage for spontaneous CTA mutations in codon 61 of H-*ras* (Anna *et al.*, 1994).

Expression of c-*myc*, and c-H-*ras* in mRNA was studied by in-situ hybridization in the livers of male B6C3F1 mice treated with dichloroacetate at 1 or 2 g/L in the drinking-water for 52 weeks. Expression of c-*myc*, corrected for the background frequency, was increased by about three times in hepatic hyperplastic nodules and hepatic carcinomas in comparison with normal liver. A similar comparison of c-H-*ras* expression showed no significant increase in hyperplastic nodules but an approximately fourfold increase in hepatic carcinomas (Nelson *et al.*, 1990). [The

Working Group considered that the changes in proto-oncogene expression could not be attributed conclusively to dichloroacetic acid.]

5. Summary and Evaluation

5.1 Exposure data

Dichloroacetic acid is produced commercially in small quantities for use as an intermediate in the production of glyoxylic acid, dialkyloxy and diaryloxy acids and sulfonamides. Human exposure may occur during the production and use of dichloroacetic acid and from drinking chlorinated water.

5.2 Human carcinogenicity data

The available data were too limited to form the basis for an evaluation of the carcinogenicity of dichloroacetic acid to humans.

5.3 Animal carcinogenicity data

Neutralized dichloroacetic acid was tested by oral administration in males of one strain of mice in four studies. Increased incidences of hepatocellular adenomas and carcinomas were observed in all of the studies.

5.4 Other relevant data

Dichloroacetic acid is metabolized in humans and experimental animals, and oxalate, thiodiacetic acid and unchanged dichloroacetic acid are excreted in urine. Clearance is decreased in humans after repeated administration. Species differences in the clearance of dichloroacetic acid are observed in rodents: clearance in rats is much slower than in mice. Dichloroacetic acid induces peroxisome proliferation in the livers of both rats and mice.

No data were available on the effects of dichloroacetic acid on human reproduction. In rats and dogs, testicular degeneration can occur after exposure to this compound. The development of the heart, major vessels and kidney of rats can be affected by exposure *in utero*.

The evidence for induction of DNA strand breaks in liver cells of rodents exposed to dichloroacetic acid *in vivo* was inconclusive. Strand breaks were not induced in human or rodent cells *in vitro*. The results of assays for mutagenesis in bacteria were inconsistent.

The spectrum of mutations in H-*ras* proto-oncogenes in hepatic tumours from mice treated with dichloroacetic acid was different from that seen in hepatic tumours from untreated mice.

5.5 Evaluation[1]

There is *inadequate evidence* in humans for the carcinogenicity of dichloroacetic acid.

There is *limited evidence* in experimental animals for the carcinogenicity of dichloroacetic acid.

Overall evaluation

Dichloroacetic acid *is not classifiable as to its carcinogenicity to humans (Group 3).*

6. References

Anna, C.H., Maronpot, R.R., Pereira, M.A., Foley, J.F., Malarkey, D.E. & Anderson, M.W. (1994) *ras* Proto-oncogene activation in dichloroacetic acid-, trichloroethylene- and tetrachloroethylene-induced liver tumors in B6C3F1 mice. *Carcinogenesis*, **15**, 2255–2261

Bhat, H.K., Kanz, M.F., Campbell, G.A.S. & Ansari, G.A.S. (1991) Ninety day toxicity study of chloroacetic acids in rats. *Fundam. appl. Toxicol.*, **17**, 240–253

Budavari, S., ed. (1989) *The Merck Index*, 11th Ed., Rahway, NJ, Merck & Co., p. 481

Bull, R.J., Sanchez, I.M., Nelson, M.A., Larson, J.L. & Lansing, A.J. (1990) Liver tumor induction in B6C3F1 mice by dichloroacetate and trichloroacetate. *Toxicology*, **63**, 341–359

Bull, R.J., Templin, M., Larson, J.L. & Stevens, D.K. (1993) The role of dichloroacetate in the hepatocarcinogenicity of trichloroethylene. *Toxicol. Lett.*, **68**, 203–211

Chang, L.W., Daniel, F.B. & DeAngelo, A.B. (1992) Analysis of DNA strand breaks induced in rodent liver *in vivo*, hepatocytes in primary culture, and a human cell line by chlorinated acetic acids and chlorinated acetaldehydes. *Environ. mol. Mutag.*, **20**, 277–288

Chemical Information Services, Inc. (1994) *Directory of World Chemical Producers 1995/1996 Standard Edition*, Dallas, TX, p. 245

Christman, R.F., Norwood, D.L., Millington, D.S. & Johnson, J.D. (1983) Identity and yields of major halogenated products of aquatic fulvic acid chlorination. *Environ. Sci. Technol.*, **17**, 625–628

Cicmanec, J.L., Condie, L.W., Olson, G.R. & Wang, S.R. (1991) 90-Day toxicity study of dichloroacetate in dogs. *Fundam. appl. Pharmacol.*, **17**, 376–389

Clemens, M. & Schöler, H.-F. (1992a) Determination of halogenated acetic acids and 2,2-dichloropropionic acid in water samples. *Fresenius J. anal. Chem.*, **344**, 47–49

Clemens, M. & Schöler, H.-F. (1992b) Halogenated organic compounds in swimming pool waters. *Zbl. Hyg.*, **193**, 91–98 (in German)

Crabb, D.W. & Harris, R.A. (1979) Mechanism responsible for the hypoglycemic actions of dichloroacetate and 2-chloropropionate. *Arch. Biochem. Biophys.*, **198**, 145–152

Curry, S.H., Chu, P.-I., Baumgartner, T.G. & Stacpoole, P.W. (1985) Plasma concentrations and metabolic effects of intravenous sodium dichloroacetate. *Clin. Pharmacol. Therap.*, **37**, 89–93

---

[1] For definition of the italicized terms, see Preamble, pp. 22–26.

Daniel, F.B., DeAngelo, A.B., Stober, J.A., Olson, G.R. & Page, N.P. (1992) Hepatocarcinogenicity of chloral hydrate, 2-chloroacetaldehyde and dichloroactic acid in B6C3F$_1$ mice. *Fundam. appl. Toxicol.*, **19**, 159–168

Davis, M.E. (1986) Effect of chloroacetic acids on the kidneys. *Environ. Health Perspectives*, **69**, 209–214

Davis, M.E. (1992) Dichloroacetic acid and trichloroacetic acid increase chloroform toxicity. *J. Toxicol. environ. Health*, **37**, 139–148

DeAngelo, A.B., Daniel, F.B., McMillan, L., Wernsing, P. & Savage, R.E., Jr (1989) Species and strain sensitivity to the induction of peroxisome proliferation by chloroacetic acids. *Toxicol. appl. Pharmacol.*, **101**, 285–298

DeAngelo, A.B., Daniel, F.B., Stober, J.A. & Olson, G.R. (1991) The carcinogenicity of dichloroacetic acid in the male B6C3F$_1$ mouse. *Fundam. appl. Toxicol.*, **16**, 337–347

DeMarini, D.M., Perry, E. & Shelton, M.L. (1994) Dichloroacetic acid and related compounds: induction of prophage in *E. coli* and mutagenicity and mutation spectra in Salmonella TA100. *Mutagenesis*, **9**, 429–437

Environmental Chemicals Data and Information Network (1993) *Dichloroacetic acid*, Ispra, JRC-CEC, last update: 02.09.1993

Epstein, D.L., Nolen, G.A., Randall, J.L., Christ, S.A., Read, E.J., Stober, J.A. & Smith, M.K. (1992) Cardiopathic effects of dichloroacetate in the fetal Long-Evans rat. *Teratology*, **46**, 225–235

Frank, H., Vital, J. & Frank, W. (1989) Oxidation of airborne C$_2$-chlorocarbons to trichloroacetic and dichloroacetic acid. *Fresenius Z. anal. Chem.*, **333**, 713

Geist, S., Lesemann, C., Schütz, C., Seif, P. & Frank, E. (1991) Determination of chloroacetic acids in surface water. In: Angeletti, G. & Bjørseth, A., eds, *Organic Micropollutants in the Aquatic Environment*, Dordrecht, Kluwer Academic Publishers, pp. 393–397

Glaze, W.H., Kenneke, J.F. & Ferry, J.L. (1993) Chlorinated byproducts from the TiO$_2$-mediated photodegradation of trichloroethylene and tetrachloroethylene in water. *Environ. Sci. Technol.*, **27**, 177–184

Hansch, C., Leo, A. & Hoekman, D.H. (1995) *Exploring QSAR*, Washington, DC, American Chemical Society

Hargesheimer, E.E. & Satchwill, T. (1989) Gas-chromatographic analysis of chlorinated acids in drinking water. *J. Water Sewage Res. Treat. Aqua*, **38**, 345–351

Herbert, V., Gerdner, A. & Colman, N. (1980) Mutagenicity of dichloroacetate, an ingredient of some formulations of pangamic acid (trade-named 'vitamin B$_{15}$'). *Am. J. clin. Nutr.*, **33**, 1179–1182

Herren-Freund, S.L., Pereira, M.A., Khoury, M.D. & Olson, G. (1987) The carcinogenicity of trichloroethylene and its metabolites, trichloroacetic acid and dichloroacetic acid, in mouse liver. *Toxicol. appl. Pharmacol.*, **90**, 183–189

Hoechst Chemicals (1990) *Chemical Information Sheet: Dichloroacetic acid*, Dallas, TX

Husain, S., Narsimha, R., Alvi, S.N. & Rao, R.N. (1993) Monitoring the levels of chloroacetic acids in the industrial environment by high-performance liquid chromatography. *J. high Resolut. Chromatogr.*, **16**, 381–383

IARC (1987a) *IARC Monographs on the Evaluation of Carcinogenic Risks to Humans*, Suppl. 7, *Overall Evaluations of Carcinogenicity: An Updating of* IARC Monographs Volumes 1–42, Lyon, p. 69

IARC (1987b) *IARC Monographs on the Evaluation of Carcinogenic Risks to Humans*, Suppl. 7, *Overall Evaluations of Carcinogenicity: An Updating of* IARC Monographs Volumes *1–42*, Lyon, pp. 354–355

IARC (1990) *IARC Monographs on the Evaluation of Carcinogenic Risks to Humans*, Vol. 50, *Pharmaceutical Drugs*, Lyon, pp. 169–193

IARC (1991a) *IARC Monographs on the Evaluation of Carcinogenic Risks to Humans*, Vol. 52, *Chlorinated Drinking-water; Chlorination By-products; Some Other Halogenated Compounds; Cobalt and Cobalt Compounds*, Lyon, pp. 55–141

IARC (1991b) *IARC Monographs on the Evaluation of the Carcinogenic Risks to Humans*, Vol. 53, *Occupational Exposures in Insecticide Application, and Some Pesticides*, Lyon, pp. 267–307

IARC (1992) *IARC Monographs on the Evaluation of Carcinogenic Risks to Humans*, Vol. 54, *Occupational Exposures to Mists and Vapours from Strong Inorganic Acids; and Other Industrial Chemicals*, Lyon, pp. 189–211

ILO (1991) *Occupational Exposure Limits for Airborne Toxic Substances: Values of Selected Countries* (Occupational Safety and Health Series No. 37), 3rd Ed., Geneva, pp. 138–139

Italia, M.P. & Uden, P.C. (1988) Multiple element emission spectral detection gas chromatographic profiles of halogenated products from chlorination of humic acid and drinking water. *J. Chromatogr.*, **438**, 35–43

Jacangelo, J.G., Patania, N.L., Reagan, K.M., Aieta, E.M., Krasner, S.W. & McGuire, M.J. (1989) Ozonation: assessing its role in the formation and control of disinfection by-products. *J. Am Water Works Assoc.*, **81**, 74–84

Katz, R., Tai, C.N., Diener, R.M., McConnell, R.F. & Semonick, D.E. (1981) Dichloroacetate, sodium: 3-month oral toxicity studies in rats and dogs. *Toxicol. appl. Pharmacol.*, **57**, 273–287

Koenig, G., Lohmar, E. & Rupprich, N. (1986) Chloroacetic acids. In: Gerhartz, W., Yamamoto, Y.S., Campbell, F.T., Pfefferkorn, R. & Rounsaville, J.F., eds, *Ullmann's Encyclopedia of Industrial Chemistry*, 5th rev. Ed., Vol. A6, New York, VCH Publishers, pp. 537–552

Krasner, S.W., McGuire, M.J., Jacangelo, J.G., Patania, N.L., Reagan, K.M. & Aieta, E.M. (1989) The occurrence of disinfection by-products in US drinking water. *J. Am. Water Works Assoc.*, **81**, 41–53

Larson, J.L. & Bull, R.J. (1992a) Metabolism and lipoperoxidative activity of trichloroacetate and dichloroacetate in rats and mice. *Toxicol. appl. Pharmacol.*, **115**, 268–277

Larson, J.L. & Bull, R.J. (1992b) Species differences in the metabolism of trichloroethylene to the carcinogenic metabolites trichloroacetate and dichloroacetate. *Toxicol. appl. Pharmacol.*, **115**, 278–285

Legube, B., Croue, J.P. & Dore, M. (1985) Chlorination of humic substances in aqueous solution: yields of volatile and major non-volatile organic halides. *Sci. total Environ.*, **47**, 217–222

Lide, D.R., ed. (1993) *CRC Handbook of Chemistry and Physics*, 74th Ed., Boca Raton, FL, CRC Press, p. 3–213

Lin, E.L.C., Mattox, J.K. & Daniel, F.B. (1993) Tissue distribution, excretion, and urinary metabolites of dichloroacetic acid in the male Fischer 344 rat. *J. Toxicol. environ. Health*, **38**, 19–32

Mather, G.G., Exon, J.H. & Koller, L.D. (1990) Subchronic 90 day toxicity of dichloroacetic and trichloroacetic acid in rats. *Toxicology*, **64**, 71–80

Mazze, R.I. & Cousins, M.J. (1974) Biotransformation of methoxyflurane. *Int. Anesthesiol. Clin.*, **12**, 93–105

Miller, J.W. & Uden, P.C. (1983) Characterization of non-volatile aqueous chlorination products of humic substances. *Environ. Sci. Technol.*, **17**, 150–157

Mohamed, M., Matayun, M. & Lim, T.S. (1989) Chlorinated organics in tropical hardwood kraft pulp and paper mill effluents and their elimination in an activated sludge treatment system. *Pertanika*, **12**, 387–394

Moore, G.W., Swift, L.L., Rabinowitz, D., Crofford, O.B., Oates, J.A. & Stacpoole, P.W. (1979) Reduction of serum cholesterol in two patients with homozygous familial hypercholesterolemia by dichloroacetate. *Artherosclerosis*, **33**, 285–293

Nair, L.M., Saari-Nordhaus, R. & Anderson, J.M., Jr (1994) Determination of haloacetic acids by ion chromatography. *J. Chromatogr. A.*, **671**, 309–313

Nelson, M.A. & Bull, R.J. (1988) Induction of strand breaks in DNA by trichloroethylene and metabolites in rat and mouse liver *in vivo*. *Toxicol. appl. Pharmacol.*, **94**, 45–54

Nelson, M.A., Lansing, A.J., Sanchez, I.M., Bull, R.J. & Springer, D.L. (1989) Dichloroacetic acid and trichloroacetic acid-induced DNA strand breaks are independent of peroxisome proliferation. *Toxicology*, **58**, 239–248

Nelson, M.A., Sanchez, I.M., Bull, R.J. & Sylvester, S.R. (1990) Increased expression of c-myc and c-H-ras in dichloroacetate and trichloroacetate-induced liver tumors in B6C3F1 mice. *Toxicology*, **64**, 47–57

Nicholson, B.C., Hayes, K.P. & Bursill, D.B. (1984) By-products of chlorination. *Water*, **September**, 11–15

Ozawa, H. (1993) Gas chromatographic–mass spectrometric determination of halogenated acetic acids in water after direct derivatization. *J. Chromatogr*, **644**, 375–382

Ozawa, H. & Tsukioka, T. (1990) Gas chromatographic separation and determination of chloroacetic acids in water by a difluoroanilide derivatization method. *Analyst*, **115**, 1343–1347

Randall, J.L., Christ, S.A., Nolen, G.A., Read, E.J. & Smith, M.K. (1991) Developmental effects of dichloroacetic acid in Long-Evans rats. II. Establishment of no adverse effect level (Abstract P123). *Teratology*, **43**, 454

Reckhow, D.A. & Singer, P.C. (1990) Chlorination by-products in drinking waters; from formation potentials to finished water concentrations. *J. Am. Water Works Assoc.*, **92**, 173–180

Reckhow, D.A., Singer, P.C. & Malcom, R.L. (1990) Chlorination of humic materials: byproduct formation and chemical interpretations. *Environ. Sci. Technol.*, **24**, 1655–1664

Roth, A.C., Crocker, W., Wessendarp, T.K., Smith, M.K. & Gordon, D.A. (1991) Dose related absorption and distribution of dichloroacetate (DCA) in pregnant Long-Evans rats (Abstract P36). *Teratology*, **43**, 428

Sadtler Research Laboratories (1980) *1980 Cumulative Index*, Philadelphia, PA

Sanchez, I.M. & Bull, R.J. (1990) Early induction of reparative hyperplasia in the liver of B6C3F1 mice treated with dichloroacetate and trichloroacetate. *Toxicology*, **64**, 33–46

Schultz, D.R., Marxmiller, R.L. & Koos, B.A. (1971) Residue determination of dichlorvos and related metabolites in animal tissues and fluids. *J. agric. Food Chem.*, **19**, 1238–1243

Smith, M.K., Christ, S.A., Randall, J.L., Nolen, G.A., Read, E.J. & Stober, J.A. (1991) Interaction study of chlorinated acetic acids in pregnant Long-Evans rats (Abstract P121). *Teratology*, **43**, 453–454

Smith, M.K., Randall, J.L., Read, E.J. & Stober, J.A. (1992) Developmental toxicity of dichloroacetate in the rat. *Teratology*, **46**, 217–223

Spectrum Chemical Mfg Corp. (1994) *1993–1994 Spectrum Chemical and Safety Products Catalog*, New Brunswick, NJ, p. C-156

Stacpoole, P.W. (1989) The pharmacology of dichloroacetate. *Metabolism*, **38**, 1124–1144

Stacpoole, P.W., Harwood, H.J., Jr, Cameron, D.F., Curry, S.H., Samuelson, D.A., Cornwell, P.E. & Sauberlich, H.E. (1990) Chronic toxicity of dichloroacetate: possible relation to thiamine deficiency in rats. *Fundam. appl. Toxicol.*, **14**, 327–337

Stevens, D.K., Eyre, R.J. & Bull, R.J. (1992) Adduction of hemoglobin and albumin *in vivo* by metabolites of trichloroethylene, trichloroacetate, and dichloroacetate in rats and mice. *Fundam. appl. Toxicol.*, **19**, 336–342

Toth, G.P., Kelty, K.C., George, E.L., Read, E.J. & Smith, M.K. (1992) Adverse male reproductive effects following subchronic exposure of rats to sodium dichloroacetate. *Fundam. appl. Toxicol.*, **19**, 57–63

Uden, P.C. & Miller, J.W. (1983) Chlorinated acids and chloral in drinking water. *J. Am. Water Works Assoc.*, **75**, 524–527

United States Environmental Protection Agency (1990) *Methods for the Determination of Organic Compounds in Drinking Water. Supplement 1* (EPA-600/4-90-020; US NTIS PB91-146027), Cincinnati, OH, Environmental Monitoring Systems Laboratory, Method 552

United States Environmental Protection Agency (1994) Drinking water; national primary drinking water regulations: disinfectants and disinfectant byproducts. *Fed. Reg.*, **59**, 38668–38710

United States National Institute for Occupational Safety and Health (1994) *National Occupational Exposure Survey (1981–1983)*, Cincinnati, OH, p. 11

Vartiainen, T., Pukkala, E., Rienoja, T., Strandman, T. & Kaksonen, K. (1993) Population exposure to tri- and tetrachloroethene and cancer risk: two cases of drinking water pollution. *Chemosphere*, **27**, 1171–1181

Waskell, L. (1978) A study of the mutagenicity of anesthetics and their metabolites. *Mutat. Res.*, **57**, 141–153

Weast, R.C. & Astle, M.J. (1985) *CRC Handbook of Data on Organic Compounds*, Vols I & II, Boca Raton, FL, CRC Press, p. 552 (I), 575 (II)

WHO (1989) *Dichlorvos* (Environmental Health Criteria 79), Geneva

WHO (1993) *Guidelines for Drinking-water Quality*, 2nd Ed., Vol. 1, *Recommendations*, Geneva, pp. 102, 177

Yount, E.A., Felten, S.Y., O'Connor, B.L., Peterson, R.G., Powell, R.S., Yum, M.N. & Harris, R.A. (1982) Comparison of metabolic and toxic effects of 2-chloroproprionate and dichloroacetate. *J. Pharmacol. exp. Ther.*, **222**, 501–508

TRICHLOROACETIC ACID

1. Exposure Data

1.1 Chemical and physical data

1.1.1 Nomenclature

Chem. Abstr. Serv. Reg. No.: 76-03-9
Chem. Abstr. Name: Trichloroacetic acid
IUPAC Systematic Name: Trichloroacetic acid
Synonyms: TCA; TCAA; TCA (acid); trichloracetic acid; trichloroethanoic acid; trichloromethanecarboxylic acid

1.1.2 Structural and molecular formulae and relative molecular mass

$$Cl-\underset{\underset{Cl}{|}}{\overset{\overset{Cl}{|}}{C}}-C\overset{\overset{O}{\parallel}}{\underset{OH}{}}$$

$C_2HCl_3O_2$ Relative molecular mass: 163.39

1.1.3 Chemical and physical properties of the pure substance

From Lide (1993), unless otherwise specified

(a) *Description*: Colourless to slightly yellowish, deliquescent crystals with a slight characteristic odour (Budavari, 1989; Hoechst Chemicals, 1990)

(b) *Boiling-point*: 197.5 °C

(c) *Melting-point*: 58 °C (α-form); 49.6 °C (β-form)

(d) *Density*: 1.62 at 25 °C/4 °C

(e) *Spectroscopy data*: Infrared (prism [36, 11 346]; grating [29 681]), ultraviolet [1–6], nuclear magnetic resonance (proton [6]; C-13 [10]) and mass [1026] spectral data have been reported (Sadtler Research Laboratories, 1980; Weast & Astle, 1985).

(f) *Solubility*: Soluble in water, acetone, methanol, ethanol and diethyl ether; slightly soluble in hydrocarbons and chlorinated hydrocarbons (Hoechst Chemicals, 1990; Lide, 1993). In aqueous solution, trichloroacetic acid and trichloroacetate exist as an equilibrium mixture, the proportions of each depending principally on the pH. The pK_a of trichloroacetic acid is 0.70.

(g) *Volatility*: Vapour pressure, 1 mm Hg [0.133 kPa] at 51 °C (Verschueren, 1983)

(h) *Stability*: Aqueous solutions undergo gradual hydrolysis, depending on the temperature, especially in the presence of alkali and alkaline earth ions; hydrolysed to chloroform, carbon dioxide and carbonate (Budavari, 1989; Hoechst Chemicals, 1990)

(i) *Reactivity*: Highly reactive with most metals (Hoechst Chemicals, 1990)

(j) *Octanol/water partition coefficient (P)*: log P, 1.33 (Hansch et al., 1995)

(k) *Conversion factor*: $mg/m^3 = 6.68 \times ppm^1$

1.1.4 Technical products and impurities

Trichloroacetic acid is available commercially in three grades: crude (purity 96%), technical (purity, 98%) and a 90% aqueous solution with the following specifications: dichloroacetic acid (see monograph, this volume), 1.2–2.5% max.; sulfuric acid, 0.3–0.5% max. and iron, 0.001–0.002% max. (Hoechst Chemicals, 1990). Trichloroacetic acid is also available in a reagent grade at a purity of 98–100%, containing maxima of 0.001% chloride, 0.002% heavy metals (as lead), 0.001% iron, 0.002% nitrate, 0.0005% phosphate and 0.02% sulfate (Sigma Chemical Co., 1992; Mallinckrodt Chemical, Inc., 1993; Aldrich Chemical Co., 1994a,b). Trade names for trichloroacetic acid include Aceto-Caustin and Amchem Grass Killer.

1.1.5 Analysis

Two ion chromatography methods have been described for the determination of haloacetic acids, including trichloroacetic acid. The first is based on anion-exchange separation with suppressed electrolytic conductivity detection; the second is based on anion-exclusion separation with ultraviolet detection. The detection limits for trichloroacetic acid are 80 µg/L for the first method and 5.1 µg/L for the second (Nair et al., 1994).

A high-performance liquid chromatography method is available for the separation and quantitative determination of trichloroacetic acid in effluents. The samples were chromatographed with aqueous ammonium sulfate as the mobile phase and quantified by ultraviolet absorption at 210 nm. The detection limit for trichloroacetic acid is 10 µg/L (Husain et al., 1992, 1993).

A method has been described for the determination of organic acids, including trichloroacetic acid, in ambient air. Ion-exchange resin was used as the adsorbent for sampling and subsequently as the catalyst for methylation of the adsorbed acids by methyl formate. The methyl esters were analysed by capillary GC–mass spectrometry. The limit of detection was 0.8 ng/L with and 56.4 ng/L without preconcentration (Sollinger et al., 1992).

In a method for the microdetermination of chloroacetic acids in water, trichloroacetic acid was converted into the difluoroanilide derivative by reaction with difluoroaniline and dicyclohexylcarbodiimide. The derivative was extracted into ethyl acetate and determined by GC with

---

[1] Calculated from: mg/m^3 = (molecular weight/24.45) × ppm, assuming normal temperature (25 °C) and pressure (101 kPa)

electron capture detection. The detection limit for trichloroacetic acid was about 0.6 µg/L (Ozawa & Tsukioka, 1990; Ozawa, 1993).

A method for the determination of trichloroacetic acid using flow injection and chemiluminescence detection is based on generation of bromine from the bromate–bromide–acid reaction *in situ*; the bromine is then reacted with hydrogen peroxide to liberate oxygen for the oxidation of luminol. The minimal detectable concentration of trichloroacetic acid was 15 mmol [2.5 g] (Shakir & Faizullah, 1990).

A multi-channel, microwave-induced plasma atomic spectroscopic GC detector has been used to characterize the profiles of chlorinated humic acid on capillary columns and the contents of carbon, chlorine and bromine in drinking-water. This technique makes it possible to estimate the empirical formulae of separated compounds with sufficient accuracy for useful peak identification. Trichloroacetic acid was among the compounds characterized by this method (Italia & Uden, 1988).

Trichloroacetic acid can be determined in water and urine by differential pulse polarography (Pergola *et al.*, 1988). The authors also discuss the advantages and disadvantages of the other methods commonly used for the determination of trichloroacetic acid in urine (spectrophotometry and chromatography) and in water (fluorimetric and mass spectroscopy).

A method has been described for the determination of trichloroethylene metabolites, including trichloroacetic acid, in rat liver homogenate, blood and urine. The method is based on selective thermal conversion of trichloroacetic acid into chloroform, which is determined by head-space GC with electron capture detection (Christensen *et al.*, 1988; Køppen *et al.*, 1988). A similar procedure was reported for human plasma (Ziglio *et al.*, 1984).

A spectophotometric method for the determination of trichloroacetic acid at the level of parts per million in air, serum and urine involves a modification of the Fujiwara reaction and measurement of a red colour developed by the reaction between a polyhalogenated compound and pyridine in the presence of alkali. The addition of acetic acid and sulfanilic–formic acid reagent produced a yellow–orange chromophore with an absorption maximum at 505 nm (Bhattacharjee *et al.*, 1991).

The United States Environmental Protection Agency (1990) reported a method for the determination of haloacetic acids, including trichloroacetic acid, in drinking-water, groundwater, raw water and water in any intermediate treatment stage. The method involves adjusting the pH to 11.5 and extraction with methyl-*tert*-butyl ether to remove neutral and basic organic compounds. The aqueous sample is then acidified to a pH of 0.5, and the acids are extracted into methyl-*tert*-butyl ether. The acids are then converted to their methyl esters with diazomethane. The methyl esters are determined by capillary GC with electron capture detection. The detection limit for this method is 0.085 µg/L.

1.2 Production and use

1.2.1 Production

Trichloroacetic acid is produced on an industrial scale by exhaustive chlorination of acetic or chloroacetic acid. Heavy metal salts have been used as chlorination catalysts (Koenig *et al.*,

1986). Trichloroacetic acid has also been manufactured by nitric acid oxidation of chloral (see monograph, this volume) and hydrolytic oxidation of tetrachloroethylene (see monograph, this volume) (Freiter, 1978).

Information on the global production and use of trichloroacetic acid are not available; however, most is converted to its sodium salt, and about 21 000–23 000 tonnes of this product are used annually as a selective herbicide (Koenig et al., 1986). Production in the Member States of the European Union was estimated to have been 750 tonnes in 1979 (Environmental Chemicals Data and Toxicology of Chemicals, 1993).

Trichloroacetic acid is produced by four companies in Japan, by three companies each in Germany and India, and by one company each in Brazil, China, Italy, Romania, the Russian Federation, Spain and the United Kingdom (Chemical Information Services, Inc., 1994).

1.2.2 Use

The main use for trichloroacetic acid is in the production of its sodium salt, which is used as a selective herbicide. The sodium salt is also used in formulations with 2,4-D and 2,4,5-T (see IARC, 1987a) as a broad-spectrum herbicide. Trichloroacetic acid has been used as an etching agent for surface treatment of metals, as a swelling agent and solvent in plastics and as an auxiliary in textile finishing (Koenig et al., 1986). It has been used as a decalcifier, a fixative in microscopy and to precipitate proteins (Budavari, 1989). It is used as a colour reagent in thin-layer and paper chromatography. The esters of trichloroacetic acid are important starting materials in organic synthesis (Neumüller, 1988). Because it is strongly corrosive, trichloroacetic acid has been used in medicine to remove genital warts, to treat extensive actinic keratosis of the face and scalp (Brodland & Roenigk, 1988; Boothby et al., 1990; Kling, 1992; Wang et al., 1992) and as an astringent and antiseptic (Heithersay & Wilson, 1988). In dentistry, trichloroacetic acid is recommended for treatment of external cervical root resorption (Heithersay & Wilson, 1988; Lewinstein & Rotstein, 1992).

1.3 Occurrence

1.3.1 Natural occurrence

Trichloroacetic acid is not known to occur as a natural product.

1.3.2 Occupational exposures

The National Occupational Exposure Survey conducted between 1981 and 1983 indicated that 35 124 employees in the United States of America were potentially exposed occupationally to trichloroacetic acid in seven industries and 1562 plants (United States National Institute for Occupational Safety and Health, 1994a).

Because trichloroacetic acid is the major end metabolite of trichloroethylene and tetrachloroethylene in humans (see pp. 112 and 186), it has been used as a biological marker of exposure to those compounds for many years. It is also a metabolite of 1,1,1-trichloroethane (see IARC, 1987b) and 1,1,2,2-tetrachloroethane (see IARC, 1987c); and chloral hydrate (see monograph this volume) is rapidly oxidized to trichloroacetic acid in humans. The levels of trichloroacetic acid reported in human blood and urine after occupational exposure to trichloroethylene, tetrachloroethylene or 1,1,1-trichloroethane are summarized in Table 1.

Table 1. Concentrations of trichloroacetic acid in human blood and urine

Job description (country)	Exposure	Concentration of trichloroacetic acid	Reference
Metal degreasing (Switzerland)	Trichloroethylene, 10–300 ppm [54–1611 mg/m^3]	57–980 mg/L (urine)	Boillat (1970)
Metal degreasing (USA)	Trichloroethylene, 170–420 mg/m^3	3–116 mg/g creatinine (urine)	Lowry et al. (1974)
Workshop (Japan)	Trichloroethylene, 3–175 ppm [16.1–940 mg/m^3]	9–297 mg/L (urine)	Ikeda et al. (1972)
	Tetrachloroethylene, 50–400 ppm [339–2712 mg/m^3]	Mean, 50 mg/L (urine)	
Printing factory (Japan)	1,1,1-Trichloroethane, 4.3–53.5 ppm [23–289 mg/m^3]	0.5–5.5 mg/L	Seki et al. (1975)
Workshop (Japan)	Trichloroethylene	Range of means, 108–133 mg/L (urine) (trichloroacetate)	Itoh (1989)
Automobile workshop (Japan)	Trichloroethylene, 1–50 ppm [5–269 mg/m^3]	Average, 136 mg/g creatinine (urine)	Ogata et al. (1987)
Dry cleaning, degreasing (former Yugoslavia)	Trichloroethylene	0.43–154.92 µmol/L (blood) [0.07–25.3 mg/L] 0.58–42.44 mmol/mol creatinine (urine) [0.84–61 mg/g]	Skender et al. (1988)
Solvent exposure (Republic of Korea)	Tetrachloroethylene, 0–61 ppm [0–414 mg/m^3]	0.6–3.5 mg/L (urine)	Jang et al. (1993)
Degreasing (Sweden)	Trichloroethylene, 3–114 mg/m^3	2–260 µmol/L (urine) [0.3–42.5 mg/L]	Ulander et al. (1992)
Dry cleaning (former Yugoslavia)	Trichloroethylene, 25–40 ppm [134–215 mg/m^3]	13.47–393.56 µmol/L (blood) [2.2–64 mg/L] 1.92–77.35 mmol/mol creatinine (urine) [2.8–112 mg/g]	Skender et al. (1991)
	Tetrachloroethylene, 33–53 ppm [224–359 mg/m^3]	1.71–20.93 µmol/L (blood) [0.3–3.4 mg/L] 0.81–15.76 mmol/mol creatinine (urine) [1.2–23 mg/g]	
Printing workshop (Japan)	1,1,1-Trichloroethane, 5–65 ppm [27–351]	2–5 mg/L (urine)	Kawai et al. (1991)
Printing and ceramics workshop (Germany)	Trichloroethylene, 5–70 ppm [26.9–376 mg/m^3]	2.0–201.0 mg/g creatinine (urine)	Triebig et al. (1982)

Table 1 (contd)

Job description (country)	Exposure	Concentration of trichloroacetic acid	Reference
Environmental levels (Japan)	Trichloroethylene: air: 1.7–26.9 µg/m³ water: 12–123 µg/L Tetrachloroethylene: air: 2.9–40 µg/m³ water: 2–68 µg/L	8.1–60.0 µg/L (plasma) 6.2–72.0 µg/g creatinine (urine)	Ziglio et al. (1985)

The average concentration of trichloroacetic acid in 177 urinary measurements made in 1986–88 at the Finnish Institute of Occupational Health was 77.1 µmol/L [12.6 mg/L], with a range of < 50–860 µmol/L [< 8.2–140.5 mg/L]. Seven samples exceeded the Finnish biological action level of 360 µmol/L [59 mg/L] (Rantala et al., 1992).

1.3.3 Air

No data were available to the Working Group.

1.3.4 Water

Trichloroacetic acid is produced as a by-product during aqueous chlorination of humic substances (Christman et al., 1983; Miller & Uden, 1983; Legube et al., 1985; Reckhow et al., 1990). Consequently, it may occur in drinking-water after chlorine disinfection of raw waters containing natural organic substances (Hargesheimer & Satchwill, 1989; see IARC, 1991). The concentrations of trichloroacetic acid measured in various water sources are summarized in Table 2. It has been identified as a major chlorinated by-product of the photocatalytic degradation of tetrachloroethylene in water but a minor by-product of the degradation of trichloroethylene (Glaze et al., 1993).

1.3.5 Food

Residues of trichloroacetic acid have been found in the seed of wheat, barley and oats after treatment with trichloroacetic acid as a postemergence herbicide (Kadis et al., 1972). Trace concentrations (0.01–0.19 ppm [mg/kg]) of trichloroacetic acid have been detected in vegetables and fruits irrigated with water containing trichloroacetic acid; slightly higher levels (0.13–0.43 mg/kg) were detected in field bean pods and seeds (Demint et al., 1975).

1.3.6 Other

Trichloroacetic acid concentrations were determined in the urine of people living in the vicinity of dry cleaning shops where tetrachloroethylene was used in Germany. The mean

Table 2. Concentrations of trichloroacetic acid in water

Water type (location)	Concentration range (µg/L)	Reference
Chlorinated tap-water (USA) (drinking-water)	33.6–161	Uden & Miller (1983)
Chlorinated drinking-water (USA)	4.23–53.8	Norwood et al. (1986)
Raw water (USA)	95–2120	
Chlorinated surface, reservoir, lake and groundwaters (USA)	4.0–6.0	Krasner et al. (1989)
Chlorinated surface water (USA)	7.4–22	Jacangelo et al. (1989)
Chlorinated drinking-water (USA)	15–64	Reckhow & Singer (1990)
Raw water (USA)	60–1630	
Chlorinated treated water (Australia)	Max 200	Nicholson et al. (1984)
Surface, ground- and drinking-waters	< 0.1–1.0	Artho et al. (1991)
Chlorinated water (Switzerland)	3.0	
Chlorinated tap-water (Germany)	Not detected–3	Lahl et al. (1984)
Chlorinated drinking-water (Japan)	7.5	Ozawa (1993)
Rainwater (Germany)	0.9	Clemens & Schöler (1992a)
Groundwater (Germany)	0.05	
Rainwater (Germany)	0.1–20	Plümacher & Renner (1993)
Swimming pool (Germany)	Indoor: 3.3–9.1 Open-air: 46.5–100.6[a]	Clemens & Schöler (1992b)
Irrigation water (USA)[b]	0–297	Comes et al. (1975)
Biologically treated kraft pulp mill effluent (Malaysia)	838–994	Mohamed et al. (1989)
Surface water (downstream from a paper mill) (Austria)	< 3–558	Geist et al. (1991)
Trichloroacetic acid process effluents (India)	0.41–4.5%	Husain et al. (1992)

[a] The higher levels found in open-air swimming pools may be due to input of organic material by swimmers.
[b] Trichloroacetic acid was added to the water from its use as a herbicide.

values were 105 µg/L for 29 neighbours and 682 µg/L for 12 workers (maximum, 1720 µg/L) (Popp et al., 1992). In Zagreb, Croatia, the levels of trichloroacetic acid in the urine of 39 people with no known exposure to solvents were 14–160 µg/L in plasma and 2–292 µg/24 h in urine (Skender et al., 1993). In 66 students in Japan, the corresponding levels ranged from not detected to 930 µg/g creatinine (urine) (Ikeda & Ohtsuji, 1969); those in 94 unexposed subjects in Germany were 5–221 µg/L in serum and 0.6–261 µg/24 h in urine (Hajimiragha et al., 1986).

Trichloroacetic acid was detected at levels of 10–150 µg/kg in spruce needles from the Black Forest in Germany and the Montafon region in Austria, both considered to be relatively non-polluted areas (Frank *et al.*, 1989; Frank, 1991). The concentrations of trichloroacetic acid in pine needles from an urban area in Germany were 0.7–175 μg/kg fresh weight (Plümacher & Renner, 1993); those in conifer needles in Finland were 3–126 μg/kg fresh weight (Frank *et al.*, 1992). In the vicinity of a pulp mill in Finland, concentrations of 2–135 μg/kg were found (Juuti *et al.*, 1993). Trichloroacetic acid was also detected in earthworms (at 150–400 μg/kg wet weight) from a contaminated forest site (Back & Süsser, 1992).

1.4 Regulations and guidelines

Occupational exposure limits for trichloroacetic acid have been recommended in many countries, either as a time-weighted average or as a short-term exposure limit. The following time-weighted averages have been set: Australia, 7 mg/m^3; Belgium, 6.7 mg/m^3; Denmark, 1 mg/m^3; France, 5 mg/m^3 (ILO, 1991); Switzerland, 7 mg/m^3 (Schweizerische Unfallversicherungsanstalt, 1994); United Kingdom, 5 mg/m^3 (United Kingdom Health and Safety Executive, 1993); and the United States, 6.7 mg/m^3 (American Conference of Governmental Industrial Hygienists, 1994) and 7 mg/m^3 (United States National Institute for Occupational Safety and Health, 1994b). In the Russian Federation, a short-term exposure limit of 5 mg/m^3 is recommended (ILO, 1991).

In the United States and Switzerland, levels of trichloroacetic acid in urine have been recommended as biological indices of exposure, at 7 mg/L at the end of the work week for exposure to tetrachloroethylene and 100 mg/g creatinine at the end of the work week for trichloroethylene. The American Conference of Governmental Industrial Hygienists (1994) noted that trichloroacetic acid in urine is a nonspecific determinant of exposure to trichloroethylene. In other countries where urinary trichloroacetic acid is the recommended biological marker for assessing exposure to trichloroethylene, there is an action level of 360 μmol/L (59 mg/L) in Finland (Aitio *et al.*, 1995); a biological tolerance value of 100 mg/L in Germany (Deutsche Forschungsgemeinschaft, 1993); and a biological tolerance value of 100 mg/g creatinine in Switzerland (Schweizerische Unfallversicherungsanstalt, 1994).

The United States Environmental Protection Agency (1994) proposed that the maximal level of haloacetic acids (the sum of the concentrations of mono-, di- and trichloroacetic acids and mono- and dibromoacetic acids) in drinking-water be 0.06 mg/L. WHO (1993) recommend a provisional guideline value of 100 μg/L for trichloroacetic acid in drinking-water.

2. Studies of Cancer in Humans

No studies were available on people exposed to trichloroacetic acid. In view of the fact that it is a metabolite of trichloroethylene and tetrachloroethylene, the results of studies on populations exposed to those compounds may be relevant (see pp. 95 and 176). In particular, urinary levels of trichloroacetic acid were measured in workers exposed to the parent compounds in two cohort studies (Axelson *et al.*, 1978, 1984 [abstract], 1994; Anttila *et al.*, 1995) and in one descriptive study (Vartiainen *et al.*, 1993).

3. Studies of Cancer in Experimental Animals[1]

3.1 Oral administration

Mouse: A group of 25 male B6C3F1 mice, four weeks of age, was given drinking-water containing 5 g/L trichloroacetic acid (purity, > 99%) neutralized with sodium hydroxide to a pH of 6.5–7.5. A control group of 27 mice received drinking-water containing 2 g/L sodium chloride. Both groups were kept for 61 weeks, at which time they were killed and necropsied. Two of 22 control mice and 8/22 treated mice had hepatic adenomas ($p < 0.01$, Fisher's exact test); and 0/22 controls and 7/22 treated mice had hepatocellular carcinomas ($p < 0.01$) (Herren-Freund et al., 1987). [The Working Group noted that the study also included groups of mice treated with 2.5 or 10 µg/kg bw N-nitrosoethylurea at 15 days of age followed four weeks later by treatment with drinking-water containing 2 or 5 g/L trichloroacetic acid; no promoting effect was demonstrated. The results of this experiment are not included, since phenobarbital, used as a positive control, was inactive.]

Groups of male and female B6C3F1 mice, 37 days old, received either 1.0 or 2.0 g/L trichloroacetic acid (neutralized to pH 6.8–7.2 with 10 N sodium hydroxide) in the drinking-water for up to 52 weeks, at which time the study was terminated. A group of 11 male mice received 1 g/L trichloroacetic acid for 52 weeks; 24 male mice received 2.0 g/L trichloroacetic acid for 52 weeks and a further 11 males received 2.0 g/L trichloroacetic acid for 37 weeks and then water alone until week 52. The last group was included in order to assess the reversibility of any hepatic effects. Two groups of 35 and 11 male control mice were kept for 52 weeks. Groups of 10 female mice received either 0 or 2.0 g/L trichloroacetic acid for 52 weeks. The study also included groups of five male mice given 2.0 g/L trichloroacetic acid and sacrificed at 15, 24 or 37 weeks, and the corresponding controls given water alone. Livers and kidneys were weighed and examined for macroscopic lesions. Microscopic examination was undertaken only of lesions of the liver found in the 35 male controls, the 11 males treated with 2.0 g/L trichloroacetic acid for 37 weeks and other groups [numbers unspecified] chosen at random. The lesions were classified histologically as hyperplastic nodules, adenomas or hepatocellular carcinomas. The results in male mice at 52 weeks are summarized in Table 3. No hyperplastic nodules or neoplastic lesions were seen at week 52 in the females receiving 2.0 g/L trichloroacetic acid. Multifocal necrosis was seen in both males and females (Bull et al., 1990). [The Working Group noted the lack of detailed reporting of the results.]

3.2 Mid-term assays

Rat: The potential initiation–promotion activity of trichloroacetic acid in the liver was investigated in two sets of mid-term studies in rats, in which enzyme-altered foci were the endpoint. Use of a standard initiation protocol followed by promotion with phenobarbital did not result in initiation of γ-glutamyl transpeptidase-positive foci, but promotion with trichloroacetic acid at doses of 50–5000 mg/L in drinking-water for 12 months after initiation with N-

---

[1] The Working Group was aware of studies in progress in male rats treated by gavage and in mice administered the compound in the drinking-water (IARC, 1994).

nitrosodiethylamine increased the size of the foci (Parnell et al., 1988). In similar study, trichloroacetic acid did not promote the formation of glutathione S-transferase placental form-positive foci (Hasegawa & Ito, 1992).

Table 3. Lesions in the livers of male B6C3F1 mice given trichloroacetic acid in the drinking-water

Dose (g/L) × no. of weeks	No. of mice	No. with macroscopic lesions	Total no. of lesions	No. of macroscopic lesions examined histologically (no. of mice)	No. of hyperplastic nodules (no. of mice)	No. of hepatic adenomas (no. of mice)	No. of hepatocellular carcinomas (no. of mice)
Control (water alone) × 52	35	2	2	2 (2)	1 (1)	0	0
1 × 52	11	5	7	7 (5)	3 (1)	2 (2)	2 (2)
2 × 52	24	19	30	16 (11)	10 (9)	1 (1)	4 (4)
2 × 37	11	4	5	5 (4)	2 (2)	0	3 (3)

From Bull et al. (1990)

4. Other Data Relevant to an Evaluation of Carcinogenicity and its Mechanisms

4.1 Absorption, distribution, metabolism and excretion

4.1.1 Humans

After administration of a single oral dose of 3 mg/kg bw trichloroacetic acid to healthy volunteers, the mean plasma half-life of the compound was about 50 h and the volume of distribution was about 115 ml/kg bw (Müller et al., 1974). The long plasma-half life of trichloroacetic acid in human subjects is consistent with its extensive binding to plasma proteins (Sellers & Koch-Weser, 1971).

4.1.2 Experimental systems

As trichloroacetic acid is a metabolite of trichloroethylene and tetrachloroethylene in both humans and experimental animals, its toxicokinetics are also described in the monographs on those compounds.

After administration of trichloroacetic acid as single doses of 5, 20 or 100 mg/kg bw in water to male Fischer 344 rats and male B6C3F1 mice, the total amount of radiolabel excreted in the urine of animals of each species was 60–70% of the administered dose, about 60% of which appeared to be unchanged trichloroacetic acid. The plasma half-life of trichloroacetic acid in rodents was about 6 h, and the concentration in blood over time was similar in the two species (Larson & Bull, 1992a,b). The volume of distribution was 365–485 ml/kg bw in rats and 335–555 ml/kg bw in mice.

The major urinary products in mice and rats after administration of trichloroacetic acid by gavage are trichloroacetic acid itself and oxalic and thiodiacetic acids (Larson & Bull, 1992a). Reductive dechlorination of trichloroacetic acid yields dichloroacetic acid (see monograph, this volume), which is metabolized by two main routes: ultimately to oxalate and carbon dioxide, probably via glucolate and glyoxylate as intermediate metabolites; or to monochloroacetate after reductive dechlorination, followed by glutathione conjugation of monochloroacetate to thiodiacetic acid (Crabb & Harris, 1979; Stacpoole *et al.*, 1990; Larson & Bull, 1992a).

4.1.3 Comparison of humans and animals

Biotransformation of trichloroacetic acid occurs by similar routes in humans and rodents; however, the apparent volume of distribution appears to be smaller in humans than in rats or mice, perhaps because of the extensive binding of trichloroacetic acid to human plasma proteins. This effect may also account for the slower rate of systemic clearance in humans than in rodents.

4.2 Toxic effects

4.2.1 Humans

Little is known about the toxicity of trichloroacetic acid, although it is corrosive to the skin and eyes.

4.2.2 Experimental systems

Since trichloroacetic acid undergoes reductive metabolism to free radicals, its ability to elicit a lipid peroxidative response in liver was investigated in male Fischer 344 rats and male B6C3F1 mice after administration of a single oral dose of 100–2000 mg/kg bw in water. A dose-dependent response was induced, which was more potent in mice than in rats. Significant increases in the formation of thiobarbituric acid reactive substances were observed at a dose of 300 mg/kg bw in mice but only at a dose of 1000 mg/kg bw in rats (Larson & Bull, 1992a).

More total radioactivity was associated with albumin and haemoglobin in male Fischer 344 rats than in male B6C3F1 mice 4–120 h after a single oral dose of 20 mg/kg bw ^{14}C-trichloroacetic acid. In contrast, incorporation of radiolabelled amino acids resulting from the metabolism of the chloroacetate (glycine from glyoxylic acid) was more extensive in mice; this result is consistent with the more extensive metabolism in this species (Stevens *et al.*, 1992).

Exposure of male and female Sprague-Dawley rats to trichloroacetic acid at 24 or 240 mg/kg bw per day in the drinking-water for 14 days resulted in reduced weight gain only in the group given the higher dose (Davis, 1986). Male Sprague-Dawley rats administered trichloroacetic acid in the drinking-water for 90 days, at concentrations providing daily doses of about 4, 35 or 350 mg/kg bw, had decreased body weights. Animals given the high dose also had increased ratios of liver and kidney weight to body weight and increased hepatic peroxisomal activity (Mather *et al.*, 1990). Male Sprague-Dawley rats were given trichloroacetic acid in the drinking-water at a concentration of 45.8 mmol/L [7.48 g/L] to provide an approximate intake of 785 mg/kg bw per day. After 90 days, body weights were decreased, and minimal histopathological changes were seen in the liver and lung (Bhat *et al.*, 1991).

Trichloroacetic acid, dichloroacetic acid and chloroform are formed during the chlorination of drinking-water (see IARC, 1991). After concomitant administration of trichloroacetic acid (at 0.92 and 2.45 mmol [150 and 400 mg]/kg bw by gavage, three times over 24 h) and chloroform (one intraperitoneal injection of 0.75 mg/kg bw once after the last dose of trichloroacetic acid) to male and female Sprague-Dawley rats, the renal toxicity of chloroform was increased (Davis, 1992).

The main effect in male and female B6C3F1 mice of exposure to trichloroacetic acid at 1000 and 2000 mg/L in drinking-water for up to 52 weeks was marked accumulation of lipid and lipofuscin in liver cells, in the virtual absence of any significant increase in cell size or in the onset of focal necrotic areas and with only marginal induction of cell proliferation and organ hypertrophy (Bull *et al.*, 1990; Sanchez & Bull, 1990; Bull *et al.*, 1993). In another study, however, a threefold increase in hepatic DNA synthesis was seen in male and female B6C3F1 mice exposed to trichloroacetic acid at 100–1000 mg/kg bw per day in corn oil by gavage for 11 days (Dees & Travis, 1994). Similarly, administration of trichloroacetic acid at 5 g/L in drinking-water for seven days resulted in an almost fourfold increase in the labelling index in livers of male B6C3F1 mice, whereas hepatic DNA synthesis was reduced to 10% of the control values in male Fischer 344 rats. Administration of trichloroacetic acid produced similar induction of peroxisomal palmitoyl-coenzyme A oxidation in rats and mice (Watson *et al.*, 1995).

Induction of peroxisome proliferation has been repeatedly associated with the chronic toxicity and carcinogenicity of trichloroacetic acid to the liver (DeAngelo *et al.*, 1989). It can induce peroxisome proliferation in the livers of both mice and rats, as indicated by increased activities of palmitoyl-coenzyme A oxidase and carnitine acetyl transferase, the appearance of a peroxisome proliferation-associated protein and increased volume-density of peroxisomes after exposure to trichloroacetic acid for 14 days. The compound induced peroxisome proliferation in mouse but not in rat kidney (Goldsworthy & Popp, 1987), and it induced peroxisome proliferation in primary cultures of hepatocytes from rats and mice but not in those from humans (Elcombe, 1985; Herren-Freund *et al.*, 1987).

4.3 Reproductive and prenatal effects

4.3.1 Humans

No data were available to the Working Group.

4.3.2 Experimental systems

Treatement of pregnant rats with trichloroacetic acid at 0, 330, 800, 1200 or 1800 mg/kg bw per day by gavage on gestation days 6–15 resulted in a dose-dependent increase in the frequency of resorptions per litter, a reduction in fetal weight and length and an increased frequency of anomalies of the heart and eyes. Maternal toxicity, as evidenced by decreased weight, was observed at all doses (Smith *et al.*, 1989).

4.4 Genetic and related effects

4.4.1 Humans

No data were available to the Working Group.

4.4.2 Experimental systems (see also Table 4 and Appendices 1 and 2)

Trichloroacetic acid did not induce λ prophage in *Escherichia coli* and was not mutagenic to *Salmonella typhimurium* strains in the presence or absence of metabolic activation. Trichloroacetic acid, however, reacts with dimethyl sulfoxide (a solvent used commonly in this assay) to form unstable, mutagenic substances, which have not been identified (Nestmann *et al.*, 1980).

The frequency of chlorophyll mutations was increased in *Arabidopsis* after treatment of seeds with trichloroacetic acid.

It did not induce DNA strand breaks in mammalian cells *in vitro*. Gap-junctional intercellular communication was observed in mouse hepatocytes *in vitro*. Chromosomal aberrations were not induced in human lymphocytes exposed *in vitro* to trichloroacetic acid neutralized to avoid the effects of low pH seen in cultured mammalian cells.

DNA strand breaks were reported in one laboratory in the livers of mice and rats treated 4 h previously with trichloroacetate; none were observed 24 h after repeated daily dosing with 500 mg/kg bw (Nelson & Bull, 1988; Nelson *et al.*, 1989). Peroxisome proliferation, as indicated by β-oxidation of palmitoyl-coenzyme A, was observed only after induction of DNA damage (Nelson *et al.*, 1989). DNA strand breakage was not observed in the livers of mice or rats (Chang *et al.*, 1992). [The reasons for the contrasting results obtained using similar techniques are unclear.]

In one study, trichloroacetic acid induced micronuclei and chromosomal aberrations in bone-marrow cells and abnormal sperm morphology after injection into Swiss mice *in vivo*. In another study, in which a 10-fold higher dose was injected into C57BL/JfBL/Alpk mice, no micronucleus induction was observed.

Mutation of proto-oncogenes in tumours induced by trichloroacetic acid

The expression of c-*myc* and c-H-*ras* in mRNA was studied by in-situ hybridization in the livers of male B6C3F1 mice treated with trichloroacetate at 1 or 2 g/L in the drinking-water for 52 weeks. Expression of c-*myc*, corrected for the background frequency, was increased by about three times in hepatic hyperplastic nodules and by almost six times in hepatic carcinomas in comparison with normal liver. A similar comparison of c-H-*ras* expression showed no significant increase in hyperplastic nodules but an approximately fourfold increase in hepatic carcinomas (Nelson *et al.*, 1990). [The Working Group considered that the changes in proto-oncogene expression could not be attributed conclusively to trichloroacetic acid.]

Table 4. Genetic and related effects of trichloroacetic acid

Test system	Result[a]		Dose[b] (LED/HID)	Reference
	Without exogenous metabolic system	With exogenous metabolic activation		
PRB, λ Prophage induction, *Escherichia coli* WP2s	–	–	10 000	DeMarini *et al.* (1994)
BSD, *Bacillus subtilis* H17 rec⁺ and M45 rec⁻	–	0	20	Shirasu *et al.* (1976)
ECR, *Escherichia coli*, Br/try WP2, reverse mutation	–	0	20	Shirasu *et al.* (1976)
SA0, *Salmonella typhimurium* TA100, reverse mutation	–	–	225	Waskell (1978)
SA0, *Salmonella typhimurium* TA100, reverse mutation	–	0	2000	Nestmann *et al.* (1980)
SA0, *Salmonella typhimurium* TA100, reverse mutation	–	0	500	Rapson *et al.* (1980)
SA0, *Salmonella typhimurium* TA100, reverse mutation	–	–	0.00	Moriya *et al.* (1983)
SA0, *Salmonella typhimurium* TA100, reverse mutation	–	–	4	DeMarini *et al.* (1994)
SA5, *Salmonella typhimurium* TA1535, reverse mutation	–	0	20	Shirasu *et al.* (1976)
SA5, *Salmonella typhimurium* TA1535, reverse mutation	–	0	2000	Nestmann *et al.* (1980)
SA7, *Salmonella typhimurium* TA1537, reverse mutation	–	0	20	Shirasu *et al.* (1976)
SA7, *Salmonella typhimurium* TA1537, reverse mutation	–	0	1000	Nestmann *et al.* (1980)
SA8, *Salmonella typhimurium* TA1538, reverse mutation	–	0	20	Shirasu *et al.* (1976)
SA8, *Salmonella typhimurium* TA1538, reverse mutation	–	–	1000	Nestmann *et al.* (1980)
SA9, *Salmonella typhimurium* TA98, reverse mutation	–	–	225	Waskell (1978)
SA9, *Salmonella typhimurium* TA98, reverse mutation	–	–	1000	Nestmann *et al.* (1980)
SA9, *Salmonella typhimurium* TA98, reverse mutation	–	–	0.00	Moriya *et al.* (1983)
SAS, *Salmonella typhimurium* TA1536, reverse mutation	–	0	20	Shirasu *et al.* (1976)
ASM, *Arabidopsis* species, mutation	+	0	1000	Plotnikov & Petrov (1976)
DIA, DNA strand breaks, B6C3F1 mouse hepatocytes *in vitro*	–	0	1630	Chang *et al.* (1992)
DIA, DNA strand breaks, Fischer 344 rat hepatocytes *in vitro*	–	0	1630	Chang *et al.* (1992)
DIH, DNA strand breaks, human CCRF-CEM cells *in vitro*	–	0	1630	Chang *et al.* (1992)
ICR, Inhibition of intercellular communication, B6C3F1 mouse hepatocytes *in vitro*	+	0	16.3	Klaunig *et al.* (1990)
CHL, Chromosomal aberrations, human lymphocytes *in vitro*	–	[c]	5000	Mackay *et al.* (1995)
DVA, DNA strand breaks, B6C3F1 mouse hepatic cells *in vivo*	+		1.0 po × 1	Nelson & Bull (1988)
DVA, DNA strand breaks, B6C3F1 mouse hepatic cells *in vivo*	+		500 po × 1	Nelson *et al.* (1989)

Table 4 (contd)

Test system	Result[a]		Dose[b]	Reference
	Without exogenous metabolic system	With exogenous metabolic activation	(LED/HID)	
DVA, DNA strand breaks, B6C3F1 mouse hepatic cells *in vivo*	–		500 po × 10	Nelson *et al.* (1989)
DVA, DNA strand breaks, B6C3F1 mouse hepatic cells and epithelial cells from stomach and duodenum *in vivo*	–		1630 po × 1	Chang *et al.* (1992)
DVA, DNA strand breaks, Sprague-Dawley rat hepatic cells *in vivo*	+		100 po × 1	Nelson & Bull (1988)
DVA, DNA strand breaks, Fischer 344 rat hepatic cells *in vivo*	–		1630 po × 1	Chang *et al.* (1992)
MVM, Micronucleus induction, Swiss mice *in vivo*	+		125 ip × 2	Bhunya & Behera (1987)
MVM, Micronucleus induction, C57Bl/JfBl 10/Alpk female mice	–		1300 ip × 2	Mackay *et al.* (1995)
MVM, Micronucleus induction, C57Bl/JfBl 10/Alpk male mice	–		1080 ip × 2	Mackay *et al.* (1995)
CBA, Chromosomal aberrations, Swiss mouse bone-marrow cells *in vivo*	+		125 ip × 1	Bhunya & Behera (1987)
CBA, Chromosomal aberrations, Swiss mouse bone-marrow cells *in vivo*	+		100 ip × 5	Bhunya & Behera (1987)
CBA, Chromosomal aberrations, Swiss mouse bone-marrow cells *in vivo*[c]	+		500 po × 1	Bhunya & Behera (1987)
SPM, Sperm morphology, Swiss mice *in vivo*	+		125 ip × 5	Bhunya & Behera (1987)

[a] +, considered to be positive; –, considered to be negative; 0, not tested
[b] LED, lowest effective dose; HID, highest ineffective dose; in-vitro tests, mg/ml; in-vivo tests, mg/kg bw; 0.00, dose not reported; ip, intraperitoneally; po, orally
[c] Neutralized trichloroacetic acid

5. Summary and Evaluation

5.1 Exposure data

Trichloroacetic acid is produced commercially in small amounts by chlorination of acetic or chloroacetic acid. It is used principally in the form of the sodium salt, as a herbicide. Most human exposure to trichloroacetic acid occurs because of its metabolic formation from tetrachloroethylene, trichloroethylene, 1,1,1-trichloroethane, 1,1,2,2-tetrachloroethane and chloral hydrate. Trichloroacetic acid can also be formed during the chlorination of drinking-water.

5.2 Human carcinogenicity data

The available data were too limited to form the basis for an evaluation of the carcinogenicity of trichloroacetic acid to humans.

5.3 Animal carcinogenicity data

Trichloroacetic acid was tested by oral administration in the drinking-water in two studies in males of one strain of mice. In both studies, the incidence of hepatocellular adenomas and carcinomas was increased.

5.4 Other relevant data

Trichloroacetic acid has a longer plasma half-life in humans than in rodents, presumably because there is more binding to plasma proteins in humans. Much of an administered dose of trichloroacetic acid is excreted unchanged in the urine of rats and mice. Reductive dechlorination and glutathione conjugation are involved in the formation of the urinary metabolites, oxalate and thiodiacetic acid.

Little is known about the toxicity of this compound to humans. Single doses of high concentrations of trichloroacetic acid induce lipid peroxidation in the livers of rats and mice. Trichloroacetic acid causes hepatic peroxisome proliferation in both rats and mice *in vivo* and in cultured hepatocytes from mice and rats, but not from humans. Short-term, repeated administrations of trichloroacetic acid induced cell proliferation in the livers of mice but reduced cell proliferation in the livers of rats.

No data were available on the effects of trichloroacetic acid on human reproduction. In rats, fetotoxicity was observed at doses that are maternally toxic.

Trichloroacetic acid induced chromosomal aberrations and abnormal sperm in mice in one study. The results of studies on the induction of DNA strand breaks and micronuclei were inconclusive.

Trichloroacetic acid did not induce chromosomal aberrations in a single study or DNA strand breaks in cultured mammalian cells. Inhibition of intercellular communication has been reported. It was not mutagenic to bacteria.

5.5 Evaluation[1]

There is *inadequate evidence* in humans for the carcinogenicity of trichloroacetic acid.

There is *limited evidence* in experimental animals for the carcinogenicity of trichloroacetic acid.

Overall evaluation

Trichloroacetic acid *is not classifiable as to its carcinogenicity to humans (Group 3)*.

6. References

Aitio, A., Luotamo, M. & Kiilunen, M., eds (1995) *Kemikaalialtistumisen Biomonitorointi* [Biological monitoring of exposure to chemicals], Helsinki, Finnish Institute of Occupational Health, pp. 284–286 (in Finnish)

Aldrich Chemical Co. (1994a) *Material Safety Data Sheet: Trichloroacetic Acid, 98%*, Milwaukee, WI

Aldrich Chemical Co. (1994b) *Material Safety Data Sheet: Trichloroacetic Acid, 99+%, ACS*, Milwaukee, WI

American Conference of Governmental Industrial Hygienists (1994) *1994–1995 Threshold Limit Values for Chemical Substances and Physical Agents and Biological Exposure Indices*, Cincinnati, OH, pp. 34, 61–62

Anttilla, A., Pukkala, E., Sallmén, M., Hernberg, S. & Hemminki, K. (1995) Cancer incidence among Finnish workers exposed to halogenated hydrocarbons. *J. occup. Med.* (in press)

Artho, A., Grob, K. & Giger, P. (1991) Trichloroacetic acid in surface, ground- and drinking-waters. *Mitt. geb. Lebensm. Hyg.*, **82**, 487–491 (in German)

Axelson, O., Andersson, K., Hogstedt, C., Holmberg, B., Molina, G. & de Verdier, A. (1978) A cohort study on trichloroethylene exposure and cancer mortality. *J. occup. Med.*, **20**, 194–196

Axelson, O., Andersson, K., Seldén, A. & Hogstedt, C. (1984) Cancer morbidity and exposure to trichloroethylene (Abstract). *Arb. Hälsa*, **29**, 126

Axelson, O., Seldén, A., Andersson, K. & Hogstedt, C. (1994) Updated and expanded Swedish cohort study on trichloroethylene and cancer risk. *J. occup. Med.*, **36**, 556–562

Back, H & Süsser, P. (1992) Concentrations of volatile chlorinated hydrocarbons and trichloroacetic acid in earthworms. *Soil Biol. Biochem.*, **24**, 1745–1748

Bhat, H.K., Kanz, M.F., Campbell, G.A. & Ansari, G.A.S. (1991) Ninety day toxicity study of chloroacetic acids in rats. *Fundam. appl. Toxicol.*, **17**, 240–253

Bhattacharjee, M., Cherian, L. & Gupta, V.K. (1991) Modified Fujiwara reaction for the determination of trichloroacetic acid. *Microchem. J.*, **43**, 109–111

Bhunya, S.P. & Behera, B.C. (1987) Relative genotoxicity of trichloroacetic acid (TCA) as revealed by different cytogenetic assays: bone marrow chromosome aberration, micronucleus and sperm-head abnormality in the mouse. *Mutat. Res.*, **188**, 215–221

---

[1] For definition of the italicized terms, see Preamble, pp. 22–26.

Boillat, M.-A. (1970) Usefulness of the determination of trichloroacetic acid in urine for detecting poisoning due to trichloroethylene and perchloroethylene. Study of several firms. *Z. Präventivmed.*, **15**, 447–456 (in French)

Boothby, R.A., Carlson, J.A., Rubin, M., Morgan, M. & Mikuta, J.J. (1990) Single application treatment of human papillomavirus infection of the cervix and vagina with trichloroacetic acid: a randomised trial. *Obstet. Gynecol.*, **76**, 278–280

Brodland, D.G. & Roenigk, R.K. (1988) Trichloroacetic acid chemexfoliation (chemical peel) for extensive premalignant actinic damage of the face and scalp. *Mayo Clin. Proc.*, **63**, 887–896

Budavari, S., ed. (1989) *The Merck Index*, 11th Ed., Rahway, NJ, Merck & Co., p. 1515

Bull, R.J., Sanchez, I.M., Nelson, M.A., Larson, J.L. & Lansing, A.J. (1990) Liver tumor induction in B6C3F1 mice by dichloroacetate and trichloroacetate. *Toxicology*, **63**, 341–359

Bull, R.J., Templin, M., Larson, J.L. & Stevens, D.K. (1993) The role of dichloroacetate in the hepatocarcinogenicity of trichloroethylene. *Toxicol. Lett.*, **68**, 203–211

Chang, L.W., Daniel, F.B. & DeAngelo, A.B. (1992) Analysis of DNA strand breaks induced in rodent liver *in vivo*, hepatocytes in primary culture, and a human cell line by chlorinated acetic acids and chlorinated acetaldehydes. *Environ. mol. Mutag.*, **20**, 277–288

Chemical Information Services, Inc. (1994) *Directory of World Chemical Producers 1995/96 Standard Edition*, Dallas, TX, p. 677

Christensen, J.M., Rasmussen, K. & Køppen, B. (1988) Automatic headspace gas chromatographic method for the simultaneous determination of trichloroethylene and metabolites in blood and urine. *J. Chromatogr.*, **442**, 317–323

Christman, R.F., Norwood, D.L., Millington, D.S., Johnson, J.D. & Stevens, A.A. (1983) Identity and yields of major halogenated products of aquatic fulvic acid chlorination. *Environ. Sci. Technol.*, **17**, 625–628

Clemens, M. & Schöler, H.F. (1992a) Determination of halogenated acetic acids and 2,2-dichloropropionic acid in water samples. *Fresenius J. anal. Chem.*, **344**, 47–49

Clemens, M. & Schöler, H.F. (1992b) Halogenated organic compounds in swimming-pool waters. *Zbl. Hyg.*, **193**, 91–98 (in German)

Comes, R.D., Frank, P.A. & Demint, R.J. (1975) TCA in irrigation water after bank treatments for weed control. *Weed Sci.*, **23**, 207–210

Crabb, D.W. & Harris, R.A. (1979) Mechanism responsible for the hypoglycemic actions of dichloroacetate and 2-chloropropionate. *Arch. Biochem. Biophys.*, **198**, 439–452

Davis, M.E. (1986) Effect of chloroacetic acids on the kidneys. *Environ. Health Perspectives*, **69**, 209–214

Davis, M.E. (1992) Dichloroacetic acid and trichloroacetic acid increase chloroform toxicity. *J. Toxicol. environ. Health*, **37**, 139–148

DeAngelo, A.B., Daniel, F.B., McMillan, L., Wernsing, P. & Savage, R.E., Jr (1989) Species and strain sensitivity to the induction of peroxisome proliferation by chloroacetic acids. *Toxicol. appl. Pharmacol.*, **101**, 285–298

Dees, C. & Travis, C. (1994) Trichloroacetate stimulation of liver DNA synthesis in male and female mice. *Toxicol. Lett.*, **70**, 343–355

DeMarini, D.M., Perry, E. & Shelton, M.L. (1994) Dichloroacetic acid and related compounds: induction of prophage in *E. coli* and mutagenicity and mutation spectra in Salmonella TA100. *Mutagenesis*, **9**, 429–437

Demint, R.J., Pringle, J.C., Jr, Hattrup, A., Bruns, V.F. & Frank, P.A. (1975) Residues in crops irrigated with water containing trichloroacetic acid. *J. agric. Food Chem.*, **23**, 81–84

Deutsche Forschungsgemeinschaft (1993) *MAK- und BAT-Werte Liste* [MAK and BAT List], Weinheim, VCH Verlagsgesellschaft, p. 129 (in German)

Elcombe, C.R. (1985) Young Scientists Award lecture 1984: Species differences in carcinogenicity and peroxisome proliferation due to trichloroethylene: a biochemical human hazard assessment. *Arch. Toxicol.*, **Suppl. 8**, 6–17

Environmental Chemicals Data and Information Network (1993) *Trichloroacetic acid*, Ispra, JRC-CEC, last update: 02.09.1993

Frank, H. (1991) Airborne chlorocarbons, photooxidants, and forest decline. *Ambio*, **20**, 13–18

Frank, H., Vital, J. & Frank, W. (1989) Oxidation of airborne C_2 chlorocarbons to trichloroacetic and dichloroacetic acid. *Fresenius Z. anal. Chem.*, **333**, 713

Frank, H., Scholl, H., Sutinen, S. & Norokorpi, Y. (1992) Trichloroacetic acid in conifer needles in Finland. *Ann. bot. fenn.*, **29**, 263–267

Freiter, E.R. (1978) Halogenated acetic acid derivatives. In: Mark, H.F., Othmer, D.F., Overberger, C.G., Seaborg, G.T. & Grayson, N., eds, *Kirk-Othmer Encyclopedia of Chemical Technology*, 3rd Ed., Vol. 1, New York, John Wiley & Sons, pp. 171–178

Geist, S., Lesemann, C., Schütz, C., Seif, P. & Frank, E. (1991) Determination of chloroacetic acids in surface water. In: Angeletti, G. & Bjørseth, A., eds, *Organic Micropollutants in the Aquatic Environment*, Dordrecht, Kluwer Academic Publishers, pp. 393–397

Glaze, W.H., Kenneke, J.F. & Ferry, J.L. (1993) Chlorinated by-products from the TiO_2-mediated photodegradation of trichloroethylene and tetrachloroethylene in water. *Environ. Sci. Technol.*, **27**, 177–184

Goldsworthy, T.L. & Popp, J.A. (1987) Chlorinated hydrocarbon-induced peroxisomal enzyme activity in relation to species and organ carcinogenicity. *Toxicol. appl. Pharmacol.*, **88**, 225–233

Hajimiragha, H., Ewers, U., Jansen-Rosseck, R. & Brockhaus, A. (1986) Human exposure to volatile halogenated hydrocarbons from the general environment. *Int. Arch. occup. environ. Health*, **58**, 141–150

Hansch, C., Leo, A. & Hoekman, D.H. (1995) *Exploring QSAR*, Washington DC, American Chemical Society

Hargesheimer, E.E. & Satchwill, T. (1989) Gas-chromatographic analysis of chlorinated acids in drinking water. *J. Water Sewage Res. Treat. Aqua*, **38**, 345–351

Hasegawa, R. & Ito, N. (1992) Liver medium-term bioassay in rats for screening of carcinogens and modifying factors in hepatocarcinogenesis. *Food chem. Toxicol.*, **30**, 979–992

Heithersay, G.S. & Wilson, D.F. (1988) Tissue responses in the rat to trichloroacetic acid—an agent used in the treatment of invasive cervical resorption. *Austr. dent. J.*, **33**, 451–461

Herren-Freund, S.L., Pereira, M.A., Khoury, M.D. & Olson, G. (1987) The carcinogenicity of trichloroethylene and its metabolites, trichloroacetic acid and dichloroacetic acid, in mouse liver. *Toxicol. appl. Pharmacol.*, **90**, 183–189

Hoechst Chemicals (1990) *Chemical Information Sheet: Trichloroacetic Acid*, Dallas, TX

Husain, S., Narsimha, R., Alvi, S.N., & Nageswara Rao, R. (1992) Monitoring the effluents of the trichloroacetic acid process by high-performance liquid chromatography. *J. Chromatogr.*, **600**, 316–319

Husain, S., Narsimha, R., Alvi, S.N. & Nageswara Rao, R. (1993) Monitoring the levels of chloroacetic acids in the industrial environment by high-performance liquid chromatography. *J. high Resolut. Chromatogr.*, **16**, 381–383

IARC (1987a) *IARC Monographs on the Evaluation of Carcinogenic Risks to Humans*, Suppl. 7, *Overall Evaluations of Carcinogenicity: An Updating of* IARC Monographs *Volumes 1–42*, Lyon, pp. 156–160

IARC (1987b) *IARC Monographs on the Evaluation of Carcinogenic Risks to Humans*, Suppl. 7, *Overall Evaluations of Carcinogenicity: An Updating of* IARC Monographs *Volumes 1–42*, Lyon, p. 73

IARC (1987c) *IARC Monographs on the Evaluation of Carcinogenic Risks to Humans*, Suppl. 7, *Overall Evaluations of Carcinogenicity: An Updating of* IARC Monographs *Volumes 1–42*, Lyon, pp. 354–355

IARC (1991) *IARC Monographs on the Evaluation of the Carcinogenic Risks to Humans*, Vol. 52, *Chlorinated Drinking-water; Chlorination By-products; Some Other Halogenated Compounds; Cobalt and Cobalt Compounds*, Lyon, pp. 55–141

IARC (1994) *Directory of Agents Being Tested for Carcinogenicity*, No. 16, Lyon

Ikeda, M. & Ohtsuji, H. (1969) Hippuric acid, phenol, and trichloroacetic acid levels in the urine of Japanese subjects with no known exposure to organic solvents. *Br. J. ind. Med.*, **26**, 126–164

Ikeda, M., Ohtsuji, H., Imamura, T. & Komoike, Y. (1972) Urinary excretion of total trichloro-compounds, trichloroethanol, and trichloroacetic acid as a measure of exposure to trichloroethylene and tetrachloroethylene. *Br. J. ind. Med.*, **29**, 328–333

ILO (1991) *Occupational Exposure Limits for Airborne Toxic Substances: Values of Selected Countries* (Occupational Safety and Health Series No. 37), 3rd Ed., Geneva, pp. 394–395

Italia, M.P. & Uden, P.C. (1988) Multiple element emission spectral detection gas chromatographic profiles of halogenated products from chlorination of humic acid and drinking water. *J. Chromatogr.*, **438**, 35–43

Itoh, H. (1989) Determination of trichloroacetate in human serum and urine by ion chromatography. *Analyst*, **114**, 1637–1640

Jacangelo, J.G., Patania, N.L., Reagan, K.M., Aieta, E.M., Krasner, S.W. & McGuire, M.J. (1989) Ozonation: assessing its role in the formation and control of disinfection by-products. *J. Am. Water Works Assoc.*, **81**, 74–84

Jang, J.-Y., Kang, S.-K. & Chung, H.K. (1993) Biological exposure indices of organic solvents for Korean workers. *Int. Arch. occup. environ. Health*, **65**, S219–S222

Juuti, S., Hirvonen, A., Tarhanen, J., Holopainen, J.K. & Ruuskanen, J. (1993) Trichloroacetic acid in pine needles in the vicinity of a pulp mill. *Chemosphere*, **26**, 1859–1868

Kadis, V.W., Yarish, W., Molberg, E.S. & Smith, A.E. (1972) Trichloroacetic acid residues in cereals and flax. *Can. J. Plant Sci.*, **52**, 674–676

Kawai, T., Yamaoka, K., Uchida, Y. & Ikeda, M. (1991) Exposure to 1,1,1-trichloroethane and dose-related excretion of metabolites in urine of printing workers. *Toxicol. Lett.*, **55**, 39–45

Klaunig, J.E., Ruch, R.J., Hampton, J.A., Weghorst, C.M. & Hartnett, J.A. (1990) Gap-junctional intercellular communication and murine hepatic carcinogenesis. *Prog. clin. biol. Res.*, **331**, 277–291

Kling, A.R. (1992) Genital warts—therapy. *Semin. Dermatol.*, **11**, 247–255

Koenig, G., Lohmar, E. & Rupprich, N. (1986) Chloroacetic acids. In: Gerhartz, W., Yamamoto, Y.S., Campbell, F.T., Pfefferkorn, R. & Rounsaville, J.F., eds, *Ullmann's Encyclopedia of Industrial Chemistry*, 5th rev. Ed., Vol. A6, New York, VCH Publishers, pp. 537–552

Køppen, B., Dalgaard, L. & Christensen, J.M. (1988) Determination of trichloroethylene metabolites in rat liver homogenate using headspace gas chromatography. *J. Chromatogr.*, **442**, 325–332

Krasner, S.W., McGuire, M.J., Jacangelo, J.G., Patania, N.L., Reagan, K.M. & Aieta, E.M. (1989) The occurrence of disinfection by-products in US drinking water. *J. Am. Water Works Assoc.*, **81**, 41–53

Lahl, U., Stachel, B., Schröer W. & Zeschmar, B. (1984) Determination of organohalogenic acids in water samples. *Z. Wasser Abwasser-Forsch.*, **17**, 45–49 (in German)

Larson, J.L. & Bull, R.J. (1992a) Metabolism and lipoperoxidative activity of trichloroacetate and dichloroacetate in rats and mice. *Toxicol. appl. Pharmacol.*, **115**, 268–277

Larson, J.L. & Bull, R.J. (1992b) Species differences in the metabolism of trichloroethylene to the carcinogenic metabolites trichloroacetate and dichloroacetate. *Toxicol. appl. Pharmacol.*, **115**, 278–285

Legube, B., Croue, J.P. & Dore, M. (1985) Chlorination of humic substances in aqueous solution: yields of volatile and major non-volatile organic halides. *Sci. total Environ.*, **47**, 217–222

Lewinstein, I. & Rotstein, I. (1992) Effect of trichloroacetic acid on the microhardness and surface morphology of human dentin and enamel. *Endodont. dent. Traumatol.*, **8**, 16–20

Lide, D.R., ed. (1993) *CRC Handbook of Chemistry and Physics*, 74th Ed., Boca Raton, FL, CRC Press, pp. 3–500, 8–46

Lowry, L.K., Vandervort, R. & Polakoff, P.L. (1974) Biological indicators of occupational exposure to trichloroethylene. *J. occup. Med.*, **16**, 98–101

Mackay, J.M., Fox, V., Griffiths, K., Fox, D.A., Howard, C.A., Coutts, C., Wyatt, I. & Styles, J.A. (1995) Trichloroacetic acid: investigation into the mechanism of chromosomal damage in the *in vitro* human lymphocyte cytogenetic assay and the mouse bone marrow micronucleus test. *Carcinogenesis* (in press)

Mallinckrodt Chemical, Inc. (1993) *Specification Sheet: Trichloroacetic Acid AR (Crystals) (ACS)*, Paris, KY

Mather, G.G., Exon, J.H. & Koller, L.D. (1990) Subchronic 90 day toxicity of dichloroacetic and trichloroacetic acid in rats. *Toxicology*, **64**, 71–80

Miller, J.W. & Uden, P.C. (1983) Characterization of nonvolatile aqueous chlorination products of humic substances. *Environ. Sci. Technol.*, **17**, 150–157

Mohamed, M., Matayun, M. & Lim, T.S. (1989) Chlorinated organics in tropical hardwood kraft pulp and paper mill effluents and their elimination in an activated sludge treatment system. *Pertanika*, **12**, 387–394

Moriya, M., Ohta, T., Watanabe, K., Miyazawa, T., Kato, K. & Shirasu, Y. (1983) Further mutagenicity studies on pesticides in bacterial reversion assay systems. *Mutat. Res.*, **116**, 185–216

Müller, G., Spassovski, M. & Henschler, D. (1974) Metabolism of trichloroethylene in man. II. Pharmacokinetics of metabolites. *Arch. Toxicol.*, **32**, 283–295

Nair, L.M., Saari-Nordhaus, R. & Anderson, J.M., Jr (1994) Determination of haloacetic acids by ion chromatography. *J. Chromatogr. A*, **671**, 309–313

Nelson, M.A. & Bull, R.J. (1988) Induction of strand breaks in DNA by trichloroethylene and metabolites in rat and mouse liver *in vivo*. *Toxicol. appl. Pharmacol.*, **94**, 45–54

Nelson, M.A., Lansing, A.J., Sanchez, I.M., Bull, R.J. & Springer, D.L. (1989) Dichloroacetic acid and trichloroacetic acid-induced DNA strand breaks are independent of peroxisome proliferation. *Toxicology*, **58**, 239–248

Nelson, M.A., Sanchez, I.M., Bull, R.J. & Sylvester, S.R. (1990) Increased expression of c-myc and c-H-ras in dichloroacetate and trichloroacetate-induced liver tumors in B6C3F1 mice. *Toxicology*, **64**, 47–57

Nestmann, E.R., Chu, I., Kowbel, D.J. & Matula, T.I. (1980) Short-lived mutagen in *Salmonella* produced by reaction of trichloroacetic acid and dimethyl sulphoxide. *Can. J. Genet. Cytol.*, **22**, 35–40

Neumüller, O.A. (1988) Trichloressigsaure [Trichloroacetic acid]. In: *Rompps Chemie-Lexikon*, 8th Ed., Vol. 6, Stuttgart, W. Keller & Co. p. 4341 (in German)

Nicholson, B.C., Hayes, K.P. & Bursill, D.B. (1984) By-products of chlorination. *Water*, **September**, 11–15

Norwood, D.L., Christman, R.F., Johnson, J.D. & Hass, J.R. (1986) Using iostope dilution mass spectrometry to determine aqueous trichloroacetic acid. *J. Am. Water Works Assoc.*, **78**, 175–180

Ogata, M., Shimada, Y. & Taguchi, T. (1987) A new microdetermination method used in an analysis of the excretion of trichloro-compounds in the urine of workers exposed to trichloroethylene vapour. *Ind. Health*, **25**, 103–112

Ozawa, H. (1993) Gas chromatographic–mass spectrometric determination of halogenated acetic acids in water after direct derivatization. *J. Chromatogr.*, **644**, 375–382

Ozawa, H. & Tsukioka, T. (1990) Gas chromatographic separation and determination of chloroacetic acids in water by a difluoroanilide derivitization method. *Analyst*, **115**, 1343–1347

Parnell, M.J., Exon, J.H. & Koller, L.D. (1988) Assessment of hepatic initiation–promotion properties of trichloroacetic acid. *Arch. environ. Contam. Toxicol.*, **17**, 429–436

Pergola, F., Pezzatini, G. & Carrai, P. (1988) Determination of trichloroacetic acid by differential pulse polarography. Application to clinical analysis. *Anal. Lett.*, **21**, 977–991

Plotnikov, V.A. & Petrov, A.P. (1976) Combined action of trichloroacetic acid and dimethyl sulfate on *Arabidopsis* in relation to the theory of the role of histones in the mutation process. *Tsitol. Genet.*, **10**, 35–39

Plümacher, J. & Renner, I. (1993) Determination of volatile chlorinated hydrocarbons and trichloroacetic acid in conifer needles by headspace gas chromatography. *Fresenius J. anal. Chem.*, **347**, 129–135

Popp, W., Müller, G., Baltes-Schmitz, B., Wehner, B., Vahrenholz, C., Schmieding, W., Benninghoff, M. & Norpoth, K. (1992) Concentrations of tetrachloroethene in blood and trichloroacetic acid in urine in workers and neighbours of dry-cleaning shops. *Int. Arch. occup. environ. Health*, **63**, 393–395

Rantala, K., Riipinen, H. & Anttila, A. (1992) *Altisteet Työssä, 22. Halogeenihiilivedyt* [Exposure at work: 22. Halogenated hydrocarbons], Helsinki, Finnish Institute of Occupational Health–Finnish Work Environment Fund, pp. 18–19 (in Finnish)

Rapson, W.H., Nazar, M.A. & Butsky,V.V. (1980) Mutagenicity produced by aqueous chlorination of organic compounds. *Bull. environ. Contam. Toxicol.*, **24**, 590–596

Reckhow, D.A. & Singer, P.C. (1990) Chlorination by-products in drinking waters: from formation potentials to finished water concentrations. *J. Am. Water Works Assoc.*, **92**, 173–180

Reckhow, D.A., Singer, P.C. & Malcom, R.L. (1990) Chlorination of humic materials: byproduct formation and chemical interpretations. *Environ. Sci. Technol.*, **24**, 1655–1664

Sadtler Research Laboratories (1980) *1980 Cumulative Index*, Philadelphia, PA

Sanchez, I.M. & Bull, R.J. (1990) Early induction of reparative hyperplasia in the liver of B6C3F1 mice treated with dichloroacetate and trichloroacetate. *Toxicology*, **64**, 33–46

Schweizerische Unfallversicherungsanstalt [Swiss Accident Insurance Company] (1994) *Grenzwerte am Arbeitsplatz* [Limit Values in the Workplace], Lucerne, pp. 101, 131

Seki, Y., Urashima, Y., Aikawa, H., Matsumura, H., Ichikawa, Y., Hiratsuka, F., Yoshioka, Y., Shimbo, S. & Ikeda, M. (1975) Trichloro-compounds in the urine of humans exposed to methyl chloroform at sub-threshold levels. *Int. Arch. Arbeitsmed.*, **34**, 39–49

Sellers, E.M. & Koch-Weser, J. (1971) Kinetics and clinical importance of displacement of warfarin from albumin by acidic drugs. *Ann. N.Y. Acad. Sci.*, **179**, 213–225

Shakir, I.M.A. & Faizullah, A.T. (1990) Determination of oxonium ion in strongly ionisable inorganic acids and determination of substituted acetic acids using flow injection and chemiluminescence detection. *Analyst*, **115**, 69–72

Shirasu, Y., Moriya, M., Kato, K., Furuhashi, A. & Kada, T. (1976) Mutagenicity screening of pesticides in the microbial system. *Mutat. Res.*, **40**, 19–30

Sigma Chemical Co. (1992) *Specification Sheet: Trichloroacetic Acid ACS Reagent*, St Louis, MO

Skender, L.J., Karacic, V. & Prpic-Majic, D. (1988) Metabolic activity of antipyrine in workers occupationally exposed to trichloroethylene. *Int. Arch. occup. environ. Health*, **61**, 189–195

Skender, L.J., Karacic, V. & Prpic-Majic, D. (1991) A comparative study of human levels of trichloroethylene and tetrachloroethylene after occupational exposure. *Arch. environ. Health*, **46**, 174–178

Skender, L., Karacic, V., Bosner, B. & Prpic-Majic, D. (1993) Assessment of exposure to trichloroethylene and tetrachloroethylene in the population of Zagreb, Croatia. *Int. Arch. occup. environ. Health*, **65**, S163–S165

Smith, M.K., Randal, J.L., Read, E.J. & Stober, J.A. (1989) Teratogenic activity of trichloroacetic acid in the rat. *Teratology*, **40**, 445–451

Sollinger, S., Levsen, K. & Emmrich, M. (1992) Determination of trace amounts of carboxylic acids in ambient air by capillary gas chromatography–mass spectrometry. *J. Chromatogr.*, **609**, 297–304

Stacpoole, P.W., Harwood, H.J., Jr, Cameron, D.F., Curry, S.H., Samuelson, D.A., Cornwell, P.E. & Sauberlich, H.E. (1990) Chronic toxicity of dichloroacetate: possible relation to thiamine deficiency in rats. *Fundam. appl. Toxicol.*, **14**, 327–337

Stevens, D.K., Eyre, R.J. & Bull, R.J. (1992) Adduction of hemoglobin and albumin *in vivo* by metabolites of trichloroethylene, trichloroacetate, and dichloroacetate in rats and mice. *Fundam. appl. Toxicol.*, **19**, 336–342

Triebig, G., Trautner, P., Weltle, D., Saure, E. & Valentin, H. (1982) Investigations on neurotoxicity of chemical substances at the workplace. III. Determination of the motor and sensory nerve conduction velocity in persons occupationally exposed to trichloroethylene. *Int. Arch. occup. environ. Health*, **51**, 25–34 (in German)

Uden, P.C. & Miller, J.W. (1983) Chlorinated acids and chloral in drinking water. *J. Am. Water Works Assoc.*, **75**, 524–527

Ulander, A., Sélden, A. & Ahlborg, G., Jr (1992) Assessment of intermittent trichloroethylene exposure in vapor degreasing. *Am. ind. Hyg. Assoc. J.*, **53**, 742–743

United Kingdom Health and Safety Executive (1993) *Occupational Exposure Limits 1993* (EH 40/93), London, Her Majesty's Stationery Office, p. 29

United States Environmental Protection Agency (1990) *Methods for the Determination of Organic Compounds in Drinking Water. Supplement 1* (EPA-600/4-90-020; US NTIS PB91-146027), Cincinnati, OH, Environmental Monitoring Systems Laboratory, Method 552

United States Environmental Protection Agency (1994) Drinking water; national primary drinking water regulations: disinfectants and disinfection byproducts. *Fed. Regist.*, **59**, 38668–38829

United States National Institute for Occupational Safety and Health (1994a) *National Occupational Exposure Survey (1981–1983)*, Cincinnati, OH, p. 9

United States National Institute for Occupational Safety and Health (1994b) *Pocket Guide to Chemical Hazards* (DHHS Publ. 94-116), Cincinnati, OH, pp. 314–315

Vartiainen, T., Pukkala, E., Rienoja, T., Strandman, T. & Kaksonen, K. (1993) Population exposure to tri- and tetrachloroethene and cancer risk: two cases of drinking water pollution. *Chemosphere*, **27**, 1171–1181

Verschueren, K. (1983) *Handbook of Environmental Data on Organic Chemicals*, 2nd Ed., New York, Van Nostrand Reinhold Co., p. 1121

Wang, A.C., Hsueh, S. & Sun, C.F. (1992) Therapeutic effect of carbon dioxide laser versus single application of trichloroacetic acid for koilocytic squamous papillae. *J. Formosan med. Assoc.*, **91**, 1054–1058

Waskell, L. (1978) A study of the mutagenicity of anesthetics and their metabolites. *Mutat. Res.*, **57**, 141–153

Watson, S.C., Foster, J.R. & Elcombe, C.R. (1995) TCA induced hepatocarcinogenesis correlates better with DNA synthesis than with peroxisome proliferation. *Carcinogenesis* (in press)

Weast, R.C. & Astle, M.J. (1985) *CRC Handbook of Data on Organic Compounds*, Vol. II, Boca Raton, FL, CRC Press, pp. 390, 759

WHO (1993) *Guidelines for Drinking-water Quality*, 2nd ed., Vol. 1, *Recommendations*, Geneva, pp. 102–103, 177

Ziglio, G., Beltramelli, G. & Pregliasco, F. (1984) A procedure for determining plasmatic trichloroacetic acid in human subjects exposed to chlorinated solvents at environmental levels. *Arch. environ. Contam. Toxicol.*, **13**, 129–134

Ziglio, G., Beltramelli, G., Pregliasco, F. & Ferrari, G. (1985) Metabolites of chlorinated solvents in blood and urine of subjects exposed at environmental level. *Sci. total Environ.*, **47**, 473–477

1-CHLORO-2-METHYLPROPENE

1. Exposure Data

1.1 Chemical and physical data

1.1.1 Nomenclature

Chem. Abstr. Serv. Reg. No.: 513-37-1
Chem. Abstr. Name: 1-Chloro-2-methyl-1-propene
IUPAC Systematic Name: 1-Chloro-2-methylpropene
Synonyms: α-Chloroisobutylene; 1-chloroisobutylene; dimethylvinyl chloride; 2,2-dimethylvinyl chloride; β,β-dimethylvinyl chloride; isocrotyl chloride; 2-methyl-1-chloropropene; 2-methyl-1-propenyl chloride

1.1.2 Structural and molecular formulae and relative molecular mass

$$\begin{array}{c}\text{Cl}\\ \text{H}\end{array}\!\!\!>\!\!\text{C}=\text{C}\!\!<\!\!\!\begin{array}{c}\text{CH}_3\\ \text{CH}_3\end{array}$$

C_4H_7Cl Relative molecular mass: 90.55

1.1.3 Chemical and physical properties of the pure substance

(a) *Description*: Colourless to brown liquid (Budavari, 1989; Aldrich Chemical Co., 1994a)
(b) *Boiling-point*: 68 °C at 754 mm Hg [100 kPa] (Lide, 1993)
(c) *Freezing-point*: –1 °C (Aldrich Chemical Co., 1994b)
(d) *Density*: 0.9186 at 20 °C/4 °C (Lide, 1993)
(e) *Spectroscopy data*: Infrared (prism [80595], grating [80595]) and nuclear magnetic resonance (proton [53505]) spectral data have been reported (Sadtler Research Laboratories, 1994).
(f) *Solubility*: Soluble in acetone, chloroform, diethyl ether and ethanol (Lide, 1993)
(g) *Conversion factor*: $mg/m^3 = 3.7 \times ppm$[1]

---

[1] Calculated from: mg/m^3 = (relative molecular mass/24.45) × ppm, assuming normal temperature (25 °C) and pressure (101 kPa)

1.1.4 Technical products and impurities

1-Chloro-2-methylpropene is available commercially at a purity of at least 98% (Fluka Chemical Corp., 1993; Aldrich Chemical Co., 1994b). 3-Chloro-2-methylpropene (see monograph, this volume) has been reported as an impurity (US National Toxicology Program, 1986).

1.1.5 Analysis

Capillary gas chromatography–mass spectrometry has been used for the analysis of emissions of organic vapours near sites for the disposal of industrial and chemical wastes. Samples of ambient air were collected with a sampler equipped with Tenax GC sorbent cartridges. For 1-chloro-2-methylpropene, the method has an estimated detection limit of 62 ng/m^3 (Krost et al., 1982; Pellizzari, 1982).

1.2 Production and use

1.2.1 Production

1-Chloro-2-methylpropene is a by-product of the production of 3-chloro-2-methylpropene (Hooper et al., 1992). 1-Chloro-2-methylpropene is produced in research quantities by one company in Germany (Chemical Information Services Inc., 1994).

1.2.2 Use

1-Chloro-2-methylpropene is not known to be used commercially other than for research purposes.

1.3 Occurrence

1.3.1 Natural occurrence

1-Chloro-2-methylpropene is not known to occur as a natural product.

1.3.2 Occupational exposure

No data were available to the Working Group.

1.3.3 Air

1-Chloro-2-methylpropene was found among other organic compounds in the vapour phase of ambient air near industrial complexes and chemical waste disposal sites in the United States of America. Levels of 90–670 µg/m^3 were found in ambient air around four of five industrial complexes near Curtis Bay, MD (Pellizzari, 1982).

1.3.4 Water

No data were available to the Working Group.

1.4 Regulations and guidelines

No occupational exposure limits have been reported for 1-chloro-2-methylpropene (ILO, 1991).

2. Studies of Cancer in Humans

No data were available to the Working Group.

3. Studies of Cancer in Experimental Animals

Oral administration

Mouse: Groups of 50 male and 50 female B6C3F1 mice, eight weeks of age, were administered 1-chloro-2-methylpropene (purity, 98%; the major impurity was 3-chloro-2-methylpropene) in corn oil by gavage at doses of 0, 100, or 200 mg/kg bw on five days per week for 102 weeks. Survival of treated male and female mice was significantly lower ($p < 0.001$) than that of vehicle controls; the numbers of survivors at the end of the experiment were: 38 male controls, eight at the low dose and two at the high dose; and 41 female controls, six at the low dose and three at the high dose. Histopathological evaluation revealed dose-related increased incidences (by life-table tests) of forestomach neoplasms in both males and females and of preputial gland neoplasms in males. In males, the incidences of squamous-cell papillomas of the forestomach were 1/48 controls, 3/47 at the low dose ($p = 0.054$) and 8/44 at the high dose ($p = 0.011$); the incidences of squamous-cell carcinomas were 0/48 controls, 42/47 at the low dose ($p < 0.001$) and 35/44 at the high dose ($p < 0.001$); and the combined incidences of squamous-cell papillomas or carcinomas were 1/48 controls, 43/47 at the low dose ($p < 0.001$) and 41/44 at the high dose ($p < 0.001$). In females, the incidences of squamous-cell papillomas of the forestomach were not significantly increased in the treated groups in comparison with controls; the incidences of squamous-cell carcinomas were 0/50 controls, 40/47 at the low dose ($p < 0.001$) and 36/43 at the high dose ($p < 0.001$); and the combined incidences of squamous-cell papillomas or carcinomas were 0/50 controls, 40/47 at the low dose ($p < 0.001$) and 38/43 at the high dose ($p < 0.001$). The incidences of squamous-cell carcinomas of the preputial gland in male mice were 1/48 controls, 3/47 at the low dose and 16/44 at the high dose ($p < 0.001$) (United States National Toxicology Program, 1986).

Rat: Groups of 50 male and 50 female Fischer 344/N rats, seven weeks of age, were administered 1-chloro-2-methylpropene (purity, 96–98%; the major impurities were 3-chloro-2-methylpropene, 1,2-dichloro-2-methylpropane, 2,2,4-trimethyl-3-hydroxypentanal and *tert*-butyl chloride) in corn oil by gavage at doses of 0, 100 or 200 mg/kg bw on five days per week for 103 weeks. Survival of treated male and female rats was significantly lower ($p < 0.001$) than that of vehicle controls; the numbers of survivors at the end of the experiment were 38 control males, nine at the low dose and none at the high dose; and 43 female controls, 11 at the low dose and none at the high dose. Histopathological evaluation revealed increased incidences (by life-table tests) of neoplasms of the nasal cavity, oral cavity, oesophagus and forestomach in males and females (Table 1). The incidences of epithelial hyperplasia of the forestomach were increased in treated males, occurring in 0/49 controls, 24/50 at the low dose and 19/50 at the high dose; and in treated females, occurring in 0/50 controls, 29/50 at the low dose and 24/49 at the high dose. The incidences of epithelial hyperplasia of the oesophagus were also increased in treated animals, occurring in 0/50 male controls, 6/50 at the low dose and 4/49 at the high dose and in 0/49 female controls, 7/50 at the low dose and 4/49 at the high dose (United States National Toxicology Program, 1986).

Table 1. Incidences of neoplastic lesions in Fischer 344/N rats in two-year studies in which 1-chloro-2-methylpropene was administered by gavage

Site and tumour type	Male			Female		
	Vehicle control	100 mg/kg bw	200 mg/kg bw	Vehicle control	100 mg/kg bw	200 mg/kg bw
Nasal cavity						
Adenocarcinoma	0/47	8/46 $p < 0.001$	4/32 $p < 0.001$	0/50	3/49 $p = 0.011$	6/41 $p < 0.001$
Squamous-cell carcinoma	0/47	3/46	0/32	0/50	2/49	2/41
Carcinoma (not otherwise specified)	0/47	12/46 $p < 0.001$	24/32 $p < 0.001$	0/50	11/49 $p < 0.001$	28/41 $p < 0.001$
Oral cavity						
Squamous-cell carcinoma	0/50	5/50 $p = 0.007$	2/50 $p=0.030$	0/50	2/50	1/50
Squamous-cell papilloma	0/50	0/50	2/50	0/50	0/50	4/50 $p = 0.004$
Papilloma or carcinoma	0/50	5/50 $p = 0.007$	4/50 $p = 0.001$	0/50	2/50	5/50 $p < 0.001$
Oesophagus						
Squamous-cell carcinoma	0/50	4/50 $p = 0.004$	1/49	0/49	3/50 $p = 0.024$	1/49
Squamous-cell papilloma	0/50	2/50 $p = 0.043$	3/49 $p = 0.011$	0/49	0/50	0/49
Papilloma or carcinoma	0/50	6/50 $p < 0.001$	4/49 $p = 0.004$			
Forestomach						
Squamous-cell carcinoma	0/49	7/50 $p < 0.001$	0/50	0/50	5/50 $p = 0.004$	1/49
Squamous-cell papilloma	0/49	7/50 $p < 0.001$	0/50	1/50	4/50 $p = 0.027$	1/49
Papilloma or carcinoma	0/49	14/50 $p < 0.001$	0/50	1/50	9/50 $p < 0.001$	2/49

From United States National Toxicology Program (1986)
p values are given for incidences that are significantly greater than those of controls, on the basis of life-table tests.

4. Other Data Relevant to an Evaluation of Carcinogenicity and its Mechanisms

4.1 Absorption, distribution, metabolism and excretion

4.1.1 Humans

No data were available to the Working Group.

4.1.2 Experimental systems

[2-^{14}C]1-Chloro-2-methylpropene (3.5 mCi/mmol; radiochemical purity, 96%) was administered by gavage to male Fischer 344 rats in single or up to four daily doses of 150 mg/kg bw in corn oil. The compound was extensively absorbed and rapidly excreted: 92% of the single dose was eliminated within 24 h after treatment. It was rapidly distributed to the tissues; liver and kidney contained higher concentrations of radiolabel than forestomach and glandular stomach, which had similar levels. The tissue concentrations increased monotonically after repeated doses but had declined significantly after four days of recovery following treatment. After a single dose, about 35% of the administered radiolabel was found in urine, 52% was exhaled and 5% was detected in the faeces. In the expired air, about 25% of the dose was ^{14}C-carbon dioxide; 30% was volatile compounds, of which 96% was unchanged compound. The main urinary metabolite of 1-chloro-2-methylpropene was *trans*-2-amino-6-methyl-4-thia-5-heptene-1,7-dioic acid, indicating oxidation before glutathione conjugation. This metabolite constituted 23% of the total urinary radiolabel, and its corresponding N-acetyl derivative accounted for 9%. The profile of the urinary metabolites was quantitatively and qualitatively similar after one, two and four treatments (Ghanayem & Burka, 1987).

[2-^{14}C]1-Chloro-2-methylpropene (3.5 mCi/mmol; radiochemical purity, 96%) was administered by gavage to male B6C3F1 mice as a single dose of 150 mg/kg bw in corn oil. The compound was extensively absorbed and rapidly excreted; urine contained about 46% of the dose after 24 h, 25% was expired as carbon dioxide and 5% as the unchanged compound, whereas 9% of the dose was found in the faeces. The highest levels of radiolabel in tissues were detected in kidney, liver, forestomach and thymus 24 h after administration. As in rats, *trans*-2-amino-6-methyl-4-thia-5-heptene-1,7-dioic acid was the major urinary metabolite, constituting 35% of the total urinary radiolabel. Its N-acetyl derivative accounted for 12% (Ghanayem & Burka, 1987).

Incubation of 1-chloro-2-methylpropene with liver microsomes from phenobarbital-pretreated male Fischer 344 rats, untreated male rats or untreated female rats in the presence of NADPH resulted in the formation of (*E*)- and (*Z*)-3-chloro-2-methylpropen-1-ol in a ratio of 2:1. No alcohol was formed in the presence of microsomes from β-naphthoflavone-treated male rats. Synthetic (*E*)- and (*Z*)-3-chloro-2-methylpropenal reacted rapidly with *N*-acetylcysteine, the *E*-adduct being the sole product in both cases. In contrast, the corresponding acids reacted very slowly with sulfur nucleophiles. These results suggest that 1-chloro-2-methylpropene is oxidized to *E*- and *Z*-alcohols and subsequently to their respective electrophilic α,β-unsaturated chloro-aldehydes, which then react instantaneously with glutathione to form the *E*-adduct (Srinivas & Burka, 1988).

4.2 Toxic effects

4.2.1 *Humans*

No data were available to the Working Group.

4.2.2 *Experimental animals*

Groups of 10 male and 10 female Fischer 344/N rats and B6C3F1 mice were administered 0, 63, 125, 250, 500 or 750 mg/kg bw 1-chloro-2-methylpropene (purity, 96%) in corn oil by gavage, on five days per week for 13 weeks. Rats developed necrosis of the crypts of the small intestine at doses of 250 mg/kg bw and higher; similar changes, but at a lower incidence, occurred in the colonic mucosa. In addition, bone-marrow hypoplasia was noted at the two highest doses. In mice, necrosis of lymphopoietic cells, leading to atrophy of the thymus, lymph nodes and spleen, was observed at doses of 125 mg/kg bw and higher. Compound-related necrotic and/or degenerative changes also occurred in the pancreatic islet cells, liver, ovary and testis at the highest doses (United States National Toxicology Program, 1986).

In a study of cell proliferation in the forestomach, 1-chloro-2-methylpropene (purity, 98%) in corn oil was administered by gavage to groups of eight male Fischer 344/N rats at doses of 0, 100 or 200 mg/kg bw on five days per week for two weeks. Multifocal epithelial cell proliferation of the forestomach occurred in two animals at the low dose and in eight at the high dose, and hyperkeratosis of the forestomach developed in eight animals at the high dose (Ghanayem *et al.*, 1986).

4.3 Reproductive and prenatal effects

No data were available to the Working Group.

4.4 Genetic and related effects

4.4.1 *Humans*

No data were available to the Working Group

4.4.2 *Experimental systems* (see also Table 2 and Appendices 1 and 2)

1-Chloro-2-methylpropene was mutagenic to *Salmonella typhimurium* strain TA100 in the presence of exogenous metabolic activation in a single study, in which bacterial cells were exposed in chambers specially devised for testing volatile compounds. No clearly positive responses were obtained in a preincubation test.

Table 2. Genetic and related effects of 1-chloro-2-methylpropene

Test system	Result[a] Without exogenous metabolic system	Result[a] With exogenous metabolic system	Dose[b] (LED/HID)	Reference
SA0, *Salmonella typhimurium* TA100, reverse mutation	–	–	0.00	Neudecker et al. (1986)
SA0, *Salmonella typhimurium* TA100, reverse mutation	–	–	3850	Zeiger et al. (1987)
SA0, *Salmonella typhimurium* TA100, reverse mutation	–	–	3850	Zeiger (1990)
SA0, *Salmonella typhimurium* TA100, reverse mutation	–	+	100[c]	Zeiger (1990)
SA5, *Salmonella typhimurium* TA1535, reverse mutation	–	–	3850	Zeiger et al. (1987)
SA7, *Salmonella typhimurium* TA1537, reverse mutation	–	–	3850	Zeiger et al. (1987)
SA7, *Salmonella typhimurium* TA1537, reverse mutation	–	–	3850	Zeiger (1990)
SA9, *Salmonella typhimurium* TA98, reverse mutation	–	–	3850	Zeiger et al. (1987)
SA9, *Salmonella typhimurium* TA98, reverse mutation	–	–	3850	Zeiger (1990)
SA9, *Salmonella typhimurium* TA98, reverse mutation	–	–	200[c]	Zeiger (1990)
DMX, *Drosophila melanogaster*, sex-linked recessive lethal mutation	+		12 750 feed	US National Toxicology Program (1986)
DMH, *Drosophila melanogaster*, heritable translocation	+		12 750 feed	US National Toxicology Program (1986)
G5T, Gene mutation, mouse lymphoma L5178Y cells, *tk* locus *in vitro*	+	0	0.4	US National Toxicology Program (1986)
SIC, Sister chromatid exchange, Chinese hamster ovary (CHO) cells *in vitro*	+	+	100	Anderson et al. (1990)
CIC, Chromosomal aberrations, Chinese hamster ovary (CHO) cells *in vitro*	–	–	1600	Anderson et al. (1990)

[a]+, considered to be positive; –, considered to be negative; 0, not tested
[b]LED, lowest effective dose; HID, highest ineffective dose; in-vitro tests, mg/ml; in-vivo tests, mg/kg bw; 0.00, dose not reported
[c]Exposure in a desiccator: atmospheric concentration (µg/ml)

1-Chloro-2-methylpropene induced a significant increase in the frequency of both sex-linked recessive lethal mutations and heritable reciprocal translocations in the germ cells of *Drosophila melanogaster* after adult males were fed for three days at a dose of 12 750 ppm [µg/g] in 5% sucrose.

1-Chloro-2-methylpropene induced gene mutation at the thymidine kinase locus of L5178Y mouse lymphoma cells in the absence of metabolic activation.

Sister chromatid exchange, but not chromosomal aberration, was induced in Chinese hamster ovary cells in both the presence and absence of metabolic activation.

5. Summary and Evaluation

5.1 Exposure data

1-Chloro-2-methylpropene occurs as an impurity in the production of 3-chloro-2-methylpropene. It has no known commercial application.

5.2 Human carcinogenicity data

No data were available to the Working Group.

5.3 Animal carcinogenicity data

1-Chloro-2-methylpropene was tested for carcinogenicity by oral administration in one experiment in mice and in one experiment in rats. It produced squamous-cell carcinomas of the preputial gland in male mice and squamous-cell carcinomas of the forestomach in animals of each sex. In rats, it produced carcinomas of the nasal cavity and papillomas and carcinomas of the oral cavity, oesophagus and forestomach in animals of each sex.

5.4 Other relevant data

No data were available on the toxicokinetics or toxic effects of 1-chloro-2-methylpropene in humans. It is rapidly absorbed and excreted after oral administration to rats and mice. In rats, more of the dose was excreted via the lungs than in the urine, whereas in mice similar proportions were excreted by the two routes. Both the unchanged compound and carbon dioxide were exhaled. The major urinary metabolite in rats and mice was formed after oxidation and glutathione conjugation.

Repeated oral administration of 1-chloro-2-methylpropene to rats resulted in tissue necrosis in a number of organs, including the small and large intestine, thymus and spleen. Repeated administration of the compound to rats by gavage induced proliferation of forestomach cells.

Gene mutation and sister chromatid exchange, but not chromosomal aberrations, were induced in cultured rodent cells in single studies. 1-Chloro-2-methylpropene induced gene and chromosomal mutation in insects (in a single study). It was mutagenic to bacteria.

5.5 Evaluation[1]

There is *inadequate evidence* in humans for the carcinogenicity of 1-chloro-2-methylpropene.

There is *sufficient evidence* in experimental animals for the carcinogenicity of 1-chloro-2-methylpropene.

Overall evaluation

1-Chloro-2-methylpropene *is possibly carcinogenic to humans (Group 2B)*.

6. References

Aldrich Chemical Co. (1994a) *Specification Sheet:1-Chloro-2-methylpropene*, Milwaukee, WI

Aldrich Chemical Co. (1994b) *Aldrich Catalog/Handbook of Fine Chemicals 1994–1995*, Milwaukee, WI, p. 338

Anderson, B.E., Zeiger, E., Shelby, M.D., Resnick, M.A., Gulati, D.K., Ivett, J.L. & Loveday, K.S. (1990) Chromosome aberration and sister chromatid exchange test results with 42 chemicals. *Environ. mol. Mutag.*, **16** (Suppl. 18), 55–137

Budavari, S., ed. (1989) *The Merck Index*, 11th Ed., Rahway, NJ, Merck & Co., p. 331

Chemical Information Services, Inc. (1994) *Directory of World Chemical Producers 1995/96 Standard Edition*, Dallas, TX, p. 179

Fluka Chemical Corp. (1993) *Fluka Chemika-BioChemika*, Buchs, p. 321

Ghanayem, B.I. & Burka, L.T. (1987) Comparative metabolism and disposition of 1-chloro- and 3-chloro-2-methylpropene in rats and mice. *Drug Metab. Disposition*, **15**, 91–96

Ghanayem, B.I., Maronpot, R.R. & Matthews, H.B. (1986) Association of chemically induced forestomach cell proliferation and carcinogenesis. *Cancer Lett.*, **32**, 271–278

Hooper, K., LaDou, J., Rosenbaum, J.S. & Book, S.A. (1992) Regulation of priority carcinogens and reproductive or developmental toxicants. *Am. J. ind. Med.*, **22**, 793–808

ILO (1991) *Occupational Exposure Limits for Airborne Toxic Substances: Values of Selected Countries* (Occupational Safety and Health Series No. 37), 3rd Ed., Geneva

Krost, K.J., Pellizzari, E.D., Walburn, S.G. & Hubbard, S.A. (1982) Collection and analysis of hazardous organic emissions. *Anal. Chem.*, **54**, 810–817

Lide, D.R., ed. (1993) *CRC Handbook of Chemistry and Physics*, 74th Ed., Boca Raton, FL, CRC Press, p. 3-146

Neudecker, T., Dekant, W., Jorns, M., Eder, E. & Henschler, D. (1986) Mutagenicity of chloroolefins in the *Salmonella*/mammalian microsome test. II. Structural requirements for the metabolic activation of non-allylic chloroprenes and methylated derivatives via epoxide formation. *Biochem. Pharmacol.*, **35**, 195–200

---

[1] For definition of the italicized terms, see Preamble, pp. 22–26.

Pellizzari, E.D. (1982) Analysis of organic vapour emissions near industrial and chemical waste disposal sites. *Environ. Sci. Technol.*, **16**, 781–785

Sadtler Research Laboratories (1994) *Sadtler Standard Spectra, 1981–1994, Supplementary Index*, Philadelphia, PA

Srinivas, P. & Burka, L.T. (1988) Metabolism of 1-chloro-2-methylpropene. Evidence for reactive chloroaldehyde intermediates. *Drug Metab. Disposition*, **16**, 449–454

United States National Toxicology Program (1986) *Toxicology and Carcinogenesis Studies of Dimethylvinyl Chloride (1-Chloro-2-methylpropene) (CAS No. 513-37-1) in F344/N Rats and B6C3F1 Mice (Gavage Studies)* (NTP Tech. Rep. No. 316; NIH Publ. No. 86-2572), Research Triangle Park, NC

Zeiger, E. (1990) Mutagenicity of 42 chemicals in *Salmonella*. *Environ. mol. Mutag.*, **16** (Suppl. 18), 32–54

Zeiger, E., Anderson, B., Haworth, S., Lawlor, T., Mortelmans, K. & Speck, W. (1987) *Salmonella* mutagenicity tests: III. Results from the testing of 255 chemicals. *Environ. mol. Mutag.*, **9** (Suppl. 9), 1–110

3-CHLORO-2-METHYLPROPENE

1. Exposure Data

1.1 Chemical and physical data

1.1.1 Nomenclature

Chem. Abstr. Serv. Reg. No.: 563-47-3
Chem. Abstr. Name: 3-Chloro-2-methyl-1-propene
IUPAC Systematic Name: 3-Chloro-2-methylpropene
Synonyms: 3-Chloroisobutene; γ-chloroisobutylene; 3-chloroisobutylene; 2-(chloromethyl)-1-propene; isobutenyl chloride; MAC; methallylchloride; β-methylallylchloride; methallyl chloride; β-methallyl chloride; 2-methallyl chloride; methylallyl chloride; 2-methylallyl chloride; 2-methyl-3-chloropropene; 2-methyl-2-propenyl chloride

1.1.2 Structural and molecular formulae and relative molecular mass

$$Cl-\underset{H}{\overset{H}{C}}-\underset{}{\overset{CH_3}{C}}=C\underset{H}{\overset{H}{\diagup}}$$

C_4H_7Cl Relative molecular mass: 90.55

1.1.3 Chemical and physical properties of the pure substance

(a) *Description*: Colourless to pale-yellow liquid with sharp irritating odour (Verschueren, 1983; FMC Corp., 1990; Aldrich Chemical Co., 1994a)

(b) *Boiling-point*: 71–72 °C (Lide, 1993)

(c) *Freezing-point*: –12 °C (Aldrich Chemical Co., 1994b)

(d) *Density*: 0.9165 at 20 °C/4 °C (Lide, 1993)

(e) *Spectroscopy data*: Infrared (prism [4689]; grating [29193]), nuclear magnetic resonance (proton [9682, 9369]; C-13 [2018]) and mass [174] spectral data have been reported (Sadtler Research Laboratories, 1980; Weast & Astle, 1985).

(f) *Solubility*: Insoluble in water; soluble in acetone, chloroform, diethyl ether and ethanol (FMC Corp., 1990; Lide, 1993)

(g) *Volatility*: Vapour pressure, 102 mm Hg [13.6 kPa] at 20 °C; relative vapour density (air = 1), 3.12 (FMC Corp., 1990)

(h) *Stability*: Lower inflammable limit (air), 3.2%; polymerizes at room temperature and in the presence of sunlight (FMC Corp., 1990)

(i) *Conversion factor*: $mg/m^3 = 3.7 \times ppm$[1]

1.1.4 Technical products and impurities

3-Chloro-2-methylpropene is available commercially at purities ranging from 95% (technical grade) to 98%, with isocrotyl chloride (1-chloro-2-methylpropene; see monograph, this volume) as an impurity (Crescent Chemical Co., 1990; FMC Corp., 1990; Aldrich Chemical Co., 1994b).

1.1.5 Analysis

Capillary gas chromatography–mass spectrometry has been used for the analysis of emissions of organic vapours near sites for the disposal of industrial and chemical wastes. Samples of ambient air were collected with a sampler equipped with Tenax GC sorbent cartridges. For 3-chloro-2-methylpropene, the method has an estimated detection limit of 62 ng/m^3 (Krost *et al.*, 1982; Pellizzari, 1982).

1.2 Production and use

1.2.1 Production

3-Chloro-2-methylpropene is produced by substitutive chlorination of isobutylene. Production of this compound in the United States of America in 1984 was 5.4–11 thousand tonnes (United States National Toxicology Program, 1986). It is produced by two companies in the United States and one each in China, Germany and Japan (Chemical Information Services Inc., 1994).

1.2.2 Use

3-Chloro-2-methylpropene is used as an insecticide and fumigant and as an intermediate in the production of plastics, pharmaceuticals and other organic chemicals (Hooper *et al.*, 1992). Its use as a fumigant in individual sacks of grain in order to control the maize weevil in developing countries was promoted in the 1970s (Taylor, 1975); this use has been claimed to have been successful (Braby, 1992). It has been used in the Russian Federation to fumigate bulk grains (Taylor, 1975) and pulses that are not for human consumption (Braby, 1992).

1.3 Occurrence

1.3.1 Natural occurrence

3-Chloro-2-methylpropene is not known to occur as a natural product.

---

[1] Calculated from: mg/m^3 = (relative molecular mass/24.45) × ppm, assuming normal temperature (25 °C) and pressure (101 kPa)

1.3.2 Occupational exposure

No data were available to the Working Group.

1.3.3 Air

3-Chloro-2-methylpropene was found among other organic compounds in the vapour phase of ambient air near industrial complexes and chemical waste disposal sites in the United States. Levels of 110–400 μg/m^3 were found in ambient air around four of five industrial complexes near Curtis Bay, MD (Pellizzari, 1982).

1.3.4 Water

No data were available to the Working Group.

1.4 Regulations and guidelines

No occupational exposure limits have been reported (ILO, 1991). When the status of 3-chloro-2-methylpropene was reviewed in Germany in 1991, no MAK value was established, and it was placed in carcinogen category IIIB (justifiably suspected of having carcinogenic potential) (Deutsche Forschungsgemeinschaft, 1993).

2. Studies of Cancer in Humans

No data were available to the Working Group.

3. Studies of Cancer in Experimental Animals

Oral administration

Mouse: Groups of 50 male and 50 female B6C3F1 mice, eight weeks of age, were administered 3-chloro-2-methylpropene (technical-grade, containing 5% dimethylvinyl chloride [1-chloro-2-methylpropene; see monograph, p. 315]) in corn oil by gavage at doses of 0, 100 or 200 mg/kg bw on five days per week for 103 weeks. Survival in the treated groups was not significantly lower than that in vehicle controls; the numbers of survivors at the end of the experiment were: 26 male controls, 37 at the low dose and 32 at the high dose; and 37 female controls, 43 at the low dose and 27 at the high dose. Histopathological evaluation revealed dose-related increased incidences (by the incidental tumour test) of forestomach neoplasms in males and females. In males, the incidences of squamous-cell papillomas were 3/49 controls, 19/49 ($p < 0.001$) at the low dose and 30/49 ($p < 0.001$) at the high dose; the incidences of squamous-cell carcinomas were 0/49 controls, 5/49 ($p = 0.031$) at the low dose and 7/49 ($p = 0.016$) at the high dose; and the combined incidences of squamous-cell papillomas or carcinomas were 3/49 controls, 24/49 ($p < 0.001$) at the low dose and 36/49 ($p < 0.001$) at the high dose. In females, the incidences of squamous-cell papillomas were 0/50 controls, 15/48 ($p < 0.001$) at the low dose and 29/44 ($p < 0.001$) at the high dose; the incidences of squamous-cell carcinomas were: 0/50 controls, 1/48 at the low dose and 2/44 at the high dose; and the combined incidences of squamous-cell papillomas or carcinomas were 0/50 controls, 16/48 ($p < 0.001$) at the low dose and 31/44 ($p < 0.001$) at the high dose. The incidences of epithelial hyperplasia were also increased in treated animals: males – 0/49 controls, 14/49 at the low dose and 15/49 at the high dose; females – 4/50 controls, 6/48 at the low dose and 13/44 at the high dose (Chan et al., 1986; United States National Toxicology Program, 1986).

Rat: Groups of 50 male and 50 female Fischer 344/N rats, eight weeks of age, were administered 3-chloro-2-methylpropene (technical grade, containing 5% dimethylvinyl chloride) in corn oil by gavage at doses of 0, 75 or 150 mg/kg bw on five days per week for 103 weeks. Survival was marginally reduced among male rats receiving the high dose; the numbers of survivors at the end of the experiment were 30 male controls, 25 at the low dose and 17 at the high dose; and 31 female controls, 32 at the low dose and 26 at the high dose. Histopathological examination revealed increased incidences (by the incidental tumour test) of squamous-cell papillomas of the forestomach in animals of each sex receiving the high dose: 1/50 male controls, 5/50 at the low dose and 30/48 ($p < 0.001$) at the high dose; 1/50 female controls, 1/50 at the low dose and 10/50 ($p = 0.006$) at the high dose. The incidences of basal-cell or epithelial hyperplasia of the forestomach were also increased in treated animals: 19/50 male controls, 41/50 at the low dose and 44/48 at the high dose; 24/50 female controls, 42/50 at the low dose and 45/50 at the high dose (Chan et al., 1986; United States National Toxicology Program, 1986).

4. Other Data Relevant to an Evaluation of Carcinogenicity and its Mechanisms

4.1 Absorption, distribution, metabolism and excretion

4.1.1 Humans

No data were available to the Working Group.

4.1.2 Experimental systems

[2-^{14}C]3-Chloro-2-methylpropene (specific activity, 2.5 mCi/mmol; radiochemical purity, 93%; 5% 1-chloro-2-methylpropene) was administered by gavage to male Fischer 344 rats as single or up to four daily doses of 150 mg/kg bw in corn oil. The compound was extensively absorbed and rapidly excreted: 82% of the single dose was eliminated within 24 h after treatment. It was rapidly distributed to tissues, and the highest concentrations were found in forestomach, liver and kidney; the concentration of radiolabel was considerably lower in glandular stomach than in forestomach. The tissue concentrations were approximately doubled after two doses, but little additional increase was observed after four doses. The concentrations decreased after cessation of treatment. After a single dose, about 58% of the administered radiolabel was found in the urine, 22% was exhaled and 2% was detected in the faeces. In the expired air, about 12% of the dose was ^{14}C-carbon dioxide and 7% was volatile compounds. The main urinary metabolite was N-acetyl-S-(2-methylpropenyl)cysteine, which constituted 45% of the total urinary radiolabel. This metabolite is presumed to arise from direct conjugation of glutathione with 3-chloro-2-methylpropene (Ghanayem & Burka, 1987).

4.2 Toxic effects

4.2.1 Humans

No data were available to the Working Group.

4.2.2 Experimental animals

Groups of 10 male and 10 female Fischer 344/N rats were administered 0, 50, 100, 200, 300 or 400 mg/kg bw 3-chloro-2-methylpropene (purity, 93%; containing 5% 1-chloro-2-methylpropene) in corn oil by gavage on five days per week for 13 weeks. Groups of 10 male and 10 female B6C3F1 mice received 0, 125, 250, 500, 750 or 1250 mg/kg bw by the same schedule. Focal areas of necrosis with inflammation were noted in the livers of rats given 300 and 400 mg/kg bw. In mice, degeneration and necrosis of the cortical tubules of the kidneys were observed at doses of 500 mg/kg bw and higher, and the incidence and severity of these lesions were greater in males than in females. Coagulative necrosis in the liver was also observed in mice at doses of 500 mg/kg bw and higher (United States National Toxicology Program, 1986).

In a study of cell proliferation in the forestomach, 3-chloro-2-methylpropene (purity, 90%; most of the remainder was 1-chloro-2-methylpropene) was administered by gavage to groups of eight male Fischer 344/N rats at doses of 0, 75 or 160 mg/kg bw on five days per week for two weeks. All treated animals had generalized epithelial cell proliferation and hyperkeratosis of the forestomach (Ghanayem *et al.*, 1986).

4.3 Reproductive and prenatal effects

No data were available to the Working Group.

4.4 Genetic and related effects

4.4.1 Humans

No data were available to the Working Group

4.4.2 Experimental systems (see also Table 1 and Appendices 1 and 2)

3-Chloro-2-methylpropene (purity, 90.7%) was mutagenic to *Salmonella typhimurium* strain TA100 both in the presence and absence of metabolic activation and when the cells were treated in a modified liquid suspension test or, as reported in an abstract, in chambers specially devised for volatile compounds. A 100% pure preparation was also mutagenic to this strain

Table 1. Genetic and related effects of 3-chloro-2-methylpropene

Test system	Result[a] Without exogenous metabolic system	Result[a] With exogenous metabolic system	Dose[b] (LED/HID)	Purity	Reference
SA0, *Salmonella typhimurium* TA100, reverse mutation	+	+	0.00	100	Eder *et al.* (1982)
SA0, *Salmonella typhimurium* TA100, reverse mutation	–	(+)	1280	90.7	Haworth *et al.* (1983)
SA0, *Salmonella typhimurium* TA100, reverse mutation (Tedlar bag technique)	+	+	0.00	–	Warner *et al.* (1988) (Abstract)
SA5, *Salmonella typhimurium* TA1535, reverse mutation	–	–	1280	90.7	Haworth *et al.* (1983)
SA7, *Salmonella typhimurium* TA1537, reverse mutation	–	+	385	90.7	Haworth *et al.* (1983)
SA9, *Salmonella typhimurium* TA98, reverse mutation	–	–	3850	90.7	Haworth *et al.* (1983)
DMG, *Drosophila melanogaster*, genetic crossing-over or recombination	+		1000 inh.	–	Vogel & Nivard (1993)
G5T, Gene mutation, mouse lymphoma L5178Y cells, *tk* locus *in vitro*	+	0	23.0	90.7	Myhr & Caspary (1991)
SIC, Sister chromatid exchange, Chinese hamster ovary (CHO) cells *in vitro*	+	+	16	90.7	Gulati *et al.* (1989)
CIC, Chromosomal aberrations, Chinese hamster ovary (CHO) cells *in vitro*	+	?	120	90.7	Gulati *et al.* (1989)
MVM, Micronucleus induction, mouse bone-marrow cells *in vivo*	–		250 ip	90.7	Shelby *et al.* (1993)

[a]+, considered to be positive; (+), considered to be weakly positive in an inadequate study; –, considered to be negative; ?, considered to be inconclusive (variable responses in several experiments within an adequate study); 0, not tested

[b]LED, lowest effective dose; HID, highest ineffective dose; in-vitro tests, mg/ml; in-vivo tests, mg/kg bw;0.00, dose not reported; inh, inhalation; ip, intraperitoneal

(Eder *et al.*, 1982). In standard assays, the compound was more weakly mutagenic to TA100; it was mutagenic to TA1537 only in the presence of an exogenous metabolic system.

Induction of somatic recombination was observed in *Drosophila melanogaster*, in the *white/white⁺* eye mosaic assay, after larvae had been exposed by inhalation to 3-chloro-2-methylpropene of unknown purity.

The frequencies of both large and small colonies in the L5178Y mouse lymphoma *tk* locus assay, indicative of intragenic changes at the thymidine kinase locus and chromosomal rearrangements, respectively, were increased by treatment with 3-chloro-2-methylpropene (purity, 90.7%) in the absence of metabolic activation.

3-Chloro-2-methylpropene (purity, 90.7%) induced chromosomal aberrations and sister chromatid exchange in Chinese hamster ovary cells both in the presence and absence of metabolic activation.

Micronuclei were not induced in bone-marrow cells of male B6C3F1 mice treated intraperitoneally with 3-chloro-2-methylpropene (purity, 90.7%) at doses up to 250 mg/kg bw.

[A quantitative comparison with results obtained with 1-chloro-2-methylpropene indicates that the genotoxicity of 3-chloro-2-methylpropene cannot be accounted for by the presence of 1-chloro-2-methylpropene.]

5. Summary and Evaluation

5.1 Exposure data

3-Chloro-2-methylpropene is produced commercially as a chemical intermediate. It has had limited use as an insecticide and grain fumigant.

5.2 Human carcinogenicity data

No data were available to the Working Group.

5.3 Animal carcinogenicity data

3-Chloro-2-methylpropene containing 5% 1-chloro-2-methylpropene (see monograph, p. 315) was tested for carcinogenicity by oral administration in one experiment in mice and in one experiment in rats. Tumours of the forestomach were induced in mice and rats of each sex.

5.4 Other relevant data

No data were available on the toxicokinetics or toxic effects of 3-chloro-2-methylpropene in humans. It is rapidly absorbed, extensively metabolized and rapidly excreted after oral administration to rats. Most of the excretory products were found in urine; a mercapturic acid was the main metabolite. Considerable amounts were exhaled, some as carbon dioxide.

After repeated oral administrations, 3-chloro-2-methylpropene induced liver necrosis in rats and mice and kidney necrosis in mice; it also induced forestomach hyperplasia in rats.

No data were available on the effects of 3-chloro-2-methylpropene on reproduction in humans or experimental animals.

Micronuclei were not induced in the bone marrow of mice treated *in vivo* in a single study. 3-Chloro-2-methylpropene induced gene mutation, sister chromatid exchange and chromosomal aberrations in rodent cells in single studies. It was mutagenic to insects and bacteria. The genotoxic effects of this compound cannot be attributed solely to the presence of 1-chloro-2-methylpropene as an impurity.

5.5 Evaluation[1]

There is *inadequate evidence* in humans for the carcinogenicity of 3-chloro-2-methylpropene.

There is *limited evidence* in experimental animals for the carcinogenicity of 3-chloro-2-methylpropene.

Overall evaluation

3-Chloro-2-methylpropene *is not classifiable as to its carcinogenicity to humans (Group 3)*.

6. References

Aldrich Chemical Co. (1994a) *Specification Sheet: 3-Chloro-2-methylpropene*, Milwaukee, WI

Aldrich Chemical Co. (1994b) *Aldrich Catalog/Handbook of Fine Chemicals 1994–1995*, Milwaukee, WI, p. 338

Braby, M.F. (1992) Sorption of methallyl chloride on wheat. *J. stored Prod. Res.*, **28**, 183–186

Chan, P.C., Haseman, J.K., Boorman, G.A., Huff, J., Manus, A.G. & Cardy, R.H. (1986) Forestomach lesions in rats and mice administered 3-chloro-2-methylpropene by gavage for two years. *Cancer Res.*, **46**, 6349–6352

Chemical Information Services, Inc. (1994) *Directory of World Chemical Producers 1995/96 Standard Edition*, Dallas, TX, p. 146

Crescent Chemical Co. (1990) *Laboratory Chemicals Catalog*, Hauppauge, NY, p. 814

Deutsche Forschungsgemeinschaft (1993) *MAK- und BAT-Werte-Liste 1993* [MAK and BAT values list] (Report 29), Weinheim, Verlagsgesellschaft, p. 30

Eder, E., Neudecker, T., Lutz, D. & Henschler, D. (1982) Correlation of alkylating and mutagenic activities of allyl and allylic compounds: standard alkylation test vs. kinetic investigation. *Chem.-biol. Interactions*, **38**, 303–315

FMC Corp. (1990) *Material Safety Data Sheet: Methallyl Chloride*, Philadelphia, PA

---

[1] For definition of the italicized terms, see Preamble, pp. 22–26.

Ghanayem, B.I. & Burka, L.T. (1987) Comparative metabolism and disposition of 1-chloro- and 3-chloro-2-methylpropene in rats and mice. *Drug Metab. Disposition*, **15**, 91–96

Ghanayem, B.I., Maronpot, R.R. & Matthew, H.B. (1986) Association of chemically induced forestomach cell proliferation and carcinogenesis. *Cancer Lett.*, **32**, 271–278

Gulati, D.K., Witt, K., Anderson, B., Zeiger, E. & Shelby, M.D. (1989) Chromosome aberration and sister chromatid exchange tests in Chinese hamster ovary cells *in vitro*. III: Results with 27 chemicals. *Environ. mol. Mutag.*, **13**, 133–193

Haworth, S., Lawlor, T., Mortelmans, K., Speck, W. & Zeiger, E. (1983) Salmonella mutagenicity test results for 250 chemicals. *Environ. Mutag.*, **Suppl. 1**, 3–142

Hooper, K., LaDou, J., Rosenbaum, J.S. & Book, S.A. (1992) Regulation of priority carcinogens and reproductive or developmental toxicants. *Am. J. ind. Med.*, **22**, 793–808

ILO (1991) *Occupational Exposure Limits for Airborne Toxic Substances: Values of Selected Countries* (Occupational Safety and Health Series No. 37), 3rd Ed., Geneva

Krost, K.J., Pellizzari, E.D., Walburn, S.G. & Hubbard, S.A. (1982) Collection and analysis of hazardous organic emissions. *Anal. Chem.*, **54**, 810–817

Lide, D.R., ed. (1993) *CRC Handbook of Chemistry and Physics*, 74th Ed., Boca Raton, FL, CRC Press, pp. 3–146

Myhr, B.C. & Caspary, W.J. (1991) Chemical mutagenesis at the thymidine kinase locus in L5178Y mouse lymphoma cells: results from 31 coded compounds in the National Toxicology Program. *Environ. mol. Mutag.*, **18**, 51–83

Pellizzari, E.D. (1982) Analysis for organic vapor emissions near industrial and chemical waste disposal sites. *Environ. Sci. Technol.*, **16**, 781–785

Sadtler Research Laboratories (1980) *1980 Cumulative Index*, Philadelphia, PA

Shelby, M.D., Erexson, G.L., Hook, G.J. & Tice, R.R. (1993) Evaluation of a three-exposure mouse bone marrow micronucleus protocol: results with 49 chemicals. *Environ. mol. Mutag.*, **21**, 160–179

Taylor, R.W.D. (1975) Fumigation of individual sacks of grain using methallyl chloride for control of maize weevil. *Int. Pest Control*, **17**, 4–8

United States National Toxicology Program (1986) *Toxicology and Carcinogenesis Studies of 3-Chloro-2-methylpropene (Technical Grade Containing 5% Dimethylvinyl Chloride) (CAS No. 563-47-3) in F344/N Rats and B6C3F1 Mice (Gavage Studies)* (NTP Tech. Rep. No. 300; NIH Publ. No. 86-2556), Research Triangle Park, NC

Verschueren, K. (1983) *Handbook of Environmental Data on Organic Chemicals*, 2nd Ed., New York, Van Nostrand Reinhold, p. 815

Vogel, E.W. & Nivard, M.J.M. (1993) Performance of 181 chemicals in a Drosophila assay predominantly monitoring interchromosomal mitotic recombination. *Mutagenesis*, **8**, 57–81

Warner, J.R., Hughes, T.J. & Claxton, L.D. (1988) Mutagenicity of 16 volatile organic chemicals in a vaporization technique with *Salmonella typhimurium* TA100 (Abstract 273). *Environ. mol. Mutag.*, **11** (Suppl. 11), 111

Weast, R.C. & Astle, M.J. (1985) *CRC Handbook of Data on Organic Compounds*, Vol. 2, Boca Raton, FL, CRC Press, pp. 182, 711

OTHER INDUSTRIAL CHEMICALS

ACROLEIN

This substance was considered by previous Working Groups, in February 1978, June 1984 and March 1987 (IARC, 1979, 1985, 1987a). Since that time, new data have become available, and these have been incorporated into the monograph and taken into consideration in the present evaluation.

1. Exposure Data

1.1 Chemical and physical data

1.1.1 Nomenclature

Chem. Abstr. Serv. Reg. No.: 107-02-8
Deleted CAS Reg. No.: 25314-61-8
Chem. Abstr. Name: 2-Propenal
IUPAC Systematic Name: Acrolein
Synonyms: Acraldehyde; acrylaldehyde; acrylic aldehyde; allyl aldehyde; ethylene aldehyde; propenal; prop-2-en-1-al

1.1.2 Structural and molecular formulae and relative molecular mass

$$\begin{array}{c} H \\ H \\ \diagdown \\ H \end{array} C=C-C \begin{array}{c} O \\ \diagup \\ H \end{array}$$

C_3H_4O Relative molecular mass: 56.06

1.1.3 Chemical and physical properties of the pure substance

(a) *Description*: Colourless to yellowish liquid with extremely acrid, irritating odour (Verschueren, 1983; Budavari, 1989)

(b) *Boiling-point*: 52.5–53.5 °C (Lide, 1993)

(c) *Melting-point*: –86.9 °C (Lide, 1993)

(d) *Density*: 0.8410 at 20 °C/4 °C (Lide, 1993)

(e) *Spectroscopy data*: Infrared (prism [6646]; grating [29776]), ultraviolet [5-8], nuclear magnetic resonance (proton [9153]; C-13 [6242]) and mass [22] spectral data have been reported (Sadtler Research Laboratories, 1980; Weast & Astle, 1985).

(f) *Solubility*: Soluble in water (206 g/L at 20 °C), ethanol, diethyl ether and acetone (WHO, 1992; Lide, 1993)

(g) *Volatility*: Vapour pressure, 210 mm Hg [27.9 kPa] at 20 °C (Budavari, 1989); relative vapour density (air = 1), 1.9 (Union Carbide, 1993)

(h) *Stability*: Unstable; polymerizes, especially under light or in the presence of alkali or strong acid, to form disacryl, a plastic solid (Budavari, 1989)

(i) *Reactivity*: Reacts with air (oxygen), oxidizers, acids, alkalis and ammonia (United States National Institute for Occupational Safety and Health, 1994a)

(j) *Octanol/water partition coefficient (P)*: log (P) = – 0.01 (Hansch et al., 1995)

(k) *Conversion factor*: $mg/m^3 = 2.29 \times ppm$[1]

1.1.4 Technical products and impurities

Acrolein is available commercially with the following typical specifications: purity, 96.5%; water, 3.0%; hydroquinone (see IARC, 1987b), 0.10%; acetaldehyde (see IARC, 1987c), 0.30%; propionaldehyde, 0.002%; acetone, 0.07%; acetic acid, 0.07%; allyl alcohol, 0.005%; allyl acrylate, 0.03%; and benzene (see IARC, 1987d), 0.0003% (Union Carbide, 1993). Trade names for acrolein include Aqualin, Magnacide B and Magnacide H.

1.1.5 Analysis

Methods for the analysis of acrolein in air, water, biological media, tissue and food have been reviewed (WHO, 1992). Selected methods for the analysis of acrolein in various media are presented in Table 1. A method similar to that of the United States Environmental Protection Agency (1988) has been described that can be used for ambient air, industrial emissions and automobile exhaust, with a limit of detection of 1.2 ppb [2.7 µg/m^3] (Lodge, 1989).

The methods generally used for the determination of acrolein are spectrophotometry, fluorimetry, liquid chromatography, gas chromatography (GC) with electron capture detection and high-performance liquid chromatography with fluorescence detection. The oxime derivatives used in determination of carbonyls by GC are methoximes, benzyloximes, *para*-nitrobenzyloximes and pentafluorobenzyloximes; in these methods, flame ionization and nitrogen-specific detection systems are used (Nishikawa et al., 1987a).

A method for identifying carbonyl compounds, including acrolein, in environmental samples involves derivatization with *O*-(2,3,4,5,6-pentafluorobenzyl)hydroxylamine hydrochloride, followed by GC–mass spectrometry (Le Lacheur et al., 1993). A similar method for the determination of low-relative molecular-mass aldehydes, including acrolein, formed by the ozonation of drinking-water involves derivatization with *O*-(2,3,4,5,6-pentafluorobenzyl)-hydroxylamine hydrochloride, followed by analysis by high-resolution capillary GC. The detection limits of methods involving GC–electron capture detection and GC–mass spectrometry with ion-selective monitoring are 3.5 and 16.4 µg/L, respectively (Glaze et al., 1989).

---

[1] Calculated from: mg/m^3 = (relative molecular mass/24.45) × ppm, assuming normal temperature (25 °C) and pressure (101 kPa)

Passive and active sampling methods for the assessment of personal exposures to airborne aldehydes, including acrolein, due to emissions from methanol-fuelled vehicles are based on derivatization of the aldehydes with 2,4-dinitrophenylhydrazine during collection. The adsorbent materials are extracted with toluene and analysed by GC with flame-ionization detection (Otson et al., 1993).

Table 1. Methods for the analysis of acrolein

Sample matrix	Sample preparation	Assay procedure	Limit of detection	Reference
Air	Adsorb on sorbent coated with 2-(hydroxymethyl)piperidine on XAD-2; desorb with toluene; analyse for oxazolidine derivative	GC/NSD	2 µg/sample (6.1 µg/m^3)	Eller (1994); US Occupational Safety and Health Administration (1990)
		GC/FID and GC/MS	2 µg/sample	Eller (1994)
	Draw air through midget impinger containing acidified DNPH and isooctane; extract DNPH derivative with hexane:dichloromethane (70:30) solution; evaporate to dryness; dissolve in methanol	Reversed-phase HPLC/UV	NR	US Environmental Protection Agency (1988)
	Draw air through bubblers in series containing 4-hexylresorcinol in an alcoholic trichloroacetic acid solvent medium with mercuric chloride	Colorimetry	10 ppb [22.9 µg/m^3]	Feldstein et al. (1989a)
	Draw air through midget impinger containing 1% sodium bisulfite; react with 4-hexylresorcinol in an alcoholic trichloroacetic acid solvent medium with mercuric chloride	Colorimetry	10 ppb [22.9 µg/m^3]	Feldstein et al. (1989b)
Moist air	Collect in DNPH-impregnated adsorbent tubes (with CaCl, tubes); extract with acetonitrile	HPLC/UV	0.3 µg/sample (0.01 mg/m^3)	Vainiotalo & Matveinen (1992)
Exhaust gas	Derivatize with O-benzyl-hydroxylamine to O-benzyloxime; brominate with sulfuric acid, potassium bromate and potassium bromide; reduce with sodium thiosulfate; extract with diethyl ether	GC/ECD	NR	Nishikawa et al. (1987a)
Aqueous solution	Derivatize with O-(2,3,4,5,6-pentafluorobenzyl)hydroxylamine	MIMS/EIMS	10 ppb (µg/L)	Choudhury et al. (1992)

Table 1 (contd)

Sample matrix	Sample preparation	Assay procedure	Limit of detection	Reference
Rain-water	Derivatize with O-methoxylamine to O-methyloxime; brominate with sulfuric acid, potassium bromate and potassium bromide; reduce with sodium thiosulfate; elute with diethyl ether	GC/ECD	0.4 µg/L	Nishikawa et al. (1987b)
Liquid and solid wastes	Purge (inert gas); trap on suitable adsorbent material; desorb as vapour onto packed gas chromatographic column	GC/FID	0.7 µg/L[a]	US Environmental Protection Agency (1986)
Biological samples	Derivatize with DNPH; extract with chloroform; wash with hydrochloric acid; dry with nitrogen; dissolve in methanol	HPLC/UV	1 ng	Boor & Ansari (1986)

GC, gas chromatography; NSD, nitrogen selective detection; FID, flame ionization detection; MS, mass spectrometry; HPLC/UV, high-performance liquid chromatography/ultraviolet detection; DNPH, 2,4-dinitrophenylhydrazine; ECD, electron capture detection; MIMS/EIMS, membrane introduction mass spectrometry/electron impact mass spectrometry; NR, not reported

[a] Practical quantification limits for other matrices: 7 µg/L for groundwater; 7 µg/kg for low-level soil samples; 350 µg/L for water-miscible liquid waste samples; 875 µg/kg for high-level soil and sludge samples; 875 µg/L for non-water-miscible waste samples

1.2 Production and use

1.2.1 Production

Acrolein was first prepared in 1843 by Redtenbacher by the dry distillation of fat (Prager et al., 1918). Commercial production of acrolein began in Germany in 1942, by a process based on the vapour-phase condensation of acetaldehyde and formaldehyde (see IARC, 1995). This method was used until 1959, when a process was introduced for producing acrolein by vapour-phase oxidation of propylene (see IARC, 1994) (Ohara et al., 1985). Several catalysts have been used in the vapour-phase oxidation of propylene, including cuprous oxide, bismuth molybdate and antimony oxide (Hess et al., 1978). All commercial production of acrolein is currently based on propylene oxidation (Ohara et al., 1985).

In 1975, global production of acrolein was about 59 000 tonnes (Hess et al., 1978). The worldwide capacity for production of refined acrolein is about 113 000 tonnes per year (Etzkorn et al., 1991).

Acrolein is produced by three companies each in Japan and the United States of America and by one company each in France and Germany (Chemical Information Services, Inc., 1994).

1.2.2 Use

The principal use of acrolein is as an intermediate in the synthesis of acrylic acid (see IARC, 1987e), which is used to make acrylates, and of DL-methionine, an essential amino acid used as an animal feed supplement. Other important derivatives of acrolein are glutaraldehyde, pyridines, tetrahydrobenzaldehyde, allyl alcohol and glycerol, 1,4-butanedial and 1,4-butenediol, 1,3-propanediol, DL-glyceraldehyde, flavours and fragrances, polyurethane and polyester resins (Ohara *et al.*, 1985; Sax & Lewis, 1987).

The most important direct use of acrolein is as a biocide: It is used as a herbicide and to control algae, aquatic weeds and molluscs in recirculating process water systems. It is further used to control the growth of microorganisms in liquid fuel, the growth of algae in oil fields and the formation of slime in paper manufacture. Acrolein has been used in leather tanning and as a tissue fixative in histological work (Hess *et al.*, 1978; Ohara *et al.*, 1985; United States Environmental Protection Agency, 1985; Etzkorn *et al.*, 1991; WHO, 1992).

Acrolein has also been used as a warning agent in methyl chloride refrigerants and other gases, in poison gas mixtures for military use, in the manufacture of colloidal forms of metals (Budavari, 1989) and as a test gas for gas masks (Neumüller, 1979).

1.3 Occurrence

1.3.1 Natural occurrence

Acrolein has been identified as a volatile component of essential oils extracted from the wood of oak trees (Egorov *et al.*, 1976), in biogenic emissions from pine (0.49 µg/m^3) and deciduous (0.27 µg/m^3) forests in Europe and in remote, high-altitude areas with scarce vegetation (e.g. Nepal; 0.08–0.25 µg/m^3) (Ciccioli *et al.*, 1993). Acrolein occurs in a wide variety of food and food components (Feron *et al.*, 1991) (see also section 1.3.5).

1.3.2 Occupational exposures

The National Occupational Exposure Survey conducted between 1981 and 1983 indicated that 1298 employees in four industries (a total of 37 plants) in the United States were potentially exposed occupationally to acrolein (United States National Institute for Occupational Safety and Health, 1994b). The estimate is based on a survey of companies and did not involve measurements of actual exposures. Exposure to acrolein can occur in a wide variety of occupations, as indicated in Table 2.

1.3.3 Air (including emissions and combustion)

The levels of acrolein in ambient air and various emission rates have been reviewed (IARC, 1985; WHO, 1992).

Acrolein can be formed *in situ* in the atmosphere by photochemical oxidation of hydrocarbons (Atkinson & Arey, 1993). Concentrations of acrolein in ambient air have been reported to be 2–7 pbb [4.58–16 µg/m^3] in the United States, 0.5 pbb [1.15 µg/m^3] in the Netherlands and 0.13–0.56 pbb [0.30–1.28 µg/m^3] in Brazil (Grosjean, 1990).

Table 2. Occupational exposure of acrolein

Country	No. of plants	Job, task or industry	No. of samples	Concentration in air (mg/m³) Mean	Concentration in air (mg/m³) Range	Reference
Finland (1980–92)		Various industries, e.g. manufacture of plastics products, pulp, paper, paperboard, metal, glass products, electronic equipment	257 (A and P)	96.9% of measurements < 0.25		Finnish Institute of Occupational Health (1994)
Finland	5 2 1	Restaurant kitchen Bakery Food factory	(A)	0.02 0.01	0.06–0.59	Vainiotalo & Matveinen (1993)
Finland	3	Bakery	11(A)	0.12	< 0.03–0.59	Linnainmaa et al. (1990)
USA		Bakery	(A)	–	0.02–0.32 mg/batch	Lane & Smathers (1991)
China		Emission from rapeseed oil		Qualitative identification		Shields et al. (1993)
Former USSR		Emission from sunflower oil (160–170 °C)	(A)	≤ 1.1		Izmerov (1984)
Finland	1	Shipyard	82(A)	0.01–0.07 (median)	0.04–1.4 (max)	Engström et al. (1990)
Denmark	3	Engine workshops	(A)		ND–0.61	Rietz (1985)
USA		Wildland fire fighters	1(P)		0.05	Materna et al. (1992)
USA	1	Truck maintenance shop		0.005		Castle & Smith (1974)
Russian Federation	1	Rubber vulcanization			0.44–1.5	Volkova & Bagdinov (1969)
Russian Federation		Workshop, welding of metals coated with anti-corrosive primers			0.11–1.0	Protsenko et al. (1973)

Table 2 (contd)

Country	No. of plants	Job, task or industry	No. of samples	Concentration in air (mg/m^3) Mean	Concentration in air (mg/m^3) Range	Reference
Former Czechoslovakia	1	Pitch-coking plant Coal-coking plant	10 20	0.27 0.05	0.1–0.6 0.002–0.55	Mašek (1972)
USA	1	Workshop, repair and service (diesel exhaust)			< 0.1	Apol (1973)
Russian Federation		Quarries, exhaust from diesel engines			2.1–7.2	Klochkovskii et al. (1981)
Russian Federation	1	Production of acrolein and methyl mercaptopropionic aldehyde	(A)		0.1–8.2	Izmerov (1984)
Russian Federation	1	Press shops in oil seed mills			2–10	WHO (1992)
Finland	14	Manufacture of thermoplastics (17 different processes)	67(A)		< 0.02	Pfäffli (1982)

A, area sample; P, personal air sample (breathing zone)

In the former USSR, acrolein was measured at concentrations of 2 mg/m^3 in the air 100 m from oil chemical plants and 0.64 mg/m^3 at 1000 m; it was also measured at 0.4 mg/m^3 in the air 100 m from an oil mill and at 0.1–0.2 mg/m^3 1000 m from the mill. The concentrations of acrolein 150–800 m from an imitation-leather cloth and oil-cloth factory were 0.088–0.02 mg/m^3, and those 50 m from a perfume factory were 0.04–0.48 mg/m^3 (Izmerov, 1984). In the Netherlands, the annual emission of acrolein in 1989–90 was estimated to be 1.4 tonnes from the production of acrylonitrile and 0.03 tonnes from the metal and metallurgical industry (Sloof et al., 1991).

Acrolein has been measured in smoky indoor air. In a tavern in the United States, the concentrations were 21–24 µg/m^3 (Löfroth et al., 1989). In Germany, concentrations of 30–100 ppb [68.7–229 µg/m^3] were measured in five cafes, 3–13 pbb [6.87–29.8 µg/m^3] in two restaurants, 20–120 ppb [45.8–275 µg/m^3] in a train, 5–18 ppb [11.5–41.2 µg/m^3] in a tavern and 1–10 pbb [2.29–22.9 µg/m^3] in a cafeteria (Triebig & Zober, 1984).

Acrolein has been detected in exhaust gases from both gasoline engines, at 0.02–12.1 ppm [0.05–27.7 mg/m^3], and diesel engines, at 0.05–0.09 ppm [0.12–0.21 mg/m^3]. It has also been measured in exhaust from a diesel truck, at 6.9 ppm [15.8 mg/m^3], and a two-stroke motorcycle, at 6.5 ppm [14.9 mg/m^3] (Kuwata et al., 1979; Lipari & Swarin, 1982; Nishikawa et al., 1987a; Sigrist, 1994). The emission rate of acrolein from gasoline-fuelled vehicles with different emission control systems varied from undetectable to 1.7 mg/km (Victorin et al., 1988). The emission rate from gasoline-fuelled light-duty vehicles operated under different driving conditions ranged from 0.004 to 0.17 mg/km (Westerholm et al., 1992) and from undetectable to 0.4 mg/mile [0.25 mg/km] (Warner-Selph, 1989). In a study of light-duty vehicles fuelled with natural gas, the emission rate of acrolein was 0.0121 g/kg fuel (Siewert et al., 1993). The emission from heavy-duty engines run with natural gas was 0.32 mg/kW h [0.09 mg/MJ], and that of diesel engines was 0.30 mg/kW h [0.08 mg/MJ] (Gambino et al., 1993). In another study of diesel engines, the acrolein emissions were 7 mg/bhp h [2.59 mg/MJ] from base fuel, 3 mg/bhp h [1.11 mg/MJ] from fuels with ethylhexyl nitrate-containing additives and 3 mg/bhp h [1.11 mg/MJ] from fuels with peroxide-containing additives (Liotta, 1993). In Japan, acrolein was measured at concentrations of 0.9–1.3 ppb [2.06–3.00 µg/m^3] in urban air, 1.4–1.8 ppb [3.21–4.12 µg/m^3] in a road tunnel and 1.5–3.6 ppb [3.44–8.24 µg/m^3] in automobile exhaust (Nishikawa et al., 1986). In the former USSR, acrolein concentrations of 0.6–22 µg/m^3 were measured on a highway and from undetectable to 13 µg/m^3 in a neighbouring residential area (Izmerov, 1984). The estimated annual emissions of acrolein from road traffic in 12 European countries (Bouscaren et al., 1987) are summarized in Table 3.

Acrolein has been determined as an odorous constituent in aircraft emissions. In the United States, it was emitted by model jet engines operating at idling power at concentrations of 0.80–2.23 ppm [1.83–5.11 mg/m^3] (Rossi, 1992). In Japan, concentrations of 0.009–0.052 ppm [0.02–0.12 mg/m^3] were measured about 50 m behind a low-smoke combustor jet engine at idling power (Miyamoto, 1986).

Acrolein concentrations associated with residences where wood stoves were used were 0.7–6.0 µg/m^3 in indoor air and 1.6–4.9% outdoors (Highsmith et al., 1988). The rate of emission of acrolein from wood-burning fireplaces varied from 0.021 to 0.132 g/kg (Lipari et al., 1984).

Acrolein was identified in oil combustion products in a hospital at a level of 166 µg/m^3 (Götze & Harke, 1989).

Table 3. Estimated annual emissions of acrolein (tonnes/year) from road traffic in 12 European countries

Country	Acrolein emission (tonnes/year)	
	Gasoline engines	Diesel engines
Belgium	70	60
Denmark	30	30
Germany	550	450
France	450	400
Greece	40	70
Ireland	20	10
Italy	400	400
Luxembourg	4	2
Netherlands[a]	90	70
Portugal	20	60
Spain	140	200
United Kingdom	500	250

From Bouscaren *et al.* (1987)
[a]Plus 30 tonnes per year from the chemical industry

Acrolein has been identified among the decomposition products of cellophane (Feron *et al.*, 1991) and polyvinyl chloride (Boettner & Ball, 1980) used for food wrapping. It has also been identified in Chinese incense smoke (Lin & Wang, 1994).

1.3.4 Water

The levels of acrolein in water from various sources have been reviewed (WHO, 1992). Acrolein was detected in surface water in an irrigation canal downstream from its application as a slimicide or herbicide at concentrations of 30–100 µg/L (WHO, 1992). It has also been detected in raw sewage in treatment plants at concentrations of 216–825 ppb [µg/L] and in municipal effluents at 20–200 ppb (United States Environmental Protection Agency, 1985).

Acrolein has not been detected in drinking-water (WHO, 1992). It was detected at concentrations of 1.5–3.1 µg/L in samples of rainwater in Japan (Nishikawa *et al.*, 1987b); it was not detected in rainwater in the Po Valley, Italy, but was found in fog at levels varying from undetectable (< 1 µmol [56 µg]/L) to 120 µg/L (Facchini *et al.*, 1986, 1990).

1.3.5 Food and beverages

Acrolein has been detected in a wide variety of fruits (apples, grapes, raspberries, strawberries, blackberries) at concentrations of < 0.01–0.05 ppm [mg/kg] and in vegetables (cabbage, carrots, potatoes, tomatoes) at ≤ 0.59 ppm. It has also been detected in caviar, lamb, hops, sour salted pork, the aroma of cooked horse mackerel and of white bread and in raw chicken breast (Feron *et al.*, 1991). It was detected in doughnuts fried at 182 °C at concentrations of 0.1–0.9 ppm and in the coating of codfish fillets fried at 182 °C and 204 °C at a concentration of 0.1 ppm (Lane & Smathers, 1991).

Acrolein has been detected in cheese at levels of 290–1300 ppb [μg/kg] (Collin *et al.*, 1993). It was found in whisky at 0.67–11.1 ppb [μg/L] (Miller & Danielson, 1988), in red wine at 3.8 ppm [mg/L] and in fresh lager beer at 0.0011–0.002 ppm [mg/L]. It has also been identified in coffee and tea (Feron *et al.*, 1991) and in the emissions from heated animal fat and vegetable oils (Umano & Shibamoto, 1987; Yasuhara *et al.*, 1989).

1.3.6 Tobacco smoke

The occurrence of acrolein in tobacco smoke has been reviewed. The acrolein concentrations in the smoke from various cigarettes were 3–220 μg/cigarette (IARC, 1985, 1986). Levels as high as 463–684 μg/cigarette were reported in Japan (Kuwata *et al.*, 1979). The mean acrolein concentrations detected in 75 brands of cigarettes in the United Kingdom during 1983–90 varied from < 10 to 140 μg/cigarette (Phillips & Waller, 1991).

1.3.7 Humans

Acrolein was detected in the urine of a bone-marrow recipient after intravenous administration of cyclosphamide (Al-Rawithi et al., 1993).

1.4 Regulations and guidelines

Occupational exposure limits and guidelines for acrolein in several countries are given in Table 4. In the Russian Federation, the maximal acceptable background concentration of acrolein in ambient air is 0.03 mg/m^3 (Environmental Chemicals Data and Information Network, 1993). In the United States, acrolein may be used as a slimicide in the manufacture of paper and paperboard that will come into contact with food (United States Food and Drug Administration, 1994). An Expert Panel of the European Commission recommended a time-weighted average occupational exposure limit of 0.12 mg/m^3, with a short-term exposure limit of 0.23 mg/m^3 (European Commission, 1994).

2. Studies of Cancer in Humans

Epidemiological studies of exposure to acrolein in complex mixtures originating from the combustion of organic products, such as indoor cooking and occupations related to motor exhaust, were available but were considered to be insufficiently specific for use in evaluating the carcinogenicity of acrolein. Only one study in which individual exposure to acrolein was assessed was available to the Working Group.

Table 4. Occupational exposure limits and guidelines for acrolein

Country	Year	Concentration (mg/m³)	Interpretation
Australia	1991	0.25	TWA
		0.8	STEL
Belgium	1991	0.23	TWA
		0.69	STEL
Canada	1994	0.25	TWA
		0.8	STEL
Denmark	1991	0.25	TWA
Finland	1993	0.25	STEL
France	1991	0.25	STEL
Germany	1993	0.25	TWA
Hungary	1991	0.25	TWA
		0.5	STEL
Italy	1994	0.25	TWA
		0.75	STEL
Japan	1991	0.23	TWA
Netherlands	1994	0.25	TWA
Poland	1991	0.5	TWA
Romania	1994	0.3	TWA
		0.5	STEL
Russian Federation	1991	0.2	STEL
Sweden	1993	0.2	TWA
		0.7	STEL
Switzerland	1994	0.25	TWA
		0.5	STEL
United Kingdom	1993	0.25	TWA
		0.8	STEL
United States			
ACGIH (TLV)	1994	0.23	TWA
		0.69	STEL
NIOSH (REL)	1994	0.25	TWA
		0.8	STEL
OSHA (PEL)	1994	0.25	TWA

From ILO (1991); United States National Institute for Occupational Safety and Health (NIOSH) (1994a); Arbetarskyddsstyrelsens (1993); Deutsche Forschungsgemeinschaft (1993); Environmental Chemicals Data and Information Network (1993); Työministeriö (1993); United Kingdom Health and Safety Executive (1993); Arbeidsinspectie (1994); American Conference of Governmental Industrial Hygienists (ACGIH) (1994); Schweizerische Unfallversicherungsanstalt (1994); United States Occupational Safety and Health Adminitration (OSHA) (1994)

TWA, time-weighted average; STEL, short-term exposure limit; PEL, permissible exposure limit; REL, recommended exposure limit; TLV, threshold limit value

In the case–control study of Ott *et al.* (1989), described in the monograph on vinyl acetate, p. 450, exposure to acrolein was reported for two men who had died with non-Hodgkin's lymphoma (odds ratio, 2.6), one with multiple myeloma (odds ratio, 1.7), three with non-lymphocytic leukaemia (odds ratio, 2.6) and none who had died with lymphocytic leukaemia.

3. Studies of Cancer in Experimental Animals

3.1 Oral administration

3.1.1 Mouse

Four groups of 70–75 male and 70–75 female Swiss albino CD-1 mice, eight weeks of age, received 0, 0.5, 2.0 or 4.5 mg/kg bw acrolein (purity, 94.9–98.5%; containing 0.25–0.3% hydroquinone as a stabilizer) per day by gavage in deionized water for 18 months, at which time all surviving mice were killed. Treated mice had decreased body weight gain, and treated males had an increased mortality rate, particularly at the high dose. All mice killed at the end of treatment and those found dead or moribund were necropsied, and tissues from major organs were examined histologically. No treatment-related increase in tumour frequency was observed (Parent *et al.*, 1991a).

3.1.2 Rat

Groups of 20 male and 20 female Fischer 344 rats, seven to eight weeks of age, received 0 or 625 mg/L acrolein ([purity unspecified] stabilized with hydroquinone) per day in the drinking-water on five days a week for 124 or 104 weeks and were killed at 132 weeks. Additional groups of 20 males received 100 or 250 mg/L acrolein for 124 weeks, and surviving rats were killed at 130 weeks. Survival was comparable in treated and control groups. No significant, dose-related increase in the frequency of tumours at any site was observed (Lijinsky & Reuber, 1987; Lijinsky, 1988). [The Working Group noted the small number of animals used.]

Groups of 50 male and 50 female Sprague-Dawley rats, about six weeks of age, received 0, 0.05, 0.5 or 2.5 mg/kg bw acrolein (purity, 94.9–98.5%; containing 0.25–0.3% hydroquinone as a stabilizer) in deionized water by gavage daily for 102 weeks, at which time all surviving rats were killed. Survival was significantly reduced among males and females at the high dose during the first year, and this trend continued among females throughout the treatment period. All rats, including those that were found dead, were necropsied, and tissues from major organs were examined histologically. There was no significant increase in the incidence of neoplastic or non-neoplastic lesions in treated rats in comparison with controls (Parent *et al.*, 1992a).

3.2 Inhalation and/or intratracheal administration

Hamster: Two groups of 18 male and 18 female Syrian golden hamsters, six weeks old, were exposed to 0 or 4 ppm (0 or 9.2 mg/m^3) acrolein vapour [purity unspecified] for 7 h per day on five days per week for 52 weeks. Six animals per group were killed at 52 weeks and the remainder at 81 weeks. Survival was similar in treated and control groups. All animals were subjected to necropsy, and all tissues from the respiratory tract and gross or suspect lesions were examined histologically. A single papilloma of the respiratory tract was found in a treated female. Inflammation and epithelial metaplasia of the respiratory tract were observed in about 20% of animals killed at 81 weeks, even after a withdrawal period of six months (Feron & Kruysse, 1977). [The Working Group noted the short exposure period and the small number of animals used.]

3.3 Skin application

Mouse: A group of 15 S strain mice [sex and age unspecified] received 10 weekly skin applications of a 0.5% solution of acrolein [purity unspecified] in acetone (total dose of acrolein, 12.6 mg per animal). Starting 25 days after the first application of acrolein, the mice received weekly skin applications of 0.17% croton oil for 18 weeks; for the second and third applications, the concentration was reduced to 0.085%. When croton oil and acrolein were administered together, each compound was given alternately at three- or four-day intervals. At the end of the treatment with croton oil, all 15 mice were still alive, and two had a total of three skin papillomas; 4/19 controls that received the croton oil treatment alone had four skin papillomas (Salaman & Roe, 1956). [The Working Group noted the small number of animals used and the short duration of the experiment.]

3.4 Administration with known carcinogens

3.4.1 Rat

In a two-stage initiation–promotion assay in rat urinary bladder, male Fischer 344 rats, five weeks of age, were divided into eight groups. During the initial phase, groups 1, 3 and 5 (30 rats per group) received intraperitoneal injections of 2 mg/kg bw acrolein (purity, 97%) in distilled water twice a week for six weeks; groups 2, 6 and 7 (30, 30 and 40 rats per group, respectively) were fed 0.2% *N*-[4-(5-nitro-2-furyl)-2-thiazolyl]formamide (FANFT) in the diet for six weeks; group 4 (30 rats) received injections of distilled water and the control diet; and group 8 (40 rats) received no treatment and the control diet. In the second phase, groups 1, 2 and 4 received 3.0% uracil in the diet for 20 weeks, followed by six weeks of control diet; group 3 received the control diet only; groups 5 and 6 received intraperitoneal injections of 2 mg/kg bw acrolein twice a week for two weeks, 1.5 mg/kg bw once in week 10, then 1.5 mg/kg bw twice a week for seven weeks (weeks 11–17) and 1.0 mg/kg bw once in week 18, twice in week 19 and once in weeks 20 and 21; group 7 received intraperitoneal injections of distilled water and the control diet; and group 8 received control diet only. In the group receiving acrolein followed by uracil (group 1), 18/30 animals developed urinary bladder papillomas, in comparison with 8/30 in the group receiving distilled water followed by uracil (group 4) ($p < 0.05$, Fisher's exact test). The group receiving FANFT followed by uracil (group 2) developed 9/30 urinary bladder papillomas and 21/30 bladder carcinomas. No bladder tumour was observed in the rats receiving acrolein alone (group 5) (Cohen *et al.*, 1992).

3.4.2 Hamster

Groups of 30 male and 30 female Syrian golden hamsters, about six weeks of age, were exposed by inhalation to 0 or 4.0 ppm [0 or 9.2 mg/m^3] acrolein [purity unspecified] for 7 h per day on five days per week for 52 weeks, together with either weekly intratracheal instillations of 0.175 or 0.35% benzo[a]pyrene (purity, > 99%) in 0.9% saline (total doses, 18.2 or 36.4 mg/animal) or subcutaneous injections of 0.0675% N-nitrosodiethylamine in saline once every three weeks (total dose, 2 µl/animal). The experiment was terminated at 81 weeks, and all survivors were killed and autopsied. Papillomas, adenomas, adenocarcinomas and squamous-cell carcinomas of the respiratory tract were found in male and female hamsters treated with benzo[a]pyrene and N-nitrosodiethylamine. Exposure to acrolein vapour did not increase the incidence of these tumours (Feron & Kruysse, 1977).

3.5 Carcinogenicity of possible metabolites

Acrolein is metabolized *in vitro* by liver and lung microsomes to glycidaldehyde, which is carcinogenic to mice after skin application and to mice and rats after subcutaneous injection, producing tumours at the site of application (IARC, 1987f).

4. Other Data Relevant to an Evaluation of Carcinogenicity and its Mechanisms

4.1 Absorption, distribution, metabolism and excretion

4.1.1 Humans

No data were available to the Working Group.

4.1.2 Experimental systems

Exposure of dogs to acrolein by inhalation results in extensive retention in the respiratory tract (Egle, 1972), probably because of its strong reactivity with tissues (Beauchamp *et al.*, 1985; WHO, 1992). There seems to be limited, if any, distribution to other organs under these circumstances. After exposure of rats to concentrations of 0.1–5 ppm [0.23–11.5 mg/m^3], the amount of reduced glutathione (GSH) in respiratory mucosa was reduced in a dose-dependent manner; even at 5 ppm, however, there was no change in the amount of GSH in liver (McNulty *et al.*, 1984). In rats and guinea-pigs, exposure to acrolein by inhalation for 24 h/day for 90 days at concentrations of 0.22, 1 and 1.8 ppm [0.51, 2.3 and 4.1 mg/m^3] resulted in some changes to the liver (Lyon *et al.*, 1970), suggesting that it may also be distributed to sites distant from the respiratory tract after exposure by this route. The effects could be due to a locally produced metabolite. Subcutaneous (Kaye, 1973) and oral (Draminski *et al.*, 1983; Sanduja *et al.*, 1989) administration resulted in urinary metabolites of acrolein, indicating absorption of the parent molecule.

Acrolein reacts rapidly with thiols such as GSH, and conjugation is complete within seconds in nonenzymatic incubations *in vitro* with millimolar concentrations of GSH (Esterbauer *et al.*, 1975; Beauchamp *et al.*, 1985). This pathway seems to dominate the metabolism of acrolein (Berhane & Mannervik, 1990). The formation of GSH adducts may be catalysed in part by glutathione *S*-transferase (Beauchamp *et al.*, 1985). Acrolein may also bind to the enzyme itself, which would result in some elimination of the parent molecule (Haenen *et al.*, 1988; Berhane & Mannervik, 1990). Acrolein also binds to other proteins at sulfydryl and amino groups (Beauchamp *et al.*, 1985; Esterbauer *et al.*, 1991).

The urinary metabolites that have been identified include *S*-(2-carboxylethyl)mercapturic acid and *S*-(3-hydroxypropyl)mercapturic acid (Draminski *et al.*, 1983). Other metabolic products have been identified *in vitro*, including acrylic acid and glyceraldehyde. Acrylic acid was formed during NAD^+- or $NADP^+$-dependent incubation of acrolein with cytosol or microsomes from rat liver, but not from rat lung (Patel *et al.*, 1980). No evidence was found for formation of acrylic acid from acrolein by rat liver mitochondrial or cytosolic aldehyde dehydrogenases, but acrolein was a potent inhibitor of these enzymes, suggesting that the GSH adduct of acrolein is the substrate (Mitchell & Petersen, 1988). The epoxide glycidaldehyde is generated from acrolein by cytochrome P450 epoxidase, which is then subjected to the action of epoxide hydrolase to form glyceraldehyde. Acrylic acid and glyceraldehyde may then be incorporated into normal cellular metabolism (WHO, 1992). [Owing to the very efficient conjugation of acrolein with glutathione, it is unlikely that this pathway is significant *in vivo*.]

The metabolism of acrolein (summarized in Figure 1) has been studied extensively because acrolein is probably the toxic metabolite of cyclophosphamide, a chemotherapeutic agent (see IARC, 1987g). The mercapturic acid adduct of acrolein has been detected in the urine of humans after intravenous administration of cyclophosphamide. Acrolein and its glutathione adduct have also been shown to produce oxygen radicals through interaction with aldehyde dehydrogenase and xanthine oxidase (Adams & Klaidman, 1993). The depletion of GSH does not, however, involve generation of active oxygen species (Grafström *et al.*, 1988).

The 1:1 acrolein:glutathione adduct *S*-(3-oxopropyl)glutathione is nephrotoxic in rats when administered intravenously; however, this activity can be inhibited by acivicin, a γ-glutamyl-transpeptidase inhibitor, suggesting that *S*-(3-oxopropyl)glutathione is further metabolized before it becomes toxic (Horvath *et al.*, 1992). In primary cultures of rat kidney proximal tubule cells, the mercapturic acid that can be derived from the adduct, *S*-(3-oxopropyl)-*N*-acetyl-L-cysteine, is cytotoxic and can release acrolein (Hashmi *et al.*, 1992).

The results of such experiments with acrolein *in vitro* must be interpreted with caution (Beauchamp *et al.*, 1985), because the main route of exposure is inhalation and the distribution of tolerable levels is likely to be negligible. Thus, while consideration of such pathways is important for understanding the toxicity of cyclophosphamide, they are of lesser importance for the effects of acrolein itself at distant sites.

Figure 1. Possible metabolic pathways of acrolein

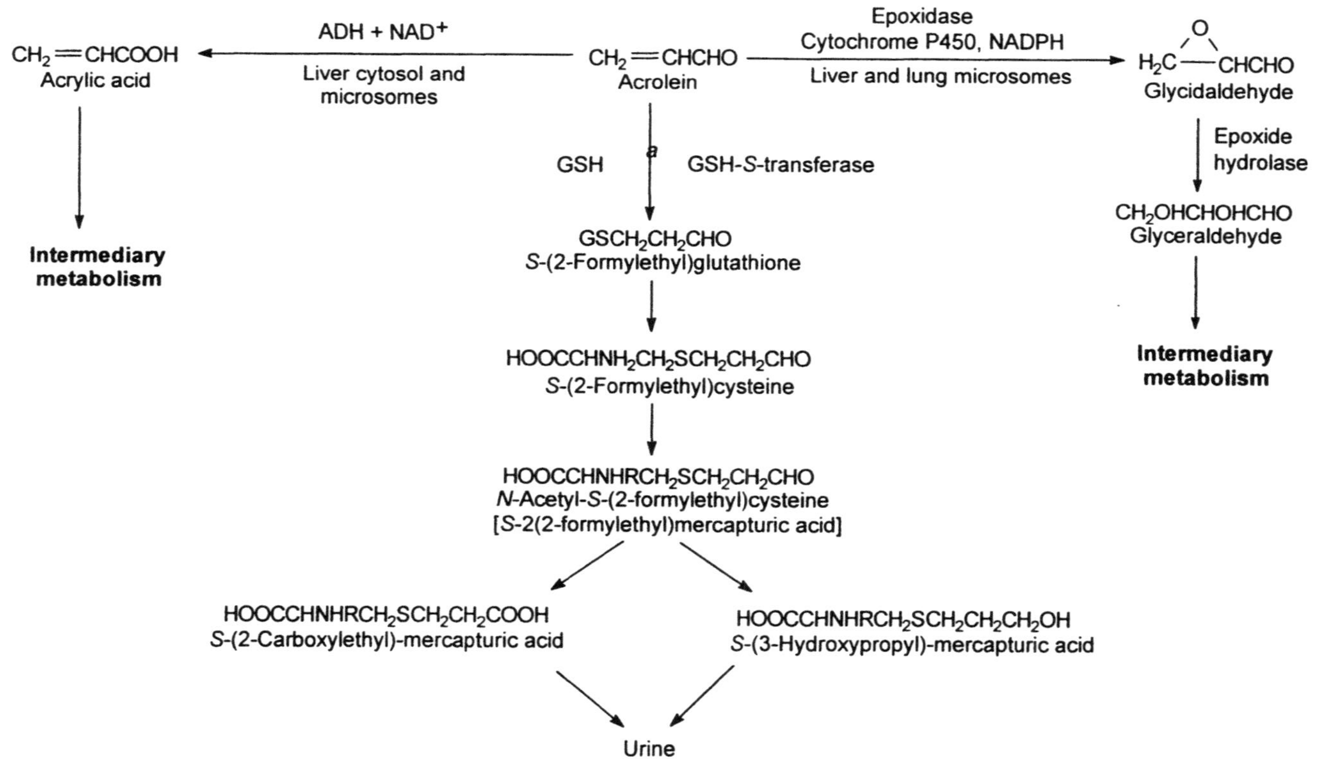

Modified from Draminski et al. (1983)
GSH, glutathione or glutamylcysteinylglycine; ADH, aldehyde dehydrogenase; R, COCH,
ᵃ A spontaneous reaction occurs rapidly

4.2 Toxic effects

4.2.1 Humans

The toxic effects of acrolein to humans have been reviewed (WHO, 1992). It causes intense eye and respiratory irritation, the irritation threshold being about 0.1–0.2 mg/m^3, with no effects reported below 0.05 mg/m^3. This intense irritation may limit the amount of exposure to acrolein, even at low concentrations. For example, exposure to 1 ppm [2.29 mg/m^3] for 5 min results in intolerable eye irritation (Beauchamp et al., 1985).

Skin burns and dermatitis occur after prolonged or repeated exposures (Beauchamp et al., 1985). Sensitization has been reported (Key et al., 1983).

4.2.2 Experimental systems

Acrolein was reported to be ciliostatic in rabbit tracheal slices exposed to > 13 mg/m^3 for 1 h (Dalhamn & Rosengren, 1971). It did not induce sensitization in the guinea-pig maximization test (Susten & Breitenstein, 1990).

Acrolein inhibited the respiratory rate in mice at a concentration as low as 0.7 ppm [1.6 mg/m^3] (Steinhagen & Barrow, 1984). A level of 6 ppm [13.8 mg/m^3] was required to inhibit respiration in rats (Babiuk et al., 1985).

The effects of repeated exposure have been reviewed (Beauchamp et al., 1985; WHO, 1992). Exposure of rats, guinea-pigs, monkeys and dogs by inhalation, the most relevant route, causes reductions in body weight gain, interference with pulmonary function and a variety of histopathological changes in the nose, airways and lungs. After groups of rats, Syrian hamsters and rabbits were exposed to acrolein at concentrations of 0, 0.4, 1.4 or 4.9 ppm [0, 0.92, 3.21 or 11.2 mg/m^3] for 6 h per day, five days per week for 13 weeks, squamous metaplasia and neutrophilic infiltration of the nasal mucosa were observed in rats at ≥ 0.9 mg/m^3; Syrian hamsters had similar nasal effects at 4.9 ppm but minimal inflammatory changes at 1.4 ppm. Rats exposed to 4.9 ppm also had squamous metaplasia in the larynx and trachea and hyperplasia in the bronchi and bronchioli. Syrian hamsters exposed to the same concentration had slight thickening of the larynx and focal hyperplasia and metaplasia of the trachea. Effects were seen on the airways of rabbits only at 4.9 ppm (Feron et al., 1978).

Groups of Fischer 344 rats were exposed to acrolein at concentrations of 0, 0.4, 1.4 or 4.0 ppm [0, 0.92, 3.21 or 9.16 mg/m^3] for 6 h per day, five days per week for 62 days. Studies of lung mechanics and diffusion suggested airway obstruction only at the highest dose, at which severe peribronchiolar and bronchiolar damage was apparent. Similar damage was observed in a few rats at 1.4 ppm, but lung function was unaffected. In contrast, air flow dynamics were significantly enhanced in the rats exposed to 0.4 ppm, with no observed changes in histological appearance or composition. The lack of an apparent effect at 1.4 ppm was considered to be due to a cancellation of the effects observed at the higher and lower levels (Costa et al., 1986).

Exposure of rats for one day or for three consecutive days to acrolein at 0.2 or 0.6 ppm [0.46 or 1.37 mg/m^3] increased cell proliferation in the respiratory tract (Roemer et al., 1993).

Groups of Sprague-Dawley rats, guinea-pigs, beagle dogs and squirrel monkeys (*Saimiri scuirea*) were exposed to acrolein at concentrations of 0, 0.22, 1.0 or 1.8 ppm [0, 0.50, 2.29 and 4.12 mg/m^3] for 24 h per day, continuously for 90 days. Both dogs and monkeys showed eye and respiratory irritation at the highest concentration. Significant histological changes considered to be related to the exposure consisted of squamous metaplasia and basal-cell hyperplasia of the trachea in monkeys and confluent bronchopneumonia in dogs at the highest concentration. While no effects were noted in guinea-pigs and rats at the lowest concentration, animals of both species showed focal liver necrosis and guinea-pigs showed pulmonary inflammation at 1.0 ppm. In dogs exposed continuously, emphysema, acute congestion and focal vacuolization of bronchiolar epithelial cells with increased secretory activity occurred at 0.50 mg/m^3 (Lyon *et al.*, 1970).

In dogs given acrolein in gelatin capsules at 0.1, 0.5 or 1.5 mg/kg bw per day (the high dose being increased to 2 mg/kg bw per day after four weeks) orally for one year, the main effect at the two higher doses was vomiting, the frequency of which decreased with time. The levels of serum albumin, calcium and total protein were decreased at the highest dose, with some variability in red blood cell parameters and coagulation times (Parent *et al.*, 1992b).

Male and female Sprague-Dawley rats treated with acrolein in water at 0.05, 0.5 or 2.5 mg/kg bw per day by gavage for two years showed depression of creatinine phosphokinase activity and a dose-related increase in the frequency of early mortality. No histopathological effects were observed (Parent *et al.*, 1992b). [The results obtained after exposure orally cannot be compared directly with those obtained after inhalation, especially as there are local effects in the respiratory tract. The breakdown of acrolein in water may contribute to the lack of local effects after treatment by gavage.]

In an effort to understand the mechanism of the pulmonary toxicity of acrolein, plasma α_1-proteinase inhibitor, which is inactivated by acrolein, probably by adduct formation with lysine and histidine, has been studied, as it would reduce protection against leukocyte elastase activity in the lungs (Gan & Ansari, 1987, 1989). It has also been suggested that acrolein-induced bronchial hyperresponsiveness is mediated by sulfidopeptide leukotrienes, since an immediate increase in leukotriene C_4 and cyclooxygenases is seen in bronchoalveolar lavage fluid from guinea-pigs before bronchial hyperreactivity. Cultured bovine airway epithelial cells exposed to 100 µmol/L [5.6 mg/L] acrolein released both cyclooxygenases and lipoxygenases (Leikauf *et al.*, 1989; Doupnik & Leikauf, 1990).

There is some evidence that acrolein interferes with aspects of the immune system, including macrophages (Sherwood *et al.*, 1986; Witz *et al.*, 1987; Jakab & Hemenway, 1993), lymphocytes (Wood *et al.*, 1992), thymocytes (Comment *et al.*, 1992) and polymorphonuclear leukocytes (Bridges *et al.*, 1980; Bridges, 1985; Witz *et al.*, 1987). Defense against pulmonary infection due to carbon black was impaired by co-exposure to acrolein, but neither substance alone had this effect (Jakab, 1993).

Acrolein has been reported to decrease cytochrome P450 enzyme activity and to inhibit microsomal cytochrome c reductase *in vitro* (Cooper *et al.*, 1987). Treatment of male Fischer 344 rats with a single intraperitoneal dose of 89 µmol /kg bw [5 mg/kg bw] did not inhibit the reductase, but both cytochrome P450 activity and ethylmorphine *N*-demethylation were decreased after 24 h, to 61 and 35% of the control levels, respectively (Cooper *et al.*, 1992). Acrolein has also been reported to inhibit aldehyde dehydrogenases (Mitchell & Petersen, 1988).

4.3 Reproductive and prenatal effects

4.3.1 Humans

No data were available to the Working Group.

4.3.2 Experimental systems

Acrolein was both embryolethal and teratogenic in rats treated by intraamnionic injection *in vivo* on day 13 of gestation, with doses as low as 0.1 µg/fetus for embryolethality and 5 µg/fetus for malformations. The defects reported included tail abnormalities, microencephaly and prosencephalic hypoplasia (Slott & Hales, 1985). No significant reproductive toxicity was seen after acrolein was administered by gavage to two generations of rats at ≤ 6 mg/kg per day (Parent *et al.*, 1992c), and no developmental toxicity was reported in rabbits administered acrolein by gavage at ≤ 2 mg/kg bw per day on days 7–19 of gestation (Parent *et al.*, 1993).

4.4 Genetic and related effects

4.4.1 Humans

Acrolein-modified DNA was detected by an antibody reaction in the peripheral blood lymphocytes of patients receiving cyclophosphamide. No untreated cancer patients had detectable adducts. There was, however, no clear association between the total drug dose and acrolein–DNA adduct formation (McDiarmid *et al.*, 1991)

4.4.2 Experimental systems (see also Table 5 and Appendices 1 and 2)

(a) Adducts

Acrolein binds to proteins, thus inhibiting critical enzymes involved in replicative DNA synthesis, RNA transcription and cell membrane integrity, although inactivation of enzymes associated with sister chromatid exchange and DNA excision repair could not be established (Wilmer *et al.*, 1986).

Acrolein reacts chemically with proteins and DNA constituents (Nelsestuen, 1980; Chung *et al.*, 1984), but DNA binding has not been demonstrated in acrolein-treated animals. Acrolein-derived adducts were observed in DNA from livers of unexposed mice, rats and humans by a very sensitive procedure (Nath & Chung, 1994).

(b) Mutagenic effects

Acrolein induced SOS repair in *Escherichia coli* PQ37, but only when ethanol was used as the solvent. DNA–histone cross-links were detected in *E. coli* HB101pUC13.

Table 5. Genetic and related effects of acrolein

Test system	Result[a]		Dose[b] LED/HID	Reference
	Without exogenous metabolic system	With exogenous metabolic system		
PRB, SOS (*umu*) induction assay, *Salmonella typhimurium* TA1535 pks 1002	–	0	5.6	Benamira & Marnett (1992)
PRB, *Escherichia coli* PQ37, SOS repair	+	0	0.00	Eder *et al.* (1993)
ECB, *Escherichia coli* HB101pUC13, DNA–histone cross-links	+		3	Kuykendall & Bogdanffy (1992)
SA0, *Salmonella typhimurium* TA100, reverse mutation (spot test)	–	–	17	Florin *et al.* (1980)
SA0, *Salmonella typhimurium* TA100, reverse mutation	–	–	43	Lijinsky & Andrews (1980)
SA0, *Salmonella typhimurium* TA100, reverse mutation	–	–	28	Loquet *et al.* (1981)
SA0, *Salmonella typhimurium* TA100, reverse mutation	–	(+)	38	Haworth *et al.* (1983)
SA0, *Salmonella typhimurium* TA100, reverse mutation	+	–	2.1	Lutz *et al.* (1982)
SA0, *Salmonella typhimurium* TA100, reverse mutation	–	–	0.00	Basu & Marnett (1984)
SA0, *Salmonella typhimurium* TA100, reverse mutation	+		224	Foiles *et al.* (1989)
SA0, *Salmonella typhimurium* TA100, reverse mutation	+	Toxic	0.00	Eder *et al.* (1993)
SA2, *Salmonella typhimurium* TA102, reverse mutation	–	0	0.00	Marnett *et al.* (1985)
SA4, *Salmonella typhimurium* TA104, reverse mutation	+	0	8	Marnett *et al.* (1985)
SA4, *Salmonella typhimurium* TA104, reverse mutation	+	0	224	Foiles *et al.* (1989)
SA5, *Salmonella typhimurium* TA1535, reverse mutation (spot test)	–	–	17	Florin *et al.* (1980)
SA5, *Salmonella typhimurium* TA1535, reverse mutation	–	–	43	Lijinsky & Andrews (1980)
SA5, *Salmonella typhimurium* TA1535, reverse mutation	–	–	28	Loquet *et al.* (1981)
SA5, *Salmonella typhimurium* TA1535, reverse mutation	–	(+)	0.005	Hales (1982)
SA5, *Salmonella typhimurium* TA1535, reverse mutation	–	–	13	Haworth *et al.* (1983)
SA7, *Salmonella typhimurium* TA1537, reverse mutation (spot test)	–	–	17	Florin *et al.* (1980)
SA7, *Salmonella typhimurium* TA1537, reverse mutation	–	–	43	Lijinsky & Andrews (1980)
SA7, *Salmonella typhimurium* TA1537, reverse mutation	–	–	13	Haworth *et al.* (1983)

Table 5 (contd)

Test system	Result[a] Without exogenous metabolic system	Result[a] With exogenous metabolic system	Dose[b] LED/HID	Reference
SA8, *Salmonella typhimurium* TA1538, reverse mutation	–	–	43	Lijinsky & Andrews (1980)
SA9, *Salmonella typhimurium* TA98, reverse mutation	–	–	17	Florin *et al.* (1980)
SA9, *Salmonella typhimurium* TA98, reverse mutation	+	–	8.4	Lijinsky & Andrews (1980)
SA9, *Salmonella typhimurium* TA98, reverse mutation	–	–	28	Loquet *et al.* (1981)
SA9, *Salmonella typhimurium* TA98, reverse mutation	–	–	13	Haworth *et al.* (1983)
SA9, *Salmonella typhimurium* TA98, reverse mutation	–	–	0.00	Basu & Marnett (1984)
SA9, *Salmonella typhimurium* TA98, reverse mutation	–	–	0.00	Claxton (1985)
SA9, *Salmonella typhimurium* TA98, reverse mutation	+	+	0.00	Basu & Marnett (1984)
SAS, *Salmonella typhimurium* hisD3052/nopKM101, reverse mutation	–	–	0.00	Hemminki *et al.* (1980)
ECW, *Escherichia coli* WP2 (*uvrA*), reverse mutation	+	0	560	Hemminki *et al.* (1980)
SSB, *Saccharomyces cerevisiae*, DNA strand breaks and interstrand cross-links	–	0	5.6	Fleer & Brendel (1982)
SCR, *Saccharomyces cerevisiae* S211 and S138, reverse mutation	–	0	100	Izard (1973)
DMM, *Drosophila melanogaster*, SMART eye spot mutation	+		280 feed	Sierra *et al.* (1991)
DMM, *Drosophila melanogaster*, SMART wing spot mutation	+		280 feed	Sierra *et al.* (1991)
DMX, *Drosophila melanogaster*, sex-linked recessive lethal mutation	–		280 feed	Sierra *et al.* (1991)
	+		168 inj	
DMX, *Drosophila melanogaster*, sex-linked recessive lethal mutation	–		3000 feed	Zimmering *et al.* (1985)
	–		200 inj	
DMX, *Drosophila melanogaster*, sex-linked recessive lethal mutation	–		800 feed	Zimmering *et al.* (1989)
DMX, *Drosophila melanogaster*, sex-linked recessive lethal mutation	+		168 inj	Barros *et al.* (1994a,b)
	–		560 feed	
DMN, *Drosophila melanogaster*, sex chromosome loss	–		280 feed	Sierra *et al.* (1991)
			280 inj	
DIA, DNA–protein cross-links, rat nasal mucosal cells *in vitro*	+	0	168	Lam *et al.* (1985)
DIA, DNA–strand breaks (alkaline elution), mouse leukaemia L1210 cells *in vitro*	+	0	56	Eder *et al.* (1993)
DIA, DNA–strand breaks (alkaline elution), Chinese hamster ovary K1 cells *in vitro*	+	0	1.2	Deaton *et al.* (1993)

Table 5 (contd)

Test system	Result[a] Without exogenous metabolic system	Result[a] With exogenous metabolic system	Dose[b] LED/HID	Reference
GCL, Gene mutation, Chinese hamster lung V79 cells, *hprt* locus *in vitro*	+	0	0.06	Smith *et al.* ((1990a)
GCO, Gene mutation, Chinese hamster ovary (CHO) cells, *hprt* locus *in vitro*	–	0	5.6	Foiles *et al.* (1990)
GCO, Gene mutation, Chinese hamster ovary (CHO) cells, *hprt* locus *in vitro*	–	–	5	Parent *et al.* (1991b)
SIC, Sister chromatid exchange, Chinese hamster ovary (CHO) cells *in vitro*	–	–	0.56	Au *et al.* (1980)
SIC, Sister chromatid exchange, Chinese hamster ovary (CHO) cells *in vitro*	(+)	–	1.0	Galloway *et al.* (1987)
CIC, Chromosomal aberrations, Chinese hamster ovary (CHO) cells *in vitro*	–	(+)	2.24	Au *et al.* (1980)
CIC, Chromosomal aberrations, Chinese hamster ovary (CHO) cells *in vitro*	–	–	1.0	Galloway *et al.* (1987)
TCM, Cell transformation, C3H 10T½ mouse cells *in vitro*	–	0	0.00	Abernethy *et al.* (1983) (abstract)
DIH, DNA cross-links, human bronchial epithelial cells *in vitro*	–	0	5.6	Grafström *et al.* (1986)
DIH, DNA strand breaks (alkaline elution), human bronchial epithelial cells *in vitro*	–	0	5.6	Grafström *et al.* (1986)
DIH, DNA single-strand breaks, human myeloid leukaemia K562 cells *in vitro*	+	0	0.3	Crook *et al.* (1986)
DIH, DNA–protein cross-links, human bronchial epithelial cells *in vitro*	+	0	5.6	Grafström *et al.* (1986)
GIH, Gene mutation, human xeroderma pigmentosum fibroblasts, 6-thioguanine resistance *in vitro*	+	0	0.1	Curren (1988)
SHL, Sister chromatid exchange, human lymphocytes *in vitro*	+	0	0.84	Wilmer *et al.* (1986)
DVA, DNA–protein cross-links in rat nasal mucosa *in vivo*	–		1.0 inh 6 h	Lam *et al.* (1985)
DLM, Dominant lethal mutation, mice *in vivo*	–		2.2 ip ×1	Epstein *et al.* (1972)
BID, Binding (covalent) to calf thymus DNA *in vitro*	+	0	58	Chung *et al.* (1984)
***, Binding to poly dC *in vitro*	+	0	1960	Smith *et al.* (1988)
***, Cyclic binding to adenine *in vitro*	+	0	1275	Sodum & Shapiro (1988)
***, Cyclic binding to cytosine *in vitro*	+	0	2125	Sodum & Shapiro (1988)
***, Binding to poly dA *in vitro*	+	0	2.0	Smith *et al.* (1990b)
BID, Binding (covalent) to DNA (dG) of *Salmonella typhimurium in vitro*	+	0	224	Foiles *et al.* (1989)

Table 5 (contd)

Test system	Result[a]		Dose[b] LED/HID	Reference
	Without exogenous metabolic system	With exogenous metabolic system		
BID, Binding (covalent) to DNA (dGuo), Chinese hamster ovary (CHO) cells *in vitro*	+	0	5.6	Foiles *et al.* (1990)
BID, Binding (covalent) to calf thymus DNA *in vitro*	+	0	5.6	Wilson *et al.* (1991)
BID, Binding (covalent) to DNA, human fibroblasts *in vitro*	+	0	5.6	Wilson *et al.* (1991)

[a]+, considered to be positive; (+), considered to be weakly positive in an inadequate study; −, considered to be negative; 0, not tested
[b]LED, lowest effective dose; HID, highest effective dose. In-vitro tests, µg/ml; in-vivo tests, mg/kg bw; 0.00, dose not reported; inh, inhalation; ip, intraperitoneally
***, Not included on the profile
dC, deoxycytosine; dA, deoxyadenine; dG, deoxyguanine; dGuo, deoxyguanosine

Acrolein caused gene mutation in *E. coli* and *Salmonella typhimurium* in the absence of an exogenous metabolic system. The high toxicity of the compound led to conflicting results when the standard protocol was not modified.

No interstrand cross-links or mutations were seen in yeast after exposure to acrolein. In *Drosophila melanogaster*, it induced somatic mutations and (after injection) sex-linked recessive lethal mutations. No sex chromosome loss was detected.

Acrolein induced DNA–protein cross-links and DNA strand breaks in cultured mammalian cells. It induced gene mutations at the *hprt* locus in one study with Chinese hamster V79 cells but not in two other studies with Chinese hamster ovary cells. Acrolein induced sister chromatid exchange and chromosomal aberrations in Chinese hamster ovary cells *in vitro*. As reported in an abstract, it did not transform C3H10T^1/$_2$ cells, but it showed some initiating activity in a two-stage assay in which treatment with acrolein was followed by application of 12-*O*-tetradecanoylphorbol 13-acetate. It induced gene mutations (6-thioguanine resistance) in human xeroderma pigmentosum fibroblasts *in vitro*.

Acrolein did not induce DNA–protein cross-links in nasal mucosa of rats exposed by inhalation *in vivo*. It did not induce dominant lethal mutations in mice after males were treated on a single occasion by intraperitoneal injection.

5. Summary and Evaluation

5.1 Exposure data

Acrolein has been produced commercially since the 1940s. It is used mainly in the production of acrylic acid, a starting material for acrylate polymers. It is also used in the production of DL-methionine and as a herbicide and slimicide.

Acrolein occurs naturally in foods and is formed during the combustion of fossil fuels (including engine exhausts), wood and tobacco and during the heating of cooking oils. Human exposure occurs from these sources and during its production and use.

5.2 Human carcinogenicity data

The available data were inadequate to form the basis for an evaluation of the carcinogenicity of acrolein to humans.

5.3 Animal carcinogenicity data

Acrolein was tested for carcinogenicity in one experiment in mice and in two experiments in rats by oral administration. No increase in tumour incidence was observed in mice or in rats in the one adequate study.

An increased incidence of urinary bladder papillomas was observed in rats receiving intraperitoneal injections of acrolein in combination with uracil in the diet.

5.4 Other relevant data

Acrolein is retained irreversibly in the respiratory tract after exposure by inhalation, probably because of its high tissue reactivity. Consequently, there is little, if any, distribution to other organs. Subcutaneous and oral exposure and long-term inhalation result in some systemic distribution and urinary excretion. Acrolein reacts readily with reduced glutathione, and this is the dominant detoxification pathway.

Acrolein is an intense irritant, and its irritancy may limit exposure to this substance. Repeated inhalation results in changes in the upper and lower respiratory tract. In dogs, acute congestion, changes in bronchiolar epithelial cells and emphysema were found after inhalation of the lowest dose tested.

No data were available on the effects of acrolein on human reproduction. No reproductive toxicity was seen in rats or rabbits treated with acrolein by gavage.

In single studies, acrolein did not induce DNA damage in rats or dominant lethal mutations in mice treated *in vivo*.

In cultured mammalian cells, acrolein induced gene mutation, sister chromatid exchange and DNA damage; weak induction of chromosomal aberrations was observed in one study.

Acrolein induced both somatic and germinal mutations in insects and DNA mutation and DNA damage in bacteria. DNA binding *in vitro* was observed in several studies.

5.5 Evaluation[1]

There is *inadequate evidence* in humans for the carcinogenicity of acrolein.

There is *inadequate evidence* in experimental animals for the carcinogenicity of acrolein.

Overall evaluation

Acrolein *is not classifiable as to its carcinogenicity to humans (Group 3)*.

6. References

Abernethy, D.J., Frazelle, J.H. & Boreiko, C.J. (1983) Relative cytotoxic and transforming potential of respiratory irritants in the C3H/10T1/2 cell transformation system (Abstract Cd-20). *Environ. Mutag.*, **5**, 419

Adams, J.D. & Klaidman, L.K. (1993) Acrolein-induced oxygen radical formation. *Free Radicals Biol. Med.*, **15**, 187–193

Al-Rawithi, S., El-Yazigi, A. & Nicholls, P.J. (1993) Determination of acrolein in urine by liquid chromatography and fluorescence detection of its quinoline derivative. *Pharm. Res.*, **10**, 1587–1590

---

[1] For definition of the italicized terms, see Preamble, pp. 22–26.

American Conference of Governmental Industrial Hygienists (1994) *1994–1995 Threshold Limit Values for Chemical Substances and Physical Agents and Biological Exposure Indices*, Cincinnati, OH, p. 12

Apol, A.G. (1973) *Health Hazard Evaluation/Toxicity Determination Report 72–32, Union Pacific Railroad, Pocatello, ID*, Cincinnati, OH, United States National Institute for Occupational Safety and Health

Arbeidsinspectie [Labour Inspection] (1994) *De Nationale MAC-Lijst 1994* [National MAC list 1994], The Hague, p. 17

Arbetarskyddsstyrelsens [National Board of Occupational Safety and Health] (1993) *Hygieniska Gränsvärden* [Hygienic limit values], Stockholm, p. 12

Atkinson, R. & Arey, J. (1993) *Lifetimes and Fates of Toxic Air Contaminants in California's Atmosphere. Final Report*, Riverside, CA, Statewide Air Pollution Research Center, University of California

Au, W., Sokova, O.I., Kopnin, B. & Arrighi, F.E. (1980) Cytogenetic toxicity of cyclophosphamide and its metabolites *in vitro. Cytogenet. Cell Genet.*, **26**, 108–116

Babiuk, C., Steinhagen, W.H. & Barrow, C.S. (1985) Sensory irritation response to inhaled aldehydes after formaldehyde pretreatment. *Toxicol. appl. Pharmacol.*, **79**, 143–149

Barros, A.R., Comendador, M.A. & Sierra, L.M. (1994a) Acrolein genotoxicity in *Drosophila melanogaster*. II. Influence of *mus201* and *mus308* mutations. *Mutat. Res.*, **306**, 1–8

Barros, A.R., Sierra, L.M. & Comendador, M.A. (1994b) Acrolein genotoxicity in *Drosophila melanogaster*. III. Effects of metabolism modification. *Mutat. Res.*, **321**, 119–126

Basu, A.K. & Marnett, L.J. (1984) Molecular requirements for the mutagenicity of malondialdehyde and related acroleins. *Cancer Res.*, **44**, 2848–2854

Beauchamp, R.O., Jr, Andjelkovich, D.A., Kligerman, A.D., Morgan, K.T. & Heck, H.d'A. (1985) A critical review of the literature on acrolein toxicity. *C.R.C. Crit. Rev. Toxicol.*, **14**, 309–378

Benamira, M. & Marnett, L.J. (1992) The lipid peroxidation product 4-hydroxynonenal is a potent inducer of the SOS response. *Mutat. Res.*, **293**, 1–10

Berhane, K. & Mannervik, B. (1990) Inactivation of the genotoxic aldehyde acrolein by human glutathione transferases of classes alpha, mu, and pi. *Mol. Pharmacol.*, **37**, 251–254

Boettner, E.A. & Ball, G.L. (1980) Thermal degradation products from PVC film in food-wrapping operations. *Am. ind. Hyg. Assoc. J.*, **41**, 513–522

Boor, P.J. & Ansari, G.A.S. (1986) High-performance liquid chromatographic method for quantitation of acrolein in biological samples. *J. Chromatogr.*, **375**, 159–164

Bouscaren, R., Frank, R. & Veldt, C. (1987) *Hydrocarbons. Identification of Air Quality Problems in Member States of the European Communities. Final Report*, Luxembourg, European Commission

Bridges, R.B. (1985) Protective action of thiols on neutrophil function. *Eur. J. respir. Dis.*, **66** (Suppl. 139), 40–48

Bridges, R.B., Hsieh, L. & Haack, D.G. (1980) Effects of cigarette smoke and its constituents on the adherence of polymorphonuclear leukocytes. *Infect. Immun.*, **29**, 1096–1101

Budavari, S. (1989) *The Merck Index*, 11th Ed., Rahway, New Jersey, Merck & Co., p. 118

Castle, C.N. & Smith, T.N. (1974) *Environmental Sampling at a Copper Smelter* (US NTIS PB82-164948), Cincinnati, OH, United States National Institute of Occupational Safety and Health, pp. 4–7, 13–14, 22, 29–30

Chemical Information Services, Inc. (1994) *Directory of World Chemical Producers 1995/96 Standard Edition*, Dallas, TX, p. 16

Choudhury, T.K., Kotiaho, T. & Cooks, R.G. (1992) Analysis of acrolein and acrylonitrile in aqueous solution by membrane introduction mass spectrometry. *Talanta*, **39**, 1113–1120

Chung, F.-L., Young, R. & Hecht, S.S. (1984) Formation of cyclic 1,N^2-propanodeoxyguanosine adducts in DNA upon reaction with acrolein or crotonaldehyde. *Cancer Res.*, **44**, 990–995

Ciccioli, P., Brancaleoni, E., Cecinato, A., Sparapani, R. & Frattoni, M. (1993) Identification and determination of biogenic and anthropogenic volatile organic compounds in forest areas of northern and southern Europe and a remote site of the Himalaya region by high-resolution gas chromatography–mass spectrometry. *J. Chromatogr.*, **643**, 55–69

Claxton, L.D. (1985) Assessment of bacterial mutagenicity methods for volatile and semivolatile compounds and mixtures. *Environ. int.*, **11**, 375–382

Cohen, S.M., Garland, E.M., St John, M., Okamura, T. & Smith, R.A. (1992) Acrolein initiates rat urinary bladder carcinogenesis. *Cancer Res.*, **52**, 3577–3581

Collin, S., Osman, M., Delcambre, S., El-Zayat, A.I. & Dufour, J.-P. (1993) Investigation of volatile flavor compounds in fresh and ripened Domiati cheeses. *J. agric. Food Chem.*, **41**, 1659–1663

Comment, C.E., Blaylock, B.L., Germolec, D.R., Pollock, P.L., Kouchi, Y., Brown, H.W., Rosenthal, G.J. & Luster, M.I. (1992) Thymocyte injury after in vitro chemical exposure: potential mechanisms for thymic atrophy. *J. Pharmacol. exp. Ther.*, **262**, 1267–1273

Cooper, K.O., Witmer, C.M. & Witz, G. (1987) Inhibition of microsomal cytochrome c reductase activity by a series of α,β-unsaturated aldehydes. *Biochem. Pharmacol.*, **36**, 627–631

Cooper, K.O., Witz, G. & Witmer, C. (1992) The effects of α,β-unsaturated aldehydes on hepatic thiols and thiol-containing enzymes. *Fundam. appl. Toxicol.*, **19**, 343–349

Costa, D.L., Kutzman, R.S., Lehmann, J.R. & Drew, R.T. (1986) Altered lung function and structure in the rat after subchronic exposure to acrolein. *Am. Rev. respir. Dis.*, **133**, 286–291

Crook, T.R., Souhami, R.L. & Mclean, A.E.M. (1986) Cytotoxicity, DNA cross-linking and single strand breaks induced by activated cyclophosphamide and acrolein in human leukemia cells. *Cancer Res.*, **46**, 5029–5034

Curren, R.D., Yang, L.L., Conklin, P.M., Grafström, R.C. & Harris, C.C. (1988) Mutagenesis of xeroderma pigmentosum fibroblasts by acrolein. *Mutat. Res.*, **209**, 17–22

Dalhamn, T. & Rosengren, A. (1971) Effect of different aldehydes on tracheal mucosa. *Arch. Otolaryngol.*, **93**, 496–500

Deaton, A.P., Dozier, M.M., Lake, R.S. & Heck, J.D. (1993) Acute DNA strand breaks (SB) induced by acrolein are distinguished from those induced by other aldehydes by modulators of active oxygen (Abstract). *Environ. mol. Mutag.*, **21** (Suppl. 22), 16

Deutsche Forschungsgemeinschaft (1993) *MAK-und-BAT-Werte Liste* [MAK and BAT values list] (Report No. 29), Weinheim, VCH Verlagsgesellschaft, p. 67

Doupnik, C.A. & Leikauf, G.D. (1990) Acrolein stimulates eicosanoid release from bovine airway epithelial cells. *Am. J. Physiol.*, **259**, L222–L229

Draminski, W., Eder, E. & Henschler, D. (1983) A new pathway of acrolein metabolism in rats (Short communication). *Arch. Toxicol.*, **52**, 243–247

Eder, E., Scheckenbach, Deininger, C. & Hoffman, C. (1993) The possible role of α,β-unsaturated carbonyl compounds in mutagenesis and carcinogenesis. *Toxicol. Lett.*, **67**, 87–103

Egle, J.L., Jr (1972) Retention of inhaled formaldehyde, propionaldehyde and acrolein in the dog. *Arch. environ. Health*, **25**, 114–124

Egorov, I.A., Pisarnitskii, A.F., Zinkevich, E.P. & Gavrilov, A.I. (1976) Study of some volatile components of oak wood. *Prikl. Biochim. Microbiol.*, **12**, 108–112 (in Russian) [*Chemical Abstracts*, 85-182256y]

Eller, P.M., ed. (1994) *NIOSH Manual of Analytical Methods*, 4th Ed., Vol. 1, (DHHS (NIOSH) Publ. No. 94-113), Washington DC, United States Government Printing Office, Methods 2501, 2539

Engström, B., Henricks-Eckerman, M.-L. & Ånäs, E. (1990) Exposure to paint degradation products when welding, flame cutting, or straightening painted steel. *Am. ind. Hyg. Assoc. J.*, **51**, 561–565

Environmental Chemicals Data and Information Network (1993) *Acrolein*, Ispra, JRC-CEC, last update: 02.09.1993

Epstein, S.S., Arnold, E., Andrea, J., Bass, W. & Bishop, Y. (1972) Detection of chemical mutagens by the dominant lethal assay in the mouse. *Toxicol. appl. Pharmacol.*, **23**, 288–325

Esterbauer, H., Zollner, H. & Scholz, N. (1975) Reaction of glutathione with conjugated carbonyls. *Z. Naturforsch.*, **30**, 466–473

Esterbauer, H., Schaur, R.J. & Zollner, H. (1991) Chemistry and biochemistry of 4-hydroxynonenal, malonaldehyde and related aldehydes. *Free Radical Biol. Med.*, **11**, 81–128

Etzkorn, W.G., Kurland, J.J. & Neilsen, W.D. (1991) Acrolein and derivatives. In: Kroschwitz, J.I. & Howe-Grant, M., eds, *Kirk-Othmer Encyclopedia of Chemical Technology*, 4th Ed., Vol. 1, New York, John Wiley & Sons, pp. 232–251

European Commission (1994) *Recommendation from a Scientific Expert Group on Occupational Exposure Limits for Acrolein* (SEG/SUN/32C), Luxembourg

Facchini, M.C., Chiavari, G. & Fuzzi, S. (1986) An improved HPLC method for carbonyl compound speciation in the atmospheric liquid phase. *Chemosphere*, **15**, 667–674

Facchini, M.C., Lind, J., Orsi, G. & Fuzzi, S. (1990) Chemistry of carbonyl compounds in Po Valley fog water. *Sci. total Environ.*, **91**, 79–86

Feldstein, M., Bryan, R.J., Hyde, D.L., Levaggi, D.A., Locke, D.C., Rasmussen, R.A. & Warner, P.O. (1989a) 114. Determination of acrolein content of the atmosphere (colorimetric). In: Lodge, J.P., Jr, ed., *Methods of Air Sampling and Analysis*, 3rd Ed., Chelsea, MI, Lewis Publishers, pp. 271–273

Feldstein, M., Bryan, R.J., Hyde, D.L., Levaggi, D.A., Locke, D.C., Rasmussen, R.A. & Warner, P.O. (1989b) 826. Determination of acrolein in air. In: Lodge, J.P., Jr, ed., *Methods of Air Sampling and Analysis*, 3rd Ed., Chelsea, MI, Lewis Publishers, pp. 646–648

Feron, V.J. & Kruysse, A. (1977) Effects of exposure to acrolein vapor in hamsters simultaneously treated with benzo[a]pyrene or diethylnitrosamine. *J. Toxicol. environ. Health*, **3**, 379–394

Feron, V.J., Kruysse, A., Til, H.P. & Immel, H.R. (1978) Repeated exposure to acrolein vapour: subacute studies in hamsters, rats and rabbits. *Toxicology*, **9**, 47–57

Feron, V.J., Til, H.P., de Vrijer, F., Woutersen, R.A., Cassee, F.R. & van Bladeren, P.J. (1991) Aldehydes: occurrence, carcinogenic potential, mechanism of action and risk assessment. *Mutat. Res.*, **259**, 363–385

Finnish Institute of Occupational Health (1994) *Finnish Occupational Exposure Database*, Helsinki (in Finnish)

Fleer, R. & Brendel, M. (1982) Toxicity, interstrand cross-links and DNA fragmentation induced by activated cyclophosphamide in yeast: comparative studies on 4-hydroperoxy-cyclophosphamide, its monofunctional analogon, acrolein, phosphoramide mustard, and nor-nitrogen mustard. *Chem.-biol. Interactions*, 39, 1–15

Florin, I., Rutberg, L., Curvall, M. & Enzell, C.R. (1980) Screening of tobacco smoke constitutents for mutagenicity using the Ames' test. *Toxicology*, 15, 219–232

Foiles, P.G., Akerkar, S.A. & Chung, F.-L. (1989) Application of an immunoassay for cyclic acrolein deoxyguanosine adducts to assess their formation in DNA of *Salmonella typhimurium* under conditions of mutation induction by acrolein. *Carcinogenesis*, 10, 87–90

Foiles, P.G., Akerkar, S.A., Miglietta, L.M. & Chung, F.-L. (1990) Formation of cyclic deoxyguanosine adducts in Chinese hamster ovary cells by acrolein and crotonaldehyde. *Carcinogenesis*, 11, 2059–2061

Galloway, S.M., Armstrong, M.J., Reuben, C., Colman, S., Brown, B., Cannon, C., Bloom, A.D., Nakamura, F., Ahmed, M., Duk, S., Rimpo, J., Margolin, B.H., Resnick, M.A., Anderson, B. & Zeiger, E. (1987) Chromosome aberrations and sister chromatid exchanges in Chinese hamster ovary cells: evaluations of 108 chemicals. *Environ. mol. Mutag.*, 10 (Suppl.), 1–175

Gambino, M., Cericola, R., Corbo, P. & Iannaccone, S. (1993) Carbonyl compounds and PAH emissions from CNG heavy-duty engine. *J. Eng. Gas Turbines Power*, 115, 747–749

Gan, J.C. & Ansari, G.A.S. (1987) Plausible mechanism of inactivation of plasma $alpha_1$-proteinase inhibitor by acrolein. *Res. Commun. chem. Pathol. Pharmacol.*, 55, 419–422

Gan, J.C. & Ansari, G.A.S. (1989) Inactivation of plasma $alpha_1$-proteinase inhibitor by acrolein: adduct formation with lysine and histidine residues. *Mol. Toxicol.*, 2, 137–145

Glaze, W.H., Koga, M. & Cancilla, D. (1989) Ozonation byproducts. 2. Improvement of an aqueous-phase derivatization method for the detection of formaldehyde and other carbonyl compounds formed by the ozonation of drinking water. *Environ. Sci. Technol.*, 23, 838–847

Götze, H.-J. & Harke, S. (1989) Determination of aldehydes and ketones in natural gas combustion in the ppb range by high-performance liquid chromatography. *Fresenius' Z. anal. Chem.*, 335, 286–288

Grafström, R.C., Curren, R.D., Yang, L.L. & Harris, C.C. (1986) Aldehyde-induced inhibition of DNA repair and potentiation of *N*-nitrosocompound-induced mutagenesis in cultured human cells. *Prog. clin. biol. Res.*, 209A, 255–264

Grafström, R.C., Dypbukt, J.M., Willey, J.C., Sundqvist, K., Edman, C., Atzori, L. & Harris, C.C. (1988) Pathological effects of acrolein in cultured human bronchial epithelial cells. *Cancer Res.*, 48, 1717–1721

Grosjean, D. (1990) Atmospheric chemistry of toxic contaminants. 3. Unsaturated aliphatics: acrolein, acrylonitrile, maleic anhydride. *J. Air Waste Manage. Assoc.*, 40, 1664–1668

Haenen, G.R.M.M., Vermeulen, N.P.E., Tai Tin Tsoi, J.N.L., Ragetli, H.M.N., Timmerman, H. & Bast, A. (1988) Activation of the microsomal glutathione-*S*-transferase and reduction of the glutathione dependent protection against lipid peroxidation by acrolein. *Biochem. Pharmacol.*, 37, 1933–1938

Hales, B.F. (1982) Comparison of the mutagenicity and teratogenicity of cyclophosphamide and its active metabolites, 4-hydroxycyclophosphamide, phosphoramide mustard, and acrolein. *Cancer Res.*, 42, 3016–3021

Hansch, C., Leo, A. & Hoekman, D.H. (1995) *Exploring QSAR*, Washington DC, American Chemical Society

Hashmi, M., Vamvakas, S. & Anders, M.W. (1992) Bioactivation mechanism of *S*-(3-oxopropyl)-*N*-acetyl-L-cysteine, the mercapturic acid of acrolein. *Chem. Res. Toxicol.*, **5**, 360–365

Haworth, S., Lawlor, T., Mortelmans, K., Speck, W. & Zeiger, E. (1983) *Salmonella* mutagenicity test results for 250 chemicals. *Environ. Mutag.*, **Suppl. 1**, 3–142

Hemminki, K., Falck, K. & Vainio, H. (1980) Comparison of alkylation rates and mutagenicity of directly acting industrial and laboratory chemicals. Epoxides, glycidyl ethers, methylating and ethylating agents, halogenated hydrocarbons, hydrazine derivatives, aldehydes, thiuram and dithiocarbamate derivatives. *Arch. Toxicol.*, **46**, 277–285

Hess, L.G., Kurtz, A.N. & Stanton, D.B. (1978) Acrolein and derivatives. In: Mark, H.F., Othmer, D.F., Overberger, C.G., Seaborg, G.T. & Grayson, M., eds, *Kirk-Othmer Encyclopedia of Chemical Technology*, 3rd Ed., Vol. 1, New York, John Wiley & Sons, pp. 277–297

Highsmith, V.R., Zweidinger, R.B. & Merrill, R.G. (1988) Characterization of indoor and outdoor air associated with residences using woodstoves: a pilot study. *Environ. int.*, **14**, 213–219

Horvath, J.J., Witmer, C.M. & Witz, G. (1992) Nephrotoxicity of the 1:1 acrolein–glutathione adduct in the rat. *Toxicol. appl. Pharmacol.*, **117**, 200–207

IARC (1979) *IARC Monographs on the Evaluation of the Carcinogenic Risk of Chemicals to Humans*, Vol. 19, *Some Monomers, Plastics and Synthetic Elastomers, and Acrolein*, Lyon, pp. 479–494

IARC (1985) *IARC Monographs on the Evaluation of the Carcinogenic Risk of Chemicals to Humans*, Vol. 36, *Allyl Compounds, Aldehydes, Epoxides and Peroxides*, Lyon, pp. 133–161

IARC (1986) *IARC Monographs on the Evaluation of the Carcinogenic Risk of Chemicals to Humans*, Vol. 38, *Tobacco Smoking*, Lyon, p. 86

IARC (1987a) *IARC Monographs on the Evaluation of Carcinogenic Risks to Humans*, Suppl. 7, *Overall Evaluations of Carcinogenicity: An Updating of* IARC Monographs *Volumes 1–42*, Lyon, p. 78

IARC (1987b) *IARC Monographs on the Evaluation of Carcinogenic Risks to Humans*, Suppl. 7, *Overall Evaluations of Carcinogenicity: An Updating of* IARC Monographs *Volumes 1–42*, Lyon, p. 64

IARC (1987c) *IARC Monographs on the Evaluation of Carcinogenic Risks to Humans*, Suppl. 7, *Overall Evaluations of Carcinogenicity: An Updating of* IARC Monographs *Volumes 1–42*, Lyon, pp. 77–78

IARC (1987d) *IARC Monographs on the Evaluation of Carcinogenic Risks to Humans*, Suppl. 7, *Overall Evaluations of Carcinogenicity: An Updating of* IARC Monographs *Volumes 1–42*, Lyon, pp. 120–122

IARC (1987e) *IARC Monographs on the Evaluation of Carcinogenic Risks to Humans*, Suppl. 7, *Overall Evaluations of Carcinogenicity: An Updating of* IARC Monographs *Volumes 1–42*, Lyon, p. 56

IARC (1987f) *IARC Monographs on the Evaluation of Carcinogenic Risks to Humans*, Suppl. 7, *Overall Evaluations of Carcinogenicity: An Updating of* IARC Monographs *Volumes 1–42*, Lyon, p. 64

IARC (1987g) *IARC Monographs on the Evaluation of Carcinogenic Risks to Humans*, Suppl. 7, *Overall Evaluations of Carcinogenicity: An Updating of* IARC Monographs *Volumes 1–42*, Lyon, pp. 182–184

IARC (1994) *IARC Monographs on the Evaluation of Carcinogenic Risks to Humans*, Vol. 60, *Some Industrial Chemicals*, Lyon, pp. 161–180

IARC (1995) *IARC Monographs on the Evaluation of Carcinogenic Risks to Humans*, Vol. 62, *Wood Dust and Formaldehyde*, Lyon, pp. 217–362

ILO (1991) *Occupational Exposure Limits for Airborne Toxic Substances: Values of Selected Countries* (Occupational Safety and Health Series No. 37), 3rd Ed., Geneva, pp. 6–7

Izard, C. (1973) Studies of the mutagenic effects of acrolein and its two epoxides: glycidol and glycidal, on *Saccharomyces cerevisiae*. *C.R. Acad. Sci. Paris*, **276**, 3037–3040

Izmerov, N.F., ed. (1984) *Acrolein* (Scientific Reviews of Soviet Literature on Toxicity and Hazards of Chemicals), Geneva, UNEP, International Register of Potentially Toxic Chemicals

Jakab, G.J. (1993) The toxicologic interactions resulting from inhalation of carbon black and acrolein on pulmonary antibacterial and antiviral defenses. *Toxicol. appl. Pharmacol.*, **121**, 167–175

Jakab, G.J. & Hemenway, D.R. (1993) Inhalation coexposure to carbon black and acrolein suppresses alveolar macrophage phagocytosis and TNF-alpha release and modulates peritoneal macrophage phagocytosis. *Inhal. Toxicol.*, **5**, 275–289

Kaye, C.M. (1973) Biosynthesis of mercapturic acids from allyl alcohol, allyl esters and acrolein. *Biochem. J.*, **134**, 1093–1101

Key, M.M., Henschel, A.F., Butler, J., Ligo, R.N. & Tabershaw, I.R. (1983) *Occupational Diseases: A Guide to Their Recognition*, Rev. Ed., Cincinnati, OH, United States National Institute for Occupational Safety and Health

Klochkovskii, S.P., Lukashenko, R.D., Podvysotsii, K.S. & Kagramanyan, N.P. (1981) Acrolein and formaldehyde content in the air of quarries. *Bezop. tr. Promsti.*, **12**, 38 [*Chemical Abstracts*, 96-128666] (in Russian)

Kuwata, K., Uebori, M. & Yamasaki, Y. (1979) Determination of aliphatic and aromatic aldehydes in polluted airs as their 2,4-dinitrophenylhydrazones by high performance liquid chromatography. *J. chromatogr. Sci.*, **17**, 264–268

Kuykendall, J.R. & Bogdanffy, M.S. (1992) Efficiency of DNA–histone crosslinking induced by saturated and unsaturated aldehydes *in vitro*. *Mutat. Res.*, **283**, 131–136

Lam, C-W., Casanova, M. & Heck, H.d'A. (1985) Depletion of nasal mucosal glutathione by acrolein and enhancement of formaldehyde-induced DNA–protein cross-linking by simultaneous exposure to acrolein. *Arch. Toxicol.*, **58**, 67–71

Lane, R.H. & Smathers, J.L. (1991) Monitoring aldehyde production during frying by reversed-phase liquid chromatography. *J. Assoc. off. anal. Chem.*, **74**, 957–960

Leikauf, G.D., Doupnik, C.A., Leming, L.M. & Wey, H.E. (1989) Sulfidopeptide leukotrienes mediate acrolein-induced bronchial hyperresponsiveness. *J. appl. Physiol.*, **66**, 1838–1845

Le Lacheur, R.M., Sonnenberg, L.B., Singer, P.C., Christman, R.F. & Charles, M.J. (1993) Identification of carbonyl compounds in environmental samples. *Environ. Sci. Technol.*, **27**, 2745–2753

Lide, D.R., ed. (1993) *CRC Handbook of Chemistry and Physics*, 74th Ed., Boca Raton, FL, CRC Press, p. 3-27

Lijinsky, W. (1988) Chronic studies in rodents of vinyl acetate and compounds related to acrolein. *Ann. N.Y. Acad. Sci.*, **554**, 246–254

Lijinsky, W. & Andrews, A.W. (1980) Mutagenicity of vinyl compounds in *Salmonella typhimurium*. *Teratog. Carcinog. Mutag.*, **1**, 259–267

Lijinsky, W. & Reuber, M.D. (1987) Chronic carcinogenesis studies of acrolein and related compounds. *Toxicol. ind. Health*, **3**, 337–345

Lin, J.M. & Wang, L.H. (1994) Gaseous aliphatic aldehydes in Chinese incense smoke. *Bull. environ. Contam. Toxicol.*, **53**, 374–381

Linnainmaa, M., Eskelinen, T., Louhelainen, K. & Piirainen, J. (1990) *Occupational Hygiene Survey in Bakeries. Final Report*, Kuopio, Kuopio Regional Institute of Occupational Health (in Finnish)

Liotta, F.J., Jr (1993) A peroxide based cetane improvement additive with favorable fuel blending properties. In: *Diesel Fuels for the Nineties: Composition and Additives to Meet Emissions and Performance Needs* (Spec. Publ. SP-994), Warrendale, PA, Society of Automotive Engineers, pp. 149–161

Lipari, F. & Swarin, S.J. (1982) Determination of formaldehyde and other aldehydes in automobile exhaust with an improved 2,4-dinitrophenylhydrazine method. *J. Chromatogr.*, **247**, 297–306

Lipari, F., Dasch, J.M. & Scruggs, W.F. (1984) Aldehyde emissions from wood-burning fireplaces. *Environ. Sci. Technol.*, **18**, 326–330

Lodge, J.P., Jr (1989) 122. Determination of C_1 through C_5 aldehydes in ambient air and source emissions as 2,4-dinitrophenylhydrazones by HPLC. In: Lodge, J.P., Jr, ed., *Methods of Air Sampling and Analysis*, 3rd Ed., Chelsea, MI, Lewis Publishers, pp. 293–295

Löfroth, G., Burton, R.M., Forehand, L., Hammond, S.K., Seila, R.L., Zweidinger, R.B. & Lewtas, J. (1989) Characterization of environmental tobacco smoke. *Environ. Sci. Technol.*, **23**, 610–614

Loquet, C., Toussaint, G. & LeTalaer, J.Y. (1981) Studies on mutagenic constituents of apple brandy and various alcoholic beverages collected in western France, a high incidence area for oesophageal cancer. *Mutat. Res.*, **88**, 155–164

Lutz, D., Eder, E., Neudecker, T. & Henschler, D. (1982) Structure–mutagenicity relationship in α,ß-unsaturated carbonylic compounds and their corresponding allylic alcohols. *Mutat. Res.*, **93**, 305–315

Lyon, J.P., Jenkins, L.J., Jr, Jones, R.A., Coon, R.A. & Siegel, J. (1970) Repeated and continuous exposure of laboratory animals to acrolein. *Toxicol. appl. Pharmacol.*, **17**, 726–732

Marnett, L.J., Hurd, H.K., Hollstein, M.C., Levin, D.E., Esterbauer, H. & Ames, B.N. (1985) Naturally occurring carbonyl compounds are mutagens in *Salmonella* tester strain TA 104. *Mutat. Res.*, **148**, 25–34

Mašek, V. (1972) Aldehydes in the air at workplaces in coal and pitch coking plants. *Staub-Reinhalt. Luft*, **32**, 26–28

Materna, B.L., Jones, J.R., Sutton, P.M., Rothman, N. & Harrison, R.J. (1992) Occupational exposures in California wildland fire fighting. *Am. ind. Hyg. Assoc. J.*, **53**, 69–76

McDiarmid, M.A., Iype, P.T., Kolodner, K., Jacobson-Kram, D. & Strickland, P.T. (1991) Evidence for acrolein-modified DNA in peripheral blood leukocytes of cancer patients treated with cyclophosphamide. *Mutat. Res.*, **248**, 93–99

McNulty, M.J., Heck, H.d'A. & Casanova-Schmitz, M. (1984) Depletion of glutathione in rat respiratory mucosa by inhaled acrolein (Abstract 1695). *Fed. Proc.*, **43**, 575

Miller, B.E. & Danielson, N.D. (1988) Derivatization of vinyl aldehydes with anthrone prior to high-performance liquid chromatography with fluorometric detection. *Anal. Chem.*, **60**, 622–626

Mitchell, D.Y. & Petersen, D.R. (1988) Inhibition of rat liver aldehyde dehydrogenases by acrolein. *Drug Metab. Disposition*, **16**, 37–42

Miyamoto, Y. (1986) Eye and respiratory irritants in jet engine exhaust. *Aviation Space environ. Med.*, **November**, 1104–1108

Nath, R.G. & Chung, F.-L. (1994) Detection of exocyclic $1,N^2$-propanodeoxyguanosine adducts as common DNA lesions in rodents and humans. *Proc. natl Acad. Sci. USA*, **91**, 7491–7495

Nelsestuen, G.L. (1980) Origin of life: consideration of alternatives to proteins and nucleic acids. *J. mol. Evol.*, **15**, 59–72

Neumüller, O.-A. (1979) *Römpps Chemie-Lexikon*, 8th Ed., Vol. 1, Stuttgart, Franckh'sche Verlagshandlung, W. Keller & Co., p. 54

Nishikawa, H., Hayakawa, T. & Sakai, T. (1986) Determination of micro amounts of acrolein in air by gas chromatography. *J. Chromatogr.*, **370**, 327–332

Nishikawa, H., Hayakawa, T. & Sakai, T. (1987a) Determination of acrolein and crotonaldehyde in automobile exhaust by gas chromatography with electron-capture detection. *Analyst*, **112**, 859–862

Nishikawa, H., Hayakawa, T. & Sakai, T. (1987b) Gas chromatographic determination of acrolein in rain water using bromination of *O*-methyloxime. *Analyst*, **112**, 45–48

Ohara, T., Sato, T., Shimizu, N., Prescher, G., Schwind, H. & Weiberg, O. (1985) Acrolein and methacrolein. In: Gerhartz, W., Yamamoto, Y.S., Campbell, F.T., Pfefferkorn, R. & Rounsaville, J.F., eds, *Ullmann's Encyclopedia of Industrial Chemistry*, 5th rev. Ed., Vol. A1, New York, VCH Publishers, pp. 149–160

Otson, R., Fellin, P., Tran, Q. & Stoyanoff, R. (1993) Examination of sampling methods for assessment of personal exposures to airborne aldehydes. *Analyst*, **118**, 1253–1259

Ott, M.G., Teta, J. & Greenberg, H.L. (1989) Lymphatic and hematopoietic tissue cancer in a chemical manufacturing environment. *Am. J. ind. Med.*, **16**, 631–643

Parent, R.A., Caravello, H.E. & Long, J.E. (1991a) Oncogenicity study of acrolein in mice. *J. Am. Coll. Toxicol.*, **10**, 647–659

Parent, R.A., Caravello, H.E. & Harbell, J.W. (1991b) Gene mutation assay of acrolein in the CHO/HGPRT test system. *J. appl. Toxicol.*, **11**, 91–95

Parent, R.A., Caravello, H.E. & Long, J.E. (1992a) Two-year toxicity and carcinogenicity study of acrolein in rats. *J. appl. Toxicol.*, **12**, 131–139

Parent, R.A., Caravello, H.E., Balmer, M.F., Shellenberger, T.E. & Long, J.E. (1992b) One-year toxicity of orally administered acrolein to the beagle dog. *J. appl. Toxicol.*, **12**, 311–316

Parent, R.A., Caravello, H.E. & Hoberman, A.M. (1992c) Reproductive study of acrolein on two generations of rats. *Fundam. appl. Toxicol.*, **19**, 228–237

Parent, R.A., Caravello, H.E., Christian, M.S. & Hoberman, A.M. (1993) Developmental toxicity of acrolein in New Zealand white rabbits. *Fundam. appl. Toxicol.*, **20**, 248–256

Patel, J.M., Wood, J.C. & Leibman, K.C. (1980) The biotransformation of allyl alcohol and acrolein in rat liver and lung preparations. *Drug Metab. Disposition*, **8**, 305–308

Pfäffli, P. (1982) III. Industrial hygiene measurements. *Scand. J. Work. Environ. Health*, **8** (Suppl. 2), 27–43

Phillips, G.F. & Waller, R.E. (1991) Yields of tar and other smoke components from UK cigarettes. *Food chem. Toxicol.*, **29**, 469–474

Prager, B., Jacobson, P., Schmidt, P. & Stern, D., eds (1918) *Beilsteins Handbuch der Organischen Chemie* [Beilsteins Handbook of Organic Chemistry], 4th Ed., Vol. 1, Syst. No. 90, Berlin, Springer, p. 725

Protsenko, G.A., Danilov, V.I., Timchenko, A.N., Nenartovich, A.V., Trubilko, V.I. & Savchenkov, V.A. (1973) Working conditions when metals to which primer has been applied are welded evaluated from the health and hygienic aspect. *Avt. Svarka.*, **2**, 65–68

Rietz, B. (1985) Determination of three aldehydes in the air of working environments. *Anal. Lett.*, **18**, 2369–2379

Roemer, E., Anton, H.J. & Kindt, R. (1993) Cell proliferation in the respiratory tract of the rat after acute inhalation of formaldehyde or acrolein. *J. appl. Toxicol.*, **13**, 103–107

Rossi, R.J. (1992) Odorous hydrocarbon emissions from aircraft. In: *Proceedings from the 85th Annual Meeting and Exhibition of Air and Waste Management Association.* Vol. 7, *Human Health and Environmental Effects*, Pittsburg, PA, Air and Waste Management Association, Paper 92-144.02

Sadtler Research Laboratories (1980) *1980 Cumulative Index*, Philadelphia, PA

Salaman, M.H. & Roe, F.J.C. (1956) Further tests for tumour-initiating activity: *N*,*N*-di-(2-chloroethyl)-*p*-aminophenylbutyric acid (CB1348) as an initiator of skin tumour formation in the mouse. *Br. J. Cancer*, **10**, 363–378

Sanduja, R., Ansari, G.A.S. & Boor, P.J. (1989) 3-Hydroxypropylmercapturic acid: a biologic marker of exposure to allylic and related compounds. *J. appl. Toxicol.*, **9**, 235–238

Sax, N.I. & Lewis, R.J. (1987) *Hawley's Condensed Chemical Dictionary*, 11th Ed., New York, Van Nostrand Reinhold, p. 18

Schweizerische Unfallversicherungsanstalt [Swiss Accident Insurance Company] (1994) *Grenzwerte am Arbeitsplaz* [Limit values in the work place], Lucerne, p. 87

Sherwood, R.L., Leach, C.L., Hatoum, N.S. & Aranyi, C. (1986) Effects of acrolein on macrophage functions in rats. *Toxicol. Lett.*, **32**, 41–49

Shields, P.G., Xu, G.X., Blot, W., Trivers, G.E., Weston, A., Pellizzari, E.D., Qu, Y.H. & Harris, C.C. (1993) Volatile emissions of wok cooking oils (Abstract 701). *Proc. Am. Assoc. Cancer Res.*, **34**, 118

Sierra, L.M., Barros, A.R., García, M., Ferreiro, J.A. & Comendador, M.A. (1991) Acrolein genotoxicity in *Drosophila melanogaster*. I. Somatic and germinal mutagenesis under proficient repair conditions. *Mutat. Res.*, **260**, 247–256

Siewert, R.M., Mitchell, P.J. & Mulawa, P.A. (1993) Environmental potential of natural gas fuel for light-duty vehicles: an engine-dynamometer study of exhaust-emission-control strategies and fuel consumption. In: *Advanced Alternative Fuels Technology* (Spec. Publ. SP-995), Warrendale, PA, Society of Automotive Engineers, pp. 1–17

Sigrist, M.W. (1994) Laser photoacoustic spectrometry for trace gas monitoring. *Analyst*, **119**, 525–531

Slooff, W., Bont, P.F.H., Janus, J.A. & Ros, J.P.M. (1991) *Exploratory Report: Acrolein* (Report No. RIVM-710401009), Bilthoven, National Institute of Public Health and Environmental Protection

Slott, V.L. & Hales, B.F. (1985) Teratogenicity and embryolethality of acrolein and structurally related compounds in rats. *Teratology*, **32**, 65–72

Smith, R.A., Sysel, I.A., Tibbels, T.S. & Cohen, S.M. (1988) Implications for the formation of abasic sites following modification of polydeoxycytidylic acid by acrolein *in vitro*. *Cancer Lett.*, **40**, 103–109

Smith, R.A., Cohen, S.M. & Lawson, T.A. (1990a) Short communication. Acrolein mutagenicity in the V79 assay. *Carcinogenesis*, **11**, 497–498

Smith, R.A., Williamson, D.S., Cerny, R.L. & Cohen, S.M. (1990b) Detection of 1,N^6-propanodeoxyadenosine in acrolein-modified polydeoxyadenylic acid and DNA by ^{32}P postlabeling. *Cancer Res.*, **50**, 3005–3012

Sodum, R.S. & Shapiro, R. (1988) Reaction of acrolein with cytosine and adenine derivatives. *Bioorganic Chem.*, **16**, 272–282

Steinhagen, W.H. & Barrow, C.S. (1984) Sensory irritation structure–activity study of inhaled aldehydes in B6C3F1 and Swiss-Webster mice. *Toxicol. appl. Pharmacol.*, **72**, 495–503

Susten, A.S. & Breitenstein, M.J. (1990) Failure of acrolein to produce sensitization in the guinea pig maximization test (Short communication). *Contact Derm.*, **22**, 299–300

Triebig, G. & Zober, M.A. (1984) Indoor air pollution by smoke constituents—a survey. *Prev. Med.*, **13**, 570–581

Työministeriö [Ministry of Labour] (1993) *HTP-Arvot 1993* [Occupational exposure limits 1993], Tampere, p. 8 (in Finnish)

Umano, K. & Shibamoto, T. (1987) Analysis of acrolein from heated cooking oils and beef fat. *J. agric. Food Chem.*, **35**, 909–912

Union Carbide (1993) *Material Safety Data Sheet: Acrolein, Inhibited*, Danbury, CT

United Kingdom Health and Safety Executive (1993) *Occupational Exposure Limits 1993* (EH 40/93), London, Her Majesty's Stationery Office, p. 13

United States Environmental Protection Agency (1985) *Health and Environmental Effects Profile for Acrolein* (EPA-600/X-85/369; US NTIS PB88-171269), Cincinnati, OH, Environmental Criteria and Assessment Office, Office of Research and Development

United States Environmental Protection Agency (1986) Method 8030. Acrolein, acrylonitrile, acetonitrile. In: *Test Methods for Evaluating Solid Waste— Physical/Chemical Methods* (US EPA No. SW-846), 3rd Ed., Vol. 1A, Washington DC, Office of Solid Waste and Emergency Response, pp. 1–11

United States Environmental Protection Agency (1988) *Compendium of Methods for the Determination of Toxic Organic Compounds in Ambient Air* (EPA Report No. EPA-600/4-89-017; US NTIS PB90-116989), Research Triangle Park, NC, Office of Research and Development, Methods T05, T011

United States Food and Drug Administration (1994) Slimicides. *US Code fed. Regul.*, **Title 21**, part 176.300, pp. 200–202

United States National Institute for Occupational Safety and Health (1994a) *NIOSH Pocket Guide to Chemical Hazards* (DHHS (NIOSH Publ. No. 94-116), Cincinnati, OH, pp. 6–7

United States National Institute for Occupational Safety and Health (1994b) *National Occupational Exposure Survey (1981–1983)*, Cincinnati, OH

United States Occupational Safety and Health Administration (1990) *OSHA Analytical Methods Manual*, Part 1, Vol. 1, Salt Lake City, UT, US Department of Labor, Method 52

United States Occupational Safety and Health Administration (1994) Air contaminants. *US Code fed. Regul.*, **Title 29**, Part 1910.1000, pp. 6–19

Vainiotalo, S. & Matveinen, K. (1992) Determination of acrolein in air with 2,4-dinitrophenylhydrazine impregnated adsorbent tubes in the presence of water. In: Brown, R.H., Curtis, M., Saunders, K.J. & Vandendriessche, S., eds, *Clean Air at Work. New Trends in Assessment and Measurement for the 1990s*, Luxembourg, European Commission, pp. 204–206

Vainiotalo, S. & Matveinen, K. (1993) Cooking fumes as a hygienic problem in the food and catering industries. *Am. ind. Hyg. Assoc. J.*, **54**, 376–382

Verschueren, K. (1983) *Handbook of Environmental Data on Organic Chemicals*, 2nd Ed., New York, Van Nostrand Reinhold, pp. 157–160

Victorin, K., Ståhlberg, M., Alsberg, T., Strandell, M., Westerholm, R. & Egebäck, K.-E. (1988) Emission of mutagenic, irritating and odorous substances from gasoline fueled vehicles with different emission control system. *Chemosphere*, **17**, 1767–1780

Volkova, Z.A. & Bagdinov, Z.M. (1969) Industrial hygiene problems in vulcanization processes of rubber production. *Gig. Sanit.*, **34**, 33–40 (in Russian) [*Chemical Abstracts*, 71-128354b]

Warner-Selph, M.A. (1989) *Measurements of Toxic Exhaust Emissions from Gasoline-powered Light-duty Vehicles. Final Report*, San Antonio, TX, Southwest Research Institute

Weast, R.C. & Astle, M.J. (1985) *CRC Handbook of Data on Organic Compounds*, Vols I & II, Boca Raton, FL, CRC Press, pp. 42 (I), 456 (II)

Westerholm, R., Almén, J., Li, H., Rannung, U. & Rosén, Å. (1992) Exhaust emissions from gasoline-fuelled light duty vehicles operated in different driving conditions: a chemical and biological characterization. *Atmos. Environ.*, **26B**, 79–90

WHO (1992) *Acrolein* (Environmental Health Criteria 127), Geneva

Wilmer, J.L., Erexson, G.L. & Kligerman, A.D. (1986) Attenuation of cytogenetic damage by 2-mercaptoethanesulphonate in cultured human lymphocytes exposed to cyclophosphamide and its reactive metabolites. *Cancer Res.*, **46**, 203–210

Wilson, V.L., Foiles, P.G., Chung, F.-L., Poves, A.C., Frank, A.A. & Harris, C.C. (1991) Detection of acrolein and crotonaldehyde DNA adducts in cultured human cells and canine peripheral blood lymphocytes by ^{32}P-postlabeling and nucleotide chromatography. *Carcinogenesis*, **12**, 1483–1490

Witz, G., Lawrie, N.J., Amoruso, M.A. & Goldstein, B.D. (1987) Inhibition by reactive aldehydes of superoxide anion radical production from stimulated polymorphonuclear leukocytes and pulmonary alveolar macrophages. *Biochem. Pharmacol.*, **36**, 721–726

Wood, S.C., Karras, J.G. & Holsapple, M.P. (1992) Integration of the human lymphocyte into immunotoxicological investigations. *Fundam. appl. Toxicol.*, **18**, 450–459

Yasuhara, A., Dennis, K.J. & Shibamoto, T. (1989) Development and validation of new analytical method for acrolein in air. *J. Assoc. off. anal. Chem.*, **72**, 749–751

Zimmering, S., Mason, J.M., Valencia, R. & Woodruff, R.C. (1985) Chemical mutagenesis testing in *Drosophila*. II. Results of 20 coded compounds tested for the National Toxicology Program. *Environ. Mutag.*, **7**, 87–100

Zimmering, S., Mason, J.M. & Valencia, R. (1989) Chemical mutagenesis testing in *Drosophila*. VII. Results of 22 coded compounds tested in larval feeding experiments. *Environ. mol. Mutag.*, **14**, 245–251

CROTONALDEHYDE

1. Exposure Data

1.1 Chemical and physical data

1.1.1 Nomenclature

Crotonaldehyde

Chem. Abstr. Serv. Reg. No.: 4170-30-3
Chem. Abstr. Name: 2-Butenal
IUPAC Systematic Name: Crotonaldehyde
Synonyms: 2-Butenaldehyde; crotonal; crotonic aldehyde; crotylaldehyde; 1-formylpropene; β-methylacrolein; propylene aldehyde

***cis* Isomer**

Chem. Abstr. Serv. Reg. No.: 15798-64-8
Chem. Abstr. Name: (Z)-2-Butenal
IUPAC Systematic Name: (Z)-Crotonaldehyde
Synonyms: *cis*-2-Butenal; *cis*-crotonaldehyde

***trans* Isomer**

Chem. Abstr. Serv. Reg. No.: 123-73-9
Chem. Abstr. Name: (E)-2-Butenal
IUPAC Systematic Name: (E)-Crotonaldehyde
Synonyms: 2(E)-Butenal; *trans*-2-butenal; *trans*-2-buten-1-al; *trans*-crotonal; *trans*-crotonaldehyde

1.1.2 Structural and molecular formulae and relative molecular mass

C_4H_6O Relative molecular mass: 70.09

1.1.3 Chemical and physical properties of the pure substance

(a) *Description*: Colourless liquid with a pungent odour (Eastman Chemical Co., 1994)

(b) *Boiling-point*: 104–105 °C (Lide, 1993)

(c) *Melting-point*: –74 °C (Lide, 1993)

(d) *Density*: 0.8495 at 20 °C/4 °C (Lide, 1993)

(e) *Spectroscopy data*: Infrared (prism [1457]; grating [29705]), ultraviolet [418], nuclear magnetic resonance (proton [11669]) and mass [57] spectral data have been reported (Sadtler Research Laboratories, 1980; Weast & Astle, 1985).

(f) *Solubility*: Soluble in water (150 g/L at 20 °C), acetone, benzene, diethyl ether and ethanol (Eastman Chemical Co., 1991; Lide, 1993)

(g) *Volatility*: Vapour pressure, 32 mm Hg [4.3 kPa] at 20 °C; relative vapour density (air = 1), 2.4 (Budavari, 1989; Eastman Chemical Co., 1994)

(h) *Stability*: Readily dimerizes when pure; slowly oxidizes to crotonic acid (Budavari, 1989); polymerizes to become inflammable and explosive (Eastman Chemical Co., 1994)

(i) *Reactivity*: Lower explosive limit, 2.15% at 24 °C; reacts violently with bases, strong oxidizing agents and polymerization initiators (Eastman Chemical Co., 1994)

(j) *Octanol/water partition coefficient (P)*: log P, 0.63 (United States National Library of Medicine, 1994)

(k) *Conversion factor*: $mg/m^3 = 2.87 \times ppm$[1]

1.1.4 Technical products and impurities

Crotonaldehyde is available commercially at a purity of 90–99%. The commercial product consists of > 95% *trans* and < 5% *cis* isomer. A typical specification for crotonaldehyde is as follows: minimal purity, 90%; maximal acidity (as crotonic acid), 0.15%; maximal water content, 8.5%; aldol, 0.1% max; butyraldehyde, 0.02% max; low-boiling-point compounds (including acetaldehyde [see IARC, 1987a] and butyraldehyde), 0.20% max; butyl alcohol, 0.15 max; and high-boiling-point compounds, 1.0% max (Blau *et al.*, 1987; Eastman Chemical Co., 1993; Spectrum Chemical Mfg Corp., 1994). The trade names for *trans*-crotonaldehyde include Topanel CA.

1.1.5 Analysis

Gas chromatography (GC) is the method most often used for the identification and determination of crotonaldehyde. Several methods have been developed for its quantitative determination in the presence of saturated aldehydes, including sulfitometric determination of double bonds and potentiometric titration with bromine in methanol (Blau *et al.*, 1987). Selected methods for the determination of crotonaldehyde in various matrices are presented in Table 1.

---

[1] Calculated from: mg/m^3 = (relative molecular mass/24.45) × ppm, assuming normal temperature (25 °C) and pressure (101 kPa)

Table 1. Methods for the analysis of crotonaldehyde

Sample matrix	Sample preparation	Assay procedure	Limit of detection	Reference
Air	Adsorb on sorbent coated with 2-(hydroxymethyl)piperidine on XAD-2; desorb with toluene	GC/FID & GC/MS	2 µg/sample	Eller (1994)
	Draw air through midget bubbler buffer; analyse as crotonaldehyde containing hydroxylamine/formate oxime derivative	DPP	ND	Eller (1994)
	Adsorb on glass-fibre filter coated with DNPH and phosphoric acid; extract with acetonitrile	HPLC/UV	93 µg/m^3 (0.56 mg/sample)	US Occupational Safety and Health Administration (1990)
	Draw air through midget impinger containing acidified DNPH and isooctane; extract DNPH derivatives with hexane:dichloromethane (70:30) solution; evaporate to dryness; dissolve in methanol	Reversed-phase HPLC/UV	NR	US Environmental Protection Agency (1988)
Engine exhaust	Absorb in aqueous acid solution of DNPH; extract hydrazone derivative with hexane	CGC	NR	Kachi et al. (1988)

GC, gas chromatography; FID, flame ionization detection; MS, mass spectrometry; DPP, differential pulse polarography; ND, not determined; DNPH, 2,4-dinitrophenylhydrazine; HPLC/UV, high-performance liquid chromatography/ultraviolet detection; NR, not reported; CGC, capillary gas chromatography

A method for identifying carbonyl compounds, including crotonaldehyde, in environmental samples involves derivatization with O-(2,3,4,5,6-pentafluorobenzyl)hydroxylamine hydrochloride, followed by GC–mass spectrometry (Le Lacheur et al., 1993). A similar method for the determination of low-relative-molecular-mass aldehydes formed by the ozonation of drinking-water, including crotonaldehyde, involves derivatization with O-(2,3,4,5,6-pentafluorobenzyl)-hydroxylamine hydrochloride and analysis by high-resolution capillary GC. The limits of detection with GC–electron capture detection and GC–mass spectrometry with ion selective monitoring were 1.2 and 11.2 µg/L, respectively (Glaze et al., 1989).

A method similar to that of the United States Environmental Protection Agency (1988) (see Table 1) has been described that can be used for ambient air, industrial emissions and automobile exhaust (Lodge, 1989).

Several sampling methods for the assessment of personal exposures to airborne aldehydes, including crotonaldehyde, due to emissions from methanol-fuelled vehicles were evaluated. Derivatization of aldehydes was used to enhance the integrity of the samples and to facilitate analysis. 2,4-Dinitrophenylhydrazine was found to provide better sensitivity and resolution of the hydrazones of crotonaldehyde than the other derivatization agents, (2-hydroxymethyl)-

piperidine and *N*-benzylethanolamine. Passive and active samplers containing 2,4-dinitrophenylhydrazine were also evaluated (Otson *et al.*, 1993).

Another method for the determination of acrolein and crotonaldehyde in automobile exhaust is based on bromination of their *O*-benzyloxime derivatives, followed by GC–electron capture detection (Nishikawa *et al.*, 1987).

1.2 Production and use

1.2.1 Production

The most widely used method for the production of crotonaldehyde is aldol condensation of acetaldehyde, followed by dehydration. A process involving direct oxidation of 1,3-butadiene (see IARC, 1992) to crotonaldehyde with palladium catalysis has also been reported (Baxter, 1979; Blau *et al.*, 1987).

Crotonaldehyde is produced by four companies in Japan and by one company each in China, Germany, India and the United States of America (Chemical Information Services, Inc., 1994).

1.2.2 Use

The main use of crotonaldehyde in the past was in the manufacture of *n*-butanol (Baxter, 1979), but this process has been largely displaced by the oxo process. Currently, the most extensive use of crotonaldehyde is in the manufacture of sorbic acid; crotonic acid is made commercially by oxidation of crotonaldehyde and 3-methoxybutanol by the reaction of methanol with crotonaldehyde, followed by reduction (Blau *et al.*, 1987).

Crotonaldehyde has been used as a warning agent in fuel gases, for locating breaks and leaks in pipes (Budavari, 1989). It also has been used in the preparation of rubber accelerators, in leather tanning, as an alcohol denaturant (Sax & Lewis, 1987) and as a stabilizer for tetraethyllead (see IARC, 1987b).

1.3 Occurrence

1.3.1 Natural occurrence

Crotonaldehyde occurs in emissions from the Chinese arbor vitae plant (Isidorov *et al.*, 1985) and in gases emitted from volcanoes (Graedel *et al.*, 1986). It has been detected in biogenic emissions from pine (0.19 $\mu g/m^3$) and deciduous (0.49 $\mu g/m^3$) forests in Europe and in remote, high-altitude areas with scarce vegetation (e.g. Nepal; 0.24–3.32 $\mu g/m^3$) (Ciccioli *et al.*, 1993). Crotonaldehyde also occurs naturally in many foods (Feron *et al.*, 1991). (See also section 1.3.4.)

1.3.2 Occupational exposure

The National Occupational Exposure Survey conducted between 1981 and 1983 indicated that 387 employees in the United States were potentially exposed occupationally to crotonaldehyde (United States National Institute of Occupational Safety and Health, 1994a). The estimate is based on a survey of companies and did not involve measurements of actual exposures.

The concentrations of crotonaldehyde measured in various occupational settings in 1982–92 by the Finnish Institute of Occupational Health (1994) were all below the occupational exposure limit of 6 mg/m^3 [range unspecified] recommended in Finland. A total of 24 industries were listed, which included manufacture of electric and electronic equipment, plastic, glass and metal products, and pulp and paper.

In a bakery in Finland, crotonaldehyde was measured at 0.23 mg/m^3 near a pan used for frying doughnuts and pastries (Linnainmaa et al., 1990). In a chemical plant in the United States, the concentration in general area air samples varied from not detected to 3.2 mg/m^3 and those in two personal samples were 1.9 and 2.1 mg/m^3 (Fannick, 1982). In an occupational survey at a printing and finish plant in the United States, crotonaldehyde was not detected in either area or personal samples (detection limit, 0.7–2.1 mg/m^3) (Rosensteel & Tanaka, 1976). It was detected at a concentration of 1–7 mg/m^3 in a factory producing aldehydes in Germany (Bittersohl, 1975).

1.3.3 Air

Crotonaldehyde has been detected in exhaust gases from both gasoline engines, at 0.09–1.33 ppm [0.26–3.82 mg/m^3] (Kuwata et al., 1979; Nishikawa et al., 1987), and diesel engines, at 0.01–0.04 ppm [0.03–0.12 mg/m^3] (Creech et al., 1982; Lipari & Swarin, 1982). In a study of light-duty vehicles with catalysts and fuelled with natural gas, the emission rate of crotonaldehyde was 1.87 mg/kg fuel (Siewert et al., 1993). The emission from heavy-duty engines run with natural gas was 0.12 mg/kW h [0.03 mg/MJ] and that of diesel engines 0.42 mg/kW h [0.12 mg/MJ] (Gambino et al., 1993). In another study with diesel engines, the crotonaldehyde emissions were 2.8 mg/bhp h [1.04 mg/MJ] from base fuel, 1.2 mg/bhp h [0.44 mg/MJ] from fuels with ethylhexyl nitrate-containing additives and 1.0 mg/bhp h [0.37 mg/MJ] from fuels with peroxide-containing additives (Liotta, 1993).

Crotonaldehyde has been determined as an odorous constituent in aircraft emissions. In the United States, it was emitted by model jet engines operating at idling power at concentrations of 0.35–0.98 ppm [1.0–2.81 mg/m^3] (Rossi, 1992). In Japan, concentrations of 0.007–0.051 ppm [0.02–0.15 mg/m^3] were measured about 50 m behind a low-smoke combustor jet engine at idling power (Miyamoto, 1986).

Crotonaldehyde was detected in cigarette smoke at 10–228 μg/cigarette (Kuwata et al., 1979; IARC, 1986), in emissions from wood-burning fireplaces at 6–116 mg/kg (Lipari et al., 1984), in smoke from the burning of 5 kg of polypropylene in a 27-m^3 room at 1.1 ppm [3.16 mg/m^3] (Woolley, 1982) and in thermal degradation products of steel-protection paints at 6 mg/m^3 (Henricks-Eckerman et al., 1990). It was also identified in off-gases of polyurethane foam at 40–80 °C and 90% relative humidity (Krzymien, 1989). Crotonaldehyde was released from seven of 11 samples of food packaging during heating in a microwave oven, at concentrations of 0.10–4.5 μg/in^2 [0.016–0.70 μg/cm^2] (McNeal & Hollifield, 1993).

1.3.4 Food

Crotonaldehyde was found in many fruits (e.g. apples, guavas, grapes, strawberries and tomatoes) at concentrations of < 0.01 ppm [mg/kg]; in cabbage, cauliflower, Brussels sprouts, carrots and celery leaves at concentrations of 0.02–0.1 ppm; in bread, cheese, milk, meat, fish and beer at concentrations of 0–0.04 ppm; and in wine at concentrations of 0–0.7 ppm (Feron *et al.*, 1991). It was detected in samples of various liquors at < 0.02–0.21 ppm (Miller & Danielson, 1988). It has also been identified in wheat flavour essence (McWilliams & Mackey, 1969) and in beef fat heated at 150 °C in nitrogen (Yamato *et al.*, 1970).

1.3.5 Other

Crotonaldehyde was detected qualitatively in one of 42 human milk samples from volunteers in four cities (Pellizzari *et al.*, 1982).

1.4 Regulations and guidelines

Occupational exposure limits and guidelines for the mixture of *cis* and *trans* isomers and the *trans* isomer of crotonaldehyde in several countries are given in Table 2.

2. Studies of Cancer in Humans

Bittersohl (1975) recorded cases of cancer diagnosed between 1967 and 1972 among 220 people employed in a factory for the production of aldehydes in the former German Democratic Republic. Air measurements showed concentrations of crotonaldehyde in the workroom in the range of 1–7 mg/m^3; other chemicals were also present. Nine malignant neoplasms were found in men: two squamous-cell carcinomas of the oral cavity, five squamous-cell carcinomas of the lung, one adenocarcinoma of the stomach and one adenocarcinoma of the colon. On the basis of crude comparisons with national figures, the author suspected an excess cancer risk among the employees. [The Working Group noted that the data were too sparse to be conclusive.]

3. Studies of Cancer in Experimental Animals

Oral administration

Rat: Groups of 23 and 27 male Fischer 344 rats, six weeks of age, were given 0, 0.6 or 6.0 mmol/L crotonaldehyde (purity, > 99%) in distilled drinking-water [0, 42 or 421 mg/L] for 113 weeks. Survival was similar in all groups; 17, 13 and 16 rats in the three groups survived to 110 weeks. Throughout the study, those rats receiving the high dose of crotonaldehyde had lower body weights than either the controls or those at the low dose. Gross lesions and representative samples from all major organs [unspecified] were taken for microscopic

Table 2. Occupational exposure limits for crotonaldehyde

Country	Year	Concentration (mg/m³)	Interpretation
Australia	1991	6 (*cis* and *trans*)	TWA
Belgium	1991	5.7 (*cis* and *trans*)	TWA
Bulgaria	1993	0.5 (*cis* and *trans*)	STEL
Denmark	1991	6 (*cis* and *trans*, and *trans*)	TWA
Finland	1993	6 (*cis* and *trans*)	TWA; skin notation
		17 (*cis* and *trans*)	STEL
France	1991	6 (*trans*)	TWA
Germany	1993	None (*trans*)	TWA; justifiably suspected of having carcinogenic potential
Italy	1993	6 (*cis* and *trans*)	TWA
Netherlands	1994	6 (*cis* and *trans*)	TWA
Russian Federation	1991	0.5 (*trans*)	STEL
Switzerland	1994	6 (*trans*)	TWA; skin notation
United Kingdom	1993	6 (*trans*)	TWA
		18 (*trans*)	STEL
USA			
ACGIH	1994	5.7 (*cis* and *trans*)	TWA
NIOSH	1994	6 (*cis* and *trans*)	TWA
OSHA	1994	6 (*cis* and *trans*)	TWA

From ILO (1991); Deutsche Forschungsgemeinschaft (1993); Environmental Chemicals Data and Information Network (1993); United Kingdom Health and Safety Executive (1993); Työministeriö (1993); American Conference of Governmental Industrial Hygienists (ACGIH) (1994); Arbeidsinspectie (1994); Schweizerische Unfallversicherungsanstalt (1994); United States National Institute of Occupational Safety and Health (1994b); United States Occupational Safety and Health Administration (1994); TWA, time-weighted average; STEL, short-term exposure limit

examination; particular attention was paid to lesions in the liver, including altered liver-cell foci. Hepatocellular carcinomas were seen in 0/23 control rats, 2/27 at the low dose and 0/23 at the high dose; and neoplastic nodules were found in 0/23 controls, 9/27 at the low dose [$p = 0.01$] and 1/23 at the high dose. Altered liver-cell foci [considered by the authors to be precursors of hepatocellular neoplasms] were observed in 1/23 controls, 23/27 at the low dose ($p < 0.001$) and 13/23 at the high dose ($p < 0.001$). Liver damage, reported as moderate to severe and including fatty metamorphosis, focal liver necrosis, fibrosis and cholestasis, was seen only in 10/23 rats given the high dose [$p < 0.001$]; none of these 10 animals had preneoplastic or neoplastic lesions. With the exception of two transitional-cell papillomas of the bladder among animals receiving the low dose, no increase in the frequency of tumours was observed (Chung *et al.*, 1986a).

4. Other Data Relevant to an Evaluation of Carcinogenicity and Its Mechanisms

4.1 Absorption, distribution, metabolism and excretion

4.1.1 Humans

No data were available to the Working Group.

4.1.2 Experimental systems

No specific information was available on the absorption and distribution of crotonaldehyde, and little information is available on its metabolism; however, it is a substrate for aldehyde dehydrogenase (Cederbaum & Dicker, 1982). It has been reported to be conjugated with glutathione with or without glutathione S-transferase (Boyland & Chasseaud, 1967; Maniara et al., 1990; Esterbauer et al., 1991; Stenberg et al., 1992; Wang et al., 1992); and 3-hydroxy-1-methyl- and 2-carboxyl-1-methyl-mercapturic acids have been reported in the urine of rats after exposure to crotonaldehyde (Gray & Barnsley, 1971).

4.2 Toxic effects

4.2.1 Humans

Crotonaldehyde is a potent eye, respiratory and skin irritant (Coenraads et al., 1975). In humans, exposure to crotonaldehyde at 4.1 ppm [11.8 mg/m^3] for 15 min was highly irritating to the nose and upper respiratory tract, with lachrymation after 30 sec. Brief exposure to 45 ppm [129 mg/m^3] was very disagreeable, and conjunctival irritation was the predominant reaction (Rinehart, 1967). A textile worker was reported to have become sensitized to crotonaldehyde (Shmunes & Kempton, 1980).

4.2.2 Experimental systems

Crotonaldehyde causes eye, respiratory and skin irritation in animals. The concentration that reduces the respiratory rate to 50% was reported to be 3.5 ppm [10.0 mg/m^3] in mice and 23.2 ppm [66.6 mg/m^3] in rats (Steinhagen & Barrow, 1984; Babiuk et al., 1985).

After crotonaldehyde was administered by gavage to male and female Fischer 344 rats and B6C3F1 mice at doses of 2.5, 5, 10, 20 or 40 mg/kg bw per day on five days per week for 13 weeks, dose-related mortality was observed in rats at doses ≥ 5 mg/kg bw per day, but no deaths were seen in mice. Dose-related lesions of the forestomach (hypertension, inflammation, hyperkeratosis and necrosis) were seen in rats at doses ≥ 10 mg/kg bw per day and in mice at 40 mg/kg bw per day. Acute inflammation of the nasal cavity was seen in rats at doses ≥ 5 (males) and 20 mg/kg bw per day (females) (Wolfe et al., 1987).

Intraperitoneal injection to rats of crotonaldehyde at 450 μmol/kg bw [31.5 mg/kg bw] decreased cytochrome P450 levels to 67% and ethylmorphine N-demethylase activity to 23% of control levels within 24 h (Cooper et al., 1992). [Figure 2 of the paper shows that cytochrome P450 reductase activity at 24 h was 70% of the control value but not significantly different.]

Crotonaldehyde inhibits chemotaxis and adherence of human polymorphonuclear leukocytes *in vitro* (Bridges *et al.*, 1977, 1980; Bridges, 1985). Exposure of human polymorphonuclear cells *in vitro* inhibited superoxide anion production (Witz *et al.*, 1985); this effect may be related to decreases in surface and soluble sulfhydryl groups (Witz *et al.*, 1987). Intraperitoneal treatment of mice with crotonaldehyde resulted in necrosis of the thymus and splenic atrophy (Warholm *et al.*, 1984).

The compound is ciliostatic *in vitro* (Dalhamn & Rosengren, 1971). It inhibits the activities of some enzymes *in vitro*, including cytochrome P450 (Cooper *et al.*, 1987a) and aldehyde dehydrogenase (Dicker & Cederbaum, 1984, 1986).

4.3 Reproductive and prenatal effects

No data were available to the Working Group.

4.4 Genetic and related effects

4.4.1 Humans

No data were available to the Working Group.

4.4.2 Experimental systems (See also Table 3 and Appendices 1 and 2)

(a) DNA adducts

Crotonaldehyde binds to DNA and induces DNA–protein cross-links *in vitro*. It modifies DNA by forming cyclic 1,N^2-propanedeoxyguanosine. Both the 1- and the N^2 positions of guanine are involved in base-pairing, hence the presence of the cyclic adduct may lead to mutation (Chung *et al.*, 1986b). With a ^{32}P-postlabelling technique, authentic 1,N^2-propanedeoxyguanosine was shown to co-chromatograph with DNA of the skin of Sencar mice that had been treated topically *in vivo* with 1.4 mmol [98 mg] crotonaldehyde (Chung *et al.*, 1989). Exocyclic 1,N^2-propanodeoxyguanosine adducts detected by ^{32}P-postlabelling, however, occur *in vivo* in the absence of exposure to either crotonaldehyde or acrolein. The estimated total numbers of adducts in DNA from liver were 1.0–1.7/10^6 guanine bases for mice, 0.2–1.0 for rats and 0.3–2.0 for humans (Nath & Chung, 1994).

Not only propano-adducts but also small quantities of etheno-adducts can be detected after incubation of crotonaldehyde with deoxyadenosine or deoxyguanosine. The reaction probably involves autoxidation to an epoxide intermediate (Chen & Chung, 1994).

(b) Mutagenicity

SOS repair functions were induced in *Escherichia coli* PQ37, and a weak SOS response was obtained in *Salmonella typhimurium* TA1535/pSK1002. Crotonaldehyde induced both forward and reverse mutations in *S. typhimurium* TA100, provided that a preincubation protocol was used.

Table 3. Genetic and related effects of crotonaldehyde

Test system	Result[a] Without exogenous metabolic activation	Result[a] With exogenous metabolic activation	Dose[b] (LED/HID)	Reference
PRB, SOS chromotest, *Escherichia coli* PQ37	–	0	53	Eder et al. (1992)
PRB, SOS chromotest, *Escherichia coli* PQ37	(+)[c]	0	0.00	Eder et al. (1993)
PRB, SOS (*umu*) induction assay, *Salmonella typhimurium* TA1535 pKS 1002	(+)	0	21	Benamira & Marnett (1992)
ECB, DNA–protein cross-linking, *Escherichia coli* HB101 pUC13, modified filter binding assay	+	0	175	Kuykendall & Bogdanffy (1992)
SAF, *Salmonella typhimurium* BA9, forward mutation, liquid assay	+[d]	0	43	Ruiz-Rubio et al. (1984)
SA0, *Salmonella typhimurium* TA100, reverse mutation	+	+	21	Neudecker et al. (1981)
SA0, *Salmonella typhimurium* TA100, reverse mutation	(+)	+	100	Haworth et al. (1983)
SA0, *Salmonella typhimurium* TA100, reverse mutation, preincubation	+	0	70	Neudecker et al. (1989)
SA0, *Salmonella typhimurium* TA100, reverse mutation, preincubation assay	–	0	75	Cooper et al. (1987b)
SA0, *Salmonella typhimurium* TA100, reverse mutation, preincubation assay	+[d]	+	5	Lijinsky & Andrews (1980)
SA0, *Salmonella typhimurium* TA100, reverse mutation, preincubation assay	+[e]	0	200	Eder et al. (1992)
SA4, *Salmonella typhimurium* TA104, reverse mutation	+	0	20	Marnett et al. (1985)
SA5, *Salmonella typhimurium* TA1535, reverse mutation	–	–	500	Lijinsky & Andrews (1980)
SA5, *Salmonella typhimurium* TA1535, reverse mutation	–	–	167	Haworth et al. (1983)
SA7, *Salmonella typhimurium* TA1537, reverse mutation	–	–	500	Lijinsky & Andrews (1980)
SA7, *Salmonella typhimurium* TA1537, reverse mutation	–	–	167	Haworth et al. (1983)
SA8, *Salmonella typhimurium* TA1538, reverse mutation	–	–	500	Lijinsky & Andrews (1980)

Table 3 (contd)

Test system	Result[a]		Dose[b] (LED/HID)	Reference
	Without exogenous metabolic activation	With exogenous metabolic activation		
SA9, *Salmonella typhimurium* TA98, reverse mutation	–	–	500	Lijinsky & Andrews (1980)
SA9, *Salmonella typhimurium* TA98, reverse mutation	–[c]	–	167	Haworth et al. (1983)
SAS, *Salmonella typhimurium* BA9, reverse mutation, liquid assay	+[d]		43	Ruiz-Rubio et al. (1984)
DMX, *Drosophila melanogaster*, sex-linked recessive lethal mutation	+		3500 inj	Woodruff et al. (1985)
DMX, *Drosophila melanogaster*, sex-linked recessive lethal mutation	–		4000 feed	Woodruff et al. (1985)
DMH, *Drosophila melanogaster*, heritable translocation	+		3500 inj	Woodruff et al. (1985)
URP, Unscheduled DNA synthesis, primary rat hepatocytes *in vitro*	–		?	Williams et al. (1989)
CGC, Chromosomal aberrations, spermatogonia treated, spermatocytes observed, mouse *in vivo*	+		300	Moutschen-Dahmen et al. (1976)
CGG, Chromosomal aberrations, spermatogonia treated, spermatogonia observed, mouse *in vivo*	+		300	Moutschen-Dahmen et al. (1976)
BID, Binding (covalent) to calf thymus DNA *in vitro*	+	0	21	Chung et al. (1984)
***, Binding (covalent) to nucleosides and 5′-mononucleotides *in vitro*	+	0	70	Eder & Hoffman (1992)
BVD, Binding (covalent) to DNA in mouse skin *in vivo*	+		300 top × 15	Chung et al. (1989)

[a] +, considered to be positive; (+), considered to be weakly positive in an inadequate study; –, considered to be negative; 0, not tested
[b] LED, lowest effective dose; HID, highest effective dose. In-vitro tests, mg/ml; in-vivo tests, mg/kg bw; 0,00, dose not reported
[c] Ethanol used as solvent in place of dimethyl sulfoxide
[d] Negative in plate incorporation assay
[e] Positive with and without metabolic activation with a threefold higher cell density at 35 μg/ml
***, Not included on profile

Sex-linked recessive lethal mutations and reciprocal translocations were induced in *Drosophila melanogaster* injected with crotonaldehyde. The compound did not induce unscheduled DNA synthesis in primary cultures of rat hepatocytes.

It induced chromosomal aberrations in the spermatogonia of mice after administration in the drinking-water or by injection.

5. Summary and Evaluation

5.1 Exposure data

Crotonaldehyde is produced principally as an intermediate for the production of sorbic acid. It was formerly used in large amounts in the production of *n*-butanol.

Crotonaldehyde occurs naturally in foods and is formed during the combustion of fossil fuels (including engine exhausts), wood and tobacco and in heated cooking oils. Human exposure occurs from these sources and may occur during its production and use.

5.2 Human carcinogenicity data

The available data were too limited to form the basis for an evaluation of the carcinogenicity of crotonaldehyde to humans.

5.3 Animal carcinogenicity data

Crotonaldehyde was tested for carcinogenicity in one study in male rats by administration in the drinking-water. Increased incidences of hepatic neoplastic nodules and altered liver-cell foci were seen, but these were not dose-related.

5.4 Other relevant data

Crotonaldehyde is a substrate for aldehyde dehydrogenase and forms conjugates with glutathione, in the presence or absence of glutathione transferase. Mercapturic acid metabolites have been identified in urine.

Crotonaldehyde is a potent irritant, and it has been reported to interfere with immune function.

Crotonaldehyde did not induce DNA damage in rat hepatocytes *in vitro* in a single study. It was mutagenic to insects and bacteria. It bound to DNA of mouse skin *in vivo* after topical application and to DNA *in vitro* and caused formation of DNA–protein cross-links.

5.5 Evaluation[1]

There is *inadequate evidence* in humans for the carcinogenicity of crotonaldehyde.

There is *inadequate evidence* in experimental animals for the carcinogenicity of crotonaldehyde.

Overall evaluation

Crotonaldehyde *is not classifiable as to its carcinogenicity to humans (Group 3)*.

6. References

American Conference of Governmental Industrial Hygienists (1994) *1994–1995 Threshold Limit Values for Chemical Substances and Physical Agents and Biological Exposure Indices*, Cincinnati, OH, p. 17

Arbeidsinspectie [Labour Inspection] (1994) *De Nationale MAC-Lijst 1986* [National MAC list 1986], The Hague, p. 22

Babiuk, C., Steinhagen, W.H. & Barrow, C.S. (1985) Sensory irritation response to inhaled aldehydes after formaldehyde pretreatment. *Toxicol. appl. Pharmacol.*, **79**, 143–149

Baxter, W.F., Jr (1979) Crotonaldehyde. In: Mark, H.F., Othmer, D.F., Overberger, C.G., Seaborg, G.T. & Grayson, N., eds, *Kirk-Othmer Encyclopedia of Chemical Technology*, 3rd Ed., Vol. 7, New York, John Wiley & Sons, pp. 207–218

Benamira, M. & Marnett, L.J. (1992) The lipid peroxidation product 4-hydroxynonenal is a potent inducer of the SOS response. *Mutat. Res.*, **293**, 1–10

Bittersohl, G. (1975) Epidemiological research on cancer risk by aldol and aliphatic aldehydes. *Environ. Qual. Saf.*, **4**, 235–238

Blau, W., Baltes, H. & Mayer, D. (1987) Crotonaldehyde and crotonic acid. In: Gerhartz, W., Yamamoto, Y.S., Kaudy, L., Pfefferkorn, R. & Rounsaville, J.F., eds, *Ullmann's Encyclopedia of Industrial Chemistry*, 5th rev. Ed., Vol. A8, New York, VCH Publishers, pp. 83–90

Boyland, E. & Chasseaud, L.F. (1967) Enzyme-catalysed conjugations of glutathione with unsaturated compounds. *Biochemistry*, **104**, 95–102

Bridges, R.B. (1985) Protective action of thiols on neutrophil function. *Eur. J. respir. Dis.*, **66** (Suppl. 139), 40–48

Bridges, R.B., Kraal, J.H., Huang, L.J.T. & Chancellor, M.B. (1977) Effects of cigarette smoke components on in vitro chemotaxis of human polymorphonuclear leukocytes. *Infect. Immun.*, **16**, 240–248

Bridges, R.B., Hsieh, L. & Haack, D.G. (1980) Effects of cigarette smoke and its constituents on the adherence of polymorphonuclear leukocytes. *Infect. Immun.*, **29**, 1096–1101

Budavari, S., ed. (1989) *The Merck Index*, 11th Ed., Rahway, NJ, Merck & Co., pp. 406–407

---

[1] For definition of the italicized terms, see Preamble, pp. 22–26.

Cederbaum, A.I. & Dicker, E. (1982) Evaluation of the role of acetaldehyde in the actions of ethanol on gluconeogenesis by comparison with the effects of crotonol and crotonaldehyde. *Alcohol. Clin. exp. Res.*, **6**, 100–109

Chemical Information Services, Inc. (1994) *Directory of World Chemical Producers 1995/96 Standard Edition*, Dallas, TX, p. 207

Chen, H.-J.C. & Chung, F.-L. (1994) Formation of etheno adducts in reactions of enals via autoxidation. *Chem. Res. Toxicol.*, **7**, 857–860

Chung, F.-L., Young, R. & Hecht, S.S. (1984) Formation of cyclic 1,N^2-propanodeoxyguanosine adducts in DNA upon reaction with acrolein or crotonaldehyde. *Cancer Res.*, **44**, 990–995

Chung, F.-L., Tanaka, T. & Hecht, S.S. (1986a) Induction of liver tumors in F334 rats by crotonaldehyde. *Cancer Res.*, **46**, 1285–1289

Chung, F.-L., Hecht, S.S. & Palladio, G. (1986b) Formation of cyclic nucleic acid adducts from some simple α,β-unsaturated carbonyl compounds and cyclic nitrosamines In: Singer, B. & Bartsch, H., eds, *The Role of Cyclic Nucleic Acid Adducts in Carcinogenesis and Mutagenesis* (IARC Scientific Publications No. 70), Lyon, IARC, pp. 207–225

Chung, F.-L., Young, R. & Hecht, S.S. (1989) Detection of cyclic 1,N^2-propanodeoxyguanosine adducts in DNA of rats treated with *N*-nitrosopyrrolidine and mice treated with crotonaldehyde. *Carcinogenesis*, **10**, 1291–1297

Ciccioli, P., Brancaleoni, E., Cecinato, A., Sparapani, R. & Frattoni, M. (1993) Identification and determination of biogenic and anthropogenic volatile organic compounds in forest areas of northern and southern Europe and a remote site of the Himalaya region by high-resolution gas chromatography–mass spectrometry. *J. Chromatogr.*, **643**, 55–69

Coenraads, P.J., Bleumink, E. & Nater, J.P. (1975) Susceptibility to primary irritants: age dependence and relation to contact allergic reactions. *Contact Derm.*, **1**, 377–381

Cooper, K.O., Witmer, C.M. & Witz, G. (1987a) Inhibition of microsomal cytochrome c reductase activity by a series of α,β-unsaturated aldehydes. *Biochem. Pharmacol.*, **36**, 627–631

Cooper, K.O., Witz, G. & Witmer, C.M. (1987b) Mutagenicity and toxicity studies of several α,β-unsaturated aldehydes in the *Salmonella typhimurium* mutagenicity assay. *Environ. Mutag.*, **9**, 289–295

Cooper, K.O., Witz, G. & Witmer, C.M. (1992) The effects of α,β-unsaturated aldehydes on hepatic thiols and thiol-containing enzymes. *Fundam. appl. Toxicol.*, **19**, 343–349

Creech, G., Johnson, R.T. & Stoffer, J.O. (1982) Part I. A comparison of three different high pressure liquid chromatography systems for the determination of aldehydes and ketones in diesel exhaust. *J. chromatogr. Sci.*, **20**, 67–72

Dalhamn, T. & Rosengren, A. (1971) Effect of different aldehydes on tracheal mucosa. *Arch. Otolaryngol.*, **93**, 496–500

Deutsche Forschungsgemeinschaft (1993) *MAK- und BAT-Werte Liste* [MAK and BAT values list 1993] (Report No. 29), Weinheim, VCH Verlagsgesellschaft, p. 26

Dicker, E. & Cederbaum, A.I. (1984) Inhibition of the oxidation of acetaldehyde and formaldehyde by hepatocytes and mitochondria by crotonaldehyde. *Arch. Biochem. Biophys.*, **234**, 187–196

Dicker, E. & Cederbaum, A.I. (1986) Inhibition of CO_2 production from aminopyrine or methanol by cyanamide or crotonaldehyde and the role of mitochondrial aldehyde dehydrogenase in formaldehyde oxidation. *Biochim. biophys. Acta*, **883**, 91–97

Eastman Chemical Co. (1991) *Technical Data Sheet: Crotonaldehyde (2-Butenal)*, Kingsport, TN

Eastman Chemical Co. (1993) *Sales Specification Sheet: Crotonaldehyde (PM 161)*, Kingsport, TN

Eastman Chemical Co. (1994) *Material Safety Data Sheet: 'Eastman' Crotonaldehyde*, Kingsport, TN

Eder, E. & Hoffman, C. (1992) Identification and characterization of deoxyguanosine–crotonaldehyde adducts. Formation of 7,8 cyclic adducts and 1,N^2,7,8 bis-cyclic adducts. *Chem. Res. Toxicol.*, **5**, 802–808

Eder, E., Deininger, C., Neudecker, T. & Deininger, D. (1992) Mutagenicity of β-alkyl substituted acrolein congeners in the *Salmonella typhimurium* strain TA100 and genotoxicity testing in the SOS chromotest. *Environ. mol. Mutag.*, **19**, 338–345

Eder, E., Scheckenbach, S., Deininger, C. & Hoffman, C. (1993) The possible role of α,β-unsaturated carbonyl compounds in mutagenesis and carcinogenesis. *Toxicol. Lett.*, **67**, 87–103

Eller, P.M., ed. (1994) *NIOSH Manual of Analytical Methods*, 4th Ed., Vol. 1, (DHHS (NIOSH) Publ. No. 94-113), Washington DC, United States Government Printing Office, Methods 2539, 3516

Environmental Chemicals Data and Information Network (1993) *Crotonaldehyde*, Ispra, JRC-CEC, last update: 02.09.1993

Esterbauer, H., Schaur, R.J. & Zollner, H. (1991) Chemistry and biochemistry of 4-hydroxynonenal, malonaldehyde and related aldehydes. *Free Radicals Biol. Med.*, **11**, 81–128

Fannick, N. (1982) *Sandoz Colors and Chemicals, East Hanover, New Jersey* (Health Hazard Evaluation Report, No. HETA-81-102-1244), Cincinnati, OH, United States National Institute for Occupational Safety and Health, Hazard Evaluations and Technical Assistance Branch

Feron, V.J., Til, H.P., de Vrijer, F., Woutersen, R.A., Cassee, F.R. & van Bladeren, P.J. (1991) Aldehydes: occurrence, carcinogenic potential, mechanism of action and risk assessment. *Mutat. Res.*, **259**, 363–385

Finnish Institute of Occupational Health (1994) *Finnish Occupational Exposure Database*, Helsinki (in Finnish)

Gambino, M., Cericola, R., Corbo, P. & Iannaccone, S. (1993) Carbonyl compounds and PAH emissions from CNG heavy-duty engine. *J. Engin. Gas Turbines Power*, **115**, 747–749

Glaze, W.H., Koga, M. & Cancilla, D. (1989) Ozonation byproducts. 2. Improvement of an aqueous-phase derivatization method for the detection of formaldehyde and other carbonyl compounds formed by the ozonation of drinking water. *Environ. Sci. Technol.*, **23**, 838–847

Graedel, T.E., Hawkins, D.T. & Claxton, L.D. (1986) *Atmospheric Chemical Compounds. Sources, Occurrence, and Bioassay,* San Diego, CA, Academic Press, pp. 295–314

Gray, J.M. & Barnsley, E.A. (1971) The metabolism of crotyl phosphate, crotyl alcohol and crotonaldehyde. *Xenobiotica*, **1**, 55–67

Haworth, S., Lawlor, T., Mortelmans, K., Speck, W. & Zeiger, E. (1983) *Salmonella* mutagenicity test results for 250 chemicals. *Environ. Mutag.*, **Suppl. 1**, 3–142

Henricks-Eckerman, M.-L., Engström, B. & Ånäs, E. (1990) Thermal degradation products of steel protective paints. *Am. ind. Hyg. Assoc. J.*, **51**, 241–244

IARC (1986) *IARC Monographs on the Evaluation of the Carcinogenic Risk of Chemicals to Humans*, Vol. 38, *Tobacco Smoking*, Lyon, p. 86

IARC (1987a) *IARC Monographs on the Evaluation of Carcinogenic Risks to Humans*, Suppl. 7, *Overall Evaluations of Carcinogenicity: An Updating of* IARC Monographs *Volumes 1–42*, Lyon, pp. 77–78

IARC (1987b) *IARC Monographs on the Evaluation of Carcinogenic Risks to Humans*, Suppl. 7, *Overall Evaluations of Carcinogenicity: An Updating of* IARC Monographs *Volumes 1–42*, Lyon, pp. 230–232

IARC (1992) *IARC Monographs on the Evaluation of Carcinogenic Risks to Humans*, Vol. 54, *Occupational Exposures to Mists and Vapours from Strong Inorganic Acids; and Other Industrial Chemicals*, Lyon, pp. 237–285

ILO (1991) *Occupational Exposure Limits for Airborne Toxic Substances: Values of Selected Countries* (Occupational Safety and Health Series No. 37), 3rd Ed., Geneva, pp. 60–61, 114–115

Isidorov, V.A., Zenkevich, I.G. & Ioffe, B.V. (1985) Volatile organic compounds in the atmosphere of forests. *Atmos. Environ.*, **19**, 1–8

Kachi, H., Akiyama, K. & Tsuruga, F. (1988) Analytical method for aldehyde emissions from methanol engines in 2,4-dinitrophenylhydrazones form using a glass capillary column. *Jpn Soc. Automotive Eng. Rev.*, **9**, 97–100

Krzymien, M.E. (1989) GC–MS analysis of organic vapours emitted from polyurethane foam insulation. *Int. J. environ. anal. Chem.*, **36**, 193–207

Kuwata, K., Uebori, M. & Yamasaki, Y. (1979) Determination of aliphatic and aromatic aldehydes in polluted airs as their 2,4-dinitrophenylhydrazones by high performance liquid chromatography. *J. chromatogr. Sci.*, **17**, 264–268

Kuykendall, J.R. & Bogdanffy, M.S. (1992) Efficiency of DNA–histone crosslinking induced by saturated and unsaturated aldehydes *in vitro*. *Mutat. Res.*, **283**, 131–136

Le Lacheur, R.M., Sonnenberg, L.B., Singer, P.C., Christman, R.F. & Charles, M.J. (1993) Identification of carbonyl compounds in environmental samples. *Environ. Sci. Technol.*, **27**, 2745–2753

Lide, D.R., ed. (1993) *CRC Handbook of Chemistry and Physics*, 74th Ed., Boca Raton, FL, CRC Press, p. 3–190

Lijinsky, W. & Andrews, A.W. (1980) Mutagenicity of vinyl compounds in *Salmonella typhimurium*. *Teratog. Carcinog. Mutag.*, **1**, 259–267

Linnainmaa, M., Eskelinen, T., Louhelainen, K. & Piirainen, J. (1990) *Työhygieeniset Mittaukset Leipomoissa* [Occupational hygiene survey in bakeries], Kuopio, Kuopio Regional Institute of Occupational Health (in Finnish)

Liotta, F.J., Jr (1993) A peroxide based cetane improvement additive with favorable fuel blending properties. In: *Diesel Fuels for the Nineties: Composition and Additives to Meet Emissions and Performance Needs* (Spec. Publ. SP-994), Warrendale, PA, Society of Automotive Engineers, pp. 149–161

Lipari, F. & Swarin, S.J. (1982) Determination of formaldehyde and other aldehydes in automobile exhaust with an improved 2,4-dinitrophenylhydrazine method. *J. Chromatogr.*, **247**, 297–306

Lipari, F., Dasch, J.M. & Scruggs, W.F. (1984) Aldehyde emissions from wood-burning fireplaces. *Environ. Sci. Technol.*, **18**, 326–330

Lodge, J.P., Jr (1989) 122. Determination of C_1 through C_5 aldehydes in ambient air and source emissions as 2,4-dinitrophenylhydrazones by HPLC. In: Lodge, J.P., Jr, ed., *Methods of Air Sampling and Analysis*, Chelsea, MI, Lewis Publishers, pp. 293–295

Maniara, W.M., Santiago, A., Jowa, L. & Witz, G. (1990) The detection of glutathione–aldehyde adducts in red blood cells incubated with acrolein and crotonaldehyde (Abstract 2802). *FASEB J.*, **4**, A749

Marnett, L.J., Hurd, H.K., Hollstein, M.C., Levin, D.E., Esterbauer, H. & Ames, B.N. (1985) Naturally occurring carbonyl compounds are mutagens in *Salmonella* tester strain TA 104. *Mutat. Res.*, **148**, 25–34

McNeal, T.P. & Hollifield, H.C. (1993) Determination of volatile chemicals released from microwave-heat-susceptor food packaging. *J. Assoc. off. anal. Chem. Int.*, **76**, 1268–1275

McWilliams, M. & Mackey, A.C. (1969) Wheat flavor components. *J. Food Sci.*, **34**, 493–496

Miller, B.E. & Danielson, N.D. (1988) Derivatization of vinyl aldehydes with anthrone prior to high-performance liquid chromatography with fluorometric detection. *Anal. Chem.*, **60**, 622–626

Miyamoto, Y. (1986) Eye and respiratory irritants in jet engine exhaust. *Aviat. Space environ. Med.*, **November**, 1104–1108

Moutschen-Dahmen, J., Moutschen-Dahmen, M., Houbrechts, N. & Colizzi, A. (1976) Cytotoxicity and mutagenicity of two aldehydes: crotonaldehyde and butyraldehyde, in the mouse. *Bull. Soc. R. Sci. Liège*, **45**, 58–72 (in French)

Nath, R.G. & Chung, F.-L. (1994) Detection of exocyclic $1,N^2$-propanodeoxyguanosine adducts as common DNA lesions in rodents and humans. *Proc. natl Acad. Sci. USA*, **91**, 7491–7495

Neudecker, T., Lutz, D., Eder, E. & Henschler, D. (1981) Crotonaldehyde is mutagenic in a modified *Salmonella typhimurium* mutagenicity testing system. *Mutat. Res.*, **91**, 27–31

Neudecker, T., Eder, E., Deininger, C. & Henschler, D. (1989) Crotonaldehyde is mutagenic in *Salmonella typhimurium* TA 100. *Environ. mol. Mutag.*, **14**, 146–148

Nishikawa, H., Hayakawa, T. & Sakai, T. (1987) Determination of acrolein and crotonaldehyde in automobile exhaust gas by gas chromatography with electron-capture detection. *Analyst*, **112**, 859–862

Otson, R., Fellin, P., Quang, T. & Stoyanoff, R. (1993) Examination of sampling methods for assessment of personal exposure to airborne aldehydes. *Analyst*, **118**, 1253–1259

Pellizzari, E.D., Hartwell, T.D., Harris, B.S.H., III, Waddell, R.D., Whitaker, D.A. & Erickson, M.D. (1982) Purgeable organic compounds in mother's milk. *Bull. environ. Contam. Toxicol.*, **28**, 322–328

Rinehart, W.E. (1967) The effect on rats of single exposures to crotonaldehyde vapor. *Am. ind. Hyg. Assoc. J.*, **28**, 561–566

Rosensteel, R.E. & Tanaka, S. (1976) *Rock Hill Printing and Finishing Company, Rock Hill, SC* (Health Hazard Evaluation Report No. HHE-75-89-344), Cincinnati, OH, United States National Institute for Occupational Safety and Health

Rossi, R.J. (1992) Odorous hydrocarbon emissions from aircraft. In: *Proceedings of the 85th Annual Meeting and Exhibition of Air and Waste Management Association, Kansas City, MO*, Vol. 7, *Human Health and Environmental Effects*, Pittsburgh, PA, Air and Waste Management Association Paper 92-114-02

Ruiz-Rubio, M., Hera, C. & Pueyo, C. (1984) Comparison of a forward and a reverse mutation assay in *Salmonella typhimurium* measuring L-arabinose resistance and histidine prototrophy. *EMBO J.*, **3**, 1435–1440

Sadtler Research Laboratories (1980) *1980 Cumulative Index*, Philadelphia, PA

Sax, N.I. & Lewis, R.J. (1987) *Hawley's Condensed Chemical Dictionary*, 11th Ed., New York, Van Nostrand Reinhold, p. 323

Schweizerische Unfallversicherungsanstalt [Swiss Accident Insurance Company] (1994) *Grenzwerte am Arbeitsplaz* [Limit values in the work place], Lucerne, p. 24

Shmunes, E. & Kempton, R.J. (1980) Allergic contact dermatitis to dimethoxane in a spin finish. *Contact Derm.*, **6**, 421–424

Siewert, R.M., Mitchell, P.J. & Mulawa, P.A. (1993) Environmental potential of natural gas fuel for light-duty vehicles: an engine-dynamometer study of exhaust-emission-control strategies and fuel consumption. In: *Advanced Alternative Fuels Technology* (Spec. Publ. SP-995), Warrendale, PA, Society of Automotive Engineers, pp. 1–17

Spectrum Chemical Mfg Corp. (1994) *1993–1994 Spectrum Chemical and Safety Products Catalog*, New Brunswick, NJ, p. C-140

Steinhagen, W.H. & Barrow, C.S. (1984) Sensory irritation structure–activity study of inhaled aldehydes in B6C3F1 and Swiss-Webster mice. *Toxicol. appl. Pharmacol.*, **72**, 495–503

Stenberg, G., Ridderström, M., Engström, Å., Pemble, S.E. & Mannervik, B. (1992) Cloning and heterologous expression of cDNA encoding class alpha rat glutathione transferase 8-8, an enzyme with high catalytic activity towards genotoxic α,β-unsaturated carbonyl compounds. *Biochem. J.*, **284**, 313–319

Työministeriö [Ministry of Labour] (1993) *HTP-Arvot 1993* [Occupational exposure limits 1993], Tampere, p. 14 (in Finnish)

United Kingdom Health and Safety Executive (1993) *Occupational Exposure Limits 1993* (EH 40/93), London, Her Majesty's Stationery Office, p. 29

United States Environmental Protection Agency (1988) *Compendium of Methods for the Determination of Toxic Organic Compounds in Ambient Air* (EPA-600/4-89-017; US NTIS PB90-116989), Research Triangle Park, NC, Office of Research and Development, Method TO5

United States National Institute for Occupational Safety and Health (1994a) *National Occupational Exposure Survey (1981–1983)*, Cincinnati, OH, p. 1

United States National Institute for Occupational Safety and Health (1994b) *NIOSH Pocket Guide to Chemical Hazards* (DHHS (NIOSH) Publ. No. 94-116), Cincinnati, OH, pp. 80–81

United States National Library of Medicine (1994) *Hazardous Substances Data Bank (HSDB)*, Bethesda, MD

United States Occupational Safety and Health Administration (1990) *OSHA Analytical Methods Manual*, 2nd Ed., Part 1, Vol. 4, Salt Lake City, UT, United States Department of Labor, Methods 81, 82

United States Occupational Safety and Health Administration (1994) Air contaminants. *US Code Fed. Regul.*, **Title 29**, Part 1910-1000, p. 10

Wang, M., Nishikawa, A. & Chung, F.-L. (1992) Differential effects of thiols on DNA modifications via alkylation and Michael addition by α-acetoxy-*N*-nitrosopyrrolidine. *Chem. Res. Toxicol.*, **5**, 528–531

Warholm, M., Holmberg, B., Högberg, J., Kronevi, T. & Götharson, A. (1984) The acute effects of single and repeated injections of acrolein and other aldehydes. *Int. J. Tissue React.*, **6**, 61–70

Weast, R.C. & Astle, M.J. (1985) *CRC Handbook of Data on Organic Compounds*, Vols I & II, Boca Raton, FL, CRC Press, pp. 336 (I), 524 (II)

Williams, G.M., Mori, H. & McQueen, C.A. (1989) Structure–activity relationships in the rat hepatocyte DNA-repair test for 300 chemicals. *Mutat. Res.*, **221**, 263–286

Witz, G., Lawrie, N.J., Amoruso, M.A. & Goldstein, B.D. (1985) Inhibition by reactive aldehydes of superoxide anion radical production in stimulated human neutrophils. *Chem.-biol. Interactions*, **53**, 13–23

Witz, G., Lawrie, N.J., Amoruso, M.A. & Goldstein, B.D. (1987) Inhibition by reactive aldehydes of superoxide anion radical production from stimulated polymorphonuclear leukocytes and pulmonary alveolar macrophages. *Biochem. Pharmacol.*, **36**, 721–726

Wolfe, G.W., Rudroin, M., French, J.E. & Parker, G.A. (1987) Thirteen week subchronic toxicity study of crotonaldehyde (CA) in F344 rats and B6C3F1 mice (Abstract). *Toxicology*, **7**, 209

Woodruff, R.C., Mason, J.M., Valencia, R. & Zimmering, S. (1985) Chemical mutagenesis testing in *Drosophila*. V. Results of 53 coded compounds tested for the National Toxicology Program. *Environ. Mutag.*, **7**, 677–702

Woolley, W.D. (1982) Smoke and toxic gas production from burning polymers. *J. macromol. Sci. Chem.*, **A17**, 1–33

Yamato, T., Kurata, T., Kato, H. & Fujimaki, M. (1970) Volatile carbonyl compounds from heated beef fat. *Agric. Biol. Chem.*, **34**, 88–94

FURAN

1. Exposure Data

1.1 Chemical and physical data

1.1.1 Nomenclature

Chem. Abstr. Serv. Reg. No.: 110-00-9
Chem. Abstr. Name: Furan
IUPAC Systematic Name: Furan
Synonyms: Divinylene oxide; 1,4-epoxy-1,3-butadiene; furfuran; furfurane; oxacyclopentadiene

1.1.2 Structural and molecular formulae and relative molecular mass

C_4H_4O Relative molecular mass: 68.08

1.1.3 Chemical and physical properties of the pure substance

(a) *Description*: Clear colourless liquid with strong ether-like odour (McKillip *et al.*, 1989; United States National Toxicology Program, 1993)

(b) *Boiling-point*: 31.4 °C (Lide, 1993)

(c) *Melting-point*: –85.6 °C (Lide, 1993)

(d) *Density*: 0.9514 at 20 °C/4 °C (Lide, 1993)

(e) *Spectroscopy data*: Infrared (prism [3664]; grating [33 024]), ultraviolet [28 593], nuclear magnetic resonance (proton [16 937], C-13 [570]) and mass [51] spectral data have been reported (Sadtler Research Laboratories, 1980; Weast & Astle, 1985).

(f) *Solubility*: Insoluble in water; soluble in acetone, benzene, diethyl ether and ethanol (Spectrum Chemical Mfg Corp., 1992; Lide, 1993)

(g) *Volatility*: Relative vapour density (air = 1), 2.3 (Spectrum Chemical Mfg Corp., 1992)

(h) *Stability*: Stable to alkalis; forms resin on evaporation and when in contact with mineral acids (Budavari, 1989); forms unstable peroxides on exposure to air; vapour–air mixtures are explosive above the flash-point (–35 °C closed cup) (Spectrum Chemical Mfg Corp., 1992)

(i) *Reactivity*: Lower inflammable/explosive limit in air, 2.3%; dangerous fire hazard and moderate explosive hazard when exposed to heat or flame (Spectrum Chemical Mfg Corp., 1992)

(j) *Octanol:water partition coefficient (P)*: log P, 1.34 (Hansch et al., 1995)

(k) *Conversion factor*: mg/m^3 = 2.78 × ppm[1]

1.1.4 Technical products and impurities

Furan is available commercially with the following typical specifications (wt%): minimal purity, 99.0; methylfuran, 0.2 max; tetrahydrofuran, 0.05 max; furfural (see monograph, this volume), 0.05 max; water, 0.2 max; peroxide, 0.015 max; butylated hydroxytoluene (see IARC, 1987), 0.025–0.040 max (McKillip et al., 1989). Trade names for furan include Axole, Oxole, Tetrol, Tetrole and U124.

1.1.5 Analysis

A method for the identification and determination of biogenic and anthropogenic volatile organic compounds, including furan, in ambient air is based on the combined use of carbon absorption traps and high-resolution gas chromatography–mass spectrometry (Ciccioli et al., 1993). Furan can be detected in water by direct aqueous injection on a glass column packed with Tenax GC and analysis by gas chromatography with flame ionization detection. The detection limit is about 1.0 µg/ml (Knuth & Hogland, 1984).

1.2 Production and use

1.2.1 Production

Furan is produced commercially by the decarbonylation of furfural over a palladium/-charcoal catalyst (McKillip et al., 1989). It is produced by one company each in Belgium, the Russian Federation and the United States of America (Chemical Information Services, Inc., 1994).

1.2.2 Use

Furan is an intermediate in the manufacture of tetrahydrofuran, pyrrole and thiophene and in the formation of lacquers and solvent for resins. It is also used in the production of pharmaceuticals, agricultural chemicals and stabilizers (McKillip & Sherman, 1980; McKillip et al., 1989).

---

[1] Calculated from: mg/m^3 = (relative molecular mass/24.45) × ppm, assuming normal temperature (25 °C) and pressure (101 kPa)

1.3 Occurrence

1.3.1 Natural occurrence

Furan occurs in oils obtained by the distillation of pine wood containing rosin (Budavari, 1989). It has also been identified in volatile emissions from sorb trees (Isodorov *et al.*, 1985).

1.3.2 Occupational exposure

No data were available to the Working Group.

1.3.3 Air

Furan is released into the air as a gas-phase component of cigarette smoke, wood smoke and exhaust gas from diesel and gasoline engines. It has been detected in the expired air of both smokers, at 0–98 µg/h, and nonsmokers, at 0–28 µg/h (Howard *et al.*, 1990). Residential burning of brown-coal briquets also led to emission of furfural, at an emission factor of 1.70 mg/kg. The estimated total amount of furfural emitted in the city of Leipzig was 550 kg/year (Engewald *et al.*, 1993). In a study of nuisance odours in Flanders, a concentration of 170 µg/m^3 furan was detected in the emission from a deep-frier (Moortgat *et al.*, 1992).

1.3.4 Water

Furan was identified qualitatively in the Niagara River, United States, and in one creek in the Niagara River watershed (Elder *et al.*, 1981; Howard *et al.*, 1990).

It was detected in one of 63 industrial effluents at a concentration of < 10 µg/L (Perry *et al.*, 1979). It was also detected in aqueous condensate samples from low-heat gasification of rosebud coal, at a concentration of 7 ± 4 ppb [µg/L]. It was not detected (detection limit, 0.1 ppb) in groundwater or coal water before coal gasification *in situ*, in water samples obtained during coal gasification *in situ*, in retort water from shale processing *in situ* or in boiler blowdown water from oil-shale processing *in situ* (Pellizzari *et al.*, 1979).

1.4 Regulations and guidelines

The occupational exposure limit for furan in the Russian Federation is 0.5 mg/m^3, with a notation of skin irritancy (ILO, 1991).

2. Studies of Cancer in Humans

No data were available to the Working Group.

3. Studies of Cancer in Experimental Animals

Oral administration

Mouse: Groups of 50 male and 50 female B6C3F1 mice, eight weeks of age, were administered 0, 8 or 15 mg/kg bw furan (purity, > 99%) by gavage in corn oil on five days per week for 104 weeks. The numbers of surviving animals at the end of the study were reduced: 33/50 male controls, 17/50 at the low dose and 16/50 at the high dose ($p = 0.009$); and 29/50 female controls, 25/50 at the low dose and 2/50 at the high dose ($p < 0.001$). The incidence of hepatocellular adenomas was increased, with 20/50 among male controls, 33/50 at the low dose ($p = 0.001$, logistic regression analysis) and 42/50 at the high dose ($p < 0.001$); and 5/50 among female controls, 31/50 at the low dose and 48/50 at the high dose groups ($p < 0.001$)). The incidence of hepatocellular carcinoma was also increased, occurring in 7/50 male controls, 32/50 at the low dose and 34/50 at the high dose ($p < 0.001$); and in 2/50 female controls, 7/50 at the low dose ($p = 0.08$) and 27/50 at the high dose ($p < 0.001$)). All treated groups had high incidences of biliary hyperplasia, cholangiofibrosis, hepatocellular necrosis and focal hyperplasia (United States National Toxicology Program, 1993).

Rat: Groups of 70 male and 70 female Fischer 344/N rats, seven to eight weeks of age, were administered 0, 2, 4 or 8 mg/kg bw furan (purity, > 99%) dissolved in corn oil by gavage on five days a week for 102 weeks. Interim evaluations were made on 10 rats from each group at 9 and 15 months. The numbers of animals still alive at the end of the study were 33/50 male controls, 28/50 at the low dose, 26/50 at the middle dose and 16/50 at the high dose ($p < 0.001$); and 34/50 female controls, 32/50 at the low dose, 28/50 at the middle dose and 19/50 at the high dose ($p = 0.006$). The incidence of hepatocellular adenomas was increased, and many rats had multiple adenomas; adenomas were found in 1/50 male controls, 4/50 at the low dose, 18/50 at the middle dose ($p < 0.001$, logistic regression analysis) and 27/50 at the high dose ($p < 0.001$); and in none of 50 female controls, 2/50 at the low dose, 4/50 at the middle dose ($p = 0.048$) and 7/50 at the high dose ($p = 0.002$). Hepatocellular carcinomas occurred only in males: 0/50 controls, 1/50 at the low dose, 6/50 at the middle dose ($p = 0.009$) and 18/50 at the high dose ($p < 0.001$). Cholangiocarcinomas were found in none of the control animals and in 43/50 males at the low dose, 48/50 at the middle dose and 49/50 at the high dose; and in 49/50 females at the low dose, 50/50 at the middle dose and 49/50 at the high dose ($p < 0.001$ for any group and either sex in comparison with controls). Biliary hyperplasia, cholangiofibrosis, hepatocellular necrosis and hyperplasia occurred in treated rats of each sex. The incidence of mononuclear-cell leukaemia increased with dose, occurring in 8/50 male controls, 11/50 at the low dose, 17/50 at the middle dose ($p = 0.016$, life-table test) and 25/50 at the high dose ($p < 0.001$); and in 8/50 female controls, 9/50 at the low dose, 17/50 at the middle dose ($p = 0.022$) and 21/50 at the high dose ($p < 0.001$). Cholangiocarcinomas, but no other tumours of the liver, were observed at 9 and 15 months, the incidence increasing with dose up to 100% at the high dose. In a 'stop-exposure' experiment, 50 males received 30 mg/kg bw furan for 13 weeks, and 10 rats were killed and examined at 13 weeks and at 9 and 15 months; all animals were observed for two years. In addition to the lesions seen previously, cholangiocarcinomas were seen in all treated rats that survived for at least nine months, and 6/40 had hepatocellular carcinomas (United States National Toxicology Program, 1993).

In a study to characterize the histopathogenesis of intrahepatic cholangiocarcinomas induced by furan, groups of 10–12 young adult male Fischer 344 rats, weighing 160–190 g, were treated with 30 mg/kg bw furan (purity, > 99%) in corn oil by gavage on five days a week for 6–13 weeks and observed up to 16 months. Lesions described as 'hepatic adenocarcinomas' occurred in the right caudate lobe of 4/9 rats treated for six weeks, 6/8 treated for nine weeks, 5/7 treated for 12 weeks and 9/10 treated for 13 weeks. Most of the tumours showed small-intestinal differentiation, as reflected by the presence of goblet, Paneth and serotonin-positive cells. Two hepatocellular carcinomas were also found in the animals treated for 13 weeks. Cells from the adenocarcinoma cells did not contain TGF-α, but hepatocellular carcinoma cells did (Elmore & Sirica, 1993).

4. Other Data Relevant to an Evaluation of Carcinogenicity and its Mechanisms

4.1 Absorption, distribution, metabolism and excretion

4.1.1 Humans

No data were available to the Working Group.

4.1.2 Experimental systems

[2,5-^{14}C]Furan (specific activity, 56 mCi/mmol; radiochemical purity, 99%) was administered by gavage to male Fischer 344 rats once a day for up to eight days at a dose of 8 mg/kg bw in corn oil. Furan was absorbed rapidly and extensively, and only about 14% of the administered dose was expired as unchanged furan, 11% within the first hour. After administration of a single dose, ^{14}C-carbon dioxide accounted for 26% of the radiolabel, 20% was found in urine and 22% was found in faeces within 24 h. Of the 19% of the administered radiolabel that remained in the tissues, 68% (13% of the dose) was found in the liver; smaller amounts were detected in the kidney and gastrointestinal tract. Only 20% of the radiolabelled material in the liver could be extracted with organic solvents, and the remaining radiolabel was assumed to be bound covalently to tissue macromolecules; none of the radiolabel was bound to liver DNA. Elimination of unbound radiolabelled material from the liver appeared to follow first-order kinetics, with a half-life of 1.8 days, whereas the kinetics of elimination from kidney and blood was more complex. The blood concentration of radiolabel remained at about 2–5 nmol equivalents of furan per gram for eight days after a single dose. The concentration of radiolabel increased with multiple doses, by about sixfold in blood and kidney and fourfold in the liver after eight doses, as compared with the levels seen after a single dose. The percentage of the administered radiolabel that was eliminated in the urine increased during multiple dosing from 20% after one day to 33% after four days. Furan appeared to be metabolized extensively, as high-performance liquid chromatography of the radiochromatogram of 24-h urine showed at least 10 peaks. The pattern of metabolites did not change appreciably after multiple doses. None of the urinary metabolites was identified. Exhalation of radiolabel in carbon dioxide after administration of furan indicates opening of the furan ring (Burka et al., 1991).

Male Fischer 344 rats were exposed to furan (purity, > 99%) by inhalation at 52, 107 or 208 ppm [145, 297 or 578 mg/m^3] for 4 h, at which time the concentrations of furan were determined in blood and liver. A physiologically based kinetic model was developed on the basis of the tissue partition coefficients for furan determined *in vitro* in a vial equilibration technique. The model adequately simulated the concentrations of furan after inhalation *in vivo* and revealed a single saturable process, with a maximal metabolic velocity (V_{max}) of 27 µmol[1.8 mg]/h per 250-g rat and a Michaelis-Menten constant (K_m) of 2 µmol[136 µg]/L. The biotransformation of furan by rat hepatocytes *in vitro* had a K_m of 0.4 µmol[27 µg]/L and a V_{max} of 0.02 µmol[1.4 µg]/h per 10^6 cells. The metabolism of this compound was increased by fivefold in hepatocytes isolated from rats treated with acetone, but not after phenobarbital treatment, and was inhibited by treatment of rats with aminobenzotriazole or after addition of ethanol and 1-phenylimidazole in vitro. The metabolism of furan *in vivo* was inhibited by pyrazole (Kedderis et al., 1993).

Incubation of [2,5-^{14}C]furan (specific activity, 56 mCi/mmol; radiochemical purity, > 99%) with liver microsomes from male Fischer 344 rats in the presence of NADPH resulted in covalent binding of radiolabel to microsomal protein. The rates of covalent binding to protein *in vitro* were increased by two- to threefold after treatment of rats with phenobarbital and by four- to fivefold after treatment with imidazole or pyrazole, but were reduced in the presence of microsomes from rats treated with β-naphthoflavone. Addition of semicarbazide, N-acetyl-cysteine or glutathione inhibited binding of ^{14}C-furan to microsomal protein. Twenty-four hours after a single oral administration of 8 or 25 mg/kg bw of furan, the liver microsomal P450 levels were reduced to 90 and 71% of those in controls, respectively, with concomitant decreases in the activities of aniline hydroxylase, 7-ethoxycoumarin-O-deethylase and 7-ethoxyresorufin-O-deethylase. The amount of radiolabel bound to microsomes was distributed almost equally between the haem and protein moieties of carbon monoxide-binding particles. The haem–furan adduct(s) was not identified (Parmar & Burka, 1993).

4.2 Toxic effects

4.2.1 Humans

No data were available to the Working Group.

4.2.2 Experimental systems

Groups of 10 male Swiss albino mice (GP strain) given furan [purity not given] at a single dose of 300 mg/kg bw in 0.9% sodium chloride intraperitoneally had centrilobular hepatic necrosis and coagulative necrosis of the proximal convoluted tubules of the outer renal cortex. Pretreatment of mice with the cytochrome P450 inhibitor piperonyl butoxide reduced the extent of liver necrosis and totally inhibited the renal necrosis (McMurtry & Mitchell, 1977).

The liver damage and hepatocyte proliferation induced by single and repeated oral doses of furan were studied in B6C3F1/CrlBR mice and Fischer 344/CrlBR rats. In the study of single doses, groups of five male mice and five male rats were administered corn oil or furan (same lot as used in the two-year bioassay of the United States National Toxicology Program) in corn oil by gavage at doses of 30 or 50 mg/kg bw. Animals were killed at various times between 12 h and eight days; 2 h before being killed they were injected intraperitoneally with 2000 mCi/kg bw of [^{3}H-methyl]thymidine so that labelling indices could be determined. In the study of repeated doses, groups of six male mice and six male and female rats were administered the vehicle or doses of furan equivalent to the highest dose used in the bioassay (8 mg/kg bw for rats, 15 mg/kg bw for mice), by gavage, daily for five days per week for up to six weeks. Animals were killed after one, three or six weeks of treatment; six days before sacrifice, they received a subcutaneously implanted osmotic pump containing ^{3}H-thymidine. The effects observed after single doses were hepatocellular necrosis, marked increases in the activities of plasma aspartate aminotransferase, alanine aminotransferase and lactate dehydrogenase and a sharp increase in the labelling index (23.9 in mice and 17.8 in rats, in comparison with < 0.5 in controls). After six weeks of treatment, male and female rats, but not mice, had bile-duct hyperplasia and metaplasia, resembling intestinal cells, in areas of liver fibrosis. The hepatocyte labelling indices in male mice were 25.1 at week 1, 12.0 at week 3 and 3.2 at week 6, in comparison with control values of 0.41 at week 1 and 0.89 at week 6 (5–39-fold increase over combined control values). The hepatocyte labelling indices in male rats were 3.2 at week 1, 9.2 at week 3 and 6.5 at week 6, in comparison with control values of 0.08 at week 1 and 0.29 at week 6 (18–51-fold increase over combined control values). The hepatocyte labelling indices in female rats were 11.7 at week 1, 9.2 at week 3 and 14.4 at week 6, in comparison with control values of 0.77 at week 1 and 0.75 at week 6 (12–19-fold increase over combined control values) (Wilson *et al.*, 1992).

Groups of five male Fischer 344 rats were administered furan (purity, 99%) in corn oil by gavage at doses of 0, 15, 30, 45 or 60 mg/kg bw daily on five days per week for three weeks; liver lesions were assessed by histology and histo- and immunochemistry. Animals at the two highest doses showed rapid development of severe cholangiofibrosis in the caudate liver lobe. Histologically, the lesion was characterized by well-differentiated hyperplastic bile ductules and numerous metaplastic intestinal glands supported by fibrous tissue. At the highest dose, cholangiolar-like structures were also observed. Phenotypic analysis of the hyperplastic ductular structures and of the metaplastic intestinal glands in the cholangiofibrotic areas suggested a precursor relationship between the ductular cells and the metaplastic cells (Elmore & Sirica, 1991).

Groups of male Fischer 344 rats were administered furan (purity, 99%) at 45 mg/kg bw by gavage daily on five days per week for up to 32 days. Animals were killed on day 1, 3, 5, 7, 9, 12, 16 or 32. Severe hepatocellular necrosis was evident in the right and caudate lobes at the time of early sacrifices, and was followed by inflammatory cell infiltration, obvious bile-duct hyperplasia and then by development of metaplastic glands, ending with cholangiofibrosis by day 32 (Elmore & Sirica, 1992).

Synergism was demonstrated between bile-duct ligation in male Fischer 344 rats and administration of furan at 45 mg/kg bw per day, five times per week for five to six weeks beginning one week after bile duct ligation. In treated animals, 72.6 ± 16.3% of the liver was replaced by well-differentiated hyperplastic bile ductules, whereas in control rats, which were ligated and dosed with corn oil, 2.0 ± 4.2% of the liver was bile-duct tissue. Treatment with furan after a sham operation resulted in only 11.9 ± 3.1% bile-duct tissue. The combined treatment also resulted in greatly increased activities of γ-glutamyl transpeptidase in liver homogenates (Sirica et al., 1994a).

Furan administered by gavage to young adult male Fischer 344 rats at a dose of 60 mg/kg bw per day five times per week for 10–14 days was highly toxic and induced marked atrophy of the right liver lobe. The cholangiolar-like structures that developed in this region consisted of biliary epithelial cells and ductular hepatocytes in various stages of maturation. Immunohistochemical and other phenotypic features of the cells were considered to be consistent with the development of rare bile ductular-like cell types from hepatocytes; in contrast, less severe treatment with furan led to the development of small intestinal-like glands (Sirica et al., 1994b).

Groups of 10 male and female Fischer 344/N rats and 10 female B6C3F1 mice received furan (purity, 99%) in corn oil at doses of 0, 4, 8, 15, 30 or 60 mg/kg bw by gavage on five days per week for 13 weeks. Groups of 10 male mice received doses of 0, 2, 4, 8, 15 or 30 mg/kg bw. In rats, dose-dependent bile-duct hyperplasia and cholangiofibrosis were observed, and animals at the high dose had renal tubular necrosis and dilatation and atrophy of the thymus, testes or ovaries. Mice had dose-dependent hepatic lesions, with cytomegaly, degeneration, necrosis and bile-duct hyperplasia (United States National Toxicology Program, 1993).

Repeated treatment of male Fischer 344 rats with furan by gavage at daily doses of 8 mg/kg bw in corn oil, five days per week, resulted in a sustained, approximately 35-fold increase in the percentage of hepatocytes in S-phase over that in controls at weeks 1, 3 and 6. Northern blot analysis of mRNA expression in the livers showed up to a doubling of Ha-*ras* expression, but no measurable expression of *fos*, at weeks 1, 3 and 6. At the end of week 6, but not earlier, an approximately 10-fold increase in the expression of *myc* was observed (Butterworth et al., 1992, 1994).

Isolated hepatocytes from male Fischer 344 rats were incubated in suspension culture with 2–100 μmol[136–6808 μg]/L of furan (vacuum-distilled) for 4 h and then in a monolayer culture for an additional 25.5 h. Concentration-dependent cytolethality (5–70%) and modest glutathione depletion were observed. These concentrations were similar to the hepatotoxic tissue concentrations predicted by dosimetry modelling. Both glutathione depletion and cytolethality were prevented by addition of 1-phenylimidazole to the cultures and were enhanced by pretreatment of the rats with acetone (Carfagna et al., 1993).

4.3 Reproductive and prenatal effects

No data were available to the Working Group.

4.4 Genetic and related effects

4.4.1 Humans

No data were available to the Working Group

4.4.2 Experimental systems (see also Table 1 and Appendices 1 and 2)

Furan was not mutagenic to *Salmonella typhimurium* after preincubation in the presence or absence of an exogenous metabolic activation system. It was reported in an abstract that furan was not mutagenic in a vapour-phase protocol.

Furan did not induce sex-linked recessive lethal mutations in *Drosophila melanogaster* when administered to adult flies by feeding or abdominal injection.

It was reported in an abstract that furan did not induce unscheduled DNA synthesis in rat or mouse hepatocytes after treatment *in vitro* or *in vivo*. It induced gene mutation at the thymidine kinase locus of L5178Y mouse lymphoma cells in the absence of metabolic activation. Furan induced sister chromatid exchange and chromosomal aberrations in Chinese hamster ovary cells both in the presence and absence of metabolic activation. In another study, a positive response for the induction of chromosomal aberrations was observed only at comparatively high doses in the presence of a liver activation system from Aroclor 1254-pretreated male Swiss rats.

No sister chromatid exchanges or chromosomal aberrations were induced in bone-marrow cells of B6C3F1 male mice injected intraperitoneally with furan at doses up to 350 mg/kg bw *in vivo*; however, chromosomal aberrations were induced at 250 mg/kg bw in an extended protocol involving a late harvest time.

Mutation of proto-oncogenes in tumours induced by furan

ras Proto-oncogene activation was studied in liver adenomas and carcinomas induced in B6C3F1 mice by furan. The frequency of activated H-*ras* and K-*ras* oncogenes was similar in hepatocellular tumours from 12/29 treated mice and in 15/27 vehicle controls, but the spectrum of activating mutations in the H-*ras* gene differed significantly: Although mutations occurred at codon 61 in tumours from both treated and untreated animals, mutations (G→T and G→C transversions) were observed at codon 117 only in animals treated with furan. The authors interpreted these findings as suggesting that the novel mutations in *ras* genes could have been due to a genotoxic effect of furan (Reynolds *et al.*, 1987).

5. Summary and Evaluation

5.1 Exposure data

Furan is produced commercially by decarbonylation of furfural. It is used mainly in the production of tetrahydrofuran, thiophene and pyrrole. It also occurs naturally in certain woods and during the combustion of coal and is found in engine exhausts, wood smoke and tobacco smoke.

5.2 Human carcinogenicity data

No data were available to the Working Group.

Table 1. Genetic and related effects of furan

Test system	Result[a] Without exogenous metabolic system	Result[a] With exogenous metabolic system	Dose[b] (LED/HID)	Reference
SA0, *Salmonella typhimurium* TA100, reverse mutation	–	–	3850	US National Toxicology Program (1993)
SA0, *Salmonella typhimurium* TA100, reverse mutation	–	–	0.00	Dillon *et al.* (1992) (Abstract)
SA2, *Salmonella typhimurium* TA102, reverse mutation	–	–	0.00	Dillon *et al.* (1992) (Abstract)
SA4, *Salmonella typhimurium* TA104, reverse mutation	–	–	0.00	Dillon *et al.* (1992) (Abstract)
SA5, *Salmonella typhimurium* TA1535, reverse mutation	–	–	3850	US National Toxicology Program (1993)
SA7, *Salmonella typhimurium* TA1537, reverse mutation	–	–	3850	US National Toxicology Program (1993)
SA9, *Salmonella typhimurium* TA98, reverse mutation	–	–	3850	US National Toxicology Program (1993)
DMX, *Drosophila melanogaster*, sex-linked recessive lethal mutation	–		10 000 feed	Foureman *et al.* (1994)
DMX, *Drosophila melanogaster*, sex-linked recessive lethal mutation	–		25 000 inj	Foureman *et al.* (1994)
URP, Unscheduled DNA synthesis, rat primary hepatocytes *in vitro*	–		680	Wilson & Butterworth (1989) (Abstract); Wilson *et al.* (1990) (Abstract)
UIA, Unscheduled DNA synthesis, mouse primary hepatocytes *in vitro*	–		680	Wilson & Butterworth (1989) (Abstract); Wilson *et al.* (1990) (Abstract)
G5T, Gene mutation, mouse lymphoma L5178Y cells, *tk* locus *in vitro*	+	0	1140	McGregor *et al.* (1988)
SIC, Sister chromatid exchange, Chinese hamster ovary (CHO) cells *in vitro*	+	+	160	US National Toxicology Program (1993)

Table 1 (contd)

Test system	Result[a]		Dose[b] (LED/HID)	Reference
	Without exogenous metabolic system	With exogenous metabolic system		
CIC, Chromosomal aberrations, Chinese hamster ovary (CHO) cells *in vitro*	-	+	4800	Stich *et al.* (1981)
CIC, Chromosomal aberrations, Chinese hamster ovary (CHO) cells *in vitro*	+	+	100	US National Toxicology Program (1993)
UPR, Unscheduled DNA synthesis, rat primary hepatocytes *in vivo*	-		100 po × 1	Wilson & Butterworth (1989) (Abstract); Wilson *et al.* (1990) (Abstract)
UVM, Unscheduled DNA synthesis, mouse primary hepatocytes *in vivo*	-		200 po × 1	Wilson & Butterworth (1989) (Abstract); Wilson *et al.* (1990) (Abstract)
SVA, Sister chromatid exchange, mouse bone-marrow cells *in vivo*	-		350 ip × 1	US National Toxicology Program (1993)
CBA, Chromosomal aberrations, mouse bone-marrow cells *in vivo*	+[c]		250 ip × 1	US National Toxicology Program (1993)

[a] +, considered to be positive; (+), considered to be weakly positive in an inadequate study; -, considered to be negative; 0, not tested
[b] LED, lowest effective dose; HID, highest ineffective dose; in-vitro tests, µg/ml; in-vivo tests, mg/kg bw; 0.00, dose not reported; ip, intraperitoneally; po, orally
[c] Extended harvest protocol (36-h harvest); negative result with standard protocol (17-h harvest)

5.3 Animal carcinogenicity data

Furan was tested for carcinogenicity by oral administration in one study in mice and in one study in rats. It produced hepatocellular adenomas and carcinomas in mice. In rats, it produced hepatocellular adenomas in animals of each sex and carcinomas in males; a high incidence of cholangiocarcinomas was seen in both males and females. The incidence of mononuclear-cell leukaemia was also increased in animals of each sex.

5.4 Other relevant data

Furan is rapidly and extensively absorbed by rats after oral administration; part of the absorbed dose becomes covalently bound to protein, mainly in the liver. No DNA binding could be demonstrated in the liver.

Repeated administration of furan to mice and rats leads to liver necrosis, liver-cell proliferation and bile-duct hyperplasia; in rats, prominent cholangiofibrosis develops.

Induction of chromosomal aberrations but not of sister chromatid exchange was observed in rodents treated *in vivo* in one study. Gene mutation, sister chromatid exchange (in single studies) and chromosomal aberrations were induced in rodent cells *in vitro*.

Furan was not mutagenic to insects or bacteria.

5.5 Evaluation[1]

There is *inadequate evidence* in humans for the carcinogenicity of furan.

There is *sufficient evidence* in experimental animals for the carcinogenicity of furan.

Overall evaluation

Furan *is possibly carcinogenic to humans (Group 2B)*.

6. References

Budavari, S., ed. (1989) *The Merck Index*, 11th Ed., Rahway, NJ, Merck & Co., p. 672

Burka, L.T., Washburn, K.D. & Irwin, R.D. (1991) Disposition of [^{14}C]furan in the male F344 rat. *J. Toxicol. environ. Health*, **34**, 245–257

Butterworth, B.E., Sprankle, C.S., Goldsworthy, S.M., Wilson, D.M. & Goldsworthy, T.L. (1992) Overexpression of *myc* independent of induced cell proliferation: a delayed response to 6 weeks furan administration (Abstract 637). *Proc. Am. Assoc. Cancer Res.*, **33**, 106

---

[1] For definition of the italicized terms, see Preamble, pp. 22–26.

Butterworth, B.E., Sprankle, C.S., Goldsworthy, S.M., Wilson, D.M. & Goldsworthy, T.L. (1994) Expression of *myc*, *fos*, and Ha-*ras* in the livers of furan-treated F344 rats and B6C3F$_1$ mice. *Mol. Carcinog.*, **9**, 24–32

Carfagna, M.A., Held, S.D. & Kedderis, G.L. (1993) Furan-induced cytolethality in isolated rat hepatocytes: correspondence with in vivo dosimetry. *Toxicol. appl. Pharmacol.*, **123**, 265–273

Chemical Information Services, Inc. (1994) *Directory of World Chemical Producers 1995/96 Standard Edition*, Dallas, TX, p. 298

Ciccioli, P., Brancaleoni, E., Cecinato, A. & Sparapani, R. (1993) Identification and determination of biogenic and anthropogenic volatile organic compounds in forest areas of northern and southern Europe and a remote site of the Himalaya region by high-resolution gas chromatography–mass spectrometry. *J. Chromatogr.*, **643**, 55–69

Dillon, D.M., McGregor, D.B., Combes, R.D. & Zeiger, E. (1992) Detection of mutagenicity in *Salmonella* of some aldehydes and peroxides (Abstract). *Environ. mol. Mutag.*, **19** (Suppl. 20), 15

Elder, V.A., Proctor, B.L. & Hites, R.A. (1981) Organic compounds found near dump sites in Niagara Falls, New York. *Environ. Sci. Technol.*, **15**, 1237–1243

Elmore, L.W. & Sirica, A.E. (1991) Phenotypic characterization of metaplastic intestinal glands and ductular hepatocytes in cholangiofibrotic lesions rapidly induced in the caudate liver lobe of rats treated with furan. *Cancer Res.*, **51**, 5752–5759

Elmore, L.W. & Sirica, A.E. (1992) Sequential appearance of intestinal mucosal cell types in the right and caudate liver lobes of furan-treated rats. *Hepatology*, **16**, 1220–1226

Elmore, L. & Sirica, A.E. (1993) 'Intestinal-type' of adenocarcinoma preferentially induced in right/caudate liver lobes of rats treated with furan. *Cancer Res.*, **53**, 254–259

Engewald, W., Knobloch, T. & Efer J. (1993) Volatile organic compounds in emissions from brown-coal-fired residential stoves. *Z. Umweltchem. Ökotox.*, **5**, 303–308 (in German)

Foureman, P., Mason, J.M., Valencia, R. & Zimmering, S. (1994) Chemical mutagenesis testing in *Drosophila*. IX. Results of 50 coded compounds tested for the National Toxicology Program. *Environ. mol. Mutag.*, **23**, 51–63

Hansch, C., Leo, A. & Hoekman, D.H. (1995) *Exploring QSAR*, Washington DC, American Chemical Society

Howard, P.H., Sage, G.W., Jarvis, W.F. & Gray, D.A., eds (1990) *Handbook of Environmental Fate and Exposure Data for Organic Chemicals*, Vol. II, *Solvents*, Chelsea, MI, Lewis Publishers

IARC (1987) *IARC Monographs on the Evaluation of Carcinogenic Risks to Humans*, Suppl. 7, *Overall Evaluations of Carcinogenicity: An Updating of* IARC Monographs *Volumes 1–42*, Lyon, p. 59

ILO (1991) *Occupational Exposure Limits for Airborne Toxic Substances: Values of Selected Countries* (Occupational Safety and Health Series No. 37), 3rd Ed., Geneva, pp. 206–207

Isidorov, V.A., Zenkevich, I.G. & Ioffe, B.V. (1985) Volatile organic compounds in the atmosphere of forests. *Atmos. Environ.*, **19**, 1–8

Kedderis, G.L., Carfagna, M.A., Held, S.D., Batra, R., Murphy, J.E. & Gargas, M.L. (1993) Kinetic analysis of furan biotransformation by F-344 rats *in vivo* and *in vitro*. *Toxicol. appl. Pharmacol.*, **123**, 274–282

Knuth, M.L. & Hogland, M.D. (1984) Quantitative analysis of 68 polar compounds from ten chemical classes by direct aqueous injection gas chromatography. *J. Chromatogr.*, **285**, 153–160

Lide, D.R., ed. (1993) *CRC Handbook of Chemistry and Physics*, 74th Ed., Boca Raton, FL, CRC Press, pp. 3–250, B-11

McGregor, D.B., Brown, A., Cattanach, P., Edwards, I., McBride, D., Riach, C. & Caspary, W.J. (1988) Responses of the L5178Y tk$^+$/tk^n mouse lymphoma cell forward mutation assay. III: 72 coded chemicals. *Environ. mol. Mutag.*, **12**, 85–154

McKillip, W.J. & Sherman, E. (1980) Furan derivatives. In: Mark, H.F., Othmer, D.F., Overberger, C.G., Seaborg, G.T. & Grayson, N., eds, *Kirk-Othmer Encyclopedia of Chemical Technology*, 3rd Ed., Vol. 11, New York, John Wiley & Sons, pp. 516–520

McKillip, W.J., Collin, G. & Höke, H. (1989) Furan and derivatives. In: Elvers, B., Hawkins, S., Ravenscroft, M., Rounsaville, J.F. & Schulz, G., eds, *Ullmann's Encyclopedia of Industrial Chemistry*, 5th rev. Ed., Vol. A12, New York, VCH Publishers, pp. 119–121

McMurtry, R.J. & Mitchell, J.R. (1977) Renal and hepatic necrosis after metabolic activation of 2-substituted furans and thiophenes, including furosemide and cephaloridine. *Toxicol. appl. Pharmacol.*, **42**, 285–300

Moortgat, M., Schamp, N. & Van Langenhove, H. (1992) Assessment of odour nuisance problems in Flanders: a practical approach. In: Dragt, A.J. & van Ham, J., eds, *Biotechniques for Air Pollution Abatement and Odour Control Policies* (Studies in Environmental Science 51), Amsterdam, Elsevier, pp. 447–452

Parmar, D. & Burka, L.T. (1993) Studies on the interaction of furan with hepatic cytochrome P-450. *J. biochem. Toxicol.*, **8**, 1–9

Pellizzari, E.D., Castillo, N.P., Willis, S., Smith, D. & Bursey, J.T. (1979) Identification of organic compounds in aqueous effluents from energy-related processes. In: Van Hall, C.E., ed., *Measurement of Organic Pollutants in Water and Wastewater* (ASTM STP 686), Philadelphia, PA, American Society for Testing and Materials, pp. 256–274

Perry, D.L., Chuang, C.C., Jungclaus, G.A. & Warner, J.S. (1979) *Identification of Organic Compounds in Industrial Effluent Discharges* (USEPA 600/4-79-016; NTIS PB-294794), Washington DC, United States Environmental Protection Agency

Reynolds, S.H., Stowers, S.J., Patterson, R.M., Maronpot, R.R., Aaronson, S.A. & Anderson, M.W. (1987) Activated oncogenes in B6C3F1 mouse liver tumors: implications for risk assessment. *Science*, **237**, 1309–1316

Sadtler Research Laboratories (1980) *1980 Cumulative Index*, Philadelphia, PA

Sirica, A.E., Cole, S.L. & Williams, T. (1994a) A unique rat model of bile ductular hyperplasia in which liver is almost totally replaced with well-differentiated bile ductules. *Am. J. Pathol.*, **144**, 1257–1268

Sirica, A.E., Gainey, T.W. & Mumaw, V.R. (1994b) Ductular hepatocytes. Evidence for a bile ductular cell origin in furan-treated rats. *Am. J. Pathol.*, **145**, 375–383

Spectrum Chemical Mfg Corp. (1992) *Material Safety Data Sheet: Furan*, Gardena, CA

Stich, H.F., Rosin, M.P., Wu, C.H. & Powrie, W.D. (1981) Clastogenicity of furans found in food. *Cancer Lett.*, **13**, 89–95

United States National Toxicology Program (1993) *Toxicology and Carcinogenesis Studies of Furan (CAS No. 110-00-9) in F344/N Rats and B6C3F1 Mice (Gavage Studies)* (NTP Tech. Rep. No. 402; NIH Publ. No. 93-2857), Research Triangle Park, NC

Weast, R.C. & Astle, M.J. (1985) *CRC Handbook of Data on Organic Compounds*, Vols I & II, Boca Raton, FL, CRC Press, pp. 647 (I), 597 (II)

Wilson, D.M. & Butterworth, B.E. (1989) Evaluation of furan-induced genotoxicity and cell proliferation in rat and mouse hepatocytes (Abstract 635). *Environ. mol. Mutag.*, **14** (Suppl. 15), 219

Wilson, D.M., Goldsworthy, T.L., Popp, J.A. & Butterworth, B.E. (1990) Evaluation of genotoxicity, cytotoxicity, and cell proliferation in hepatocytes from rats and mice treated with furan (Abstract 613). *Proc. Am. Assoc. Cancer Res.*, **31**, 103

Wilson, D.M., Goldsworthy, T.L., Popp, J.A. & Butterworth, B.E. (1992) Evaluation of genotoxicity, pathological lesions, and cell proliferation in livers of rats and mice treated with furan. *Environ. mol. Mutag.*, **19**, 209–222

FURFURAL

1. Exposure Data

1.1 Chemical and physical data

1.1.1 Nomenclature

Chem. Abstr. Serv. Reg. No.: 98-01-1
Chem. Abstr. Name: 2-Furancarboxaldehyde
IUPAC Systematic Name: 2-Furaldehyde
Synonyms: Artificial ant oil; 2-formylfuran; fural; furaldehyde; 2-furanaldehyde; 2-furancarbaldehyde; furancarbonal; 2-furancarbonal; 2-furfural; furfuraldehyde; 2-furfuraldehyde; furfurol; furfurole; furfurylaldehyde; furole; α-furole; 2-furylaldehyde; 2-furylcarboxaldehyde; pyromucic aldehyde

1.1.2 Structural and molecular formulae and relative molecular mass

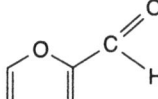

$C_5H_4O_2$ Relative molecular mass: 96.09

1.1.3 Chemical and physical properties of the pure substance

(a) *Description*: Clear, colourless oily liquid with a benzaldehyde-like odour (Budavari, 1989; United States National Toxicology Program, 1990)

(b) *Boiling-point*: 161.7 °C (Lide, 1993)

(c) *Melting-point*: –38.7 °C (Lide, 1993)

(d) *Density*: 1.1594 at 20 °C/4 °C (Lide, 1993)

(e) *Spectroscopy data*: Infrared (prism [113]; grating [35]), ultraviolet [42], nuclear magnetic resonance (proton [10203]; C-13 [1618]) and mass [200] spectral data have been reported (Sadtler Research Laboratories, 1980; Weast & Astle, 1985).

(f) *Solubility*: Moderately soluble in water (83 g/L at 20 °C); soluble in acetone, benzene, chloroform, diethyl ether and ethanol (Verschueren, 1983; Lide, 1993)

(g) *Volatility*: Vapour pressure, 1 mm Hg [0.13 kPa] at 20 °C; relative vapour density (air = 1), 3.31 (Verschueren, 1983)

(h) *Stability*: Turns yellow to brown on exposure to air and light (Budavari, 1989)

(i) *Reactivity*: Vapour forms inflammable mixture with air; reacts with oxidizing materials, strong alkalis and strong acids (Sittig, 1985; Sax & Lewis, 1989; United States National Institute for Occupational Safety and Health, 1994a)

(j) *Octanol/water partition coefficient (P)*: log (P), 0.41 (Hansch *et al.*, 1995)

(k) *Conversion factor*: mg/m^3 = 3.93 × ppm[1]

1.1.4 Technical products and impurities

Furfural is available commercially at a purity > 98% (Aldrich Chemical Co., 1994; Lancaster Synthesis Ltd, 1994; TCI America, 1994).

1.1.5 Analysis

Selected methods for the analysis of furfural in various matrices are presented in Table 1.

Table 1. Methods for the analysis of furfural

Sample matrix	Sample preparation	Assay procedure	Limit of detection	Reference
Air	Adsorb on sorbent coated with 2-(hydroxymethyl)piperidine on XAD-2; desorb with toluene; analyse for oxazolidine derivative	GC/FID	5 µg/sample	Eller (1994)
		GC/FID & GC/MS	2 µg/sample	Eller (1994)
	Adsorb on petroleum-based charcoal; desorb with carbon disulfide containing 1% dimethylformamide; remove water with magnesium sulfate	GC/FID	29.9 µg/sample (166 µg/m^3)	US Occupational Safety and Health Administration (1990)
	Adsorb on charcoal; desorb with carbon disulfide	GC/FID	About 25 ppb [98 µg/m^3]	Feldstein *et al.* (1989)
Urine	Treat urine with sodium hydroxide, acidify, extract with ethyl acetate; esterify (furoic acid) with diazomethane	GC/FID	NR	Šedivec & Flek (1978)
Distilled liquors	Steam distil; measure at 277 nm; compare with standard curve	UV	NR	Helrich (1990)

GC/FID, gas chromatography/flame ionization detection; MS, mass spectrometry; UV, ultraviolet spectrophotometry; NR, not reported

A method for identifying carbonyl compounds, including furfural, in environmental samples involves derivatization with *O*-(2,3,4,5,6-pentafluorobenzyl)hydroxylamine hydro-

---

[1] Calculated from: mg/m^3 = (relative molecular mass/24.45) × ppm, assuming normal temperature (25 °C) and pressure (101 kPa)

chloride, followed by gas chromatography (GC)–mass spectrometry (Le Lacheur et al., 1993). A similar method was used for the quantitative analysis of carbonyl compounds in cognac at a level of parts per billion (10^{-9}) (Vidal et al., 1993).

Traces of hydrophilic and volatile organic compounds, including furfural, can be determined in water after preconcentration with activated carbon. Organic substances adsorbed on the activated carbon are eluted with acetone and dichloromethane, and quantitative analysis is carried out by GC–mass spectrometry (Kadokami et al., 1990). In a method for determining picomolar concentrations of carbonyl compounds, including furfural, in natural waters such as seawater, the samples were derivatized with 2,4-dinitrophenylhydrazine and analysed by liquid chromatography. The detection limit for aldehydes after direct injection of derivatized water samples was 0.5 nmol [48 ng] (Kieber & Mopper, 1990).

Several methods have been described for the determination of furfural in processed citrus juices (orange, grapefruit, apple), including high-performance liquid chromatography (HPLC) of the 2,4-dinitrophenylhydrazine derivative; solid–liquid extraction of the juice before reversed-phase separation, followed by HPLC, with detection at 280 nm; micellar electrokinetic capillary chromatography after direct sample injection, with sodium dodecyl sulfate as the anionic surfactant; and colorimetry (Dinsmore & Nagy, 1974; Marcy & Rouseff, 1984; Lee et al., 1986; Beeman, 1987; Li et al., 1988; Blanco Gomis et al., 1991; Corradini & Corradini, 1992; Lo Coco et al., 1994).

Furfural was determined in commercial, aged brandies and in aqueous alcohol solutions of caramels by direct injection of the samples and HPLC (Villalón Mir et al., 1992). A complexometric method has also been used to determine furfural in cognacs and cognac brandies (Maltabar & Girenko, 1969; Girenko et al., 1970).

Dissolved gases and furan-related compounds, including furfural, were determined in transformer insulating oils by static headspace capillary GC with porous-layer open tubular columns. The detection limit for furfural was 0.5 ppm [mg/L] (Leblanc et al., 1993). A reversed-phase liquid chromatographic method was developed for determining furfural in lubricating oils. After a simple extraction step to remove interfering hydrocarbons, residual furfural was determined with an ultraviolet detector (DiSanzo et al., 1988).

The furan components, including furfural, in the aqueous condensate from wood smoke were extracted with dichloromethane, and the extract was analysed by GC–mass spectrometry (Edye & Richards, 1991).

A fluorimetric method has been described for the determination of aromatic aldehydes, including furfural, based on their reaction in dilute acid with 4,5-dimethoxy-1,2-diaminobenzene, to give a compound that fluoresces in alkaline solution (Nakamura et al., 1982). A similar method is based on the reaction of furfural in dilute sulfuric acid with 2,2'-dithiobis(1-aminonaphthalene) in the presence of tri-n-butyl phosphine, sodium sulfite and sodium phosphite (Ohkura et al., 1978).

Biological monitoring is best done by determining the total concentration of furoic acid in urine taken after an entire work shift and expressed in milligrams per unit of time (Šedivec & Flek, 1978; Dutch Expert Committee on Occupational Standards, 1992).

1.2 Production and use

1.2.1 Production

Furfural has been produced since 1832 from renewable agricultural sources, such as non-food residues of food crops and wood wastes (Hitchcock, & Duffey, 1948). The pentosan polysaccharides of these vegetable materials (xylan and arabinan) are almost as widely distributed in nature as is cellulose and are the major precursors of furfural. All pentosan-containing substances are potentially usable, but only a relatively few (corn cobs, oat hulls and bagasse), which are available in large tonnages within a limited radius of furfural production facilities, are commercially significant. Furfural is produced commercially in batch or continuous digesters, in which the pentosans are hydrolysed with strong inorganic acids to pentoses and the pentoses are subsequently cyclodehydrated to furfural (McKillip & Sherman, 1980). Furfural is also produced by continuous hydrolysis of bark at 190 °C and 12 atmospheres [1012 kPa], with acetic acid as a catalyst (Dutch Expert Committee on Occupational Standards, 1992).

Worldwide consumption of furfural was estimated to have been 91 thousand tonnes in 1965, 136 thousand tonnes in 1973 and 159 thousand tonnes in 1978 (McKillip & Sherman, 1980). It is produced by 31 companies in 15 countries (Chemical Information Services, Inc., 1994), with eight producers in China and six in the Russian Federation. In the United States of America, 45–57 thousand tonnes were used annually between 1965 and 1989 (Mannsville Chemical Products Corp., 1984).

1.2.2 Use

Large quantities of furfural are used in solvent extraction in the petroleum refining industry. It is also used as a solvent (for nitrated cotton, cellulose acetate and gums), to accelerate vulcanization, as an ingredient of phenolic resins (Durite), as an intermediate in the synthesis of furan derivatives, as a weed killer, as a fungicide and as a flavouring agent (McKillip & Sherman, 1980; Budavari, 1989). In the Netherlands, furfural is used as a component of a diesel fuel colourant at 50 g/L (Dutch Expert Committee on Occupational Standards, 1992).

1.3 Occurrence

1.3.1 Natural occurrence

Furfural is a volatile component of a wide range of fruits and vegetables (Dutch Expert Committee on Occupational Standards, 1992).

1.3.2 Occupational exposure

In the National Occupational Exposure Survey conducted between 1981 and 1983 it was estimated that 135 914 workers were potentially exposed to furfural in 6447 plants in the United States (United States National Institute for Occupational Safety and Health, 1994b). The estimate is based on a survey of companies and did not involve measurements of actual exposures. In the Harvard job–exposure matrix (Hoar *et al.*, 1980), potential exposure to furfural was linked to work in several industries, including construction, metal, food, tobacco, textiles, chemicals, drugs, paints, rubber, plastics, paints and leather, and to several jobs in agriculture, medicine, science, art and even to work as a homemaker or chambermaid.

Furfural production is either semicontinuous or batch, and few workers are involved in either case (Mutchler & Proskie, 1982). The main steps in the production of furfural are: unloading and storage of raw materials; mixing and feeding raw materials to a reactor; digestion, reaction, steam distillation and emptying the reactor; finishing by condensation, stripping, separation and distillation and storage and loading. Exposure to furfural may occur at the end of the reaction stage in the batch operation, and furfural vapours can escape into the workroom when the reactors are emptied of residue. Intermittent exposure to furfural can occur if the seals of the reaction vessel lose their integrity, especially in the packing glands and the mechanical shaft seals of rotating, pressurized equipment.

The warm, damp residue from the reactor passes through a hopper and, depending on the workplace configuration, vapour and gases may be carried by convection into the workplace. The part of the process in which the furfural produced during the reaction phase is recovered, enriched and refined is totally enclosed and includes several condensate-level tanks, decanters, stripping columns and final distillation. Exposure to furfural during this stage of the production process is unlikely. Storage and shipping of furfural takes place in a pump house, drum-filling stations and railcar or truck-loading facilities. In all of these operations, exposure to furfural is incidental and usually occurs only if there is a significant spill, leak or equipment malfunction. The pump station and drum-filling operations may have local exhaust ventilation. Transfers during storage and shipping usually take place out-of-doors or in semi-enclosed spaces where there is also natural ventilation (Mutchler & Proskie, 1982).

In two reports in which worker response was related to measured air concentrations, a time-weighted threshold limit value of 5 ppm [19.7 mg/m^3] was determined on the basis of 'comfort'. A study in a grinding-wheel plant showed a widespread incidence of eye and respiratory tract irritation, which was attributed to furfural vapours, found at concentrations of 5–16 ppm [19.7–62.9 mg/m^3] (American Conference of Governmental Industrial Hygienists, 1983). A study in a refinery yielded an 8–10-h time-weighted average exposure of 1.9 mg/m^3 (range, 1.2–2.6 mg/m^3) over six samples (Herwin & Froneberg, 1982). Four samples from selected work areas in a plant where furfural was produced by wood hydrolysis contained 4–16 mg/m^3 (Troshina *et al.*, 1986).

In a plant for the manufacture of arc carbon electrodes and dry-cell electrodes in India, furfural was used as a solvent for a phenol–formaldehyde resin, and the resulting mass was used as a binder with petroleum coke. The concentrations of furfural in air were 11–70 mg/m^3 (Ghose, 1991).

The exposure of California wildland firefighters was assessed during three consecutive fire seasons (1986–89). The mean time-weighted average personal exposure to furfural was 0.13 mg/m^3 (25 observations) and ranged from undetectable to 0.23 mg/m^3. Furfural was found in 83% of the samples taken (Materna *et al.*, 1992).

Environmental monitoring was conducted at 17 locations in Italy in the section of a chemical plant in which furfural was extracted from wood. At four locations, the threshold limit value of 8 mg/m^3 was exceeded. Twelve employees were investigated for the presence of furoic acid (expressed in mg/g creatinine) in urine, while 18 'blue collar' workers at a paper mill were used as matched controls. Urine was collected during the same 8-h period on two occasions: one at work and the other at home. The mean concentration of furoic acid in the urine of the blue-collar workers was significantly higher (27.1 ± 16.6 mg/g creatinine) than that in urine taken at home from workers exposed to furfural (21.4 ± 15.6 mg/g creatinine) ($p < 0.01$) but significantly lower than that in urine of furfural-exposed workers taken at work (34.5 ± 13.3 mg/g creatinine) ($p < 0.0001$) (Di Pede et al., 1991).

1.3.3 Air

Furfural was measured at a concentration of 0.19 µg/m^3 at the foot of Mount Everest in Nepal (Ciccioli et al., 1993). It was identified as one of the main components of smoke condensates of pine and cottonwood (Edye & Richards, 1991). It was also a major constituent of glowing fires of conifer logs (Mallory & Van Meter, 1976). Residential burning of brown-coal briquets led to emission of furfural, at an emission factor of 1.63 mg/kg. The estimated total amount of furfural emitted in the city of Leipzig was 530 kg/year (Engewald et al., 1993).

Aldehydes were measured both indoors and outdoors for six days at six houses in suburban New Jersey (United States) during the summer of 1992 under controlled conditions of ventilation and gas combustion. The mean concentration of furfural was 0.27 ppb [1.06 µg/m^3] in 36 samples taken indoors and 0.17 ppb [0.67 µg/m^3] in 36 samples taken outdoors. The presence of furfural indoors may be due to emissions during cooking (Zhang et al., 1994).

Furfural was detected in two of five houses that had been damaged by fire and refurbished, at concentrations of 0.02 and 0.23 ppb [0.08 and 0.90 µg/m^3] (Tsuchiya, 1992).

1.3.4 Water

Information on actual concentrations in water is very limited. Low levels of furfural (up to 30 mg/L) were reported in wastewater from a furfural production plant; however, derivatives of furfural occurred at concentrations up to 300 mg/L (Dodolina et al., 1977).

1.3.5 Food

Furfural has been identified in 150 foods, including fruits, vegetables, beverages, bread and bread products. The highest reported concentrations were found in wheat bread (0.8–14 ppm) [mg/kg], cognac (0.6–33 ppm), rum (22 ppm), malt whisky (10–37 ppm), port wine (2–34 ppm) and coffee (55–255 ppm). The concentrations of furfural in juices were 0.01–4.93 ppm. Furfural is formed during the reaction of amino acids with sugars at 100 °C for 4 h (Dutch Expert Committee on Occupational Standards, 1992). The occurrence of furfural in stored citrus products is recognized as an indication of deterioration; the concentrations in grapefruit juice samples stored at ambient temperature for over one year were 2.1–9.0 ppm (Lee et al., 1986).

1.4 Regulations and guidelines

Occupational exposure limits and guidelines for furfural in a number of countries are given in Table 2.

Table 2. Occupational exposure limits and guidelines for furfural

Country	Year	Concentration (mg/m^3)	Interpretation
Australia	1991	8	TWA; skin notation
Belgium	1991	7.9	TWA; skin notation
Denmark	1991	20	TWA; skin notation
Finland	1993	20	TWA; skin notation
		40	STEL
France	1991	8	STEL
Germany	1993	20	TWA; skin notation
Japan	1991	9.8	TWA; skin notation
Netherlands	1994	8	TWA
Poland	1991	10	TWA
Russian Federation	1991	10	STEL; skin notation
Sweden	1991	20	TWA; skin notation
		40	STEL
Switzerland	1991	8	TWA; skin notation
United Kingdom	1993	8	TWA; skin notation
		40	STEL; skin notation
USA			
ACGIH	1994	7.9	TWA; skin notation
OSHA	1994	20	TWA; skin notation

From ILO (1991); Dutch Expert Committee on Occupational Standards (1992); American Conference of Governmental Industrial Hygienists (ACGIH) (1994); Deutsche Forschungsgemeinschaft (1993); Työministeriö (1993); United Kingdom Health and Safety Executive (1993); Arbeidsinspectie (1994); United States Occupational Safety and Health Administration (1994)
TWA, time-weighted average; STEL, short-term exposure limit

2. Studies of Cancer in Humans

No data were available to the Working Group.

3. Studies of Cancer in Experimental Animals

3.1 Oral administration

3.1.1 Mouse

Groups of 50 male and 50 female B6C3F1 mice, aged nine weeks, were administered 0, 50, 100 or 175 mg/kg bw furfural (purity, 99%) dissolved in corn oil by gavage on five days a week for 103 weeks. Survival at the end of the study was 35/50 male controls, 28/50 at the low dose, 24/50 at the middle dose and 27/50 at the high dose; and 33/50 female controls, 28/50 at the low dose, 29/50 at the middle dose and 32/50 at the high dose. There was a dose-related increase in the incidence of chronic inflammation of the liver. In males, the incidences of hepatocellular adenomas were 9/50 controls, 13/50 at the low dose, 11/49 at the middle dose and 19/50 at the high dose ($p = 0.008$, logistic regression analysis); the incidences of hepatocellular carcinoma were 7/50 controls, 12/50 at the low dose, 6/49 at the middle dose and 21/50 at the high dose ($p = 0.001$). Female mice also had a higher incidence of hepatocellular adenomas, with 1/50 in controls, 3/50 at the low dose, 5/50 at the middle dose and 8/50 at the high dose ($p = 0.017$); the incidences of hepatocellular carcinoma (4/50, 0/50, 2/50, 4/50) were not increased. The combined incidences of hepatocellular adenomas and carcinomas were: 16/50 male controls, 22/50 at the low dose, 17/49 at the middle dose and 32/50 at the high dose ($p < 0.001$); and 5/50 female controls, 3/50 at the low dose, 7/50 at the middle dose and 12/50 at the high dose ($p = 0.051$). There was a marginal increase in the incidence of forestomach papillomas in females at the high dose: 6/50 in comparison with 1/50 in controls ($p = 0.058$) (United States National Toxicology Program, 1990).

3.1.2 Rat

Groups of 50 male and 50 female Fischer 344 rats, seven to eight weeks of age, were administered 0, 30 or 60 mg/kg bw furfural (purity, 99%) dissolved in corn oil by gavage on five days per week for 103 weeks. Survival at the end of the study was: 31/50 male controls, 28/50 at the low dose and 24/50 at the high dose; and 28/50 female controls, 32/50 at the low dose and 18/50 at the high dose (not significant). A dose-related increase in the frequency of centrilobular necrosis of the liver was seen in males: 3/50 controls, 9/50 at the low dose and 12/50 at the high dose. Two of 50 males given the high dose had bile-duct dysplasia, and two had rarely occurring cholangiocarcinomas. No such lesions were found in the other groups of males or among female rats. There were no other treatment-related lesions in the liver or other organs. The historical incidence of cholangiocarcinoma in control rats at the testing laboratory was 1/449 (United States National Toxicology Program, 1990).

In a study of enzyme-altered foci in the liver, six groups of six male Wistar rats, five weeks of age, were administered furfural [purity unspecified] in the diet at a concentration of 20 ml/kg of diet for 15–30 days and then at 30 ml/kg of diet for up to 150 days. The exposure of the six groups ceased on days 15, 30, 60, 90, 120 and 150, respectively. Six groups of four male controls were available. The rats were sacrificed 15 days after the end of exposure. Fibrosis was seen in the liver after 30 days of treatment and progressed with the length of exposure, resulting in pseudolobule formation after 150 days of treatment. Foci positive for glutathione S-transferase placental form were seen in 4/6 rats after 30 days of treatment and in 6/6 after 150 days. No such foci were seen in the controls. No cancers or neoplastic nodules occurred in any of the groups (Shimizu et al., 1989).

3.2 Skin application

Mouse: Groups of 20 female CD-1 mice, seven weeks of age, received topical applications of 50 µmol [4.8 mg] furfural ('special grade') dissolved in 0.1 ml dimethyl sulfoxide on the back twice a week for five weeks. One week after the last treatment, the mice were treated twice a week with 2.5 µg of the promoter 12-*O*-tetradecanoylphorbol 13-acetate (TPA) in 0.1 ml acetone for 47 weeks. One control group was treated with furfural and acetone, a second with dimethyl sulfoxide (vehicle control) and TPA, a third with dimethyl sulfoxide and acetone and a fourth with a total dose of 100 µg 7,12-dimethylbenz[*a*]anthracene (DMBA) and TPA (positive control). Five of 19 mice given furfural and TPA developed seven skin papillomas and one squamous-cell cancer, whereas only one of 20 mice given DMSO and TPA had a papilloma [$p = 0.08$, Fisher's exact test]. None of the other negative controls developed tumours, but all 20 mice in the positive control group developed skin tumours (Miyakawa et al., 1991).

3.3 Administration with known carcinogens

Hamster: The co-carcinogenic effect of furfural and the known carcinogens benzo[*a*]pyrene and *N*-nitrosodiethylamine on the respiratory tract of hamsters was studied in two experiments. In one study, long-term exposure to furfural vapour and repeated intratracheal instillations of benzo[*a*]pyrene or *N*-nitrosodiethylamine did not significantly affect the tumour incidence in hamster respiratory tissues (Feron & Kruysse, 1978). In the other study, repeated, simultaneous intratracheal instillations of furfural and benzo[*a*]pyrene solutions had a slight co-carcinogenic effect in the respiratory tract (Feron, 1972).

4. Other Data Relevant to an Evaluation of Carcinogenicity and its Mechanisms

4.1 Absorption, distribution, metabolism and excretion

4.1.1 Humans

Pulmonary uptake and exhalation and urinary excretion of furfural were studied in six healthy male volunteers, aged 30–55 years, exposed by inhalation in a laboratory measuring $5.4 \times 3.4 \times 3.5$ m [64 m^3], where the concentration of furfural in the atmosphere was kept constant by automatically controlled dispersion, within an accuracy of ± 5% at preselected levels of approximately 7–30 mg/m^3. [Data were presented for concentrations of 15, 20 and 31 mg/m^3.] The total duration of exposure was 8 h, interrupted by two 5-min breaks after the second and sixth hour and by one 20-min break in the middle of the experiment, to give an actual exposure of 7.5 h. Two to four individuals were exposed simultaneously and were seated throughout the exposure. An average of 78% of the furfural was absorbed; the rate was not related to the level or duration of exposure. Only a small proportion of free furfural (< 1%) was excreted by the lungs [exhalation of carbon dioxide was not determined], and none was found in urine. The biological half-life of furfural was 2–2.5 h. Free furoic acid occurred in only negligible amounts in the urine, and a conjugated form of furoic acid [presumably furoylglycine] appeared to be the major urinary metabolite. The recovery of 'total furoic acid' (free furoic acid measured after alkaline hydrolysis of urine samples) indicated that the volunteers had also been exposed dermally. In experiments in which furfural was inhaled directly, through a mask, thus avoiding dermal absorption, 93–103% of a dose of 30 mg/m^3 was recovered as 'total furoic acid'. 'Total furanacrylic acid', measured as free furanacrylic acid after hydrolysis of urine samples, amounted to only 0.5–5% of the inhaled dose of furfural. When the volunteers were exposed dermally to furfural while breathing pure air, there was considerable but variable absorption. After volunteers submerged their hands up to the wrist in a vessel containing liquid furfural for 15 min, the total amount of 'total furoic acid' excreted indicated that about 27 mg furfural had been absorbed through the hand surface. Recalculation of this amount indicated that 1 cm^2 skin absorbed approximately 3 µg furfural per min (Flek & Šedivec, 1978).

4.1.2 Experimental systems

After [carbonyl-^{14}C]furfural (specific activity, 4.1 mCi/mmol; radiochemical purity, 95%) was administered by gavage to male Fischer 344 rats at single doses of 0.127, 1.15 or 12.5 mg/kg bw in corn oil, 86–89% of the dose was absorbed, and more than 60% was excreted after 12 h, reaching a plateau after 24 h. After 72 h, high concentrations of radiolabel were found in liver and kidney; brain had the lowest concentration. The concentrations in liver and kidney were approximately proportional to the dose. The major route of excretion was urine, which contained 83–88% of the dose; about 7% of a dose of 12.5 mg/kg bw was exhaled as ^{14}C-carbon dioxide, and 2–4% of the dose was detected in the faeces. Furoylglycine was the major urinary metabolite (73–80% of dose), and furanacrylic acid (3–8%) and furoic acid (1–6%) were minor metabolites. The extent and rate of excretion of furfural metabolites were unaffected by dose. Furoic acid is an oxidation product of furfural, which may be excreted unchanged or conjugated with glycine. Furanacrylic acid is presumably formed *via* condensation with acetyl coenzyme A (Nomeir *et al.*, 1992).

[Carbonyl-^{14}C]furfural (specific activity, 135.75 µCi/mg; radiochemical purity, 98.1%) was administered by gavage to male and female Fischer 344 rats at single doses of 1, 10 or 60 mg/kg bw and to male and female CD1 mice at single doses of 1, 20 or 200 mg/kg bw in trioctanoin. More than 90% of the dose was recovered in all animals after 72 h. The main route of elimination was the urine, and more than 60% of the dose was eliminated by this route within the first 24 h. Within 72 h, faecal elimination comprised 3–7% of the dose, and mice expired 5% of the radiolabel; the carcasses of both species contained < 1% of the dose by that time. Furoylglycine was the major urinary metabolite in both species: urine collected at 0–24 h contained 76–84% (rats) and 65–89% (mice) of this metabolite. Furanacryloylglycine was the second most abundant metabolite, constituting 15–24% of the radiolabel in 0–24-h urine of rats and 11–35% of that of mice. Only subtle differences in the metabolic profile were seen as a function of dose, sex and species. The parent acids of the glycine conjugates were excreted at the higher doses (Parkash & Caldwell, 1994).

4.1.3 Comparison of humans and animals

Furfural is extensively absorbed and rapidly eliminated in humans after inhalation and in rats after oral administration. The pattern of metabolites appears to be qualitatively similar, involving oxidation of furfural to furanoic acid with subsequent conjugation, primarily with glycine. Because of limitations in the reporting of the study of humans (Flek & Šedivec, 1978), a closer, quantitative comparison of the toxicokinetic profiles of humans and rats is not possible.

4.2 Toxic effects

4.2.1 Humans

In a review, the main effect of furfural in humans was reported to be skin and mucous membrane irritation. Irritant dermatitis has in some cases led to eczema, and there have been reports of allergic skin sensitization and photosensitization (Mishra, 1992).

4.2.2 Experimental systems

In a short-term study, groups of six adult male albino rats [strain not specified] were exposed by inhalation to furfural [purity not specified] at 95 ppm [370 mg/mg^3] in a single exposure or at 38 ppm [150 mg/m^3] for 1 h daily on five days per week over 7, 15 or 30 days. Six rats served as controls. The single exposure resulted in moderate congestion and perivascular oedema in the lungs; these changes were more pronounced after repeated exposure. Alkaline phosphatase activity in the lung was substantially increased at all times, whereas lung glutamic pyruvic transaminase activity was increased after 15 and 30 days. Furfural was also reported to induce liver necrosis and to increase the activities of hepatic alkaline phosphatase, glutamic oxaloacetic transaminase and glutamic pyruvic transaminase [details not given] (Mishra et al., 1991).

In a short-term study, six adult male albino rats [strain not specified] were exposed to 40 ppm [155 mg/m^3] furfural (purity, 99%) for 1 h per day on five days per week for 7, 15 or 30 days. Six rats served as controls. Exposure to furfural increased the activities of acid and alkaline phosphatases and glutamic pyruvic transaminase, inhibited the activities of arginine and succinic dehydrogenases and increased the concentration of lactic acid in lung homogenates. [The Working Group noted that histopatological examination was not performed.] The exposure regimen was not lethal, whereas separate experiments determined an LC_{50} value of 189 ppm [735 mg/m^3] after a single exposure (Gupta et al., 1991).

Groups of 10 male and 10 female Fischer 344/N rats were administered 0, 11, 22, 45, 90 or 180 mg/kg bw furfural (purity, 99%) in corn oil by gavage on five days per week for 13 weeks, and groups of 10 male and 10 female B6C3F1 mice were administered 0, 75, 150, 300, 600 or 1200 mg/kg bw on the same schedule. The incidence of cytoplasmic vacuolization in hepatocytes was increased in male rats. Severe, dose-dependent centrilobular coagulative necrosis of hepatocytes, leading to death, was seen in mice at the high dose (United States National Toxicology Program, 1990).

In a 90–120-day study, furfural [purity not specified] was administered in the diet of four male Wistar/Slc rats at concentrations of 20 ml/kg [23 mg/kg] of diet for the first seven days, 30 ml/kg [35 mg/kg] for the next seven days and 40 ml/kg [46 mg/kg] from day 15 to day 90; from day 91 to day 120, rats were given repeated cycles of two days on basal diet and five days on 40 ml/kg [46 mg/kg] furfural [doses not calculated]. Four controls were available. Hepatic cirrhosis was seen in treated animals, and the fibrotic changes were more prominent in animals killed at 120 days than in those killed after 90 days. Massive liver necrosis was not observed after administration of a single dose of 50 mg/kg bw by gavage (Shimizu & Kanisawa, 1986).

Groups of 18 male and 18 female Syrian golden hamsters were exposed to distilled furfural at concentrations of 400 ppm [1550 mg/m^3] for 7 h per day on five days per week during the first nine weeks, 300 ppm [1280 mg/m^3] during weeks 10–20 and 250 ppm [970 mg/m^3] during weeks 21–52; 18 animals of each sex were exposed to air only. Severe damage of the olfactory epithelium of the nasal cavity was seen, with nearly complete atrophy of the sensory cells in the olfactory mucosa, large cyst-like, glandular structures in the lamina propria beneath the olfactory epithelium and karyo- and cytomegaly of Bowman's glands. There was neither evidence of recovery of the nasal changes after a period of six months nor any progression of the lesions. No changes in other parts of the respiratory tract or outside the airway system were seen that could be ascribed to treatment (Feron & Kruysse, 1978; Feron *et al.*, 1979).

Exposure of adult male albino rats [strain not specified] to 40 ppm [155 mg/m^3] furfural for 1 h daily, five days per week, for 7, 15 or 30 days led to increases in the activities of aminopyrine *N*-demethylase, aniline hydroxylase and glutathione *S*-transferase (1-chloro-2,4-dinitrobenzene as substrate) in the 9000 × *g* supernatant of lung, whereas benzo[*a*]pyrene hydroxylase activity was decreased (Gupta *et al.*, 1991).

4.3 Reproductive and prenatal effects

No data were available to the Working Group.

4.4 Genetic and related effects

4.4.1 Humans

Six workers exposed to furfural and furfuryl alcohol in a furoic resin plant showed no significant difference in sister chromatid exchange frequency in peripheral blood lymphocytes in comparison with six control individuals (Gomez-Arroyo & Souza, 1985). [The Working Group noted the small number of individuals studied and the presence of both smokers and non-smokers; moreover, the furfural concentrations in the atmosphere of the plant were not reported.]

4.4.2 Experimental systems (see also Table 3 and Appendices 1 and 2)

Furfural reacts with DNA *in vitro*, primarily at AT base pairs, leading to destabilization of the secondary structure of DNA and to single-strand breaks.

Table 3. Genetic and related effects of furfural

Test system	Result[a] Without exogenous metabolic system	Result[a] With exogenous metabolic system	Dose[b] (LED/HID)	Reference
***, Destabilization of DNA secondary structure *in vitro*	+		1:2	Uddin (1993)
***, DNA strand breaks, calf thymus DNA *in vitro*	+		1:4	Uddin et al. (1991)
***, DNA strand breaks, calf thymus DNA *in vitro*	+		1:1	Hadi et al. (1989)
PRB, SOS repair, *Salmonella typhimurium* umu	–		1932	Nakamura et al. (1987)
SA0, *Salmonella typhimurium* TA100, reverse mutation	+	+	2200	Zdzienicka et al. (1978)
SA0, *Salmonella typhimurium* TA100, reverse mutation	–	–	2100	Loquet et al. (1981)
SA0, *Salmonella typhimurium* TA100, reverse mutation	–	–	1280	US National Toxicology Program (1990)
SA0, *Salmonella typhimurium* TA100, reverse mutation	–	–	0.00	Dillon et al. (1992) (Abstract)
SA2, *Salmonella typhimurium* TA102, reverse mutation	–	–	0.00	Dillon et al. (1992) (Abstract)
SA4, *Salmonella typhimurium* TA104, reverse mutation	–	–	0.00	Dillon et al. (1992) (Abstract)
SA5, *Salmonella typhimurium* TA1535, reverse mutation	–	–	1800	Loquet et al. (1981)
SA5, *Salmonella typhimurium* TA1535, reverse mutation	–	–	1280	US National Toxicology Program (1990)
SA7, *Salmonella typhimurium* TA1537, reverse mutation	–	–	1280	US National Toxicology Program (1990)
SA9, *Salmonella typhimurium* TA98, reverse mutation	–	–	4400	Zdzienicka et al. (1978)
SA9, *Salmonella typhimurium* TA98, reverse mutation	–	–	1800	Loquet et al. (1981)
SA9, *Salmonella typhimurium* TA98, reverse mutation	–	–	1280	US National Toxicology Program (1990)
DMX, *Drosophila melanogaster*, sex-linked recessive lethal mutation	–		1000 feed	Woodruff et al. (1985)
DMX, *Drosophila melanogaster*, sex-linked recessive lethal mutation	+		100 inj	Woodruff et al. (1985)
DMH, *Drosophila melanogaster*, heritable translocation	–		100 inj	Woodruff et al. (1985)
G5T, Gene mutation, mouse lymphoma L5178Y cells, *tk* locus *in vitro*	+	0	200	McGregor et al. (1988)
SIC, Sister chromatid exchange, Chinese hamster ovary (CHO) cells *in vitro*	+	+	12	US National Toxicology Program (1990)
CIC, Chromosomal aberrations, Chinese hamster ovary (CHO) cells *in vitro*	+	+	240	Stich et al. (1981)

Table 3 (contd)

Test system	Result[a]		Dose[b] (LED/HID)	Reference
	Without exogenous metabolic system	With exogenous metabolic system		
CIC, Chromosomal aberrations, Chinese hamster ovary (CHO) cells in vitro	+	+	400	US National Toxicology Program (1990)
CIC, Chromosomal aberrations, Chinese hamster V79 cells in vitro	+	0	1000	Nishi et al. (1989)
SHL, Sister chromatid exchange, human lymphocytes in vitro	+	0	7	Gomez-Arroyo & Souza (1985)
SVA, Sister chromatid exchange, mouse bone-marrow cells in vivo	–		200.0 ip × 1	US National Toxicology Program (1990)
CBA, Chromosomal aberrations, mouse bone-marrow cells in vivo	–		200.0 ip × 1	US National Toxicology Program (1990)

[a] +, considered to be positive; –, considered to be negative; 0, not tested
[b] LED, lowest effective dose; HID, highest ineffective dose; in vitro tests, μg/ml; in vivo tests, mg/kg bw; 0.00, dose not reported; ip, intraperitoneally
***, Not included on the profile

Furfural did not induce *umu c'* gene expression, a function related to SOS DNA repair, in *Salmonella typhimurium* TA1535/pSK1002. It was reported to be mutagenic to *S. typhimurium* TA100 in the presence and absence of metabolic activation in one study, but this result was not confirmed in three subsequent studies, which gave equivocal or negative results. Furfural was also reported to be nonmutagenic in *S. typhimurium* strains G46, TA100, TA1535, C3076, TA1537, D3052, TA1538 and TA98 and in *Escherichia coli* strains WP2 and WP2 *uvr*A with a concentration gradient protocol (MacMahon *et al.*, 1979).

Injection, but not feeding, of furfural to adult flies *Drosophila melanogaster* induced sex-linked recessive lethal mutation. Furfural did not induce heritable reciprocal translocations in *D. melanogaster*.

Furfural induced gene mutation at the thymidine kinase locus of L5178Y mouse lymphoma cells in the absence of metabolic activation. It induced sister chromatid exchange in Chinese hamster ovary cells and human lymphocytes and chromosomal aberrations in Chinese hamster ovary and V79 lung cells in the absence of metabolic activation.

The frequencies of sister chromatid exchange and chromosomal aberrations were not increased in the bone-marrow cells of B6C3F1 male mice injected intraperitoneally with single doses of furfural up to 200 mg/kg bw.

Mutation of proto-oncogenes in tumours induced by furfural

ras Proto-oncogene activation was studied in liver adenomas and carcinomas of B6C3F1 mice treated with furfural. The frequency of activated H-*ras* and K-*ras* oncogenes in hepatocellular tumours was no different in furfural-treated (10/16) and vehicle-treated (15/27) mice; however, the spectrum of activating mutations in the H-*ras* gene in tumours from the furfural-treated mice differed significantly from that in tumours of untreated animals. Mutations at codon 61 occurred in tumours from both furfural-treated and untreated animals, but mutations (G→T and G→C transversions) were observed at codons 13 and 117 only in furfural-treated animals. The authors interpreted their findings as suggesting that novel mutations in *ras* genes could have resulted from a genotoxic effect of furfural (Reynolds *et al.*, 1987).

5. Summary and Evaluation

5.1 Exposure data

Furfural is produced commercially by the acid hydrolysis of pentosan polysaccharides from non-food residues of food crops and wood wastes. It is used widely as a solvent in petroleum refining, in the production of phenolic resins and in a variety of other applications. Human exposure to furfural occurs during its production and use, as a result of its natural occurrence in many foods and from the combustion of coal and wood.

5.2 Human carcinogenicity data

No data were available to the Working Group.

5.3 Animal carcinogenicity data

Furfural was tested for carcinogenicity by oral administration in one study in mice and one study in rats. In mice, it increased the incidence of hepatocellular adenomas and carcinomas in males and of hepatocellular adenomas and forestomach papillomas in females. Male rats had a low incidence of cholangiocarcinomas, which occur rarely. In a two-stage assay on mouse skin, furfural had weak initiating activity.

5.4 Other relevant data

Furfural is extensively absorbed and rapidly eliminated after inhalation by humans and rats. Furfural in air is also absorbed dermally by humans. Repeated exposure of hamsters to furfural by inhalation severely damages the olfactory epithelium. Repeated oral administration to rats causes liver necrosis and cirrhosis.

Neither chromosomal aberrations nor sister chromatid exchanges were observed in rodents treated with furfural *in vivo* in a single study.

Gene mutation (in a single study), sister chromatid exchange and chromosomal aberrations were induced in mammalian cells *in vitro*. Sex-linked recessive lethal mutations were induced in insects. Furfural induced weak or no mutagenicity in bacteria but damaged DNA *in vitro*.

5.5 Evaluation[1]

There is *inadequate evidence* in humans for the carcinogenicity of furfural.

There is *limited evidence* in experimental animals for the carcinogenicity of furfural.

Overall evaluation

Furfural *is not classifiable as to its carcinogenicity to humans (Group 3)*.

6. References

Aldrich Chemical Co. (1994) *Specification Sheet: 2-Furaldehyde*, Milwaukee, WI

American Conference of Governmental Industrial Hygienists (1983) *Documentation on the Threshold Limit Values*, 4th Ed., Cincinnati, OH, pp. 694–695

American Conference of Governmental Industrial Hygienists (1994) *1994–1995 Threshold Limit Values for Chemical Substances and Physical Agents and Biological Exposure Indices*, Cincinnati, OH, p. 22

---

[1] For definition of the italicized terms, see Preamble, pp. 22–26.

Arbeidsinspectie (1994) *De Nationale MAC-Lijst* [National MAC list], The Hague, p. 28

Beeman, C.P. (1987) Improved efficiency for colorimetric determination of furfural in citrus juices. *J. Assoc. off. anal. Chem.*, **70**, 601–602

Blanco Gomis, D., Gutiérrez Alvarez, M.D., Sopeña Naredo, L. & Mangas Alonso, J.J. (1991) High-performance liquid chromatographic determination of furfural and hydroxymethylfurfural in apple juices and concentrates. *Chromatographia*, **32**, 45–48

Budavari, S., ed. (1989) *The Merck Index*, 11th Ed., Rahway, NJ, Merck & Co., p. 673

Chemical Information Services, Inc. (1994) *Directory of World Chemical Producers 1995/96 Standard Edition*, Dallas, TX, p. 368

Ciccioli, P., Brancaleoni, E., Cecinato, A., Sparapani, R. & Frattoni, M. (1993) Identification and determination of biogenic and anthropogenic volatile organic compounds in forest areas of northern and southern Europe and a remote site of the Himalaya region by high-resolution gas chromatography–mass spectrometry. *J. Chromatogr.*, **643**, 55–69

Corradini, D. & Corradini, C. (1992) Separation and determination of 5-hydroxymethyl-2-furaldehyde and 2-furaldehyde in fruit juices by micellar electrokinetic capillary chromatography with direct sample injection. *J. Chromatogr.*, **624**, 503–509

Deutsche Forschungsgemeinschaft (1993) *MAK- und BAT-Werte-Liste 1993* [MAK and BAT values list 1993] (Report 29), Weinheim, VCH Verlagsgesellschaft, p. 48 (in German)

Dillon, D.M., McGregor, D.B., Combes, R.D. & Zeiger, E. (1992) Detection of mutagenicity in *Salmonella* of some aldehydes and peroxides (Abstract). *Environ. mol. Mutag.*, **19** (Suppl. 20), 15

Dinsmore, H.L. & Nagy, S. (1974) Improved colorimetric determination for furfural in citrus juices. *J. Assoc. off. anal. Chem.*, **57**, 332–335

Di Pede, C., Viegi, G., Taddeucci, R., Landucci, C., Settmi, L. & Paggiaro, P.O. (1991) Biological monitoring of work exposure to furfural (Abstract). *Arch. environ. Health*, **46**, 125–126

DiSanzo, F.P., Johnson, G.L., Giarrocco, V.J. & Sutton, P. (1988) Determination of furfural in petroleum oils by liquid chromatography. *J. chromatogr. Sci.*, **26**, 77–79

Dodolina, V.T., Zhirnov, B.F. & Kutepov, L.Y. (1977) Permissible concentration of alcohols and furfural in waste water used for irrigation. *Gidrotekhn. Melior.*, **4**, 101–104 (in Russian)

Dutch Expert Committee on Occupational Standards (1992) *Health-based Recommended Occupational Exposure Limit for Furfural*, 3rd draft, Zeist, Directorate-General of Labour

Edye, L.A. & Richards, G.N. (1991) Analysis of condensates from wood smoke: components derived from polysaccharides and lignins. *Environ. Sci. Technol.*, **25**, 1133–1137

Eller, P.M., ed. (1994) *NIOSH Manual of Analytical Methods*, 4th Ed., Vol. 1 (DHHS (NIOSH) Publ. No. 94-113), Washington DC, US Government Printing Office, Methods 2529, 2539

Engewald, W., Knobloch, T. & Efer, J. (1993) Volatile organic compounds in emissions from brown-coal-fired residential stoves. *Z. Umweltchem. Ökotox.*, **5**, 303–308 (in German)

Feldstein, M., Bryan, R.J., Hyde, D.L., Levaggi, D.A., Locke, D.C., Rasmussen, R.A. & Warner, P.O. (1989) 834. Determination of organic solvent vapors in air. In: Lodge, J.P., Jr, ed., *Methods of Air Sampling and Analysis*, 3rd Ed., Chelsea, MI, Lewis Publishers, pp. 678–685

Feron, V.J. (1972) Respiratory tract tumors in hamsters after intratracheal instillations of benzo(*a*)pyrene alone and with furfural. *Cancer Res.*, **32**, 28–36

Feron, V.J. & Kruysse, A. (1978) Effects of exposure to furfural vapour in hamsters simultaneously treated with benzo(*a*)pyrene or diethylnitrosamine. *Toxicology*, **11**, 127–144

Feron, V.J., Kruysse, A. & Dreef-Van der Meulen, H.C. (1979) Repeated exposure to furfural vapour: 13-week study in Syrian golden hamsters. *Zbl. Bakt. Hyg. I. Abt. Orig. B*, **168**, 442–451

Flek, J. & Šedivec, V. (1978) The absorption, metabolism and excretion of furfural in man. *Int. Arch. occup. environ. Health*, **41**, 159–168

Ghose, M.K. (1991) Evaluation of air pollution due to fugitive emissions. *Indian J. environ. Health*, **33**, 66–69

Girenko, E.K., Maltabar, V.M. & Moiseenko, L.F. (1970) Determination of furfural in cognacs and cognac brandies. *Sadovod. Vinograd. Vinodel. Mold.*, **25**, 32–34 [*Chemical Abstracts*, 74:1865]

Gomez-Arroyo, S. & Souza, V.S. (1985) In vitro and occupational induction of sister chromatid exchanges in human lymphocytes with furfuryl alcohol and furfural. *Mutat. Res.*, **156**, 233–238

Gupta, G.D., Misra, A. & Agarwal, D.K. (1991) Inhalation toxicity of furfural vapours: an assessment of biochemical response in rat lungs. *J. appl. Toxicol.*, **11**, 343–347

Hadi, S.M., Uddin, S. & Rehman, A. (1989) Specificity of the interaction of furfural with DNA. *Mutat. Res.*, **225**, 101–106

Hansch, C., Leo, A. & Hoekman, D.H. (1995) *Exploring QSAR*, Washington DC, American Chemical Society

Helrich, K., ed. (1990) *Official Methods of Analysis of the Association of Official Analytical Chemists*, 15th Ed., Vol. 2, Arlington, VA, Association of Official Analytical Chemists, p. 701

Herwin, R.L. & Froneberg, B. (1982) *Cities Services Co. Lake Charles, LA* (Health Hazard Evaluation Report HETA 80-020-1054), Cincinnati, OH, United States National Institute for Occupational Safety and Health

Hitchcock, L.B. & Duffey, H.R. (1948) Commercial production of furfural in its twenty-fifth year. *Chem. Eng. Progr.*, **44**, 669–674

Hoar, S.K., Morrison, A.S., Cole, P. & Silverman, D.T. (1980) An occupation and exposure linkage system for the study of occupational carcinogenesis. *J. occup. Med.*, **22**, 722–726

ILO (1991) *Occupational Exposure Limits for Airborne Toxic Substances: Values of Selected Countries* (Occupational Safety and Health Series No. 37), 3rd Ed., Geneva, pp. 206–207

Kadokami, K., Koga, M. & Otsuki, A. (1990) Gas chromatography/mass spectrometric determination of traces of hydrophilic and volatile organic compounds in water after preconcentration with activated carbon. *Anal. Sci.*, **6**, 843–849

Kieber, R.J. & Mopper, K. (1990) Determination of picomolar concentrations of carbonyl compounds in natural waters, including seawater, by liquid chromatography. *Environ. Sci. Technol.*, **24**, 1477–1481

Lancaster Synthesis Ltd (1994) *Product Data Sheet: 2-Furaldehyde*, Morecambe, Lancashire

Leblanc, Y., Gilbert, R., Jalbert, J., Duval, M. & Hubert, J. (1993) Determination of dissolved gases and furan-related compounds in transformer insulating oils in a single chromatographic run by headspace/capillary gas chromatography. *J. Chromatogr. A*, **657**, 111–118

Lee, H.S., Rouseff, R.L. & Nagy, S. (1986) HPLC determination of furfural and 5-hydroxymethylfurfural in citrus juices. *J. Food Sci.*, **51**, 1075–1076

Le Lacheur, R.M., Sonnenberg, L.B., Singer, P.C., Christman, R.F. & Charles, M.J. (1993) Identification of carbonyl compounds in environmental samples. *Environ. Sci. Technol.*, **27**, 2745–2753

Li, Z.-F., Sawamura, M. & Kusunose, H. (1988) Rapid determination of furfural and 5-hydroxymethylfurfural in processed citrus juices by HPLC. *Agric. Biol. Chem.*, **52**, 2231–2234

Lide, D.R., ed. (1993) *CRC Handbook of Chemistry and Physics*, 74th Ed., Boca Raton, FL, CRC Press, pp. 3-252, B-11

Lo Coco, F., Valentini, C., Novelli, V. & Ceccon, L. (1994) High performance liquid chromatographic determination of 2-furaldehyde and 5-hydroxymethyl-2-furaldehyde in processed citrus juices. *J. liq. Chromatogr.*, **17**, 603-617

Loquet, C., Toussaint, G. & LeTalaer, J.Y. (1981) Studies on mutagenic constituents of apple brandy and various alcoholic beverages collected in western France, a high incidence area for oesophageal cancer. *Mutat Res.*, **88**, 155-164

Mallory, W.R. & Van Meter, W.P. (1976) Chemicals in vapors of smoldering wood fires. *Forest Sci.*, **22**, 81-83

Maltabar, V.M. & Girenko, E.K. (1969) Determination of furfural in young colorless brandies. *Sadovod. Vinograd. Vinodel. Mold.*, **24**, 26-30 [*Chemical Abstracts*, 71-11 716]

Mannsville Chemical Products Corp. (1984) *Furfural. Chemical Products Synopsis*, Cortland, NY

Marcy, J.E. & Rouseff, R.L. (1984) High-performance liquid chromatographic determination of furfural in orange juice. *J. agric. Food Chem.*, **32**, 979-981

Materna, B.L., Jones, J.R., Sutton, P.M., Rothman, N. & Harrison, R.J. (1992) Occupational exposures in California wildland fire fighting. *Am. ind. Hyg. Assoc. J.*, **53**, 69-76

McGregor, D.B., Brown, A., Cattanach, P., Edwards, I., McBride, D. & Caspary W.J. (1988) Responses of the L5178Y tk$^+$/tk$^-$ mouse lymphoma cell forward mutation assay. II: 18 coded chemicals. *Environ. mol. Mutag.*, **11**, 91-118

McKillip, W.J. & Sherman, E. (1980) Furan derivatives. In: Mark, H.F., Othmer, D.F., Overberger, C.G., Seaborg, G.T. & Grayson, M., eds, *Kirk-Othmer Encyclopedia Of Chemical Technology*, 3rd Ed., Vol. 11, New York, John Wiley & Sons, pp. 501-510

McMahon, R.E., Cline, J.C. & Thompson, C.Z. (1989) Assay of 855 test chemicals in ten tester strains using a new modification of the Ames test for bacterial mutagens. *Cancer Res.*, **39**, 682-693

Mishra, A. (1992) Furfural: a toxic chemical. *Agric. biol. Res.*, **8**, 93-104

Mishra, A., Dwivedi, P.D., Verma, A.S., Sinha, M., Mishra, J., Lal, K., Pandya, K.P. & Dutta, K.K. (1991) Pathological and biochemical alterations induced by inhalation to furfural vapor in the lung. *Bull. environ. Contam. Toxicol.*, **47**, 668-674

Miyakawa, Y., Nishi, Y., Kato, K., Sato, H., Takahashi, M. & Hayashi, Y. (1991) Initiating activity of eight pyrolysates of carbohydrates in a two-stage mouse skin tumorigenesis model. *Carcinogenesis*, **12**, 1169-1173

Mutchler, J.E. & Proskie, K.G. (1982) Furfural. In: Cralley, L.V. & Cralley, L.J., eds, *Industrial Hygiene Aspects of Plant Operations*, Vol. 1, *Process Flows*, New York, Macmillan Publishing, pp. 257-262

Nakamura, M., Toda, M., Saito, H. & Ohkura, Y. (1982) Fluorimetric determination of aromatic aldehydes with 4,5-dimethoxy-1,2-diaminobenzene. *Anal. chim. Acta*, **134**, 39-45

Nakamura, S.-I., Oda, Y., Shimada, T., Oki, I. & Sugimoto, K. (1987) SOS-inducing activity of chemical carcinogens and mutagens in *Salmonella typhimurium* TA1535/pSK1002: examination with 151 chemicals. *Mutat. Res.*, **192**, 239-246

Nishi, Y., Miyakawa, Y. & Kato, K. (1989) Chromosome aberrations induced by pyrolysates of carbohydrates in Chinese hamster V79 cells. *Mutat. Res.*, **227**, 117-123

Nomeir, A.A., Silveira, D.M., McComish, M.F. & Chadwick, M. (1992) Comparative metabolism and disposition of furfural and furfuryl alcohol in rats. *Drug Metab. Disposition*, **20**, 198-204

Ohkura, Y., Ohtsubo, K., Zaitsu, K. & Kohashi, K. (1978) Fluorimetric determination of aromatic aldehydes with 2,2'-dithiobis(1-aminonaphthalene). *Anal. chim. Acta*, **99**, 317–324

Parkash, M.K. & Caldwell, J. (1994) Metabolism and excretion of [^{14}C]furfural in the rat and mouse. *Food chem. Toxicol.*, **32**, 887–895

Reynolds, S.H., Stowers, S.J., Patterson, R.M., Maronpot, R.R., Aaronson, S.A. & Anderson, M.W. (1987) Activated oncogenes in B6C3F1 mouse liver tumors: implications for risk assessment. *Science*, **237**, 1309–1316

Sadtler Research Laboratories (1980) *1980 Cumulative Index*, Philadelphia, PA

Sax, N.I. & Lewis, R.J. (1989) *Dangerous Properties of Industrial Materials*, 7th Ed., New York, Van Nostrand Reinhold

Šedivec, V. & Flek, J. (1978) Biologic monitoring of persons exposed to furfural vapors. *Int. Arch. occup. environ. Health*, **42**, 41–49

Shimuzu, A. & Kanisawa, M. (1986) Experimental studies on hepatic cirrhosis and hepatocarcinogenesis. I. Production of hepatic cirrhosis by furfural administration. *Acta pathol. jpn.*, **36**, 1027–1038

Shimizu, A., Nakamura, Y., Harada, M., Ono, T., Sato, K., Inoue, T. & Kanisawa, K. (1989) Positive foci of glutathione S-transferase placental form in the liver of rats given furfural by oral administration. *Jpn. J. Cancer Res.*, **80**, 608–611

Sittig, M. (1985) *Handbook of Toxic and Hazardous Chemicals and Carcinogens*, 2nd Ed., Park Ridge, NJ, Noyes Publications, p. 467

Stich, H.F., Rosin, M.P., Wu, C.H. & Powrie, W.D. (1981) Clastogenicity of furans found in food. *Cancer Lett.*, **13**, 89–95

TCI America (1994) *Certificate of Analysis: Furfural*, Portland, OR

Troshina, Z.P., Kulikova, N.P. & Andreev, E.L. (1986) Gas chromatographic determination of methanol and furfural in the air of the work area and in vapor–gas emissions of hydrolysis-yeast plant. *Gidroliz. Lesokhim. Promyshl.*, **4**, 23–25 (in Russian)

Tsuchiya, Y. (1992) Air quality problems inside a house following a fire. *J. Fire Sci.*, **10**, 58–71

Työministeriö [Ministry of Labour] (1993) *HTP-Arvot 1993* [Limit Values 1993], Tampere, p. 12

Uddin, S. (1993) Effect of furfural on the secondary structure of DNA. *Med. Sci. Res.*, **21**, 545–546

Uddin, S., Rahman, H. & Hadi, S.M. (1991) Reaction of furfural and methylfurfural with DNA: use of single-strand specific nucleases. *Food chem. Toxicol.*, **29**, 719–721

United Kingdom Health and Safety Executive (1993) *Occupational Exposure Limits* (EH40/93), London, Her Majesty's Stationery Office

United States National Institute for Occupational Safety and Health (1994a) *NIOSH Pocket Guide to Chemical Hazards* (DHHS (NIOSH) Publ. No. 94-116), Cincinnati, OH, pp. 150–151

United States National Institute for Occupational Safety and Health (1994b) *National Occupational Exposure Survey (1981–83)*, Cincinnati, OH

United States National Toxicology Program (1990) *Toxicology and Carcinogenesis Studies of Furfural (CAS No. 98-01-1) in F344/N Rats and B5C3F1 Mice (Gavage Studies)* (NTP Tech. Rep. No. 382; NIH Publ. No. 90-2837), Research Triangle Park, NC

United States Occupational Safety and Health Administration (1990) *OSHA Analytical Methods Manual*, Part 1, Vol. 3, Salt Lake City, UT, Department of Labor, Method 72

United States Occupational Safety and Health Administration (1994) *Air contaminants. US Code fed. Regul.*, **Title 29**, Part 1910.1000, p. 12

Verschueren, K. (1983) *Handbook of Environmental Data on Organic Chemicals*, 2nd Ed., New York, Van Nostrand Reinhold Co., p. 687

Vidal, J.P., Estreguil, S. & Cantagrel, R. (1993) Quantitative analysis of cognac carbonyl compounds at the ppb level by GC–MS of their O-(pentafluorobenzyl amine) derivatives. *Chromatographia*, **36**, 183–186

Villalón Mir, M., Quesada Granados, J., López Garcia de la Serrana, H. & López Martinez, M.C. (1992) High performance liquid chromatography determination of furanic compounds in commercial brandies and caramels. *J. Liq. Chromatogr.*, **15**, 513–524

Weast, R.C. & Astle, M.J. (1985) *CRC Handbook of Data on Organic Compounds*, Vols I & II, Boca Raton, FL, CRC Press, pp. 652 (I), 598 (II)

Woodruff, R.C., Mason, J.M., Valencia, R. & Zimmering, S. (1985) Chemical mutagenesis testing in *Drosophila*. V. Results of 53 coded compounds tested for the National Toxicology Program. *Environ. Mutag.*, **7**, 677–702

Zdzienicka, M., Tudek, B., Zielenska, M. & Szymczyk, T. (1978) Mutagenic activity of furfural in *Salmonella typhimurium* TA100. *Mutat. Res.*, **58**, 205–209

Zhang, J., He, Q. & Lioy, P.J. (1994) Characteristics of aldehydes: concentrations, sources, and exposures for indoor and outdoor residential microenvironments. *Environ. Sci. Technol.*, **28**, 146–152

BENZOFURAN

1. Exposure Data

1.1 Chemical and physical data

1.1.1 Nomenclature

Chem. Abstr. Serv. Reg. No.: 271-89-6
Chem. Abstr. Name: Benzofuran
IUPAC Systematic Name: Benzofuran
Synonyms: Benzo[b]furan; 2,3-benzofuran; benzofurfuran; 2,3-benzofurfuran; coumarone; cumarone; 1-oxindene

1.1.2 Structural and molecular formulae and relative molecular mass

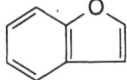

C_8H_6O

Relative molecular mass: 118.14

1.1.3 Chemical and physical properties of the pure substance

(a) *Description*: Clear, yellowish, oily liquid with an aromatic odour (Budavari, 1989; Henley Chemicals, Inc., 1990)

(b) *Boiling-point*: 174 °C (Lide, 1993)

(c) *Melting-point*: < –18 °C (Lide, 1993)

(d) *Density*: 1.0913 at 25 °C (Lide, 1993)

(e) *Spectroscopy data*: Infrared (prism [3739]; grating [33 363]), ultraviolet [20 800], nuclear magnetic resonance (proton [19 610]; C-13 [576]) and mass [448] spectral data have been reported (Sadtler Research Laboratories, 1980; Weast & Astle, 1985).

(f) *Solubility*: Insoluble in water and aqueous alkaline solutions; soluble in benzene, petroleum ether, diethyl ether and ethanol (Budavari, 1989; Lide, 1993)

(g) *Volatility*: Vapour pressure, 0.44 mm Hg [0.06 kPa] at 25 °C (Chao et al., 1983)

(h) *Stability*: Polymerizes at elevated temperatures or upon contact with acids (Budavari, 1989; Henley Chemicals, Inc., 1990)

(i) *Octanol:water partition coefficient (P)*: log P, 2.67 (Hansch et al., 1995)

(j) *Conversion factor*: mg/m^3 = 4.83 × ppm[1]

1.1.4 Technical products and impurities

Benzofuran is available commercially at a minimal purity of 99% (Henley Chemicals, Inc., 1990). Trade names for benzofuran include AT 33852 and R 7204.

1.1.5 Analysis

Benzofuran has been determined in air samples after collection on Tenax or Chromosorb, thermal desorption and high-resolution gas chromatography/mass spectrometry. It has been concentrated from water samples by the purge-and-trap method or by extraction with dichloromethane. It has also been extracted from particulate samples with dichloromethane (Ferretti & Flanagan, 1971; Erikson & Pellizzari, 1978; Pellizzari *et al.*, 1979; Hunt *et al.*, 1982; Rostad *et al.*, 1985; Juttner, 1986; van Netten *et al.*, 1988).

Benzofuran has been detected, but not quantified, in samples of blood and breast milk by purging them with an inert gas at an elevated temperature, trapping on Tenax, thermal desorption and high-resolution gas chromatography/mass spectrometry (Anderson & Harland, 1980; Pellizzari *et al.*, 1982).

1.2 Production and use

1.2.1 Production

Benzofuran is a constituent of coal-tar (see IARC, 1987a) and can be isolated from coal-tar oils. It is produced by three companies in China, one in Armenia and one in Switzerland (Chemical Information Services, Inc., 1994).

1.2.2 Use

Benzofuran derived from the crude, heavy solvent naphtha fraction of coal-tar light oil, obtained as a by-product in the coking of bituminous coal (see IARC, 1987b), is used in the manufacture of coumarone–indene resin (Budavari, 1989). The fraction of coal-tar oil that distils at 167–184 °C contains small quantities (probably < 10%) of benzofuran and about 30% indene, indan, substituted benzenes and related compounds. Polymerization is initiated by addition of an acid catalyst such as boron trifluoride or sulfuric acid (United States Agency for Toxic Substances and Disease Registry, 1992).

Coumarone–indene resins harden when heated and have been used to make floor tiles and other products. They are also used as a coating on grapefruit, lemons, limes, oranges, tangelos and tangerines and in the production of corrosion-resistant paints and varnishes, in water-resistant coatings on paper products and fabrics and as adhesives in food containers (United States Agency for Toxic Substances and Disease Registry, 1992).

---

[1] Calculated from: mg/m^3 = (relative molecular mass/24.45) × ppm, assuming normal temperature (25 °C) and pressure (101 kPa)

1.3 Occurrence

1.3.1 Natural occurrence

Benzofuran is not known to occur as a natural product.

1.3.2 Occupational exposure

No data were available to the Working Group, although occupational exposure to benzofuran may occur in several industries, including coke production and coal gasification facilities (see IARC, 1987c) and the polymerization process used to produce coumarone–indene resin. Benzofuran was not found in two surveys in the United States of America (United States National Institute for Occupational Safety and Health, 1994).

The naphtha fraction of coal-tar was included among the occupational hazards associated with coal gasification (United States National Institute for Occupational Safety and Health, 1978). Individuals occupationally exposed to coal-tars, including road-pavement workers, or to the naphtha fraction of coal-tar distillate are potentially exposed to benzofuran (United States Agency for Toxic Substances and Disease Registry, 1992). In a study of laser cutting of fibre-reinforced plastics, benzofuran was produced at a rate of 0.03–0.25 mg/g, the highest levels occurring during the cutting of epoxy resins reinforced with aramide fibres (Busch et al., 1989; Levsen et al., 1991).

1.3.3 Air

Benzofuran may be released into the environment during the production and use of benzofuran-containing products and in coke production, coal gasification and oil-shale facilities (see IARC, 1987d). Coke oven emissions were composed on average of 0.29% benzofuran (Kirton et al., 1991). Benzofuran was also detected in emissions from a Swedish floor finish used on domestic flooring (van Netten et al., 1988), in emissions from the pyrolysis of silk (Junk & Ford, 1980), in combustor flue gas emissions from fluidized-bed coal combustion at a concentration of 900 ng/g (Hunt et al., 1982) and from burning scrap tyres at a rate of 25 mg/kg tyre (Lemieux & Ryan, 1993).

Exhaust produced by an automobile burning simple hydrocarbon fuels contained benzofuran at concentrations of < 0.1–2.8 ppm [0.48–13.5 mg/m^3] (Seizinger & Dimitriades, 1972), but an analysis of air in a highway tunnel used by both diesel- and gasoline-powered vehicles showed no benzofuran (Hampton et al., 1982). Residential burning of brown-coal briquets led to an estimated emission factor of 6.85 mg/kg benzofuran. The estimated total amount of benzofuran emitted in the city of Leipzig, Germany, was 2220 kg/year (Engewald et al., 1993). Benzofuran was detected but not quantified in one of 10 samples of ambient air taken in an industrial area in the Kanawha Valley, West Virginia, United States (Erickson & Pellizzari, 1978); it was also detected among pollutants in the air of the southern Black Forest in Germany (Juttner, 1986). Benzofuran was identified as a minor component of smoke condensates of cottonwood in an experimental setting (Edye & Richards, 1991).

1.3.4 Water

Benzofuran was detected in the effluents from a coal gasification facility at concentrations of 6–267 ppb [μg/L], but it was not detected (detection limit, 0.1 ppb) in effluents from oil-shale processing facilities (Pellizzari *et al.*, 1979). It was also detected in one of 18 wastewater concentrates and at a concentration of 770 ppb in a groundwater sample, but not in surface water samples, taken at a hazardous waste site in the United States (United States Agency for Toxic Substances and Disease Registry, 1992). It was detected in contaminated groundwater at a coal-tar distillation and wood-preserving facility in Minnesota, United States (Rostad *et al.*, 1985).

1.3.5 Soil

Benzofuran was found at a concentration of 60 ppb [μg/kg] in one soil and sediment sample taken at a hazardous waste site in the United States (United States Agency for Toxic Substances and Disease Registry, 1992).

1.3.6 Other media

Benzofuran has not usually been found in foods, but it was one of the volatile constituents of freeze-dried whey powder subjected to accelerated browning (Ferreti & Flanagan, 1971). It was also detected in three samples of human milk (Pellizzari *et al.*, 1982) and in the blood of fire victims (Anderson & Harland, 1980). It is reportedly a constituent of cigarette smoke (Florin *et al.*, 1980; Curvall *et al.*, 1984; IARC, 1986; Schlotzhauer & Chortyk, 1987).

1.4 Regulations and guidelines

No guidelines or regulations have been reported for benzofuran (ILO, 1991).

2. Studies of Cancer in Humans

No data were available to the Working Group.

3. Studies of Cancer in Experimental Animals

Oral administration

Mouse: Groups of 50 male and 50 female B6C3F1 mice, seven to eight weeks of age, were administered benzofuran (purity, 99%) dissolved in corn oil at 0, 60 or 120 mg/kg bw (males) and 0, 120 or 240 mg/kg bw (females) by gavage on five days a week for 104 weeks. The numbers of male mice still alive at the end of the experiment were 33/50 controls, 20/50 at the low dose and 28/50 at the high dose; survival of treated females was significantly lower than that of controls ($p = 0.005$), with 37/50 controls, 19/50 at the low dose and 21/50 at the high dose still alive at the end of the experiment. The incidences of hepatocellular adenomas were increased in treated animals: 4/49 male controls, 24/39 at the low dose ($p < 0.001$, incidental tumour test) and 34/48 at the high dose ($p < 0.001$); and 1/50 female controls, 22/48 at the low dose ($p < 0.001$) and 21/47 at the high dose ($p < 0.001$). The incidences of hepatoblastoma, which had not been observed in historical controls in the study laboratory, were increased in male animals: 0/49 male controls, 3/39 at the low dose ($p = 0.083$) and 18/48 at the high dose ($p < 0.001$); one hepatoblastoma was found in a female at the low dose and in two females at the high dose. The incidence of hepatocellular carcinomas was not increased in treated animals. Male mice had increased incidences of squamous-cell papillomas of the forestomach, which occurred in 2/49 controls, 7/39 at the low dose ($p = 0.018$) and 10/48 at the high dose ($p = 0.007$); and of squamous-cell carcinomas, seen in 0/49 controls, 4/39 at the low dose ($p = 0.05$) and 3/48 at the high dose. The combined incidences of forestomach papillomas and carcinomas in males were 2/49 controls, 11/39 at the low dose ($p = 0.001$) and 13/48 at the high dose ($p = 0.001$). Squamous-cell papillomas of the forestomach occurred in 2/50 female controls, 8/50 at the low dose ($p = 0.035$) and 5/50 at the high dose; the incidence of carcinomas of the forestomach was not increased in females. Treated animals had increased incidences of pulmonary alveolar–bronchiolar adenomas, which occurred in 4/49 male controls, 7/39 at the low dose and 15/48 at the high dose ($p = 0.003$); and in 1/50 female controls, 5/48 at the low dose ($p = 0.040$) and 13/47 at the high dose ($p < 0.001$). Females had an increased incidence of alveolar–bronchiolar carcinomas, seen in 1/50 controls, 4/48 at the low dose ($p = 0.038$) and 3/47 at the high dose. The frequency of epithelial hyperplasia of the forestomach was increased in all treated mice except males at the low dose. Bronchiolar epithelial hyperplasia was observed in all treated animals (United States National Toxicology Program, 1989).

Rat: Groups of 50 male and 50 female Fischer 344/N rats, six to seven weeks of age, were administered benzofuran (purity, 99%) dissolved in corn oil at doses of 0, 30 or 60 mg/kg bw (males) and 0, 60 or 120 mg/kg bw (females) by gavage on five days a week for 103 weeks. The numbers of animals that survived to the end of the study were 33/50 male controls, 12/50 at the low dose ($p < 0.001$) and 18/50 at the high dose ($p = 0.003$); and 27/50 female controls, 23/50 at the low dose and 25/50 at the high dose. Nephropathy was seen in almost all aged male rats, including controls, but was more severe in treated animals; nephropathy was more frequent and more severe in treated females. The incidences of renal-cell adenocarcinomas were increased in female, but not in male, rats, occurring in 0/50 controls, 1/50 at the low dose and 4/50 at the high dose ($p = 0.032$, incidental tumour test). This tumour is rare in female rats: the incidence in historical controls at the study laboratory was 1/149 and that in the United States National Toxicology Testing Program was 2/2094 (United States National Toxicology Program, 1989).

4. Other Data Relevant to an Evaluation of Carcinogenicity and its Mechanisms

4.1 Absorption, distribution, metabolism and excretion

No data were available to the Working Group.

4.2 Toxic effects

4.2.1 Humans

No data were available to the Working Group.

4.2.2 Experimental systems

Groups of 10 male and 10 female Fischer 344/N rats and B6C3F1 mice were administered 0, 31, 63, 125, 250 or 500 mg/kg bw benzofuran (purity, 99%) in corn oil by gavage on five days per week for 13 weeks. Minimal centrilobular degeneration and necrosis of individual hepatocytes throughout the liver occurred in rats of each sex at the two highest doses and in males at 125 mg/kg bw. Severe nephropathy with foci of tubular regeneration and dilated tubules occurred in male rats at the two highest doses. Cytoplasmic vacuolization was observed in all animals at the highest dose. Of the mice, 7/10 males at the high dose died, and 7/10 males at 250 mg/kg bw had nephrosis, with tubular-cell necrosis, inflammation, fibrosis, regeneration and focal mineralization (United States National Toxicology Program, 1989).

Female CD-1 mice administered benzofuran at 5 mmol/kg bw [590 mg/kg bw] in corn oil by gavage for 10 days had liver microsomal P450 levels that were decreased by 70% in comparison with controls, 7-ethoxycoumarin O-deethylase activity increased by 150% and no significant changes in the activities of aniline hydroxylase and aminopyrine N-demethylase. Treatment also increased the activities of microsomal epoxide hydrolase (218% of the control level), UDP-glucuronosyl transferase (237% of control levels), cytosolic glutathione S-transferase (631% of control levels, with 1-chloro-2,4-dinitrobenzene as the substrate) and NADH-quinone reductase (230% of control levels) (Cha *et al.*, 1985). The same data were subsequently published elsewhere (Heine *et al.*, 1986).

4.3 Reproductive and prenatal effects

No data were available to the Working Group.

4.4 Genetic and related effects

4.4.1 Humans

No data were available to the Working Group

4.4.2 Experimental systems (see also Table 1 and Appendices 1 and 2)

Benzofuran was not mutagenic to *Salmonella typhimurium* when tested in either the standard plate incorporation assay or in a preincubation protocol in the presence or absence of an exogenous metabolic activation system.

Benzofuran induced gene mutation at the thymidine kinase locus of L5178Y mouse lymphoma cells in the absence of metabolic activation. It induced sister chromatid exchange and chromosomal aberrations in Chinese hamster ovary cells both in the presence and absence of metabolic activation.

Table 1. Genetic and related effects of benzofuran

Test system	Result[a]		Dose[b] (LED/HID)	Reference
	Without exogenous metabolic system	With exogenous metabolic system		
SA0, *Salmonella typhimurium* TA100, reverse mutation	–	–	350	Florin et al. (1980)
SA0, *Salmonella typhimurium* TA100, reverse mutation	–	–	0.00	Weill-Thevenet et al. (1981)
SA0, *Salmonella typhimurium* TA100, reverse mutation	–	–	385	US National Toxicology Program (1989)
SA5, *Salmonella typhimurium* TA1535, reverse mutation	–	–	350	Florin et al. (1980)
SA5, *Salmonella typhimurium* TA1535, reverse mutation	–	–	0.00	Weill-Thevenet et al. (1981)
SA5, *Salmonella typhimurium* TA1535, reverse mutation	–	–	385	US National Toxicology Program (1989)
SA7, *Salmonella typhimurium* TA1537, reverse mutation	–	–	350	Florin et al. (1980)
SA7, *Salmonella typhimurium* TA1537, reverse mutation	–	–	0.00	Weill-Thevenet et al. (1981)
SA7, *Salmonella typhimurium* TA1537, reverse mutation	–	–	385	US National Toxicology Program (1989)
SA8, *Salmonella typhimurium* TA1538, reverse mutation	–	–	0.00	Weill-Thevenet et al. (1981)
SA9, *Salmonella typhimurium* TA98, reverse mutation (spot test)	–	–	350	Florin et al. (1980)
SA9, *Salmonella typhimurium* TA98, reverse mutation	–	–	0.00	Weill-Thevenet et al. (1981)
SA9, *Salmonella typhimurium* TA98, reverse mutation	–	–	385	US National Toxicology Program (1989)
SAS, *Salmonella typhimurium* TA98NR, reverse mutation	–	–	0.00	Weill-Thevenet et al. (1981)
G5T, Gene mutation, mouse lymphoma L5178Y cells, *tk* locus *in vitro*	+	0	100	McGregor et al. (1988)
SIC, Sister chromatid exchange, Chinese hamster ovary (CHO) cells *in vitro*	+	+	37	US National Toxicology Program (1989)
CIC, Chromosomal aberrations, Chinese hamster ovary (CHO) cells *in vitro*	+	+	280	US National Toxicology Program (1989)

[a] +, considered to be positive; –, considered to be negative; 0, not tested
[b] LED, lowest effective dose; HID, highest ineffective dose; in-vitro tests, mg/ml; in-vivo tests, mg/kg bw; 0.00, dose not given

5. Summary and Evaluation

5.1 Exposure data

Benzofuran is produced by isolation from coal-tar oils, which are obtained as by-products of coking coal. Its major use is in the production of coumarone–indene resins. Human exposure can occur during coke production, coal gasification, the production of coumarone–indene resins or the combustion of coal and from tobacco smoke.

5.2 Human carcinogenicity data

No data were available to the Working Group.

5.3 Animal carcinogenicity data

Benzofuran was tested for carcinogenicity by oral administration in one study in mice and in one study in rats. In mice, benzofuran increased the incidence of hepatocellular adenomas in animals of each sex and of hepatoblastomas and forestomach papillomas and carcinomas in males; the incidence of alveolar–bronchiolar adenomas was increased in both males and females. Female rats had an increased incidence of renal-cell adenocarcinomas, which occur rarely in animals of this sex.

5.4 Other relevant data

No data were available on the toxicokinetics of benzofuran. Repeated administration of benzofuran to rats and mice caused renal toxicity; rats also developed slight hepatic toxicity.

Induction of gene mutation, sister chromatid exchange and chromosomal aberrations was seen in cultured rodent cells treated with benzofuran in single studies. Benzofuran was not mutagenic to bacteria.

5.5 Evaluation[1]

There is *inadequate evidence* in humans for the carcinogenicity of benzofuran.

There is *sufficient evidence* in experimental animals for the carcinogenicity of benzofuran.

Overall evaluation

Benzofuran *is possibly carcinogenic to humans (Group 2B)*.

---

[1] For definition of the italicized terms, see Preamble, pp. 22–26.

6. References

Anderson, R.A. & Harland, W.A. (1980) The analysis of volatiles in blood from fire fatalities. In: Oliver, J.S., ed., *Proceedings of the European Meeting of the International Association of Forensic Toxicologists, 1979*, London, Croom Helm Ltd, pp. 279–292

Budavari, S., ed. (1989) *The Merck Index*, 11th Ed., Rahway, NJ, Merck & Co., pp. 169–170

Busch, H., Holländer, W., Levsen, K., Schilhabel, J., Trasser, F.J. & Neder, L. (1989) Aerosol formation during laser cutting of fibre reinforced plastics. *J. aerosol Sci.*, **20**, 1473–1476

Cha, Y.-N., Thompson, D.C., Heine, H.S. & Chung, J.-H. (1985) Differential effects of indole, indole-3-carbinol and benzofuran on several microsomal and cytosolic enzyme activities in mouse liver. *Kor. J. Pharmacol.*, **21**, 1–11

Chao, J., Lin, L.T. & Chung, T.H. (1983) Vapor pressure of coal chemicals. *J. phys. chem. Ref. Data*, **12**, 1033–1063

Chemical Information Services, Inc. (1994) *Directory of World Chemical Producers 1995/96 Standard Edition*, Dallas, TX, p. 82

Curvall, M., Enzell, C.R. & Pettersson, B. (1984) An evaluation of the utility of four in vitro short term tests for predicting the cytotoxicity of individual compounds derived from tobacco smoke. *Cell Biol. Toxicol.*, **1**, 173–193

Edye, L.A. & Richards, G.N. (1991) Analysis of condensates from wood smoke: components derived from polysaccharides and lignins. *Environ. Sci. Technol.*, **25**, 1133–1137

Engewald, W., Knobloch, T. & Efer J. (1993) Volatile organic compounds in emissions from brown-coal-fired residential stoves. *Z. Umweltchem. Ökotox.*, **5**, 303–308 (in German)

Erickson, M.D. & Pellizzari, E.D. (1978) *Analysis of Organic Air Pollutants in the Kanawha Valley, WV and the Shenandoah Valley, VA* (EPA-903/9-78-007; NTIS No. PB-286 141), Philiedelphia, PA, United States Environmental Protection Agency

Ferretti, A. & Flanagan, V.P. (1971) Volatile constituents of whey powder subjected to accelerated browning. *J. Dairy Sci.*, **54**, 1764–1768

Florin, I., Rutberg, L., Curvall, M. & Enzell, C.R. (1980) Screening of tobacco smoke constituents for mutagenicity using the Ames test. *Toxicology*, **18**, 219–232

Hampton, C.V., Pierson, W.R., Harvey, T.M., Updegrave, W.S. & Marano, R.S. (1982) Hydrocarbon gases emitted from vehicles on the road. 1. A qualitative gas chromatography/mass spectrometry survey. *Environ. Sci. Technol.*, **16**, 287–298

Hansch, C., Leo, A. & Hoekman, D.H. (1995) *Exploring QSAR*, Washington DC, American Chemical Society

Heine, H.S., Stoskopf, M.K., Thompson, D.C. & Cha, Y.-N. (1986) Enhancement of epoxide hydrolase activity in hepatic microsomes of mice given heterocyclic compounds. *Chem.-biol. Interactions*, **59**, 219–230

Henley Chemicals, Inc. (1990) *Specification Sheet: 2,3-Benzofuran*, Montvale, NJ

Hunt, G.T., Kindya, R.J., Hall, R.R., Fennelly, P.F. & Hoyt, M. (1982) The polycyclic aromatic environment of the fluidized bed coal combustion process—an investigation of chemical and biological activity. In: Cooke, M., Dennis, A.J. & Fisher, G.L., eds, *Proceedings of the 6th International Physical, Biological and Chemical Symposium on Polynuclear Aromatic Hydrocarbons*, Columbus, OH, Battelle Press, pp. 367–381

IARC (1986) *IARC Monographs on the Evaluation of the Carcinogenic Risk of Chemicals to Humans*, Vol. 38, *Tobacco Smoking*, Lyon, p. 92

IARC (1987a) *IARC Monographs on the Evaluation of Carcinogenic Risks to Humans*, Suppl. 7, *Overall Evaluations of Carcinogenicity: An Updating of* IARC Monographs *Volumes 1–42*, Lyon, pp. 175–176

IARC (1987b) *IARC Monographs on the Evaluation of Carcinogenic Risks to Humans*, Suppl. 7, *Overall Evaluations of Carcinogenicity: An Updating of* IARC Monographs *Volumes 1–42*, Lyon, pp. 176–177

IARC (1987c) *IARC Monographs on the Evaluation of Carcinogenic Risks to Humans*, Suppl. 7, *Overall Evaluations of Carcinogenicity: An Updating of* IARC Monographs *Volumes 1–42*, Lyon, pp. 173–174

IARC (1987d) *IARC Monographs on the Evaluation of Carcinogenic Risks to Humans*, Suppl. 7, *Overall Evaluations of Carcinogenicity: An Updating of* IARC Monographs *Volumes 1–42*, Lyon, pp. 339–341

ILO (1991) *Occupational Exposure Limits for Airborne Toxic Substances: Values of Selected Countries* (Occupational Safety and Health Series No. 37), 3rd Ed., Geneva

Junk, G.A. & Ford, C.S. (1980) A review of organic emissions from selected combustion processes. *Chemosphere*, **9**, 187–230

Juttner, F. (1986) Analysis of organic compounds (VOC) in the forest air of the southern Black Forest. *Chemosphere*, **15**, 985–992

Kirton, P.J., Ellis, J. & Crisp, P.T. (1991) The analysis of organic matter in coke oven emissions. *Fuel*, **70**, 1383–1389

Lemieux, P.M. & Ryan, J.V. (1993) Characterization of air pollutants emitted from a simulated scrap tire fire. *J. Air Waste Manage. Assoc.*, **43**, 1106–1115

Levsen, K., Emmrich, M., Kock, H., Priess, B., Sollinger S. & Trasser, F.-J. (1991) Organic emissions during laser cutting of fibre-reinforced plastics. *Staub. Reinh. Luft*, **51**, 365–372

Lide, D.R., ed. (1993) *CRC Handbook of Chemistry and Physics*, 74th Ed., Boca Raton, FL, CRC Press, p. 3-96

McGregor, D.B., Brown, A., Cattanach, P., Edwards, I., McBride, D. & Caspary, W.J. (1988) Responses of the L5178Y tk$^+$/tk$^-$ mouse lymphoma cell forward mutation assay. II: 18 coded chemicals. *Environ. mol. Mutag.*, **11**, 91–118

van Netten, C., Shirtliffe, C. & Svec, J. (1988) Formaldehyde release characteristics from a Swedish floor finish. *Bull. environ. Contam. Toxicol.*, **40**, 672–677

Pellizzari, E.D., Castillo, N.P., Willis, S., Smith, D. & Bursey, J.T. (1979) Identification of organic compounds in aqueous effluents from energy-related processes. In: Van Hall, C.E., ed., *Measurement of Organic Pollutants in Water and Wastewater* (ASTM STP 686), Philidelphia, PA, American Society for Testing and Materials, pp. 256–274

Pellizzari, E.D., Hartwell, T.D., Harris, B.S.H., III, Waddell, R.D., Whitaker, D.A. & Erickson, M.D. (1982) Purgeable organic compounds in mother's milk. *Bull. environ. Contam. Toxicol.*, **28**, 322–328

Rostad, C.E., Pereira, W.E. & Hult, M.F. (1985) Partitioning studies of coal-tar constituents in a two-phase contaminated ground-water system. *Chemosphere*, **14**, 1023–1036

Sadtler Research Laboratories (1980) *1980 Cumulative Index*, Philadelphia, PA

Schlotzhauer, W.S. & Chortyk, O.T. (1987) Recent advances in studies on the pyrosynthesis of cigarette smoke constituents. *J. anal. appl. Pyrolysis*, **12**, 193–222

Seizinger, D.E. & Dimitriades, B. (1972) Oxygenates in exhaust from simple hydrocarbon fuels. *J. Air Pollut. Control Assoc.*, **22**, 47–51

United States Agency for Toxic Substances and Disease Registry (1992) *Toxicological Profile for 2,3-Benzofuran*, Washington DC, US Department of Health and Human Services

United States National Institute for Occupational Safety and Health (1978) *Occupational Exposures in Coal Gasification Plants* (DHEW (NIOSH) Publ. No. 78-191), Cincinnati, OH

United States National Institute for Occupational Safety and Health (1994) *National Occupational Exposure Survey (NOES)*, Cincinnati, OH

United States National Toxicology Program (1989) *Toxicology and Carcinogenesis Studies of Benzofuran (CAS No. 271-89-6) in F344/N Rats and B6C3F1 Mice (Gavage Studies)* (Tech. Rep. No. 370; NIH Publ. No. 90-2825), Research Triangle Park, NC

Weast, R.C. & Astle, M.J. (1985) *CRC Handbook of Data on Organic Compounds*, Vols I & II, Boca Raton, FL, CRC Press, pp. 209 (I), 495 (II)

Weill-Thevenet, N., Buisson, J.-P., Royer, R. & Hofnung, M. (1981) Mutagenic activity of benzofurans and naphthofurans in the *Salmonella*/microsome assay: 2-nitro-7-methoxynaphtho[2,1-*b*]furan (R7000), a new highly potent mutagenic agent. *Mutat. Res.*, **88**, 355–362

VINYL ACETATE

This substance was considered by previous Working Groups, in February 1978, in June 1985 and in March 1987 (IARC, 1979, 1986, 1987a). Since that time, new data have become available and these have been incorporated into the monograph and taken into consideration in the present evaluation.

1. Exposure Data

1.1 Chemical and physical data

1.1.1 Nomenclature

Chem. Abstr. Serv. Reg. No.: 108-05-4
Deleted CAS Reg. Nos.: 61891-42-7; 82041-23-4
Chem. Abstr. Name: Acetic acid, ethenyl ester
IUPAC Systematic Name: Vinyl acetate
Synonyms: Acetoxyethene; acetoxyethylene; 1-acetoxyethylene; ethenyl acetate; ethenyl ethanoate; VA; VAC; VAM; vinyl A monomer

1.1.2 Structural and molecular formulae and relative molecular mass

$$\begin{array}{c} \text{H} \quad \text{O} \quad \text{H} \\ | \quad \; \| \quad \; | \\ \text{H}-\text{C}-\text{C}-\text{O}-\text{C}=\text{C} \diagdown \text{H} \\ | \qquad\qquad\qquad \diagup \text{H} \\ \text{H} \end{array}$$

$C_4H_6O_2$ Relative molecular mass: 86.09

1.1.3 Chemical and physical properties of the pure substance

(a) *Description*: Colourless liquid with an ether-like odour (Union Carbide Corp., 1981; Hoechst Celanese Corp., 1993)

(b) *Boiling-point*: 72.2 °C (Lide, 1993)

(c) *Melting-point*: –93.2 °C (Lide, 1993)

(d) *Density*: 0.9317 at 20 °C/4 °C (Lide, 1993)

(e) *Spectroscopy data*: Infrared (prism [982]; grating [8112]), ultraviolet [1-29], nuclear magnetic resonance (proton [10245]; C-13 [1841]) and mass spectral data [136] have been reported (Sadtler Research Laboratories, 1980; Weast & Astle, 1985).

(f) *Solubility*: Slightly soluble in water (20 g/L at 20 °C); soluble in acetone, benzene, diethyl ether, ethanol, chloroform and most organic solvents (Budavari, 1989; Lewis, 1993; Lide, 1993)

(g) *Volatility*: Vapour pressure, 100 mm Hg [13.3 kPa] at 23.3 °C (Lide, 1993); relative vapour density (air = 1), 3.0 (Hoechst Celanese Corp., 1993)

(h) *Stability*: Polymerizes in light to a colourless, transparent mass (Budavari, 1989)

(i) *Reactivity*: Vapour may react vigorously in contact with silica gel or alumina (Bretherick, 1985); can react with oxidizing materials (United States National Institute for Occupational Safety and Health, 1994a)

(j) *Octanol:water partition coefficient (P)*: log P, 0.73 (Hansch et al., 1995)

(k) *Conversion factor*: mg/m^3 = 3.52 × ppm[1]

1.1.4 Technical products and impurities

Various commercial grades of vinyl acetate are available, which differ in the amount and type of added inhibitor. Typically, 3–5 mg/L hydroquinone (see IARC, 1987b) are added if the chemical is to be stored for less than two months; for longer storage periods, 14–17 mg/L or even 20–30 mg/L hydroquinone are added. Typical specifications for vinyl acetate are: minimal purity, 99.9%; acidity (as acetic acid), 0.005% max.; acetaldehyde (see IARC, 1987c), 0.010% max. by weight; and water, 0.04% max. by weight (Union Carbide Corp., 1981; Hoechst Celanese Corp., 1993; Union Carbide Corp., 1994). Trade names for vinyl acetate include Vyac, Zeset T and RP 251.

1.1.5 Analysis

Selected methods for the analysis of vinyl acetate in various matrices are presented in Table 1.

1.2 Production and use

1.2.1 Production

The oldest process for producing vinyl acetate is the addition of acetic acid to acetylene, which was first accomplished by Klatte in 1912 in a liquid-phase process with a mercuric salt as a catalyst. The liquid-phase process was converted to a vapour-phase process, with a zinc salt as a catalyst, in Germany in the 1920s (Roscher et al., 1983).

Production of vinyl acetate by the addition of acetic acid to ethylene (see IARC, 1994a) came into widespread use in the 1970s. Because ethylene is an inexpensive feed stock, essentially all manufacture since 1980 has been based on that compound (Daniels, 1983).

Both liquid-phase and vapour-phase processes are used. Liquid-phase processes, which were developed in Germany, Japan and the United Kingdom, involve palladium and copper salts as catalysts and production of vinyl acetate and acetaldehyde. The products are separated, and the acetaldehyde can be converted to acetic acid, which is fed back into the process. The vapour-phase process was developed independently by companies in Germany and the United States of America, and currently about 75% of world production is made by this technique (Roscher et al.,

---

[1] Calculated from: mg/m^3 = (relative molecular mass/24.45) × ppm, assuming normal temperature (25 °C) and pressure (101 kPa)

1983). The vapour-phase process differs from the liquid-phase in that very little acetaldehyde is formed. In Japan, most of the vinyl acetate produced by the vapour-phase process is used in the production of polyvinyl alcohol (see IARC, 1987d) via alcoholysis of polyvinyl acetate (see IARC, 1987e). The liberated acetic acid is then recycled for synthesis of the monomer (Daniels, 1983).

Table 1. Methods for the analysis of vinyl acetate

Sample matrix	Sample preparation	Assay procedure	Limit of detection	Reference
Air	Adsorb on Carboxen-564; desorb with dichloromethane:methanol (95:5) solution	GC/FID	1.0 µg/sample	Eller (1994)
	Adsorb on Ambersorb XE-347; desorb with dichloromethane:methanol (95:5) solution	GC/FID	0.04 mg/m^3	US Occupational Safety and Health Administration (1990)
Water[a]	Purge (nitrogen); trap (Tenax-GC/OV-1/-silica gel); desorb as vapour (heat purging with nitrogen) directly onto gas chromatographic column	GC/MS	PQL	US Environmental Protection Agency (1986)

GC/FID, gas chromatography/flame ionization detection; MS, mass spectrometry; PQL, practical quantification limit: 50 µg/L in groundwater; 50 µg/kg in soil and sediment samples with low levels of vinyl acetate; 2.5 mg/L in water-miscible liquid waste samples; 6.25 mg/kg in soil and sediment samples with high levels of vinyl acetate; 25 mg/L in non-water-miscible waste samples

[a] This method is used to determine volatile organic compounds in a variety of solid wastes, regardless of water content, including groundwater, aqueous sludges, caustic liquors, acid liquors, waste solvents, oily wastes, mousses, tars, fibrous wastes, polymeric emulsions, filter cakes, spent carbons, spent catalysts, soils and sediments.

A third technique for producing vinyl acetate involves the reaction of acetaldehyde with acetic anhydride to yield ethylidene diacetate, which is pyrolytically cleaved to vinyl acetate and acetic acid (Leonard, 1970; Daniels, 1983).

Other methods for producing vinyl acetate include the reaction of vinyl chloride (see IARC, 1987f) with sodium acetate in solution in the presence of palladium chloride (Daniels, 1983), conversion of methyl acetate with carbon monoxide and hydrogen in the presence of catalysts to ethylidene diacetate and subsequent cleavage to vinyl acetate and acetic acid (Roscher et al., 1983). Synthetic gas has been used as a feed for vinyl acetate (Mannsville Chemical Products Corp., 1982).

The total world capacity for production of vinyl acetate was about 1000 thousand tonnes in 1965. World production rose to 3000 thousand tonnes in 1982 (Roscher et al., 1983) but had decreased to 2500 thousand tonnes by 1988. In that year, 530 thousand tonnes were produced in the Member States of the European Union, the main producers being France, Germany, Italy, Spain and the United Kingdom, and 460 thousand tonnes in Japan (Environmental Chemicals

Data and Information Network, 1993). In 1992, 1240 thousand tonnes were produced in the United States (Anon., 1993a).

Commercial production of vinyl acetate in the United States was first reported in 1928 (United States Agency for Toxic Substances and Disease Registry, 1992). Between 1960 and 1980, United States production rose from 114 to 850 thousand tonnes (Daniels, 1983). In 1992, 1200 thousand tonnes were produced, so that vinyl acetate ranked 40th among the 50 chemicals produced in the largest volumes (Anon., 1993b). Production of vinyl acetate in Japan in 1993 was about 530 thousand tonnes, about 44% of which was used to produce polyvinyl alcohol (Anon., 1993c).

Vinyl acetate is produced by six companies in Japan, five each in China and the United States, two companies each in Germany and the Russian Federation, and one company each in Brazil, Canada, France, Mexico, Poland, Romania, Spain, Thailand, the United Kingdom and Venezuela (Chemical Information Services, Inc., 1994).

1.2.2 Use

The only commercial use for vinyl acetate is in the preparation of polymers and copolymers (Daniels, 1983). In 1987, 55% of the vinyl acetate used in the United States was to produce polyvinyl acetate emulsions and resins for paints and adhesives; 20% was used in the production of polyvinyl alcohol, 5% for polyvinyl butyral, 10% for ethylene–vinyl acetate resins, 4% for polyvinyl chloride copolymers (see IARC, 1987g) and 5% for miscellaneous uses (United States Agency for Toxic Substances and Disease Registry, 1992).

Polyvinyl acetate is used as an intermediate in conversion to polyvinyl alcohol and acetals. The principal use of polyvinyl acetate is in adhesives for paper, wood, glass, metals and porcelain. It is also used in latex water paints, as a coating for paper, for textile and leather finishing, as a base for inks and lacquers, in heat-sealing films and in shatterproof photographic bulbs (Sax & Lewis, 1987; Rosato, 1993). Polyvinyl acetate has been used as an emulsifying agent in cosmetics, pesticide formulations and pharmaceuticals and as a food additive (Anon., 1992).

Polyvinyl alcohol is the synthetic, water-soluble plastic produced in the largest volume in the world. It is used in sizing for textile warp and yarn, in laminating adhesives, photosensitive films and cements and as a binder for cosmetics, ceramics, leather, nonwoven fabrics and paper. It is also used as an emulsifying agent, thickener and stabilizer (Sax & Lewis, 1987; Rosato, 1993).

Polyvinyl acetals result from the condensation of polyvinyl alcohol with an aldehyde. Commonly used aldehydes are formaldehyde (see IARC, 1995), acetaldehyde and butyraldehyde. Polyvinyl formal, polyvinyl acetal and polyvinyl butyral are mainly used in adhesives, paints, laquers and films. Polyvinyl butyral is used in sheet form as an interlayer in safety glasses and shatter-resistant acrylic protection in aircraft (Sax & Lewis, 1987).

Ethylene–vinyl acetate copolymers improve the adhesion properties of hot-melt and pressure-sensitive adhesives. They are also used in medical tubing, milk packaging and beer-dispensing equipment (Sax & Lewis, 1987; Rosato, 1993). Plastic containers with barrier layers of ethylene–vinyl alcohol copolymers are replacing many glass and metal containers for packaging food (Rosato, 1993).

Polyvinyl chloride–acetate copolymers, compounded with plasticizers, are used for cable and wire coverings, in chemical plants and in protective garments (Rosato, 1993).

1.3 Occurrence

1.3.1 Natural occurrence

Vinyl acetate is not known to occur in nature.

1.3.2 Occupational exposures

The National Occupational Exposure Survey conducted between 1981 and 1983 indicated that 129 024 employees in the United States are potentially exposed to vinyl acetate. Vinyl acetate was found to be used in 27 industries with a total of 6264 facilities. The estimate is based on a survey of companies and did not involve actual measurements of exposure (United States National Institute for Occupational Safety and Health, 1994b). Data on occupational exposure to vinyl acetate are given in Table 2.

1.3.3 Air

Concentrations of 0.07–0.57 ppm [0.25–2 mg/m^3] were reported in ambient air in an area where several vinyl acetate manufacturers or process facilities were located (United States Agency for Toxic Substances and Disease Registry, 1992). An ambient air concentration of 0.5 µg/m^3 was detected near a chemical waste disposal site (Pellizzari, 1982). Vinyl acetate was monitored, but not detected, in the United States in 1990–91, in ambient air close to a furniture factory, in a paint incineration plant and in indoor air in an office and a machine shop (limit of detection, 2 ppb [7 µg/m^3] (Kelly *et al.*, 1993). At least 8.8 million pounds (4000 tonnes) of vinyl acetate were released to the environment from manufacturing and processing facilities in the United States in 1987 (United States Agency for Toxic Substances and Disease Registry, 1992).

1.3.4 Water

Vinyl acetate was detected at 50 mg/L in wastewater effluents from a polyvinyl acetate plant (Stepanyan *et al.*, 1970).

1.3.5 Other

Vinyl acetate was among the volatile chemicals released from food packaging during heating in a microwave oven, at concentrations of 0.01–0.88 µg/in^2 [0.002–0.14 µg/cm^2] (McNeal & Hollifield, 1993). A high rate of emission of vinyl acetate was found from new carpets that had a secondary backing of polyvinyl chloride (Hodgson *et al.*, 1993). Vinyl acetate has been detected in cigarette smoke at concentrations of 400 ng/cigarette (Guerin, 1980) and 0.5 µg/puff (Battista, 1976). Emissions of vinyl acetate from the combustion of pulverized coal were estimated to vary from 2.081×10^{-8} to 4.274×10^{-7} lb/10^6 Btu [0.009–0.18 mg/kJ] under different test conditions (Miller *et al.*, 1994).

Table 2. Occupational exposure to vinyl acetate

Country	No. of plants	Job, task or industry	No. of samples	Air concentration (mg/m³) Mean	Air concentration (mg/m³) Range	Reference
USA (1969)	1	Vinyl acetate production	P (TWA)	NR	1.4–17	US National Institute for Occupational Safety and Health (1978)
USA	1	Vinyl acetate production and polymerization	40 (P)	0.7–106 (range of means)	0–173	Deese & Joyner (1969)
USA (1982)	1	Polymer adhesive manufacture	5 (P)	10.6	< 0.4–18	Boxer & Reed (1983)
USA (1980)	1	Latex paint manufacture	8 (P) 8 (A)		[< 6.7]–126 [< 4.2]–36.6	Belanger & Coy (1981)
Poland	1	Vinyl copolymer production Vinyl acetate production Polyvinyl acetate production	(A) (A) (A)		9.7–11.5 0.6–4.3 1.2–1.4	Jedrychowski et al. (1979)
USA (1979)	1	Polymerization	(P) (A)		0.4–4.3 0.35–19.4	Kimble et al. (1982)
Former Czechoslovakia	1	Vinyl acetate production	(A) (P)		2.5–416 3.3–388	Kollár et al. (1988)
Finland (1980–92)	NR	Various industries	119 (A and P)	All samples	< 35	Finnish Institute of Occupational Health (1994)
USA	1	Polyvinyl acetate painters (airless spray equipment)	(A) (P)	< [35.2 µg/m³] < [774 µg/m³]		International Technology Corporation (1992)

P, personal sample; A, area sample; NR, not reported; TWA, time-weighted average

1.4 Regulations and guidelines

Occupational exposure limits and guidelines for vinyl acetate in several countries are given in Table 3. Polyvinyl acetate, ethylene–vinyl acetate–vinyl alcohol copolymers, ethylene–vinyl acetate copolymers, vinyl acetate–vinyl chloride copolymers and vinyl acetate–crotonic acid copolymers have been approved for use by the United States Food and Drug Administration (1984) as components of surfaces in contact with food and beverages.

Table 3. Occupational exposure limits for vinyl acetate

Country	Year	Concentration (mg/m³)	Interpretation
Australia	1991	30	TWA
		60	STEL
Belgium	1991	35	TWA
		70	STEL
Canada	1993	30	TWA
		60	STEL
Denmark	1991	30	TWA
Finland	1993	35	TWA
		70	STEL
France	1991	30	TWA
Germany	1993	35	TWA; justifiably suspected of having carcinogenic potential
Netherlands	1994	30	TWA
Poland	1991	10	TWA
Russian Federation	1991	10	STEL
Sweden	1993	18	TWA
		35	STEL
Switzerland	1994	35	TWA
		70	STEL
United Kingdom	1993	30	TWA
		60	STEL
USA			
ACGIH	1994	35	TWA; animal carcinogen
NIOSH	1994	15	TWA; ceiling

From ILO (1991); Arbetarskyddsstyrelsens (1993); Deutsche Forschungsgemeinschaft (1993); Environmental Chemicals Data and Information Network (1993); Työministeriö (1993); United Kingdom Health and Safety Executive (1993); American Conference of Governmental Industrial Hygienists (ACGIH) (1994); Arbeidsinspectie (1994); Schweizerische Unfallversicherungsanstalt (1994); United States National Institute for Occupational Safety and Health NIOSH) (1994a). TWA, time-weighted average; STEL, short-term exposure limit

2. Studies of Cancer in Humans

2.1 Cohort study

Waxweiler *et al.* (1981) undertook a cohort study of 4806 men employed at a plant for the manufacture of synthetic chemicals in the United States between 1942 and 1973. Death certificates could not be found for 16 (3%) of the deceased. In comparison with national rates, the cohort had an excess risk for cancer of the respiratory system (42 observed; standardized mortality ratio, 1.5 [95% confidence interval, 1.1–2.0]). In a review of medical records, 45 men were found to have had lung cancer. Histological specimens were found for 27 of these men, and eight showed undifferentiated large-cell lung cancer. The exposure of the patients with lung cancer to 19 chemicals, including vinyl acetate, was examined. The group had an estimated cumulative dose of exposure to vinyl acetate below the estimated weighted mean expected for the members of the cohort with the same year of birth and age at commencement of work in the plant. The subgroup with undifferentiated large-cell lung cancer had had slightly higher cumulative exposure to vinyl acetate.

2.2 Case–control study

Ott *et al.* (1989) undertook a nested case–control study in a cohort of 29 139 men employed in two large chemical manufacturing facilities and a research and development centre in the United States, who had died in 1940–78 with non-Hodgkin's lymphoma, multiple myeloma, lymphocytic leukaemia or nonlymphocytic leukaemia as the underlying or contributing cause of death. Controls were frequency matched to the dead cases by date of hire and length of survival in a 5:1 ratio. Exposure to 21 chemicals was assessed on the basis of information about work activity, work area and production over time. Potential exposure to vinyl acetate was reported for seven of 52 men who had died with non-Hodgkin's lymphoma (odds ratio, 1.2), three of 20 with multiple myeloma (odds ratio, 1.6), two of 39 with non-lymphocytic leukaemia (odds ratio, 0.5) and two of 18 with lymphocytic leukaemia (odds ratio, 1.8).

3. Studies of Cancer in Experimental Animals[1]

3.1 Oral administration

Rat: Three groups of 20 male and 20 female Fischer 344 rats, seven to eight weeks of age, received 0, 1000 or 2500 mg/L vinyl acetate [purity unspecified] in the drinking-water on five days a week for 100 weeks and were observed for their lifetime (maximum, 130 weeks). The vinyl acetate solutions were prepared weekly; however, owing to the instability of vinyl acetate in water, the rats were estimated to have received only about half of the nominal dose. Survival did not differ significantly between the treated and control groups: the numbers of survivors at

---

[1] The Working Group was aware of studies in progress in which mice and rats were administered vinyl acetate in the drinking-water (IARC, 1994b).

130 weeks were 7/20 male controls, 8/20 at the low dose and 6/20 at the high dose; and 5/20 female controls, 11/20 at the low dose and 11/20 at the high dose. All animals were necropsied, and all major organs and lesions were examined histologically. Treated animals had increased incidences of liver neoplastic nodules (none in male controls, four at the low dose and two at the high dose; none in female controls or at the low dose and six at the high dose [$p = 0.01$, Fisher's exact test]); increased incidences of uterine adenocarcinomas (none in female controls, one at the low dose and five at the high dose [$p = 0.024$]); increased incidences of uterine polyps (none in controls, three at the low dose and five at the high dose) [$p = 0.024$]; and increased incidences of thyroid C-cell adenomas (none in female controls, two at the low dose and five at the high dose [$p = 0.024$]). No malignant neoplasm was found in the liver (Lijinsky & Reuber, 1983; Lijinsky, 1988). [The Working Group noted the small number of animals used and that the animals received less than the nominal doses of the compound.]

Newborn Wistar rats given 200 or 400 mg/kg bw vinyl acetate per day orally for three weeks had not developed hepatic enzyme-altered foci by 14 weeks, whereas newborn rats similarly treated with vinyl chloride or vinyl carbamate did (Laib & Bolt, 1986).

3.2 Inhalation

3.2.1 Mouse

Groups of 60 male and 60 female Swiss-derived Crl:Cd-1(1CR)BR mice, about seven weeks of age and weighing 22.4–35.5 and 16.0–28.5 g, respectively, were exposed by inhalation to 0, 50, 200 or 600 ppm [0, 176, 704 or 2112 mg/m^3] vinyl acetate (purity, > 99%) for 6 h per day on five days a week for about 104 weeks. The body weight gain of treated mice was consistently depressed, but the survival rates of treated and control groups were similar. All animals were subjected to complete necropsy, and all tissues from controls and animals receiving the high dose and respiratory tissues from animals at all doses were examined histologically. One squamous-cell carcinoma of the lung was found in a male at the high dose, and a single adenocarcinoma of the lung occurred in a male control. Non-neoplastic lesions related to treatment were seen in the respiratory tract and included atrophy of the olfactory epithelium accompanied by respiratory metaplasia and nonkeratinizing squamous metaplasia of the respiratory epithelium in the nasal cavities; tracheal epithelial hyperplasia was also seen (Bogdanffy *et al.*, 1994a).

3.2.2 Rat

Groups of 60 male and 60 female Sprague-Dawley-derived Crl:Cd(SD)BR rats, about seven weeks of age and weighing 116–243 and 128–189 g, respectively, were exposed by inhalation to 0, 50, 200 or 600 ppm [0, 176, 704 or 2112 mg/m^3] vinyl acetate (purity, > 99%) for 6 h per day on five days a week for about 104 weeks. Body weight gain was consistently depressed in males and females at the high dose, but survival rates were similar in treated and control groups. Four of 59 females given the high dose had a squamous-cell carcinoma of the nasal cavity; no such tumour was seen in the other treated female rats or in 60 female controls [$p = 0.06$]. One of 59 males given the middle dose developed a nasal papilloma, two at the high dose developed squamous-cell carcinomas of the nasal cavity and one at the high dose had a carcinoma *in situ*;

four males at the high dose had nasal papillomas. The total number of nasal tumours (benign and malignant) in males at the high dose (7/59) was significantly greater than that in controls (0/59) ($p < 0.01$, Fisher's exact test). Non-neoplastic lesions related to treatment were seen in the respiratory tract and especially in the nasal cavity; the most consistent was thinning of the olfactory epithelium accompanied by basal-cell hyperplasia. No effects were seen in the nasal respiratory epithelium (Bogdanffy et al., 1994a).

3.3 Exposure *in utero*

Rat: A total of 72 male and 144 female Sprague-Dawley-derived Crl:CD(SD)BR rats [age unspecified] were divided into four groups and received vinyl acetate (purity, > 99%) at 0, 200, 1000 or 5000 mg/L in the drinking-water daily. The solutions were prepared each day, and 5% was added to each dose in order to compensate for an observed degradation of 6–14% over 24 h. The average concentrations achieved over the entire study were 0, 208, 1019 and 5166 mg/L. Treatment commenced 10 weeks before mating; treatment of males was continued for an additional four weeks and that of females throughout mating, gestation and lactation. Two males of the F_0 generation in each group were paired with one female from the same group for up to 15 days. After weaning, groups of 60 male and 60 female F_1 pups were selected and were administered 0, 200, 1000 or 5000 mg/L vinyl acetate in the drinking-water daily (average daily consumption: 10, 47 or 202 mg/kg bw for males and 16, 76 or 302 mg/kg bw for females) for 104 weeks. Water consumption was decreased among rats given 1000 or 5000 mg/L, and both food consumption and body weight gain were decreased in the latter group. The percentages of rats still alive at the end of the study were 42% of male controls, 57% of those at 200 mg/L, 50% of those at 1000 mg/L and 64% of those at 5000 mg/L; and 55% of female controls, 49% at 200 mg/L, 40% at 1000 mg/L and 57% at 5000 mg/L. No increase in tumour incidence was observed that was related to treatment (Bogdanffy et al., 1994b).

3.4 Carcinogenicity of metabolites

A previous working group concluded that there is *sufficient evidence* for the carcinogenicity of acetaldehyde to experimental animals (IARC, 1987b).

4. Other Data Relevant to an Evaluation of Carcinogenicity and its Mechanisms

4.1 Absorption, distribution, metabolism and excretion

4.1.1 Humans

No data were available to the Working Group.

4.1.2 Experimental systems

The elimination of vinyl acetate and formation of acetaldehyde were studied *in vitro* in a closed system comprising heparinized human blood and separated plasma and erythrocytes. The half-lives of vinyl acetate were 4.1 min in blood, 62 min in plasma and 5.5 min in erythrocytes. The esterase activity was shown to be sensitive to eserine and thus of the cholinesterase type; when carboxylesterase activity was inhibited, less effect was seen. Other reactions of vinyl acetate in blood, such as conjugation with reduced glutathione in erythrocytes, appeared to be of minor importance (Fedtke & Wiegand, 1990). In human lymphocytes cultured in whole blood and treated with 5.4 mmol[465 mg]/L vinyl acetate, the disappearance of parent compound and the formation of acetaldehyde were completed within 20 min (Norppa *et al.*, 1985).

The major metabolites of vinyl acetate *in vivo* are acetaldehyde (through an unstable vinyl alcohol) and acetic acid, which are formed by esterases, mostly in blood (Simon *et al.*, 1985a; Laib & Bolt, 1986). Disappearance of atmospheric vinyl acetate due to uptake and metabolism by male Wistar rats was measured in closed exposure chambers (Simon *et al.*, 1985a). The kinetics indicated that some aspects of the process must be saturable, since individual curves were composed of an apparent zero-order section at a higher concentration and a first-order section at a lower concentration. First-order kinetics applied up to about 200 ppm [704 mg/m^3]. Under these conditions, the total clearance (dose/area under the curve) from the gas phase was 30 000 ml/h per kg bw. This value is almost identical to the maximal ventilation rate in rats, 32 000 ml/h per kg bw (Guyton, 1947). The calculated rates of elimination due to metabolism indicated that vinyl acetate is metabolized in linear correlation with the atmospheric concentration up to 650 ppm [2288 mg/m^3], above which a saturation phenomenon occurs. In the closed exposure system, the concentration of acetaldehyde increased transiently as the vinyl acetate concentration declined. The kinetics of vinyl acetate appeared to be unaffected by prior treatment of the rats with diethyldithiocarbamate, indicating that monooxygenases may not be important in its metabolism; however, rat hepatic microsomes and plasma are particularly efficient in hydrolysing vinyl acetate to acetic acid. In a similar system, Fedtke and Wiegand (1990) found a half-life for vinyl acetate of < 1 min in rat blood and 1.2 min in rat plasma. The plasma hydrolytic activity was sensitive to phosphoric acid bis-4-nitrophenyl ester, a typical carboxylesterase inhibitor. It has been suggested (Norppa *et al.*, 1985) that some effects of vinyl acetate are due to acetaldehyde. No epoxide metabolite has been demonstrated (Simon *et al.*, 1985a).

The results of studies on DNA–protein cross-linking after exposure of rats to vinyl acetate and acetaldehyde have been presented in a series of reports (Kuykendall & Bogdanffy, 1992; Kuykendall *et al.*, 1993a,b [abstract]). The results, which include those for nasal tissues (which have high carboxylesterase levels), indicate that acetaldehyde is responsible for the cross-links; the other metabolite, acetic acid, was considered to be responsible for the cytotoxic response.

The kinetics of the hydrolysis of vinyl acetate in respiratory and olfactory nasal mucosa in rats and mice of each sex are similar (Bogdanffy & Taylor, 1993). The finding that metabolism occurs faster in olfactory then in respiratory mucosa may partially explain the distribution of nasal lesions induced by vinyl acetate, but these results do not explain the species difference in its carcinogenic effects.

Vinyl acetate is conjugated with glutathione *in vitro* in the presence of a rat liver supernatant (Boyland & Chasseaud, 1967), providing an explanation for the decreased levels of reduced glutathione seen after exposure to vinyl acetate *in vivo* (Boyland & Chasseaud, 1970; Holub & Tarkowski, 1982).

4.2 Toxic effects

4.2.1 Humans

Irritation of the eye and nose occurs at a threshold of 10–22 ppm [35–77 mg/m^3] (Deese & Joyner, 1969). Throat irritation occurred at 200 ppm (704 mg/m^3) and eye irritation at 72 ppm (253 mg/m^3) (United States National Institute for Occupational Safety and Health, 1978). Dermal contact may cause skin irritation and blister formation.

4.2.2 Experimental systems

Groups of four male and four female rats were exposed by inhalation to various concentrations of vinyl acetate for 6 h per day for 15 days. Rats exposed to 2000 ppm [7040 mg/m^3] experienced eye and nose irritation and respiratory difficulty and had reduced body weight gain. At autopsy, large numbers of macrophages were seen in the lung. Rats exposed to 630 or 250 ppm [2220 or 880 mg/m^3] had reduced weight gain. The lowest dose, 100 ppm [350 mg/m^3] induced no signs of toxicity (Gage, 1970).

Mice exposed to vinyl acetate at 1500 ppm [5280 mg/m^3] for 6 h per day on five days per week for 28 days had hyperplasia and metaplasia of the upper and lower respiratory tract (Clary, 1988).

As reported in an abstract, groups of 10 male and 10 female rats and mice were exposed by inhalation to 50, 200 or 1000 ppm [176, 704 or 3520 mg/m^3] for 6 h per day on five days per week for three months. Rats at 1000 ppm showed respiratory distress and reduced body weight. Mice showed respiratory distress at 200 and 1000 ppm, reduced body weight at 1000 ppm and rhinitis, nasal mucosal metaplasia, tracheal metaplasia and bronchitis/bronchiolitis, mainly at 1000 ppm. Animals of both species had increased lung weights [presumably at all doses] (Owen, 1983).

Rats exposed to 10, 100 or 500 mg/m^3 vinyl acetate for 5 h per day on five days per week for 10 months showed dose-related reductions in body weight and reticulopenia. Squamous metaplasia of the bronchi was seen at all doses, and fatty degeneration of the hepatic parenchyma, proliferation of smooth endoplasmic reticulum and changes in biliary canaliculi were noted at the two higher doses (Czajkowska *et al.*, 1986).

Groups of 60 male and 60 female rats and mice were exposed to vinyl acetate by inhalation for 104 weeks at doses of 0, 50, 200 or 600 ppm [0, 176, 704 or 2112 mg/m^3] (Bogdanffy *et al.*, 1994b). Ten animals per group were allowed to recover for 15 (males) or 16 (females) weeks after termination of exposure at 70 weeks. Weight gain was depressed in all groups receiving 600 ppm and in mice at 200 ppm. Animals that were allowed to recover showed improvement in weight gain, except for female rats at 600 ppm. No haematological changes were consistently related to treatment; a decrease in blood glucose in females at 600 ppm was the only change in clinical chemistry observed. Increased lung weights were observed in both species, especially at 600 ppm, and all treated rats [not clear whether males and females or only females] had increases at the time of terminal sacrifice; the changes in lung weights were associated with bronchial exfoliation, macrophage accumulation, fibrous plaques and buds and bronchial/bronchiolar epithelial disorganization. Histopathological changes were observed in the nasal cavity; the main non-neoplastic effects in the olfactory epithelium of both species were epithelial atrophy, regeneration, basal-cell hyperplasia and epithelial nest-like infolds. No non-neoplastic changes were seen in the respiratory epithelium of rats, but squamous metaplasia was seen in mice.

In rats and mice that received vinyl acetate in the drinking-water at concentrations of 200, 1000 or 5000 ppm [mg/L], no obvious toxic effects were noted (Clary, 1988).

Groups of 60 male and 60 female rats were exposed to vinyl acetate in drinking-water at concentrations of 0, 200, 1000 or 5000 ppm [mg/L] (v/v) for 104 weeks, after exposure of their F_0 parents to vinyl acetate for 10 weeks before mating. Decreased body weights and decreased food and water consumption were recorded in animals at 5000 ppm; water consumption was reduced in rats at 1000 ppm. No other compound-related effects on chemical, haematological or pathological end-points were observed (Bogdanffy *et al.*, 1994a).

4.3 Reproductive and prenatal effects

4.3.1 Humans

No data were available to the Working Group.

4.3.2 Experimental systems

Male and female rats were given 0, 200, 1000 or 5000 ppm [mg/L] vinyl acetate in the drinking-water over two generations. The F_1 generation of male and female controls and those at 5000 ppm were then cross-mated. Signs of general toxicity were observed in the F_0 and F_1 generations at 5000 ppm, and several reproductive paramaters were altered (decreased fertility and mating performance); however, the responses were inconsistent across generations. The no-observed adverse effect level was considered to be 1000 ppm (Mebus *et al.*, 1995).

Developmental toxicity was studied in rats exposed in the drinking-water to up to 5000 ppm [mg/L] or by inhalation to up to 1000 ppm [3520 mg/m^3] for 6 h per day on days 6–15 of gestation. Administration in the drinking-water induced no maternal or developmental toxicity. After exposure by inhalation, evidence was found of maternal toxicity and retarded embryonic growth, and a significant increase in the incidence of minor skeletal alterations (mainly delayed ossification) was seen in fetuses of dams exposed to 1000 ppm, but not at lower doses (Hurtt *et al.*, 1995).

4.4 Genetic and related effects

4.4.1 Humans

People exposed occupationally to vinyl acetate were reported to have an increased frequency of chromosomal aberrations in cultured lymphocytes (Shirinian & Arutyunyan, 1980). [The Working Group noted the inadequate reporting of the study.]

4.4.2 Experimental systems (See also Table 4 and Appendices 1 and 2)

Vinyl acetate did not induce mutations in *Salmonella typhimurium* or SOS repair functions in *Escherichia coli*. It induced DNA–protein cross-links in *Escherichia coli* HB 101 in the presence of rat liver microsomes.

It induced DNA cross-links in isolated human lymphocytes and in isolated rat nasal epithelial cells. It induced sister chromatid exchange in cultured Chinese hamster ovary cells in the presence or absence of exogenous metabolic activation and enhanced the viral transformation of Syrian hamster embryo cells. It also induced sister chromatid exchange, chromosomal aberrations and micronuclei in human whole blood and in isolated lymphocyte cultures *in vitro*.

In mice treated *in vivo* with vinyl acetate, the frequencies of sister chromatid exchange, of micronuclei in bone-marrow cells and of sperm with abnormal morphology were increased, but there was no increase in the frequency of micronucleated meiotic cells. The increased frequency of sister chromatid exchange was not affected by hepatectomy of the mice before treatment (Takeshita *et al.*, 1986).

Vinyl acetate did not produce DNA adducts in the livers of male and female rats exposed either orally or by inhalation (Simon *et al.*, 1985b).

4.4.3 Genotoxicity of metabolites

Acetaldehyde, the primary metabolite of vinyl acetate, is genotoxic in a wide range of assays *in vitro* and *in vivo* (IARC, 1987b).

5. Summary and Evaluation

5.1 Exposure data

Vinyl acetate is used in the production of a wide range of polymers, including polyvinyl acetate, polyvinyl alcohol, polyvinyl acetals, ethylene–vinyl acetate copolymers and polyvinyl chloride–vinyl acetate copolymers, which are widely used in the production of adhesives, paints and food packaging.

Human exposure to vinyl acetate occurs mainly by inhalation or dermal contact during production of the monomer or during production of polymers and water-based paints.

Table 4. Genetic and related effects of vinyl acetate

Test system	Result[a] Without exogenous metabolic activation	Result[a] With exogenous metabolic activation	Dose[b] (LED/HID)	Reference
PRB, SOS chromotest (commercial kit), *Escherichia coli* PQ 37	–	–	8600	Brams *et al.* (1987)
***, DNA–protein cross-links, *Escherichia coli* HB 101 pUC13, filter binding assay	–	+	860	Kuykendall & Bogdanffy (1992)
SA0, *Salmonella typhimurium* TA100, reverse mutation	–	–	500	Lijinsky & Andrews (1980)
SA0, *Salmonella typhimurium* TA100, reverse mutation	–	–	5000	McCann *et al.* (1975)
SA0, *Salmonella typhimurium* TA100, reverse mutation	–	–	500	Brams *et al.* (1987)
SA0, *Salmonella typhimurium* TA100, reverse mutation	–	–	260	Florin *et al.* (1980)
SA0, *Salmonella typhimurium* TA100, reverse mutation	–	–	340[c]	Bartsch *et al.* (1979)
SA3, *Salmonella typhimurium* TA1530, reverse mutation	–	–	340[c]	Bartsch *et al.* (1979)
SA5, *Salmonella typhimurium* TA1535, reverse mutation	–	–	500	Lijinsky & Andrews (1980)
SA5, *Salmonella typhimurium* TA 1535, reverse mutation	–	–	5000	McCann *et al.* (1975)
SA5, *Salmonella typhimurium* TA1535, reverse mutation	–	–	260	Florin *et al.* (1980)
SA7, *Salmonella typhimurium* TA1537, reverse mutation	–	–	500	Lijinsky & Andrews (1980)
SA7, *Salmonella typhimurium* TA1537, reverse mutation	–	–	5000	McCann *et al.* (1975)
SA7, *Salmonella typhimurium* TA1537, reverse mutation	–	–	260	Florin *et al.* (1980)
SA8, *Salmonella typhimurium* TA1538, reverse mutation	–	–	500	Lijinsky & Andrews (1980)
SA9, *Salmonella typhimurium* TA98, reverse mutation	–	–	500	Linjinsky & Andrews (1980)
SA9, *Salmonella typhimurium* TA98, reverse mutation	–	–	5000	McCann *et al.* (1975)
SA9, *Salmonella typhimurium* TA98, reverse mutation	–	–	500	Brams *et al.* (1987)
SA9, *Salmonella typhimurium* TA98, reverse mutation	–	–	260	Florin *et al.* (1980)
SAS, *Salmonella typhimurium* TA97, reverse mutation	–	–	500	Brams *et al.* (1987)
DIA, DNA cross-links, rat olfactory epithelial cells *in vitro*	+	0	860	Kuykendall *et al.* (1993a)
DIA, DNA cross-links, rat nasal respiratory epithelial cells *in vitro*	+	0	860	Kuykendall *et al.* (1993a)
SIC, Sister chromatid exchange, Chinese hamster ovary (CHO) cells *in vitro*	+	+	11	Norppa *et al.* (1985)
T7S, Cell transformation, SA7/Syrian hamster embryo (SHE) cells *in vitro*	+	0	500	Casto (1981)

Table 4 (contd)

Test system	Result[a] Without exogenous metabolic activation	Result[a] With exogenous metabolic activation	Dose[b] (LED/HID)	Reference
DIH, DNA cross-link formation, alkaline elution, human lymphocytes in vitro	+[d]	0	860	Lambert et al. (1985)
SHL, Sister chromatid exchange, human lymphocytes in vitro	+	0	4	Norppa et al. (1985)
SHL, Sister chromatid exchange, human lymphocytes in vitro	+	0	8.6	He & Lambert (1985)
SHL, Sister chromatid exchange, human lymphocytes in vitro	+	0	43	Sipi et al. (1992)
MIH, Micronucleus induction, human lymphocytes in vitro	+	0	43	Mäki-Paakkanen & Norppa (1987)
CHL, Chromosomal aberrations, human lymphocytes in vitro	+	0	17	Norppa et al. (1985)
CHL, Chromosomal aberrations, human lymphocytes in vitro	+	0	22	Jantunen et al. (1986)
SVA, Sister chromatid exchange, mouse cells in vivo	+		370 ip × 1	Takeshita et al. (1986)
MVM, Micronucleus induction, mouse bone marrow in vivo	+		1000 ip × 1	Mäki-Paakkanen & Norppa (1987)
MVM, Meiotic micronucleus induction, mice in vivo	–		1000 ip × 5	Lähdetie (1988)
BVD, DNA binding (covalent), rat hepatocytes in vivo (^{14}C label)	–		45 po × 1	Simon et al. (1985b)
BVD, DNA binding (covalent), rat hepatocytes in vivo (^{14}C label)	–		175 inh 4 h[c]	Simon et al. (1985b)
SPM, Sperm morphology, F$_1$ mice in vivo	+		500 ip × 5	Lähdetie (1988)

[a] +, considered to be positive; –, considered to be negative
[b] LED, lowest effective dose; HID, highest effective dose. In-vitro tests, μg/ml; in-vivo tests, mg/kg bw; ip, intraperitoneally; inh, inhalation
[c] Exposure to vapour phase also negative
[d] Negative for single-strand breaks at 1700 μg/ml
[e] Assuming 100% absorption based on mean body weight and vinyl acetate concentration from reported range
***, Not included on the profile

5.2 Human carcinogenicity data

The available data were too limited to form the basis for an evaluation of the carcinogenicity of vinyl acetate to humans.

5.3 Animal carcinogenicity data

Vinyl acetate was tested in one experiment in mice and in one experiment in rats by inhalation. No treatment-related increase in tumour incidence was observed in mice; in rats, an increased incidence of nasal cavity tumours was found in animals of each sex. No increase in tumour incidence was found in rats administered vinyl acetate in the drinking-water *in utero* and then for life.

5.4 Other relevant data

Vinyl acetate is rapidly metabolized by esterases in human blood and animal tissues to acetaldehyde and acetic acid.

Vinyl acetate irritates the eye and respiratory system. Respiratory distress is seen after subchronic exposure by inhalation. Other effects included nasal irritation, nasal mucosal metaplasia, tracheal metaplasia and bronchitis or bronchiolitis. After chronic exposure by inhalation, changes were observed in the lung. Non-neoplastic effects, atrophic and regenerative changes, were seen in the nasal cavity. After chronic exposure via the drinking-water, the only effects observed were decrements in body weight at high doses.

There are no data on the effects of vinyl acetate on human reproduction. A two-generation study in rats showed evidence of parental toxicity and decreased fertility at the highest dose tested. Oral administration of vinyl acetate to rats during pregnancy did not result in maternal or developmental toxicity, whereas exposure by inhalation induced maternal toxicity, retarded embryonic growth and minor skeletal alterations at the highest dose tested.

Vinyl acetate induced sperm abnormalities and sister chromatid exchange in rodents exposed *in vivo*; micronuclei were induced in bone marrow but not in meiotic cells. No DNA binding was seen in rat hepatocytes. In human lymphocytes *in vitro*, vinyl acetate produced chromosomal aberrations, micronuclei, sister chromatid exchange and DNA cross-links. It enhanced viral transformation and sister chromatid exchange in mammalian cells *in vitro*, and it induced DNA–protein cross-links in rat nasal epithelial cells *in vitro*. Vinyl acetate did not induce mutation in bacteria but induced DNA–protein cross-links in plasmid DNA. The primary metabolite of vinyl acetate, acetaldehyde, is genotoxic in a wide range of assays.

5.5 Evaluation[1]

There is *inadequate evidence* in humans for the carcinogenicity of vinyl acetate.

There is *limited evidence* in experimental animals for the carcinogenicity of vinyl acetate.

---

[1] For definition of the italicized terms, see Preamble, pp. 22–26.

Overall evaluation

Vinyl acetate *is possibly carcinogenic to humans (Group 2B)*.

In making the overall evaluation, the Working Group took into account the following evidence:

(i) Vinyl acetate is rapidly transformed into acetaldehyde in human blood and animal tissues.

(ii) There is *sufficient evidence* in experimental animals for the carcinogenicity of acetaldehyde (IARC, 1987b). Both vinyl acetate and acetaldehyde induce nasal cancer in rats after administration by inhalation.

(iii) Vinyl acetate and acetaldehyde are genotoxic in human cells *in vitro* and in animals *in vivo*.

6. References

American Conference of Governmental Industrial Hygienists (1994) *1994–1995 Threshold Limit Values for Chemical Substances and Physical Agents and Biological Exposure Indices*, Cincinnati, OH, p. 35

Anon. (1992) Chemical profile: vinyl acetate. *Chem. Marketing Rep.*, **241**, 42

Anon. (1993a) Facts & figures for the chemical industry. *Chem. Eng. News*, **71**, 38–83

Anon. (1993b) Top 50 list of chemical products. *Chem. Eng. News*, **71**, 10–20

Anon. (1993c) Production of plastics materials in Japan. *Plast. Ind. News*, **39**

Arbeidsinspectie [Labour Inspection] (1994) *De Nationale MAC-Lijst 1994* [National MAC list 1994], The Hague, p. 41

Arbetarskyddsstyrelsens [National Swedish Board of Occupational Safety and Health] (1993) *Hygieniska Gränsvärden* [Occupational exposure limits 1993] (AFS 1993:9), Stockholm, p. 58 (in Swedish)

Bartsch, H., Malaveille, C., Barbin, A. & Planche, G. (1979) Mutagenic and alkylating metabolites of halo-ethylenes, chlorobutadienes and dichlorobutenes produced by rodent or human liver tissues. Evidence for oxirane formation by P450-linked microsomal mono-oxygenases. *Arch. Toxicol.*, **41**, 249–277

Battista, S.P. (1976) Cilia toxic components of cigarette smoke. In: Wynder, E.L., Hofmann, D. & Gori, G.B., eds, *Proceedings of the 3rd World Conference on Smoking and Health* (DHEW(NIH) Publication No. 76-1221), Bethesda, MD, US Department of Health, Education, and Welfare, National Institutes of Health, pp. 517–534

Belanger, P.L. & Coy, M.Y. (1981) *Sinclair Paint Company, Los Angeles, CA* (Health Hazard Evaluation Report No. HHE-80-68-871) (US NTIS PB82-214396), Cincinnati, OH, United States National Institute for Occupational Safety and Health

Bogdanffy, M.S. & Taylor, M.L. (1993) Kinetics of nasal carboxylesterase-mediated metabolism of vinyl acetate. *Drug Metab. Disposition*, **21**, 1107–1111

Bogdanffy, M.S., Dreef-van der Meulen, H.C., Beems, R.B., Feron, V.J., Cascieri, T.C., Tyler, T.R., Vinegar, M.B. & Rickard, R.W. (1994a) Chronic toxicity and oncogenicity inhalation study with vinyl acetate in the rat and mouse. *Fundam. appl. Toxicol.*, **23**, 215–229

Bogdanffy, M.S., Tyler, T.R., Vinegar, M.B., Rickard, R.W., Carpanini, F.M.B. & Cascieri, T.C. (1994b) Chronic toxicity and oncogenicity study with vinyl acetate in the rat: in utero exposure in drinking-water. *Fundam. appl. Toxicol.*, **23**, 206–214

Boyland, E. & Chasseaud, L.F. (1967) Enzyme-catalysed conjugations of glutathione with unsaturated compounds. *Biochem J.*, **104**, 95–102

Boyland, E. & Chasseaud, L.F. (1970) The effect of some carbonyl compounds on rat liver glutathione levels (Short communication). *Biochem. Pharmacol.*, **19**, 1526–1528

Boxer, P.A. & Reed, L.D. (1983) *National Starch & Chemical, Meredosia, IL* (Health Hazard Evaluation Report No. HETA-82-051-1269) (US NTIS PB84-209923), Cincinnati, OH, United States National Institute for Occupational Safety and Health

Brams, A., Buchet, J.P., Crutzen-Fayt, M.C., de Meester, C., Lauwerys, R. & Léonard, A. (1987) A comparative study, with 40 chemicals, of the efficiency of the *Salmonella* assay and the SOS chromotest (kit procedure). *Toxicol. Lett.*, **38**, 123–133

Bretherick, L. (1985) *Handbook of Reactive Chemical Hazards*, 3rd Ed., London, Butterworths, pp. 430–431

Budavari, S., ed. (1989) *The Merck Index*, 11th Ed., Rahway, NJ, Merck & Co., p. 1572

Casto, B.C. (1981) Effect of chemical carcinogens and mutagens on the transformation of mammalian cells by DNA viruses. In: *Antiviral Chemotherapy—Design and Inhibition of Viral Functions*, New York, Academic Press, pp. 261–278

Chemical Information Services, Inc. (1994) *Directory of World Chemical Producers 1995/1996 Standard Edition*, Dallas, TX, p. 706

Clary, J.J. (1988) Chronic and reproduction toxicologic studies on vinyl acetate. Status report. *Ann. N.Y. Acad. Sci.*, **534**, 255–260

Czajkowska, T., Sokal, J., Knobloch, K., Górny, R., Kolakowski, J., Lao, I., Stetkiewicz, J. & Binkowski, J. (1986) Experimental study on chronic toxic effect of vinyl acetate. *Med. Prac.*, **37**, 26–35

Daniels, W. (1983) Vinyl acetate monomer. In: Mark, H.F., Othmer, D.F., Overberger, C.G., Seaborg, G.T. & Grayson, N., eds, *Kirk-Othmer Encyclopedia of Chemical Technology*, 3rd Ed., Vol. 23, New York, John Wiley & Sons, pp. 817–847

Deese, D.E. & Joyner, R.E. (1969) Vinyl acetate: a study of chronic human exposure. *Am. ind. Hyg. Assoc. J.*, **30**, 449–457

Deutsche Forschungsgemeinschaft (1993) *MAK- und BAT-Werte Liste* [MAK and BAT values list], (Report No. 29), Weinheim, VCH Verlagsgesellschaft, p. 77

Eller, P.M., ed. (1994) *NIOSH Manual of Analytical Methods*, 4th Ed., Vol. 3, (DHHS (NIOSH) Publ. No. 94-113), Washington DC, United States Government Printing Office, Method 1453

Environmental Chemicals Data and Information Network (1993) *Vinyl Acetate*, Ispra, JRC-CEC, last update: 02.09.1993

Fedtke, N. & Wiegand, H.-J. (1990) Hydrolysis of vinyl acetate in human blood (Letter to the Editor). *Arch. Toxicol.*, **64**, 428–429

Finnish Institute of Occupational Health (1994) *Työhygieenisten mittausten rekisteri* (Finnish occupational exposure database), Helsinki (in Finnish)

Florin, I., Rutberg, L., Curvall, M. & Enzell, C.R. (1980) Screening of tobacco smoke constituents for mutagenicity using the Ames' test. *Toxicology*, **15**, 219–232

Gage, J.C. (1970) The subacute inhalation toxicity of 109 industrial chemicals. *Br. J. ind. Med.*, **27**, 1–18

Guerin, M.R. (1980) Chemical composition of cigarette smoke. In: Gori, G.B. & Bock, F.G., eds, *Banbury Report: A Safe Cigarette?* Vol. 3, Cold Spring Harbor, NY, CSH Press, pp. 191–204

Guyton, A.C. (1947) Measurement of the respiratory volumes of laboratory animals. *Am. J. Physiol.*, **150**, 70–77

Hansch, C., Leo, A. & Hoekman, D.H. (1995) *Exploring QSAR*, Washington DC, American Chemical Society

He, S.-M. & Lambert, B. (1985) Induction and persistence of SCE-inducing damage in human lymphocytes exposed to vinyl acetate and acetaldehyde *in vitro*. *Mutat. Res.*, **158**, 201–208

Hodgson, A.T., Wooley, J.D. & Daisey, J.M (1993) Emissions of volatile organic compounds from new carpets measured in large-scale environmental chamber. *J. Air Waste Manage. Assoc.*, **43**, 316–324

Hoechst Celanese Corp. (1993) *Material Safety Data Sheet: Vinyl Acetate*, Dallas, TX

Holub, I. & Tarkowski, S. (1982) Hepatic content of free sulfhydryl compounds in animals exposed to vinyl acetate. *Int. Arch. occup. environ. Health*, **51**, 185–189

Hurtt, M.E., Vinegar, M.B., Rickard, R.W., Cascieri, T. & Tyler, T.R. (1995) Developmental toxicity of oral and inhaled vinyl acetate in the rat. *Fundam. appl. Toxicol.*, **24**, 198–205

IARC (1979) *IARC Monographs on the Evaluation of the Carcinogenic Risk of Chemicals to Humans*, Vol. 19, *Some Monomers, Plastics and Synthetic Elastomers, and Acrolein*, Lyon, pp. 341–366

IARC (1986) *IARC Monographs on the Evaluation of the Carcinogenic Risk of Chemicals to Humans*, Vol. 39, *Some Chemicals Used in Plastics and Elastomers*, Lyon, pp. 113–131

IARC (1987a) *IARC Monographs on the Evaluation of Carcinogenic Risks to Humans*, Suppl. 7, *Overall Evaluations of Carcinogenicity: An Uptating of* IARC Monographs *Volumes 1–42*, Lyon, p. 73

IARC (1987b) *IARC Monographs on the Evaluation of Carcinogenic Risks to Humans*, Suppl. 7, *Overall Evaluations of Carcinogenicity: An Updating of* IARC Monographs *Volumes 1–42*, Lyon, p. 64

IARC (1987c) *IARC Monographs on the Evaluation of Carcinogenic Risks to Humans*, Suppl. 7, *Overall Evaluations of Carcinogenicity: An Updating of* IARC Monographs *Volumes 1–42*, Lyon, pp. 77–78

IARC (1987d) *IARC Monographs on the Evaluation of Carcinogenic Risks to Humans*, Suppl. 7, *Overall Evaluations of Carcinogenicity: An Updating of* IARC Monographs *Volumes 1–42*, Lyon, p. 70

IARC (1987e) *IARC Monographs on the Evaluation of Carcinogenic Risks to Humans*, Suppl. 7, *Overall Evaluations of Carcinogenicity: An Updating of* IARC Monographs *Volumes 1–42*, Lyon, p. 70

IARC (1987f) *IARC Monographs on the Evaluation of Carcinogenic Risks to Humans*, Suppl. 7, *Overall Evaluations of Carcinogenicity: An Updating of* IARC Monographs *Volumes 1–42*, Lyon, pp. 373–376

IARC (1987g) *IARC Monographs on the Evaluation of Carcinogenic Risks to Humans*, Suppl. 7, *Overall Evaluations of Carcinogenicity: An Updating of* IARC Monographs *Volumes 1–42*, Lyon, p. 70

IARC (1994a) *IARC Monographs on the Evaluation of Carcinogenic Risks to Humans*, Vol. 60, *Some Industrial Chemicals*, Lyon, pp. 45–71

IARC (1994b) *Directory of Agents Being Tested for Carcinogenicity*, No. 16, Lyon, p. 72

IARC (1995) *IARC Monographs on the Evaluation of Carcinogenic Risks to Humans*, Vol. 62, Lyon, pp. 217–362

ILO (1991) *Occupational Exposure Limits for Airborne Toxic Substances: Values of Selected Countries* (Occupational Safety and Health Series No. 37), 3rd Ed., Geneva, pp. 418–419

International Technology Corporation (1992) *Exposure to Volatile Components of Polyvinyl Acetate (PVA) Emulsion Paints During Application and Drying*, Washington DC, National Paint and Coatings Association, Inc., Vinyl Acetate Toxicology Group

Jantunen, K., Mäki-Paakkanen, J. & Norppa, H. (1986) Induction of chromosome aberrations by styrene and vinylacetate in cultured human lymphocytes: dependence on erythrocytes. *Mutat. Res.*, **159**, 109–116

Jedrychowski, W., Prochowska, K., Garlinska, J. & Bruzgielewicz, J. (1979) The occurrence of chronic nonspecific diseases of respiratory tract in workers of vinyl resins establishment. *Przegl. Lék*, **36**, 679–682 (in Polish)

Kelly, T.J., Callahan, P.J., Plell, J. & Evans, G.F. (1993) Method development and field measurements for polar organic compounds in ambient air. *Environ. Sci. Technol.*, **27**, 1146–1153

Kimble, H.J., Ketcham, N.H., Kuryla, W.C., Neff, J.E. & Patel, M.A. (1982) A solid sorbent tube for vinyl acetate monomer that eliminates the effect of moisture in environmental sampling. *Am. ind. Hyg. Assoc. J.*, **43**, 137–144

Kollár, V., Kemka, R., Mišianik, J. & Tölgyessy, J. (1988) Determination of vinyl chloride and vinyl acetate in working atmosphere. *Chem. Pap.*, **42**, 147–160

Kuykendall, J.R. & Bogdanffy, M.S. (1992) Reaction kinetics of DNA–histone crosslinking by vinyl acetate and acetaldehyde. *Carcinogenesis*, **13**, 2095–2100

Kuykendall, J.R., Taylor, M.L. & Bogdanffy, M.S. (1993a) Cytotoxicity and DNA–protein crosslink formation in rat nasal tissues exposed to vinyl acetate are carboxylesterase-mediated. *Toxicol. appl. Pharmacol.*, **123**, 283–292

Kuykendall, J.R., Taylor, M.L. & Bogdanffy, M.S. (1993b) DNA–protein crosslink formation in isolated rat nasal epithelial cells exposed to vinyl acetate and acetaldehyde (Abstract 929). *Proc. Am. Assoc. Cancer Res.*, **34**, 156

Lähdetie, J. (1988) Effects of vinyl acetate and acetaldehyde on sperm morphology and meiotic micronuclei in mice. *Mutat. Res.*, **202**, 171–178

Laib, R.J. & Bolt, H.M. (1986) Vinyl acetate, a structural analog of vinyl carbamate, fails to induce enzyme-altered foci in rat liver. *Carcinogenesis*, **7**, 841–843

Lambert, B., Chen, Y., He, S.-M. & Sten, M. (1985) DNA cross-links in human leucocytes treated with vinyl acetate and acetaldehyde *in vitro*. *Mutat. Res.*, **146**, 301–303

Leonard, E.C. (1970) Vinyl acetate. In: Leonard, E.C., ed., *Vinyl and Diene Monomers*, Part 1, New York, Wiley Interscience, pp. 263–328

Lewis, R.J. (1993) *Hawley's Condensed Chemical Dictionary*, 12th Ed., New York, Van Nostrand Reinhold Co., p. 1215

Lide, D.R. (1993) *CRC Handbook of Chemistry and Physics*, 74th Ed., Boca Raton, FL, CRC Press, p. 3–517

Lijinsky, W. (1988) Chronic studies in rodents of vinyl acetate and compounds related to acrolein. *Ann. N.Y. Acad. Sci.*, **534**, 246–254

Lijinsky, W. & Andrews, A.W. (1980) Mutagenicity of vinyl compounds in *Salmonella typhimurium*. *Teratog. Carcinog. Mutag.*, **1**, 259–267

Lijinsky, W. & Reuber, M.D. (1983) Chronic toxicity studies of vinyl acetate in Fischer rats. *Toxicol. appl. Pharmacol.*, **68**, 43–53

Mäki-Paakkanen, J. & Norppa, H. (1987) Induction of micronuclei by vinyl acetate in mouse bone marrow cells and cultured human lymphocytes. *Mutat. Res.*, **190**, 41–45

Mannsville Chemical Products Corp. (1982) *Vinyl Acetate (Chemical Products Synopsis)*, Cortland, NY

McCann, J., Choi, E., Yamasaki, E. & Ames, B.N. (1975) Detection of carcinogens as mutagens in the *Salmonella*/microsome test: assay of 300 chemicals. *Proc. natl Acad. Sci. USA*, **72**, 5135–5139

McNeal, T.P. & Hollifield, H.C. (1993) Determination of volatile chemicals released from microwave-heat-susceptor food packaging. *J. Assoc. off. anal. Chem. Int.*, **76**, 1268–1275

Mebus, C.A., Carpanini, F.M.B., Rickard, R.W., Cascieri, T.C., Tyler, T.R. & Vinegar, M.B. (1995) A two-generation reproduction study in rats receiving drinking water containing vinyl acetate. *Fundam. appl. Toxicol.*, **24**, 206–216

Miller, C.A., Srivastava, R.K. & Ryan, J.V. (1994) Emmissions of organic hazardous air pollutants from the combustion of pulverized coal in a small-scale combustor. *Environ. Sci. Technol.*, **28**, 1150–1158

Norppa, H., Tursi, F., Pfäffli, P., Mäki-Paakkanen, J. & Järventaus, H. (1985) Chromosome damage induced by vinyl acetate through in vitro formation of acetaldehyde in human lymphocytes and Chinese hamster ovary cells. *Cancer Res.*, **45**, 4816–4821

Ott, M.G., Teta, M.J. & Greenberg, H.L. (1989) Lymphatic and hematopoietic tissue cancer in a chemical manufacturing environment. *Am. J. ind. Med.*, **16**, 631–643

Owen, P.E. (1983) Vinyl acetate—three month inhalation toxicity study in the rat and mouse (Abstract). *Hum. Toxicol.*, **2**, 416

Pellizzari, E.D. (1982) Analysis for organic vapor emissions near industrial and chemical waste disposal sites. *Environ. Sci. Technol.*, **16**, 781–785

Rosato, D.V. (1993) *Rosato's Plastics Encyclopedia and Dictionary*, New York, Hanser, pp. 235–236, 573

Roscher, G., Hofmann, E., Armstrong Adey, K., Jeblick, W., Klimisch, H.-J. & Kieczka, H. (1983) Vinyl acetate. In: Gerhartz, W., Mayer, C., Moegling, D., Monkhouse, P. & Pfefferkorn, R., eds, *Ullmann's Encyclopedia of Industrial Chemistry*, 4th Ed., Vol. 23, New York, VCH Publishers, pp. 597–618

Sadtler Research Laboratories (1980) *1980 Cumulative Index*, Philadelphia, PA

Sax, N.I. & Lewis, R.J. (1987) *Hawley's Condensed Chemical Dictionary*, 11th Ed., New York, Van Nostrand Reinhold, pp. 491, 945–946, 1222

Schweizerische Unfallversicherungsanstalt [Swiss Accident Insurance Company] (1994) *Grenzwerte am Arbeitsplaz* [Limit values in the work place], Lucerne, p. 104

Shirinian, G.S. & Arutyunyan, R.H. (1980) Study of cytogenetic change levels under PVA production. *Biol. Z. Armenii*, **33**, 748–752

Simon, P., Filser, J.G. & Bolt, H.M. (1985a) Metabolism and pharmacokinetics of vinyl acetate. *Arch. Toxicol.*, **57**, 191–195

Simon, P., Ottenwälder, H. & Bolt, H.M. (1985b) Vinyl acetate: DNA–binding assay in vivo. *Toxicol. Lett.*, **27**, 115–120

Sipi, P., Järventaus, H. & Norppa, H. (1992) Sister-chromatid exchanges induced by vinyl esters and respective carboxylic acids in cultured human lymphocytes. *Mutat. Res.*, **279**, 75–82

Stepanyan, I.S., Padaryan, G.M., Airapetyan, L.K. & Maslyukova, D.F. (1970) Ionization–chromatographic method for determining some components of waste waters from the plant 'Polivinilatsetat'. *Prom. Arm.*, **9**, 76–78 [*Chem. Abstr.*, **74**, 90928p] (in Russian)

Takeshita, T., Iijima, S. & Higurashi, M. (1986) Vinyl acetate-induced sister chromatid exchanges in murine bone marrow cells. *Proc. Jpn Acad.*, **62**, 239–242

Työministeriö [Ministry of Labour] (1993) *HTP-Arvot 1993* [Occupational exposure limits 1993] Tampere, p. 19 (in Finnish)

Union Carbide Corp. (1981) *Product Information Sheet: Vinyl Acetate Monomer*, Danbury, CT

Union Carbide Corp. (1994) *Product Specification Sheet: Vinyl Acetate*, South Charleston, WV

United Kingdom Health and Safety Executive (1993) *Occupational Exposure Limits 1993* (EH 40/93), London, Her Majesty's Stationery Office, p. 27

United States Agency for Toxic Substances and Disease Registry (1992) *Toxicological Profile for Vinyl Acetate* (Clement International Corp., Contract No. 205-88-0608), Washington DC, United States Department of Health and Human Services

United States Environmental Protection Agency (1986) Method 8240. Gas chromatography/mass spectrometry for volatile organics. In: *Test Methods for Evaluating Solid Waste—Physical/-Chemical Methods* (US EPA No. SW-846), 3rd Ed., Vol. 1A, Washington DC, Office of Solid Waste and Emergency Response, pp. 1–43

United States Food and Drug Administration (1984) Food and Drugs. *US Code fed. Regul.*, **Title 21**, Parts 175.105, 175.300, 175.350, 176.170, 176.180, 177.1010, 177.1200, 177.1350, 177.1360, 177.1670, 177.2260, 177.2800, 178.3910, pp. 133, 145, 160, 180, 189, 199, 211, 225, 255, 271, 289, 348

United States National Institute for Occupational Safety and Health (1978) *Criteria for a Recommended Standard for Occupational Exposure to Vinyl Acetate* (DHEW (NIOSH) Pub. No. 78-205), Cincinnati, OH

United States National Institute for Occupational Safety and Health (1994a) *NIOSH Pocket Guide to Chemical Hazards* (DHHS (NIOSH) Publ. No. 94-116), Cincinnati, OH, pp. 328–329

United States National Institute for Occupational Safety and Health (1994b) *National Occupational Exposure Survey (1981–83)*, Cincinnati, OH

United States Occupational Safety and Health Administration (1990) *OSHA Analytical Methods Manual*, 2nd Ed., Part 1, Vol. 2, Salt Lake City, UT, United States Department of Labor, pp. 51-1–51-14

Waxweiler, R.J., Smith, A.H., Falk, H. & Tyroler, H.A. (1981) Excess lung cancer risk in a synthetic chemicals plant. *Environ. Health Perspectives*, **41**, 159–165

Weast, R.C. & Astle, M.J. (1985) *CRC Handbook of Data on Organic Compounds*, Vols I & II, Boca Raton, FL, CRC Press, pp. 22 (I), 451 (II)

VINYL FLUORIDE

This substance was considered by previous working groups, in June 1985 and March 1987 (IARC, 1986, 1987a). Since that time, new data have become available, and these have been incorporated into the monograph and taken into consideration in the present evaluation.

1. Exposure Data

1.1 Chemical and physical data

1.1.1 Nomenclature

Chem. Abstr. Serv. Reg. No.: 75-02-5
Chem. Abstr. Name: Fluoroethene
IUPAC Systematic Name: Fluoroethylene
Synonyms: 1-Fluoroethene; 1-fluoroethylene; monofluoroethene; monofluoroethylene

1.1.2 Structural and molecular formulae and relative molecular mass

$$\begin{array}{c}H\\H\end{array}\!\!\!>\!\!C\!=\!C\!\!<\!\!\begin{array}{c}H\\F\end{array}$$

C_2H_3F Relative molecular mass: 46.04

1.1.3 Chemical and physical properties of the pure substance

(a) *Description*: Colourless gas (Ebnesajjad & Snow, 1994)
(b) *Boiling-point*: –72.2 °C (Lide, 1993)
(c) *Melting-point*: –160.5 °C (Lide, 1993)
(d) *Spectroscopy data*: Infrared (prism [30864]; grating [48458P]) and mass [15] spectral data have been reported (Sadtler Research Laboratories, 1980; Weast & Astle, 1985).
(e) *Solubility*: Slightly soluble in water (1.1% by weight) (PCR Inc., 1994); soluble in acetone, ethanol and diethyl ether (Sax & Lewis, 1989; Lide, 1993)
(f) *Volatility*: Vapour pressure, 370 psi [2553 kPa] at 21 °C (Ebnesajjad & Snow, 1994); relative vapour density (air = 1), 1.6 (PCR Inc., 1994)
(g) *Stability*: Lower explosive limit (in air), 2.6% (PCR Inc., 1994); polymerizes freely; forms explosive mixtures with air (Buckingham, 1982); ignites in the presence of heat or sources of ignition (Sax & Lewis, 1989)

(h) *Reactivity*: Reacts with alkali and alkaline earth metals, powdered aluminium, zinc and beryllium (PCR Inc., 1994)

(i) *Conversion factor*: $mg/m^3 = 1.88 \times ppm$[1]

1.1.4 Technical products and impurities

Vinyl fluoride is available commercially at a purity of 99.9%; 0.1% *d*-limonene (see IARC, 1993) is added as a stabilizer (PCR Inc., 1994).

1.1.5 Analysis

Vinyl fluoride has been determined in workplace air collected in poly(tetrafluoroethylene) bags and analysed by gas chromatography (Oser, 1980). Nonspecific methods involving fluorescence spectrophotometry and chemiluminescence have been reported (Quickert *et al.*, 1975; Sutton *et al.*, 1979).

1.2 Production and use

1.2.1 Production

Vinyl fluoride was first prepared from 1,1-difluoro-2-bromoethane, with zinc as the catalyst, in Belgium in the early 1900s. Modern production involves direct addition of hydrogen fluoride to acetylene in the presence of mercury- or aluminium-based catalysts (Brasure, 1980). In the past, vinyl fluoride was produced by dehydrofluorination of 1,1-difluoroethane or by direct conversion of ethylene (see IARC, 1994) in the presence of a carbon catalyst impregnated with palladium and copper chloride (Brasure, 1980; Siegmund *et al.*, 1988). Dehydrochlorination of 1-chloro-1-fluoroethane and 1-chloro-2-fluoroethane is used commercially (Siegmund *et al.*, 1988).

Vinyl fluoride is produced by one company each in Japan and the United States of America (Chemical Information Services, Inc., 1994).

1.2.2 Use

The main use of vinyl fluoride is in the production of polyvinyl fluoride and other fluoropolymers. Polyvinyl fluoride was first prepared in the 1930s (Brasure, 1980) and was introduced commercially in the United States in 1961 (Ebnesajjad & Snow, 1994). It is converted into a thin film by plasticized melt extrusion (Brasure, 1980) and is sold under the trademarks Tedlar PVF film and Dalvor (Siegemund *et al.*, 1988).

Polymers of vinyl fluoride are characterized by strong resistance to weather, great strength, chemical inertness and low permeability to air and water. Polyvinyl fluoride laminated with aluminium, galvanized steel and cellulosic materials has been used as a protective surfacing for the exteriors of residential and industrial buildings. Laminations with various plastics have also

---

[1] Calculated from: mg/m^3 = (relative molecular mass/24.45) × ppm, assuming normal temperature (25 °C) and pressure (101 kPa)

been used as protective coverings on walls, pipes and electric equipment and inside aircraft cabins (Cohen & Kraft, 1971; Brasure, 1980).

Use of vinyl fluoride in the Member States of the European Union in 1991 was estimated at about 3600 tonnes (Environmental Chemicals Data and Information Network, 1993).

1.3 Occurrence

1.3.1 Natural occurrence

Vinyl fluoride is not known to occur as a natural product.

1.3.2 Occupational exposures

Vinyl fluoride was determined in a manufacturing and a polymerization plant in the United States. The concentrations in eight samples taken at the manufacturing plant were generally < 2 ppm [3.76 mg/m^3], but a level of 21 ppm [39.5 mg/m^3] was reported in one personal sample. The concentrations in seven personal samples taken in the polymerization plant were 1–4 ppm [1.88–7.52 mg/m^3], and those in four general area samples were 1–5 ppm [1.88–9.4 mg/m^3] (Oser, 1980).

1.4 Regulations and guidelines

In most countries, exposure limits have not been recommended (ILO, 1991). The United States National Institute for Occupational Safety and Health (1994) recommended an exposure limit of 1 ppm [1.88 mg/m^3] as an 8-h time-weighted average, with a ceiling value of 5 ppm [9.4 mg/m^3] for short-term (15-min) exposure.

2. Studies of Cancer in Humans

No data were available to the Working Group.

3. Studies of Cancer in Experimental Animals

Inhalation

Mouse: Groups of 95 male and 95 female Swiss-derived mice, 47 days of age, were exposed to 0, 25, 250 or 2500 ppm [0, 47, 470 or 4700 mg/m^3] vinyl fluoride (purity, > 99.94%) for 6 h per day on five days per week for up to 550 days. Ten mice per group were killed at six months for interim evaluation. All surviving mice were killed at various times between 375 and 550 days because of early, dose-related mortality. Animals were necropsied, and all organs were preserved for histological examination. All organs of control animals and those at the high dose were examined; only the nose, lungs, liver, kidneys, gross lesions and target organs of animals in the other groups were examined microscopically. The overall incidences of primary lung tumours (predominantly alveolar–bronchiolar adenomas) were 11/81 male controls, 45/80 at 25 ppm, 52/80 at 250 ppm and 56/81 at 2500 ppm; and 9/81 female controls, 24/80 at 25 ppm, 47/80 at 250 ppm and 53/81 at 2500 ppm. Fatal hepatic haemangiosarcomas occurred in 1/81, 16/80, 42/80 and 42/81 males and 0/81, 13/81, 25/80 and 32/81 females in the four groups, respectively. Mammary gland neoplasms (adenoma, adenocarcinoma and fibroadenoma) occurred in 0/77 control females, 22/76 at 25 ppm, 20/78 at 250 ppm and 20/77 at 2500 ppm. Harderian gland adenomas occurred in 3/66 control males, 13/68 at 25 ppm, 12/66 at 250 ppm and 31/62 at 2500 ppm (killed between 7 and 18 months). The authors did not perform statistical analyses because of the differences in the times of killing in the various groups (Bogdanffy et al., 1995).

Rat: Groups of 95 male and 95 female Sprague-Dawley-derived rats, 40 days of age, were exposed to 0, 25, 250 or 2500 ppm [0, 47, 470 or 4700 mg/m^3] vinyl fluoride (purity, > 99.94%) for 6 h per day on five days per week for 725 days. Ten rats per group were killed at 271–276 days (although the interim kill had been scheduled for 12 months), and surviving rats were killed at various times between 552 and 725 days. Animals were necropsied, and organs were preserved for histological examination. All organs of control animals and those at the high dose were examined; only the nose, lungs, liver, kidneys, gross lesions and target organs of animals in the other groups were examined microscopically. Increased incidences of tumours were seen at several sites (Table 1). The authors did not perform statistical analyses because of the differences in the times of killing in the various groups (Bogdanffy et al., 1995).

Table 1. Tumour incidences in Fischer 344 rats exposed to vinyl fluoride by inhalation for two years

Sex	Target organ	Control	25 ppm [47 mg/m^3]	250 ppm [470 mg/m^3]	2500 ppm [4700 mg/m^3]
Males	Hepatic haemangiosarcomas	0/81	5/80	30/80	20/80
	Zymbal gland tumours	0/80	2/80	3/80	11/80
	Hepatocellular adenomas and carcinomas	5/80	10/80	10/80	7/80
Females	Hepatic haemangiosarcomas	0/80	8/80	19/80	15/80
	Zymbal gland tumours	0/80	0/80	1/80	12/80
	Hepatocellular adenomas and carcinomas	0/80	4/80	9/80	8/10

From Bogdanffy et al. (1995)

4. Other Data Relevant to an Evaluation of Carcinogenicity and its Mechanisms

4.1 Absorption, distribution, metabolism and excretion

4.1.1 Humans

No data were available to the Working Group.

4.1.2 Experimental systems

The metabolism and toxicity of vinyl fluoride have been reviewed (Kennedy, 1990).

Vinyl fluoride is readily absorbed after its inhalation and reaches a tissue concentration in rats that is 90% of that in the inhaled air (Filser & Bolt, 1979, 1981). The metabolism is saturable and dose-dependent, saturation occurring at concentrations > 75 ppm (> 140 mg/m^3) (Filser & Bolt, 1979). Pharmacokinetic data imply that the rate of biotransformation of vinyl fluoride is about one-fifth that of vinyl chloride (Bolt *et al.*, 1981).

Fluoride appears to be a metabolite of vinyl fluoride as it is found in the urine six days after exposure (Dilley *et al.*, 1974). The fluoride concentrations in the urine of rats were found to be increased 45 and 90 days after exposure by inhalation to 0, 200, 2000 or 20 000 ppm [373, 3733 or 37 333 mg/m^3] vinyl fluoride for 6 h per day, five days per week for about 90 days. A plateau was observed at about 2000 ppm [3733 mg/m^3], suggesting saturation of vinyl fluoride metabolism (Bogdanffy *et al.*, 1990). When rats and mice were exposed to 0, 25, 250 or 2500 ppm [0, 47, 470 or 4700 mg/m^3] vinyl fluoride for two years and 18 months, respectively, a plateau of urinary fluoride excretion was seen at ≥ 250 ppm (Bogdanffy *et al.*, 1995).

4.2 Toxic effects

4.2.1 Humans

Vinyl fluoride may burn the skin and eyes and may cause headache or dizziness (United States National Library of Medicine, 1994).

4.2.2 Experimental systems

The acute lethality of vinyl fluoride is so low that concentrations can be increased to the point at which oxygen becomes limiting (Lester & Greenberg, 1950; Kopecný *et al.*, 1964; Clayton, 1967).

Rats were exposed to vinyl fluoride at 3000 ppm [5640 mg/m^3] for 30 min and followed for seven days thereafter. Increased urine output and potassium excretion were observed at various times (Dilley *et al.*, 1974). [The Working Group noted that the control values on some days were similar to those that were significantly different from control values on other days.]

Groups of 15 male and 15 female rats and mice were exposed to vinyl fluoride at concentrations of 0, 200, 2000 or 20 000 ppm [0, 375, 3750 or 37 500 mg/m^3] for 6 h per day, on five days per week for 90 days. No significant changes in histopathological or haematological parameters or in clinical chemistry were found at 45 days or at the end of the 90-day exposure

(Bogdanffy et al., 1990). Rats and mice exposed to vinyl fluoride by inhalation at 0, 25, 250 or 2500 ppm [47, 470 or 4700 mg/m^3] also showed no changes in these measures (Bogdanffy et al., 1995). The survival of the exposed mice and rats was decreased, however, such that, after early sacrifices, only control animals and those receiving the lowest dose were available for clinical evaluation. Liver-cell proliferation, measured as the labelling index in animals implanted with osmotic minipumps containing ^{3}H-thymidine, was increased in rats and mice of each sex at all concentrations in one study (Bogdanffy et al., 1990) but not in a subsequent study with interim sampling for cell proliferation (Bogdanffy et al., 1995), owing perhaps to use of a less sensitive technique in the later study.

Increased exhalation of acetone has been reported to be due to the effects of vinyl fluoride on intermediary metabolism (Filser et al., 1982). Intermediate metabolites of vinyl fluoride have been shown to alkylate the haem group of cytochrome P450 enzymes *in vitro*, suggesting potential inhibition of the metabolism of other substances (Ortiz de Montellano et al., 1982).

Exposure of rats to vinyl fluoride at 10 000 ppm [18 800 mg/m^3] after pretreatment with Aroclor 1254 resulted in liver damage (Conolly et al., 1978). Treatment of newborn rats with 2000 ppm [3760 mg/m^3] vinyl fluoride for 8 h per day, on five days per week for 14 weeks resulted in increased numbers of preneoplastic foci in their livers (Bolt et al., 1981).

4.3 Reproductive and prenatal effects

No data were available to the Working Group.

4.4 Genetic and related effects

As reported in an abstract, vinyl fluoride induced gene mutations and chromosomal aberrations in Chinese hamster ovary cells in the presence of an exogenous metabolic system, sex-linked recessive lethal mutations in *Drosophila melanogaster* and micronuclei in polychromatic erythrocytes from the bone marrow of female mice exposed to 19.1 or 38.8% vinyl fluoride for 6 h. No unscheduled DNA synthesis was seen in pachytene spermatocytes, no single strand breaks or cross-links in testicular DNA and no dominant lethal mutations in male rats (Bentley et al., 1992).

4.5 Structure–activity relationship

Vinyl fluoride bears a close structural relationship to vinyl chloride (see IARC, 1987b).

5. Summary and Evaluation

5.1 Exposure data

Vinyl fluoride has been produced commercially since the 1960s for use in the production of polyvinylfluoride and fluoropolymers. Human exposure may occur during its production and use.

5.2 Human carcinogenicity data

No data were available to the Working Group.

5.3 Animal carcinogenicity data

Vinyl fluoride was tested for carcinogenicity in one experiment in mice and one experiment in rats by inhalation. It produced haemangiosarcomas in the liver and alveolar–bronchiolar adenomas in mice of each sex, mammary tumours in females and Harderian gland adenomas in males. In rats, it produced haemangiosarcomas of the liver and Zymbal gland tumours in animals of each sex and an increased incidence of hepatocellular adenomas and carcinomas in females.

5.4 Other relevant data

Vinyl fluoride is readily absorbed after administration by inhalation. Its metabolism is saturable and dose-dependent. Vinyl fluoride has very low acute toxicity. High doses produced no measurable toxic effects after subchronic exposure. Survival was decreased after chronic exposure, but no other toxic effects were seen in surviving animals.

5.5 Evaluation[1]

There is *inadequate evidence* in humans for the carcinogenicity of vinyl fluoride.

There is *sufficient evidence* in experimental animals for the carcinogenicity of vinyl fluoride.

Overall evaluation

Vinyl fluoride *is probably carcinogenic to humans (Group 2A)*.

In making the overall evaluation, the Working Group took into account the following evidence: Vinyl fluoride is closely related structurally to the known human carcinogen, vinyl chloride. The two chemicals cause the same rare tumour (hepatic haemangiosarcoma) in experimental animals, which is also a tumour caused by vinyl chloride in humans.

6. References

Bentley, K.S., Mullin, L.S., Rickard, L.B., Vlachos, D.A., Sarrif, A.M., Sernau, R.C., Brusick, D.J. & Curren, R.D. (1992) Vinyl fluoride: mutagenic in mammalian somatic cells in vitro and in vivo and in Drosophila germ cells but nongenotoxic in germ cells of mammals (Abstract). *Environ. mol. Mutag.*, **19** (Suppl. 20), 5

---

[1] For definition of the italicized terms, see Preamble, pp. 22–26.

Bogdanffy, M.S., Kee, C.R., Kelly, D.P., Carakostas, M.C. & Sykes, G.P. (1990) Subchronic inhalation study with vinyl fluoride: effects on hepatic cell proliferation and urinary fluoride excretion. *Fundam. appl. Toxicol.*, **15**, 394–406

Bogdanffy, M.S., Makovec, G.T. & Frame, S.R. (1995) Inhalation oncogenicity bioassay in rats and mice with vinyl fluoride. *Fundam. appl. Toxicol.* (in press)

Bolt, H.M., Laib, R.J. & Klein, K.-P. (1981) Formation of pre-neoplastic hepatocellular foci by vinyl fluoride in newborn rats. *Arch. Toxicol.*, **47**, 71–73

Brasure, D.E. (1980) Poly(vinyl fluoride). In: Mark, H.F., Othmer, D.F., Overberger, C.G., Seaborg, G.T. & Grayson, N., eds, *Kirk-Othmer Encyclopaedia of Chemical Technology*, 3rd Ed., Vol. 11, New York, John Wiley & Sons, pp. 57–64

Buckingham, J., ed. (1982) *Dictionary of Organic Compounds*, 5th Ed., Vol. 3, New York, Chapman and Hall, p. 2649

Chemical Information Services, Inc. (1994) *Directory of World Chemical Producers 1995/1996 Standard Edition*, Dallas, TX, p.706

Clayton, J.W., Jr (1967) Fluorocarbon toxicity and biological action. *Fluorine Chem. Rev.*, **1**, 197–252

Cohen, F.S. & Kraft, P. (1971) Vinyl fluoride polymers. In: Bikales, N.M., ed., *Encyclopedia of Polymer Science and Technology*, Vol. 14, New York, Wiley Interscience, pp. 522–540

Conolly, R.B., Jaeger, R.J. & Szabo, S. (1978) Acute hepatotoxicity of ethylene, vinyl fluoride, vinyl chloride, and vinyl bromide after Aroclor 1254 pretreatment. *Exp. mol. Pathol.*, **28**, 25–33

Dilley, J.V., Carter, V.L., Jr & Harris, E.S. (1974) Fluoride ion excretion by male rats after inhalation of one of several fluoroethylenes or hexafluoropropene. *Toxicol. appl. Pharmacol.*, **27**, 582–590

Ebnesajjad, S. & Snow, L.G. (1994) Poly(vinyl fluoride). In: Kroschwitz, J.I. & Howe-Grant, M., eds, *Kirk-Othmer Encyclopedia of Chemical Technology*, 4th Ed., Vol. 11, New York, John Wiley & Sons, pp. 683–694

Environmental Chemicals Data and Information Network (1993) *Fluoroethene*, Ispra, JRC-CEC, last update: 02.09.1993

Filser, J.G. & Bolt, H.M. (1979) Pharmacokinetics of halogenated ethylenes in rats. *Arch. Toxicol.*, **42**, 123–136

Filser, J.G. & Bolt, H.M. (1981) Inhalation pharmacokinetics based on gas uptake studies. I. Improvement of kinetic models. *Arch. Toxicol.*, **47**, 279–292

Filser, J.G., Jung, P. & Bolt, H.M. (1982) Increased acetone exhalation induced by metabolites of halogenated C_1 and C_2 compounds. *Arch. Toxicol.*, **49**, 107–116

IARC (1986) *IARC Monographs on the Evaluation of the Carcinogenic Risk of Chemicals to Humans*, Vol. 39, *Some Chemicals used in Plastics and Elastomers*, Lyon, pp. 147–154

IARC (1987a) *IARC Monographs on the Evaluation of Carcinogenic Risks to Humans*, Suppl. 7, *Overall Evaluations of Carcinogenicity: An Updating of* IARC Monographs *Volumes 1–42*, Lyon, p. 73

IARC (1987b) *IARC Monographs on the Evaluation of Carcinogenic Risks to Humans*, Suppl. 7, *Overall Evaluations of Carcinogenicity: An Updating of* IARC Monographs *Volumes 1–42*, Lyon, pp. 373–376

IARC (1993) *IARC Monographs on the Evaluation of Carcinogenic Risks to Humans*, Vol. 56, *Some Naturally Occurring Substances: Food Items and Constituents, Heterocyclic Aromatic Amines and Mycotoxins*, Lyon, pp. 135–162

IARC (1994) *IARC Monographs on the Evaluation of Carcinogenic Risks to Humans*, Vol. 60, *Some Industrial Chemicals*, Lyon, pp. 45–71

ILO (1991) *Occupational Exposure Limits for Airborne Toxic Substances: Values of Selected Countries* (Occupational Safety and Health Series No. 37), 3rd Ed., Geneva

Kennedy, G.L., Jr (1990) Toxicology of fluorine-containing monomers. *Crit. Rev. Toxicol.*, **21**, 149–170

Kopecný, J., Lúcanská, N., Šipka, F., Cerný, E. & Ambros, D. (1964) Toxicity of vinyl fluoride. *Pracov. Lék.*, **16**, 310–311 (in Polish)

Lester, D. & Greenberg, L.A. (1950) Acute and chronic toxicity of some halogenated derivatives of methane and ethane. *Arch. ind. Hyg. occup. Med.*, **2**, 335–344

Lide, D.R., ed. (1993) *CRC Handbook of Chemistry and Physics*, 74th Ed., Boca Raton, FL, CRC Press, pp. 3-518, B-25

Ortiz de Montellano, P.R., Kunze, K.L., Beilan, H.S. & Wheeler, C. (1982) Destruction of cytochrome P-450 by vinyl fluoride, fluroxene, and acetylene. Evidence for a radical intermediate in olefin oxidation. *Biochemistry*, **21**, 1331–1339

Oser, J.L. (1980) Extent of industrial exposure to epichlorohydrin, vinyl fluoride, vinyl bromide and ethylene dibromide. *Am. ind. Hyg. Assoc. J.*, **41**, 463–468

PCR Inc. (1994) *Material Safety Data Sheet: Vinyl Fluoride*, Gainesville, FL

Quickert, N., Findlay, W.J. & Monkman, J.L. (1975) Modification of a chemiluminescent ozone monitor for the measurement of gaseous unsaturated hydrocarbons. *Sci. total Environ.*, **3**, 323–328

Sadtler Research Laboratories (1980) *1980 Cumulative Index*, Philadelphia, PA

Sax, N.I. & Lewis, R.J. (1989) *Dangerous Properties of Industrial Materials*, 7th Ed., New York, Van Nostrand Reinhold, p. 3476

Siegmund, G., Schwertfeger, W., Feiring, A., Smart, B., Behr, F., Vogel, H. & McKusick, B. (1988) Fluorine compounds. In: Gerhartz, W., Yamamoto, Y.S., Elvers, B., Rounsaville, J.F. & Schulz, G., eds, *Ullmann's Encyclopedia of Industrial Chemistry*, 5th Ed., Vol. A11, New York, VCH Publishers, pp. 349–392

Sutton, D.G., Westberg, K.R. & Melzer, J.E. (1979) Chemiluminescence detector based on active nitrogen for gas chromatography of hydrocarbons. *Anal. Chem.*, **51**, 1399–1401

United States National Institute for Occupational Safety and Health (1994) *NIOSH Pocket Guide to Chemical Hazards* (DHHS (NIOSH) Publ. No. 94-116), Cincinnati, OH, pp. 330–331

United States National Library of Medicine (1994) *Hazardous Substances Data Bank (HSDB)*, Bethesda, MD

Weast, R.C. & Astle, M.J. (1985) *CRC Handbook of Data on Organic Compounds*, Vols I & II, Boca Raton, FL, CRC Press, pp. 433, 624 (I), 591, 770 (II)

SUMMARY OF FINAL EVALUATIONS

Agent (occupation)	Degree of evidence of carcinogenicity		Overall evaluation of carcinogenicity to humans
	Human	Animal	
Acrolein	I	I	3
Benzofuran	I[a]	S	2B
Chloral	I[a]	I	3
Chloral hydrate	I[a]	L	3
1-Chloro-2-methylpropene	I[a]	S	2B
3-Chloro-2-methylpropene	I[a]	L	3
Crotonaldehyde	I	I	3
Dichloroacetic acid	I	L	3
Dry cleaning (occupational exposures in)	L		2B
Furan	I[a]	S	2B
Furfural	I[a]	L	3
Tetrachloroethylene	L	S	2A
Trichloroacetic acid	I	L	3
Trichloroethylene	L	S	2A
1,2,3-Trichloropropane	I[a]	S	2A[b]
Vinyl acetate	I	L	2B[b]
Vinyl fluoride	I[a]	S	2A[b]

S, sufficient evidence; L, limited evidence; I, inadequate evidence; for definitions of criteria for degrees of evidence and groups, see preamble, pp. 23–27

[a] No data available

[b] Other relevant data taken into account in making the overall evaluation

APPENDIX 1

SUMMARY TABLES OF
GENETIC AND RELATED EFFECTS

APPENDIX 1

Summary table of genetic and related effects of trichloroethylene without mutagenic stabilizers

Non-mammalian systems													Mammalian systems																						
Pro-karyotes		Lower eukaryotes				Plants			Insects				In vitro																	In vivo					
													Animal cells								Human cells									Animal							Humans				
D	G	D	R	G	A	D	G	C	R	G	C	A	D	G	S	M	C	A	T	I	D	G	S	M	C	A	T	I	D	G	S	M	C	DL	A	D	S	M	C	A
−¹	?	+	+	+¹								−¹	−¹	+¹	+¹	−	−	+¹		+¹						−¹			+	−¹	−¹	+	−¹	−¹	−¹					

A, aneuploidy; C, chromosomal aberrations; D, DNA damage; DL, dominant lethal mutation; G, gene mutation; I, inhibition of intercellular communication; M, micronuclei; R, mitotic recombination and gene conversion; S, sister chromatid exchange; T, cell transformation

In completing the table, the following symbols indicate the consensus of the Working Group with regard to the results for each end-point:

+   considered to be positive for the specific end-point and level of biological complexity
+¹ considered to be positive, but only one valid study was available to the Working Group
−   considered to be negative
−¹ considered to be negative, but only one valid study was available to the Working Group
?   considered to be equivocal or inconclusive (e.g. there were contradictory results from different laboratories; there were confounding exposures; the results were equivocal)

Summary table of genetic and related effects of trichloroethylene containing mutagenic stabilizers, or of uncertain purity

Non-mammalian systems				Mammalian systems			
Pro-karyotes	Lower eukaryotes	Plants	Insects	In vitro		Human cells	In vivo
				Animal cells			Animal	Humans
D G	D R G A	A D G C	R G C A	D G S M C A T I	D G S M C A T I	D G S M C DL A	D S M C A
–¹ +	+ + +¹	+¹		+ –¹ –¹ +	?¹	–¹ + –	

A, aneuploidy; C, chromosomal aberrations; D, DNA damage; DL, dominant lethal mutation; G, gene mutation; I, inhibition of intercellular communication; M, micronuclei; R, mitotic recombination and gene conversion; S, sister chromatid exchange; T, cell transformation

In completing the table, the following symbols indicate the consensus of the Working Group with regard to the results for each end-point:

+   considered to be positive for the specific end-point and level of biological complexity
+¹ considered to be positive, but only one valid study was available to the Working Group
–   considered to be negative
–¹ considered to be negative, but only one valid study was available to the Working Group
?   considered to be equivocal or inconclusive (e.g. there were contradictory results from different laboratories; there were confounding exposures; the results were equivocal)

APPENDIX 1

Summary table of genetic and related effects of tetrachloroethylene

Non-mammalian systems				Mammalian systems			
Pro-karyotes	Lower eukaryotes	Plants	Insects	In vitro			In vivo	
				Animal cells	Human cells	Animal	Humans
D G	D R G A	A D G C	R G C A	D G S M C A T I	D G S M C A T I	D G S M C DL A	D S M C A
– –	? – ? –	+¹ ?¹	–¹	– –¹ –¹ –		?	–¹ –¹

A, aneuploidy; C, chromosomal aberrations; D, DNA damage; DL, dominant lethal mutation; G, gene mutation; I, inhibition of intercellular communication; M, micronuclei; R, mitotic recombination and gene conversion; S, sister chromatid exchange; T, cell transformation

In completing the table, the following symbols indicate the consensus of the Working Group with regard to the results for each end-point:

+   considered to be positive for the specific end-point and level of biological complexity
+¹ considered to be positive, but only one valid study was available to the Working Group
–   considered to be negative
–¹ considered to be negative, but only one valid study was available to the Working Group
?   considered to be equivocal or inconclusive (e.g. there were contradictory results from different laboratories; there were confounding exposures; the results were equivocal)

Summary table of genetic and related effects of 1,2,3-trichloropropane

Non-mammalian systems									Mammalian systems																						
Pro-karyotes		Lower eukaryotes				Plants				Insects					In vitro										In vivo						
															Animal cells							Human cells					Animal				Humans
D	G	D	R	G	A	D	G	C	R	G	C	A			D	G	S	M	C	A	T	I	D	G	S	M	C	A	T	I	D	G	S	M	C	DL	A	D	S	M	C	A
$-^!$	+														$-^!$	$+^!$	+		$+^!$		$+^!$										+					$-^!$						

A, aneuploidy; C, chromosomal aberrations; D, DNA damage; DL, dominant lethal mutation; G, gene mutation; I, inhibition of intercellular communication; M, micronuclei; R, mitotic recombination and gene conversion; S, sister chromatid exchange; T, cell transformation

In completing the table, the following symbols indicate the consensus of the Working Group with regard to the results for each end-point:

+ considered to be positive for the specific end-point and level of biological complexity
+! considered to be positive, but only one valid study was available to the Working Group
− considered to be negative
−! considered to be negative, but only one valid study was available to the Working Group
? considered to be equivocal or inconclusive (e.g. there were contradictory results from different laboratories; there were confounding exposures; the results were equivocal)

Summary table of genetic and related effects of chloral hydrate

Non-mammalian systems					Mammalian systems				
Pro-karyotes	Lower eukaryotes	Plants	Insects		In vitro				In vivo
					Animal cells		Human cells		Animal	Humans
D G	D R G A	A D G C	R G C A		D G S M C A T I		D G S M C A T I		D G S M C DL A	D S M C A
? ?	? –¹ +		+¹		– + +¹ + + +		–¹ – ?¹ + +		? + + +	

A, aneuploidy; C, chromosomal aberrations; D, DNA damage; DL, dominant lethal mutation; G, gene mutation; I, inhibition of intercellular communication; M, micronuclei; R, mitotic recombination and gene conversion; S, sister chromatid exchange; T, cell transformation

In completing the table, the following symbols indicate the consensus of the Working Group with regard to the results for each end-point:

+   considered to be positive for the specific end-point and level of biological complexity
+¹ considered to be positive, but only one valid study was available to the Working Group
–   considered to be negative
–¹ considered to be negative, but only one valid study was available to the Working Group
?   considered to be equivocal or inconclusive (e.g. there were contradictory results from different laboratories; there were confounding exposures; the results were equivocal)

Summary table of genetic and related effects of dichloroacetic acid

Non-mammalian systems				Mammalian systems			
				In vitro		In vivo	
Pro-karyotes	Lower eukaryotes	Plants	Insects	Animal cells	Human cells	Animal	Humans
D G	D R G A	D G C	R G C A	D G S M C A T I	D G S M C A T I	D G S M C DL A	D S M C A
? ?				–¹	–¹	?	

A, aneuploidy; C, chromosomal aberrations; D, DNA damage; DL, dominant lethal mutation; G, gene mutation; I, inhibition of intercellular communication; M, micronuclei; R, mitotic recombination and gene conversion; S, sister chromatid exchange; T, cell transformation

In completing the table, the following symbols indicate the consensus of the Working Group with regard to the results for each end-point:

+ considered to be positive for the specific end-point and level of biological complexity
+¹ considered to be positive, but only one valid study was available to the Working Group
– considered to be negative
–¹ considered to be negative, but only one valid study was available to the Working Group
? considered to be equivocal or inconclusive (e.g. there were contradictory results from different laboratories; there were confounding exposures; the results were equivocal)

Summary table of genetic and related effects of trichloroacetic acid

Non-mammalian systems													Mammalian systems																									
Pro-karyotes		Lower eukaryotes				Plants			Insects				In vitro																		In vivo							
													Animal cells								Human cells										Animal						Humans				
D	G	D	R	G	A	D	G	C	R	G	C	A	D	G	S	M	C	A	T	I	D	G	S	M	C	A	T	I	D	G	S	M	C	DLA	D	S	M	C	A
–	–						+¹						–¹							+¹	–¹						–¹		?	?	?	?	+¹						

A, aneuploidy; C, chromosomal aberrations; D, DNA damage; DL, dominant lethal mutation; G, gene mutation; I, inhibition of intercellular communication; M, micronuclei; R, mitotic recombination and gene conversion; S, sister chromatid exchange; T, cell transformation

In completing the table, the following symbols indicate the consensus of the Working Group with regard to the results for each end-point:

+ considered to be positive for the specific end-point and level of biological complexity
+¹ considered to be positive, but only one valid study was available to the Working Group
– considered to be negative
–¹ considered to be negative, but only one valid study was available to the Working Group
? considered to be equivocal or inconclusive (e.g. there were contradictory results from different laboratories; there were confounding exposures; the results were equivocal)

Summary table of genetic and related effects of 1-chloro-2-methylpropene

Non-mammalian systems						Mammalian systems			
Prokaryotes	Lower eukaryotes			Plants	Insects	In vitro			In vivo
						Animal cells	Human cells	Animal	Humans
D G	D R G A			D R G C	R G C A	D G S M C A T I	D G S M C A T I	D G S M C DL A	D S M C A
+[1]						+[1] +[1] –[1]			

A, aneuploidy; C, chromosomal aberrations; D, DNA damage; DL, dominant lethal mutation; G, gene mutation; I, inhibition of intercellular communication; M, micronuclei; R, mitotic recombination and gene conversion; S, sister chromatid exchange; T, cell transformation

In completing the table, the following symbols indicate the consensus of the Working Group with regard to the results for each end-point:

+ considered to be positive for the specific end-point and level of biological complexity
+[1] considered to be positive, but only one valid study was available to the Working Group
– considered to be negative
–[1] considered to be negative, but only one valid study was available to the Working Group
? considered to be equivocal or inconclusive (e.g. there were contradictory results from different laboratories; there were confounding exposures; the results were equivocal)

APPENDIX 1

Summary table of genetic and related effects of 3-chloro-2-methylpropene

Non-mammalian systems																Mammalian systems																		
Pro-karyotes		Lower eukaryotes				Plants			Insects				In vitro															In vivo						
													Animal cells								Human cells								Animal							Humans					
D	G	D	R	G	A	D	G	C	R	G	C	A	D	G	S	M	C	A	T	I	D	G	S	M	C	A	T	I	D	G	S	M	C	DL	A	D	S	M	C	A
	+									+¹					+¹	+¹																	−¹							

A, aneuploidy; C, chromosomal aberrations; D, DNA damage; DL, dominant lethal mutation; G, gene mutation; I, inhibition of intercellular communication; M, micronuclei; R, mitotic recombination and gene conversion; S, sister chromatid exchange; T, cell transformation

In completing the table, the following symbols indicate the consensus of the Working Group with regard to the results for each end-point:

+ considered to be positive for the specific end-point and level of biological complexity
+¹ considered to be positive, but only one valid study was available to the Working Group
− considered to be negative
−¹ considered to be negative, but only one valid study was available to the Working Group
? considered to be equivocal or inconclusive (e.g. there were contradictory results from different laboratories; there were confounding exposures; the results were equivocal)

Summary table of genetic and related effects of acrolein

Non-mammalian systems													Mammalian systems																					
Pro-karyotes		Lower eukaryotes				Plants				Insects				In vitro													In vivo							
														Animal cells								Human cells					Animal						Humans
D	G	D	R	G	A	D	G	C	A	R	G	C	A	D	G	S	M	C	A	T	I	D	G	S	M	C	A	T	I	D	G	S	M	C	DL	A	D	S	M	C	A
+	+	–¹	–¹							+¹	–¹			+	?	?	?	?			–¹	?	+¹	+¹						–¹					–¹						

A, aneuploidy; C, chromosomal aberrations; D, DNA damage; DL, dominant lethal mutation; G, gene mutation; I, inhibition of intercellular communication; M, micronuclei; R, mitotic recombination and gene conversion; S, sister chromatid exchange; T, cell transformation

In completing the table, the following symbols indicate the consensus of the Working Group with regard to the results for each end-point:

+ considered to be positive for the specific end-point and level of biological complexity
+¹ considered to be positive, but only one valid study was available to the Working Group
– considered to be negative
–¹ considered to be negative, but only one valid study was available to the Working Group
? considered to be equivocal or inconclusive (e.g. there were contradictory results from different laboratories; there were confounding exposures; the results were equivocal)

APPENDIX 1

Summary table of genetic and related effects of crotonaldehyde

Non-mammalian systems				Mammalian systems			
Pro-karyotes	Lower eukaryotes	Plants	Insects	In vitro		In vivo	
				Animal cells	Human cells	Animal	Humans
D G	D R G A	D G C R	R G C A	D G S M C A T I	D G S M C A T I	D G S M C DL A	D G S M C A
+ +			+¹ +¹	–¹		+¹	

A, aneuploidy; C, chromosomal aberrations; D, DNA damage; DL, dominant lethal mutation; G, gene mutation; I, inhibition of intercellular communication; M, micronuclei; R, mitotic recombination and gene conversion; S, sister chromatid exchange; T, cell transformation

In completing the table, the following symbols indicate the consensus of the Working Group with regard to the results for each end-point:

+ considered to be positive for the specific end-point and level of biological complexity
+¹ considered to be positive, but only one valid study was available to the Working Group
– considered to be negative
–¹ considered to be negative, but only one valid study was available to the Working Group
? considered to be equivocal or inconclusive (e.g. there were contradictory results from different laboratories; there were confounding exposures; the results were equivocal)

Summary table of genetic and related effects of furan

Non-mammalian systems												Mammalian systems																					
Pro-karyotes		Lower eukaryotes				Plants				Insects		In vitro													In vivo								
												Animal cells							Human cells						Animal							Humans	
D	G	D	R	G	A	D	G	C	R	G	C	A	D	G	S	M	C	A	T	I	D	G	S	M	C	A	T	I	D	G	S	M	C	DL	A	D	S	M	C	A
−		−								−			−	+[1]	+[1]	−	+[1]												−[1]	−[1]		+[1]								

A, aneuploidy; C, chromosomal aberrations; D, DNA damage; DL, dominant lethal mutation; G, gene mutation; I, inhibition of intercellular communication; M, micronuclei; R, mitotic recombination and gene conversion; S, sister chromatid exchange; T, cell transformation

In completing the table, the following symbols indicate the consensus of the Working Group with regard to the results for each end-point:

+ considered to be positive for the specific end-point and level of biological complexity
+[1] considered to be positive, but only one valid study was available to the Working Group
− considered to be negative
−[1] considered to be negative, but only one valid study was available to the Working Group
? considered to be equivocal or inconclusive (e.g. there were contradictory results from different laboratories; there were confounding exposures; the results were equivocal)

APPENDIX 1

Summary table of genetic and related effects of furfural

Non-mammalian systems																		Mammalian systems																						
Prokaryotes		Lower eukaryotes				Plants			Insects				*In vitro*																		*In vivo*							
													Animal cells								Human cells									Animal							Humans				
D	G	D	R	G	A	D	G	C	R	G	C	A	D	G	S	M	C	A	T	I	D	G	S	M	C	A	T	I	D	G	S	M	C	DL	A	D	S	M	C	A
−¹	−								+¹	−¹			+¹	+¹	+	+					+¹										−¹	−¹								

A, aneuploidy; C, chromosomal aberrations; D, DNA damage; DL, dominant lethal mutation; G, gene mutation; I, inhibition of intercellular communication; M, micronuclei; R, mitotic recombination and gene conversion; S, sister chromatid exchange; T, cell transformation

In completing the table, the following symbols indicate the consensus of the Working Group with regard to the results for each end-point:

+ considered to be positive for the specific end-point and level of biological complexity
+¹ considered to be positive, but only one valid study was available to the Working Group
− considered to be negative
−¹ considered to be negative, but only one valid study was available to the Working Group
? considered to be equivocal or inconclusive (e.g. there were contradictory results from different laboratories; there were confounding exposures; the results were equivocal)

Summary table of genetic and related effects of benzofuran

Non-mammalian systems				Mammalian systems			
Pro-karyotes	Lower eukaryotes	Plants	Insects	In vitro			In vivo
				Animal cells	Human cells	Animal	Humans
D G	D R G A	D G C A	D R G C A	D G S M C A T I	D G S M C A T I	D G S M C DL A	D G S M C A
–				+¹ +¹ +¹			

A, aneuploidy; C, chromosomal aberrations; D, DNA damage; DL, dominant lethal mutation; G, gene mutation; I, inhibition of intercellular communication; M, micronuclei; R, mitotic recombination and gene conversion; S, sister chromatid exchange; T, cell transformation

In completing the table, the following symbols indicate the consensus of the Working Group with regard to the results for each end-point:
+ considered to be positive for the specific end-point and level of biological complexity
+¹ considered to be positive, but only one valid study was available to the Working Group
– considered to be negative
–¹ considered to be negative, but only one valid study was available to the Working Group
? considered to be equivocal or inconclusive (e.g. there were contradictory results from different laboratories; there were confounding exposures; the results were equivocal)

APPENDIX 1

Summary table of genetic and related effects of vinyl acetate

Non-mammalian systems				Mammalian systems			
Pro-karyotes	Lower eukaryotes	Plants	Insects	In vitro			In vivo
				Animal cells	Human cells		Animal	Humans
D G	D R G A	D G C	R G C A	D G S M C A T I	D G S M C A T I		D G S M C DL A	D S M C A
+¹ −				+¹ +¹	+¹ + +¹ +		+¹ +¹	

A, aneuploidy; C, chromosomal aberrations; D, DNA damage; DL, dominant lethal mutation; G, gene mutation; I, inhibition of intercellular communication; M, micronuclei; R, mitotic recombination and gene conversion; S, sister chromatid exchange; T, cell transformation

In completing the table, the following symbols indicate the consensus of the Working Group with regard to the results for each end-point:

+ considered to be positive for the specific end-point and level of biological complexity
+¹ considered to be positive, but only one valid study was available to the Working Group
− considered to be negative
−¹ considered to be negative, but only one valid study was available to the Working Group
? considered to be equivocal or inconclusive (e.g. there were contradictory results from different laboratories; there were confounding exposures; the results were equivocal)

APPENDIX 2

ACTIVITY PROFILES FOR
GENETIC AND RELATED EFFECTS

APPENDIX 2

ACTIVITY PROFILES FOR GENETIC AND RELATED EFFECTS

Methods

The x-axis of the activity profile (Waters *et al.*, 1987, 1988) represents the bioassays in phylogenetic sequence by end-point, and the values on the y-axis represent the logarithmically transformed lowest effective doses (LED) and highest ineffective doses (HID) tested. The term 'dose', as used in this report, does not take into consideration length of treatment or exposure and may therefore be considered synonymous with concentration. In practice, the concentrations used in all the in-vitro tests were converted to µg/ml, and those for in-vivo tests were expressed as mg/kg bw. Because dose units are plotted on a log scale, differences in the relative molecular masses of compounds do not, in most cases, greatly influence comparisons of their activity profiles. Conventions for dose conversions are given below.

Profile-line height (the magnitude of each bar) is a function of the LED or HID, which is associated with the characteristics of each individual test system—such as population size, cell-cycle kinetics and metabolic competence. Thus, the detection limit of each test system is different, and, across a given activity profile, responses will vary substantially. No attempt is made to adjust or relate responses in one test system to those of another.

Line heights are derived as follows: for negative test results, the highest dose tested without appreciable toxicity is defined as the HID. If there was evidence of extreme toxicity, the next highest dose is used. A single dose tested with a negative result is considered to be equivalent to the HID. Similarly, for positive results, the LED is recorded. If the original data were analysed statistically by the author, the dose recorded is that at which the response was significant ($p < 0.05$). If the available data were not analysed statistically, the dose required to produce an effect is estimated as follows: when a dose-related positive response is observed with two or more doses, the lower of the doses is taken as the LED; a single dose resulting in a positive response is considered to be equivalent to the LED.

In order to accommodate both the wide range of doses encountered and positive and negative responses on a continuous scale, doses are transformed logarithmically, so that effective (LED) and ineffective (HID) doses are represented by positive and negative numbers, respectively. The response, or logarithmic dose unit (LDUij), for a given test system i and chemical j is represented by the expressions

LDU$_{ij}$ = $-\log_{10}$ (dose), for HID values; LDU ≤ 0

and (1)

LDU$_{ij}$ = $-\log_{10}$ (dose × 10^{-5}), for LED values; LDU ≥ 0.

These simple relationships define a dose range of 0 to -5 logarithmic units for ineffective doses (1–100 000 µg/ml or mg/kg bw) and 0 to +8 logarithmic units for effective doses (100 000–0.001 µg/ml or mg/kg bw). A scale illustrating the LDU values is shown in Figure 1. Negative responses at doses less than 1 µg/ml (mg/kg bw) are set equal to 1. Effectively, an LED value $\geq$ 100 000 or an HID value $\leq$ 1 produces an LDU = 0; no quantitative information is gained from such extreme values. The dotted lines at the levels of log dose units 1 and -1 define a 'zone of uncertainty' in which positive results are reported at such high doses (between 10 000 and 100 000 mg/ml or mg/kg bw) or negative results are reported at such low doses (1 to 10 mg/ml or mg/kg bw) as to call into question the adequacy of the test.

Fig. 1. Scale of log dose units used on the y-axis of activity profiles

Positive (µg/ml or mg/kg bw)		Log dose units
0.001		8 ----
0.01		7 --
0.1		6 --
1.0		5 --
10		4 --
100		3 --
1000		2 --
10 000		1 --
100 000	1	0 ----
	10	-1 --
	100	-2 --
	1000	-3 --
	10 000	-4 --
	100 000	-5 ----

Negative
(µg/ml or mg/kg bw)

In practice, an activity profile is computer generated. A data entry programme is used to store abstracted data from published reports. A sequential file (in ASCII) is created for each compound, and a record within that file consists of the name and Chemical Abstracts Service number of the compound, a three-letter code for the test system (see below), the qualitative test result (with and without an exogenous metabolic system), dose (LED or HID), citation number and additional source information. An abbreviated citation for each publication is stored in a segment of a record accessing both the test data file and the citation file. During processing of the data file, an average of the logarithmic values of the data subset is calculated, and the length of the profile line represents this average value. All dose values are plotted for each profile line, regardless of whether results are positive or negative. Results obtained in the absence of an exogenous metabolic system are indicated by a bar (−), and results obtained in the presence of an exogenous metabolic system are indicated by an upward-directed arrow (↑). When all results for a given assay are either positive or negative, the mean of the LDU values is plotted as a solid line; when conflicting data are reported for the same assay (i.e. both positive and negative results), the majority data are shown by a solid line and the minority data by a dashed line (drawn to the extreme conflicting response). In the few cases in which the numbers of positive and negative results are equal, the solid line is drawn in the positive direction and the maximal negative response is indicated with a dashed line. Profile lines are identified by three-letter code words representing the commonly used tests. Code words for most of the test systems in current use in genetic toxicology were defined for the US Environmental Protection Agency's GENE-TOX Program (Waters, 1979; Waters & Auletta, 1981). For *IARC Monographs* Supplement 6, Volume 44 and subsequent volumes, including this publication, codes were redefined in a manner that should facilitate inclusion of additional tests. Naming conventions are described below.

Data listings are presented in the text and include end-point and test codes, a short test code definition, results, either with (M) or without (NM) an exogenous activation system, the associated LED or HID value and a short citation. Test codes are organized phylogenetically and by end-point from left to right across each activity profile and from top to bottom of the corresponding data listing. End-points are defined as follows: A, aneuploidy; C, chromosomal aberrations; D, DNA damage; F, assays of body fluids; G, gene mutation; H, host-mediated assays; I, inhibition of intercellular communication; M, micronuclei; P, sperm morphology; R, mitotic recombination or gene conversion; S, sister chromatid exchange; and T, cell transformation.

Dose conversions for activity profiles

Doses are converted to µg/ml for in-vitro tests and to mg/kg bw per day for in-vivo experiments.

1. In-vitro test systems
    (a) Weight/volume converts directly to µg/ml.
    (b) Molar (M) concentration × molecular weight = mg/ml = 10^3 mg/ml; mM concentration × molecular weight = µg/ml.
    (c) Soluble solids expressed as % concentration are assumed to be in units of mass per volume (i.e. 1% = 0.01 g/ml = 10 000 µg/ml; also, 1 ppm = 1 µg/ml).
    (d) Liquids and gases expressed as % concentration are assumed to be given in units of volume per volume. Liquids are converted to weight per volume using the density (D) of the solution (D = g/ml). Gases are converted from volume to mass using the ideal gas law, PV = nRT. For exposure at 20–37 °C at standard atmospheric pressure, 1% (v/v) = 0.4 µg/ml × molecular weight of the gas. Also, 1 ppm (v/v) = 4×10^5 µg/ml × molecular weight.

(e) In microbial plate tests, it is usual for the doses to be reported as weight/plate, whereas concentrations are required to enter data on the activity profile chart. While remaining cognisant of the errors involved in the process, it is assumed that a 2-ml volume of top agar is delivered to each plate and that the test substance remains in solution within it; concentrations are derived from the reported weight/plate values by dividing by this arbitrary volume. For spot tests, a 1-ml volume is used in the calculation.

(f) Conversion of particulate concentrations given in $\mu g/cm^2$ is based on the area (A) of the dish and the volume of medium per dish; i.e. for a 100-mm dish: $A = \pi R^2 = \pi \times (5 \text{ cm})^2 = 78.5 \text{ cm}^2$. If the volume of medium is 10 ml, then 78.5 cm^2 = 10 ml and 1 cm^2 = 0.13 ml.

2. In-vitro systems using in-vivo activation

For the body fluid-urine (BF-) test, the concentration used is the dose (in mg/kg bw) of the compound administered to test animals or patients.

3. In-vivo test systems

(a) Doses are converted to mg/kg bw per day of exposure, assuming 100% absorption. Standard values are used for each sex and species of rodent, including body weight and average intake per day, as reported by Gold et al. (1984). For example, in a test using male mice fed 50 ppm of the agent in the diet, the standard food intake per day is 12% of body weight, and the conversion is dose = 50 ppm × 12% = 6 mg/kg bw per day.

Standard values used for humans are: weight—males, 70 kg; females, 55 kg; surface area, 1.7 m^2; inhalation rate, 20 L/min for light work, 30 L/min for mild exercise.

(b) When reported, the dose at the target site is used. For example, doses given in studies of lymphocytes of humans exposed *in vivo* are the measured blood concentrations in µg/ml.

Codes for test systems

For specific nonmammalian test systems, the first two letters of the three-symbol code word define the test organism (e.g. SA- for *Salmonella typhimurium*, EC- for *Escherichia coli*). If the species is not known, the convention used is -S-. The third symbol may be used to define the tester strain (e.g. SA8 for *S. typhimurium* TA1538, ECW for *E. coli* WP2*uvr*A). When strain designation is not indicated, the third letter is used to define the specific genetic end-point under investigation (e.g. --D for differential toxicity, --F for forward mutation, --G for gene conversion or genetic crossing-over, --N for aneuploidy, --R for reverse mutation, --U for unscheduled DNA synthesis). The third letter may also be used to define the general end-point under investigation when a more complete definition is not possible or relevant (e.g. --M for mutation, --C for chromosomal aberration). For mammalian test systems, the first letter of the three-letter code word defines the genetic end-point under investigation: A-- for aneuploidy, B-- for binding, C-- for chromosomal aberration, D-- for DNA strand breaks, G-- for gene mutation, I-- for inhibition of intercellular communication, M-- for micronucleus formation, R-- for DNA repair, S-- for sister chromatid exchange, T-- for cell transformation and U-- for unscheduled DNA synthesis.

For animal (i.e. non-human) test systems *in vitro*, when the cell type is not specified, the code letters -IA are used. For such assays *in vivo*, when the animal species is not specified, the code letters -VA are used. Commonly used animal species are identified by the third letter (e.g. --C for Chinese hamster, --M for mouse, --R for rat, --S for Syrian hamster).

For test systems using human cells *in vitro*, when the cell type is not specified, the code letters -IH are used. For assays on humans *in vivo*, when the cell type is not specified, the code letters -VH are used. Otherwise, the second letter specifies the cell type under investigation (e.g. -BH for bone marrow, -LH for lymphocytes).

Some other specific coding conventions used for mammalian systems are as follows: BF- for body fluids, HM- for host-mediated, --L for leukocytes or lymphocytes *in vitro* (-AL, animals; -HL, humans), -L- for leukocytes *in vivo* (-LA, animals; -LH, humans), --T for transformed cells.

Note that these are examples of major conventions used to define the assay code words. The alphabetized listing of codes must be examined to confirm a specific code word. As might be expected from the limitation to three symbols, some codes do not fit the naming conventions precisely. In a few cases, test systems are defined by first-letter code words, for example: MST, mouse spot test; SLP, mouse specific locus mutation, postspermatogonia; SLO, mouse specific locus mutation, other stages; DLM, dominant lethal mutation in mice; DLR, dominant lethal mutation in rats; MHT, mouse heritable translocation.

The genetic activity profiles and listings were prepared in collaboration with Environmental Health Research and Testing Inc. (EHRT) under contract to the United States Environmental Protection Agency; EHRT also determined the doses used. The references cited in each genetic activity profile listing can be found in the list of references in the appropriate monograph.

References

Garrett, N.E., Stack, H.F., Gross, M.R. & Waters, M.D. (1984) An analysis of the spectra of genetic activity produced by known or suspected human carcinogens. *Mutat. Res.*, **134**, 89–111

Gold, L.S., Sawyer, C.B., Magaw, R., Backman, G.M., de Veciana, M., Levinson, R., Hooper, N.K., Havender, W.R., Bernstein, L., Peto, R., Pike, M.C. & Ames, B.N. (1984) A carcinogenic potency database of the standardized results of animal bioassays. *Environ. Health Perspect.*, **58**, 9–319

Waters, M.D. (1979) *The GENE-TOX program*. In: Hsie, A.W., O'Neill, J.P. & McElheny, V.K., eds, *Mammalian Cell Mutagenesis: The Maturation of Test Systems* (Banbury Report 2), Cold Spring Harbor, NY, CSH Press, pp. 449–467

Waters, M.D. & Auletta, A. (1981) The GENE-TOX program: genetic activity evaluation. *J. chem. Inf. comput. Sci.*, **21**, 35–38

Waters, M.D., Stack, H.F., Brady, A.L., Lohman, P.H.M., Haroun, L. & Vainio, H. (1987) Appendix 1: Activity profiles for genetic and related tests. In: *IARC Monographs on the Evaluation of the Carcinogenic Risk of Chemicals to Humans*, Suppl. 6, *Genetic and Related Effects: An Updating of Selected* IARC Monographs *from Volumes 1 to 42*, Lyon, IARC, pp. 687–696

Waters, M.D., Stack, H.F., Brady, A.L., Lohman, P.H.M., Haroun, L. & Vainio, H. (1988) Use of computerized data listings and activity profiles of genetic and related effects in the review of 195 compounds. *Mutat. Res.*, **205**, 295–312

APPENDIX 2

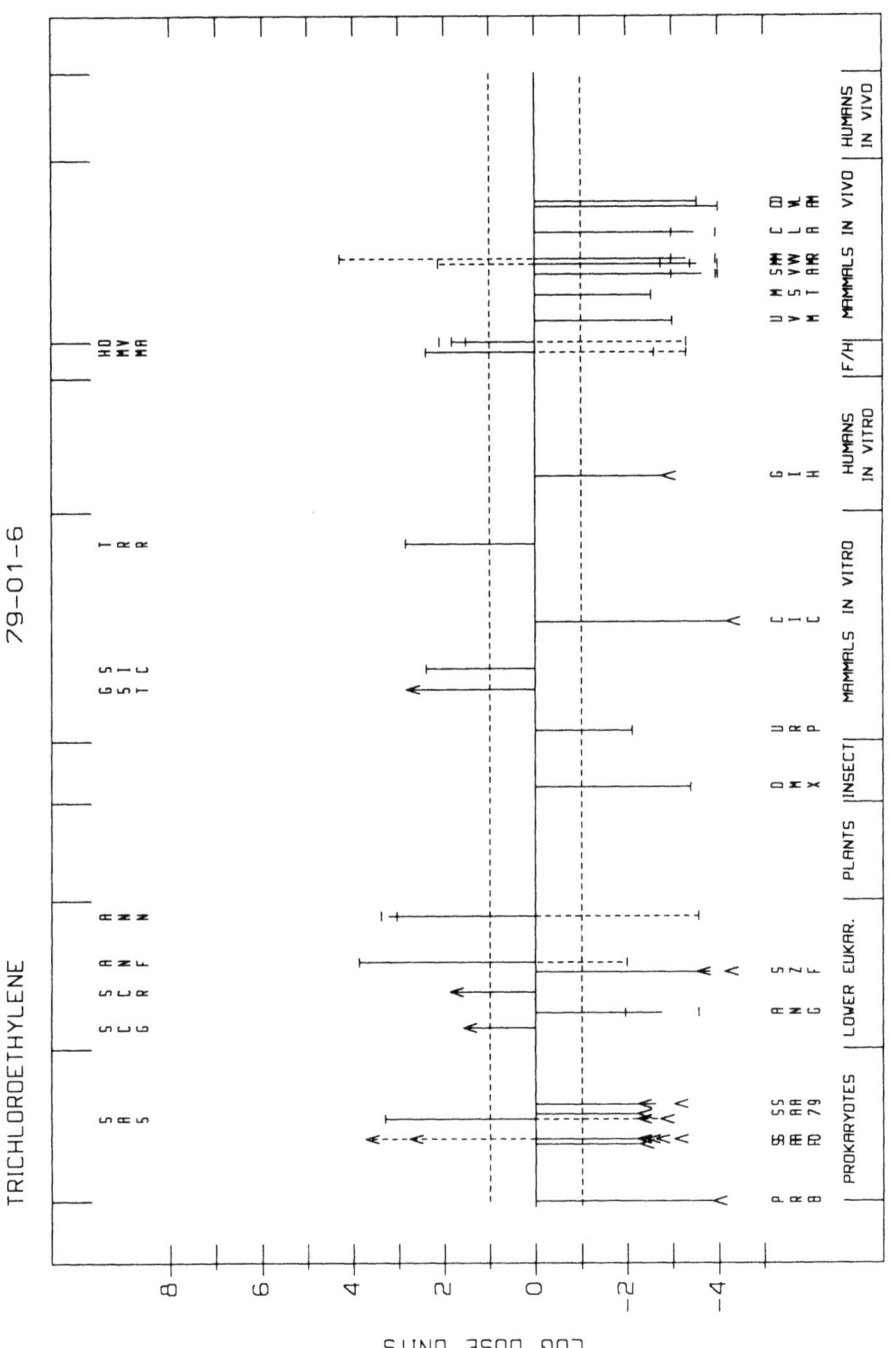

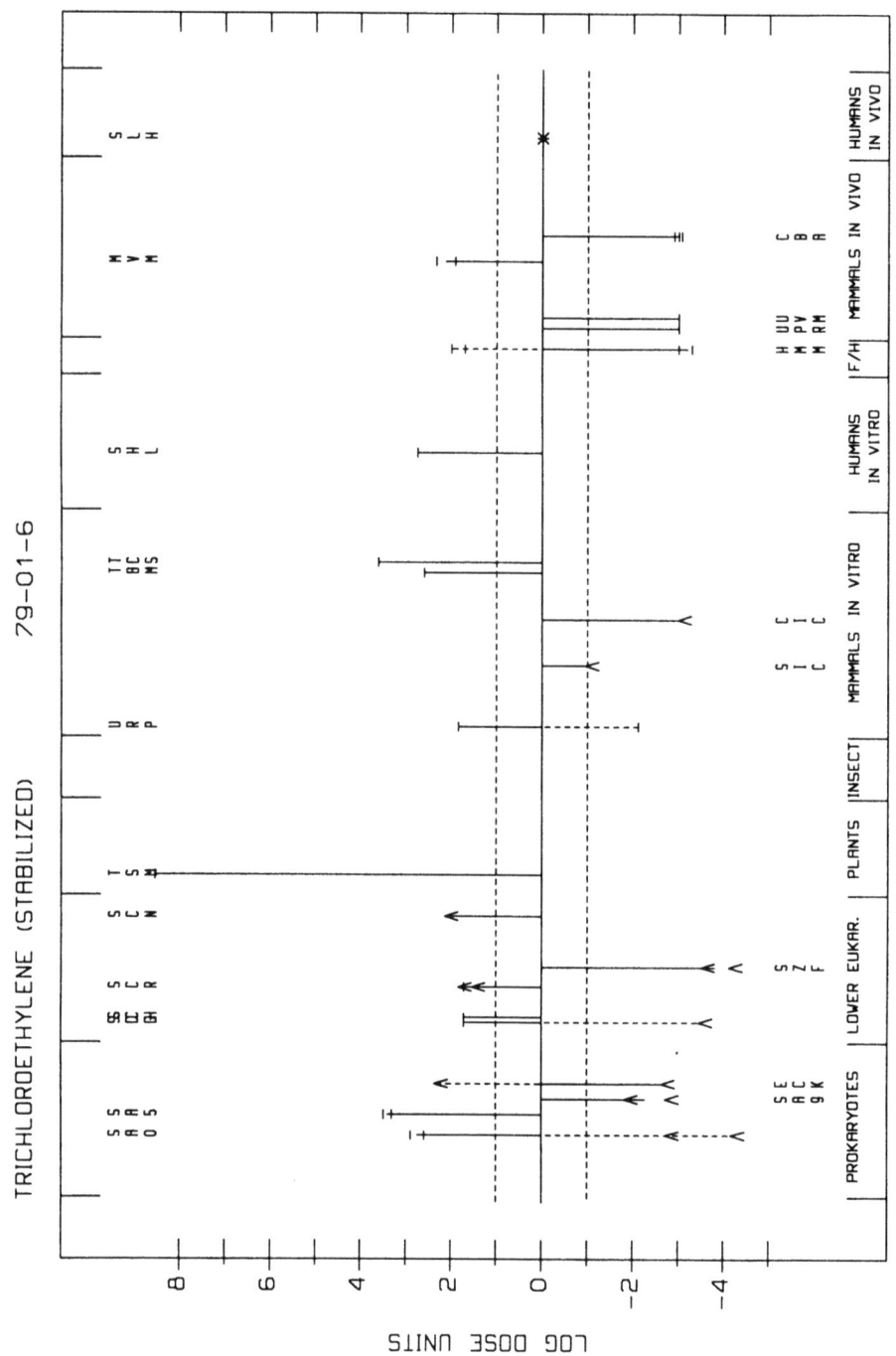

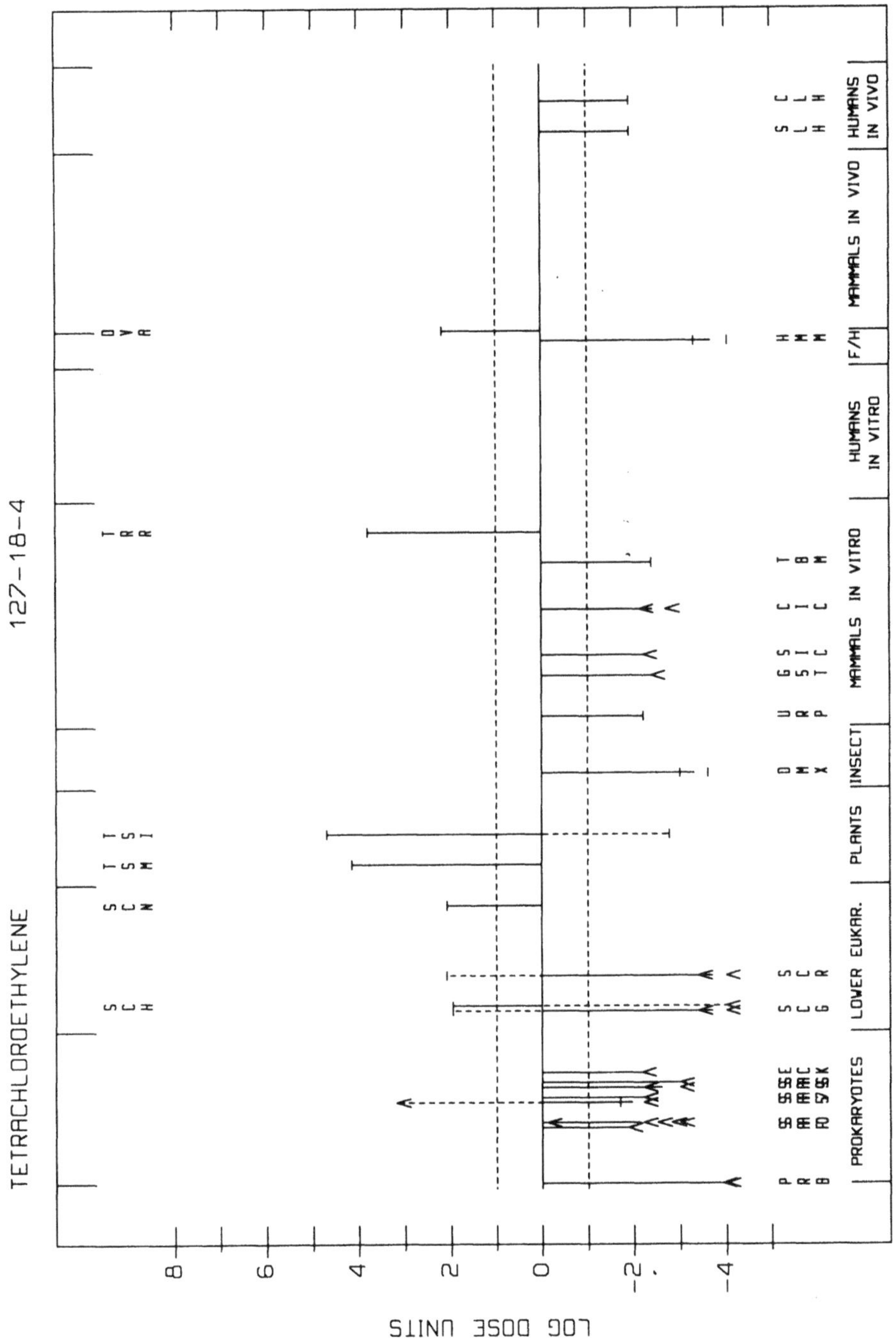
TETRACHLOROETHYLENE
127-18-4
LOG DOSE UNITS
8
6
4
2
0
-2
-4
PROKARYOTES
LOWER EUKAR.
PLANTS
INSECT
MAMMALS IN VITRO
HUMANS IN VITRO
F/H
MAMMALS IN VIVO
HUMANS IN VIVO

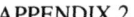

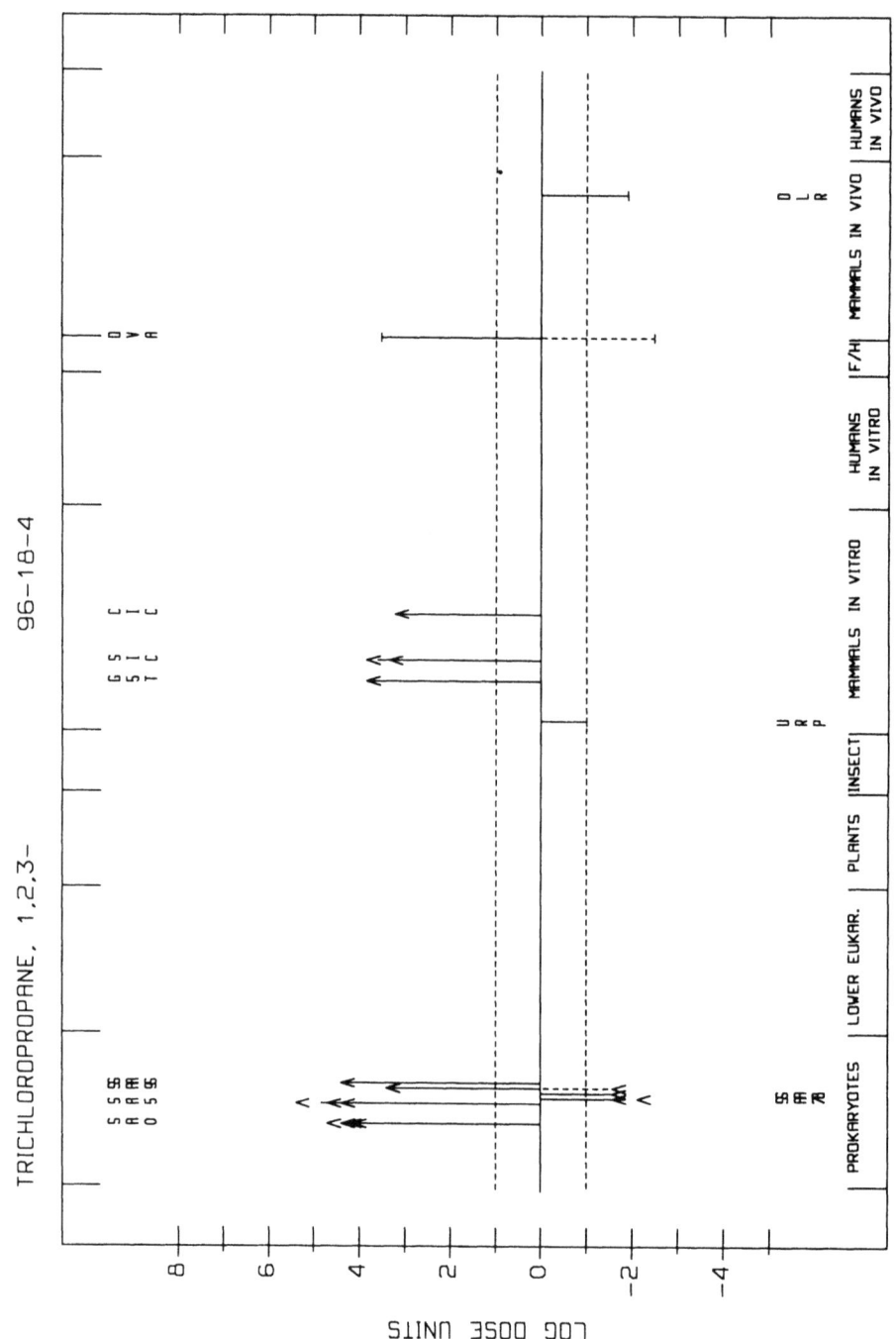

APPENDIX 2

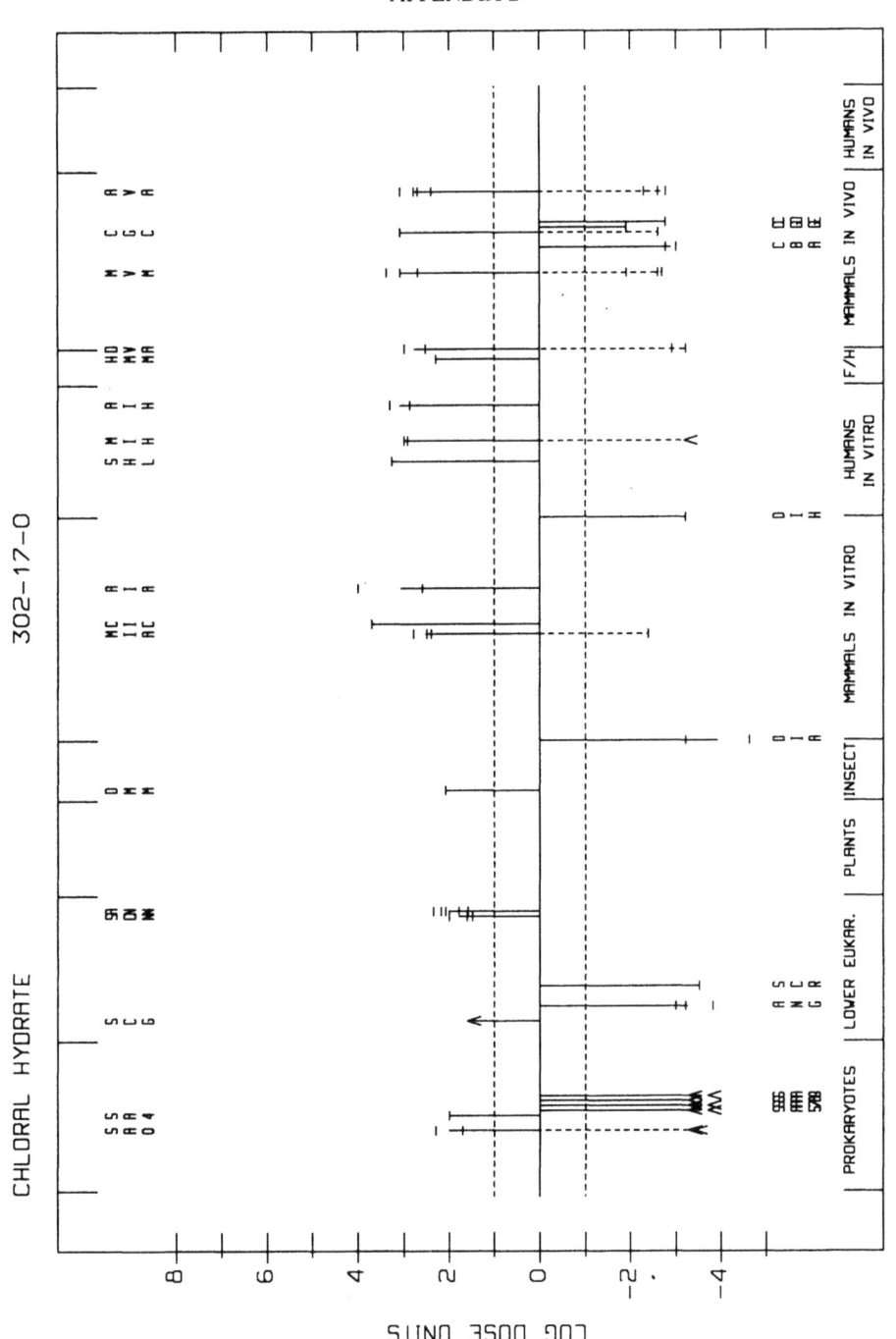

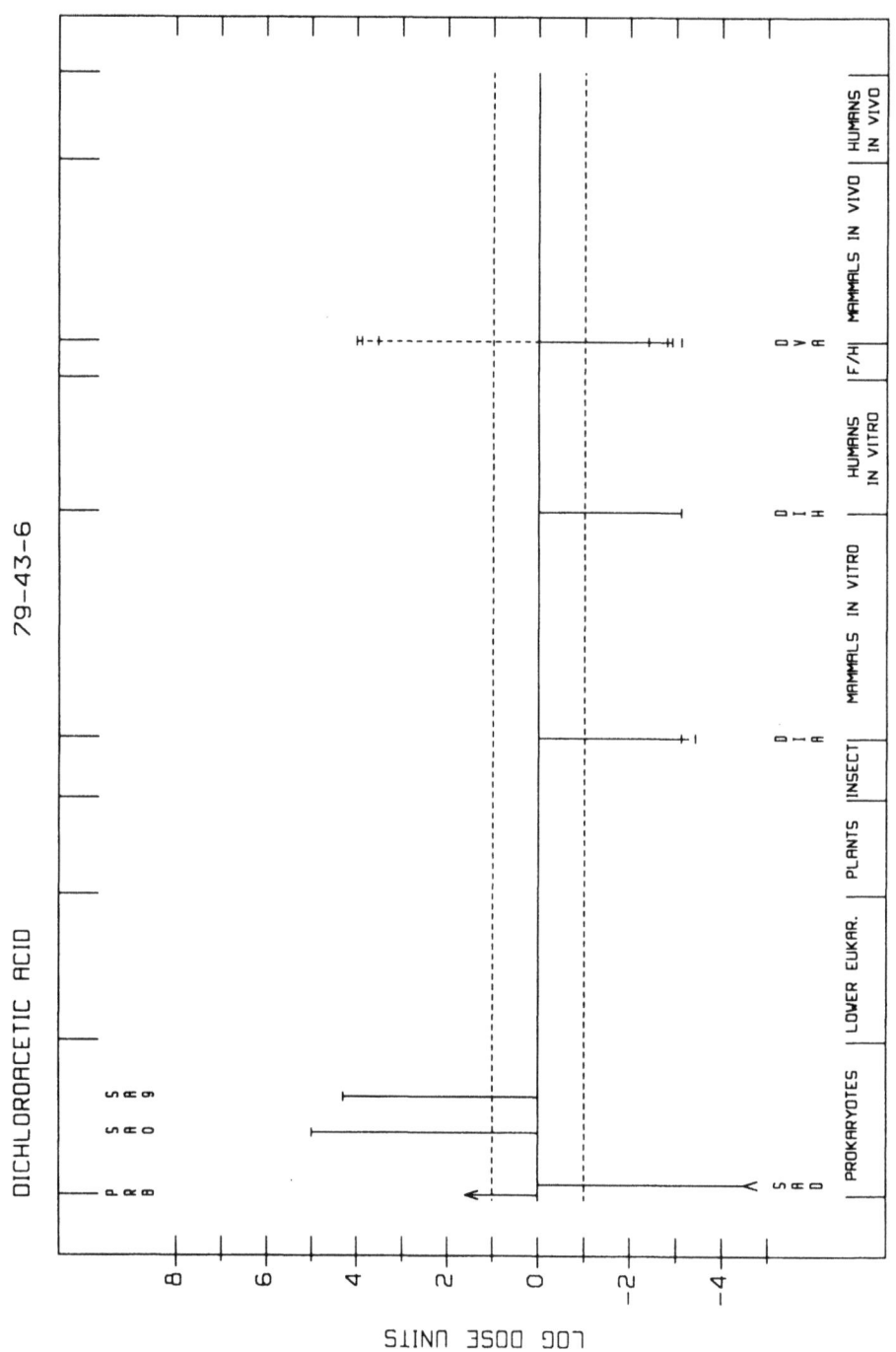

APPENDIX 2

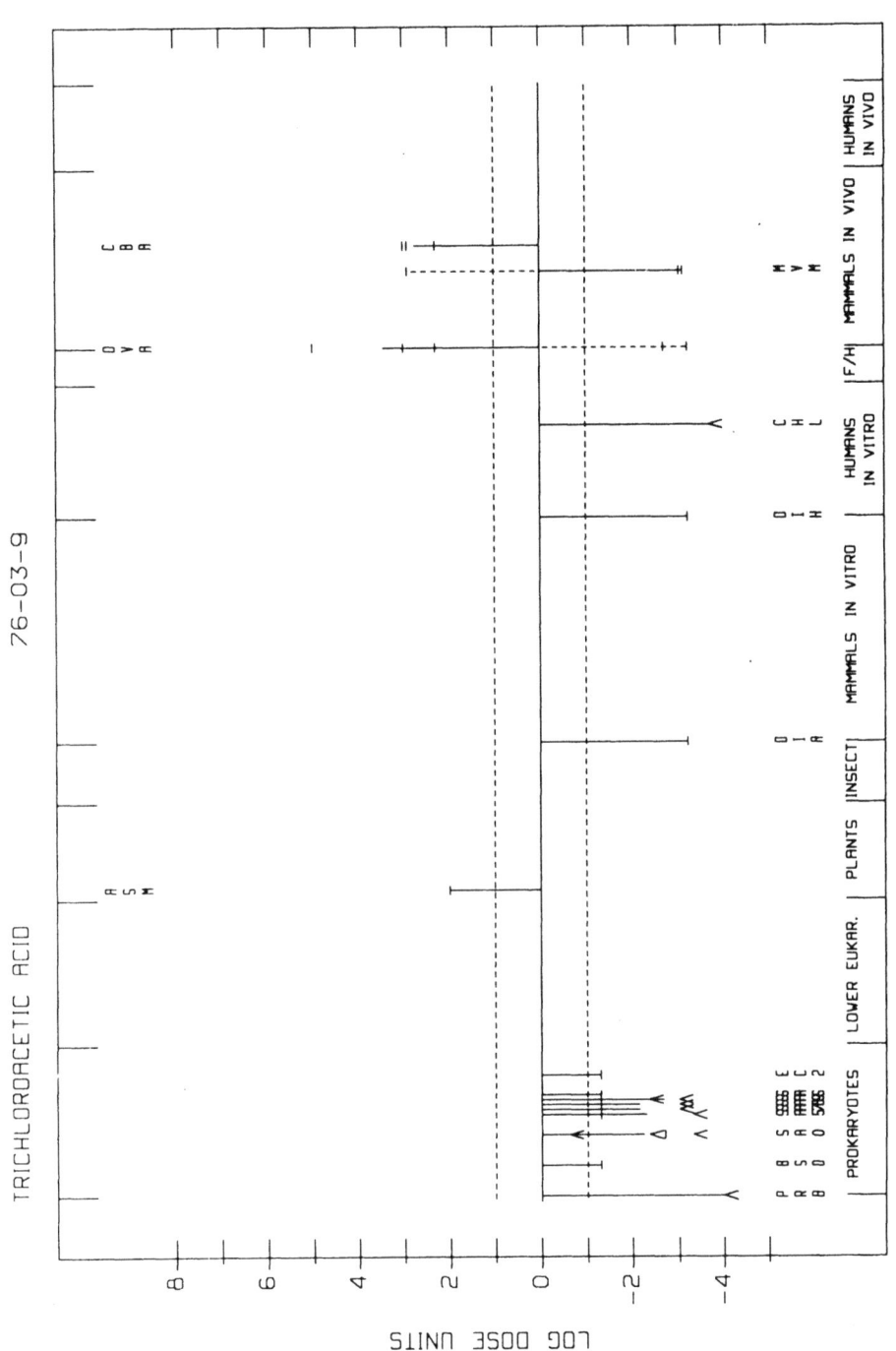

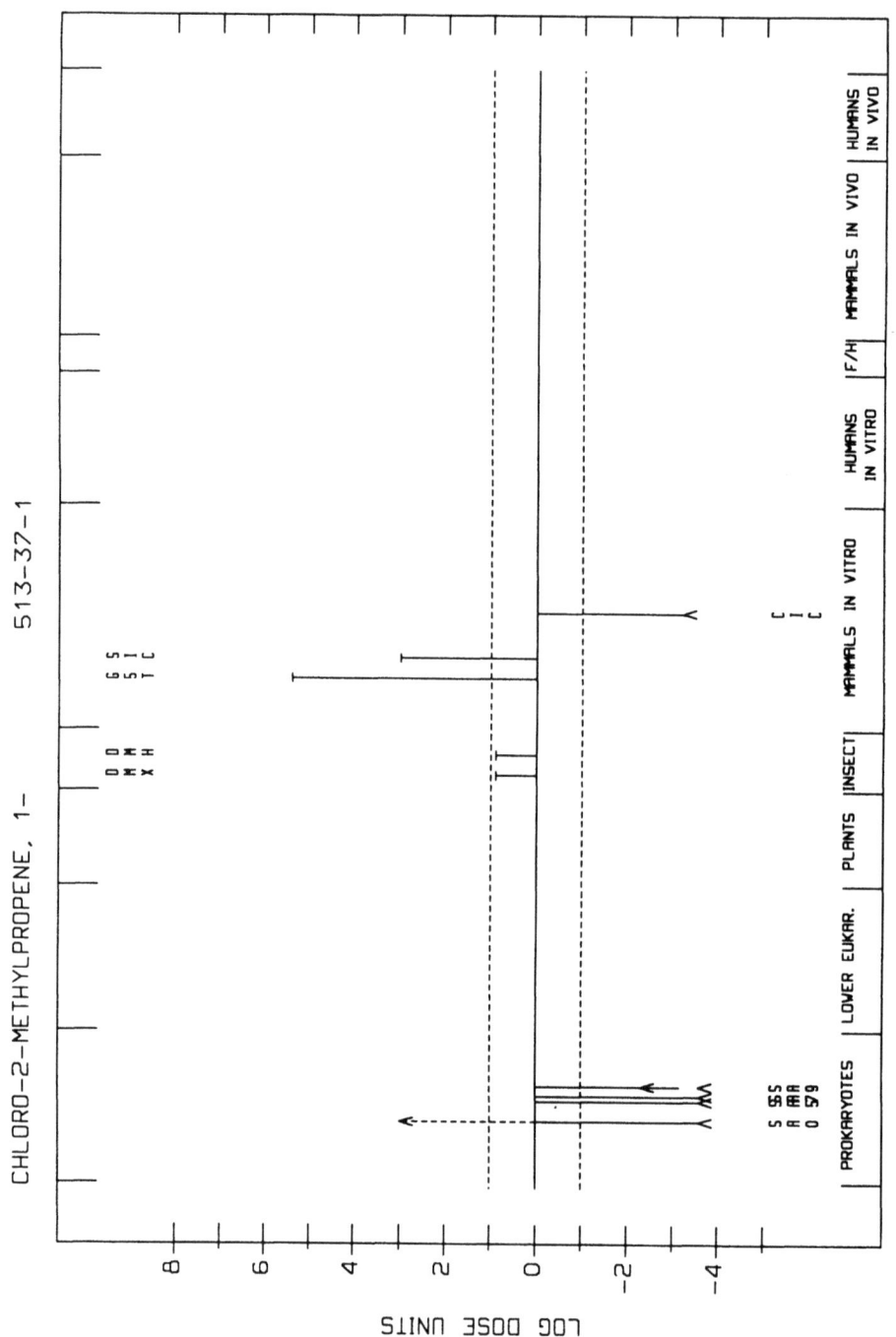

APPENDIX 2

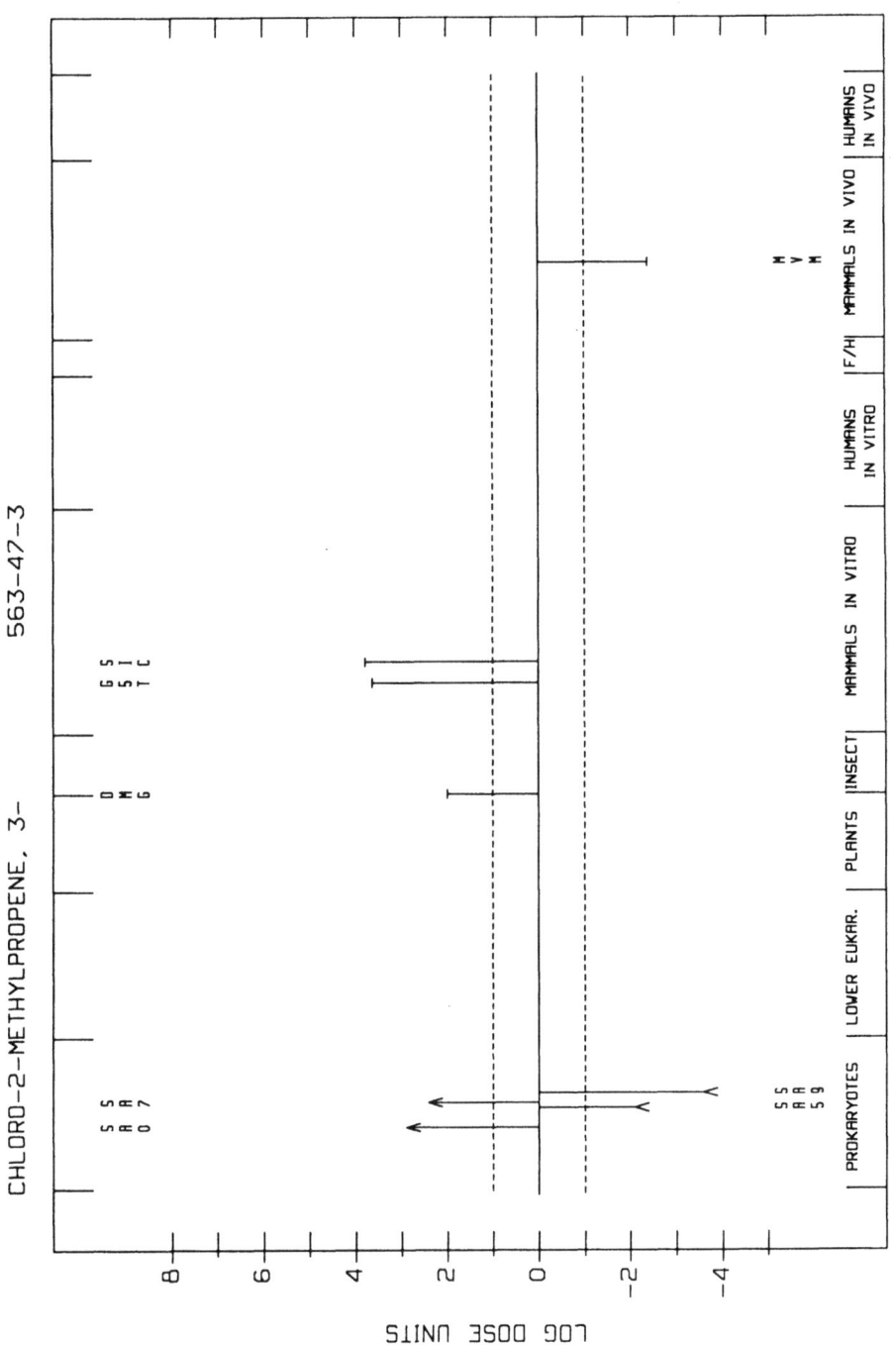

514 IARC MONOGRAPHS VOLUME 63

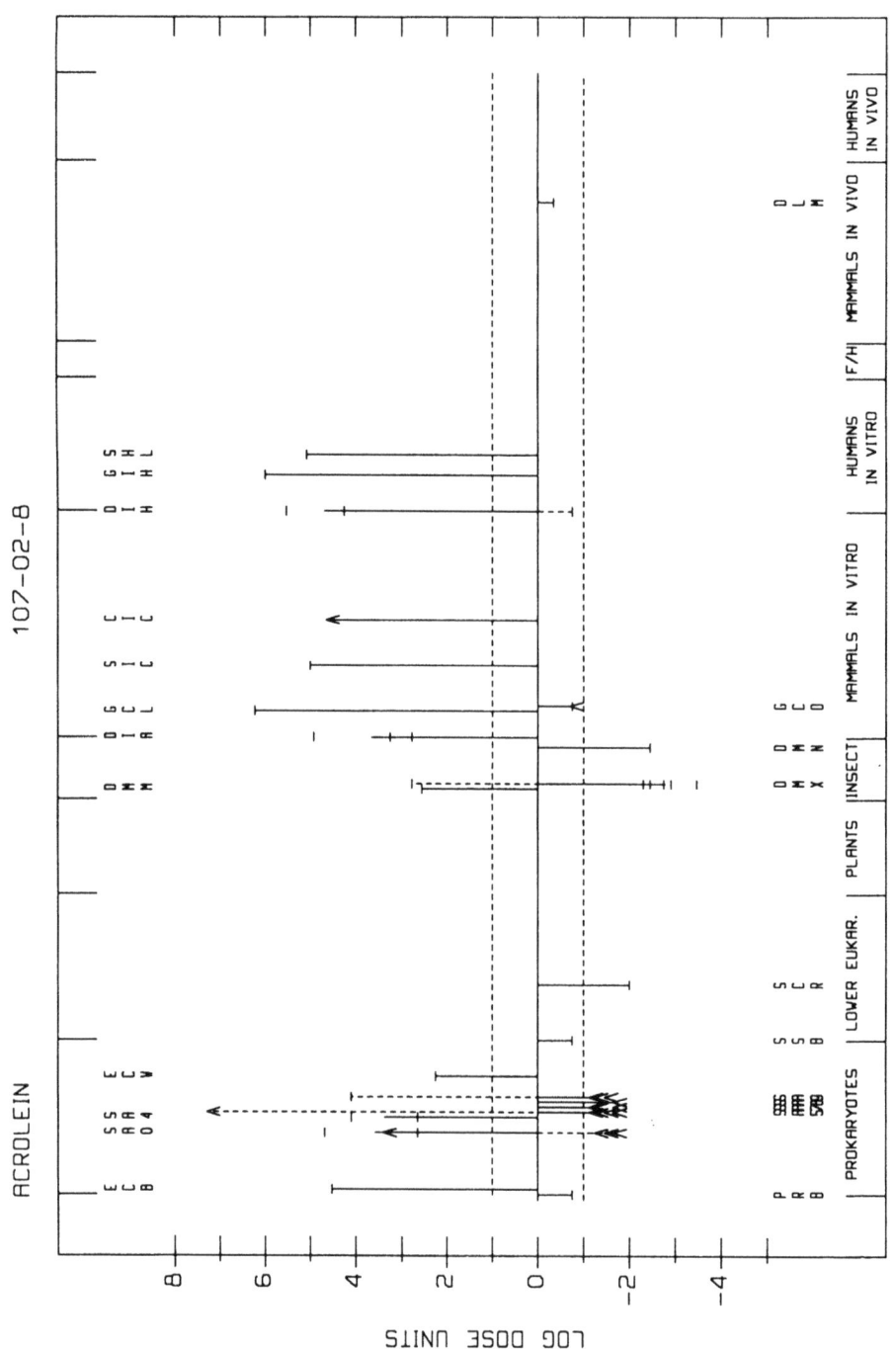

APPENDIX 2

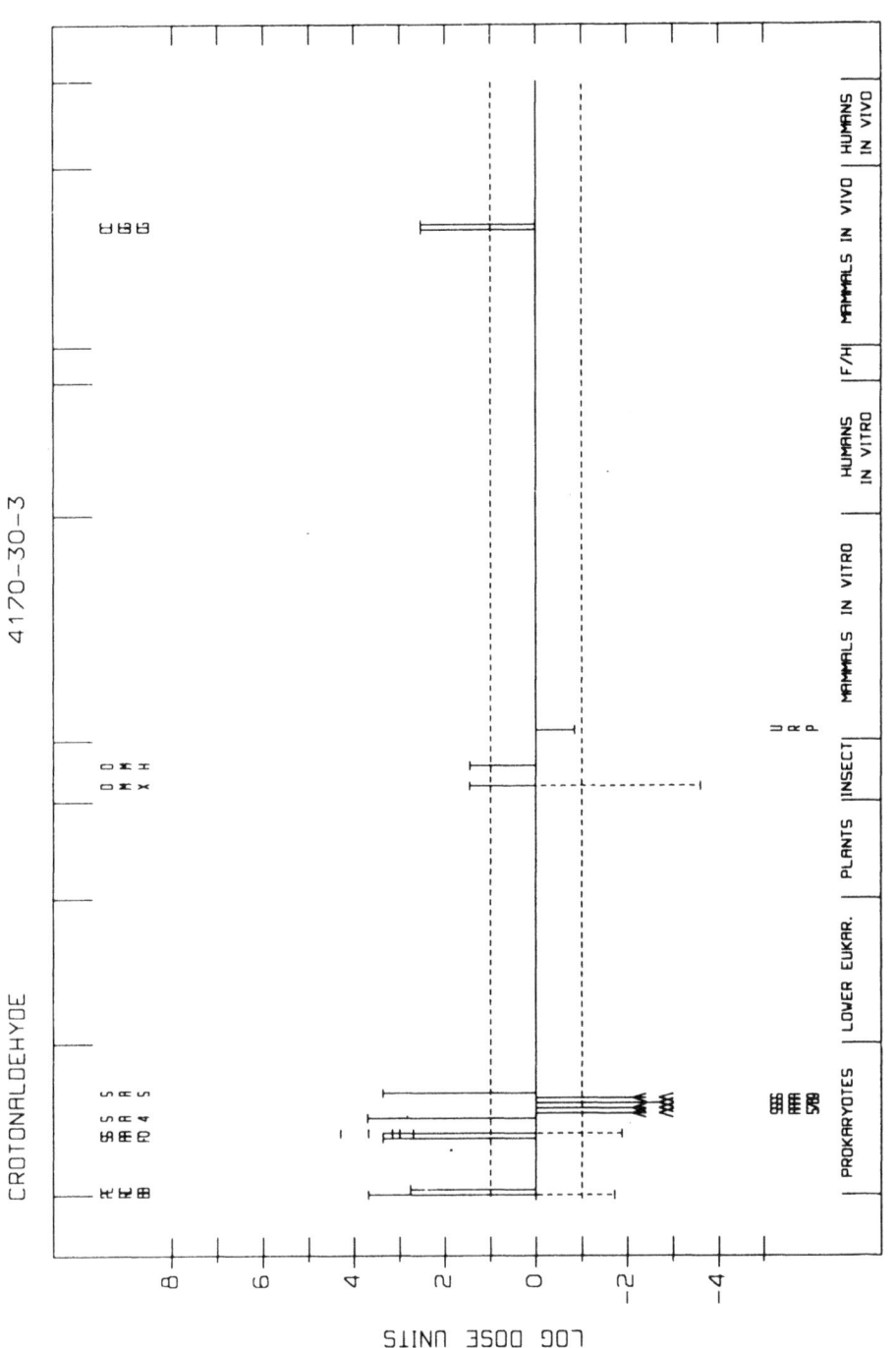

516

IARC MONOGRAPHS VOLUME 63

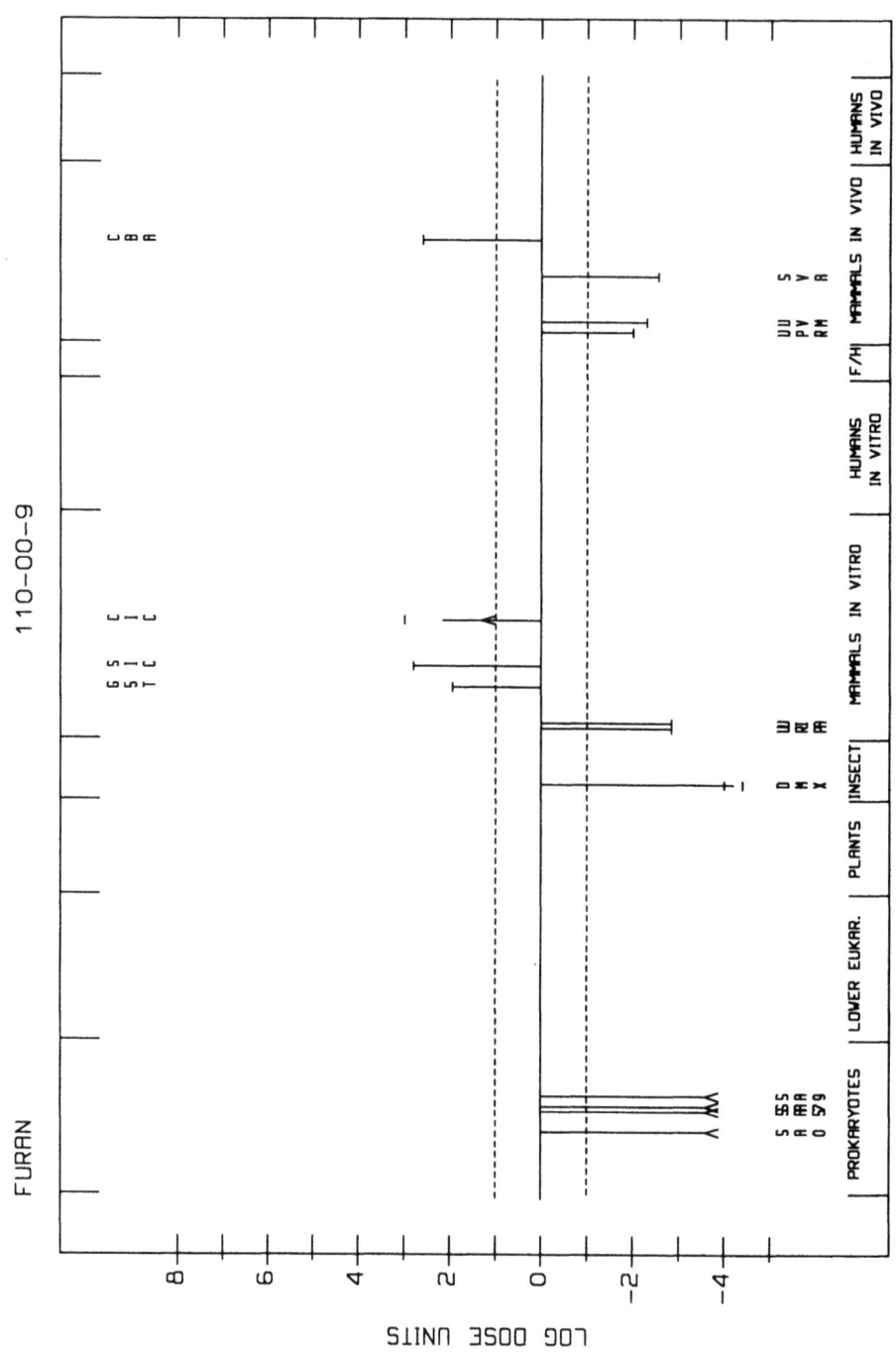

APPENDIX 2

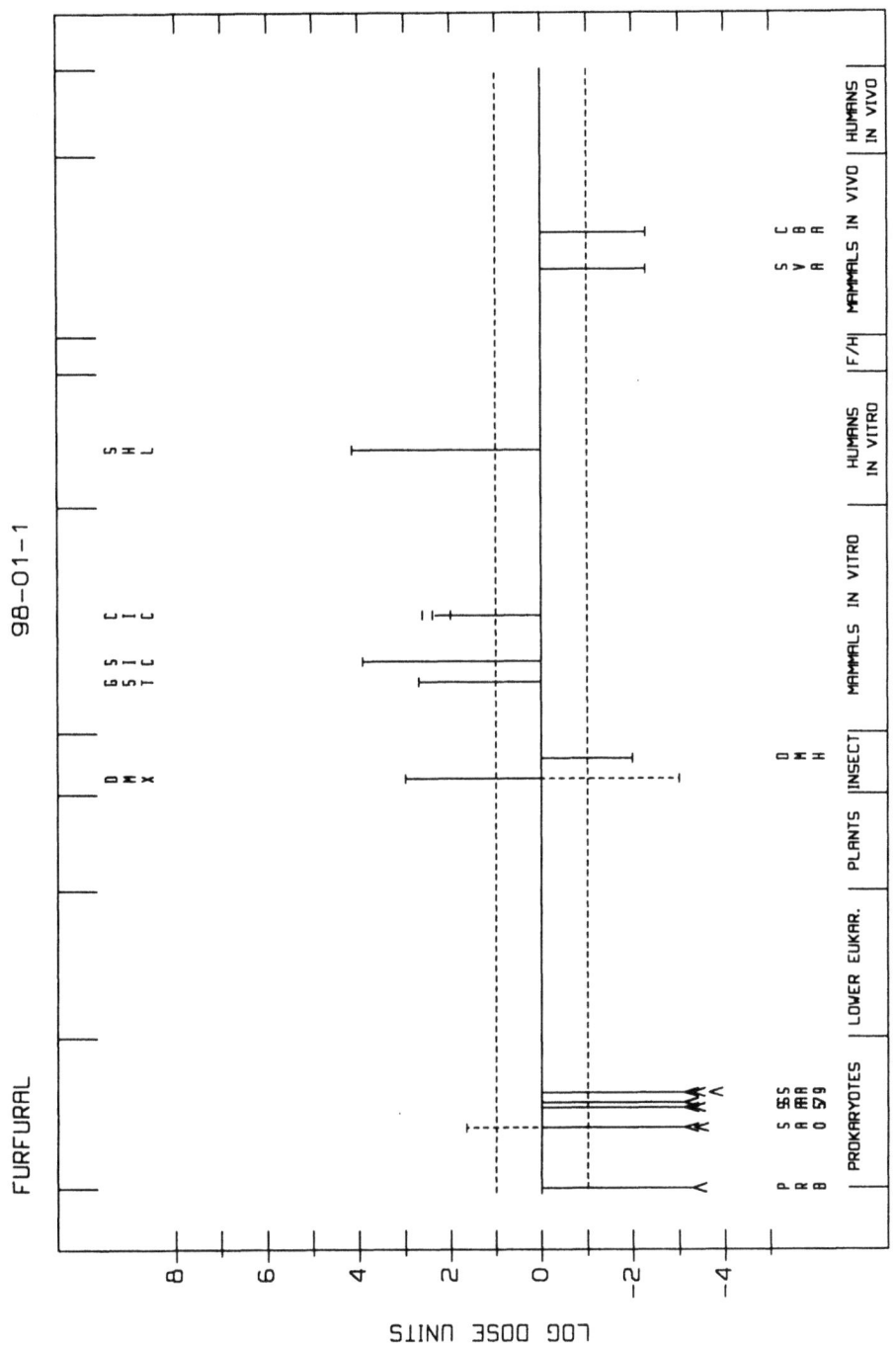

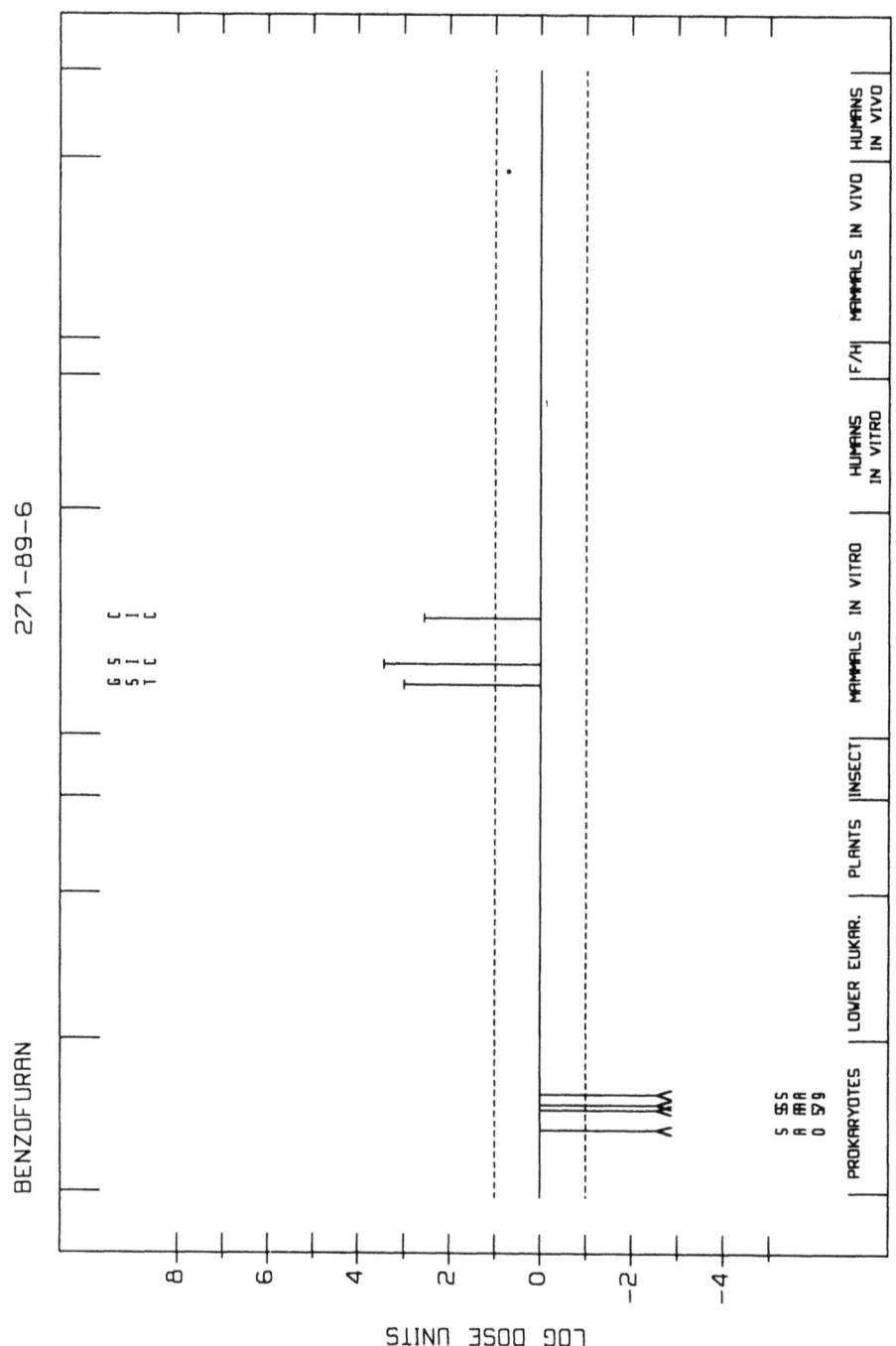

APPENDIX 2

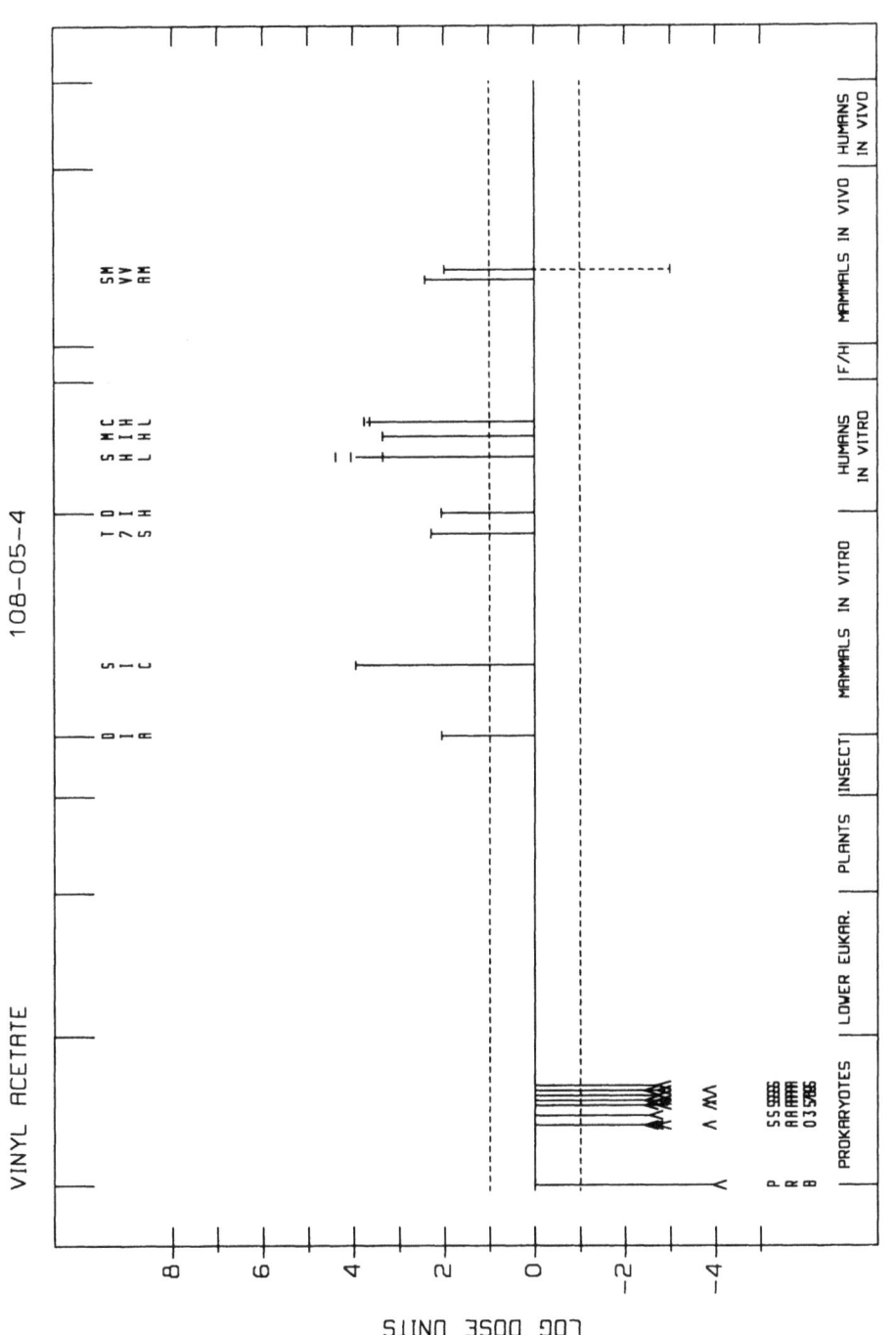

SUPPLEMENTARY CORRIGENDA TO VOLUMES 1–62

Volume 57 p. 358 *Under* Prokaryotes, *move + from* D *to* G
 p. 367 *Move – from* Plants, G *to* Insects, G
 p. 370 *Under* Human cells, *delete + under* S; *move +*[1] *under* C *to* Animal cells, C

CUMULATIVE CROSS INDEX TO *IARC MONOGRAPHS ON THE EVALUATION OF CARCINOGENIC RISKS TO HUMANS*

The volume, page and year of publication are given. References to corrigenda are given in parentheses.

A

A-α-C	*40*, 245 (1986); *Suppl. 7*, 56 (1987)
Acetaldehyde	*36*, 101 (1985) (*corr. 42*, 263); *Suppl. 7*, 77 (1987)
Acetaldehyde formylmethylhydrazone (*see* Gyromitrin)	
Acetamide	*7*, 197 (1974); *Suppl. 7*, 389 (1987)
Acetaminophen (*see* Paracetamol)	
Acridine orange	*16*, 145 (1978); *Suppl. 7*, 56 (1987)
Acriflavinium chloride	*13*, 31 (1977); *Suppl. 7*, 56 (1987)
Acrolein	*19*, 479 (1979); *36*, 133 (1985); *Suppl. 7*, 78 (1987); *63*, 337 (1995)
Acrylamide	*39*, 41 (1986); *Suppl. 7*, 56 (1987); *60*, 389 (1994)
Acrylic acid	*19*, 47 (1979); *Suppl. 7*, 56 (1987)
Acrylic fibres	*19*, 86 (1979); *Suppl. 7*, 56 (1987)
Acrylonitrile	*19*, 73 (1979); *Suppl. 7*, 79 (1987)
Acrylonitrile-butadiene-styrene copolymers	*19*, 91 (1979); *Suppl. 7*, 56 (1987)
Actinolite (*see* Asbestos)	
Actinomycins	*10*, 29 (1976) (*corr. 42*, 255); *Suppl. 7*, 80 (1987)
Adriamycin	*10*, 43 (1976); *Suppl. 7*, 82 (1987)
AF-2	*31*, 47 (1983); *Suppl. 7*, 56 (1987)
Aflatoxins	*1*, 145 (1972) (*corr. 42*, 251); *10*, 51 (1976); *Suppl. 7*, 83 (1987); *56*, 245 (1993)
Aflatoxin B_1 (*see* Aflatoxins)	
Aflatoxin B_2 (*see* Aflatoxins)	
Aflatoxin G_1 (*see* Aflatoxins)	
Aflatoxin G_2 (*see* Aflatoxins)	
Aflatoxin M_1 (*see* Aflatoxins)	
Agaritine	*31*, 63 (1983); *Suppl. 7*, 56 (1987)
Alcohol drinking	*44* (1988)
Aldicarb	*53*, 93 (1991)
Aldrin	*5*, 25 (1974); *Suppl. 7*, 88 (1987)
Allyl chloride	*36*, 39 (1985); *Suppl. 7*, 56 (1987)
Allyl isothiocyanate	*36*, 55 (1985); *Suppl. 7*, 56 (1987)
Allyl isovalerate	*36*, 69 (1985); *Suppl. 7*, 56 (1987)
Aluminium production	*34*, 37 (1984); *Suppl. 7*, 89 (1987)

Amaranth	8, 41 (1975); *Suppl. 7*, 56 (1987)
5-Aminoacenaphthene	16, 243 (1978); *Suppl. 7*, 56 (1987)
2-Aminoanthraquinone	27, 191 (1982); *Suppl. 7*, 56 (1987)
para-Aminoazobenzene	8, 53 (1975); *Suppl. 7*, 390 (1987)
ortho-Aminoazotoluene	8, 61 (1975) (*corr. 42*, 254); *Suppl. 7*, 56 (1987)
para-Aminobenzoic acid	16, 249 (1978); *Suppl. 7*, 56 (1987)
4-Aminobiphenyl	1, 74 (1972) (*corr. 42*, 251); *Suppl. 7*, 91 (1987)
2-Amino-3,4-dimethylimidazo[4,5-*f*]quinoline (*see* MeIQ)	
2-Amino-3,8-dimethylimidazo[4,5-*f*]quinoxaline (*see* MeIQx)	
3-Amino-1,4-dimethyl-5*H*-pyrido[4,3-*b*]indole (*see* Trp-P-1)	
2-Aminodipyrido[1,2-*a*:3′,2′-*d*]imidazole (*see* Glu-P-2)	
1-Amino-2-methylanthraquinone	27, 199 (1982); *Suppl. 7*, 57 (1987)
2-Amino-3-methylimidazo[4,5-*f*]quinoline (*see* IQ)	
2-Amino-6-methyldipyrido[1,2-*a*:3′,2′-*d*]imidazole (*see* Glu-P-1)	
2-Amino-1-methyl-6-phenylimidazo[4,5-*b*]pyridine (*see* PhIP)	
2-Amino-3-methyl-9*H*-pyrido[2,3-*b*]indole (*see* MeA-α-C)	
3-Amino-1-methyl-5*H*-pyrido[4,3-*b*]indole (*see* Trp-P-2)	
2-Amino-5-(5-nitro-2-furyl)-1,3,4-thiadiazole	7, 143 (1974); *Suppl. 7*, 57 (1987)
4-Amino-2-nitrophenol	16, 43 (1978); *Suppl. 7*, 57 (1987)
2-Amino-4-nitrophenol	57, 167 (1993)
2-Amino-5-nitrophenol	57, 177 (1993)
2-Amino-5-nitrothiazole	31, 71 (1983); *Suppl. 7*, 57 (1987)
2-Amino-9*H*-pyrido[2,3-*b*]indole (*see* A-α-C)	
11-Aminoundecanoic acid	39, 239 (1986); *Suppl. 7*, 57 (1987)
Amitrole	7, 31 (1974); 41, 293 (1986) (*corr. 52*, 513; *Suppl. 7*, 92 (1987)
Ammonium potassium selenide (*see* Selenium and selenium compounds)	
Amorphous silica (*see also* Silica)	42, 39 (1987); *Suppl. 7*, 341 (1987)
Amosite (*see* Asbestos)	
Ampicillin	50, 153 (1990)
Anabolic steroids (*see* Androgenic (anabolic) steroids)	
Anaesthetics, volatile	11, 285 (1976); *Suppl. 7*, 93 (1987)
Analgesic mixtures containing phenacetin (*see also* Phenacetin)	*Suppl. 7*, 310 (1987)
Androgenic (anabolic) steroids	*Suppl. 7*, 96 (1987)
Angelicin and some synthetic derivatives (*see also* Angelicins)	40, 291 (1986)
Angelicin plus ultraviolet radiation (*see also* Angelicin and some synthetic derivatives)	*Suppl. 7*, 57 (1987)
Angelicins	*Suppl. 7*, 57 (1987)
Aniline	4, 27 (1974) (*corr. 42*, 252); 27, 39 (1982); *Suppl. 7*, 99 (1987)
ortho-Anisidine	27, 63 (1982); *Suppl. 7*, 57 (1987)
para-Anisidine	27, 65 (1982); *Suppl. 7*, 57 (1987)
Anthanthrene	32, 95 (1983); *Suppl. 7*, 57 (1987)
Anthophyllite (*see* Asbestos)	
Anthracene	32, 105 (1983); *Suppl. 7*, 57 (1987)
Anthranilic acid	16, 265 (1978); *Suppl. 7*, 57 (1987)
Antimony trioxide	47, 291 (1989)
Antimony trisulfide	47, 291 (1989)
ANTU (*see* 1-Naphthylthiourea)	

Apholate	9, 31 (1975); *Suppl. 7*, 57 (1987)
Aramite®	5, 39 (1974); *Suppl. 7*, 57 (1987)
Areca nut (*see* Betel quid)	
Arsanilic acid (*see* Arsenic and arsenic compounds)	
Arsenic and arsenic compounds	1, 41 (1972); 2, 48 (1973); 23, 39 (1980); *Suppl. 7*, 100 (1987)
Arsenic pentoxide (*see* Arsenic and arsenic compounds)	
Arsenic sulfide (*see* Arsenic and arsenic compounds)	
Arsenic trioxide (*see* Arsenic and arsenic compounds)	
Arsine (*see* Arsenic and arsenic compounds)	
Asbestos	2, 17 (1973) (*corr. 42*, 252); 14 (1977) (*corr. 42*, 256); *Suppl. 7*, 106 (1987) (*corr. 45*, 283)
Atrazine	53, 441 (1991)
Attapulgite	42, 159 (1987); *Suppl. 7*, 117 (1987)
Auramine (technical-grade)	1, 69 (1972) (*corr. 42*, 251); *Suppl. 7*, 118 (1987)
Auramine, manufacture of (*see also* Auramine, technical-grade)	*Suppl. 7*, 118 (1987)
Aurothioglucose	13, 39 (1977); *Suppl. 7*, 57 (1987)
Azacitidine	26, 37 (1981); *Suppl. 7*, 57 (1987); 50, 47 (1990)
5-Azacytidine (*see* Azacitidine)	
Azaserine	10, 73 (1976) (*corr. 42*, 255); *Suppl. 7*, 57 (1987)
Azathioprine	26, 47 (1981); *Suppl. 7*, 119 (1987)
Aziridine	9, 37 (1975); *Suppl. 7*, 58 (1987)
2-(1-Aziridinyl)ethanol	9, 47 (1975); *Suppl. 7*, 58 (1987)
Aziridyl benzoquinone	9, 51 (1975); *Suppl. 7*, 58 (1987)
Azobenzene	8, 75 (1975); *Suppl. 7*, 58 (1987)

B

Barium chromate (*see* Chromium and chromium compounds)	
Basic chromic sulfate (*see* Chromium and chromium compounds)	
BCNU (*see* Bischloroethyl nitrosourea)	
Benz[*a*]acridine	32, 123 (1983); *Suppl. 7*, 58 (1987)
Benz[*c*]acridine	3, 241 (1973); 32, 129 (1983); *Suppl. 7*, 58 (1987)
Benzal chloride (*see also* α-Chlorinated toluenes)	29, 65 (1982); *Suppl. 7*, 148 (1987)
Benz[*a*]anthracene	3, 45 (1973); 32, 135 (1983); *Suppl. 7*, 58 (1987)
Benzene	7, 203 (1974) (*corr. 42*, 254); 29, 93, 391 (1982); *Suppl. 7*, 120 (1987)
Benzidine	1, 80 (1972); 29, 149, 391 (1982); *Suppl. 7*, 123 (1987)
Benzidine-based dyes	*Suppl. 7*, 125 (1987)
Benzo[*b*]fluoranthene	3, 69 (1973); 32, 147 (1983); *Suppl. 7*, 58 (1987)
Benzo[*j*]fluoranthene	3, 82 (1973); 32, 155 (1983); *Suppl. 7*, 58 (1987)

Benzo[k]fluoranthene	32, 163 (1983); Suppl. 7, 58 (1987)
Benzo[ghi]fluoranthene	32, 171 (1983); Suppl. 7, 58 (1987)
Benzo[a]fluorene	32, 177 (1983); Suppl. 7, 58 (1987)
Benzo[b]fluorene	32, 183 (1983); Suppl. 7, 58 (1987)
Benzo[c]fluorene	32, 189 (1983); Suppl. 7, 58 (1987)
Benzofuran	63, 431 (1995)
Benzo[ghi]perylene	32, 195 (1983); Suppl. 7, 58 (1987)
Benzo[c]phenanthrene	32, 205 (1983); Suppl. 7, 58 (1987)
Benzo[a]pyrene	3, 91 (1973); 32, 211 (1983); Suppl. 7, 58 (1987)
Benzo[e]pyrene	3, 137 (1973); 32, 225 (1983); Suppl. 7, 58 (1987)
para-Benzoquinone dioxime	29, 185 (1982); Suppl. 7, 58 (1987)
Benzotrichloride (see also α-Chlorinated toluenes)	29, 73 (1982); Suppl. 7, 148 (1987)
Benzoyl chloride	29, 83 (1982) (corr. 42, 261); Suppl. 7, 126 (1987)
Benzoyl peroxide	36, 267 (1985); Suppl. 7, 58 (1987)
Benzyl acetate	40, 109 (1986); Suppl. 7, 58 (1987)
Benzyl chloride (see also α-Chlorinated toluenes)	11, 217 (1976) (corr. 42, 256); 29, 49 (1982); Suppl. 7, 148 (1987)
Benzyl violet 4B	16, 153 (1978); Suppl. 7, 58 (1987)
Bertrandite (see Beryllium and beryllium compounds)	
Beryllium and beryllium compounds	1, 17 (1972); 23, 143 (1980) (corr. 42, 260); Suppl. 7, 127 (1987); 58, 41 (1993)
Beryllium acetate (see Beryllium and beryllium compounds)	
Beryllium acetate, basic (see Beryllium and beryllium compounds)	
Beryllium-aluminium alloy (see Beryllium and beryllium compounds)	
Beryllium carbonate (see Beryllium and beryllium compounds)	
Beryllium chloride (see Beryllium and beryllium compounds)	
Beryllium-copper alloy (see Beryllium and beryllium compounds)	
Beryllium-copper-cobalt alloy (see Beryllium and beryllium compounds)	
Beryllium fluoride (see Beryllium and beryllium compounds)	
Beryllium hydroxide (see Beryllium and beryllium compounds)	
Beryllium-nickel alloy (see Beryllium and beryllium compounds)	
Beryllium oxide (see Beryllium and beryllium compounds)	
Beryllium phosphate (see Beryllium and beryllium compounds)	
Beryllium silicate (see Beryllium and beryllium compounds)	
Beryllium sulfate (see Beryllium and beryllium compounds)	
Beryl ore (see Beryllium and beryllium compounds)	
Betel quid	37, 141 (1985); Suppl. 7, 128 (1987)
Betel-quid chewing (see Betel quid)	
BHA (see Butylated hydroxyanisole)	
BHT (see Butylated hydroxytoluene)	
Bis(1-aziridinyl)morpholinophosphine sulfide	9, 55 (1975); Suppl. 7, 58 (1987)
Bis(2-chloroethyl)ether	9, 117 (1975); Suppl. 7, 58 (1987)
N,N-Bis(2-chloroethyl)-2-naphthylamine	4, 119 (1974) (corr. 42, 253); Suppl. 7, 130 (1987)
Bischloroethyl nitrosourea (see also Chloroethyl nitrosoureas)	26, 79 (1981); Suppl. 7, 150 (1987)
1,2-Bis(chloromethoxy)ethane	15, 31 (1977); Suppl. 7, 58 (1987)
1,4-Bis(chloromethoxymethyl)benzene	15, 37 (1977); Suppl. 7, 58 (1987)

Bis(chloromethyl)ether	4, 231 (1974) (corr. 42, 253); Suppl. 7, 131 (1987)
Bis(2-chloro-1-methylethyl)ether	41, 149 (1986); Suppl. 7, 59 (1987)
Bis(2,3-epoxycyclopentyl)ether	47, 231 (1989)
Bisphenol A diglycidyl ether (see Glycidyl ethers)	
Bisulfites (see Sulfur dioxide and some sulfites, bisulfites and metabisulfites)	
Bitumens	35, 39 (1985); Suppl. 7, 133 (1987)
Bleomycins	26, 97 (1981); Suppl. 7, 134 (1987)
Blue VRS	16, 163 (1978); Suppl. 7, 59 (1987)
Boot and shoe manufacture and repair	25, 249 (1981); Suppl. 7, 232 (1987)
Bracken fern	40, 47 (1986); Suppl. 7, 135 (1987)
Brilliant Blue FCF, disodium salt	16, 171 (1978) (corr. 42, 257); Suppl. 7, 59 (1987)
Bromochloroacetonitrile (see Halogenated acetonitriles)	
Bromodichloromethane	52, 179 (1991)
Bromoethane	52, 299 (1991)
Bromoform	52, 213 (1991)
1,3-Butadiene	39, 155 (1986) (corr. 42, 264 Suppl. 7, 136 (1987); 54, 237 (1992)
1,4-Butanediol dimethanesulfonate	4, 247 (1974); Suppl. 7, 137 (1987)
n-Butyl acrylate	39, 67 (1986); Suppl. 7, 59 (1987)
Butylated hydroxyanisole	40, 123 (1986); Suppl. 7, 59 (1987)
Butylated hydroxytoluene	40, 161 (1986); Suppl. 7, 59 (1987)
Butyl benzyl phthalate	29, 193 (1982) (corr. 42, 261); Suppl. 7, 59 (1987)
β-Butyrolactone	11, 225 (1976); Suppl. 7, 59 (1987)
γ-Butyrolactone	11, 231 (1976); Suppl. 7, 59 (1987)

C

Cabinet-making (see Furniture and cabinet-making)	
Cadmium acetate (see Cadmium and cadmium compounds)	
Cadmium and cadmium compounds	2, 74 (1973); 11, 39 (1976) (corr. 42, 255); Suppl. 7, 139 (1987); 58, 119 (1993)
Cadmium chloride (see Cadmium and cadmium compounds)	
Cadmium oxide (see Cadmium and cadmium compounds)	
Cadmium sulfate (see Cadmium and cadmium compounds)	
Cadmium sulfide (see Cadmium and cadmium compounds)	
Caffeic acid	56, 115 (1993)
Caffeine	51, 291 (1991)
Calcium arsenate (see Arsenic and arsenic compounds)	
Calcium chromate (see Chromium and chromium compounds)	
Calcium cyclamate (see Cyclamates)	
Calcium saccharin (see Saccharin)	
Cantharidin	10, 79 (1976); Suppl. 7, 59 (1987)
Caprolactam	19, 115 (1979) (corr. 42, 258); 39, 247 (1986) (corr. 42, 264); Suppl. 7, 390 (1987)
Captafol	53, 353 (1991)

Captan	30, 295 (1983); Suppl. 7, 59 (1987)
Carbaryl	12, 37 (1976); Suppl. 7, 59 (1987)
Carbazole	32, 239 (1983); Suppl. 7, 59 (1987)
3-Carbethoxypsoralen	40, 317 (1986); Suppl. 7, 59 (1987)
Carbon blacks	3, 22 (1973); 33, 35 (1984); Suppl. 7, 142 (1987)
Carbon tetrachloride	1, 53 (1972); 20, 371 (1979); Suppl. 7, 143 (1987)
Carmoisine	8, 83 (1975); Suppl. 7, 59 (1987)
Carpentry and joinery	25, 139 (1981); Suppl. 7, 378 (1987)
Carrageenan	10, 181 (1976) (corr. 42, 255); 31, 79 (1983); Suppl. 7, 59 (1987)
Catechol	15, 155 (1977); Suppl. 7, 59 (1987)
CCNU (see 1-(2-Chloroethyl)-3-cyclohexyl-1-nitrosourea)	
Ceramic fibres (see Man-made mineral fibres)	
Chemotherapy, combined, including alkylating agents (see MOPP and other combined chemotherapy including alkylating agents)	
Chloral	63, 245 (1995)
Chloral hydrate	63, 245 (1995)
Chlorambucil	9, 125 (1975); 26, 115 (1981) Suppl. 7, 144 (1987)
Chloramphenicol	10, 85 (1976); Suppl. 7, 145 (1987); 50, 169 (1990)
Chlordane (see also Chlordane/Heptachlor)	20, 45 (1979) (corr. 42, 258)
Chlordane/Heptachlor	Suppl. 7, 146 (1987); 53, 115 (1991)
Chlordecone	20, 67 (1979); Suppl. 7, 59 (1987)
Chlordimeform	30, 61 (1983); Suppl. 7, 59 (1987)
Chlorendic acid	48, 45 (1990)
Chlorinated dibenzodioxins (other than TCDD)	15, 41 (1977); Suppl. 7, 59 (1987)
Chlorinated drinking-water	52, 45 (1991)
Chlorinated paraffins	48, 55 (1990)
α-Chlorinated toluenes	Suppl. 7, 148 (1987)
Chlormadinone acetate (see also Progestins; Combined oral contraceptives)	6, 149 (1974); 21, 365 (1979)
Chlornaphazine (see N,N-Bis(2-chloroethyl)-2-naphthylamine)	
Chloroacetonitrile (see Halogenated acetonitriles)	
para-Chloroaniline	57, 305 (1993)
Chlorobenzilate	5, 75 (1974); 30, 73 (1983); Suppl. 7, 60 (1987)
Chlorodibromomethane	52, 243 (1991)
Chlorodifluoromethane	41, 237 (1986) (corr. 51, 483); Suppl. 7, 149 (1987)
Chloroethane	52, 315 (1991)
1-(2-Chloroethyl)-3-cyclohexyl-1-nitrosourea (see also Chloroethyl nitrosoureas)	26, 137 (1981) (corr. 42, 260); Suppl. 7, 150 (1987)
1-(2-Chloroethyl)-3-(4-methylcyclohexyl)-1-nitrosourea (see also Chloroethyl nitrosoureas)	Suppl. 7, 150 (1987)
Chloroethyl nitrosoureas	Suppl. 7, 150 (1987)
Chlorofluoromethane	41, 229 (1986); Suppl. 7, 60 (1987)

Chloroform	*1*, 61 (1972); *20*, 401 (1979)
	Suppl. 7, 152 (1987)
Chloromethyl methyl ether (technical-grade) (*see also* Bis(chloromethyl)ether)	*4*, 239 (1974); *Suppl. 7*, 131 (1987)
(4-Chloro-2-methylphenoxy)acetic acid (*see* MCPA)	
1-Chloro-2-methylpropene	*63*, 315 (1995)
3-Chloro-2-methylpropene	*63*, 325 (1995)
Chlorophenols	*Suppl. 7*, 154 (1987)
Chlorophenols (occupational exposures to)	*41*, 319 (1986)
Chlorophenoxy herbicides	*Suppl. 7*, 156 (1987)
Chlorophenoxy herbicides (occupational exposures to)	*41*, 357 (1986)
4-Chloro-*ortho*-phenylenediamine	*27*, 81 (1982); *Suppl. 7*, 60 (1987)
4-Chloro-*meta*-phenylenediamine	*27*, 82 (1982); *Suppl. 7*, 60 (1987)
Chloroprene	*19*, 131 (1979); *Suppl. 7*, 160 (1987)
Chloropropham	*12*, 55 (1976); *Suppl. 7*, 60 (1987)
Chloroquine	*13*, 47 (1977); *Suppl. 7*, 60 (1987)
Chlorothalonil	*30*, 319 (1983); *Suppl. 7*, 60 (1987)
para-Chloro-*ortho*-toluidine and its strong acid salts (*see also* Chlordimeform)	*16*, 277 (1978); *30*, 65 (1983); *Suppl. 7*, 60 (1987); *48*, 123 (1990)
Chlorotrianisene (*see also* Nonsteroidal oestrogens)	*21*, 139 (1979)
2-Chloro-1,1,1-trifluoroethane	*41*, 253 (1986); *Suppl. 7*, 60 (1987)
Chlorozotocin	*50*, 65 (1990)
Cholesterol	*10*, 99 (1976); *31*, 95 (1983);
	Suppl. 7, 161 (1987)
Chromic acetate (*see* Chromium and chromium compounds)	
Chromic chloride (*see* Chromium and chromium compounds)	
Chromic oxide (*see* Chromium and chromium compounds)	
Chromic phosphate (*see* Chromium and chromium compounds)	
Chromite ore (*see* Chromium and chromium compounds)	
Chromium and chromium compounds	*2*, 100 (1973); *23*, 205 (1980); *Suppl. 7*, 165 (1987); *49*, 49 (1990) (*corr. 51*, 483)
Chromium carbonyl (*see* Chromium and chromium compounds)	
Chromium potassium sulfate (*see* Chromium and chromium compounds)	
Chromium sulfate (*see* Chromium and chromium compounds)	
Chromium trioxide (*see* Chromium and chromium compounds)	
Chrysazin (*see* Dantron)	
Chrysene	*3*, 159 (1973); *32*, 247 (1983); *Suppl. 7*, 60 (1987)
Chrysoidine	*8*, 91 (1975); *Suppl. 7*, 169 (1987)
Chrysotile (*see* Asbestos)	
CI Acid Orange 3	*57*, 121 (1993)
CI Acid Red 114	*57*, 247 (1993)
CI Basic Red 9	*57*, 215 (1993)
Ciclosporin	*50*, 77 (1990)
CI Direct Blue 15	*57*, 235 (1993)
CI Disperse Yellow 3 (see Disperse Yellow 3)	
Cimetidine	*50*, 235 (1990)
Cinnamyl anthranilate	*16*, 287 (1978); *31*, 133 (1983); *Suppl. 7*, 60 (1987)
CI Pigment Red 3	*57*, 259 (1993)

CI Pigment Red 53:1 (see D&C Red No. 9)
Cisplatin   26, 151 (1981); Suppl. 7, 170 (1987)
Citrinin   40, 67 (1986); Suppl. 7, 60 (1987)
Citrus Red No. 2   8, 101 (1975) (corr. 42, 254)
    Suppl. 7, 60 (1987)
Clofibrate   24, 39 (1980); Suppl. 7, 171 (1987)
Clomiphene citrate   21, 551 (1979); Suppl. 7, 172 (1987)
Clonorchis sinensis (infection with)   61, 121 (1994)
Coal gasification   34, 65 (1984); Suppl. 7, 173 (1987)
Coal-tar pitches (see also Coal-tars)   35, 83 (1985); Suppl. 7, 174 (1987)
Coal-tars   35, 83 (1985); Suppl. 7, 175 (1987)
Cobalt[III] acetate (see Cobalt and cobalt compounds)
Cobalt-aluminium-chromium spinel (see Cobalt and cobalt compounds)
Cobalt and cobalt compounds   52, 363 (1991)
Cobalt[II] chloride (see Cobalt and cobalt compounds)
Cobalt-chromium alloy (see Chromium and chromium compounds)
Cobalt-chromium-molybdenum alloys (see Cobalt and cobalt compounds)
Cobalt metal powder (see Cobalt and cobalt compounds)
Cobalt naphthenate (see Cobalt and cobalt compounds)
Cobalt[II] oxide (see Cobalt and cobalt compounds)
Cobalt[II,III] oxide (see Cobalt and cobalt compounds)
Cobalt[II] sulfide (see Cobalt and cobalt compounds)
Coffee   51, 41 (1991) (corr. 52, 513)
Coke production   34, 101 (1984); Suppl. 7, 176 (1987)
Combined oral contraceptives (see also Oestrogens, progestins   Suppl. 7, 297 (1987)
    and combinations)
Conjugated oestrogens (see also Steroidal oestrogens)   21, 147 (1979)
Contraceptives, oral (see Combined oral contraceptives;
    Sequential oral contraceptives)
Copper 8-hydroxyquinoline   15, 103 (1977); Suppl. 7, 61 (1987)
Coronene   32, 263 (1983); Suppl. 7, 61 (1987)
Coumarin   10, 113 (1976); Suppl. 7, 61 (1987)
Creosotes (see also Coal-tars)   35, 83 (1985); Suppl. 7, 177 (1987)
meta-Cresidine   27, 91 (1982); Suppl. 7, 61 (1987)
para-Cresidine   27, 92 (1982); Suppl. 7, 61 (1987)
Crocidolite (see Asbestos)
Crotonaldehyde   63, 373 (1995)
Crude oil   45, 119 (1989)
Crystalline silica (see also Silica)   42, 39 (1987); Suppl. 7, 341 (1987)
Cycasin   1, 157 (1972) (corr. 42, 251); 10,
    121 (1976); Suppl. 7, 61 (1987)
Cyclamates   22, 55 (1980); Suppl. 7, 178 (1987)
Cyclamic acid (see Cyclamates)
Cyclochlorotine   10, 139 (1976); Suppl. 7, 61 (1987)
Cyclohexanone   47, 157 (1989)
Cyclohexylamine (see Cyclamates)
Cyclopenta[cd]pyrene   32, 269 (1983); Suppl. 7, 61 (1987)
Cyclopropane (see Anaesthetics, volatile)
Cyclophosphamide   9, 135 (1975); 26, 165 (1981);
    Suppl. 7, 182 (1987)

D

2,4-D (*see also* Chlorophenoxy herbicides; Chlorophenoxy herbicides, occupational exposures to)	*15*, 111 (1977)
Dacarbazine	*26*, 203 (1981); *Suppl. 7*, 184 (1987)
Dantron	*50*, 265 (1990) (*corr. 59*, 257)
D&C Red No. 9	*8*, 107 (1975); *Suppl. 7*, 61 (1987); *57*, 203 (1993)
Dapsone	*24*, 59 (1980); *Suppl. 7*, 185 (1987)
Daunomycin	*10*, 145 (1976); *Suppl. 7*, 61 (1987)
DDD (*see* DDT)	
DDE (*see* DDT)	
DDT	*5*, 83 (1974) (*corr. 42*, 253); *Suppl. 7*, 186 (1987); *53*, 179 (1991)
Decabromodiphenyl oxide	*48*, 73 (1990)
Deltamethrin	*53*, 251 (1991)
Deoxynivalenol (*see* Toxins derived from *Fusarium graminearum*, *F. culmorum* and *F. crookwellense*)	
Diacetylaminoazotoluene	*8*, 113 (1975); *Suppl. 7*, 61 (1987)
N,N'-Diacetylbenzidine	*16*, 293 (1978); *Suppl. 7*, 61 (1987)
Diallate	*12*, 69 (1976); *30*, 235 (1983); *Suppl. 7*, 61 (1987)
2,4-Diaminoanisole	*16*, 51 (1978); *27*, 103 (1982); *Suppl. 7*, 61 (1987)
4,4'-Diaminodiphenyl ether	*16*, 301 (1978); *29*, 203 (1982); *Suppl. 7*, 61 (1987)
1,2-Diamino-4-nitrobenzene	*16*, 63 (1978); *Suppl. 7*, 61 (1987)
1,4-Diamino-2-nitrobenzene	*16*, 73 (1978); *Suppl. 7*, 61 (1987); *57*, 185 (1993)
2,6-Diamino-3-(phenylazo)pyridine (*see* Phenazopyridine hydrochloride)	
2,4-Diaminotoluene (*see also* Toluene diisocyanates)	*16*, 83 (1978); *Suppl. 7*, 61 (1987)
2,5-Diaminotoluene (*see also* Toluene diisocyanates)	*16*, 97 (1978); *Suppl. 7*, 61 (1987)
ortho-Dianisidine (*see* 3,3'-Dimethoxybenzidine)	
Diazepam	*13*, 57 (1977); *Suppl. 7*, 189 (1987)
Diazomethane	*7*, 223 (1974); *Suppl. 7*, 61 (1987)
Dibenz[*a,h*]acridine	*3*, 247 (1973); *32*, 277 (1983); *Suppl. 7*, 61 (1987)
Dibenz[*a,j*]acridine	*3*, 254 (1973); *32*, 283 (1983); *Suppl. 7*, 61 (1987)
Dibenz[*a,c*]anthracene	*32*, 289 (1983) (*corr. 42*, 262); *Suppl. 7*, 61 (1987)
Dibenz[*a,h*]anthracene	*3*, 178 (1973) (*corr. 43*, 261); *32*, 299 (1983); *Suppl. 7*, 61 (1987)
Dibenz[*a,j*]anthracene	*32*, 309 (1983); *Suppl. 7*, 61 (1987)
7*H*-Dibenzo[*c,g*]carbazole	*3*, 260 (1973); *32*, 315 (1983); *Suppl. 7*, 61 (1987)
Dibenzodioxins, chlorinated (other than TCDD) [*see* Chlorinated dibenzodioxins (other than TCDD)]	
Dibenzo[*a,e*]fluoranthene	*32*, 321 (1983); *Suppl. 7*, 61 (1987)
Dibenzo[*h,rst*]pentaphene	*3*, 197 (1973); *Suppl. 7*, 62 (1987)

Dibenzo[a,e]pyrene	3, 201 (1973); 32, 327 (1983); Suppl. 7, 62 (1987)
Dibenzo[a,h]pyrene	3, 207 (1973); 32, 331 (1983); Suppl. 7, 62 (1987)
Dibenzo[a,i]pyrene	3, 215 (1973); 32, 337 (1983); Suppl. 7, 62 (1987)
Dibenzo[a,l]pyrene	3, 224 (1973); 32, 343 (1983); Suppl. 7, 62 (1987)
Dibromoacetonitrile (see Halogenated acetonitriles)	
1,2-Dibromo-3-chloropropane	15, 139 (1977); 20, 83 (1979); Suppl. 7, 191 (1987)
Dichloroacetic acid	63, 271 (1995)
Dichloroacetonitrile (see Halogenated acetonitriles)	
Dichloroacetylene	39, 369 (1986); Suppl. 7, 62 (1987)
ortho-Dichlorobenzene	7, 231 (1974); 29, 213 (1982); Suppl. 7, 192 (1987)
para-Dichlorobenzene	7, 231 (1974); 29, 215 (1982); Suppl. 7, 192 (1987)
3,3'-Dichlorobenzidine	4, 49 (1974); 29, 239 (1982); Suppl. 7, 193 (1987)
trans-1,4-Dichlorobutene	15, 149 (1977); Suppl. 7, 62 (1987)
3,3'-Dichloro-4,4'-diaminodiphenyl ether	16, 309 (1978); Suppl. 7, 62 (1987)
1,2-Dichloroethane	20, 429 (1979); Suppl. 7, 62 (1987)
Dichloromethane	20, 449 (1979); 41, 43 (1986); Suppl. 7, 194 (1987)
2,4-Dichlorophenol (see Chlorophenols; Chlorophenols, occupational exposures to)	
(2,4-Dichlorophenoxy)acetic acid (see 2,4-D)	
2,6-Dichloro-*para*-phenylenediamine	39, 325 (1986); Suppl. 7, 62 (1987)
1,2-Dichloropropane	41, 131 (1986); Suppl. 7, 62 (1987)
1,3-Dichloropropene (technical-grade)	41, 113 (1986); Suppl. 7, 195 (1987)
Dichlorvos	20, 97 (1979); Suppl. 7, 62 (1987); 53, 267 (1991)
Dicofol	30, 87 (1983); Suppl. 7, 62 (1987)
Dicyclohexylamine (see Cyclamates)	
Dieldrin	5, 125 (1974); Suppl. 7, 196 (1987)
Dienoestrol (see also Nonsteroidal oestrogens)	21, 161 (1979)
Diepoxybutane	11, 115 (1976) (corr. 42, 255); Suppl. 7, 62 (1987)
Diesel and gasoline engine exhausts	46, 41 (1989)
Diesel fuels	45, 219 (1989) (corr. 47, 505)
Diethyl ether (see Anaesthetics, volatile)	
Di(2-ethylhexyl)adipate	29, 257 (1982); Suppl. 7, 62 (1987)
Di(2-ethylhexyl)phthalate	29, 269 (1982) (corr. 42, 261); Suppl. 7, 62 (1987)
1,2-Diethylhydrazine	4, 153 (1974); Suppl. 7, 62 (1987)
Diethylstilboestrol	6, 55 (1974); 21, 173 (1979) (corr. 42, 259); Suppl. 7, 273 (1987)
Diethylstilboestrol dipropionate (see Diethylstilboestrol)	
Diethyl sulfate	4, 277 (1974); Suppl. 7, 198 (1987); 54, 213 (1992)

Diglycidyl resorcinol ether	*11*, 125 (1976); *36*, 181 (1985); Suppl. 7, 62 (1987)
Dihydrosafrole	*1*, 170 (1972); *10*, 233 (1976) Suppl. 7, 62 (1987)
1,8-Dihydroxyanthraquinone (*see* Dantron)	
Dihydroxybenzenes (*see* Catechol; Hydroquinone; Resorcinol)	
Dihydroxymethylfuratrizine	*24*, 77 (1980); Suppl. 7, 62 (1987)
Diisopropyl sulfate	*54*, 229 (1992)
Dimethisterone (*see also* Progestins; Sequential oral contraceptives)	*6*, 167 (1974); *21*, 377 (1979))
Dimethoxane	*15*, 177 (1977); Suppl. 7, 62 (1987)
3,3'-Dimethoxybenzidine	*4*, 41 (1974); Suppl. 7, 198 (1987)
3,3'-Dimethoxybenzidine-4,4'-diisocyanate	*39*, 279 (1986); Suppl. 7, 62 (1987)
para-Dimethylaminoazobenzene	*8*, 125 (1975); Suppl. 7, 62 (1987)
para-Dimethylaminoazobenzenediazo sodium sulfonate	*8*, 147 (1975); Suppl. 7, 62 (1987)
trans-2-[(Dimethylamino)methylimino]-5-[2-(5-nitro-2-furyl)-vinyl]-1,3,4-oxadiazole	*7*, 147 (1974) (*corr. 42*, 253); Suppl. 7, 62 (1987)
4,4'-Dimethylangelicin plus ultraviolet radiation (*see also* Angelicin and some synthetic derivatives)	Suppl. 7, 57 (1987)
4,5'-Dimethylangelicin plus ultraviolet radiation (*see also* Angelicin and some synthetic derivatives)	Suppl. 7, 57 (1987)
2,6-Dimethylaniline	*57*, 323 (1993)
N,N-Dimethylaniline	*57*, 337 (1993)
Dimethylarsinic acid (*see* Arsenic and arsenic compounds)	
3,3'-Dimethylbenzidine	*1*, 87 (1972); Suppl. 7, 62 (1987)
Dimethylcarbamoyl chloride	*12*, 77 (1976); Suppl. 7, 199 (1987)
Dimethylformamide	*47*, 171 (1989)
1,1-Dimethylhydrazine	*4*, 137 (1974); Suppl. 7, 62 (1987)
1,2-Dimethylhydrazine	*4*, 145 (1974) (*corr. 42*, 253); Suppl. 7, 62 (1987)
Dimethyl hydrogen phosphite	*48*, 85 (1990)
1,4-Dimethylphenanthrene	*32*, 349 (1983); Suppl. 7, 62 (1987)
Dimethyl sulfate	*4*, 271 (1974); Suppl. 7, 200 (1987)
3,7-Dinitrofluoranthene	*46*, 189 (1989)
3,9-Dinitrofluoranthene	*46*, 195 (1989)
1,3-Dinitropyrene	*46*, 201 (1989)
1,6-Dinitropyrene	*46*, 215 (1989)
1,8-Dinitropyrene	*33*, 171 (1984); Suppl. 7, 63 (1987); *46*, 231 (1989)
Dinitrosopentamethylenetetramine	*11*, 241 (1976); Suppl. 7, 63 (1987)
1,4-Dioxane	*11*, 247 (1976); Suppl. 7, 201 (1987)
2,4'-Diphenyldiamine	*16*, 313 (1978); Suppl. 7, 63 (1987)
Direct Black 38 (*see also* Benzidine-based dyes)	*29*, 295 (1982) (*corr. 42*, 261)
Direct Blue 6 (*see also* Benzidine-based dyes)	*29*, 311 (1982)
Direct Brown 95 (*see also* Benzidine-based dyes)	*29*, 321 (1982)
Disperse Blue 1	*48*, 139 (1990)
Disperse Yellow 3	*8*, 97 (1975); Suppl. 7, 60 (1987); *48*, 149 (1990)
Disulfiram	*12*, 85 (1976); Suppl. 7, 63 (1987)
Dithranol	*13*, 75 (1977); Suppl. 7, 63 (1987)
Divinyl ether (*see* Anaesthetics, volatile)	
Dry cleaning	*63*, 33 (1995)

Dulcin   *12*, 97 (1976); *Suppl. 7*, 63 (1987)

E

Endrin   *5*, 157 (1974); *Suppl. 7*, 63 (1987)
Enflurane (*see* Anaesthetics, volatile)
Eosin   *15*, 183 (1977); *Suppl. 7*, 63 (1987)
Epichlorohydrin   *11*, 131 (1976) (*corr. 42*, 256);
                                                    Suppl. 7, 202 (1987)
1,2-Epoxybutane   *47*, 217 (1989)
1-Epoxyethyl-3,4-epoxycyclohexane (*see* 4-Vinylcyclohexene diepoxide)
3,4-Epoxy-6-methylcyclohexylmethyl-3,4-epoxy-6-methyl- *11*, 147 (1976); *Suppl. 7*, 63 (1987)
  cyclohexane carboxylate
cis-9,10-Epoxystearic acid   *11*, 153 (1976); *Suppl. 7*, 63 (1987)
Erionite   *42*, 225 (1987); *Suppl. 7*, 203 (1987)
Ethinyloestradiol (*see also* Steroidal oestrogens) *6*, 77 (1974); *21*, 233 (1979)
Ethionamide   *13*, 83 (1977); *Suppl. 7*, 63 (1987)
Ethyl acrylate   *19*, 57 (1979); *39*, 81 (1986);
                                                    Suppl. 7, 63 (1987)
Ethylene   *19*, 157 (1979); *Suppl. 7*, 63 (1987);
                                                    60, 45 (1994)
Ethylene dibromide   *15*, 195 (1977); *Suppl. 7*, 204 (1987)
Ethylene oxide   *11*, 157 (1976); *36*, 189 (1985)
                                                    (*corr. 42*, 263); *Suppl. 7*, 205
                                                    (1987); *60*, 73 (1994)
Ethylene sulfide   *11*, 257 (1976); *Suppl. 7*, 63 (1987)
Ethylene thiourea   *7*, 45 (1974); *Suppl. 7*, 207 (1987)
2-Ethylhexyl acrylate   *60*, 475 (1994)
Ethyl methanesulfonate   *7*, 245 (1974); *Suppl. 7*, 63 (1987)
N-Ethyl-*N*-nitrosourea   *1*, 135 (1972); *17*, 191 (1978);
                                                    Suppl. 7, 63 (1987)
Ethyl selenac (*see also* Selenium and selenium compounds) *12*, 107 (1976); *Suppl. 7*, 63 (1987)
Ethyl tellurac   *12*, 115 (1976); *Suppl. 7*, 63 (1987)
Ethynodiol diacetate (*see also* Progestins; Combined oral *6*, 173 (1974); *21*, 387 (1979)
  contraceptives)
Eugenol   *36*, 75 (1985); *Suppl. 7*, 63 (1987)
Evans blue   *8*, 151 (1975); *Suppl. 7*, 63 (1987)

F

Fast Green FCF   *16*, 187 (1978); *Suppl. 7*, 63 (1987)
Fenvalerate   *53*, 309 (1991)
Ferbam   *12*, 121 (1976) (*corr. 42*, 256);
                                                    Suppl. 7, 63 (1987)
Ferric oxide   *1*, 29 (1972); *Suppl. 7*, 216 (1987)
Ferrochromium (*see* Chromium and chromium compounds)
Fluometuron   *30*, 245 (1983); *Suppl. 7*, 63 (1987)
Fluoranthene   *32*, 355 (1983); *Suppl. 7*, 63 (1987)
Fluorene   *32*, 365 (1983); *Suppl. 7*, 63 (1987)

Fluorescent lighting (exposure to) (*see* Ultraviolet radiation)	
Fluorides (inorganic, used in drinking-water)	27, 237 (1982); *Suppl. 7*, 208 (1987)
5-Fluorouracil	26, 217 (1981); *Suppl. 7*, 210 (1987)
Fluorspar (*see* Fluorides)	
Fluosilicic acid (*see* Fluorides)	
Fluroxene (*see* Anaesthetics, volatile)	
Formaldehyde	29, 345 (1982); *Suppl. 7*, 211 (1987); 62, 217 (1995)
2-(2-Formylhydrazino)-4-(5-nitro-2-furyl)thiazole	7, 151 (1974) (*corr. 42*, 253); *Suppl. 7*, 63 (1987)
Frusemide (*see* Furosemide)	
Fuel oils (heating oils)	45, 239 (1989) (*corr. 47*, 505)
Fumonisin B_1 (*see* Toxins derived from Fusarium moniliforme)	
Fumonisin B_2 (*see* Toxins derived from Fusarium moniliforme)	
Furan	63, 393 (1995)
Furazolidone	31, 141 (1983); *Suppl. 7*, 63 (1987)
Furfural	63, 409 (1995)
Furniture and cabinet-making	25, 99 (1981); *Suppl. 7*, 380 (1987)
Furosemide	50, 277 (1990)
2-(2-Furyl)-3-(5-nitro-2-furyl)acrylamide (*see* AF-2)	
Fusarenon-X (*see* Toxins derived from *Fusarium graminearum*, *F. culmorum* and *F. crookwellense*)	
Fusarenone-X (*see* Toxins derived from *Fusarium graminearum*, *F. culmorum* and *F. crookwellense*)	
Fusarin C (*see* Toxins derived from *Fusarium moniliforme*)	

G

Gasoline	45, 159 (1989) (corr. 47, 505)
Gasoline engine exhaust (see Diesel and gasoline engine exhausts)	
Glass fibres (see Man-made mineral fibres)	
Glass manufacturing industry, occupational exposures in	58, 347 (1993)
Glasswool (*see* Man-made mineral fibres)	
Glass filaments (*see* Man-made mineral fibres)	
Glu-P-1	40, 223 (1986); *Suppl. 7*, 64 (1987)
Glu-P-2	40, 235 (1986); *Suppl. 7*, 64 (1987)
L-Glutamic acid, 5-[2-(4-hydroxymethyl)phenylhydrazide] (*see* Agaritine)	
Glycidaldehyde	11, 175 (1976); *Suppl. 7*, 64 (1987)
Glycidyl ethers	47, 237 (1989)
Glycidyl oleate	11, 183 (1976); *Suppl. 7*, 64 (1987)
Glycidyl stearate	11, 187 (1976); *Suppl. 7*, 64 (1987)
Griseofulvin	10, 153 (1976); *Suppl. 7*, 391 (1987)
Guinea Green B	16, 199 (1978); *Suppl. 7*, 64 (1987)
Gyromitrin	31, 163 (1983); *Suppl. 7*, 391 (1987)

H

Haematite	1, 29 (1972); *Suppl. 7*, 216 (1987)

Haematite and ferric oxide	*Suppl. 7*, 216 (1987)
Haematite mining, underground, with exposure to radon	*1*, 29 (1972); *Suppl. 7*, 216 (1987)
Hairdressers and barbers (occupational exposure as)	*57*, 43 (1993)
Hair dyes, epidemiology of	*16*, 29 (1978); *27*, 307 (1982);
Halogenated acetonitriles	*52*, 269 (1991)
Halothane (*see* Anaesthetics, volatile)	
HC Blue No. 1	*57*, 129 (1993)
HC Blue No. 2	*57*, 143 (1993)
α-HCH (*see* Hexachlorocyclohexanes)	
β-HCH (*see* Hexachlorocyclohexanes)	
γ-HCH (*see* Hexachlorocyclohexanes)	
HC Red No. 3	*57*, 153 (1993)
HC Yellow No. 4	*57*, 159 (1993)
Heating oils (*see* Fuel oils)	
Helicobacter pylori (infection with)	*61*, 177 (1994)
Hepatitis B virus	*59*, 45 (1994)
Hepatitis C virus	*59*, 165 (1994)
Hepatitis D virus	*59*, 223 (1994)
Heptachlor (*see also* Chlordane/Heptachlor)	*5*, 173 (1974); *20*, 129 (1979)
Hexachlorobenzene	*20*, 155 (1979); *Suppl. 7*, 219 (1987)
Hexachlorobutadiene	*20*, 179 (1979); *Suppl. 7*, 64 (1987)
Hexachlorocyclohexanes	*5*, 47 (1974); *20*, 195 (1979)
	(*corr. 42*, 258); *Suppl. 7*, 220 (1987)
Hexachlorocyclohexane, technical-grade (*see* Hexachlorocyclohexanes)	
Hexachloroethane	*20*, 467 (1979); *Suppl. 7*, 64 (1987)
Hexachlorophene	*20*, 241 (1979); *Suppl. 7*, 64 (1987)
Hexamethylphosphoramide	*15*, 211 (1977); *Suppl. 7*, 64 (1987)
Hexoestrol (*see* Nonsteroidal oestrogens)	
Hycanthone mesylate	*13*, 91 (1977); *Suppl. 7*, 64 (1987)
Hydralazine	*24*, 85 (1980); *Suppl. 7*, 222 (1987)
Hydrazine	*4*, 127 (1974); *Suppl. 7*, 223 (1987)
Hydrochloric acid	*54*, 189 (1992)
Hydrochlorothiazide	*50*, 293 (1990)
Hydrogen peroxide	*36*, 285 (1985); *Suppl. 7*, 64 (1987)
Hydroquinone	*15*, 155 (1977); *Suppl. 7*, 64 (1987)
4-Hydroxyazobenzene	*8*, 157 (1975); *Suppl. 7*, 64 (1987)
17α-Hydroxyprogesterone caproate (*see also* Progestins)	*21*, 399 (1979) (*corr. 42*, 259)
8-Hydroxyquinoline	*13*, 101 (1977); *Suppl. 7*, 64 (1987)
8-Hydroxysenkirkine	*10*, 265 (1976); *Suppl. 7*, 64 (1987)
Hypochlorite salts	*52*, 159 (1991)

I

Indeno[1,2,3-*cd*]pyrene	*3*, 229 (1973); *32*, 373 (1983);
	Suppl. 7, 64 (1987)
Inorganic acids (see Sulfuric acid and other strong inorganic acids, occupational exposures to mists and vapours from)	
Insecticides, occupational exposures in spraying and application of	*53*, 45 (1991)
IQ	*40*, 261 (1986); *Suppl. 7*, 64 (1987);
	56, 165 (1993)

CUMULATIVE INDEX 537

Iron and steel founding	*34*, 133 (1984); *Suppl. 7*, 224 (1987)
Iron-dextran complex	*2*, 161 (1973); *Suppl. 7*, 226 (1987)
Iron-dextrin complex	*2*, 161 (1973) (*corr. 42*, 252); *Suppl. 7*, 64 (1987)
Iron oxide (*see* Ferric oxide)	
Iron oxide, saccharated (*see* Saccharated iron oxide)	
Iron sorbitol-citric acid complex	*2*, 161 (1973); *Suppl. 7*, 64 (1987)
Isatidine	*10*, 269 (1976); *Suppl. 7*, 65 (1987)
Isoflurane (*see* Anaesthetics, volatile)	
Isoniazid (*see* Isonicotinic acid hydrazide)	
Isonicotinic acid hydrazide	*4*, 159 (1974); *Suppl. 7*, 227 (1987)
Isophosphamide	*26*, 237 (1981); *Suppl. 7*, 65 (1987)
Isoprene	*60*, 215 (1994)
Isopropanol	*5*, 223 (1977); *Suppl. 7*, 229 (1987)
Isopropanol manufacture (strong-acid process)	*Suppl. 7*, 229 (1987)
(*see* also Isopropyl alcohol; Sulfuric acid and other strong inorganic acids, occupational exposures to mists and vapours from)	
Isopropyl oils	*15*, 223 (1977); *Suppl. 7*, 229 (1987)
Isosafrole	*1*, 169 (1972); *10*, 232 (1976); *Suppl. 7*, 65 (1987)

J

Jacobine	*10*, 275 (1976); *Suppl. 7*, 65 (1987)
Jet fuel	*45*, 203 (1989)
Joinery (*see* Carpentry and joinery	

K

Kaempferol	*31*, 171 (1983); *Suppl. 7*, 65 (1987)
Kepone (*see* Chlordecone)	

L

Lasiocarpine	*10*, 281 (1976); *Suppl. 7*, 65 (1987)
Lauroyl peroxide	*36*, 315 (1985); *Suppl. 7*, 65 (1987)
Lead acetate (*see* Lead and lead compounds)	
Lead and lead compounds	*1*, 40 (1972) (*corr. 42*, 251); *2*, 52, 150 (1973); *12*, 131 (1976); *23*, 40, 208, 209, 325 (1980); *Suppl. 7*, 230 (1987)
Lead arsenate (*see* Arsenic and arsenic compounds)	
Lead carbonate (*see* Lead and lead compounds)	
Lead chloride (*see* Lead and lead compounds)	
Lead chromate (*see* Chromium and chromium compounds)	
Lead chromate oxide (*see* Chromium and chromium compounds)	
Lead naphthenate (*see* Lead and lead compounds)	
Lead nitrate (*see* Lead and lead compounds)	

Lead oxide (*see* Lead and lead compounds)	
Lead phosphate (*see* Lead and lead compounds)	
Lead subacetate (*see* Lead and lead compounds)	
Lead tetroxide (*see* Lead and lead compounds)	
Leather goods manufacture	25, 279 (1981); *Suppl. 7*, 235 (1987)
Leather industries	25, 199 (1981); *Suppl. 7*, 232 (1987)
Leather tanning and processing	25, 201 (1981); *Suppl. 7*, 236 (1987)
Ledate (*see also* Lead and lead compounds)	12, 131 (1976)
Light Green SF	16, 209 (1978); *Suppl. 7*, 65 (1987)
d-Limonene	56, 135 (1993)
Lindane (*see* Hexachlorocyclohexanes)	
Liver flukes (*see Clonorchis sinensis, Opisthorchis felineus* and *Opisthorchis viverrini*)	
The lumber and sawmill industries (including logging)	25, 49 (1981); *Suppl. 7*, 383 (1987)
Luteoskyrin	10, 163 (1976); *Suppl. 7*, 65 (1987)
Lynoestrenol (*see also* Progestins; Combined oral contraceptives)	21, 407 (1979)

M

Magenta	4, 57 (1974) (*corr. 42*, 252); *Suppl. 7*, 238 (1987); 57, 215 (1993)
Magenta, manufacture of (*see also* Magenta)	*Suppl. 7*, 238 (1987)
Malathion	30, 103 (1983); *Suppl. 7*, 65 (1987)
Maleic hydrazide	4, 173 (1974) (*corr. 42*, 253); *Suppl. 7*, 65 (1987)
Malonaldehyde	36, 163 (1985); *Suppl. 7*, 65 (1987)
Maneb	12, 137 (1976); *Suppl. 7*, 65 (1987)
Man-made mineral fibres	43, 39 (1988)
Mannomustine	9, 157 (1975); *Suppl. 7*, 65 (1987)
Mate	51, 273 (1991)
MCPA (*see also* Chlorophenoxy herbicides; Chlorophenoxy herbicides, occupational exposures to)	30, 255 (1983)
MeA-α-C	40, 253 (1986); *Suppl. 7*, 65 (1987)
Medphalan	9, 168 (1975); *Suppl. 7*, 65 (1987)
Medroxyprogesterone acetate	6, 157 (1974); 21, 417 (1979) (*corr. 42*, 259); *Suppl. 7*, 289 (1987)
Megestrol acetate (*see also* Progestins; Combined oral contraceptives)	
MeIQ	40, 275 (1986); *Suppl. 7*, 65 (1987); 56, 197 (1993)
MeIQx	40, 283 (1986); *Suppl. 7*, 65 (1987); 56, 211 (1993)
Melamine	39, 333 (1986); *Suppl. 7*, 65 (1987)
Melphalan	9, 167 (1975); *Suppl. 7*, 239 (1987)
6-Mercaptopurine	26, 249 (1981); *Suppl. 7*, 240 (1987)
Mercuric chloride (*see* Mercury and mercury compounds)	
Mercury and mercury compounds	58, 239 (1993)
Merphalan	9, 169 (1975); *Suppl. 7*, 65 (1987)
Mestranol (*see also* Steroidal oestrogens)	6, 87 (1974); 21, 257 (1979) (*corr. 42*, 259)

CUMULATIVE INDEX

Metabisulfites (*see* Sulfur dioxide and some sulfites, bisulfites and metabisulfites)
Metallic mercury (*see* Mercury and mercury compounds)
Methanearsonic acid, disodium salt (*see* Arsenic and arsenic compounds)
Methanearsonic acid, monosodium salt (*see* Arsenic and arsenic compounds)
Methotrexate 26, 267 (1981); *Suppl. 7*, 241 (1987)
Methoxsalen (*see* 8-Methoxypsoralen)
Methoxychlor 5, 193 (1974); 20, 259 (1979); *Suppl. 7*, 66 (1987)

Methoxyflurane (*see* Anaesthetics, volatile)
5-Methoxypsoralen 40, 327 (1986); *Suppl. 7*, 242 (1987)
8-Methoxypsoralen (*see also* 8-Methoxypsoralen plus ultraviolet radiation) 24, 101 (1980)
8-Methoxypsoralen plus ultraviolet radiation *Suppl. 7*, 243 (1987)
Methyl acrylate 19, 52 (1979); 39, 99 (1986); *Suppl. 7*, 66 (1987)

5-Methylangelicin plus ultraviolet radiation (*see also* Angelicin and some synthetic derivatives) *Suppl. 7*, 57 (1987)
2-Methylaziridine 9, 61 (1975); *Suppl. 7*, 66 (1987)
Methylazoxymethanol acetate 1, 164 (1972); 10, 131 (1976); *Suppl. 7*, 66 (1987)

Methyl bromide 41, 187 (1986) (*corr. 45*, 283); *Suppl. 7*, 245 (1987)

Methyl carbamate 12, 151 (1976); *Suppl. 7*, 66 (1987)
Methyl-CCNU [*see* 1-(2-Chloroethyl)-3-(4-methylcyclohexyl)-1-nitrosourea]
Methyl chloride 41, 161 (1986); *Suppl. 7*, 246 (1987)
1-, 2-, 3-, 4-, 5- and 6-Methylchrysenes 32, 379 (1983); *Suppl. 7*, 66 (1987)
N-Methyl-*N*,4-dinitrosoaniline 1, 141 (1972); *Suppl. 7*, 66 (1987)
4,4′-Methylene bis(2-chloroaniline) 4, 65 (1974) (*corr. 42*, 252); *Suppl. 7*, 246 (1987); 57, 271 (1993)
4,4′-Methylene bis(*N*,*N*-dimethyl)benzenamine 27, 119 (1982); *Suppl. 7*, 66 (1987)
4,4′-Methylene bis(2-methylaniline) 4, 73 (1974); *Suppl. 7*, 248 (1987)
4,4′-Methylenedianiline 4, 79 (1974) (*corr. 42*, 252); 39, 347 (1986); *Suppl. 7*, 66 (1987)
4,4′-Methylenediphenyl diisocyanate 19, 314 (1979); *Suppl. 7*, 66 (1987)
2-Methylfluoranthene 32, 399 (1983); *Suppl. 7*, 66 (1987)
3-Methylfluoranthene 32, 399 (1983); *Suppl. 7*, 66 (1987)
Methylglyoxal 51, 443 (1991)
Methyl iodide 15, 245 (1977); 41, 213 (1986); *Suppl. 7*, 66 (1987)

Methylmercury chloride (*see* Mercury and mercury compounds)
Methylmercury compounds (*see* Mercury and mercury compounds)
Methyl methacrylate 19, 187 (1979); *Suppl. 7*, 66 (1987); 60, 445 (1994)

Methyl methanesulfonate 7, 253 (1974); *Suppl. 7*, 66 (1987)
2-Methyl-1-nitroanthraquinone 27, 205 (1982); *Suppl. 7*, 66 (1987)
N-Methyl-*N*′-nitro-*N*-nitrosoguanidine 4, 183 (1974); *Suppl. 7*, 248 (1987)
3-Methylnitrosaminopropionaldehyde [*see* 3-(*N*-Nitrosomethylamino)-propionaldehyde]

3-Methylnitrosaminopropionitrile [*see* 3-(*N*-Nitrosomethylamino)-
 propionitrile]
4-(Methylnitrosamino)-4-(3-pyridyl)-1-butanal-[*see* 4-(*N*-Nitrosomethyl-
 amino)-4-(3-pyridyl)-1-butanal]
4-(Methylnitrosamino)-1-(3-pyridyl)-1-butanone [*see* 4-(-Nitrosomethyl-
 amino)-1-(3-pyridyl)-1-butanone]
N-Methyl-*N*-nitrosourea	*1*, 125 (1972); *17*, 227 (1978); *Suppl. 7*, 66 (1987)
N-Methyl-*N*-nitrosourethane	*4*, 211 (1974); *Suppl. 7*, 66 (1987)
N-Methylolacrylamide	*60*, 435 (1994)
Methyl parathion	*30*, 131 (1983); *Suppl. 7*, 392 (1987)
1-Methylphenanthrene	*32*, 405 (1983); *Suppl. 7*, 66 (1987)
7-Methylpyrido[3,4-*c*]psoralen	*40*, 349 (1986); *Suppl. 7*, 71 (1987)
Methyl red	*8*, 161 (1975); *Suppl. 7*, 66 (1987)
Methyl selenac (*see also* Selenium and selenium compounds)	*12*, 161 (1976); *Suppl. 7*, 66 (1987)
Methylthiouracil	*7*, 53 (1974); *Suppl. 7*, 66 (1987)
Metronidazole	*13*, 113 (1977); *Suppl. 7*, 250 (1987)
Mineral oils	*3*, 30 (1973); *33*, 87 (1984) (*corr. 42*, 262); *Suppl. 7*, 252 (1987)
Mirex	*5*, 203 (1974); *20*, 283 (1979) (*corr. 42*, 258); *Suppl. 7*, 66 (1987)
Mitomycin C	*10*, 171 (1976); *Suppl. 7*, 67 (1987)

MNNG [*see N*-Methyl-*N*'-nitro-*N*-nitrosoguanidine]
MOCA [*see* 4,4'-Methylene bis(2-chloroaniline)]
Modacrylic fibres	*19*, 86 (1979); *Suppl. 7*, 67 (1987)
Monocrotaline	*10*, 291 (1976); *Suppl. 7*, 67 (1987)
Monuron	*12*, 167 (1976); *Suppl. 7*, 67 (1987); *53*, 467 (1991)
MOPP and other combined chemotherapy including alkylating agents	*Suppl. 7*, 254 (1987)
Morpholine	*47*, 199 (1989)
5-(Morpholinomethyl)-3-[(5-nitrofurfurylidene)amino]-2-oxazolidinone	*7*, 161 (1974); *Suppl. 7*, 67 (1987)
Mustard gas	*9*, 181 (1975) (*corr. 42*, 254);

 Suppl. 7, 259 (1987)
Myleran (*see* 1,4-Butanediol dimethanesulfonate)

N

Nafenopin	*24*, 125 (1980); *Suppl. 7*, 67 (1987)
1,5-Naphthalenediamine	*27*, 127 (1982); *Suppl. 7*, 67 (1987)
1,5-Naphthalene diisocyanate	*19*, 311 (1979); *Suppl. 7*, 67 (1987)
1-Naphthylamine	*4*, 87 (1974) (*corr. 42*, 253); *Suppl. 7*, 260 (1987)
2-Naphthylamine	*4*, 97 (1974); *Suppl. 7*, 261 (1987)
1-Naphthylthiourea	*30*, 347 (1983); *Suppl. 7*, 263 (1987)

Nickel acetate (*see* Nickel and nickel compounds)
Nickel ammonium sulfate (*see* Nickel and nickel compounds)

Nickel and nickel compounds	2, 126 (1973) (*corr. 42*, 252); *11*, 75 (1976); *Suppl. 7*, 264 (1987) (*corr. 45*, 283); *49*, 257 (1990)
Nickel carbonate (*see* Nickel and nickel compounds)	
Nickel carbonyl (*see* Nickel and nickel compounds)	
Nickel chloride (*see* Nickel and nickel compounds)	
Nickel-gallium alloy (*see* Nickel and nickel compounds)	
Nickel hydroxide (*see* Nickel and nickel compounds)	
Nickelocene (*see* Nickel and nickel compounds)	
Nickel oxide (*see* Nickel and nickel compounds)	
Nickel subsulfide (*see* Nickel and nickel compounds)	
Nickel sulfate (*see* Nickel and nickel compounds)	
Niridazole	*13*, 123 (1977); *Suppl. 7*, 67 (1987)
Nithiazide	*31*, 179 (1983); *Suppl. 7*, 67 (1987)
Nitrilotriacetic acid and its salts	*48*, 181 (1990)
5-Nitroacenaphthene	*16*, 319 (1978); *Suppl. 7*, 67 (1987)
5-Nitro-*ortho*-anisidine	*27*, 133 (1982); *Suppl. 7*, 67 (1987)
9-Nitroanthracene	*33*, 179 (1984); *Suppl. 7*, 67 (1987)
7-Nitrobenz[*a*]anthracene	*46*, 247 (1989)
6-Nitrobenzo[*a*]pyrene	*33*, 187 (1984); *Suppl. 7*, 67 (1987); *46*, 255 (1989)
4-Nitrobiphenyl	*4*, 113 (1974); *Suppl. 7*, 67 (1987)
6-Nitrochrysene	*33*, 195 (1984); *Suppl. 7*, 67 (1987); *46*, 267 (1989)
Nitrofen (technical-grade)	*30*, 271 (1983); *Suppl. 7*, 67 (1987)
3-Nitrofluoranthene	*33*, 201 (1984); *Suppl. 7*, 67 (1987)
2-Nitrofluorene	*46*, 277 (1989)
Nitrofural	*7*, 171 (1974); *Suppl. 7*, 67 (1987); *50*, 195 (1990)
5-Nitro-2-furaldehyde semicarbazone (*see* Nitrofural)	
Nitrofurantoin	*50*, 211 (1990)
Nitrofurazone (*see* Nitrofural)	
1-[(5-Nitrofurfurylidene)amino]-2-imidazolidinone	*7*, 181 (1974); *Suppl. 7*, 67 (1987)
N-[4-(5-Nitro-2-furyl)-2-thiazolyl]acetamide	*1*, 181 (1972); *7*, 185 (1974); *Suppl. 7*, 67 (1987)
Nitrogen mustard	*9*, 193 (1975); *Suppl. 7*, 269 (1987)
Nitrogen mustard *N*-oxide	*9*, 209 (1975); *Suppl. 7*, 67 (1987)
1-Nitronaphthalene	*46*, 291 (1989)
2-Nitronaphthalene	*46*, 303 (1989)
3-Nitroperylene	*46*, 313 (1989)
2-Nitro-*para*-phenylenediamine (*see* 1,4-Diamino-2-nitrobenzene)	
2-Nitropropane	*29*, 331 (1982); *Suppl. 7*, 67 (1987)
1-Nitropyrene	*33*, 209 (1984); *Suppl. 7*, 67 (1987); *46*, 321 (1989)
2-Nitropyrene	*46*, 359 (1989)
4-Nitropyrene	*46*, 367 (1989)
N-Nitrosatable drugs	*24*, 297 (1980) (*corr. 42*, 260)
N-Nitrosatable pesticides	*30*, 359 (1983)
N'-Nitrosoanabasine	*37*, 225 (1985); *Suppl. 7*, 67 (1987)
N'-Nitrosoanatabine	*37*, 233 (1985); *Suppl. 7*, 67 (1987)

N-Nitrosodi-n-butylamine	4, 197 (1974); 17, 51 (1978); Suppl. 7, 67 (1987)
N-Nitrosodiethanolamine	17, 77 (1978); Suppl. 7, 67 (1987)
N-Nitrosodiethylamine	1, 107 (1972) (corr. 42, 251); 17, 83 (1978) (corr. 42, 257); Suppl. 7, 67 (1987)
N-Nitrosodimethylamine	1, 95 (1972); 17, 125 (1978) (corr. 42, 257); Suppl. 7, 67 (1987)
N-Nitrosodiphenylamine	27, 213 (1982); Suppl. 7, 67 (1987)
para-Nitrosodiphenylamine	27, 227 (1982) (corr. 42, 261); Suppl. 7, 68 (1987)
N-Nitrosodi-n-propylamine	17, 177 (1978); Suppl. 7, 68 (1987)
N-Nitroso-N-ethylurea (see N-Ethyl-N-nitrosourea)	
N-Nitrosofolic acid	17, 217 (1978); Suppl. 7, 68 (1987)
N-Nitrosoguvacine	37, 263 (1985); Suppl. 7, 68 (1987)
N-Nitrosoguvacoline	37, 263 (1985); Suppl. 7, 68 (1987)
N-Nitrosohydroxyproline	17, 304 (1978); Suppl. 7, 68 (1987)
3-(N-Nitrosomethylamino)propionaldehyde	37, 263 (1985); Suppl. 7, 68 (1987)
3-(N-Nitrosomethylamino)propionitrile	37, 263 (1985); Suppl. 7, 68 (1987)
4-(N-Nitrosomethylamino)-4-(3-pyridyl)-1-butanal	37, 205 (1985); Suppl. 7, 68 (1987)
4-(N-Nitrosomethylamino)-1-(3-pyridyl)-1-butanone	37, 209 (1985); Suppl. 7, 68 (1987)
N-Nitrosomethylethylamine	17, 221 (1978); Suppl. 7, 68 (1987)
N-Nitroso-N-methylurea (see N-Methyl-N-nitrosourea)	
N-Nitroso-N-methylurethane (see N-Methyl-N-nitrosourethane)	
N-Nitrosomethylvinylamine	17, 257 (1978); Suppl. 7, 68 (1987)
N-Nitrosomorpholine	17, 263 (1978); Suppl. 7, 68 (1987)
N'-Nitrosonornicotine	17, 281 (1978); 37, 241 (1985); Suppl. 7, 68 (1987)
N-Nitrosopiperidine	17, 287 (1978); Suppl. 7, 68 (1987)
N-Nitrosoproline	17, 303 (1978); Suppl. 7, 68 (1987)
N-Nitrosopyrrolidine	17, 313 (1978); Suppl. 7, 68 (1987)
N-Nitrososarcosine	17, 327 (1978); Suppl. 7, 68 (1987)
Nitrosoureas, chloroethyl (see Chloroethyl nitrosoureas)	
5-Nitro-ortho-toluidine	48, 169 (1990)
Nitrous oxide (see Anaesthetics, volatile)	
Nitrovin	31, 185 (1983); Suppl. 7, 68 (1987)
Nivalenol (see Toxins derived from Fusarium graminearum, F. culmorum and F. crookwellense)	
NNA [see 4-(N-Nitrosomethylamino)-4-(3-pyridyl)-1-butanal]	
NNK [see 4-(N-Nitrosomethylamino)-1-(3-pyridyl)-1-butanone]	
Nonsteroidal oestrogens (see also Oestrogens, progestins and combinations)	Suppl. 7, 272 (1987)
Norethisterone (see also Progestins; Combined oral contraceptives)	6, 179 (1974); 21, 461 (1979)
Norethynodrel (see also Progestins; Combined oral contraceptives	6, 191 (1974); 21, 461 (1979) (corr. 42, 259)
Norgestrel (see also Progestins, Combined oral contraceptives)	6, 201 (1974); 21, 479 (1979)
Nylon 6	19, 120 (1979); Suppl. 7, 68 (1987)

O

Ochratoxin A	*10*, 191 (1976); *31*, 191 (1983) (*corr. 42*, 262); *Suppl. 7*, 271 (1987); *56*, 489 (1993)
Oestradiol-17β (*see also* Steroidal oestrogens)	*6*, 99 (1974); *21*, 279 (1979)
Oestradiol 3-benzoate (*see* Oestradiol-17β)	
Oestradiol dipropionate (*see* Oestradiol-17β)	
Oestradiol mustard	*9*, 217 (1975)
Oestradiol-17β-valerate (*see* Oestradiol-17β)	
Oestriol (*see also* Steroidal oestrogens)	*6*, 117 (1974); *21*, 327 (1979)
Oestrogen-progestin combinations (*see* Oestrogens, progestins and combinations)	
Oestrogen-progestin replacement therapy (*see also* Oestrogens, progestins and combinations)	*Suppl. 7*, 308 (1987)
Oestrogen replacement therapy (*see also* Oestrogens, progestins and combinations)	*Suppl. 7*, 280 (1987)
Oestrogens (*see* Oestrogens, progestins and combinations)	
Oestrogens, conjugated (*see* Conjugated oestrogens)	
Oestrogens, nonsteroidal (*see* Nonsteroidal oestrogens)	
Oestrogens, progestins and combinations	*6* (1974); *21* (1979); *Suppl. 7*, 272 (1987)
Oestrogens, steroidal (*see* Steroidal oestrogens)	
Oestrone (*see* also Steroidal oestrogens)	*6*, 123 (1974); *21*, 343 (1979) (*corr. 42*, 259)
Oestrone benzoate (*see* Oestrone)	
Oil Orange SS	*8*, 165 (1975); *Suppl. 7*, 69 (1987)
Opisthorchis felineus (infection with)	*61*, 121 (1994)
Opisthorchis viverrini (infection with)	*61*, 121 (1994)
Oral contraceptives, combined (*see* Combined oral contraceptives)	
Oral contraceptives, investigational (*see* Combined oral contraceptives)	
Oral contraceptives, sequential (*see* Sequential oral contraceptives)	
Orange I	*8*, 173 (1975); *Suppl. 7*, 69 (1987)
Orange G	*8*, 181 (1975); *Suppl. 7*, 69 (1987)
Organolead compounds (*see also* Lead and lead compounds)	*Suppl. 7*, 230 (1987)
Oxazepam	*13*, 58 (1977); *Suppl. 7*, 69 (1987)
Oxymetholone [*see also* Androgenic (anabolic) steroids]	*13*, 131 (1977)
Oxyphenbutazone	*13*, 185 (1977); *Suppl. 7*, 69 (1987)

P

Paint manufacture and painting (occupational exposures in)	*47*, 329 (1989)
Panfuran S (*see also* Dihydroxymethylfuratrizine)	*24*, 77 (1980); *Suppl. 7*, 69 (1987)
Paper manufacture (*see* Pulp and paper manufacture)	
Paracetamol	*50*, 307 (1990)
Parasorbic acid	*10*, 199 (1976) (*corr. 42*, 255); *Suppl. 7*, 69 (1987)
Parathion	*30*, 153 (1983); *Suppl. 7*, 69 (1987)
Patulin	*10*, 205 (1976); *40*, 83 (1986); *Suppl. 7*, 69 (1987)

Penicillic acid 10, 211 (1976); *Suppl. 7*, 69 (1987)
Pentachloroethane 41, 99 (1986); *Suppl. 7*, 69 (1987)
Pentachloronitrobenzene (see Quintozene)
Pentachlorophenol (*see also* Chlorophenols; Chlorophenols, occupational exposures to) 20, 303 (1979); *53*, 371 (1991)
Permethrin 53, 329 (1991)
Perylene 32, 411 (1983); *Suppl. 7*, 69 (1987)
Petasitenine 31, 207 (1983); *Suppl. 7*, 69 (1987)
Petasites japonicus (*see* Pyrrolizidine alkaloids)
Petroleum refining (occupational exposures in) 45, 39 (1989)
Some petroleum solvents 47, 43 (1989)
Phenacetin 13, 141 (1977); *24*, 135 (1980); *Suppl. 7*, 310 (1987)
Phenanthrene 32, 419 (1983); *Suppl. 7*, 69 (1987)
Phenazopyridine hydrochloride 8, 117 (1975); *24*, 163 (1980) (*corr. 42*, 260); *Suppl. 7*, 312 (1987)
Phenelzine sulfate 24, 175 (1980); *Suppl. 7*, 312 (1987)
Phenicarbazide 12, 177 (1976); *Suppl. 7*, 70 (1987)
Phenobarbital 13, 157 (1977); *Suppl. 7*, 313 (1987)
Phenol 47, 263 (1989) (*corr. 50*, 385)
Phenoxyacetic acid herbicides (*see* Chlorophenoxy herbicides)
Phenoxybenzamine hydrochloride 9, 223 (1975); *24*, 185 (1980); *Suppl. 7*, 70 (1987)
Phenylbutazone 13, 183 (1977); *Suppl. 7*, 316 (1987)
meta-Phenylenediamine 16, 111 (1978); *Suppl. 7*, 70 (1987)
para-Phenylenediamine 16, 125 (1978); *Suppl. 7*, 70 (1987)
Phenyl glycidyl ether (*see* Glycidyl ethers)
N-Phenyl-2-naphthylamine 16, 325 (1978) (*corr. 42*, 257); *Suppl. 7*, 318 (1987)
ortho-Phenylphenol 30, 329 (1983); *Suppl. 7*, 70 (1987)
Phenytoin 13, 201 (1977); *Suppl. 7*, 319 (1987)
PhIP 56, 229 (1993)
Pickled vegetables 56, 83 (1993)
Picloram 53, 481 (1991)
Piperazine oestrone sulfate (*see* Conjugated oestrogens)
Piperonyl butoxide 30, 183 (1983); *Suppl. 7*, 70 (1987)
Pitches, coal-tar (*see* Coal-tar pitches)
Polyacrylic acid 19, 62 (1979); *Suppl. 7*, 70 (1987)
Polybrominated biphenyls 18, 107 (1978); *41*, 261 (1986); *Suppl. 7*, 321 (1987)
Polychlorinated biphenyls 7, 261 (1974); *18*, 43 (1978) (*corr. 42*, 258); *Suppl. 7*, 322 (1987)
Polychlorinated camphenes (*see* Toxaphene)
Polychloroprene 19, 141 (1979); *Suppl. 7*, 70 (1987)
Polyethylene 19, 164 (1979); *Suppl. 7*, 70 (1987)
Polymethylene polyphenyl isocyanate 19, 314 (1979); *Suppl. 7*, 70 (1987)
Polymethyl methacrylate 19, 195 (1979); *Suppl. 7*, 70 (1987)
Polyoestradiol phosphate (*see* Oestradiol-17β)
Polypropylene 19, 218 (1979); *Suppl. 7*, 70 (1987)
Polystyrene 19, 245 (1979); *Suppl. 7*, 70 (1987)
Polytetrafluoroethylene 19, 288 (1979); *Suppl. 7*, 70 (1987)

Polyurethane foams	19, 320 (1979); Suppl. 7, 70 (1987)
Polyvinyl acetate	19, 346 (1979); Suppl. 7, 70 (1987)
Polyvinyl alcohol	19, 351 (1979); Suppl. 7, 70 (1987)
Polyvinyl chloride	7, 306 (1974); 19, 402 (1979); Suppl. 7, 70 (1987)
Polyvinyl pyrrolidone	19, 463 (1979); Suppl. 7, 70 (1987)
Ponceau MX	8, 189 (1975); Suppl. 7, 70 (1987)
Ponceau 3R	8, 199 (1975); Suppl. 7, 70 (1987)
Ponceau SX	8, 207 (1975); Suppl. 7, 70 (1987)
Potassium arsenate (see Arsenic and arsenic compounds)	
Potassium arsenite (see Arsenic and arsenic compounds)	
Potassium bis(2-hydroxyethyl)dithiocarbamate	12, 183 (1976); Suppl. 7, 70 (1987)
Potassium bromate	40, 207 (1986); Suppl. 7, 70 (1987)
Potassium chromate (see Chromium and chromium compounds)	
Potassium dichromate (see Chromium and chromium compounds)	
Prednimustine	50, 115 (1990)
Prednisone	26, 293 (1981); Suppl. 7, 326 (1987)
Procarbazine hydrochloride	26, 311 (1981); Suppl. 7, 327 (1987)
Proflavine salts	24, 195 (1980); Suppl. 7, 70 (1987)
Progesterone (see also Progestins; Combined oral contraceptives)	6, 135 (1974); 21, 491 (1979) (corr. 42, 259)
Progestins (see also Oestrogens, progestins and combinations)	Suppl. 7, 289 (1987)
Pronetalol hydrochloride	13, 227 (1977) (corr. 42, 256); Suppl. 7, 70 (1987)
1,3-Propane sultone	4, 253 (1974) (corr. 42, 253); Suppl. 7, 70 (1987)
Propham	12, 189 (1976); Suppl. 7, 70 (1987)
β-Propiolactone	4, 259 (1974) (corr. 42, 253); Suppl. 7, 70 (1987)
n-Propyl carbamate	12, 201 (1976); Suppl. 7, 70 (1987)
Propylene	19, 213 (1979); Suppl. 7, 71 (1987); 60, 161 (1994)
Propylene oxide	11, 191 (1976); 36, 227 (1985) (corr. 42, 263); Suppl. 7, 328 (1987); 60, 181 (1994)
Propylthiouracil	7, 67 (1974); Suppl. 7, 329 (1987)
Ptaquiloside (see also Bracken fern)	40, 55 (1986); Suppl. 7, 71 (1987)
Pulp and paper manufacture	25, 157 (1981); Suppl. 7, 385 (1987)
Pyrene	32, 431 (1983); Suppl. 7, 71 (1987)
Pyrido[3,4-c]psoralen	40, 349 (1986); Suppl. 7, 71 (1987)
Pyrimethamine	13, 233 (1977); Suppl. 7, 71 (1987)
Pyrrolizidine alkaloids (see Hydroxysenkirkine; Isatidine; Jacobine; Lasiocarpine; Monocrotaline; Retrorsine; Riddelliine; Seneciphylline; Senkirkine)	

Q

Quercetin (see also Bracken fern)	31, 213 (1983); Suppl. 7, 71 (1987)
para-Quinone	15, 255 (1977); Suppl. 7, 71 (1987)

Quintozene	5, 211 (1974); *Suppl. 7*, 71 (1987)

R

Radon	*43*, 173 (1988) (*corr. 45*, 283)
Reserpine	*10*, 217 (1976); *24*, 211 (1980) (*corr. 42*, 260); *Suppl. 7*, 330 (1987)
Resorcinol	*15*, 155 (1977); *Suppl. 7*, 71 (1987)
Retrorsine	*10*, 303 (1976); *Suppl. 7*, 71 (1987)
Rhodamine B	*16*, 221 (1978); *Suppl. 7*, 71 (1987)
Rhodamine 6G	*16*, 233 (1978); *Suppl. 7*, 71 (1987)
Riddelliine	*10*, 313 (1976); *Suppl. 7*, 71 (1987)
Rifampicin	*24*, 243 (1980); *Suppl. 7*, 71 (1987)
Rockwool (*see* Man-made mineral fibres)	
The rubber industry	*28* (1982) (*corr. 42*, 261); *Suppl. 7*, 332 (1987)
Rugulosin	*40*, 99 (1986); *Suppl. 7*, 71 (1987)

S

Saccharated iron oxide	*2*, 161 (1973); *Suppl. 7*, 71 (1987)
Saccharin	*22*, 111 (1980) (*corr. 42*, 259); *Suppl. 7*, 334 (1987)
Safrole	*1*, 169 (1972); *10*, 231 (1976); *Suppl. 7*, 71 (1987)
Salted fish	*56*, 41 (1993)
The sawmill industry (including logging) [*see* The lumber and sawmill industry (including logging)]	
Scarlet Red	*8*, 217 (1975); *Suppl. 7*, 71 (1987)
Schistosoma haematobium (infection with)	*61*, 45 (1994)
Schistosoma japonicum (infection with)	*61*, 45 (1994)
Schistosoma mansoni (infection with)	*61*, 45 (1994)
Selenium and selenium compounds	*9*, 245 (1975) (*corr. 42*, 255); *Suppl. 7*, 71 (1987)
Selenium dioxide (*see* Selenium and selenium compounds)	
Selenium oxide (*see* Selenium and selenium compounds)	
Semicarbazide hydrochloride	*12*, 209 (1976) (*corr. 42*, 256); *Suppl. 7*, 71 (1987)
Senecio jacobaea L. (*see* Pyrrolizidine alkaloids)	
Senecio longilobus (*see* Pyrrolizidine alkaloids)	
Seneciphylline	*10*, 319, 335 (1976); *Suppl. 7*, 71 (1987)
Senkirkine	*10*, 327 (1976); *31*, 231 (1983); *Suppl. 7*, 71 (1987)
Sepiolite	*42*, 175 (1987); *Suppl. 7*, 71 (1987)
Sequential oral contraceptives (*see also* Oestrogens, progestins and combinations)	*Suppl. 7*, 296 (1987)
Shale-oils	*35*, 161 (1985); *Suppl. 7*, 339 (1987)
Shikimic acid (*see also* Bracken fern)	*40*, 55 (1986); *Suppl. 7*, 71 (1987)

Shoe manufacture and repair (*see* Boot and shoe manufacture and repair)	
Silica (*see also* Amorphous silica; Crystalline silica)	*42*, 39 (1987)
Simazine	*53*, 495 (1991)
Slagwool (*see* Man-made mineral fibres)	
Sodium arsenate (*see* Arsenic and arsenic compounds)	
Sodium arsenite (*see* Arsenic and arsenic compounds)	
Sodium cacodylate (*see* Arsenic and arsenic compounds)	
Sodium chlorite	*52*, 145 (1991)
Sodium chromate (*see* Chromium and chromium compounds)	
Sodium cyclamate (*see* Cyclamates)	
Sodium dichromate (*see* Chromium and chromium compounds)	
Sodium diethyldithiocarbamate	*12*, 217 (1976); *Suppl. 7*, 71 (1987)
Sodium equilin sulfate (*see* Conjugated oestrogens)	
Sodium fluoride (*see* Fluorides)	
Sodium monofluorophosphate (*see* Fluorides)	
Sodium oestrone sulfate (*see* Conjugated oestrogens)	
Sodium *ortho*-phenylphenate (*see also* ortho-Phenylphenol)	*30*, 329 (1983); *Suppl. 7*, 392 (1987)
Sodium saccharin (*see* Saccharin)	
Sodium selenate (*see* Selenium and selenium compounds)	
Sodium selenite (*see* Selenium and selenium compounds)	
Sodium silicofluoride (*see* Fluorides)	
Solar radiation	*55* (1992)
Soots	*3*, 22 (1973); *35*, 219 (1985); *Suppl. 7*, 343 (1987)
Spironolactone	*24*, 259 (1980); *Suppl. 7*, 344 (1987)
Stannous fluoride (*see* Fluorides)	
Steel founding (*see* Iron and steel founding)	
Sterigmatocystin	*1*, 175 (1972); *10*, 245 (1976); *Suppl. 7*, 72 (1987)
Steroidal oestrogens (*see also* Oestrogens, progestins and combinations)	*Suppl. 7*, 280 (1987)
Streptozotocin	*4*, 221 (1974); *17*, 337 (1978); *Suppl. 7*, 72 (1987)
Strobaner (*see* Terpene polychlorinates)	
Strontium chromate (*see* Chromium and chromium compounds)	
Styrene	*19*, 231 (1979) (*corr. 42*, 258); *Suppl. 7*, 345 (1987); *60*, 233 (1994)
Styrene-acrylonitrile-copolymers	*19*, 97 (1979); *Suppl. 7*, 72 (1987)
Styrene-butadiene copolymers	*19*, 252 (1979); *Suppl. 7*, 72 (1987)
Styrene-7,8-oxide	*11*, 201 (1976); *19*, 275 (1979); *36*, 245 (1985); *Suppl. 7*, 72 (1987); *60*, 321 (1994)
Succinic anhydride	*15*, 265 (1977); *Suppl. 7*, 72 (1987)
Sudan I	*8*, 225 (1975); *Suppl. 7*, 72 (1987)
Sudan II	*8*, 233 (1975); *Suppl. 7*, 72 (1987)
Sudan III	*8*, 241 (1975); *Suppl. 7*, 72 (1987)
Sudan Brown RR	*8*, 249 (1975); *Suppl. 7*, 72 (1987)
Sudan Red 7B	*8*, 253 (1975); *Suppl. 7*, 72 (1987)
Sulfafurazole	*24*, 275 (1980); *Suppl. 7*, 347 (1987)
Sulfallate	*30*, 283 (1983); *Suppl. 7*, 72 (1987)

Sulfamethoxazole	*24*, 285 (1980); *Suppl. 7*, 348 (1987)
Sulfites (*see* Sulfur dioxide and some sulfites, bisulfites and metabisulfites)	
Sulfur dioxide and some sulfites, bisulfites and metabisulfites	*54*, 131 (1992)
Sulfur mustard (*see* Mustard gas)	
Sulfuric acid and other strong inorganic acids, occupational exposures to mists and vapours from	*54*, 41 (1992)
Sulfur trioxide	*54*, 121 (1992)
Sulphisoxazole (*see* Sulfafurazole)	
Sunset Yellow FCF	*8*, 257 (1975); *Suppl. 7*, 72 (1987)
Symphytine	*31*, 239 (1983); *Suppl. 7*, 72 (1987)

T

2,4,5-T (*see also* Chlorophenoxy herbicides; Chlorophenoxy herbicides, occupational exposures to)	*15*, 273 (1977)
Talc	*42*, 185 (1987); Suppl. 7, 349 (1987)
Tannic acid	*10*, 253 (1976) (*corr. 42*, 255); *Suppl. 7*, 72 (1987)
Tannins (*see also* Tannic acid)	*10*, 254 (1976); *Suppl. 7*, 72 (1987)
TCDD (*see* 2,3,7,8-Tetrachlorodibenzo-*para*-dioxin)	
TDE (*see* DDT)	
Tea	*51*, 207 (1991)
Terpene polychlorinates	*5*, 219 (1974); *Suppl. 7*, 72 (1987)
Testosterone (*see also* Androgenic (anabolic) steroids)	*6*, 209 (1974); *21*, 519 (1979)
Testosterone oenanthate (*see* Testosterone)	
Testosterone propionate (*see* Testosterone)	
2,2',5,5'-Tetrachlorobenzidine	*27*, 141 (1982); *Suppl. 7*, 72 (1987)
2,3,7,8-Tetrachlorodibenzo-*para*-dioxin	*15*, 41 (1977); *Suppl. 7*, 350 (1987)
1,1,1,2-Tetrachloroethane	*41*, 87 (1986); *Suppl. 7*, 72 (1987)
1,1,2,2-Tetrachloroethane	*20*, 477 (1979); *Suppl. 7*, 354 (1987)
Tetrachloroethylene	*20*, 491 (1979); *Suppl. 7*, 355 (1987); *63*, 159 (1995)
2,3,4,6-Tetrachlorophenol (*see* Chlorophenols; Chlorophenols, occupational exposures to)	
Tetrachlorvinphos	*30*, 197 (1983); *Suppl. 7*, 72 (1987)
Tetraethyllead (*see* Lead and lead compounds)	
Tetrafluoroethylene	*19*, 285 (1979); *Suppl. 7*, 72 (1987)
Tetrakis(hydroxymethyl) phosphonium salts	*48*, 95 (1990)
Tetramethyllead (*see* Lead and lead compounds)	
Textile manufacturing industry, exposures in	*48*, 215 (1990) (*corr. 51*, 483)
Theobromine	*51*, 421 (1991)
Theophylline	*51*, 391 (1991)
Thioacetamide	*7*, 77 (1974); *Suppl. 7*, 72 (1987)
4,4'-Thiodianiline	*16*, 343 (1978); *27*, 147 (1982); *Suppl. 7*, 72 (1987)
Thiotepa	*9*, 85 (1975); *Suppl. 7*, 368 (1987); *50*, 123 (1990)
Thiouracil	*7*, 85 (1974); *Suppl. 7*, 72 (1987)
Thiourea	*7*, 95 (1974); *Suppl. 7*, 72 (1987)

Thiram	*12*, 225 (1976); *Suppl. 7*, 72 (1987); *53*, 403 (1991)
Titanium dioxide	*47*, 307 (1989)
Tobacco habits other than smoking (*see* Tobacco products, smokeless)	
Tobacco products, smokeless	*37* (1985) (*corr. 42*, 263; *52*, 513); *Suppl. 7*, 357 (1987)
Tobacco smoke	*38* (1986) (*corr. 42*, 263); *Suppl. 7*, 357 (1987)
Tobacco smoking (*see* Tobacco smoke)	
ortho-Tolidine (*see* 3,3′-Dimethylbenzidine)	
2,4-Toluene diisocyanate (*see also* Toluene diisocyanates)	*19*, 303 (1979); *39*, 287 (1986)
2,6-Toluene diisocyanate (*see also* Toluene diisocyanates)	*19*, 303 (1979); *39*, 289 (1986)
Toluene	*47*, 79 (1989)
Toluene diisocyanates	*39*, 287 (1986) (*corr. 42*, 264); *Suppl. 7*, 72 (1987)
Toluenes, α-chlorinated (*see* α-Chlorinated toluenes)	
ortho-Toluenesulfonamide (*see* Saccharin)	
ortho-Toluidine	*16*, 349 (1978); *27*, 155 (1982); *Suppl. 7*, 362 (1987)
Toxaphene	*20*, 327 (1979); *Suppl. 7*, 72 (1987)
T-2 Toxin (*see* Toxins derived from *Fusarium sporotrichioides*)	
Toxins derived from *Fusarium graminearum*, *F. culmorum* and *F. crookwellense*	*11*, 169 (1976); *31*, 153, 279 (1983); *Suppl. 7*, 64, 74 (1987); *56*, 397 (1993)
Toxins derived from *Fusarium moniliforme*	*56*, 445 (1993)
Toxins derived from *Fusarium sporotrichioides*	*31*, 265 (1983); *Suppl. 7*, 73 (1987); *56*, 467 (1993)
Tremolite (*see* Asbestos)	
Treosulfan	*26*, 341 (1981); *Suppl. 7*, 363 (1987)
Triaziquone [*see* Tris(aziridinyl)-*para*-benzoquinone]	
Trichlorfon	*30*, 207 (1983); *Suppl. 7*, 73 (1987)
Trichlormethine	*9*, 229 (1975); *Suppl. 7*, 73 (1987); *50*, 143 (1990)
Trichloroacetic acid	*63*, 291 (1995)
Trichloroacetonitrile (*see* Halogenated acetonitriles)	
1,1,1-Trichloroethane	*20*, 515 (1979); *Suppl. 7*, 73 (1987)
1,1,2-Trichloroethane	*20*, 533 (1979); *Suppl. 7*, 73 (1987); *52*, 337 (1991)
Trichloroethylene	*11*, 263 (1976); *20*, 545 (1979); *Suppl. 7*, 364 (1987); *63*, 75 (1995)
2,4,5-Trichlorophenol (*see also* Chlorophenols; Chlorophenols occupational exposures to)	*20*, 349 (1979)
2,4,6-Trichlorophenol (*see also* Chlorophenols; Chlorophenols, occupational exposures to)	*20*, 349 (1979)
(2,4,5-Trichlorophenoxy)acetic acid (*see* 2,4,5-T)	
1,2,3-Trichloropropane	*63*, 223 (1995)
Trichlorotriethylamine-hydrochloride (*see* Trichlormethine)	
T_2-Trichothecene (*see* Toxins derived from *Fusarium sporotrichioides*)	
Triethylene glycol diglycidyl ether	*11*, 209 (1976); *Suppl. 7*, 73 (1987)
Trifluralin	*53*, 515 (1991)

4,4',6-Trimethylangelicin plus ultraviolet radiation (see also Suppl. 7, 57 (1987)
   Angelicin and some synthetic derivatives)
2,4,5-Trimethylaniline   27, 177 (1982); Suppl. 7, 73 (1987)
2,4,6-Trimethylaniline   27, 178 (1982); Suppl. 7, 73 (1987)
4,5',8-Trimethylpsoralen   40, 357 (1986); Suppl. 7, 366 (1987)
Trimustine hydrochloride (see Trichlormethine)
Triphenylene   32, 447 (1983); Suppl. 7, 73 (1987)
Tris(aziridinyl)-para-benzoquinone   9, 67 (1975); Suppl. 7, 367 (1987)
Tris(1-aziridinyl)phosphine-oxide   9, 75 (1975); Suppl. 7, 73 (1987)
Tris(1-aziridinyl)phosphine-sulphide (see Thiotepa)
2,4,6-Tris(1-aziridinyl)-s-triazine   9, 95 (1975); Suppl. 7, 73 (1987)
Tris(2-chloroethyl) phosphate   48, 109 (1990)
1,2,3-Tris(chloromethoxy)propane   15, 301 (1977); Suppl. 7, 73 (1987)
Tris(2,3-dibromopropyl)phosphate   20, 575 (1979); Suppl. 7, 369 (1987)
Tris(2-methyl-1-aziridinyl)phosphine-oxide   9, 107 (1975); Suppl. 7, 73 (1987)
Trp-P-1   31, 247 (1983); Suppl. 7, 73 (1987)
Trp-P-2   31, 255 (1983); Suppl. 7, 73 (1987)
Trypan blue   8, 267 (1975); Suppl. 7, 73 (1987)
Tussilago farfara L. (see Pyrrolizidine alkaloids)

U

Ultraviolet radiation   40, 379 (1986); 55 (1992)
Underground haematite mining with exposure to radon   1, 29 (1972); Suppl. 7, 216 (1987)
Uracil mustard   9, 235 (1975); Suppl. 7, 370 (1987)
Urethane   7, 111 (1974); Suppl. 7, 73 (1987)

V

Vat Yellow 4   48, 161 (1990)
Vinblastine sulfate   26, 349 (1981) (corr. 42, 261);
                                                                Suppl. 7, 371 (1987)
Vincristine sulfate   26, 365 (1981); Suppl. 7, 372 (1987)
Vinyl acetate   19, 341 (1979); 39, 113 (1986);
                                                                Suppl. 7, 73 (1987); 63, 443 (1995)
Vinyl bromide   19, 367 (1979); 39, 133 (1986);
                                                                Suppl. 7, 73 (1987)
Vinyl chloride   7, 291 (1974); 19, 377 (1979)
                                                                (corr. 42, 258); Suppl. 7, 373 (1987)
Vinyl chloride-vinyl acetate copolymers   7, 311 (1976); 19, 412 (1979)
                                                                (corr. 42, 258); Suppl. 7, 73 (1987)
4-Vinylcyclohexene   11, 277 (1976); 39, 181 (1986)
                                                                Suppl. 7, 73 (1987); 60, 347 (1994)
4-Vinylcyclohexene diepoxide   11, 141 (1976); Suppl. 7, 63 (1987);
                                                                60, 361 (1994)
Vinyl fluoride   39, 147 (1986); Suppl. 7, 73 (1987);
                                                                63, 467 (1995)
Vinylidene chloride   19, 439 (1979); 39, 195 (1986);
                                                                Suppl. 7, 376 (1987)

Vinylidene chloride-vinyl chloride copolymers	*19*, 448 (1979) (*corr. 42*, 258); *Suppl. 7*, 73 (1987)
Vinylidene fluoride	*39*, 227 (1986); *Suppl. 7*, 73 (1987)
N-Vinyl-2-pyrrolidone	*19*, 461 (1979); *Suppl. 7*, 73 (1987)
Vinyl toluene	*60*, 373 (1994)

W

Welding	*49*, 447 (1990) (*corr. 52*, 513)
Wollastonite	*42*, 145 (1987); *Suppl. 7*, 377 (1987)
Wood dust	*62*, 35 (1995)
Wood industries	*25* (1981); *Suppl. 7*, 378 (1987)

X

Xylene	*47*, 125 (1989)
2,4-Xylidine	*16*, 367 (1978); *Suppl. 7*, 74 (1987)
2,5-Xylidine	*16*, 377 (1978); *Suppl. 7*, 74 (1987)
2,6-Xylidine (*see* 2,6-Dimethylaniline)	

Y

Yellow AB	*8*, 279 (1975); *Suppl. 7*, 74 (1987)
Yellow OB	*8*, 287 (1975); *Suppl. 7*, 74 (1987)

Z

Zearalenone (*see* Toxins derived from *Fusarium graminearum*, *F. culmorum* and *F. crookwellense*)	
Zectran	*12*, 237 (1976); *Suppl. 7*, 74 (1987)
Zinc beryllium silicate (*see* Beryllium and beryllium compounds)	
Zinc chromate (*see* Chromium and chromium compounds)	
Zinc chromate hydroxide (*see* Chromium and chromium compounds)	
Zinc potassium chromate (*see* Chromium and chromium compounds)	
Zinc yellow (*see* Chromium and chromium compounds)	
Zineb	*12*, 245 (1976); *Suppl. 7*, 74 (1987)
Ziram	*12*, 259 (1976); *Suppl. 7*, 74 (1987); *53*, 423 (1991)

PUBLICATIONS OF THE INTERNATIONAL AGENCY FOR RESEARCH ON CANCER

Scientific Publications Series

No. 1 Liver Cancer
1971; 176 pages (*out of print*)

No. 2 Oncogenesis and Herpesviruses
Edited by P.M. Biggs, G. de-Thé and L.N. Payne
1972; 515 pages (*out of print*)

No. 3 N-Nitroso Compounds: Analysis and Formation
Edited by P. Bogovski, R. Preussman and E.A. Walker
1972; 140 pages (*out of print*)

No. 4 Transplacental Carcinogenesis
Edited by L. Tomatis and U. Mohr
1973; 181 pages (*out of print*)

No. 5/6 Pathology of Tumours in Laboratory Animals, Volume 1, Tumours of the Rat
Edited by V.S. Turusov
1973/1976; 533 pages (*out of print*)

No. 7 Host Environment Interactions in the Etiology of Cancer in Man
Edited by R. Doll and I. Vodopija
1973; 464 pages (*out of print*)

No. 8 Biological Effects of Asbestos
Edited by P. Bogovski, J.C. Gilson, V. Timbrell and J.C. Wagner
1973; 346 pages (*out of print*)

No. 9 N-Nitroso Compounds in the Environment
Edited by P. Bogovski and E.A. Walker
1974; 243 pages (*out of print*)

No. 10 Chemical Carcinogenesis Essays
Edited by R. Montesano and L. Tomatis
1974; 230 pages (*out of print*)

No. 11 Oncogenesis and Herpesviruses II
Edited by G. de-Thé, M.A. Epstein and H. zur Hausen
1975; Part I: 511 pages
Part II: 403 pages (*out of print*)

No. 12 Screening Tests in Chemical Carcinogenesis
Edited by R. Montesano, H. Bartsch and L. Tomatis
1976; 666 pages (*out of print*)

No. 13 Environmental Pollution and Carcinogenic Risks
Edited by C. Rosenfeld and W. Davis
1975; 441 pages (*out of print*)

No. 14 Environmental N-Nitroso Compounds. Analysis and Formation
Edited by E.A. Walker, P. Bogovski and L. Griciute
1976; 512 pages (*out of print*)

No. 15 Cancer Incidence in Five Continents, Volume III
Edited by J.A.H. Waterhouse, C. Muir, P. Correa and J. Powell
1976; 584 pages (*out of print*)

No. 16 Air Pollution and Cancer in Man
Edited by U. Mohr, D. Schmähl and L. Tomatis
1977; 328 pages (*out of print*)

No. 17 Directory of On-going Research in Cancer Epidemiology 1977
Edited by C.S. Muir and G. Wagner
1977; 599 pages (*out of print*)

No. 18 Environmental Carcinogens. Selected Methods of Analysis. Volume 1: Analysis of Volatile Nitrosamines in Food
Editor-in-Chief: H. Egan
1978; 212 pages (*out of print*)

No. 19 Environmental Aspects of N-Nitroso Compounds
Edited by E.A. Walker, M. Castegnaro, L. Griciute and R.E. Lyle
1978; 561 pages (*out of print*)

No. 20 Nasopharyngeal Carcinoma: Etiology and Control
Edited by G. de-Thé and Y. Ito
1978; 606 pages (*out of print*)

No. 21 Cancer Registration and its Techniques
Edited by R. MacLennan, C. Muir, R. Steinitz and A. Winkler
1978; 235 pages (*out of print*)

No. 22 Environmental Carcinogens. Selected Methods of Analysis. Volume 2: Methods for the Measurement of Vinyl Chloride in Poly(vinyl chloride), Air, Water and Foodstuffs
Editor-in-Chief: H. Egan
1978; 142 pages (*out of print*)

No. 23 Pathology of Tumours in Laboratory Animals. Volume II: Tumours of the Mouse
Editor-in-Chief: V.S. Turusov
1979; 669 pages (*out of print*)

No. 24 Oncogenesis and Herpesviruses III
Edited by G. de-Thé, W. Henle and F. Rapp
1978; Part I: 580 pages, Part II: 512 pages (*out of print*)

Prices are subject to change without notice. Limited supplies of certain books marked '*out of print*' are available directly from IARC.

List of IARC Publications

No. 25 Carcinogenic Risk. Strategies for Intervention
Edited by W. Davis and C. Rosenfeld
1979; 280 pages (*out of print*)

No. 26 Directory of On-going Research in Cancer Epidemiology 1978
Edited by C.S. Muir and G. Wagner
1978; 550 pages (*out of print*)

No. 27 Molecular and Cellular Aspects of Carcinogen Screening Tests
Edited by R. Montesano, H. Bartsch and L. Tomatis
1980; 372 pages £30.00

No. 28 Directory of On-going Research in Cancer Epidemiology 1979
Edited by C.S. Muir and G. Wagner
1979; 672 pages (*out of print*)

No. 29 Environmental Carcinogens. Selected Methods of Analysis. Volume 3: Analysis of Polycyclic Aromatic Hydrocarbons in Environmental Samples
Editor-in-Chief: H. Egan
1979; 240 pages (*out of print*)

No. 30 Biological Effects of Mineral Fibres
Editor-in-Chief: J.C. Wagner
1980; Volume 1: 494 pages Volume 2: 513 pages (*out of print*)

No. 31 N-Nitroso Compounds: Analysis, Formation and Occurrence
Edited by E.A. Walker, L. Griciute, M. Castegnaro and M. Börzsönyi
1980; 835 pages (*out of print*)

No. 32 Statistical Methods in Cancer Research. Volume 1. The Analysis of Case-control Studies
By N.E. Breslow and N.E. Day
1980; 338 pages £18.00

No. 33 Handling Chemical Carcinogens in the Laboratory
Edited by R. Montesano *et al.*
1979; 32 pages (*out of print*)

No. 34 Pathology of Tumours in Laboratory Animals. Volume III. Tumours of the Hamster
Editor-in-Chief: V.S. Turusov
1982; 461 pages (*out of print*)

No. 35 Directory of On-going Research in Cancer Epidemiology 1980
Edited by C.S. Muir and G. Wagner
1980; 660 pages (*out of print*)

No. 36 Cancer Mortality by Occupation and Social Class 1851-1971
Edited by W.P.D. Logan
1982; 253 pages (*out of print*)

No. 37 Laboratory Decontamination and Destruction of Aflatoxins B_1, B_2, G_1, G_2 in Laboratory Wastes
Edited by M. Castegnaro *et al.*
1980; 56 pages (*out of print*)

No. 38 Directory of On-going Research in Cancer Epidemiology 1981
Edited by C.S. Muir and G. Wagner
1981; 696 pages (*out of print*)

No. 39 Host Factors in Human Carcinogenesis
Edited by H. Bartsch and B. Armstrong
1982; 583 pages (*out of print*)

No. 40 Environmental Carcinogens. Selected Methods of Analysis. Volume 4: Some Aromatic Amines and Azo Dyes in the General and Industrial Environment
Edited by L. Fishbein, M. Castegnaro, I.K. O'Neill and H. Bartsch
1981; 347 pages (*out of print*)

No. 41 N-Nitroso Compounds: Occurrence and Biological Effects
Edited by H. Bartsch, I.K. O'Neill, M. Castegnaro and M. Okada
1982; 755 pages (*out of print*)

No. 42 Cancer Incidence in Five Continents, Volume IV
Edited by J. Waterhouse, C. Muir, K. Shanmugaratnam and J. Powell
1982; 811 pages (*out of print*)

No. 43 Laboratory Decontamination and Destruction of Carcinogens in Laboratory Wastes: Some N-Nitrosamines
Edited by M. Castegnaro *et al.*
1982; 73 pages £7.50

No. 44 Environmental Carcinogens. Selected Methods of Analysis. Volume 5: Some Mycotoxins
Edited by L. Stoloff, M. Castegnaro, P. Scott, I.K. O'Neill and H. Bartsch
1983; 455 pages (*out of print*)

No. 45 Environmental Carcinogens. Selected Methods of Analysis. Volume 6: N-Nitroso Compounds
Edited by R. Preussmann, I.K. O'Neill, G. Eisenbrand, B. Spiegelhalder and H. Bartsch
1983; 508 pages (*out of print*)

No. 46 Directory of On-going Research in Cancer Epidemiology 1982
Edited by C.S. Muir and G. Wagner
1982; 722 pages (*out of print*)

No. 47 Cancer Incidence in Singapore 1968–1977
Edited by K. Shanmugaratnam, H.P. Lee and N.E. Day
1983; 171 pages (*out of print*)

No. 48 Cancer Incidence in the USSR (2nd Revised Edition)
Edited by N.P. Napalkov, G.F. Tserkovny, V.M. Merabishvili, D.M. Parkin, M. Smans and C.S. Muir
1983; 75 pages (*out of print*)

No. 49 Laboratory Decontamination and Destruction of Carcinogens in Laboratory Wastes: Some Polycyclic Aromatic Hydrocarbons
Edited by M. Castegnaro *et al.*
1983; 87 pages (*out of print*)

No. 50 Directory of On-going Research in Cancer Epidemiology 1983
Edited by C.S. Muir and G. Wagner
1983; 731 pages (*out of print*)

No. 51 Modulators of Experimental Carcinogenesis
Edited by V. Turusov and R. Montesano
1983; 307 pages (*out of print*)

List of IARC Publications

No. 52 Second Cancers in Relation to Radiation Treatment for Cervical Cancer: Results of a Cancer Registry Collaboration
Edited by N.E. Day and J.C. Boice, Jr
1984; 207 pages (*out of print*)

No. 53 Nickel in the Human Environment
Editor-in-Chief: F.W. Sunderman, Jr
1984; 529 pages (*out of print*)

No. 54 Laboratory Decontamination and Destruction of Carcinogens in Laboratory Wastes: Some Hydrazines
Edited by M. Castegnaro et al.
1983; 87 pages (*out of print*)

No. 55 Laboratory Decontamination and Destruction of Carcinogens in Laboratory Wastes: Some N-Nitrosamides
Edited by M. Castegnaro et al.
1984; 66 pages (*out of print*)

No. 56 Models, Mechanisms and Etiology of Tumour Promotion
Edited by M. Börzsönyi, N.E. Day, K. Lapis and H. Yamasaki
1984; 532 pages (*out of print*)

No. 57 N-Nitroso Compounds: Occurrence, Biological Effects and Relevance to Human Cancer
Edited by I.K. O'Neill, R.C. von Borstel, C.T. Miller, J. Long and H. Bartsch
1984; 1013 pages (*out of print*)

No. 58 Age-related Factors in Carcinogenesis
Edited by A. Likhachev, V. Anisimov and R. Montesano
1985; 288 pages (*out of print*)

No. 59 Monitoring Human Exposure to Carcinogenic and Mutagenic Agents
Edited by A. Berlin, M. Draper, K. Hemminki and H. Vainio
1984; 457 pages (*out of print*)

No. 60 Burkitt's Lymphoma: A Human Cancer Model
Edited by G. Lenoir, G. O'Conor and C.L.M. Olweny
1985; 484 pages (*out of print*)

No. 61 Laboratory Decontamination and Destruction of Carcinogens in Laboratory Wastes: Some Haloethers
Edited by M. Castegnaro et al.
1985; 55 pages (*out of print*)

No. 62 Directory of On-going Research in Cancer Epidemiology 1984
Edited by C.S. Muir and G. Wagner
1984; 717 pages (*out of print*)

No. 63 Virus-associated Cancers in Africa
Edited by A.O. Williams, G.T. O'Conor, G.B. de-Thé and C.A. Johnson
1984; 773 pages (*out of print*)

No. 64 Laboratory Decontamination and Destruction of Carcinogens in Laboratory Wastes: Some Aromatic Amines and 4-Nitrobiphenyl
Edited by M. Castegnaro et al.
1985; 84 pages (*out of print*)

No. 65 Interpretation of Negative Epidemiological Evidence for Carcinogenicity
Edited by N.J. Wald and R. Doll
1985; 232 pages (*out of print*)

No. 66 The Role of the Registry in Cancer Control
Edited by D.M. Parkin, G. Wagner and C.S. Muir
1985; 152 pages £10.00

No. 67 Transformation Assay of Established Cell Lines: Mechanisms and Application
Edited by T. Kakunaga and H. Yamasaki
1985; 225 pages (*out of print*)

No. 68 Environmental Carcinogens. Selected Methods of Analysis. Volume 7. Some Volatile Halogenated Hydrocarbons
Edited by L. Fishbein and I.K. O'Neill
1985; 479 pages (*out of print*)

No. 69 Directory of On-going Research in Cancer Epidemiology 1985
Edited by C.S. Muir and G. Wagner
1985; 745 pages (*out of print*)

No. 70 The Role of Cyclic Nucleic Acid Adducts in Carcinogenesis and Mutagenesis
Edited by B. Singer and H. Bartsch
1986; 467 pages (*out of print*)

No. 71 Environmental Carcinogens. Selected Methods of Analysis. Volume 8: Some Metals: As, Be, Cd, Cr, Ni, Pb, Se, Zn
Edited by I.K. O'Neill, P. Schuller and L. Fishbein
1986; 485 pages (*out of print*)

No. 72 Atlas of Cancer in Scotland, 1975–1980. Incidence and Epidemiological Perspective
Edited by I. Kemp, P. Boyle, M. Smans and C.S. Muir
1985; 285 pages (*out of print*)

No. 73 Laboratory Decontamination and Destruction of Carcinogens in Laboratory Wastes: Some Antineoplastic Agents
Edited by M. Castegnaro et al.
1985; 163 pages £13.50

No. 74 Tobacco: A Major International Health Hazard
Edited by D. Zaridze and R. Peto
1986; 324 pages £24.00

No. 75 Cancer Occurrence in Developing Countries
Edited by D.M. Parkin
1986; 339 pages £24.00

No. 76 Screening for Cancer of the Uterine Cervix
Edited by M. Hakama, A.B. Miller and N.E. Day
1986; 315 pages £31.50

No. 77 Hexachlorobenzene: Proceedings of an International Symposium
Edited by C.R. Morris and J.R.P. Cabral
1986; 668 pages (*out of print*)

No. 78 Carcinogenicity of Alkylating Cytostatic Drugs
Edited by D. Schmähl and J.M. Kaldor
1986; 337 pages (*out of print*)

No. 79 Statistical Methods in Cancer Research. Volume III: The Design and Analysis of Long-term Animal Experiments
By J.J. Gart, D. Krewski, P.N. Lee, R.E. Tarone and J. Wahrendorf
1986; 213 pages £23.50

List of IARC Publications

No. 80 **Directory of On-going Research in Cancer Epidemiology 1986**
Edited by C.S. Muir and G. Wagner
1986; 805 pages (*out of print*)

No. 81 **Environmental Carcinogens: Methods of Analysis and Exposure Measurement. Volume 9: Passive Smoking**
Edited by I.K. O'Neill, K.D. Brunnemann, B. Dodet and D. Hoffmann
1987; 383 pages £37.00

No. 82 **Statistical Methods in Cancer Research. Volume II: The Design and Analysis of Cohort Studies**
By N.E. Breslow and N.E. Day
1987; 404 pages £25.00

No. 83 **Long-term and Short-term Assays for Carcinogens: A Critical Appraisal**
Edited by R. Montesano, H. Bartsch, H. Vainio, J. Wilbourn and H. Yamasaki
1986; 575 pages £37.00

No. 84 **The Relevance of N-Nitroso Compounds to Human Cancer: Exposure and Mechanisms**
Edited by H. Bartsch, I.K. O'Neill and R. Schulte-Hermann
1987; 671 pages (*out of print*)

No. 85 **Environmental Carcinogens: Methods of Analysis and Exposure Measurement. Volume 10: Benzene and Alkylated Benzenes**
Edited by L. Fishbein and I.K. O'Neill
1988; 327 pages £42.00

No. 86 **Directory of On-going Research in Cancer Epidemiology 1987**
Edited by D.M. Parkin and J. Wahrendorf
1987; 676 pages (*out of print*)

No. 87 **International Incidence of Childhood Cancer**
Edited by D.M. Parkin, C.A. Stiller, C.A. Bieber, G.J. Draper, B. Terracini and J.L. Young
1988; 401 pages £35.00

No. 88 **Cancer Incidence in Five Continents Volume V**
Edited by C. Muir, J. Waterhouse, T. Mack, J. Powell and S. Whelan
1987; 1004 pages £58.00

No. 89 **Method for Detecting DNA Damaging Agents in Humans: Applications in Cancer Epidemiology and Prevention**
Edited by H. Bartsch, K. Hemminki and I.K. O'Neill
1988; 518 pages £50.00

No. 90 **Non-occupational Exposure to Mineral Fibres**
Edited by J. Bignon, J. Peto and R. Saracci
1989; 500 pages £52.50

No. 91 **Trends in Cancer Incidence in Singapore 1968–1982**
Edited by H.P. Lee, N.E. Day and K. Shanmugaratnam
1988; 160 pages (*out of print*)

No. 92 **Cell Differentiation, Genes and Cancer**
Edited by T. Kakunaga, T. Sugimura, L. Tomatis and H. Yamasaki
1988; 204 pages £29.00

No. 93 **Directory of On-going Research in Cancer Epidemiology 1988**
Edited by M. Coleman and J. Wahrendorf
1988; 662 pages (*out of print*)

No. 94 **Human Papillomavirus and Cervical Cancer**
Edited by N. Muñoz, F.X. Bosch and O.M. Jensen
1989; 154 pages £22.50

No. 95 **Cancer Registration: Principles and Methods**
Edited by O.M. Jensen, D.M. Parkin, R. MacLennan, C.S. Muir and R. Skeet
1991; 288 pages £28.00

No. 96 **Perinatal and Multigeneration Carcinogenesis**
Edited by N.P. Napalkov, J.M. Rice, L. Tomatis and H. Yamasaki
1989; 436 pages £52.50

No. 97 **Occupational Exposure to Silica and Cancer Risk**
Edited by L. Simonato, A.C. Fletcher, R. Saracci and T. Thomas
1990; 124 pages £24.00

No. 98 **Cancer Incidence in Jewish Migrants to Israel, 1961–1981**
Edited by R. Steinitz, D.M. Parkin, J.L. Young, C.A. Bieber and L. Katz
1989; 320 pages £37.00

No. 99 **Pathology of Tumours in Laboratory Animals, Second Edition, Volume 1, Tumours of the Rat**
Edited by V.S. Turusov and U. Mohr
740 pages £90.00

No. 100 **Cancer: Causes, Occurrence and Control**
Editor-in-Chief L. Tomatis
1990; 352 pages £25.50

No. 101 **Directory of On-going Research in Cancer Epidemiology 1989/90**
Edited by M. Coleman and J. Wahrendorf
1989; 818 pages £42.00

No. 102 **Patterns of Cancer in Five Continents**
Edited by S.L. Whelan, D.M. Parkin & E. Masuyer
1990; 162 pages £26.50

No. 103 **Evaluating Effectiveness of Primary Prevention of Cancer**
Edited by M. Hakama, V. Beral, J.W. Cullen and D.M. Parkin
1990; 250 pages £34.00

No. 104 **Complex Mixtures and Cancer Risk**
Edited by H. Vainio, M. Sorsa and A.J. McMichael
1990; 442 pages £40.00

No. 105 **Relevance to Human Cancer of N-Nitroso Compounds, Tobacco Smoke and Mycotoxins**
Edited by I.K. O'Neill, J. Chen and H. Bartsch
1991; 614 pages £74.00

No. 106 **Atlas of Cancer Incidence in the Former German Democratic Republic**
Edited by W.H. Mehnert, M. Smans, C.S. Muir, M. Möhner & D. Schön
1992; 384 pages £52.50

List of IARC Publications

No. 107 Atlas of Cancer Mortality in the European Economic Community
Edited by M. Smans, C.S. Muir and P. Boyle
1992; 280 pages £35.00

No. 108 Environmental Carcinogens: Methods of Analysis and Exposure Measurement. Volume 11: Polychlorinated Dioxins and Dibenzofurans
Edited by C. Rappe, H.R. Buser, B. Dodet and I.K. O'Neill
1991; 426 pages £47.50

No. 109 Environmental Carcinogens: Methods of Analysis and Exposure Measurement. Volume 12: Indoor Air Contaminants
Edited by B. Seifert, H. van de Wiel, B. Dodet and I.K. O'Neill
1993; 384 pages £45.00

No. 110 Directory of On-going Research in Cancer Epidemiology 1991
Edited by M. Coleman and J. Wahrendorf
1991; 753 pages £40.00

No. 111 Pathology of Tumours in Laboratory Animals, Second Edition, Volume 2, Tumours of the Mouse
Edited by V.S. Turusov and U. Mohr
1993; 776 pages; £90.00

No. 112 Autopsy in Epidemiology and Medical Research
Edited by E. Riboli and M. Delendi
1991; 288 pages £26.50

No. 113 Laboratory Decontamination and Destruction of Carcinogens in Laboratory Wastes: Some Mycotoxins
Edited by M. Castegnaro, J. Barek, J.-M. Frémy, M. Lafontaine, M. Miraglia, E.B. Sansone and G.M. Telling
1991; 64 pages £12.00

No. 114 Laboratory Decontamination and Destruction of Carcinogens in Laboratory Wastes: Some Polycyclic Heterocyclic Hydrocarbons
Edited by M. Castegnaro, J. Barek J. Jacob, U. Kirso, M. Lafontaine, E.B. Sansone, G.M. Telling and T. Vu Duc
1991; 50 pages £8.00

No. 115 Mycotoxins, Endemic Nephropathy and Urinary Tract Tumours
Edited by M. Castegnaro, R. Plestina, G. Dirheimer, I.N. Chernozemsky and H Bartsch
1991; 340 pages £47.50

No. 116 Mechanisms of Carcinogenesis in Risk Identification
Edited by H. Vainio, P.N. Magee, D.B. McGregor & A.J. McMichael
1992; 616 pages £69.00

No. 117 Directory of On-going Research in Cancer Epidemiology 1992
Edited by M. Coleman, J. Wahrendorf & E. Démaret
1992; 773 pages £44.50

No. 118 Cadmium in the Human Environment: Toxicity and Carcinogenicity
Edited by G.F. Nordberg, R.F.M. Herber & L. Alessio
1992; 470 pages £60.00

No. 119 The Epidemiology of Cervical Cancer and Human Papillomavirus
Edited by N. Muñoz, F.X. Bosch, K.V. Shah & A. Meheus
1992; 288 pages £29.50

No. 120 Cancer Incidence in Five Continents, Volume VI
Edited by D.M. Parkin, C.S. Muir, S.L. Whelan, Y.T. Gao, J. Ferlay & J.Powell
1992; 1080 pages £120.00

No. 121 Trends in Cancer Incidence and Mortality
M.P. Coleman, J. Estève, P. Damiecki, A. Arslan and H. Renard
1993; 806 pages, £120.00

No. 122 International Classification of Rodent Tumours. Part 1. The Rat
Editor-in-Chief: U. Mohr
1992/95; 10 fascicles of 60–100 pages, £120.00

No. 123 Cancer in Italian Migrant Populations
Edited by M. Geddes, D.M. Parkin, M. Khlat, D. Balzi and E. Buiatti
1993; 292 pages, £40.00

No. 124 Postlabelling Methods for Detection of DNA Adducts
Edited by D.H. Phillips, M. Castegnaro and H. Bartsch
1993; 392 pages; £46.00

No. 125 DNA Adducts: Identification and Biological Significance
Edited by K. Hemminki, A. Dipple, D. Shuker, F.F. Kadlubar, D. Segerbäck and H. Bartsch
1994; 480 pages; £52.00

No. 127 Butadiene and Styrene: Assessment of Health Hazards
Edited by M. Sorsa, K. Peltonen, H. Vainio and K. Hemminki
1993; 412 pages; £54.00

No. 128 Statistical Methods in Cancer Research. Volume IV. Descriptive Epidemiology
By J. Estève, E. Benhamou & L.Raymond
1994; 302 pages; £25.00

No. 129 Occupational Cancer in Developing Countries
Edited by N. Pearce, E. Matos, H. Vainio, P. Boffetta & M. Kogevinas
1994; 192 pages £20.00

No. 130 Directory of On-going Research in Cancer Epidemiology 1994
Edited by R. Sankaranarayanan, J. Wahrendorf and E. Démaret
1994; 792 pages, £46.00

No. 132 Survival of Cancer Patients in Europe. The EUROCARE Study
Edited by F. Berrino, M. Sant, A. Verdecchia, R. Capocaccia, T. Hakulinen and J. Estève
1994; 463 pages; £45.00

List of IARC Publications

IARC MONOGRAPHS ON THE EVALUATION OF CARCINOGENIC RISKS TO HUMANS

Volume 1 **Some Inorganic Substances, Chlorinated Hydrocarbons, Aromatic Amines, *N*-Nitroso Compounds, and Natural Products**
1972; 184 pages (*out of print*)

Volume 2 **Some Inorganic and Organometallic Compounds**
1973; 181 pages (*out of print*)

Volume 3 **Certain Polycyclic Aromatic Hydrocarbons and Heterocyclic Compounds**
1973; 271 pages (*out of print*)

Volume 4 **Some Aromatic Amines, Hydrazine and Related Substances, *N*-Nitroso Compounds and Miscellaneous Alkylating Agents**
1974; 286 pages Sw. fr. 18.—

Volume 5 **Some Organochlorine Pesticides**
1974; 241 pages (*out of print*)

Volume 6 **Sex Hormones**
1974; 243 pages (*out of print*)

Volume 7 **Some Anti-Thyroid and Related Substances, Nitrofurans and Industrial Chemicals**
1974; 326 pages (*out of print*)

Volume 8 **Some Aromatic Azo Compounds**
1975; 357 pages Sw. fr. 44.—

Volume 9 **Some Aziridines, *N*-, *S*- and *O*-Mustards and Selenium**
1975; 268 pages Sw.fr. 33.—

Volume 10 **Some Naturally Occurring Substances**
1976; 353 pages (*out of print*)

Volume 11 **Cadmium, Nickel, Some Epoxides, Miscellaneous Industrial Chemicals and General Considerations on Volatile Anaesthetics**
1976; 306 pages (*out of print*)

Volume 12 **Some Carbamates, Thiocarbamates and Carbazides**
1976; 282 pages Sw. fr. 41.—

Volume 13 **Some Miscellaneous Pharmaceutical Substances**
1977; 255 pages Sw. fr. 36.—

Volume 14 **Asbestos**
1977; 106 pages (*out of print*)

Volume 15 **Some Fumigants, The Herbicides 2,4-D and 2,4,5-T, Chlorinated Dibenzodioxins and Miscellaneous Industrial Chemicals**
1977; 354 pages (*out of print*)

Volume 16 **Some Aromatic Amines and Related Nitro Compounds - Hair Dyes, Colouring Agents and Miscellaneous Industrial Chemicals**
1978; 400 pages Sw. fr. 60.—

Volume 17 **Some *N*-Nitroso Compounds**
1978; 365 pages Sw. fr. 60.—

Volume 18 **Polychlorinated Biphenyls and Polybrominated Biphenyls**
1978; 140 pages Sw. fr. 24.—

Volume 19 **Some Monomers, Plastics and Synthetic Elastomers, and Acrolein**
1979; 513 pages (*out of print*)

Volume 20 **Some Halogenated Hydrocarbons**
1979; 609 pages (*out of print*)

Volume 21 **Sex Hormones (II)**
1979; 583 pages Sw. fr. 72.—

Volume 22 **Some Non-Nutritive Sweetening Agents**
1980; 208 pages Sw. fr. 30.—

Volume 23 **Some Metals and Metallic Compounds**
1980; 438 pages (*out of print*)

Volume 24 **Some Pharmaceutical Drugs**
1980; 337 pages Sw. fr. 48.—

Volume 25 **Wood, Leather and Some Associated Industries**
1981; 412 pages Sw. fr. 72.—

Volume 26 **Some Antineoplastic and Immunosuppressive Agents**
1981; 411 pages Sw. fr. 75.—

Volume 27 **Some Aromatic Amines, Anthraquinones and Nitroso Compounds, and Inorganic Fluorides Used in Drinking Water and Dental Preparations**
1982; 341 pages Sw. fr. 48.—

Volume 28 **The Rubber Industry**
1982; 486 pages Sw. fr. 84.—

Volume 29 **Some Industrial Chemicals and Dyestuffs**
1982; 416 pages Sw. fr. 72.—

Volume 30 **Miscellaneous Pesticides**
1983; 424 pages Sw. fr. 72.—

Volume 31 **Some Food Additives, Feed Additives and Naturally Occurring Substances**
1983; 314 pages Sw. fr. 66.—

Volume 32 **Polynuclear Aromatic Compounds, Part 1: Chemical, Environmental and Experimental Data**
1983; 477 pages Sw. fr. 88.—

Volume 33 **Polynuclear Aromatic Compounds, Part 2: Carbon Blacks, Mineral Oils and Some Nitroarenes**
1984; 245 pages (*out of print*)

Volume 34 **Polynuclear Aromatic Compounds, Part 3: Industrial Exposures in Aluminium Production, Coal Gasification, Coke Production, and Iron and Steel Founding**
1984; 219 pages Sw. fr. 53.—

Volume 35 **Polynuclear Aromatic Compounds, Part 4: Bitumens, Coal-tars and Derived Products, Shale-oils and Soots**
1985; 271 pages Sw. fr. 77.—

List of IARC Publications

Volume 36 **Allyl Compounds, Aldehydes, Epoxides and Peroxides**
1985; 369 pages Sw. fr. 77.–

Volume 37 **Tobacco Habits Other than Smoking: Betel-quid and Areca-nut Chewing; and some Related Nitrosamines**
1985; 291 pages Sw. fr. 77.–

Volume 38 **Tobacco Smoking**
1986; 421 pages Sw. fr. 83.–

Volume 39 **Some Chemicals Used in Plastics and Elastomers**
1986; 403 pages Sw. fr. 83.–

Volume 40 **Some Naturally Occurring and Synthetic Food Components, Furocoumarins and Ultraviolet Radiation**
1986; 444 pages Sw. fr. 83.–

Volume 41 **Some Halogenated Hydrocarbons and Pesticide Exposures**
1986; 434 pages Sw. fr. 83.–

Volume 42 **Silica and Some Silicates**
1987; 289 pages Sw. fr. 72.

Volume 43 **Man-Made Mineral Fibres and Radon**
1988; 300 pages Sw. fr. 72.–

Volume 44 **Alcohol Drinking**
1988; 416 pages Sw. fr. 83.

Volume 45 **Occupational Exposures in Petroleum Refining; Crude Oil and Major Petroleum Fuels**
1989; 322 pages Sw. fr. 72.–

Volume 46 **Diesel and Gasoline Engine Exhausts and Some Nitroarenes**
1989; 458 pages Sw. fr. 83.–

Volume 47 **Some Organic Solvents, Resin Monomers and Related Compounds, Pigments and Occupational Exposures in Paint Manufacture and Painting**
1989; 535 pages Sw. fr. 94.–

Volume 48 **Some Flame Retardants and Textile Chemicals, and Exposures in the Textile Manufacturing Industry**
1990; 345 pages Sw. fr. 72.–

Volume 49 **Chromium, Nickel and Welding**
1990; 677 pages Sw. fr. 105.–

Volume 50 **Pharmaceutical Drugs**
1990; 415 pages Sw. fr. 93.–

Volume 51 **Coffee, Tea, Mate, Methylxanthines and Methylglyoxal**
1991; 513 pages Sw. fr. 88.–

Volume 52 **Chlorinated Drinking-water; Chlorination By-products; Some Other Halogenated Compounds; Cobalt and Cobalt Compounds**
1991; 544 pages Sw. fr. 88.–

Volume 53 **Occupational Exposures in Insecticide Application and some Pesticides**
1991; 612 pages Sw. fr. 105.–

Volume 54 **Occupational Exposures to Mists and Vapours from Strong Inorganic Acids; and Other Industrial Chemicals**
1992; 336 pages Sw. fr. 72.–

Volume 55 **Solar and Ultraviolet Radiation**
1992; 316 pages Sw. fr. 65.–

Volume 56 **Some Naturally Occurring Substances: Food Items and Constituents, Heterocyclic Aromatic Amines and Mycotoxins**
1993; 600 pages Sw. fr. 95.–

Volume 57 **Occupational Exposures of Hairdressers and Barbers and Personal Use of Hair Colourants; Some Hair Dyes, Cosmetic Colourants, Industrial Dyestuffs and Aromatic Amines**
1993; 428 pages Sw. fr. 75.–

Volume 58 **Beryllium, Cadmium, Mercury and Exposures in the Glass Manufacturing Industry**
1993; 426 pages Sw. fr. 75.–

Volume 59 **Hepatitis Viruses**
1994; 286 pages Sw. fr. 65.–

Volume 60 **Some Industrial Chemicals**
1994; 560 pages Sw. fr. 90.–

Volume 61 **Schistosomes, Liver Flukes and Helicobacter pylori**
1994; 270 pages Sw. fr. 70.–

Volume 62 **Wood Dust and Formaldehyde**
1995; 406 pages Sw. fr. 80.–

Supplement No. 1
Chemicals and Industrial Processes Associated with Cancer in Humans (IARC Monographs, Volumes 1 to 20)
1979; 71 pages (*out of print*)

Supplement No. 2
Long-term and Short-term Screening Assays for Carcinogens: A Critical Appraisal
1980; 426 pages Sw. fr. 40.–

Supplement No. 3
Cross Index of Synonyms and Trade Names in Volumes 1 to 26
1982; 199 pages (*out of print*)

Supplement No. 4
Chemicals, Industrial Processes and Industries Associated with Cancer in Humans (IARC Monographs, Volumes 1 to 29)
1982; 292 pages (*out of print*)

Supplement No. 5
Cross Index of Synonyms and Trade Names in Volumes 1 to 36
1985; 259 pages (*out of print*)

Supplement No. 6
Genetic and Related Effects: An Updating of Selected IARC Monographs from Volumes 1 to 42
1987; 729 pages Sw. fr. 80.–

Supplement No. 7
Overall Evaluations of Carcinogenicity: An Updating of IARC Monographs Volumes 1-42
1987; 440 pages Sw. fr. 65.–

Supplement No. 8
Cross Index of Synonyms and Trade Names in Volumes 1 to 46
1990; 346 pages Sw. fr. 60.–

List of IARC Publications

IARC TECHNICAL REPORTS

No. 1 **Cancer in Costa Rica**
Edited by R. Sierra,
R. Barrantes, G. Muñoz Leiva, D.M. Parkin, C.A. Bieber and
N. Muñoz Calero
1988; 124 pages Sw. fr. 30.-

No. 2 **SEARCH: A Computer Package to Assist the Statistical Analysis of Case-control Studies**
Edited by G.J. Macfarlane,
P. Boyle and P. Maisonneuve
1991; 80 pages (*out of print*)

No. 3 **Cancer Registration in the European Economic Community**
Edited by M.P. Coleman and
E. Démaret
1988; 188 pages Sw. fr. 30.-

No. 4 **Diet, Hormones and Cancer: Methodological Issues for Prospective Studies**
Edited by E. Riboli and
R. Saracci
1988; 156 pages Sw. fr. 30.-

No. 5 **Cancer in the Philippines**
Edited by A.V. Laudico,
D. Esteban and D.M. Parkin
1989; 186 pages Sw. fr. 30.-

No. 6 **La genèse du Centre International de Recherche sur le Cancer**
Par R. Sohier et A.G.B. Sutherland
1990; 104 pages Sw. fr. 30.-

No. 7 Epidémiologie du cancer dans les pays de langue latine
1990; 310 pages Sw. fr. 30.-

No. 8 **Comparative Study of Antismoking Legislation in Countries of the European Economic Community**
Edited by A. Sasco, P. Dalla Vorgia and P. Van der Elst
1992; 82 pages Sw. fr. 30.-

No. 9 Epidemiologie du cancer dans les pays de langue latine
1991 346 pages Sw. fr. 30.-

No. 11 **Nitroso Compounds: Biological Mechanisms, Exposures and Cancer Etiology**
Edited by I.K. O'Neill & H. Bartsch
1992; 149 pages Sw. fr. 30.-

No. 12 Epidémiologie du cancer dans les pays de langue latine
1992; 375 pages Sw. fr. 30.-

No. 13 **Health, Solar UV Radiation and Environmental Change**
By A. Kricker, B.K. Armstrong, M.E. Jones and R.C. Burton
1993; 216 pages Sw.fr. 30.-

No. 14 Epidémiologie du cancer dans les pays de langue latine
1993; 385 pages Sw. fr. 30.-

No. 15 **Cancer in the African Population of Bulawayo, Zimbabwe, 1963-1977: Incidence, Time Trends and Risk Factors**
By M.E.G. Skinner, D.M. Parkin, A.P. Vizcaino and A. Ndhlovu
1993; 123 pages Sw. fr. 30.-

No. 16 **Cancer in Thailand, 1988-1991**
By V. Vatanasapt, N. Martin, H. Sriplung, K. Vindavijak, S. Sontipong, S. Sriamporn, D.M. Parkin and J. Ferlay
1993; 164 pages Sw. fr. 30.-

No. 18 **Intervention Trials for Cancer Prevention**
By E. Buiatti
1994; 52 pages Sw. fr. 30.-

No. 19 **Comparability and Quality Control in Cancer Registration**
By D.M. Parkin, V.W. Chen, J. Ferlay, J. Galceran, H.H. Storm and S.L. Whelan
1994; 110 pages plus diskette Sw. fr. 40.-

No. 20 Epidémiologie du cancer dans les pays de langue latine
1994; 346 pages Sw. fr. 30.-

No. 21 **ICD Conversion Programs for Cancer**
By J. Ferlay
1994; 24 pages plus diskette Sw. fr. 30.-

No. 22 **Cancer Incidence by Occupation and Industry in Tianjin, China, 1981-1987**
By Q.S. Wang, P. Boffetta, M. Kogevinas and D.M. Parkin
1994; 96 pages Sw. fr. 30.-

DIRECTORY OF AGENTS BEING TESTED FOR CARCINOGENICITY (Until Vol. 13 Information Bulletin on the Survey of Chemicals Being Tested for Carcinogenicity)

No. 8 Edited by M.-J. Ghess, H. Bartsch and L. Tomatis
1979; 604 pages Sw. fr. 40.-

No. 9 Edited by M.-J. Ghess, J.D. Wilbourn, H. Bartsch and L. Tomatis
1981; 294 pages Sw. fr. 41.-

No. 10 Edited by M.-J. Ghess, J.D. Wilbourn and H. Bartsch
1982; 362 pages Sw. fr. 42.-

No. 11 Edited by M.-J. Ghess, J.D. Wilbourn, H. Vainio and H. Bartsch
1984; 362 pages Sw. fr. 50.-

No. 12 Edited by M.-J. Ghess, J.D. Wilbourn, A. Tossavainen and H. Vainio
1986; 385 pages Sw. fr. 50.-

No. 13 Edited by M.-J. Ghess, J.D. Wilbourn and A. Aitio 1988; 404 pages Sw. fr. 43.-

No. 14 Edited by M.-J. Ghess, J.D. Wilbourn and H. Vainio
1990; 370 pages Sw. fr. 45.-

No. 15 Edited by M.-J. Ghess, J.D. Wilbourn and H. Vainio
1992; 318 pages Sw. fr. 45.-

No. 16 Edited by M.-J. Ghess, J.D. Wilbourn and H. Vainio
1994; 294 pages Sw. fr. 50.-

NON-SERIAL PUBLICATIONS

Alcool et Cancer
By A. Tuyns (in French only)
1978; 42 pages Fr. fr. 35.-

Cancer Morbidity and Causes of Death Among Danish Brewery Workers
By O.M. Jensen
1980; 143 pages Fr. fr. 75.-

Directory of Computer Systems Used in Cancer Registries
By H.R. Menck and D.M. Parkin
1986; 236 pages Fr. fr. 50.-

Facts and Figures of Cancer in the European Community
Edited by J. Estève, A. Kricker, J. Ferlay and D.M. Parkin
1993; 52 pages Sw. fr. 10.-

IARC Monographs and Technical Reports are available from:

World Health Organization,
Distribution and Sales Service
1211 Geneva 27, Switzerland
Fax: + 41 22 791 4857
and from WHO Sales Agents.

IARC Scientific Publications are available from:

Oxford University Press
Medical Marketing Manager
Walton Street
Oxford UK OX2 6DP
Fax: +44 1865 267782

All IARC Publications can also be ordered directly from:

IARC*Press*
CIRC/IARC
150 cours Albert Thomas
69372 Lyon cedex 08
France
Fax: +33 72 73 83 02

www.ingramcontent.com/pod-product-compliance
Ingram Content Group UK Ltd.
Pitfield, Milton Keynes, MK11 3LW, UK
UKHW051257180426
11947UKWH00020B/1767